Principles and Techniques in Plant Virology

Edited by

CLARENCE I. KADO

Department of Plant Pathology
College of Agricultural and Environmental Sciences
University of California, Davis

and

HARI O. AGRAWAL

Indian Council of Agricultural Research
Central Potato Research Institute
Simla, India

VNR **Van Nostrand Reinhold Company**

New York Cincinnati Toronto London Melbourne

Van Nostrand Reinhold Company Regional Offices:
New York Cincinnati Chicago Millbrae Dallas

Van Nostrand Reinhold Company International Offices:
London Toronto Melbourne

Library of Congress Catalog Card Number: 70-154805

Manufactured in the United States of America

Published by Van Nostrand Reinhold Company
450 West 33rd Street, New York, N.Y. 10001

Published simultaneously in Canada by Van Nostrand Reinhold Ltd.

15 14 13 12 11 10 9 8 7 6 5 4 3 2 1

Principles and Techniques in Plant Virology

Contributors

Hari O. Agrawal*
Central Potato Research Institute
Simla 1, India

R. Bercks
Biologische Bundesanstalt für Land-und Forstwirtschaft
Institut für Virusserologie
Messeweg 11/12
Braunschweig, Germany

George E. Bruening
Department of Biochemistry and Biophysics
University of California
Davis, California

Meyer Chessin
Department of Botany
University of Montana
Missoula, Montana

R.I.B. Francki
Department of Plant Pathology
Waite Agricultural Research Institute
University of Adelaide
South Australia

Mary G. Hamilton
Sloan-Kettering Cancer Research Institute
New York, New York

B. Hennig
Max Planck Institut für Molekulare Genetik
Berlin-Dahlem, Germany

Akikazu Hirashima
Laboratory of Molecular Genetics
Medical School
University of Osaka
Osaka, Japan

Clarence I. Kado
Department of Plant Pathology
University of California
Davis, California

Renate Koenig
Biologische Bundesanstalt für Land-und Forstwirtschaft
Institut für Virusserologie
Messeweg 11/12
Braunschweig, Germany

G. Loebenstein
Virus Laboratory
Volcani Institute of Agricultural Research
Bet Dagan, Israel

Robert G. Milne
Clinical Research Center
Northwick Park
Harrow, Middlesex
England

---

*Present address:
Virus Chemistry Section
Canada Department of Agriculture
6660 N.W. Marine Drive
Vancouver, Canada

Gertrud Querfurth
Biologische Bundesanstalt für Land-und Forstwirtschaft
Institut für Virusserologie
Messeweg 11/12
Braunschweig, Germany

Alan Rees
University of Texas
Medical School
San Antonio, Texas

M. Rubio-Huertos
Instituto Jaime Ferran de Microbiologia
Madrid, Spain

Verne Schumaker
Department of Chemistry and
The Molecular Biology Institute
University of California
Los Angeles, California

R. J. Shepherd
Department of Plant Pathology
University of California
Davis, California

J. T. Slykhuis
Ottawa Research Station
Canada Department of Agriculture
Ottawa, Ontario
Canada

C. E. Taylor
Scottish Horticultural Research Institute
Dundee, Invergowrie
Scotland, U.K.

D. S. Teakle
Department of Microbiology
University of Queensland
Medical School, Herston
Brisbane, Queensland
Australia

Rodes Trautman
Plum Island Animal Disease Laboratory
Veterinary Sciences Research Division
Agricultural Research Service
United States Department of Agriculture
Greenport, Long Island, New York

Akira Tsugita
Laboratory of Molecular Genetics
Medical School
University of Osaka
Osaka, Japan

Marc H. V. van Regenmortel
Department of Microbiology and Virology
University of Stellenbosch
Stellenbosch, South Africa

J. H. Venekamp
Department of Biochemistry
Institute of Phytopathological Research
Binnenhaven 12, Wageningen
The Netherlands

Marion A. Watson
Department of Plant Pathology
Rothamsted Experimental Station
Harpenden, Herts
England

Robert F. Whitcomb
Entomology Research Division
Agricultural Research Service
United States Department of Agriculture
Beltsville, Maryland

H. G. Wittmann
Max Planck Institut für Molekulare Genetik
Berlin-Dahlem, Germany

Contents

CHAPTER 6
Transmission of Viruses and Mycoplasma by Auchenorrhynchous Homoptera
R. F. Whitcomb

CHAPTER 7
Transmission of Plant Viruses by Eriophyid Mites
J. T. Slykhuis

CHAPTER 8
Transmission of Viruses by Nematodes
C. E. Taylor

CHAPTER 9
Transmission of Plant Viruses by Fungi
D. S. Teakle

CHAPTER 13
Chromatographic Purification of Plant Viruses
J. H. Venekamp

CHAPTER 14
Electrophoresis
Marc H. V. van Regenmortel

CHAPTER 15
Isolation and Properties of Virus Proteins
Akira Tsugita and Akikazu Hirashima

CHAPTER 16
Virus Degradation and Nucleic Acid Isolation
George E. Bruening

CHAPTER 17
Plant Virus Serology
R. Bercks, Renate Koenig, and Gertrud Querfurth

CHAPTER 18
Analytical Ultracentrifugation
Rodes Trautman and Mary G. Hamilton

PART FOUR
PHOTOBIOLOGY AND MUTATION

CHAPTER 19
Effect of Radiation on Viruses
Meyer Chessin

CHAPTER 20
Tobacco Mosaic Virus: Mutants and Strains
B. Hennig and H. G. Wittmann

PART ONE

BIOLOGY OF PLANT VIRUSES

1

Mechanical and Biological Inoculation Principles

CLARENCE I. KADO

Department of Plant Pathology
University of California
Davis, California

I. INTRODUCTION

All viruses can be transmitted to any plant by mechanical inoculation,* but not all can effectively initiate and establish infection. In theory, virus particles (or their free nucleic acid) can be introduced into living cells by standard methods of mechanical inoculation, but whether the virus becomes functional depends on a variety of conditions: degree of susceptibility of the recipient cell, functioning of the viral genome within the cell, conditions necessary for virus maturation, etc.

It is usually desirable to transfer viruses to herbaceous rather than woody hosts because the transfer is a prelude to purification of the infecting virus. Chemical, physical, and immunochemical studies are facilitated when the virus is capable of being mass cultured in an easily and rapidly growing host rather than in less suitable hosts, such as trees and vines. Furthermore, mechanical inoculation provides a more suitable method of assaying the infectivity of the virus. Hence, a means for studying the biology of the virus becomes clearly feasible.

A number of viruses, such as barley yellow dwarf, potato leaf roll, phony peach, peach mosaic, and citrus tristeza (which may later turn out to be other disease agents; see Chapter 6) have failed to infect host plants by mechanical

*Mechanical inoculation is defined as the process by which infectious virus is manually introduced into a living organism by mechanical trauma to facilitate initiation of infection in potentially infectible cells.

inoculation. In such instances, experimental infection is elicited by other means of inoculation: use of dodder, pollen, graft, or insect and other biological vectors. In these cases, experimental inoculation of the recipient host is usually mediated through biological (active) rather than mechanical processes, although transmission* of the virus is purely mechanical.

This chapter develops some principles involved in mechanical inoculation and provides selected examples of the subject. Various aspects of mechanical inoculation have recently been reviewed (Bawden, 1964*a*; Fulton, 1964, 1966; Siegel and Zaitlin, 1964; Yarwood, 1964; Yarwood and Fulton, 1967).

II. PRINCIPLES OF MECHANICAL INOCULATION

Essentially the basic premise of mechanical inoculation is the manual transfer and deposition of biologically active virus to receptors within living cells (the infectible site) by sublethal mechanical trauma. Mechanical inoculation differs from biological inoculation in that the inoculating process is performed manually rather than by some active agent such as the penetrating haustoria of dodders, the probing of the stylets of aphids, or the union of tissues in grafts.

In most instances of mechanical inoculation, wounding or abrasion of cells during inoculation is prerequisite to success in establishing infection (Siegel and Zaitlin, 1964). The sequence of events leading to this success is as follows: (1) effective extraction of infectious virus from the infected donor tissue; (2) administration of the extracted virus to infectible sites by mechanical means; and (3) migration of the administered virus to the receptor (or receptors). Mechanical inoculation does not guarantee successful infection. Virus placed in infectible sites may still remain inactive or may be inactivated. Activation of inoculated virus leading to the formation of progeny virus depends on whether the parental viral genome was first exposed, translated, and fully replicated. If these and other virus-specific functions (e.g., the maturation process) have taken place, infection may still be only partially successful. The progeny particles may remain confined to initial recipient cells unless the mechanism and avenues for its spread to adjoining cells are present and operating.

The failure of the virus to become invasive may limit initiation of general infection so that these micro-regions of infection escape detection. If infection is invasive and the virus spreads circumferentially to other cells, it is then possible to see infection as localized lesions or as other expressions of the disease. Necrotic localized lesions will appear in hypersensitive hosts, that is, in hosts containing cells which become "hyperactive" and are killed shortly after infection by virus. Other hosts respond by producing chlorotic lesions,

---

*Transmission is here defined as the mechanical process by which virus is manually transferred from the donor either as extracts or as carriers of virus in various plant tissues and administered to the site to be inoculated in order to elicit infection of the recipient.

indicating that some of the machinery responsible for the maintenance of chlorophyll is malfunctioning because of infection by virus. Also, the plant may eventually become systemically infected, producing a variety of symptoms.

A. The Virus Inoculum

To achieve success in transmitting virus to host plants, the properties of the virus inoculum must be considered first. Obviously, without proper inoculum, transmission of virus by mechanical inoculation will not succeed. Suitability of inocula may depend on parameters such as: (1) the source of the inoculum; (2) the method of preparing the inoculum; (3) the constituents in the inoculum; (4) the concentration of virus in the inoculum; and (5) the stability of the extracted virus.

1. SOURCE OF INOCULUM

Inoculum may be obtained from various parts of the infected plant, but leaves displaying symptoms are usually preferable and should be tested first. Woody tissues are usually poorer sources of infectious virus than are leaves. Where the choice is limited to woody tissues, as in dormant trees, the inner-bark regions and cambium have been used successfully (Hilborn and Bonde, 1956; Yarwood, 1957*a*; Berg, 1962). With leaves, care must be taken to select areas that contain the greatest amount of virus. Certain areas of the infected leaves may contain little or no virus. Leaves of Chinese cabbage (*Brassica pekinensis*, Rupr. cv. Wong Bok) infected by turnip yellow mosaic virus display dark-green patches as well as the usual mosaic pattern. Such dark-green areas contain little if any virus (Reid and Matthews, 1966). The tendency of certain plant viruses to be localized in definite regions of infected leaves has been demonstrated also in tobacco leaves infected with potato virus Y. The best source of inoculum for aphid transmission of the virus was the interveinal epidermis at tips of leaves situated at the middle of the tobacco plant (Bradley, 1962*a*). In addition, leaves of infected plants in the field may have very mild or no obvious symptoms, in which case selection of a good inoculum source is obviously fortuitous. For example, a strain of tomato ringspot virus (=peach yellow bud mosaic virus) occurs naturally in the sea daisy (*Erigeron glaucus* L), a plant indigenous to the shoreline of Northern California that bears no outward symptoms or any obvious suggestion when infected by the virus (Frazier et al., 1961). Thus, a large number of samples had to be collected and tested before the virus was recovered. There are cases where it was thought that a "new virus" was successfully transmitted whereas the transmission was of a latent virus in the plant. For instance, various *Chenopodium* species (e.g. *C. amaranticolor* Coste and Reyn., *C. quinoa* Willd., *C. murale* L.) commonly used as test hosts possess the semilatent virus sowbane mosaic virus (Dias and Waterworth, 1967; Kado, 1967), with symptoms developing after symptomless leaves are abraded. Hence, special precaution should be taken that latent viruses are not mistakenly assumed to be of a transmitted uncharacterized virus. For example, the so-called apple latent virus was eventually discovered to be a strain of sowbane mosaic virus associated with the *Chenopodium* test host (Bancroft and Tolin, 1967; Kado and van Regenmortel, unpublished).

As stated above, leaves are the usual source of inoculum but in some cases the virus may be restricted to certain other plant parts. For example, tobacco necrosis virus was detected only in the roots of different plant species (Smith, 1937). Very few plants have been observed to be infected systemically by tobacco necrosis virus in the aerial portions; among them are tulip (Kassanis, 1949); primrose (Bawden and Kassanis, 1947*a*); *Cleome spinosa* (Yarwood, 1960 *b*); soybean (Yarwood, 1960 *b*; Resconich, 1963); and beans (Natti, 1959; Roberts and Price, 1967). Another example of viruses that are restricted to certain plant parts are those that are restricted principally or entirely to the phloem. Two viruses transmitted persistently by aphids, barley yellow dwarf virus and potato leaf roll virus, have not been transmitted by mechanical inoculation but have been partially purified (Rochow and Brakke, 1964; Peters, 1967). This also applies to the leafhopper-transmitted wound tumor virus (Brakke, et al., 1954). The reasons why these viruses cannot be effectively transmitted mechanically are unknown, but since they can be purified and thus concentrated, the limitations are not one of inoculum dose. More likely, one of the limitations is inability to implant the virus in the requisite site for infection. Such is the case with another phloem-localized virus, sugar beet curly top virus, which also has been partially purified (Gold and Duffus, personal communication) but can be transmitted by mechanical inoculation to a very limited extent by repeated stab inoculations with a needle into the crown of sugar beet (Bennett, 1934) and tobacco plants (Fulton, 1955). Potato yellow dwarf virus and wound tumor virus also behave like sugar beet curly top virus by being difficult to transmit (Brakke et al., 1954).

In addition to these examples showing the importance of selecting the proper inoculum, the physiological status of the virus source plant can influence the infectivity of virus because of the amount of substances that are potentially inhibitory to infection. With fruit-tree viruses, the most consistent source of inoculum is very young leaves or leaves that have just emerged (Fulton, 1966; Nyland, personal communication). Young infected tissues apparently possess more infectious virus, and perhaps lower concentrations of inhibitory substances, than mature leaves. Tomato bushy stunt virus (Kassanis, 1952*b*) and turnip yellow mosaic virus (Matthews and Lyttleton, 1959) tend to decrease in specific infectivity with time in vivo. Similar findings have been made with alfalfa mosaic virus (Kuhn and Brancroft, 1961), broad bean mottle virus (Kodama and Bancroft, 1964), cowpea chlorotic mottle virus (Kuhn, 1965*a*), lucerne mosaic virus (Havranek, 1967), and tobacco ringspot virus (Schneider and Diener, 1968). Reasons have been hypothesized for the decrease of infectivity in vivo, but little is known about the mechanism of irreversible decrease of specific infectivity of viruses within the infected plant. With tobacco ringspot virus, at least, a shift of infected host plants to cooler temperature has restored symptoms and increased overall infectivity.

2. ON PREPARING INOCULUM

Infected tissue may contain numerous infectious virus particles, but adequate recovery of infectivity will depend on various conditions and techniques that can be employed. The phase of preparing the virus inoculum is therefore the second most important factor that can influence the effectiveness of the inoculum. Inoculum is prepared by extracting virus from diseased tissue, during which or shortly thereafter the virus becomes exposed to an environment which can reduce infectivity. Various substances such as host metabolities, cellular debris, and virus-modified host aromatic compounds (Gill, 1965) are released simultaneously with the virus. Some of these cellular components, when released from the cell, possess inhibitory properties. In addition, infected tissues may be homogenized in the presence of an extracting medium without releasing much if any virus into solution. In such cases, a suitable extracting solvent must be found for successful extraction. In other instances, concentrations of the virus may be so low that even with an adequate extracting procedure, the amount of virus in the extract is lower than is required for successful mechanical inoculation. A case in point is mechanical inoculation with hop mosaic virus (Legg, 1965). Although inoculation with crude extracts was eventually accomplished, infection was first achieved with inocula produced by concentrating the virus from crude extracts by high-speed centrifugation. In other instances, a virus may be highly labile when exposed to in vitro extracts. Ease of infecting with and purifying a virus is correlated fairly well with degree of virus stability. Such correlations should not be attributed to virus stability alone, however, since relatively stable viruses such as sugar beet curly top virus (Severin and Freitag, 1933; Bennett, 1935) and wound tumor virus (Brakke et al., 1954) are difficult to purify and are difficult to infect hosts by mechanical inoculation. Furthermore, such correlations are not valid for phloem-specific viruses such as potato leaf roll virus (Peters, 1967), barley yellow dwarf virus (Rochow and Brakke, 1964), maize rough dwarf virus (Lovisolo et al., 1967), and rice tungro virus (Galvez, 1967), which have not been infectious by mechanical inoculation and yet have been purified.

The effectiveness of the inoculum is usually increased by freeing a virus of contaminating host components. Virus can easily be freed of most host components by simple purification methods (see Chapter 11). Even a partially purified inoculum has a distinct advantage over crude extracts in being placed in an environment that is less likely to be deleterious. Without proper conditions, however, in some exceptional cases purification can decrease the infectivity of a given virus. For example, potato yellow dwarf virus and barley stripe mosaic virus tend to lose infectivity with purification (Brakke, 1956, 1962). The stability and aggregation problems encountered by these viruses can be alleviated by introducing various stabilizing agents such as magnesium-glycinate buffers for potato yellow dwarf virus (Brakke, 1956) and Tris buffers at neutral pH conditions for barley stripe mosaic virus (Brakke, 1962). Purification is known to increase the stability of a number of viruses such as sour cherry necrotic ringspot virus (Hampton and Fulton, 1961), prune dwarf virus (Fulton, 1959), and tobacco streak virus (Fulton, 1967*a*). Possible reasons for increased infectivity are the elimination of inhibitors and increase in virus concentration. Most inhibitors are deterrents of infection rather than of infectivity (Bawden, 1954, 1964*a*; Zaitlin and Siegel, 1963). In instances where inhibitors persist even after partial purification (perhaps by absorbing to the virus particles), more drastic purification steps might be undertaken. For example, the infected material can be directly homogenized and extracted in the presence of phenol. Phenol will denature most host proteins and will solubilize others, including ribonucleases (Kickhofen and Burger, 1962; Rushizky et al., 1963), while the nucleic acids of the host and of the virus will remain in solution in the aqueous phase. The aqueous phase can then be freed of traces of phenol, polysaccharides and other water-soluble low-molecular-weight cellular components. The aqueous phase can then be used directly as the inoculum. Such techniques, or modifications thereof, have been employed (Schlegel, 1960*a*) and have been successful in recovering highly infectious preparations of cucumber mosaic virus (Schlegel, 1960*b*). Extraction of infected leaf material with phenol has been used with other viruses such as the poorly multiplying strain of tobacco rattle virus (Sanger and Brandenburg 1961; Cadman, 1962), the unstable variant of tobacco necrosis virus (Babos and Kassanis, 1962), and potato spindle tuber virus (Diener and Raymer, 1967; Raymer and Diener, 1969; Diener, 1971). Obviously, if purification methods are unsuccessful, the alternative is to introduce various chemical additives to the inoculum in order to sustain infectivity (Yarwood, 1957*b*, 1964, 1968, 1969*a, b*; Yarwood and Fulton, 1967).

3. INOCULUM CONSTITUENTS

As stated above, various host components are released along with the virus during extraction. Most of the components usually have no effect on viral infectivity and may be removed by various methods (see Section III). Unfortunately, extracts from a number of well known test plants possess inhibitory substances (Bawden, 1954; Hollings, 1966). Some of the inhibitors are reported to be tannins and oxidized polyphenols which appear during and after extraction of the virus from the infected tissues. The degree of inhibition by reduced polyphenols is less than that of the oxidized compounds, and the amount and degree of oxidation of these compounds depends on the source of the inoculum. Phenolases (known also as polyphenol oxidases) usually require a metallic cofactor such as cupric and ferric ions (Keilin and Mann, 1938; Mallett et al., 1948) and the presence of these metals and polyphenols greatly decreases infectivity (Pierpoint and Harrison, 1963). The highly pigmented material which develops shortly after the virus extract is exposed to the air is thought to be due to the activation of phenolases on catalyzing the oxidation of various phenolic precursors. Thus, one of the reasons why extracts become colored immediately or upon standing is due to the formation of colored oxidized phenolic compounds. Removal of active chloroplasts often enhances the rate of pigment formation through the removal of particle-bound reductases. Extracts also become pigmented from an excess of copper or iron introduced unintentionally as a contaminant from the apparatus used in extracting the virus (as when meat grinders or metal spatulas are used). Therefore, both reducing and chelating agents are used to reduce the oxidation of polyphenols.

4. VIRUS CONCENTRATION IN THE INOCULUM

As mentioned previously in Section 2, the concentration of virus in crude extracts will influence the outcome of mechanical transmission tests. Also, tissue extracts may contain enough virus but infection by mechanical inoculation is hampered by interfering substances. Tissue extracts may contain very little virus yet infect successfully if the virus can be concentrated enough by high-speed centrifugation. When virus cannot be concentrated from infected tissue, it can possibly be extracted and semipurified from the insect vector instead (Peters, 1967). This method may be applicable also to viruses transmitted by nematodes (Sanger et al., 1962) and *Olpidium.*

5. VIRUS STABILITY

Infectivity of highly "unstable" viruses is difficult to demonstrate by mechanical inoculation since infectivity decreases rapidly during and after extraction or is lost in an unknown manner. Furthermore, to complicate the situation the amount of infectious virus in the tissue may be low. Thus, probability of effective infection is low and such viruses are usually infectious by other methods, such as graft unions or biological inoculation. On the other hand, the so-called "unstable" virus may not be a virus at all but another agent such as mycoplasma or naked nucleic acid, and some special trick may be necessary to transmit the unknown agent.

B. Transmission and Deposition of Virus

The method of mechanical inoculation, whether with a finger or some implement, is of minimal concern as long as infection can be established and the method of application is convenient, efficient, and reproducible. Mechanical inoculation is usually carried out by rubbing virus solutions onto leaf surfaces with sterile gauze pads, brushes, glass spatulas, or other implements (see Yarwood, 1964; Fulton, 1964; Yarwood and Fulton, 1967). The transfer of virus into cells and to "receptors" is fortuitous since nonlethal inflicted micro-injuries of cells occur at random. Such injuries are usually a prerequisite to the earliest phases of infection. On the surface of leaves the uneven epicuticle, consisting of waxes and cutinized and pectinized layers (Eglington and Hamilton, 1967), must be penetrated. The quantities of wax on the surface of leaves are sufficiently abundant (about 0.01-0.20 mg/cm^2 of leaf surface of legumes [Baker and Martin, 1967]) to prevent any natural penetration of the cuticle. Virus may be able to penetrate into the epidermal cells through the ectodesmata, which are just beneath the cutical and become exposed when the cuticular waxes are scraped off. Ectodesmata are most likely exposed to the virus during mechanical inoculation since the process involves scarification of the leaf surface, usually with the aid of an abrasive such as carborundum or celite. The role of ectodesmata in mechanical inoculation, however, is still uncertain. It has been shown that tobacco mosaic virus particles accumulate at the ectodesmata (Brants, 1965), but they also accumulate at the base of broken trichomes (Herridge and Schlegel, 1962). More recently, Thomas and Fulton (1968) concluded that ectodesmata function as infection sites, and that their number or condition may determine the degree of susceptibility of tobacco to tobacco mosaic virus. Similar conclusions were reached earlier (Zech and Vogt-Kohne, 1956). Although there are few exceptions where infection has been initiated in plants without direct injury (e.g. Schneider, 1965; Roberts and Price, 1967), successful mechanical inoculation requires rubbing.

Since abrasion of leaf surfaces by rubbing is probably not uniform, and since the abrasive and wounding action may be too severe on some cells and insufficient on others, the percentage of total number of cells infected is small and the percentage of infected cells varies from one leaf to another. In addition, the susceptibility of a leaf varies from one area to another although all may be infected. The model of one particle infecting one cell, as with coliphages, is probably not attainable with present plant-virus inoculations.

Contact of the protoplast with the virus does not seem to be enough to initiate infection (Mundry, 1963). However, single cells prepared from tomato or tobacco callus tissue can be infected with tobacco mosaic virus or its RNA following mechanical dispersion of the cells from the callus tissue (Motoyoshi and Oshima, 1968; Murakishi, 1968; Murakishi et al., 1970; Murakishi et al., 1971). Protoplasts of *Escherichia coli* are used routinely for infectivity assays of MS2-RNA (Strauss, 1964; Strauss and Sinsheimer, 1967). Protoplasts of tobacco mesophyll tissue likewise can be infected by tobacco mosaic virus RNA (Aoki and Takebe, 1969). Therefore it is becoming apparent that protoplasts of plant cells will be used in the assay of various plant viruses (e.g. Takebe et al., 1968). Tobacco mosaic virus RNA has been reported to infect the eucaryotic alga *Chlorella* (Süss et al., 1965). If this is substantiated, *Chlorella* will be ideal for studying the biology of a plant virus since much is known about the alga which can be grown synchronously.

C. The Establishment of Infection

Although various aspects of the initial phases of infection have been reviewed and postulated (Mundry, 1963; Bawden, 1964*a*; Siegel and Zaitlin, 1964; Takahashi, 1964; Levintow, 1965; Wittmann and Sanger, 1965), it remains to be seen whether plant viruses behave similarly to their animal and bacterial counterparts or whether they establish infection somewhat differently. It is highly possible that the overall biosynthetic processes leading toward the synthesis of viruses in plant cells are fundamentally the same as in other eucaryotic cells. That is, the viral messenger RNA is converted to the replicative form (RF), as with tobacco mosaic virus RNA (Shipp and Haselkorn, 1964; Burdon et al., 1964; Weissmann et al., 1966), which serves as the template for progeny RNA. The replication of progeny RNA is probably catalyzed by one or more RNA polymerases by the semiconservative mechanism (Fenwick et al., 1967), and a portion of these RNA strands is probably translated during self-replication into virus-specific proteins. Ultimately, assembly of the progeny virus nucleoprotein particles takes place during and at the end of the infection cycle within a given cell.

It should be reemphasized that the mere introduction of infectious virus into living cells does not fully ensure that the virus will infect. By mechanisms still unknown, the polycistronic viral genome must first be exposed after shedding its protective protein coat, so that its genes* can be translated and the nucleic acid replicated thereafter. The mechanism by which the virus is stripped of its protein jacket is still unknown. In tobacco mosaic virus, hydrogen, hydrophobic, and salt linkages cumulatively provide strong bonds between the protein subunits and the RNA (Caspar, 1963), and substantial activation energies are apparently required to strip off the coat protein in vitro. The rate and the mechanism by which this is achieved in vivo is still a mystery. It has been observed, however, that some of the coat protein of tobacco mosaic virus is released in tobacco leaves within a few minutes of inoculation (Shaw, 1967, 1969; Machida et al., 1969). It is speculated that, after or even during the process of shedding the coat protein, genes coding for RNA-dependent, RNA polymerase or polymerases are translated and that one or more of these virus-specific enzymes catalyze the synthesis of progeny RNA via the replicative intermediate form of the RNA (Nillsson-Tillgren, 1969). These biosynthetic steps have not yet been established for any plant virus, but the mechanisms of synthesis are probably similar to those of the RNA of bacteriophages since a "replicative intermediate" RNA was isolated from tobacco infected with alfalfa mosaic virus (Pinck et al., 1968). With the synthesis of progeny RNA strands, various virus-specific proteins such as the coat protein, maturation proteins and other enzymes are translated from these strands. The double-stranded RNA, which serves as the template for synthesis of copies of progeny RNA, contains one strand which is complementary in bases to the other. The strand that is complementary (the minus strand) is uninfectious, as in R17 (Iglewski and Franklin, 1967) and Qβ (Feix et al., 1968). It seems that the plus strand contains the gene for at least the coat protein since nitrous-acid-induced

*The term "gene" is here assigned to polynucleotide segments of viral RNA which, at the translational level, code for a single polypeptide chain (see Kado and Knight, 1968).

mutants that were derived from tobacco mosaic virus possessed amino acid replacements in their coat protein which are predictable only if the parental genome is altered by nitrous acid deamination (Wittmann and Wittmann-Liebold, 1966). Furthermore, the RNA from alfalfa mosaic virus (van Ravenswaay-Claasen et al., 1967) and from the satellite virus of tobacco necrosis virus (Clark et al., 1965; Reichmann et al., 1966) both apparently specify the formation of their respective virus-coat protein in the cell-free *Escherichia coli* system of Nirenberg and Matthaei (1961). Lastly, the progeny strands, together with the coat protein subunits, become oriented in a thermodynamically stable form, maturing into virus particles. Whether the virus is then spread from cell to cell as intact units or as RNA is not clear (Schneider, 1965). Virus particles have been seen, however, in plasmodesmata (Esau et al., 1967; Davison, 1969; deZoeten and Gaard, 1969; Kitajima and Lauritis, 1969). In localized infections (e.g. local lesions) virus movement may be retarded or stopped by the formation and deposition of callus in the surrounding uninfected tissues (Wu and Dimitman, 1970).

III. METHODS OF MECHANICAL AND BIOLOGICAL INOCULATION

The techniques used in mechanical inoculation differ from those in biological inoculation. Mechanical methods require manual manipulation for preparing, transmitting, and depositing inoculum to susceptible sites on the host plant, whereas biological inoculation requires *no preparation* of inoculum and requires only the initial transport of a virus-containing or transmitting organism (insect, dodder, etc.) or organelle (piece of tissue, pollen, etc.) to the infectible site. For example, transmission is elicited biologically when dodder stems are allowed to grow from infected plants to healthy plants, or when a piece of infected tissue is implanted into healthy tissue such as during the construction of a graft, or when infected pollen is transferred to healthy plants. Inoculation occurs when the virus is thereafter deposited on "receptors," purely by biologically active means such as during the penetration of the haustoria of dodder into healthy tissues, during the formation of the graft union, or during the probing of viruliferous aphids.

A. Methods of Preparing Inoculum

Virus inoculum is prepared from various plant tissues by homogenizing the infected source in the presence of a suitable extracting solvent. The homogenization procedure simply ruptures cells and allows virus from the infected cells to be extracted. A common procedure is to grind a piece of freshly picked leaf tissue (about 0.2 g) in a chilled mortar in the presence of 0.05 M tris-hydroxymethylaminomethane buffer (Tris), pH 7.8. Yarwood (1952) routinely prepared inoculum by vigorously grinding a piece of donor leaf tissue (about 1 cm^2) together with 3 drops of tap water in a mortar. The piece of tissue is thoroughly triturated in the mortar and then appropriately diluted with water or phosphate buffer. Others have modified the extracting procedure to enhance (or prevent the loss of) the infectivity of the constituent virus (for principles of inoculum preparation see Section II). In cases where it is believed that the virus may be susceptible to nucleases and/or without a protective protein coat, more refined techniques have been employed. For example, better infectivity than with other methods was obtained by extracting infected material directly with phenol for such viruses as: cucumber mosaic virus (Schlegel, 1960*b*; Diener et al., 1964; Francki, 1964), defective strains of tobacco mosaic virus (Siegel et al., 1962), the NM isolate of tobacco rattle virus (Cadman, 1962), and the unstable variants of tobacco necrosis virus (Babos and Kassanis, 1962; Kassanis and Welkie, 1963). Since most plant ribonucleases, if not all, that have been characterized to date possess a pH optimum between pH 5 and 6 (Kado, 1968), pH values above this range usually minimize nuclease activity. Plant phenolase activity also is reduced at relatively high pH values. Thus, relatively high pH values (e.g. pH 8-9) have been used successfully for the acquisition of infectivity of various unstable viruses: some *Prunus* viruses (Fulton, 1957), raspberry ringspot virus (Cadman, 1959), the PM4 strain of tobacco mosaic virus (Zaitlin and Keswani, 1964); variants of tobacco necrosis virus (Kassanis and Welkie, 1963), grapevine fanleaf virus (Cadman et al., 1960), lettus necrotic yellows virus (Crowley, 1967), and others.

Table 1-1 summarizes some examples of deterrents of infectivity and infection. The reasons for the lack of successful transmission by mechanical inoculation are many and quite diverse. Those that influence infectivity affect the virus in some manner (such as the inability of the viral genome to shed its capsid protein[s] or the destruction of some of its phosphodiester linkages) that prevents faithful reading of the genetic messages.

B. Methods of Inoculation

There are many methods and minor variations for inoculating plants mechanically, and no attempt is made to describe them here since literally hundreds can be or have been utilized. Some of the more elaborate techniques have been described, such as the use of microsyringes (Harpaz, 1959), ultrasonic devices (Lamborn, 1964; Lamborn et al., 1969), a ground-glass roller attached to a turntable (Selman and Milne, 1961), and spray devices (Lindner and Kirkpatrick, 1959; Toler and Hebert, 1964), but most of them have not been commonly used because conventional methods are more expedient and convenient. One of the most consistently used techniques is rubbing, using various instruments. Brushes seem to be better inoculating instruments than gauze pads, fingers, or glass spatulas. Comparative studies between glass spatulas and brushes, for instance, showed that brushes consistently induced more infection (Table 1-2). When less than 0.2 ml of inoculum is used, however, glass spatulas are preferred since the bristles of the brushes tend to retain most of the inoculum. A portion of a suspension of virus is placed on the surface of

TABLE 1-1
FACTORS CONTRIBUTING TO THE SUCCESS AND FAILURE OF TRANSMISSION OF PLANT VIRUSES BY MECHANICAL INOCULATION

Virus	*Most Probable Contributing Deterrent in Crude Extracts*	*Alteration of the Inoculum*	*References*
Grapevine fanleaf	Low pH and tannins	Buffers at pH 8.0; 2.5% nicotine	Cadman et al. (1960)
Tobacco stunt	Unknown	Reducing and chelating agents (e.g., .001 M phenyl-thiosemicarbazide)	Hiruki (1964)
Prunus necrotic ringspot	Cucumber-ribonuclease	0.5% caffeine	Diener and Weaver (1959*a*)
Raspberry ringspot	Oxidized poly-phenols	Buffers at pH 8.0	Cadman (1959)
Apple mosaic	Oxidized phenolic compounds	Direct tissue inoculation and apple petals	Yarwood (1955); DeSequeira (1967)
Cocoa swollen-shoot	Tannins	Caffeine and nicotine sulfate, proteins	Brunt and Kenten (1960); Wessel-Riemens (1965)
Hop mosaic	Low virus concentration	Concentration and partial purification	Legg (1965)
Citrus infectious variegation	Unknown inhibitor	20% sucrose and activated charcoal	Grant and Corbett (1960)
Barley yellow dwarf	Low concentration and tissue specificity	Purification of the virus from aphid vector	Rochow and Brakke (1964)
Potato leafroll	Low virus concentration and tissue specificity	Purification of the virus from the aphid vector	Peters (1967)
Cucumber mosaic	Aggregation and nonspecific adsorption	Phenol extraction	Schlegel (1960*a*)
	Also oxidized poly-phenols	.01 M diethyldithio-carbamate	Harrison and Pierpoint (1963)
Lettuce necrotic yellows	Unknown (not phenol oxidases)	0.1 M glycine buffer pH 8.6	Crowley (1967)
Tomato spotted wilt	Oxidation	Cysteine-HCl, sodium sulfite	Bald and Samuel (1934); Best (1939)

TABLE 1-1 (Continued)

Virus	*Most Probable Contributing Deterrent in Crude Extracts*	*Alteration of the Inoculum*	*References*
Rose mosaic	Oxidation and other unknown factors	Cysteine-HCl, sodium sulfite	Fulton (1952)
Tobacco streak	Oxidized poly-phenols	.02 M mercaptoethanol and partial purification	Fulton (1967*a*)
Potato yellow dwarf	Particle instability	0.01 M cysteine-HCl 0.1 M glycine + 0.01 M $MgCl_2$	Hougas (1951); Brakke (1956)
Prunus viruses and prune dwarf	Unknown	.01 M sodium phosphate pH 8 and partial purification	Fulton (1957, 1959)
Wheat striate	Aggregation	.01 M $MgCl_2$ + 0.1 M glycine	Lee (1963)
Tobacco rattle	Unknown	Liquid nitrogen	Sänger and Gold (1962)

TABLE 1-2

EFFECTIVENESS OF POSTER BRUSHES AND GLASS SPATULAS IN THE INOCULATION OF TMV ON TOBACCO LEAF DISCS

TMV (*μg/ml*)	*LESIONS PER DISC* [a]		*Relative Efficiency of Brushes*
	Glass Spatulas	*Poster Brush*	
0.001	1.8± 0.7 [b]	8.2± 2.5	4.5 [c]
0.01	11.3± 5.5	32.0± 7.7	2.8
0.10	100.7±18.1	168.0±22.1	1.7
1.00	366.0±66.1	610.0±42.0	1.7
0.001	3.3± 0.8	5.8± 0.8	1.8
0.01	27.0± 2.6	44.8± 4.5	1.7
0.10	131.2± 9.7	176.8±13.5	1.3

[a] Lesions on 50-cm^2 leaf discs of Xanthi-nc tobacco. The first set of dilutions was assayed by operator no. 1 and the second by operator no. 2.

[b] Figures represent average lesion numbers of 6-12 leaf discs per sample and the standard error of the mean.

[c] Figures represent the quotients obtained from the lesion numbers for a given concentration of TMV.

optimum was pH 8.5 when 0.1 M phosphate was used with tobacco mosaic virus on Scotia bean leaves (Thornberry, 1935*a*; Beraha et al., 1955). Experiments were performed to separate the effect of phosphate from that of pH. The pH for optimum lesion numbers depended on the phosphate concentration but lesion numbers increased greatly as the concentration of the potassium ortho-phosphate increased (Table 1-3). When the phosphate concentration was varied in a standard concentration of 0.05 M Tris buffer (tris-hydroxymethylaminomethane-PO_4) set at various pH values with phosphoric acid, the influence of pH on increasing infection was, again, less striking than that of phosphate concentration (Table 1-4). Thus, it appears from these data that the "phosphate effect" is brought about by the cooperative influence of both pH and phosphate concentration. Buffers besides phosphate (e.g. acetate, borate, glycine, Tris-HCl, etc.) also enhance infection, though the degree of the

TABLE 1-3
EFFECT OF CONCENTRATION AND pH OF PHOSPHATE ON INFECTION

Phosphate	*pH*					
(*mM*)	*7.0* [a]	*7.5*	*8.0*	*8.5*	*9.0*	*Average*
0.1	13.0 [b]	4.8	6.3	2.9	14.8	8.4
1.0	9.7	14.9	16.2	18.2	7.8	12.2
10.0	66.9	145.6	211.8	249.0	195.0	173.7
100.0	520.6	577.6	510.0	603.3	352.4	512.8
Average	152.6	185.7	184.6	218.4	142.5	

[a] Final pH of the inoculum.
[b] Tobacco mosaic virus lesions per leaf (pinto bean).

TABLE 1-4
EFFECT OF CONCENTRATION AND pH OF TRIS-PHOSPHATE ON INFECTION

Phosphate	*pH*					
(*mM*)	*7.0* [a]	*7.5*	*8.0*	*8.5*	*9.0*	*Average*
0.1	128.6 [b]	152.2	165.8	143.0	154.5	148.8
1.0	117.2	203.4	185.9	144.3	168.6	163.9
10.0	205.0	296.4	432.3	371.1	384.3	337.8
100.0	208.1	459.6	472.5	473.6	474.8	417.7
Average	164.7	277.9	314.1	283.0	395.6	

[a] Final pH of inoculum containing a buffer mixture of tris-hydroxymethylaminomethane-PO_4.
[b] Tobacco mosaic virus lesions per leaf (pinto bean).

increase in infection is less than that elicited by phosphate buffers. The reasons for the "phosphate effect" on the susceptibility of the host are still unclear. Nevertheless, the resulting increase in potential of the inoculum is sufficient reason for routine use of the buffer. It should be pointed out, however, that certain viruses seem to favor other buffers. For example, phosphate buffer is unsuitable for the preparation of lettuce mosaic virus (Tomlinson, 1963; 1964); the infectivity of this virus was destroyed when phosphate buffer at pH 7.5 was used in making the preparations. Conversely, infectivity was preserved and enhanced when borate buffer at the same pH was used instead. Also, lettuce necrotic yellows virus causes fewer lesions with phosphate than with glycine buffer at pH 8.6 (Crowley, 1967). Buffers used beyond the effective limits of their pKa are limiting in their effectiveness in buffering the pH of the inoculum. Thus phosphate buffer was less effective than sodium acetate buffer for extracting pea enation mosaic virus since the extraction of this virus was best achieved at pH 6.0 (Izadpanah and Shepherd, 1966).

In summary, buffers with ionic strengths and pH values that are physiologically most suited (i.e. nontoxic, isotonic, etc., to the cell) are preferable when they are to be applied on the leaves of test plants. Since buffers containing ions such as borate are toxic when left on the inoculated surfaces of test plants, such buffers are usually rinsed off immediately after inoculation. The major influences of buffers are both on the virus (by preserving infectivity) and on the test plant (by enhancing susceptibility), and the influence is due partly to the action of pH and due to the ionic strength of the buffer.

2. *ABRASIVES*

Wounding the cells of the test plant is essential for establishing infection by the mechanical method (Siegel and Zaitlin, 1964). Wounds can be elicited by simply rubbing the surface of leaves, though more wounds are made by rubbing with a finely divided abrasive, such as powdered diatoms (known under the trade name of Celite and Hyflo-Super-Cel), silicon carbide (known under the trade name of Carborundum), and aluminum oxide (common name is alumina). Recently, the abrasive action of magnesium silicates has been reported to be equally effective (Joshi and Thornberry, 1968; Yarwood and E. Hecht-Poinar, 1970). Abrasives seem to be an effective means of increasing the probability of infection by increasing the number of infectible sites. The increase in infection ranges from several hundredfold to over a thousandfold, depending on the conditions and method of inoculation. The size of Carborundum that will evoke the least amount of serious damage has been studied. On *Nicotiana glutinosa* L., the most suitable size was 400-mesh Carborundum (Kalmus and Kassanis, 1945), whereas on Scotia bean leaves 800-mesh was better than 400-mesh for obtaining the greatest infection (Beraha et al., 1955). Infection can be increased further by employing abrasives together with phosphate buffers (Yarwood, 1952; Beraha et al., 1955) and inoculating with poster brushes (Takahashi, 1956).

3. pH

Although the effect of pH appears synonymous with the effect of buffers, the effect of pH and ionic strength are two independent, though not mutually exclusive, effects conferred by the buffer. Since the effect of pH, especially in crude virus extracts, can severely hinder infection by mechanical inoculation, some of the salient features of the effect will be emphasized.

Most plant viruses possess isoelectric points in the acidic region, and inoculum with pH values near such isoelectric points can thus cause the virus particles to precipitate out of solution or form aggregates. The pH of crude extracts must generally therefore be kept slightly above pH 7.0 in order to prevent losses mediated by aggregation, degradation, etc., of most plant viruses. For example, crude extracts from young grape leaves (which contain little or no tannin) have pH values about 3.5, so such extracts were ineffectual as inoculum of grapevine fanleaf virus (Cadman et al., 1960). The virus was transmitted, however, when the extracts were prepared with phosphate buffer at pH 8.0.

The susceptibility of the test host is also affected by the pH of the inoculum. There are differences also in the effect of inoculum pH on the susceptibility of host plants within the same plant family. For example, the primary leaves of cowpea (*Vigna unguiculata* cv. Blackeye) produce more lesions at pH 7.0 than at pH 8.0; and conversely, the primary leaves of pinto bean (*Phaseolus vlugaris* L. cv. Pinto) produce more lesions at pH 8.0 than at pH 7.0 (Table 1-5). The

TABLE 1-5
EFFECT OF pH ON VIRUS INFECTION

Buffer			*Relative Infectivity (pH)* [b]				
(0.05 M)	*Virus* [a]	*Test Host*	*6*	*7*	*8*	*9*	*10*
Tris	TMV	Pinto bean	7.2	15.5	18.2	11.9	4.1
	TNV	Cowpea	4.0	4.4	1.9	0.8	0.2
	CMV	Cowpea	3.1	6.2	1.7	1.4	—
Borate	TMV	Pinto bean	0.6	0.8	7.9	6.0	1.4
	CMV	Cowpea	0.6	1.7	1.3	0.7	0.5
Phosphate	TNV	Pinto bean	4.1	8.9	29.8	6.6	1.6
	TNV	Cowpea	4.7	13.2	7.6	5.4	0.1
	TNV	Cucumber	2.3	3.7	1.2	0.4	0.2
Composite	TNV	Pinto bean	—	3.8	6.4	3.5	1.6
(Acetate-	TNV	Cowpea	3.0	3.6	0.9	0.5	0.2
Phosphate-	PYV	Cucumber	0.5	1.1	0.7	0.1	—
Tris)	TNV	Cucumber	1.3	1.5	0.3	0.2	—

[a] Abbreviations: TMV, tobacco mosaic virus; TNV, tobacco necrosis virus; CMV, cucumber mosaic virus; and PYV, peach yellow bud mosaic virus.

[b] Infectivity represents the ratio of the lesions obtained from test half-leaves to lesions obtained from control half-leaves; 10-14 half-leaves were used per treatment. Control inoculum (ca. pH 7.4, Cf. text) was without buffer. pH measured with microglass electrodes using a Leeds Northrup pH meter.

phenomenon is also found between tobacco (*N. tabacum* L. cv. Xanthi-nc) and pinto bean, where tobacco is more susceptible at pH 7.0 than at pH 8.0. Also, the differential action of pH on the test host does not change with a change in buffer, indicating that the pH effect on the host is an independent effect caused by pH per se (Table 1-5). In all cases, infection was optimum on cowpea at pH 7.0, and on pinto bean at pH 8.0. The reasons for the differences need further study, but it may be conjectured that the influence is on the site of inoculation and also on decreasing the effect of nucleases that can be released by rubbing (Kado, 1964). Most plant ribonucleases examined thus far display an activity maximum between pH 5 and pH 6 (Kado, 1968). Hence, pH values in the alkaline region may be best for inoculation because of a decrease in the activity of such enzymes.

The effect of alkaline pH may have an additional advantage. For instance, Helms and Zaitlin (1970) found that upon incubating tobacco mosaic virus for a short time (24 min) at pH 10 in 0.0025 M borate buffer followed by inoculation in 0.1 M neutral phosphate buffer to *Phaseolus vulgaris* cv. Pinto enhanced viral infectivity. Since treatment of virus with alkali generally strips off capsid protein, they suggested that the failure of virus to uncoat in vivo could contribute to the usually low infectivity of tobacco mosaic virus inocula.

4. RIBONUCLEASES

Plant ribonucleases are present in most if not all crude virus extracts. These enzymes can significantly influence the infectivity of the virus and the susceptibility of the test plant. For instance, Diener and Weaver (1959*b*) showed that the infectivity of prunus necrotic ringspot virus and the susceptibility of cucumber plants were respectively destroyed and inhibited by the action of pancreatic ribonuclease. They proposed that the enzyme acts on the nucleic acid of the virus in vitro and that the relative lability of the virus in extracts from infected cucumber leaves is due to relatively high concentrations of the enzyme. Although a number of inhibitors occur naturally in plants (Bawden, 1954), very few have been characterized completely. The virus inhibitor in cucumber plants reported by Sill and Walker (1952) and partially characterized by Francki (1964) has properties similar to those of cucumber ribonuclease (Kado, 1968). The virus inhibitor (like cucumber ribonuclease) is abundant in the plant, heat stable by being destroyed only partially at 100° C, and sedimented only partially by centrifugation for 1 hr at 144,000 *g* (the sedimentation coefficient of cucumber ribonuclease is 1.5 *S*). Cucumber ribonuclease can appreciably reduce infection (Fig. 1-2), as does pancreatic ribonuclease (Fig. 1-3). The effect of these enzymes is on the susceptibility of the host since the inhibition can be overcome by diluting the inoculum, by using bentonite in the inoculum, and by washing off nuclease-treated leaves prior to inoculation.

Several methods can be used to prevent the action of nucleases on the virus under investigation. One of the most effective methods used Wyoming or Utah

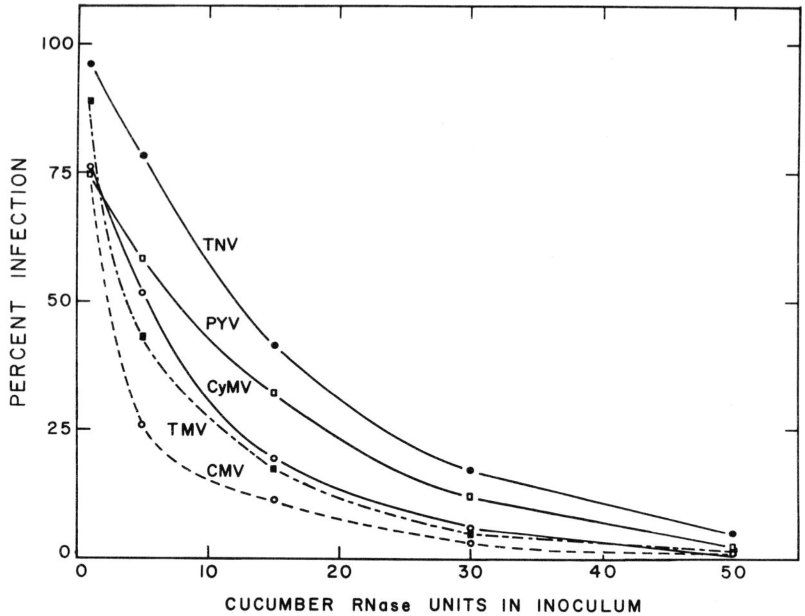

Fig. 1-2. Inhibition of infection on different test plants by cucumber ribonuclease. Plots summarize individual experiments conducted at different times. Cucumber mosaic virus (CMV), cymbidium mosaic virus (CyMV), peach yellow bud mosaic virus (PYV), tobacco mosaic virus (TMV), and tobacco necrosis virus (TNV) were respectively assayed on cowpea, *Chenopodium,* cucumber, and pinto bean.

bentonite, which is sold as the raw product and requires further purification in order to avoid unforeseen complications. This method, designed by Fraenkel-Conrat et al. (1961), has been used in preparing tobacco mosaic virus RNA and various other plant-virus nucleic acids. Using an alkaline buffer (e.g. pH 9.0) and bentonite minimizes ribonuclease activity and improves recovery of infectious RNA as in the case of cucumber virus 4 (Kado and Knight, 1970). Since bentonite is usually in the sodium form, others have purified the clay and added magnesium cations (Dunn and Hitchborn, 1965); thus, the so-called magnesium bentonite has been used in stabilizing and purifying several plant viruses including some known to lose their infectivity in vitro. Bentonite in its unpurified form was shown to enhance infection by tomato ringspot virus to pinto bean but not to cowpea or cucumber (Yarwood, 1966). The mechanism of the inhibition by bentonite is not solely attributable to its cation exchange capacity (Jacoli, 1968*a*).

A second method, discussed in Section IIA2, involves removing the enzyme by phenol extraction. Since phenol is an excellent solvent for extracting ribonucleases (Kickhofen and Burger, 1962) and has in fact been used to purify the enzyme (Rushizky et al., 1963; Glitz and Dekker, 1964), it is likely that phenol extracts made from diseased leaves remove most of the nucleases from

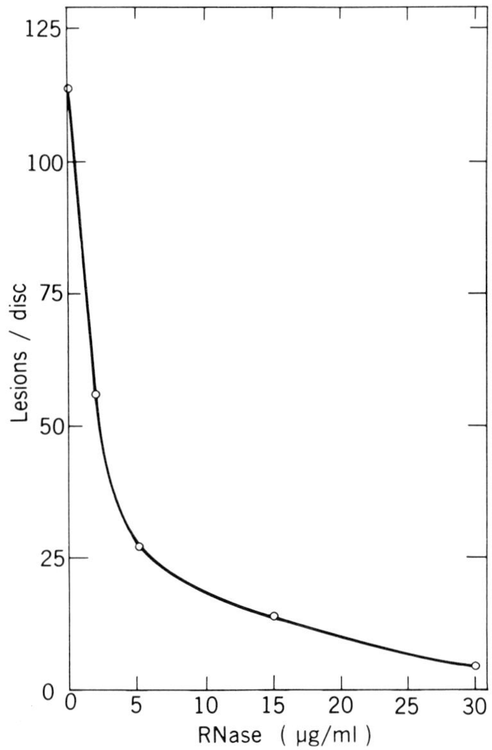

Fig. 1-3. Inhibition of infection by pancreatic ribonuclease. Test made with tobacco mosaic virus (1.0 μ g/ml) in 0.1 M phosphate, pH 7.2, using *Nicotiana tabacum* cv. Xanthi-nc as the assay host. Replicas of 10 leaf discs per inoculum were employed.

the aqueous phase containing the virus nucleic acid. On the other hand, traces of tobacco leaf nucleases can still be detected in purified tobacco mosaic virus preparations (Whitfeld and Williams, 1963).

In addition to these methods of removing nucleases from the inoculum, a third method has been used to prevent the action of nucleases: Since certain chemicals are known ribonuclease inhibitors (Philipson and Kaufman, 1964) and certain heavy-metal ions such as Ag^+ have been shown to protect tobacco mosaic virus RNA from the enzyme, infectivity of the RNA is preserved through interaction of the metal ion with the RNA, not through any inactivating action on the enzyme (Singer and Fraenkel-Conrat, 1962). Polymers of basic amino acids (e.g. polyornithine) can also be used to protect virus RNA.

5. OTHER CHEMICAL ADDITIVES

Various chemical reagents other than those mentioned above have been used in an exploratory manner to determine whether they would affect infection (for review, see Yarwood, 1957*b*; and Yarwood and Fulton, 1967). The nature and the mechanism by which they act on the infectivity of the virus and/or on the susceptibility of the host remains unclear, greatly limiting their use in plant virology. Sucrose deserves a special mention, however, because viruses are frequently purified in density gradients of sucrose (see Chapters 11 and 14). Virus collected from sucrose gradients can be applied to leaves directly, for in the case of potato yellow dwarf virus, infection of *Nicotiana rustica* was *increased* by sucrose (Whitcomb and Sinha, 1964).

6. THE TEST PLANT

The procedure for isolating the virus obtained from the field requires various test plants, the most widely used being: tobacco (*Nicotiana tabacum* L.), bean and cowpea (*Phaseolus vulgaris* L. and *Vigna unguiculata*), Chenopodium (*Chenopodium amaranticolor* and *C. quinoa*), and cucumber (*Cucumis sativus* L.). Whenever possible, it is wise to attempt first to transmit the virus to healthy plants of the same species or same family. Viruses of woody plants, however, are usually tested on a number of herbaceous plants, for reasons already stated. Once a test host suitable for routine work has been found, the initial problem of acquiring the virus from the field is removed; and, in the case of certain stone fruit viruses, the acquisition of the virus becomes easier when herbaceous rather than woody plants are used. The susceptibility of newly acquired hosts may be increased by special treatments prior to inoculation. Various factors which affect the physiology of the host will, in turn, obviously affect susceptibility. Of these factors, age, nutrition, and the surrounding environment are most influential in regulating the degree of susceptibility. Such conditions were recently discussed and reviewed (Yarwood and Fulton, 1967).

D. Biological Methods: Dodder and Graft Inoculation

Biological inoculation is mediated by active action of some organism or organelle (see p. 12). The definition thereby includes inoculation by various insects, mites, nematodes, and fungal vectors, as well as by pollen, dodder, and grafting. Vector transmission is discussed in other chapters of this book. Biological inoculation by dodder and by grafting has been among the most successful methods in experimental attempts to establish an unknown virus in a suitable test host. The mechanics of both methods were recently reviewed and discussed in detail (Fulton, 1964; Bennett, 1967*a*; Bos, 1967).

1. INOCULATION BY DODDER

Dodder is a parasitic phanerogam, lacking both chlorophyll and leaves, in the genus *Cuscuta* of the Convolvulaceae (Kuijt, 1968). Its use became apparent when it was found that dodder parasitizes a wide range of plants and can be used for transporting viruses through its vines (Bennett, 1940; Johnson, 1941). A number of virus diseases were thus shown to be infectious by using dodder (Fulton, 1964). Dodder is used when grafting is restricted to graft-amenable plants unsuited for experimental work and when extension of the host range of the virus is desired. With viruses of woody plants, dodder may be the only means of demonstrating infectiousness of the disease. Viruses reported to be inoculated successfully by various species of *Cuscuta* have been listed by Fulton (1964). Some 17 species of *Cuscuta* have been used for experimental inoculation. The most commonly used have been *C. subinclusa* and *C. campestris*.

Viruses taken up by dodder can either multiply in the dodder itself or move passively through the parasitic vines. According to Bawden (1964), the dodder method is most reliable when dodder itself is a host. Hence, inoculation can be effected by placing pieces of infected dodder onto the test plants and allowing the dodder pieces to parasitize them. Transmission of viruses that do not infect dodder can be accomplished as well, although in this instance transmission and inoculation are effected by simply placing a vine of the parasite from the virus-infected plant to the test plant and allowing the latter plant to become parasitized. The dodder method is useful for extending the host range of a given virus previously thought to be limited to very few plants. Once inoculated, the new host may become a potential donor host for use as a better source of inoculum in mechanical inoculation. Also, dodder infected with virus may itself serve as the source of inoculum for mechanical inoculation. This method is largely self-limiting since certain types of inhibitors of infection have been found in a number of *Cuscuta* species (Miyakawa and Yoshii, 1951; Schmelzer, 1956*a*). Dodder is also useful in separating one virus from another when both are infecting the same host plant. For instance, pea mottle virus could be separated from pea wilt virus since it can elicit infection through dodder whereas pea wilt virus cannot (Johnson, 1942). Also the dodder method may be limited by the species of dodder employed. Of various *Cuscuta* species tested, only *C. lupuliformis* could establish infection by tobacco etch virus (Schmelzer, 1956*b*).

The use of dodder has certain disadvantages. First, the dodder itself is a natural host of several other plant viruses (Bennett, 1944) and may be infected by a latent virus that does not produce obvious symptoms. Second, this method of transmission consumes time and is an all-or-none exploratory method of experimental transmission. And last, various species of dodder are required during the initial investigation since viruses may be transmitted only by a specific *Cuscuta* species.

To exclude possible dodder-borne viruses, the dodder used for experimental work should be started from seeds. Since dodder latent mosaic virus is the only virus known to be transmitted through dodder seeds (Bennett, 1944), starting dodder from seeds eliminates other viruses commonly infecting dodder (such as cucumber mosaic virus).

2. INOCULATION BY GRAFTING

Inoculation by grafting is mediated when an organic union is established between scion and stock and when the virus is able to pass freely between them.

Grafts can be made with various parts of the plant, including roots, tubers, corms, etc., but most grafts are made with stems and buds. The mechanics of grafting were recently described in detail (Bos, 1967), and the various ramifications of the methods of grafting are too numerous to be elaborated here. Grafting becomes relevant when other methods of inoculation fail. Inoculation by grafting is usually successful if the virus can move through the graft union.

A prerequisite for successful graft inoculation is apparently the development of plasmodesmata (Bennett, 1967*a*). This occurs during the union of the cambium layers between scion and stock or between adjacent meristematic cells. Inoculation is effective whether the graft is from diseased stock to healthy scion or conversely from healthy stock to diseased scion. Transmission can be enhanced by placing the graft recipient (whether the recipient is scion or stock) in the dark so as to direct the flow of metabolites into the recipient. Symptom expression may be enhanced by using young growing recipients rather than old ones that have lost their vigor. Like the dodder method, graft transmission also has its inherent limitations. First, if more than one virus is infecting the donor host, the recipient (with few exceptions) will receive both viruses through the graft union. Second, the graft method is time-consuming, involving several weeks to establish grafts and then a further delay in symptom expression; and third, the method is limited to compatibilities of stock and scion. In most cases, interspecific and intergeneric grafts are possible; grafts between distinct families of plants are less likely.

IV. BIOLOGICAL ASSAYS

Quantitative assays of infectivity can be designed by comparing the infectivity of a virus preparation with that of another of known concentration (either in terms of absolute nucleoprotein content or number of infectious units). Infectivity assays are limited to measuring only the virus particles that are biologically active, giving no indication of the total number of virus particles present in a preparation. In addition, since there are variations in the susceptibility of various types of indicator hosts, and since the present methods of assay are inefficient, the degree of infectiousness of a preparation reflects only a relative concentration of virus. Meaningful and highly reproducible results are obtained when the bioassay provides exacting information, that is, when there is direct relationship between the numbers of virus particles to the numbers of infectious units. The quantity of particles may be determined by electron microscopy or by chemical methods. The bioassay is usually expressed as specific infectivity (i.e. infectious units per total number of particles; or infectious units per weight; or better, infectious units per mole). The multiplicity of infection can be calculated by this assay when the number of surface cells of a leaf or half-leaf are known. The usual assay is performed by comparing unknowns on a standard curve obtained from known concentrations of virus. Very crude assays simply employ relative comparisons of lesion numbers of unknown versus "known" and the results are expressed as the ratio of unknown:known or as "percent" of control.

Studies on the biology of viruses require highly purified virus preparations, but in many instances (e.g. when the virus has not been purified), the use of crude virus preparations is unavoidable. As stressed earlier, crude virus preparations can give misleading results unless they too are assessed since various components in the preparation can affect either virus infectivity, assay host susceptibility, or both.

The subject has been reviewed by Roberts (1964); Bawden (1964*a*); and Yarwood (1964). Special emphasis will be placed here on some principles outlining the methods of infectivity assay.

A. One-Dimensional Assay

One-dimensional assay is so termed because it provides information only on the presence of virus and does not specify the amount of virus in a preparation. It is useful, however, for making surveys of the distribution of viruses in the field.

To test for virus, this simple assay is made by transmitting virus by one of several methods to plants that will express infection by the virus in question. Systemic symptoms are anticipated in the indicator test plant when inoculation is mediated by dodder or by grafting, whereas systemic and/or local lesions are expected when inoculation is mechanical.

B. Titration Assays

Quantitation can be developed from the one-dimensional assay by making serial dilutions of virus and noting the dilution that induces no infection. This type of assay is termed the end-point-dilution method. The error is great in this assay since it depends on an all-or-none response. A better assay of this type is the dose-response assay, which involves noting the dilution at which 50% of the plants become infected. This percentile gives the most precisely measurable unit and is the most sensitive region since only a small increment in concentration of virus produces the greatest difference in response. Dilution end-points can be determined graphically by plotting the percentage of plants infected against the log dilution of the virus, or can be obtained by calculations such as those described by Reed and Muench (1938). The disadvantages of end-point-dilution methods are: (1) wide variations in virus titer occur among any given assay; (2) larger numbers of plants are required; (3) accurate statistical methods are applied to an assay with uncontrolled variables. Because the titration assay is low in reproducibility and high in error, a better method was sought for quantitating virus preparations. Enumeration assays became possible when Holmes (1929*a*) found that the numbers of hypersensitive necrotic areas (local lesions) initiated on inoculated leaves could be used as a measure of virus concentration. Since then the local-lesion method has become the standard assay method in plant virology, so one of the first exploratory examinations of a new plant virus is to look for a suitable local lesion host. The local-lesion method has been refined and is presently the most precise method of evaluating plant-virus infectivity quantitatively. A serious drawback of the local-lesion method, however, is its inherent error and relatively low degree of efficiency. When compared with quantitative assays of bacteriophages where the multiplicity of infection and infection parameters can be adjusted and controlled precisely, the local-lesion method suffers since a number of parameters remain uncontrolled (e.g. irregularities in the susceptibility between test plants, variations in the physiology and size of test leaves, differences in the relative infectivities of virus preparations, and inefficient use of virus applied). The efficiency of the local-lesion assay, though still low when compared with phage assays, is nevertheless higher than previously contended, since it is possible to wash off most of the virus that is applied to surface leaf cells without appreciable loss in the number of local lesions (Harrison, 1956; Fraenkel-Conrat et al., 1964).

Upon reexamination of the specific infectivity of tobacco mosaic virus, Furumoto and Wildman (1963*a, b*) calculated that 1 of 10 tobacco mosaic virus particles was infectious and that attachment of the virus particle to a susceptible cell during rubbing is an instantaneous process.

Infectivity dilution curves are obtained by plotting lesion numbers against increasing dilutions of virus. Usually lesion numbers versus virus concentrations are plotted on log-log or semilog paper. At lower concentrations, the number of lesions is proportional to the virus concentration, whereas the proportionality is no longer faithful at high concentrations of virus. Various statistical evaluations have been made of data obtained from dilution assays, and various equations that best fit the data were proposed (e.g. Youden et al., 1935; Bald, 1937*a, b, c;* Price, 1945; Lauffer and Price, 1945; Kleczkowski, 1949, 1950; Furumoto and Mickey, 1967*a, b,* 1969).

At low virus concentrations, most experimental data fit the mathematical expression

$$Y = N(1 - e^{-cv}) \qquad (1)$$

where Y = expected lesion number; N = total number of susceptible sites; v = virus concentration; and c = a constant representing the relative infectivity of the virus. But at high concentrations (e.g. > 1 mg/ml of tobacco mosaic virus on *Nicotiana glutinosa* half-leaves) this equation can no longer "best fit" with the experimental data, for the proportionality ceases between lesion numbers and virus concentrations. The reasons for this behavior are still unclear, and various interpretations of infectivity-dilution curves have been presented on the basis of one of several assumptions, that is (1) a local lesion develops as a result of at least one virus particle in the inoculum which has been introduced into susceptible host tissue or "region"; (2) susceptible "regions" vary in susceptibility so that the minimum virus concentration required to initiate a local lesion varies from one "region" to another; (3) one virus particle is required to initiate infection and the susceptibility of one cell varies from another randomly (see also Roberts, 1964). The validity of the assumptions made previously has been examined (Lauffer and Price, 1949; Kleczkowski, 1950) and reexamined (Furumoto and Mickey, 1967*a, b*). Using statistical analyses of the data Kleczkowski (1950) argued against the hypothesis of Lauffer and Price (1945) that a local lesion is caused by single virus particles entering uniformly susceptible regions. He proposed a hypothesis based on the assumption of variation in susceptibility of susceptible regions with logarithms of minimal effective concentration distributed normally. Equation (1) was thus modified into:

$$Y = \frac{N}{\lambda\sqrt{2\pi}} \int_{-\infty}^{t} \exp\left\{-1/2\left(\frac{t-\zeta}{\lambda}\right)\right\} dt \qquad (2)$$

where Y = expected number of lesions per half-leaf; N = mean number of "susceptible regions" per half-leaf; $t = \log_{10} x$ (x = virus concentrations); $\zeta = \log_{10} x_0$ (x_0 = virus concentration when 50% of the susceptible regions develop lesions); and λ = standard deviation.

Using a computerized program for their data analyses, Furumoto and Mickey (1967*a, b*) formulated a mathematical model that best explains the infectivity-dilution curve. Their model was based on the assumption that: (1) a "region" (or site) of the leaf is an epidermal cell (differing from other one-particle one- "region" [or site] hypotheses of Youden et al., 1935; Bald, 1937*a, b*; and Lauffer and Price, 1945); and (2) the probability of penetration and infection of the cell by a given virus particle is the same for all particles but this probability may vary from cell to cell. More simply: A single virus particle inside a cell can initiate infection and the probabilities of cells becoming infected are not uniform but distributed randomly, that is, cells vary in their susceptibility. Their model thus takes into account differences in the "ability" of cells to become infected and proposes the following equation:

$$\text{Average } Y = N\left\{1 - e^{-cv}\left[1 + \frac{\beta}{(\alpha+\beta)}\, cv + \frac{(\beta)(\beta+1)}{(\alpha+\beta)(\alpha+\beta+1)}\, \frac{(cv)^2}{2!} + \cdots\right]\right\} \qquad (3)$$

or the simple approximation:

$$\text{Average } Y = N\alpha \log_e \left(1 + \frac{cv}{\beta}\right) \qquad (4)$$

where Y = average number of lesions; N = number of susceptible cells per half-leaf; v = virus concentration; c = a constant relating the virus concentration to the average number of virus particles outside a cell; and α, β = constants determining the distribution of cell penetration probabilities.

Equation (4) can be changed to the logarithm to the base 10 with the parameters being called A and B:

$$\text{Av.}(Y) = A \log_{10}(1 - Bv) \tag{5}$$

The validity of the mathematical model proposed by Furumoto and Mickey (1967*a, b*) has been questioned by Kleczkowski (1968*a*) since different equations based on different assumptions can also be fitted "satisfactorily" to a given set of experimental results, and *fewer* cells are made possible sites of infection rather than total numbers of cells.

The value of parameter N, representing the total number of susceptible cells per half-leaf, has been estimated to be 400,000 (Furumoto and Mickey, 1967*b*) rather than 500, as proposed earlier by others (Bald, 1937*a, b;* Lauffer and Price, 1945; Price, 1945; Kleczkowski, 1950). According to Furumoto and Mickey (1967*a, b*), a larger value such as 400,000 is more satisfactory than 500 since the former is consistent with various averaged and compiled data obtained from infectivity-dilution curve experiments. This implies that each cell is potentially infectible, but that most of them have an extremely small probability of being infected. From examining infectivity-dilution curve data, Furumoto and Mickey (1967*b*) suggested that an asymptotic limit of lesion numbers obtained at virus concentrations cannot be obtained experimentally because extremely high concentrations of virus are necessary (e.g. 1 g/ml of tobacco mosaic virus) and that only a small fraction of the total number of cells would have been "hit inside" with a virus particle immediately after inoculation. The mathematical model of Furumoto and Mickey (1967*a, b*) takes into account additional parameters and thus seems to be the best choice.

Computations for fitting the theoretical curve to experimental data are extremely laborious; thus high-speed computers are employed. Where computer facilities are still unavailable, alternate, quite adequate, procedures have been suggested (Furumoto and Mickey, 1969). One of the procedures employs constructing a transparent overlay on which is plotted the curve $y = \log_{10}(1 + x)$, distinguishing the vertical line, $x = 1$, and the horizontal line, $y = 1$, as the axes, and using the same scale as that used in plotting the experimental data. Furumoto and Mickey (1969) prepared the following table of points that can be plotted on the overlay and used for fitting parameters of equation (5).

x	*y*	*x*	*y*
.003	.00130	9	1.000
.01	.00432	10	1.041
.03	.0128	30	1.491
.1	.0414	100	2.004
.3	.1139	300	2.479
1.0	.3010	1000	3.000
3.0	.6021		

(From Furumoto and Mickey, 1969.)

The transparent overlay is then positioned and fitted over a plot of experimental data, keeping the axis of the overlay parallel to those of the plot. The value of *A* of equation (5) is obtained at the intersection of the horizontal axis of the overlay with the lesion count scale of the experimental plot; *B* is obtained as the reciprocal value of the intersection of the vertical axis of the overlay with the concentration scale of the experimental plot.

A slightly more tedious second procedure of Furumoto and Mickey (1969) involves determining the parameters by solving two simultaneous equations. Again a list of precalculated values, various standardized concentrations, and conversion factors are listed in a table to simplify the task of calculation (for details of the second method, see Furumoto and Mickey, 1969).

The application of various mathematical models depends on various basic assumptions. The assumption that all cells are susceptible should be qualified since, in actuality, several other parameters have not been considered. First, during mechanical inoculation, rubbing is not uniform and varies in a nonrandom manner from leaf to leaf with areas being accidentally omitted, for example areas near the midrib, tip, or edges. Second, the regions on a given leaf are not equally wounded by rubbing. Third, the susceptibility of one leaf differs from another. Fourth, the number of surface cells varies in numbers, sizes, and shapes. Fifth, the areas differ between leaves.

Also unaccounted is the parameter of the presence of defective virus particles, of satellite particles, and of multiple component viruses.

2

Inhibition, Interference and Acquired Resistance During Infection*

G. LOEBENSTEIN

The Volcani Institute of Agricultural Research
Bet Dagan, Israel

The term inhibition refers to instances whereby virus infection or replication is prevented by a nonmultiplying substance that does not inactivate the virus in vitro. Interference is defined as that phenomenon whereby the infection process of one virus interferes with the multiplication of another virus or strain; or whereby a nonmultiplying substance interferes via a host-mediated process, requiring the transcription mechanism of the cell, with the multiplication of a virus. Acquired resistance is a distinct type of interference, developing in noninvaded host tissue after viral or fungal infection of other parts of the plant; these resistant tissues are less able to support virus multiplication than nonresistant controls.

I. INHIBITION

Inhibitors have been divided by Bawden (1954) into two categories: inhibitors of infection and inhibitors of virus replication. This definition depends on the mode of action and not on the chemical nature of the inhibitor. Thiouracil, for example, may both inhibit infection and affect multiplication; substances

*Contributions from the Volcani Institute of Agricultural Research (N.U.I.A.), Bet Dagan, Israel. 1968 Series, No. 1304-E.

Supported in part by Grant FG-IS-187 from the U.S. Department of Agriculture under P.L. 480. The assistance of Miss Tirzah van Praagh is gratefully acknowledged.

such as yeast-RNA, may also inhibit infection and induce interference (Gicherman and Loebenstein, 1968).

Inhibitors generally do not inactivate viruses irreversibly, and an infectious virus can be recovered upon separation of the virus-inhibitor mixture by dilution, by sedimenting the virus by ultracentrifugation, etc.

A. Inhibitors of Infection and Mechanisms of Inhibition

Inhibitors of infection have been found in extracts of higher plants, bacteria, insects, and lichens (for lists of inhibitors see Bartels, 1955).

The effect of these inhibitors is usually assayed by comparing the effects of a mixture of purified virus and inhibitor with that of an equal concentration of a virus solution alone. A series of concentrations is tested by the half-leaf method. The duration of the inhibitory effect can be tested by timely applications of inhibitor or virus.

Inhibitors of infection are not virus specific, and the same substance may inhibit very different viruses. They generally inhibit infection when applied to the leaf tissue before or at the time of inoculation with the virus, but have little or no effect when mixed even shortly after virus inoculation. The inhibitor from carnation, for example, fully inhibits infection on *Nicotiana glutinosa* when applied before inoculation with tobacco mosaic virus. However, no significant inhibition was obtained when the inhibitor was applied later than 30 min after inoculation. Ragetli and Weintraub (1962) purified this substance and presented evidence that it is a protein. It has been suggested that this inhibitor acts by competing with the virus for receptor sites at the beginning of the infection process (van Kammen et al., 1961). The interaction between virus and receptor leading to infection or the interaction between inhibitor and receptor leading to inhibition is in dynamic equilibrium and depends on the concentrations of virus and inhibitor. Increasing the virus concentration at a given inhibitor concentration decreases the inhibition, and vice versa—increasing the amount of inhibitor at a given virus concentration increases the degree of inhibition (van Kammen et al., 1961; Gicherman and Loebenstein, (1968). Van Kammen et al. (1961), using Langmuir's adsorption theory, presented an equation describing the relation between the number of local lesions, and virus and inhibitor concentrations. They obtained a good correlation between theory and experimental results and suggested using the following equation to differentiate between inhibition of infection and inhibition of virus multiplication, by comparing the infectivity of virus inocula to which different substances have been added, and using virus in distilled water as a standard.

$$\frac{e}{E} = \frac{1}{1 + \dfrac{K_V}{c}}$$

where e = number of lesions obtained, E = maximum number of lesions at very high virus concentrations, K_V = dissociation constant of the virus receptor complex, and c = concentration of the inoculated virus.

If the value of K_V decreases, the substance acts by inhibiting the absorption of the virus to receptor sites, as is the case with carnation sap. Inhibitors of multiplication, on the other hand, should lower the value of E, i.e., the maximum number of lesions, even at very high concentrations of virus.

When the relations between the number of local lesions and virus concentrations are plotted logarithmically as a function of inhibitor concentration, the same maximum number of lesions is attained, but at a higher virus concentration (Fig. 2-1). The curves have the same course but are shifted to higher virus concentrations.

Substances belonging to different chemical groups have been reported to inhibit infection. Various proteins, such as bovine lactoglobulin, rabbit serum, and tobacco mosaic virus-protein, have been reported to inhibit infection. It is not known if in every case inhibition results from competition for receptors. It has been postulated that the reduction of infectivity in the presence of tobacco mosaic virus-A-protein, bovine serum albumin, or reconstituted noninfectious tobacco mosaic virus is due to aggregation of virus particles in the inoculum (Santilli et al., 1961). Probably only one particle participates in the conversion of an infectible site to an infective center, the other particles of the aggregate remaining inactive (Siegel and Zaitlin, 1964). The aggregation of virus particles can therefore reduce infectivity of the inoculum. Viruses with different isoelectric points may co-precipitate; for example, in salt-free solution, bromegrass mosaic virus, with an isoelectric point at about pH 8.0,

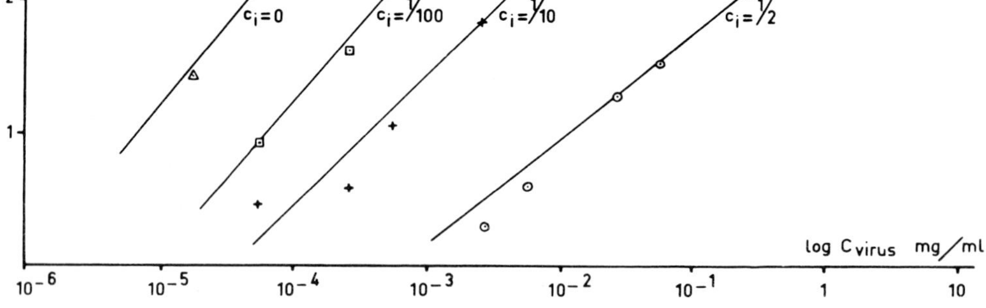

Fig. 2-1. The relation between the number of local lesions and virus concentrations, at different concentration of carnation inhibitor C_i. (From van Kammen et al., 1961.)

precipitates with tobacco mosaic virus having an isoelectric point near pH 4.5 (Kassanis and Kleczkowski, 1965).

Stahmann and co-workers showed that various polyelectrolytes, such as polylysine and polyglutamic acid, inhibit infection when mixed with tobacco mosaic virus. They suggested that inhibition by polyelectrolytes, and possibly also by other inhibitors, could be best explained on the basis of ionic combination of the inhibitor with either the virus particles or the receptor of the plant cell. Either combination may block or prevent the initial binding of the virus by the host cell. Polyelectrolytes with free basic groups, such as polylysine, apparently combine with the virus particles, because inhibition resulted only when it was mixed with the virus and not when it was applied to the plant before inoculation of the virus. Acidic polyelectrolytes such as polyglutamic acid, and the acidic polysaccharides, seem to act on the cells of the host plant (Stahmann and Gothoskhar, 1958).

Enzymes such as trypsin, papain, and ribonuclease (RNase) are also known to reduce infectivity by other means than by enzymatic degradation of the virus. Tobacco mosaic virus, for example, resists attack by enzymes such as chymotrypsin or papain. Furthermore, even its protein, when aggregated into rod-shaped particles, resists these enzymes, whereas the disaggregated protein is susceptible to proteolysis (Kleczkowski and van Kammen, 1961). When appropriate proportions of RNase are mixed with tobacco mosaic virus, a paracrystalline precipitate is formed (Kleczkowski, 1946). However, tobacco mosaic virus with full infectivity can be recovered from a mixture with trypsin or pancreatic RNase by diluting the mixture or by sedimenting the virus by ultracentrifugation. Nene (1960) concluded that, in the case of steapsin and trypsin, the enzymes complex with the virus, thereby preventing attachment of the virus to the receptor; RNase, furthermore, also complexes with the plant receptor (see Chapter 1).

Growth regulators also have been reported to inhibit infection (Upreti et al., 1964). For example, α-naphthaleneacetic acid (NAA) reduces infectivity of potato viruses X and Y and of sunhemp mosaic virus. Gibberellic acid inhibits chili mosaic virus. Gibberellic acid at a concentration of 0.001% inhibits infection when sprayed on plants up to 4 hr before inoculation. It is ineffective when applied later than 4 hr after inoculation. Karas et al. (1964) reported that the growth retardant N-dimethylaminosuccinamic acid (B-9) inhibits tobacco ringspot virus. The substance was effective when applied 45 min after inoculation or within the inoculum.

Divalent cations, such as Ba^{2+}, Ca^{2+}, and Mg^{2+}, inhibit infectivity of tobacco necrosis virus, cucumber mosaic virus, tobacco ringspot virus, and southern bean mosaic virus but are relatively less inhibitory to tobacco mosaic virus and turnip mosaic virus (rod-shaped viruses). Monovalent cations, such as K^+ or Na^+, were either stimulatory or had little effect according to Rossouw and Fulton (1963). As suggested earlier these authors also postulated that the ions may affect the ability of the virus to attach itself to receptors. This attachment is probably an electrostatic one between points of opposite polarity distributed in a complementary fashion over the virus and host surfaces. Any substance in the inoculum that influences the infection process may therefore react either with the receptor or with sites on the virus, or with both. As no inhibition was obtained by infiltrating Ca^{2+} into the leaves, immediately or 3 hr before inoculation, Rossouw and Fulton concluded that divalent cations probably affect the virus and not the host cell.

Inhibitors usually have little effect when applied several hours after inoculation. However, unique in preventing infection when applied even one day after inoculation is trichothecin. Trichothecin obtained from the culture filtrate of *Trichothecium roseum* also inhibits infection on the opposite surface of the leaf to which it is applied, apparently because the molecule is small enough to diffuse into the leaf. Its effect seems to be due to a severe alteration of the physiology of the host cells so that they are no longer able to support virus multiplication (Bawden, 1954). The inhibition is correlated with injuries to the leaves.

Applying the inhibitory substances several days before inoculation is also generally ineffective. RNase (Nene, 1960), however, and heat-killed cells of *Pseudomonas syringae* (Loebenstein and Lovrekovich, 1966) strongly inhibit infection when applied even seven days before inoculation. An inhibitor from rice has been reported to be effective when applied to bean leaves one to three days prior to TMV inoculation (Jones et al., 1959). Regarding heat-killed bacteria, however, it has been suggested that an induced interference mechanism is involved (Loebenstein and Lovrekovich, 1966). Heat-killed bacteria also decrease the size of those lesions which escaped inhibition. This fact does not seem to have been reported for other inhibitors.

The current opinions regarding possible mechanisms by which inhibitors prevent infection can be summarized as follows: (1) competition of the inhibitory substance with the virus for receptors; (2) aggregation of virus particles, thereby decreasing the number of separate particles; (3) prevention of attachment of the virus to the receptor, due to complexing of the inhibitor with the virus particle; (4) alterations in the physiology of the host, causing even visible damage in extreme cases.

1. PRACTICAL APPLICATIONS

Few research publications are available on the practical use of inhibitors to prevent virus diseases. Bawden (1954) suggested that inhibitors might be used effectively for virus diseases transmitted by contact.

Milk and its components have been studied quite intensively as possible inhibitors for practical use. In greenhouse experiments, Hare and Lucas (1959) succeeded in preventing or reducing the contact transmission of tobacco mosaic virus to peppers, tomatoes, and tobacco plants that were sprayed with milk before inoculation; spraying after inoculation gave little control. Spraying tomato transplants with milk one hour before handling may be an effective way of reducing the virus spread. Dipping the transplants in milk is not recommended, as it causes excessive wilting and loss of transplants. The surfactant dioctyl sodium sulfo-succinate (DOSS), which apparently destroys infectivity, was as effective as milk in reducing the spread of tobacco mosaic in tomatoes, but it also delayed growth and flowering (Denby and Wilks, 1963). Protective milk sprays were also effective against the sugarcane mosaic virus on sugarcane plants; and milk and whey sprays protected barley plants from barley stripe mosaic.

Milk sprays failed to prevent aphids from transmitting potato virus Y to tobacco, but did protect lettuce from aphid-transmitted lettuce mosaic (Hein, 1964). This might be due to the fat content of whole milk, because skimmed milk did not prevent aphid transmission.

Oil coverings have been shown to impede aphid transmission of stylet-borne viruses (Bradley et al., 1962; Loebenstein et al., 1964) and of a semipersistent one (Van der Veken and Semal, 1966). Control of viruses such as cucumber mosaic, potato virus A, Y, and iris yellow mosaic has been achieved in greenhouse and field experiments, using oil emulsions and concentrations which are not phytotoxic. Although oil does not inactivate tobacco mosaic virus or cucumber mosaic virus, it is premature to classify oil as an inhibitor, as its mode of action is not yet known.

B. Inhibitors of Virus Replication

Included in this group are substances that delay or inhibit virus multiplication or disease development, after the infectious particle has converted an infectible site to an infective center. To show that the effect is on virus multiplication and not on infection per se, such substances should be applied 12-24 hr after inoculation. Methods used for testing inhibitors of multiplication include floating leaf discs or detached leaves on the test solution; spraying or dipping the leaves of intact plants; infiltrating the substance into the leaves of intact plants by vacuum (Kirkpatrick and Lindner, 1961) or injecting it with a hypodermic needle into the intercellular spaces (Loebenstein and Lovrekovich, 1966), and adsorption via the roots. Attempts to use virus-infected tissue cultures have also been made (Kutsky and Rawlins, 1950). Although floating of leaf discs or detached leaves on the inhibitor solution is widely used, due to its convenience, it should probably be employed only as a preliminary screening assay. Marked physiological changes occur in the plant tissue after detaching leaves or cutting out discs. The effect of the inhibitor may therefore be quite different in detached tissue than on the intact plant. Thiouracil, for example, has no effect on tobacco leaf discs at a concentration of 4×10^{-5} M (Commoner and Mercer, 1951), but inhibits growth when sprayed on tobacco plants.

To assay the effect of the inhibitor, assays are made of the multiplication of virus in a systemic host as a function of time. A virus can be assayed by the local-lesion method, by a micromethod (Taniguchi, 1962*a*), or by a serological-chromatographic procedure (Matthews, 1957).

Compared with inhibitors of infection, the search for compounds which inhibit virus multiplication has, in many cases, been based on a rationalized approach. In particular, substances with a chemical structure similar to that of the components of viral RNA have been studied: that is, analogues of pyrimidines and purines.

1. PYRIMIDINE AND PURINE ANALOGUES

Commoner and Mercer (1951) were the first to report that the pyrimidine analogue 2-thiouracil (TU) is a powerful inhibitor of tobacco mosaic virus synthesis in tobacco leaf discs. Almost complete inhibiton of virus synthesis was obtained when the discs were floated on media containing 4.3×10^{-5} M TU. These authors suggested that TU blocks virus synthesis by interfering with some aspect of uracil metabolism which is normally essential for formation of tobacco mosaic virus. The inhibitory activity of TU could be reversed by uracil but not by cytosine or thymine.

Working with S^{35}-labeled TU, Jeener and Rosseels (1953) showed that the analogue is incorporated into the virus particle. Incorporation is not more than 10%, and the analogue is present in the viral RNA as 2-thiouridylic acid. Tobacco mosaic virus that is synthesized in the presence of the analogue is substantially less infectious than normal virus. There was much less infectivity when the virus was assayed in freshly extracted sap—about 80% of the particles from TU samples being noninfectious—than when it was assayed from purified virus preparations. In the latter case, reduction in infectivity due to treatment with TU was significantly less (Francki, 1962). Francki suggested that TU decreases infectivity of tobacco mosaic virus due to the inability of the viral RNA, containing the analogue, to initiate virus replication.

Although significant intercorrelations were found between the incorporation of TU into tobacco mosaic virus-RNA, the loss of infectivity, and the reduction in virus yield—which lends to support to the idea that the formation of "sterile" TU-containing particles leads to a reduction in the rate of virus synthesis—the question does not seem to be resolved. When TU is removed from infected leaves, or when uracil is applied together with the analogue, the virus begins to multiply. This observation has been used by Bawden (1964*a*) as evidence that the effect of TU is in altering the host plant's metabolism. Francki (1962), however, showed that uracil did not completely prevent the inhibition of virus synthesis by TU.

There is also no explanation why the large number of infectious virus particles, which are present in TU-treated leaves, do not multiply to produce "yields" of virus equivalent to those in control plants. Perhaps the strongest point, that inhibition of virus multiplication by TU is not solely through its incorporation into the viral RNA, can be deduced from the experiments of Francki and Matthews (1962) with turnip yellow mosaic virus. The multiplication of this virus is inhibited by TU, and the analogue is not incorporated into the viral RNA and has no effect on the infectivity of the virus that is formed in its presence. The formation of "sterile" tobacco mosaic virus particles, due to incorporation of the analogue, may therefore be secondary, whereas inhibition of viral multiplication is mediated via a different process.

Treatment with the analogue five days after inoculation with turnip yellow mosaic virus induces a marked increase in the production of empty protein shells that contain no RNA (Francki and Matthews, 1962). This shows that that part of the viral genome which codes for the viral protein is active in the presence of the analogue. The viral messenger may, however, be changed—due to incorporation of the analogue—because changes of the viral protein, produced in the presence of the analogue, have been reported. Turnip yellow mosaic virus protein shells from TU-treated leaves differed in their heat stability from those accumulating in untreated leaves (Ralph et al., 1965); and minor changes in the antigenic determinants of tobacco mosaic virus protein following TU treatment of infected tobacco plants have been observed (Jeener, 1965).

Studying the effect of TU in cell-free extracts was also unsuccessful in resolving the question of whether the analogue inhibits virus multiplication by being incorporated into the viral RNA. TU inhibited incorporation of labeled adenosine-triphosphate (ATP) into the double-stranded viral RNA, if the cell extract was prepared from turnip yellow mosaic virus-infected Chinese cabbage leaves that had been previously sprayed for several days with the analogue. However, if the cell-free extract was prepared from nonsprayed plants, the addition of TU stimulated ATP incorporation into the double-stranded viral RNA (Ralph and Wojcik, 1966). Although no incorporation of labeled TU into the double-stranded RNA could be detected, the incorporation of small amounts into the double-stranded RNA cannot, according to the authors, be excluded. This may explain the inhibition of ATP incorporation into the double-stranded viral RNA, after prolonged spraying of the infected leaves with the analogue.

Lwoff (1965) has discussed the mechanism by which guanidine affects the biosynthesis of polio virus. Guanidine prevents viral RNA replication, apparently by inhibiting the RNA polymerase. This inhibition may be due to allosteric changes in the polymerase in the presence of guanidine. Ralph and Wojcik (1966) suggested that TU in the turnip yellow mosaic virus-Chinese cabbage system may act in a manner similar to guanidine in the polio system.

Attempts to use TU as a therapeutic agent on virus-infected plants have so far been unsuccessful. Although prolonged application of TU may significantly reduce the virus content in the plant by slowing down new synthesis, the analogue also affects the metabolism of the plant itself: The young leaves become pale and apical growth is arrested. However, virus-free meristem or callus tissues may be obtained from infected plants by maintaining them for 3 weeks or more on media containing 100 mg/ml of TU. Virus-free callus tissue was thus obtained from potato virus Y-infected tobacco plants (Kassanis and Tinsley, 1958).

Among the substituted purines, 8-azaguanine (AZA), the analogue of guanine, has been studied most. AZA has been reported to inhibit the multiplication of a number of plant viruses, including a virus causing tumors on rumex, alfalfa mosaic, cucumber mosaic, tobacco mosaic, stone fruit ringspot, turnip yellow mosaic, sugar beet yellows, potato X and Y, and bromegrass mosaic. Compared to other analogues, AZA had the greatest latitude between the concentration required for virus inhibition and phytotoxicity (Linder et al., 1960)—only slight phytotoxic effects were noted when concentrations were around 0.01 M. The analogue markedly delays invasiveness, especially with alfalfa mosaic virus in *N. glutinosa* plants, and there is much less virus production. Systemic development may be prevented if AZA is applied close to the time of inoculation, but not when applied after systemic movement of the virus has started. The inhibitory activity of AZA can be partially or completely annulled if plants are sprayed with naturally occuring purines or their ribosides and ribonucleotides (Matthews and Smith, 1955). Matthews, in his previously cited studies, showed that up to 3% of the guanine in the viral nucleic acid was replaced by AZA, and that the inhibitory effects on virus replication were related to the substitution. The hypothesis that AZA inhibits virus multiplication by being incorporated into the viral RNA is not supported by others (Bawden, 1954).

The effect of AZA varies also with the host. *Physalis floridana* plants were found to be much more responsive to the analogue than were tobacco plants (Lindner et al., 1960). Treatment of *Physalis* with AZA resulted in virus that varied in infectivity, sedimentation time, and nucleic acid content. Treated plants contained a fraction of tobacco mosaic virus with high infectivity, and also one with very low infectivity. The mechanism of action, however, does not appear to be a simple substitution of AZA for guanine during virus RNA synthesis, because the ratio between uninfective and infective tobacco mosaic virus particles produced depends on the time of AZA application. This ratio was highest when AZA was applied during the 24-hr period after inoculation, and decreased when AZA was applied either before or after that time. The authors suggest that AZA alters directly the replication site of the virus, resulting in the production of modified virus. The variations in host response may be explained by a more ready attachment of the analogue to the replication site in one host than in another. However, the effect of AZA in inhibiting systemic virus development appears greater than can be accounted for by the inhibition of replication sites. Differences in the rate of enzymatic deamination of AZA, or other changes of the analogue in the host, have also been offered as an explanation of host variation. As with TU, the mechanism of action is probably governed by more than one factor. No correlation between the chemical structure of other analogues and their inhibitory activity has been found. Relatively few of the substituted purines and pyrimidines inhibit virus multiplication, and some, such as caffeine, even stimulate multiplication, probably by affecting the physiology of the host tissue.

The incorporation of an analogue, as with 5-fluorouracil (FU), into tobacco mosaic virus may exert a number of biological changes in the virus, although the ability to initiate local lesions is not significantly different when compared with normal virus. In a systemic host, however, less virus is produced when FU-tobacco mosaic virus is used as an inoculum. The FU-virus is more sensitive than the normal to ultraviolet irradiation, due to the alteration of the virus nucleic acid. The coding ability for viral protein is unaffected by the substitution in the viral RNA, since no alterations occur in the antigenic properties (Sutic and Djordjevic, 1964) and in the amino acid composition of the viral protein.

2. *ANTIBIOTICS*

Schlegel and Rawlins (1954) found that noformicin, an antibiotic from the actinomycete *Nocardia formicans,* reduced the amounts of tobacco mosaic virus in leaf discs, and southern bean mosaic virus as well as tobacco mosaic virus in intact plants (Gray, 1955). The antibiotic was translocated in the leaves. Cytovirin, an antibiotic from *Streptomyces* sp., protected tomato and tobacco plants from systemic infection by tomato spotted wilt virus and tobacco mosaic virus when plants were sprayed twice—2 hr and 12 days—after inoculation. Plants were protected up to 7 weeks. Spraying with cytovirin had no effect once infection was established, and it did not cure infected plants.

Cycloheximide, an antibiotic produced by *Streptomyces griseus,* inhibits tobacco mosaic virus, Wisconsin pea streak virus, sour cherry necrotic ringspot virus in cucumbers, and tobacco necrosis virus in beans. It was ineffective against two strawberry viruses and the peach rosette virus. Hirai and Shimomura (1960), studying the effect of naramycin (cycloheximide analogue) showed that the antibiotic and its acetate, oxim and semicarbazone derivatives inhibited tobacco mosaic virus multiplication; the antibiotic also inhibited P^{32} incorporation into tobacco leaf discs, without decreasing total RNA concentration. These authors suggested that naramycin suppresses the decomposition of leaf RNA. As RNA degradation products may serve as substrates for virus synthesis, a deficiency of such substrates due to naramycin retards virus synthesis. However, cycloheximide is known to be a general inhibitor of cellular protein synthesis, apparently involving the transfer of amino acids from amino acyl-RNA into protein. Therefore virus-specific proteins needed for virus replication are not synthesized. Practical application of using cycloheximide seems unlikely because at higher concentrations and prolonged treatments necessary for control, the antibiotic is highly phytotoxic.

Mitomycin C also inhibits tobacco mosaic virus multiplication in detached leaf discs (Hirai and Shimomura, 1960). Like cycloheximide, mitomycin inhibited P^{32} incorporation into leaf RNA, but in contrast, it also reduced RNA synthesis in uninfected controls. The authors suggested that inhibition of virus multiplication by mitomycin is related to the reduction in total leaf RNA synthesis. Protein-synthesis inhibitors, puromycin (Langenberg and Schlegel,

1965) and blasticidin S, isolated from *Streptomyces griseochlomogenes*, also inhibit tobacco mosaic virus synthesis (Hirai et al., 1966). Blasticidin S also reduces infection of rice by rice stripe virus. Both antibiotics are more inhibitory when applied immediately after inoculation than when applied later. They do not affect established virus but affect virus and host protein synthesis. Hirai et al. (1966) showed that blasticidin S inhibits the synthesis of 30-*S* RNA including tobacco mosaic virus, but does not inhibit the 16-*S* ribosomal RNA. These antibiotics and laurusin and flavensomycin are not virus-specific. They inhibit virus multiplication and also affect host functions. Antibiotics of potential practical therapeutic value should specifically inhibit virus multiplication, for example, by binding viral RNA synthetases without affecting normal cell processes.

3. *PLANT GROWTH REGULATORS*

Limasset et al. (1948) reported that 2-methyl-4-chloro-phenoxyacetic acid inhibited the multiplication of potato viruses X and Y in tobacco plants. Naphthalene acetic acid applied to tobacco tissue cultures reduced the amount of tobacco mosaic virus (Kutsky and Rawlins, 1950), and Augier de Montgremier and Morel (1948) suggested that less tobacco mosaic virus develops in tobacco callus cultures with high naphthalene acetic acid content. It has been suggested that the absence or low concentration of virus in meristems is due to high auxin concentration. Auxin content is less in tobacco mosaic virus-infected tobacco plants than in healthy control plants (Pavillard, 1954).

Growth regulators are known to cause masking of symptoms; 2,4-D masks potatoes of potato leaf roll and naphthalene- and indole-acetic acids mask tobacco mosaic. In the latter case symptoms were masked without reduction in virus concentration (Nichols, 1952).

A striking reversal of virus-caused stunting was obtained by three applications of gibberellic acid at 100 μg/ml on clover infected with wound tumor virus (Maramorosch, 1957). Virus could be recovered from the infected plants.

The growth retardant 2-chloroethyltrimethylammonium chloride (CCC) might be further explored for inhibitory activity. In preliminary tests with floated leaf discs, tobacco mosaic virus multiplication was reduced by more than 80%. No injuries on these discs were detected with a concentration of 0.5% (Rawlins, 1962).

Kinetin (6-furfuryl-amino purine), which stimulates protein synthesis and delays senescence, reduces the number and size of local lesions on petunias inoculated with tomato spotted wilt virus. Treatment with 50 μg/ml after inoculation reduced lesion production but had no effect when present in the inoculum. Kinetin therefore does not inhibit infection but affects virus multiplication at a later stage. Selman (1964) proposed that inhibition might be associated with an increased activity of leaf ribonuclease. Santilli et al. (1962), however, found that increasing leaf ribonuclease activity to abnormally high levels did not inhibit but rather stimulated tobacco mosaic virus multiplication. Kinetin also delayed the appearance and reduced the size and number of tobacco mosaic virus lesions on *N. glutinosa* and virus multiplication in tobacco (Kiraly and Pozsar, 1964). The authors advanced the idea that virus synthesis is enhanced when cellular host protein and nucleic acid syntheses are suppressed, as occurs in mature tissue. Vice versa, in young or kinetin-treated tissues, virus multiplication is inhibited because metabolites are used by the host plant and are less available for viral synthesis. In contrast, caffeine, which stimulates virus multiplication, enhances maturation and senescence of plant tissues (Kessler et al., 1958). However, kinetin (at 10 μg/ml) has also been reported to stimulate tobacco mosaic virus production in tobacco leaves. This is not surprising if one considers that kinetin affects the equilibrium between protein synthesis and its degradation, and may favor, especially at lower concentrations, the synthesis of one protein. Thus, the synthesis of leaf ribonuclease may be stimulated more than other leaf proteins. In several kinetin treatments, Daft (1965) found an increase of leaf ribonuclease correlated with virus increase. These nucleases may make precursors available for viral synthesis by degrading leaf RNA (Reddi, 1959*b*).

None of these growth substances is virus-specific. However, in changing the physiology of the host, cellular conditions result which are generally less favorable for virus multiplication.

4. MISCELLANEOUS COMPOUNDS

Kooistra (1959) found that p-nitrosophenol, 1-nitroso-2-naphthol and the copper sulphate complex of o-nitroso-p-chlorophenol displayed antiviral activity, and were not too phytotoxic. The compounds were applied either by spraying leaves or by soaking plant roots for 24 hr. Inhibition was determined by an increase in the incubation period from inoculation until the appearance of virus symptoms. When 1-nitroso-2-naphthol was tried in field experiments, an increase in the yield of virus-infected bean and sugar beet plants was observed. Kooistra suggests that these chemicals may react with uracil, thereby depleting its availability for virus synthesis. The nitroso compounds are decomposed in the plant, after which uracil again becomes available for virus synthesis. Therefore, only the incubation period is prolonged by the treatment and the plant is not cured of disease. Since uracil is needed for regular plant metabolism, phytotoxicity is encountered after two or three applications of these compounds.

Tomato aucuba mosaic virus multiplication is inhibited by 2,4-dinitrophenol (Merrett, 1962). Dinitrophenol uncouples respiration from phosphorylation, thereby reducing the availability of compounds with high-energy bonds, that is, adenosine triphosphate (ATP). ATP seems to be necessary for plant virus multiplication. The fact that virus synthesis is less in young tissues than in mature tissues might also be due to the need for ATP for virus multiplication, as it is used up by rapid synthesis of cellular components in young leaves. Two other compounds which inhibit respiration, sodium cyanide and malachite green, also inhibit virus multiplication.

The infiltration of photosensitizing dyes, for example, neutral red, reduces virus multiplication when leaves are placed in the light, but not when kept in the dark. However, photosensitizing dyes are probably not inhibitors *sensu stricto*, but rather viral inactivators. In vitro, tobacco mosaic virus-RNA is rapidly inactivated by the dye in the light following first-order kinetics, the dye probably binding with the viral RNA (Orlob, 1967).

Future studies on the mechanism of plant virus replication and virus-specific proteins may provide clues toward developing specific virus inhibitors. Thus studies with *Escherichia coli* phage Q_β showed that its replicase is specifically inhibited by synthetic polyribonucleotides, composed principally of either adenine (A) or uracil (U). Other polynucleotides, containing mainly cytidine (C) or guanine (G), were inert. The synthetic poly A or poly U apparently complex with the replicase, thereby preventing replication of the viral RNA. The specific inhibition by poly-A or U is related to the base composition of the RNA template (Haruna and Spiegelman, 1966).

The development of assay methods in cell-free extracts, or the incorporation of labeled precursors into viral RNA in leaf discs, may provide an advanced test method for evaluating such specific inhibitors (Semal, 1966).

II. INTERFERENCE

A. Interference Between Strains

The following phenomena, whereby one strain of a virus interferes with another, are included in this section: cross protection, interference during simultaneous inoculation, and apparent recovery. This subject has been reviewed several times, most recently by Kassanis (1963) and Price (1964).

1. CROSS PROTECTION

A plant already infected with one strain of virus is commonly protected from infection by other related strains; and this phenomenon known as cross protection is commonly used to determine rough relationships between viruses.

The classic procedure, based on Kunkel's system (Kunkel, 1934) uses a combination of two strains, one which spreads systemically and another which produces local lesions. This offers the advantage that quantitative data on the degree of protection may be obtained, which is not possible when two systemic strains are used. It is generally accepted that high or complete inhibition occurs between closely related strains, provided that the interfering strain is systemically established in the plant before challenge inoculation with the local-lesion-forming strain. Unrelated strains generally are not cross-protective. Thus, relationships between tobacco mosaic virus strains may be determined on *N. sylvestris*, where the type strain is systemic, and the aucuba mosaic strain produces local lesions. In this case the half-leaf method may be employed,

Fig. 2-2. Cross protection in a leaf of *Nicotiana sylvestris*. The common (systemic) strain of tobacco mosaic virus rubbed on the right half of the leaf, protected it from infection by the U2 (necrotic) strain, inoculated on the entire leaf three days later.

where half of the leaf is inoculated two or three days before the whole leaf is challenge inoculated with the local-lesion-producing strain. Very few lesions develop in the protected half of the leaf which was inoculated earlier with the systemic strain.

Half-leaf and opposite-leaf methods cannot be employed when systemic spread is slow. Whole plants have to be used. Cucumber mosaic virus strains, for example, can be identified on *Zinnia elegans* with the aid of an indicator local-lesion-forming strain, while most other strains of cucumber mosaic virus cause mottle mosaic symptoms on this species. It is important to challenge inoculate young leaves because local-lesion formation on older leaves is erratic. Therefore, it is advisable to inoculate very young plants with the systemic cucumber mosaic virus strain because systemic invasion is slow, and leaves which are to be subsequently challenged with the local-lesion strain, 12-14 days later, might otherwise be too old. Relationships may also be determined when both strains are systemic, as long as symptoms between the two differ markedly. Thus, a mild systemic strain of potato virus X protects *Datura stramonium* plants from severe strains, causing system; necrotic mottle mosaic symptoms. Cross-protection tests may also be carried out when the viruses cannot be transmitted mechanically, but have to be inoculated either by grafting or by insects. Thus it could be shown that four different symptoms occurring in sweet potatoes are caused by variants of the sweet potato vein-clearing virus, whereas a ringspot disease is caused by a different, unrelated virus (Loebenstein and Harpaz, 1960).

The use of intercellular changes to determine cross protection has been suggested for cases where differences in symptoms are not distinct (Mueller and Koenig, 1965). Bean yellow mosaic virus produces intranuclear inclusions in epidermal strips of leaves of *Vicia faba*. These inclusions are readily stained by trypan blue, or they may be observed unstained with a phase contrast microscope. All strains of this virus produced inclusions, but they differed in size and occurrence depending on the strain. After one strain became established it inhibited formation of crystals of another strain. Such protection was not afforded by two other viruses which are not related to bean yellow mosaic virus.

However, absolute dependence on cross-protection tests for determining relationships between viruses can lead to erroneous conclusions. Tobacco (potato) veinal necrosis virus, for example, does not protect tobacco or potato plants from other serologically related strains (Klinkowski and Schmelzer, 1960). Some, but not all strains of potato virus Y, protect against each other, and several may afford protection also against tobacco veinal necrosis virus. In the latter instance, the appearance of symptoms of tobacco veinal necrosis is delayed, and symptoms are milder.

Various strains of sugar beet curly top virus do not give any cross protection in sugar beet and in water pimpernel (*Samolus parviflorus*).

To obtain reliable results it is advisable to use reciprocal inoculation procedures. In such tests one virus is first introduced into a series of plants followed by the second virus several days later. The order of inoculation is reversed in a second series of plants. Cross protection depends on the completeness of invasion of the tissue by the first virus, on the order and method by which the two viruses are introduced, and on the relatedness of the virus isolates. Absence, or a low order of cross protection, should not be regarded as conclusive whereas positive results—a high or complete protection—can be regarded as good evidence for a close relation between two viruses.

There are two prevailing theories to explain cross protection, neither of which, however, meets the demand of all observed facts:

a. Precursor Exhaustion. After the protecting strain invades the tissue completely, available precursor materials are depleted and become limited in supporting infection of a second strain. Hageman (1964) obtained a good correlation between the decrease in the amount of RNA in the microsomal fraction and the degree of protection between two strains of tobacco mosaic virus. The U2 strain fully protected tobacco leaves from superinfection by the green aucuba strain, but transferring discs from fully protected mature U2-infected leaves to a nutrient solution (containing salts, malic acid, and uracil) decreased protection. These discs were then able to support multiplication of the second aucuba strain, whereas discs placed on tap water remained fully protected. In the discs placed on nutrient solution, the RNA content of the microsomal fraction increased markedly. The author suggested that "floating of virus-infected leaf discs on nutrient solution diminished the protection as precursor exhaustion was replenished," thus enabling further virus synthesis.

However, the results of Bawden and Kassanis (1945) with strains of tobacco etch virus are difficult to explain according to the precursor-exhaustion theory. The mild etch strain reaches only low concentrations in tobacco plants, but protects them completely against the severe etch strain, which when inoculated alone achieves a much higher concentration than the mild strain. It may be that in such cases the protecting strain is concentrated in the epidermis whereas other regions of the leaf contain little virus.

Furthermore, it is difficult to explain protection in "recovered" plants (see below) by the precursor-exhaustion theory. Leaves which have recovered from the tobacco ringspot disease contain only 10-20% the amount of virus that is present in leaves with acute symptoms, but are still protected from superinfection by another strain.

b. Competition for Sites of Infection in the Cell. It has been suggested (Siegel, 1959) that two strains inoculated simultaneously compete for the same sites, the result being that either one or the other, but not both, could initiate infection. In cross-protection tests where the first strain is allowed to multiply alone it would have the advantage of occupying all available sites, thereby preventing multiplication of another strain, introduced later, and requiring the same sites. Different viruses may multiply in the same cell, since each requires different "sites."

2. SIMULTANEOUS INOCULATION WITH STRAINS

Siegel (1959) inoculated leaves of *Nicotiana sylvestris* with a mixture of a localized necrotic (U2) and the common systemic mottling (U1) strain of tobacco mosaic virus. (See Fig. 2-2.) Lesion numbers produced by U2 are decreased by the presence of the systemic strain U1 in the inoculum, but not when U1 is inoculated one hour after U2. The degree of interference, as measured by the decrease in the number of local lesions produced by U2, depends on the concentration of the systemic strain. Interference reached 70-90% when the concentration of U1 was 25 times that of U2. A similar interference can be demonstrated in primary leaves of pinto beans, (*Phaseolus vulgaris* cv. Pinto) where the U1 strain produced local necrotic lesions, whereas the U2, VM, and Y-TAMV strains do not cause macroscopically visible local or systemic symptoms. In mixed inocula, U2, VM, or Y-TAMV interfered with U1 lesion production.

When infection is initiated by a particle of one strain, particles of another strain are "excluded" from participating in the infection; and when intact particles of two strains are present in the inoculum, they compete with each other for the same sites; either strain can convert it to an infective center. This competition occurs apparently at the nucleic acid level after the virus particles are stripped of their capsid protein, because in an inoculum containing infectious RNA of one strain and whole particles of another strain of tobacco mosaic virus, the RNA possessed a decisive competitive advantage (Wu and Rappaport, 1962). Ultraviolet-inactivated tobacco mosaic virus or its protein subunits also interfere with viable tobacco mosaic virus when present in a mixed inoculum. However, 100-1000 times of the noninfectious components are necessary to produce the same degree of interference with lesion production that occurs when one strain of the virus interferes with another (Wu et al., 1962). The mechanism in this case seems to be similar to that described for competitive inhibition with virus infection.

However, the "either/or" hypothesis, whereby two virus particles compete for an infectible site, one initiating the infection and the other being excluded, is inadequate to account for all the observed data. Evidence shows that two strains can infect the same cell. It has therefore been suggested that interference occurs at a later stage in the infection process as a consequence of multiplication of the interfering strain. Evidence for the absence of an exclusion mechanism has been obtained by Benda (1956), who succeeded, by micromanipulation, in inoculating individual hair cells of *N. glutinosa* leaves with a mixture of two tobacco mosaic virus strains. Both strains could be isolated from 5 of the 20 lesions that developed. However, virus can move to several adjacent cells, where it continues to initiate infection. The lesion (consisting of more than one cell—see Chapter 1) that contains more than one strain of virus could result from the fusion of two independent infection centers.

Woods and DuBuy (1951) and others (see Chapter 4) presented cytological data suggesting that one cell can be infected simultaneously with two strains of the same virus. A yellow strain of tobacco mosaic virus arising by mutation from a green mosaic strain, affected chloroplast development in a different way from the green one. Some cells were found in which some of the plastids were modified in a manner characteristic for yellow mosaic, the remainder showing symptoms of the parent green strain. Furthermore, some of the cells that contained yellow mosaic-type plastids, and yellow mosaic noncrystalline inclusions, also contained crystalline inclusions of the green mosaic type.

Further indications that more than one particle is present in the infective site, in contrast to the "either/or" hypothesis, have been provided by ultraviolet (UV) rescue experiments (Rappaport and Wu, 1962). Two strains of tobacco mosaic virus were used differing in their sensitivity to UV irradiation. When the UV-sensitive strain VM is inoculated together with the UV-resistant strain U1, VM interferes with lesion development by U1. However, if the leaves are UV-irradiated 2-4 min after inoculating the mixture of these two strains on *N. glutinosa,* a rise in the number of U1 lesions is observed. The authors suggested that two particles, one of each strain, could make the initial attachment, and that when one is inactivated by the UV irradiation, the other multiplies. However, it is possible that UV irradiation inactivates the sensitive strain before it has become attached to the site, thereby increasing the ratio in favor of the UV-resistant strain.

To resolve this question as to whether interference between strains is through a competitive exclusion mechanism, either at the site or at the infective center, it should be shown unequivocally that the interfering strain does not multiply and yet reduces the number of lesions formed by another strain. If, on the other hand, the interfering strain is capable of starting infection and multiplication, then the reduction in lesion number of the second strain is probably caused not only by competition for sites, but also by interference at a later stage of infection. Thus U2, which interferes with U1 when present in mixed inocula, has been shown to form microlesions on pinto bean leaves under favorable environmental conditions. These microlesions can be detected as necrotic starch lesions on bleached, iodine-stained leaves. In mixed inocula of U1 and U2, Helms (1965) found that increasing concentrations of U2 reduces the number of visible macrolesions only, whereas the number of necrotic starch lesions increases. This decrease in the number of macrolesions by U1 in the presence of U2 was an expression of the reduced size of the U1 lesions (Fig. 2-3.)

Size reduction of U1 lesions with increasing U2 concentration in the inoculum indicates that interference with lesion development occurs at a later stage subsequent to multiplication rather than by earlier competition for sites. This decrease in size of U1 lesions might have been caused by the earlier multiplication of U2, which thereby invades large numbers of cells leaving only a few uninfected cells available for multiplication of U1. Helms (1965), however, calculated that even in leaves inoculated with high concentrations of U2, less than 2% of the leaf area is occupied by microlesions. If U2 is thought to interfere with U1 directly, one has to assume that most cells, which do not show any signs of infection, still contain virus. The author therefore favors the alternative hypothesis that multiplication of U1 was inhibited as a result of metabolic changes in cells surrounding U2 infections. Such a mechanism may have similarities to localized acquired resistance, reported by Yarwood (1960*a*) and Ross (1961*a*), whereby tissue surrounding a lesion of tobacco mosaic virus develops resistance to challenge inoculation.

Current postulated mechanisms of interference by two strains present in the inoculum can be summarized as follows: (a) competition for sites; (b) changes in the physiology of host cells adjacent to the infection center of the interfering strain; and (c) existence of several independent sites within one cell. Two strains may multiply independently in the cell until they compete for some limiting precursor, and the strain with the faster rate of multiplication will dominate. It is also possible that interference results through the interaction or combination of more than one mechanism. Further studies on the biology of plant viruses will also advance our knowledge on the mechanism of interference between strains. It may be worthwhile initiating studies on the cellular mechanism of releasing inhibiting substances after the first virus particle has started replicating. The possibility that interference between strains might be mediated

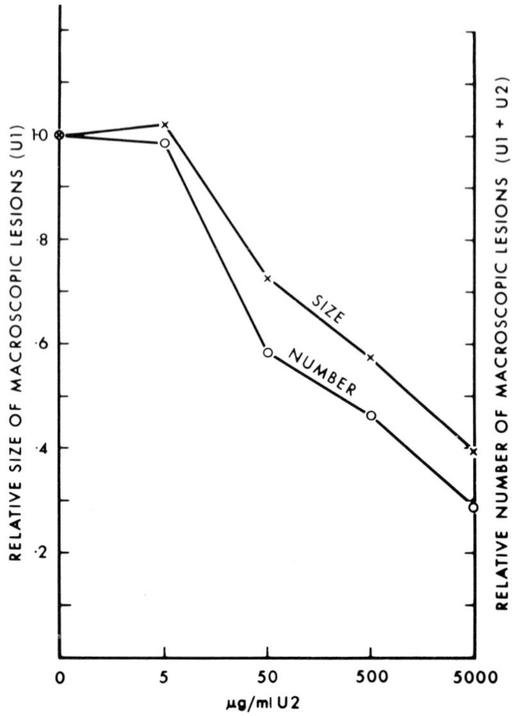

Fig. 2-3. The size of U1 lesions in pinto bean leaves inoculated with a mixed inoculum containing U1 and U2, relative to the size of U1 lesions in leaves inoculated with U2 alone. For comparison, the number of macrolesions in leaves inoculated with mixed inoculum is plotted relative to the number of macrolesions in leaves inoculated with U1 alone. (From Helms, 1965.)

via such a mechanism has been suggested by Helms (1965). This possibility has some analogy to animal virology, where interference is suggested to be mediated by interferon formation (Wagner and Levy, 1960).

3. APPARENT RECOVERY AND RESISTANCE TO REINFECTION

Tobacco plants infected by tobacco ringspot virus after showing the ringspot patterns ("shock phase") seem to recover. New leaves which form 2-3 weeks after inoculation show no typical ringspot patterns. Price (1936) observed that the virus concentration markedly decreased in such plants and recovery was followed by an "acquired immunity" to reinfection. After reinoculation, these plants did not again pass through the "shock phase."

Many infected plants, including fruit trees, pass to a later chronic stage after the acute initial reaction, but recovery is especially marked with tobacco ringspot virus, sugar beet curly top virus in tomato, tobacco and water pimpernel, and dodder latent mosaic virus and clover stunt virus in subterranean clover. With tomatoes infected with sugar beet curly top virus, recovery from the disease sometimes takes place with the production of new shoots from axillary buds. Recovery is often associated with a reduction in the infectivity of the sap, but virus is not eliminated entirely. For example, alfalfa mosaic virus in tobacco plants increases rapidly from the fourth to the tenth day after inoculation, but then decreases rapidly (Fig. 2-4). When infectivity is low, symptoms are milder than at high infectivity.

The specific infectivity of alfalfa mosaic virus, turnip yellow mosaic virus, cowpea chlorotic mottle virus, etc. also decreases with time in vivo. Specific infectivity is determined by adjusting the purified virus samples equally spectrophotometrically and serologically and then assayed on a local-lesion host (curve C, Fig. 2-4).

The decrease in specific infectivity may be caused by cleavage of the RNA within virus particles. The decrease in infectivity of the sap with time due to lower concentration of virus seems to be a common process with many virus infections, although it is not always associated with a decline in specific infectivity. Thus, with tobacco mosaic, specific infectivity levels remain constant.

Plants have been observed to recover from the disease and no longer contain the causal virus yet still remain protected. Lemon seedlings regularly recover from symptoms of seedling yellows, which apparently are caused directly by a

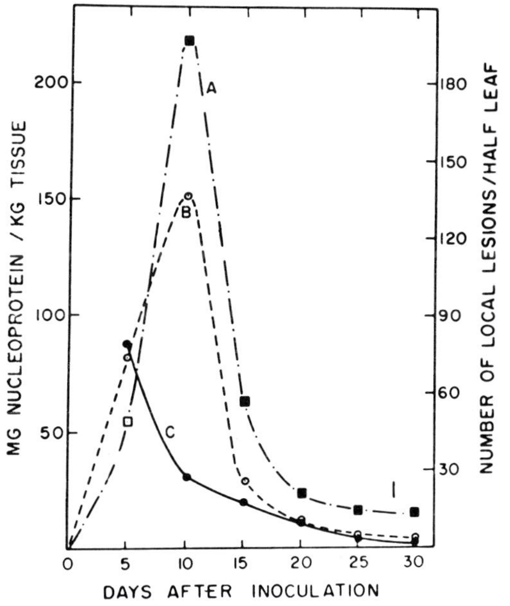

Fig. 2-4. The concentration and specific infectivity of alfalfa mosaic virus from tobacco plants at different times after inoculation. Curve A, amount of purified nucleoprotein (mg/kg total leaf wet weight); curve B, sap local-lesion curve; curve C, number of local lesions when purified samples are equaled spectrophotometrically and serologically. (From Kuhn and Bancroft, 1961.)

distinct yellow-inducing virus, or by a combination of that virus and tristeza virus. After recovery from seedling yellows, the plants retain the tristeza virus, which affects lemons only slightly, and no longer contain the yellows-inducing component. These plants, however, show a high degree of protection when reinoculated with the yellows-tristeza virus complex, the protection not resulting from the remaining tristeza component (Wallace, 1965).

Several suggestions have been made regarding the mechanism of recovery and subsequent resistance to reinfection. Wallace (1944), from his studies of sugar beet curly top in tobacco and tomato, advanced the idea that "protective substances" may be the cause of the recovery and resistance. When recovered tobacco plants were grafted on healthy tobacco or tomato plants only mild symptoms developed and the plants were protected from reinfection. However, leafhoppers were able to transmit a severe disease from recovered plants. Wallace suggested that, by grafting, "protective substances" were also transferred whereas the vector transmitted only the virus. However, it may be that the curly top cultures were mixed, and that "antagonism" between strains, leading to the multiplication of a mild strain, could account for the phenomenon. Furthermore, Price (1964) found that the severity of the disease after inoculation by leafhoppers was dependent upon the portion of the plant on which the vectors fed. Severe symptoms were produced when the vector fed on tissues near the growing point, whereas mild symptoms resulted when the insect fed on leaves farther down the stem. Insect transmission to mature tissues was thus comparable to transmission by grafting.

That new leaves which develop 2-3 weeks after inoculation with tobacco ringspot virus show only mild symptoms or remain symptomless, and are in the recovered phase, suggests that the recovery phase is initiated when the virus infects "embryonic" tissue (Price, 1936). "Embryonic" tissue apparently means young tissue, because meristems do not seem to become infected. It has also been shown that by inoculating very young tobacco seedlings with a virulent strain of curly top virus, a certain percentage of the infected plants grow to maturity without passing through a symptom phase of the disease (Benda and Bennett, 1964). Benda and Bennett propose that "recovery may result when tissue susceptible to virus multiplication and spread is invaded before it is sufficiently mature to be susceptible to injury."

B. Interference Between Unrelated Viruses

Interference is more or less a general rule with strains of viruses and an exception with unrelated viruses. Additive or synergistic effects on symptoms have often been noted by compound infection with two or more viruses, (for example, dodder latent virus with tobacco mosaic virus or tobacco etch virus; tobacco and potato plants infected by potato viruses X and Y). These plants also contain considerably more potato virus X than plants infected by only one virus. Absence of interference has been based on symptom expressions of the doubly infected host and not on virus concentrations. Interference between unrelated viruses might therefore prove to be of more common occurrence as quantitative data become available. The degree of interference between unrelated viruses is generally of a lower order of magnitude when compared to interference between strains, which is high or even complete.

Interference occurs when potato viruses X and Y are inoculated simultaneously into potato seedlings (USDA No. 41956). This seedling is a local-lesion host for virus Y and is considered immune to virus X. However, inoculation with both viruses simultaneously permits fewer virus Y lesions than with virus Y alone. Inhibition of virus Y increases as the amount of virus X in the inoculum is increased. In other hosts, such as on *Physalis floridana*, virus X does not interfere with infection by virus Y although the rate of lesion development is somewhat retarded (Ross, 1959).

The most striking example of interference between unrelated viruses has been described by Bawden and Kassanis (1945). Multiplication of virus Y was suppressed when tobacco White Burly plants were inoculated with tobacco severe etch virus together with virus Y or when virus Y-infected plants were reinoculated one month later with severe etch virus. Sap taken from young leaves of such plants failed to react with virus Y antiserum, and reacted only with severe etch virus antiserum. Severe etch virus interfered in a similar manner with multiplication of henbane mosaic virus. Based on serological properties, severe etch is unrelated to virus Y or to henbane mosaic virus. Interference takes place only in one direction, and severe etch virus was able to supplant potato virus Y or henbane mosaic virus in tissues in which the latter viruses were established. Mild etch virus reduced the content of virus Y, but not completely. Such marked suppression of potato virus Y or henbane mosaic virus by severe etch virus did not occur with Samsun tobacco (Schmelzer et al., 1960). Although the presence of severe etch reduced the concentration of the other two viruses, complete protection was not achieved. Other cases of interference between unrelated viruses have been reported (Price, 1964).

It has been suggested that the presence of the protecting virus reduces available sites necessary for infection by the second virus (Ross, 1959). To obtain a high degree of interference, the tissue has to be invaded by the first virus. In *Nicotiana repanda* plants, for example, interference increased from 0% on the third day after inoculation with cucumber mosaic virus-LL, to 22% on the fourth and 58% on the fifth day. Interference dropped subsequently, to 0% on the tenth day. However, after the fifth day the protecting virus spread to the opposite half-leaves, which served as controls. The calculated "decrease" in the interference on the protected half-leaves may therefore be due to the development of interference in the control halves. Nevertheless, the reported data to support this suggestion are not conclusive, and it might be reasoned that this type of interference, in contrast with cross protection between strains, lasts for only a short period. Cross protection also is not restricted to special strains, and is generally efective when reciprocal inoculations are employed. Interference between unrelated viruses, however, often occurs in only one direction. For example, alfalfa mosaic virus in *Chenopodium amaranticolor* reduces by 60-80% the number of lesions resulting from inoculation with cucumber mosaic virus. In contrast, cucumber mosaic virus did not reduce the number of local lesions caused by alfalfa mosaic virus in *N. glutinosa* (Nitzany and Cohen, 1960).

As stated previously, prior infection by one virus may interfere with the multiplication of a second virus delaying lesion development or reducing their size. The development of primary lesions of tomato aucuba virus was retarded in *N. sylvestris* infected with tobacco ringspot virus (Kunkel, 1934). A reduction in tobacco mosaic virus lesion size was observed in tobacco (Samsun NN) systemically infected with potato virus Y. The diameters of the lesions on these plants averaged 25-50% of those on control plants. Similarly, the size of local lesions of alfalfa mosaic virus on bean yellow mosaic virus-infected bean plants was markedly smaller than controls (Ford, 1967). In pea plants, bean yellow mosaic virus probably interferes with alfalfa mosaic virus multiplication. Interference between tobacco mosaic virus and tobacco etch virus is observed at the cellular level (Fujisawa et al., 1967). Leaf tissue infected simultaneously by these viruses contains cells with particles of both viruses. However, the challenge virus particles were unobserved when the first virus infection was already fully established. The authors suggest that interference is due to "precursor exhaustion." Interference between different viruses can also be due to virus-induced physiological changes in the plant, thereby increasing the host's resistance to the challenge virus. For instance, this resistance could occur by an inhibitor induced in the host by the first virus. Damirdagh and Ross (1967) have suggested that virus synthesis may be accompanied by the formation of an inhibitory substance. This proposed inhibitory substance, accompanying infection by potato virus X, limits further virus synthesis. Such an inhibitor may in certain cases also affect the multiplication of another unrelated virus. On the other hand, the second virus might stop the production or induce destruction of the inhibitor, thereby resulting in a synergistic interaction between the two viruses.

III. ACQUIRED RESISTANCE

A. Induction by Viruses

This section emphasizes resistance developing in host tissue not invaded by virus after infection of other parts of the plant. Prior sections discussed interference phenomena which are associated with the presence of virus in the challenged tissue.

The first observation that infection in one part of the plant confers resistance in other noninvaded tissues was reported by Gilpatrick and Weintraub (1952). The lower leaves of *Dianthus barbatus* were inoculated with carnation mosaic virus causing primary local lesions to develop and leaving the upper noninoculated leaves virus-free. These upper virus-free leaves were resistant to infection as only 5-25% as many lesions developed after inoculation with carnation mosaic virus as compared with control plants. The protecting factor, which was not due to the presence of a masked virus, could be transferred only by grafting to unprotected plants. It was not transferable by mechanical inoculation or by short contact in grafting. After 30 months of continuous transfer the protective mechanism ceased abruptly.

Acquired resistance has been reported in various hypersensitive hosts, both in tissues near the primary local lesions and in more distant parts of the plant. In the first case the term "localized acquired resistance" is being used, whereas the resistance developing in other parts is termed "systemic acquired resistance." This distinction seems to be rather one of convenience and not a basic one, as it depends only on the distance of the resistant tissue from the primary infection.

Localized acquired resistance has been shown by Yarwood (1960*a*). Tobacco mosaic virus is inoculated on one half of each primary leaf of pinto beans in two strips about 8 mm apart. On the opposite half-leaf similar strips are brushed with the appropriate control solution without the virus. Two days after the strip inoculation the whole leaf is challenge inoculated with more of the same virus. The reduction in lesion number induced by rubbing between strips is used as a criterion of acquired resistance. *Nicotiana tabacum* cv. Samsun NN (Samsun NN tobacco) plants, a variety hypersensitive to tobacco mosaic virus, can also be employed, but the challenge should be inoculated 5-7 days after the strip inoculation. The reduction in lesion numbers between strips varies between 40% and 80% of those between control strips.

The tissues around primary local lesions also become resistant. When Samsun-NN tobacco leaves with tobacco mosaic lesions are challenged after 7 days with a concentrated tobacco mosaic virus inoculum, numerous lesions develop in all areas of the inoculated leaves except for a zone of 1-2 mm surrounding each original lesion (Ross, 1961*a*). In this peripheral region few or no lesions developed and the size of this region and the degree of protection increased for 6-8 days. Resistance of the zone remained unaffected at 100% humidity, thereby reducing the argument that resistance is due to a drying-out effect resulting from abnormal loss of water from the tissues adjacent to necrotic areas. Resistance was highest when tests were carried out at 20-24° C, but could not be detected at temperatures around 30° C. The resistance is not virus-specific. Zones around tobacco mosaic virus lesions are resistant to tobacco necrosis virus, tobacco ringspot virus, and tomato ringspot virus, but not to turnip mosaic virus. Tobacco necrosis virus also induces localized resistance to tobacco mosaic virus. However, in plants infected by potato virus X no resistance occurs around tobacco mosaic virus lesions. Further studies showed that this resistance is not restricted strictly to areas around lesions, but can also develop in more distant areas. Thus inoculation of half-leaves of Samsun-NN tobacco with tobacco mosaic virus induced a high level of resistance to tobacco mosaic virus in the opposite half-leaves (Ross, 1961*b*). Similarly, resistance could be induced in the basal part of a leaf by inoculating the apical half. Challenge inoculation with tobacco mosaic virus of such leaves 7 days after the first inoculation resulted in lesions that were consistently smaller (one-fifth to one-third as large in diameter) than lesions in susceptible half-leaves. They were also usually fewer in number. Inoculation of the lower leaves on a plant induced resistance in the upper noninoculated leaves. These upper resistant leaves, and opposite noninoculated half-leaves are free of virus. A convenient test can be made using plants trimmed to five expanded leaves. In one group the three lower leaves are inoculated with tobacco mosaic virus to produce a density of about 200-300 lesions per leaf. In a second group, the control, the lower leaves are brushed with the appropriate control solution without the virus. Seven days later the two upper leaves of both series are challenge inoculated. Lesion sizes are measured and resistance, detected 2 or 3 days after the inducing inoculation, reaches a maximum in about 7 days and persists for at least 20 days. Highest resistance takes place when the plants have been kept at 20° C. No resistance was induced by mechanical rubbing or chemical injury.

Systemic acquired resistance is also inducible in the opposite primary leaves of pinto beans and cowpeas (*Vigna unguiculata* cv. Blackeye), 7 days after inoculation (Ross, 1966). The following viruses, in various combinations were used for the resistance-inducing inoculation followed by challenge inoculation: alfalfa mosaic, tobacco mosaic, tobacco necrosis, and southern bean mosaic with pinto beans; and alfalfa mosaic, cucumber mosaic, and tobacco necrosis with cowpeas. As in tobacco, lesions in the induced resistant leaves were consistently smaller (about one-half the size, in diameter), than lesions in the appropriate control leaves. Lesion numbers were also reduced, but the effect on size was more consistent. At very high levels of resistance, lesion numbers were reduced further. Systemic acquired resistance is also inducible in *Datura stramonium* by tobacco mosaic virus and in *Gomphrena globosa* by potato virus X (Loebenstein, 1963). The reduction of size and numbers of the lesions may be caused by the same factor and in highly resistant tissues infection may not progress enough to result in visible lesions (as with localized acquired resistance around primary local lesions).

To elucidate the basis of acquired resistance, metabolic pathways have been investigated. The yellow ring region surrounding virus-induced lesions possesses more respiratory enzyme activity than do normal green areas of the same leaf. More activity of the hexose monophosphate (HMP) shunt dehydrogenases (e.g. 6-phosphogluconic dehydrogenase) was found and sugars, mostly pentoses, accumulated. The lower C_6/C_1 ratio in leaves of *D. stramonium* with tobacco mosaic virus-induced local lesions also suggest an activation of the HMP shunt enzymes (Loebenstein and Linsey, 1963). As with the HMP enzymes, an increase in phenolase activity also occurs in local-lesion hosts. This increase may lead to a greater production of quinones, which could inhibit virus from spreading from the local lesion (Farkas et al., 1960). Although increases in phenolase activity in resistant tissue have been observed, its induction took place in 7 hr whereas the induction of resistance required at least 2 or 3 days. It seems therefore that phenolases play no direct role in the resistance phenomenon.

Whether other enzymes are associated with acquired resistance is still an open question. Peroxidase, for example, increases significantly in tissues with necrotic lesions (Loebenstein and Linsey, 1966). Injection of commercial peroxidase into tobacco leaves protected them against the wildfire disease, caused by *Pseudomonas tabaci* (Lovrekovich et al., 1967), and it may be worthwhile to see if peroxidase induces a similar resistance against viruses.

It has been demonstrated that resistance is associated with the movement of substances from or to the resistant tissue (Ross, 1966). Cutting the midvein of an upper leaf of Samsun-NN tobacco inoculated on the three lower leaves with tobacco mosaic virus prevented development of resistance distal to the cut. Killing a section of the petiole of each inoculated leaf with boiling water also prevented development of resistance in upper leaves.

Two hypotheses have been advanced to explain acquired resistance:

(1) *Depletion theory*. Due to hypermetabolic activity of infected cells metabolites are depleted from adjacent uninfected tissues. Thus these tissues adjacent to the infected area are less able to support virus synthesis, and become "resistant." According to Silberschmidt and Caner (1967), resistance is due to a depletion of sugars in the leaf as a consequence of the preceding infection. In their experiments systemic resistance was induced in the basal parts of *G. globosa* leaves after inoculating the apical part with potato virus X. "Ringing" the shoots, which presumably prevents the downward movement of sugars from the leaves, prevented the development of acquired resistance. They suggested that the leaves of the ringed plants retained a sugar supply sufficient for lesion development. Although ringing may prevent the sugar transport, the evidence does not necessarily suggest that resistance is due to a depletion of sugars. It may be that "ringing" prevented the production or movement of one or more precursors necessary for the development of resistance.

(2) *Resistance-inducing substance(s) theory*. Initial virus infection induces a substance(s) which confers resistance in uninfected tissues. When the inoculated Samsun-NN tobacco leaves are removed 7 days after inoculation, high resistance develops in new leaves that grew during a period when no inciting infection occurred elsewhere in the plant (Bozarth and Ross, 1964). Thus it is difficult to explain the depletion theory whereby metabolites are depleted some 19-25 days after removal of the lower inoculated leaves.

Recently, actinomycin D was shown to inhibit the development of acquired resistance when it was applied near the time of inducing inoculation (Loebenstein et al., 1968). To show this, tobacco mosaic virus was inoculated on one half of each leaf in two strips, whereas for control purposes similar strips were brushed with water on the opposite half-leaves, as done by Yarwood (1960*a*). The plants were challenged by inoculating the entire leaf and acquired resistance was then measured by the reduction in lesion numbers between strips. In Samsun-NN tobacco, challenged 5 days after the inducing inoculation, acquired resistance was almost completely abolished when actinomycin D had been injected intercellularly 3 days before the challenge. When Samsun-NN tobacco was challenged 6 days after the inducing inoculation, acquired resistance was partially inhibited by actinomycin D, which had been injected 3 days before the challenge. No inhibition resulted when actinomycin D was injected near the time of the challenges. In pinto bean acquired resistance was also significantly reduced in the presence of actinomycin D.

Except for DNA-dependent RNA synthesis in the nucleolus, actinomycin D inhibits most other DNA-dependent RNA syntheses but not viral RNA synthesis in eucaryotic cells. Development of acquired resistance depends largely on transcription and translation of the plant genome, that is, the normal protein-synthesizing machinery (see Chapters 1 and 16). If actinomycin D is injected near the time of strip inoculation, when a resistance-inducing substance is beginning to be "produced," the development of acquired resistance is inhibited almost completely. However, when actinomycin D is introduced at a later stage, near the time of the challenge, when such a substance has been produced already in larger quantities, acquired resistance is not affected. It therefore seems improbable that acquired resistance, which is sensitive to actinomycin D, is caused by drainage of metabolites as proposed by the depletion theory.

It has been suggested that acquired resistance may be caused by an interferon-like mechanism (Loebenstein, 1963). Interferon is a protein developing in animal tissues after inoculation with active or inactivated viruses and synthetic polynucleotides. It is nonspecific in its action, that is, although induced by one virus it can act against other viruses. Interferon may also be induced by other nonviral agents such as nucleic acids, poly I:C and phytohaemagglutinin. It prevents the synthesis of new viral nucleic acid. Interferon production by animal cells is inhibited in the presence of actinomycin D and puromycin; thus cellular RNA synthesis and protein synthesis are also required. For further data on interferon refer to Finter (1966). Like interferon, acquired resistance in plants is also not specific; it also is induced by various agents such as viruses, fungi (Hecht and Bateman, 1964), tobacco mosaic virus protein (Loebenstein, 1962), and yeast-RNA (Gicherman and Loebenstein, 1968), and it is sensitive to actinomycin D, which in turn can increase lesion size (Loebenstein et al., 1969).

B. Extraction of Interfering Agents from Resistant and Infected Tissues

Sap extracted from uninfected apical halves of leaves of *Datura stramonium* previously inoculated on their basal halves with tobacco mosaic virus or tobacco necrosis virus consistently interfered with infection by tobacco mosaic virus (when mixed in the inoculum), to a much greater extent than did juice from apical halves of control leaves (Loebenstein and Ross, 1963). Sap from the apical uninfected halves was extracted by squeezing, or by homogenizing in 0.1 *M* borate buffer, pH 9.0, 7-10 days after the inducing inoculation. The agent, which seems to be a protein, could be partially purified by precipitation with zinc acetate (Loebenstein and van Praagh, 1964), based on a method used for interferon purification. A similar agent could be induced by systemic infection of *Datura stramonium* with potato virus X.

This agent resembles interferon by being stable at pH 2.5, and in perchloric acid, by its molecular weight (20,000-30,000) and by its ability to precipitate with zinc acetate, acetone, and ammonium sulphate. It differs on host specificity (Loebenstein et al., 1966). Conclusions, however, should be drawn only after further purification and complete characterization and by learning the mode of action of this agent, especially since its effect has been assayed as an inhibitor of infection. It has yet to be seen if this agent interferes with virus multiplication or affects lesion size when applied after virus inoculation. The difficulties in extracting a similar agent from other hosts may be due to its adsorption to glass or plastic surfaces during the extraction procedure. Thus, the adsorption of the agent from *Datura* could be reduced by the addition of Tween 80 or bovine serum albumin (unpublished).

A substance which resembled the one induced by viruses was extracted from tobacco and *Datura* leaves 6-8 days after treatment with the sodium salt of phytic acid (inositol hexaphosphoric acid) (Maïa, 1966). This factor, which also seems to be a protein, inhibits tobacco mosaic virus lesions and slightly reduces the virus content in systemically infected discs, when applied 48 hr after inoculation.

Another agent in infected plants has been reported by Sela and co-workers (1966). Sap extracted from potato virus Y or tobacco mosaic virus-infected plants was treated with hydrated calcium phosphate, resulting in a noninfective solution. This solution, when mixed with either virus was more inhibitory than juice extracted from healthy plants. The active principle purified from *N. glutinosa* 2 days after inoculation with tobacco mosaic virus is considered to be RNA. It is not virus-specific and apparently does not inactivate the virus in vitro. Its association with acquired resistance is unknown as it is extracted and purified from infected tissues, and not from resistant uninfected parts. A suppression of infectivity was observed when this antiviral agent was applied to leaves previously infected with tobacco mosaic virus or cucumber mosaic virus (Sela et al., 1965). The authors suggest that the reduced infectivity provides evidence that the site of action is inside the cell. However, it is difficult to visualize this agent affecting virus particles already present. On the other hand, if this agent affects virus multiplication, it would seem advisable to test it under controlled conditions, shortly after inoculation of healthy tissue during the "exponential period" of virus multiplication, and not in young leaves of old infected plants, where variation may be considerable.

A substance which inhibited potato virus X was found in stems and leaves of potato plants, following inoculation with the fungus *Phytophthora infestans*

(Hodgson and Munro, 1966). Sap from within or around the blight lesions markedly inhibited virus X local-lesion production, but no inhibition was noted when sap from beyond the lesion was used. It seems reasonable to assume that the inhibition is caused by a substance in the fungal mycelium, acting as an inhibitor of infection, and it is not known if this substance affects potato virus X in vivo. However, when spores were injected into potato stems, and after necrotic areas caused by the fungus had spread up the stem, the virus content of side shoots was much reduced. It is therefore possible that the fungus evokes a host reaction which suppresses virus multiplication.

C. Induction of Acquired Resistance and Interference by Other Agents

Resistance to virus infection can be induced by other widely differing agents such as fungi, viral capsid protein, yeast-RNA, and heat-killed bacteria.

Systemic resistance to tobacco mosaic virus was induced in Xanthi-nc tobacco leaves by injecting spore suspensions of the fungus *Peronospora tabacina* into the stems of the plants (Mandryk, 1963). The plants were challenge inoculated three weeks later after allowing the fungus mycelium to develop. Virus lesions in these plants were smaller and fewer than in control tobacco. Hecht and Bateman (1964) have also demonstrated the induction of systemic resistance against tobacco mosaic virus and tobacco necrosis virus by infection of the stem or leaf tissues with the lesion-forming fungus *Thielaviopsis basicola.* Resistance was detectable within 3 days and reached a maximum within 6-8 days. Virus lesions in the resistant tissue were markedly smaller than those in the comparable control. *T. basicola* also induced systemic resistance in *D. stramonium* plants (unpublished).

Development of resistance has been associated with necrosis induced by either lesion-producing viruses or fungi. Resistance can be also induced without visible necrosis of the host with virus protein (Loebenstein, 1962) and yeast-RNA (Gicherman and Loebenstein, 1968). Partial protection was achieved by treating leaves of *N. glutinosa* or *D. stramonium* with dilute tobacco mosaic virus capsid-protein solutions and then inoculating with tobacco mosaic virus 4 days later. The treatment of leaves with capsid protein, one day or less before the challenge inoculation, did not reduce the number of lesions. Therefore, a direct inhibiting effect of the protein does not seem to be the cause. Reduced susceptibility of upper untreated leaves of plants whose lower leaves had been treated with tobacco mosaic virus protein showed that this type of protection was systemic. Tobacco mosaic virus capsid-protein protected against potato virus X but not against cucumber mosaic virus. No similar effect was obtained by several other proteins.

A much higher degree of resistance or interference could be induced when yeast-RNA (5 mg/ml) was injected intercellularly into leaves of Samsun-NN tobacco. Yeast-RNA, when applied together with tobacco mosaic virus acts as an inhibitor of infection, but it also induces interference when injected into tobacco at least 3 days before inoculation. Interference was highest when tobacco mosaic virus was inoculated 5 or 6 days after RNA injection and remained high even after 20 days. In such resistant tissue, tobacco mosaic virus lesion numbers decreased by 80-90%. However, no decrease in lesion size was observed. The development of this localized interference was significantly inhibited if actinomycin D (15 μg/ml) or puromycin (50 μg/ml) was injected shortly after the RNA induction. The fact that a time interval of at least 3 days between RNA injection and the challenge inoculation is needed for detecting resistance, and that its development is sensitive to metabolic inhibitors, strongly suggests that a resistance-inducing agent is formed in the plant by injecting yeast RNA. If actinomycin D is injected 2 days after RNA application, interference is reduced significantly. Our results therefore suggest that for the development of yeast-RNA-induced interference the transcription and translation mechanisms of the cell have to operate in order to produce a resistance-inducing substance, probably a protein.

Interference of lesion formation caused by heat-killed bacteria may have similarities to the induced-resistance phenomenon. Heat-killed cells of *Pseudomonas syringae* (Loebenstein and Lovrekovich, 1966) or *P. fluorescens* (Klement et al., 1966) were injected into the intercellular spaces of Samsun-NN or Xanthi-nc tobacco leaves, respectively. Inoculation up to 7 days thereafter resulted in a marked decrease of tobacco mosaic virus lesions. Furthermore, heat-killed bacteria suppressed lesion formation even when applied 5-72 hr after inoculation with the virus. Injection of a bacterial preparation 24 hr after inoculation or later resulted in a decreased effect on lesion numbers, but a marked reduction in lesion size persisted. The size of virus lesions in Samsun-NN tobacco leaves decreased by 87%, 58%, and 33% when the bacterial cells were injected 24, 48, and 72 hr, respectively, after virus inoculation. Injection with casein or egg albumin after inoculation had no effect. Preliminary experiments with actinomycin D showed that interference was inhibited in the presence of the antibiotic. The effect on lesion size, together with sensitivity to actinomycin D, may suggest that the bacterial cell, or one of its components, activates or induces a resistance mechanism in the plant. However, further evidence is required to support this suggestion. It is also possible that more than one mechanism is involved, a direct and an induced one, possibly brought about by different components of the bacterial preparations.

That resistance can be induced by different agents such as viruses, fungi, and yeast-RNA might also suggest that the activation of resistance involves derepression, anologous to the Jacob-Monod model. It is speculated that for induction of resistance or interference, certain macromolecules combine or remove repressors from cellular DNA regions responsible for coding the resistance substance(s). Confirming or negating such a hypothesis will be possible only when more data become available. It might be interesting to study such a proposal also for the following resistance phenomena: (a) localization of the virus in the lesion; (b) apparent recovery after infection; (c) acquired resistance or induced interference.

3

Inclusion Bodies

M. RUBIO-HUERTOS

Instituto Jaime Ferran de Microbiologia
Madrid, Spain

I. INTRODUCTION

One of the most specific changes induced by virus infection of cells is the production of microscopic bodies differing from other cellular structures. These bodies vary in morphology and composition. They are located in either the cytoplasm or the nucleus. For animal and plant viruses, inclusion bodies can be of diagnostic value.

Intracellular inclusions were first reported by Iwanowski (1903) in tobacco in plants infected with tobacco mosaic virus. He had described two kinds of inclusions, one amorphous or amoeboid-type and the other crystalline. Subsequently, inclusions induced by other viruses have also been described. Their composition, formation, and significance were not known. They were earlier misinterpreted as living organisms or developmental stages of some organisms such as protozoa (Kunkel, 1921; Lyon, 1921; and others). Better microscopes, improved fixation, staining methods, and microscopic techniques, the discovery of phase optics and fluorescence microscopy, the use of the electron microscope, and ultrathin sectioning of the whole cell, have given a fairly clear picture of the formation, composition, and diversity of the inclusion bodies associated with different viruses. There are still many gaps to be filled in understanding the significance of these bodies particularly with respect to the site of virus multiplication or assembly, and the different kinds of ultrastructures found in them depending on the infective virus.

II. TYPES OF INCLUSION BODIES

A. Amorphous Inclusion Bodies

Tobacco mosaic virus amorphous bodies were called X-bodies by Goldstein (1924) and this term is still widely used for amorphous inclusion bodies induced by virus. McWhorter (1965) called them viroplasts. In general, they are round or oval, frequently vacuolated, and have a size of 1-30 μ. They are located in the cytoplasm and rarely in the nucleus. They give the usual reactions for proteins and stain well with acid dyes. Their structure and mode of formation as seen under the light microscope vary for different viruses, the influence of the host plant being not very significant. According to their structure they can be classified into following groups:

(1) *Amorphous inclusions in the cytoplasm*

(a) Granular inclusions, round, homogeneous, occasionally vacuolate, sometimes dispersed, depending on the age of the inclusion. Type—petunia virus 1, broad bean mottle virus.

(b) Amorphous or fibrillar, oval, well developed with needle-like bodies, containing also mitochondria, plastids, and lipid droplets. Type—aucuba mosaic (Sheffield 1931, 1934), amaranthus virus, phaseolus virus 2, brassica virus 1, and others (Rubio-Huertos, 1950; Miličić, 1963).

(c) Amoeboid-type with well defined boundaries, vacuolated. Type—tobacco mosaic virus (Iwanowski, 1903; Goldstein, 1924).

(d) Round, oval, or elongated, very refringent, nonvacuolated, sharply defined from the cytoplasm. Type—brassica virus 3 (Rubio-Huertos, 1950, 1956), dahlia mosaic virus (Robb, 1964).

(e) Unusual types of inclusions which can be associated with special virus infections. Type-wound tumor virus (Black, 1947; Littau and Black, 1952).

(2) *Amorphous inclusions in the nucleus*

(f) Big round, granular X-bodies. Type—cotton leaf crumple virus (Tsao, 1963).

(g) Granular inclusions restricted to the nucleoli. Type—some strains of legume viruses (Rubio-Huertos, unpublished).

B. Crystalline Inclusion Bodies

(1) *Crystalline inclusions in the cytoplasm*

(a) Numerous crystals, with curved sides, sometimes perfect hexagons. Type—red clover vein mosaic (Porter and McWhorter, 1952).

(b) Large plates, sometimes hexagonal and paracrystalline needles. Type—tobacco mosaic virus—common strain (Iwanowski, 1903).

(c) Spindle-shaped masses or long fibers or paracrystalline fibrils. Type—cactus viruses (Amelunxen, 1958; Miličić and Plavsic, 1956).

(d) Pyramids with hexagonal bases, rhomboids, one or two in each epidermal cell. Type—petunia ringspot virus, some strains of phaseolus virus 2 (bean yellow mosaic virus) (Rubio-Huertos, 1950, 1956).

(2) *Crystalline inclusions in the nucleus*

(e) Thin, rectangular plates 1-20 in one nucleus. Type—tobacco etch viruses (Kassanis, 1939; Matsui and Yamaguchi, 1964).

(f) Isometric crystals restricted to the nucleoli. Type—strains of bean yellow mosaic virus, and pisum virus 2 (McWhorter, 1941*a*).

(g) Other crystalline inclusions. Type—strains of tobacco mosaic virus (Woods and Eck, 1948; Goldin, 1963).

The above grouping of inclusion bodies according to their structure and location as seen under the light microscope is an oversimplification since the inclusion bodies are changeable; they grow and attain a maximum size and a definite structure with age. Sometimes X-bodies of some viruses which are granular at the beginning turn out fibrillar and form a crystalline inclusion at the end. Thus, when we refer to a type of X-body, we mean its "mature stage." On the other hand, strains of the same virus could produce different types of inclusion bodies (Rubio-Huertos, 1962).

III. DEVELOPMENT OF INCLUSION BODIES

The mode of formation of X-bodies appears to vary with the virus. Most of the X-bodies due to different viruses are formed by accumulation of granular material at one place in the cytoplasm as described by Sheffield (1931) for aucuba strain of tobacco mosaic virus, cabbage ringspot (normal strain), broad bean mottle virus, bean yellow mosaic virus, and others (Rubio-Huertos, 1950).

Inclusion bodies formed by coalescence in a vacuolate body of degenerated plastids and chloroplasts were described for turnip yellow mosaic virus (Rubio-Huertos, 1950, 1956), and it has recently been confirmed by electron microscopy. The same type has been described for apple mosaic virus and for a strain of bean yellow mosaic virus (Harder and Kirkpatrick, 1966).

IV. COMPOSITION AND ULTRASTRUCTURE

By cytochemical tests and specific stains for light microscopy, inclusion bodies have been found to consist of nucleoproteins. Some amorphous inclusions also contain lipids and cellular organelles such as mitochondria, plastids, and chloroplasts.

Electron microscopy has been a useful tool in the study of the ultrastructure and composition of inclusion bodies. Normal techniques of ultrathin sectioning and specific techniques of electron microscopy are described in Chapter 4.

Sheffield (1939) was the first to attempt the isolation of crystalline inclusions of tobacco mosaic virus and tobacco etch viruses for observation in the electron microscope. She proved by infectivity tests that crystalline inclusions from both viruses had infective virus particles. Tobacco mosaic virus inclusions were very unstable and she was unable to observe them intact in the electron microscope. Tobacco etch virus inclusions were stable and she could observe very opaque crystals and some rod-shaped particles associated with them. Rubio-Huertos (1950) extracted inclusion bodies associated with different viruses from the infected cells for electron microscopy and studied their composition. He obtained evidence that the crystalline inclusions of tobacco mosaic virus aucuba strain were formed by virus particles. This was later confirmed by Steere and Williams (1953) using a very efficient technique. Amorphous inclusions studied by Rubio-Huertos (1950) were formed by cytoplasmic elements and virus particles in most cases.

The technique of ultrathin sectioning has confirmed most of the early findings and has added many new interesting facts. Masses of virus particles can be found within inclusion bodies or in any other place in the cytoplasm.

Observation by light microscopy of thick sections of methacrylate or araldite embedded tissues for electron microscopy for locating inclusion bodies, and subsequent observation of ultrathin sections by electron microscopy, give a useful correlation of both light and electron microscopy.

Most amorphous inclusions induced by different viruses are just an overgrowth consisting of a mass of cytoplasm protruding into the central vacuole of the cell. Within these inclusions are masses of virus particles arranged in different patterns depending on the virus. The inclusions are surrounded by cytoplasmic-membrane and tonoplast, and most of them also have vacuoles. The X-bodies formed by amaranthus virus are like a sponge with many ducts communicating with the central vacuole of the cell. The protrusions of the cytoplasm into the central vacuole seem to be induced by the dense bands corresponding to the crystalline needles seen in the light microscope.

Amorphous inclusions from tobacco mosaic virus have been described by several authors. Matsui (1959) divided the X-bodies into two main categories according to their structural characteristics. The first one included bodies which appear elliptical and are almost entirely composed of dense granules which clearly differ from ribosomes. The other types are usually spherical and consist of a narrow peripheral zone and large internal vacuole-like space. The peripheral zone appears in the form of a dense granular structure, the granules being of variable size. The tonoplast is frequently associated with the peripheral zone. Frequently masses of virus particles are found associated with the external outline of the X-body.

Shalla (1964) detected another type of X-body in tomato plants infected with the common strain of tobacco mosaic virus. Such a structure has also been found by Kolehmainen et al. (1965) later in tobacco mosaic virus-infected *N. tabacum* cells. This X-body appears as a cytoplasmic matrix having a loose tangle of tubules. These tubular formations are much bigger than tobacco mosaic virus particles and their role in virus infection is not known.

X-bodies associated with *Brassica* virus 3 are different from other X-bodies. They are round, oval, or elongated, and are formed by a dense matrix which may be vacuolated. Within the matrix and vacuoles, abundant spherical virus particles forming a very compact body are found. However, no membrane separates the X-body from the cytoplasm. These X-bodies are formed by the aggregation of virus particles and dense material which appears at the beginning of infection and is scattered in small patches in the cytoplasm (Rubio-Huertos et al., 1968).

X-bodies associated with petunia ringspot virus were observed in infected *Vicia faba* leaf cells. They are mainly composed of tubular elements, morphologically identical with those described in some animal cells as smooth endoplasmic reticulum, forming a compact mass in which the crystalline inclusions were embedded. These tubular elements are responsible for the positive Golgi specific staining of the X-body and probably originate by hypertrophy of Golgi apparatus. These X-bodies also contain normal cytoplasmic material such as ribosomes and mitochondria. The whole body is enclosed by tonoplasm and communicates with the normal cytoplasm by cytoplasmic strands. A similar tubular structure has been reported by Fujisawa et al., (1967) in the case of X-bodies associated with sugar beet mosaic. These X-bodies were denser than the nucleus and morphologically distinct from the nucleus, plastids, mitochondria, and ctyoplasmic ground substances. No envelopes were found around them. They are formed exclusively by the tubular elements which are like the smooth endoplasmic reticulum.

Beet yellow virus inclusions in leaf cells of beet, as seen with the light microscope, are contained in mesophyll cells. They are short, compact, and appear as bands forming a herringbone pattern. Under the electron microscope they appear to consist of filamentous particles that conform in size and structure with the virus particles. The fibrous inclusions are aggregated into strands. In the more compact banded form of inclusion the particles form crystalline arrays (Esau et al., 1966).

Some X-bodies associated with spherical viruses consist entirely of virus particles as in the case of broad bean mottle virus (Rubio-Huertos and van Slogteren, 1956; de Zoeten and Schlegel, 1967*a*) and rice dwarf virus (Fukushi et al., 1962).

Electron microscopy of X-bodies associated with turnip yellow mosaic virus infection in *Brassica pekinensis* showed that they are formed by masses of degenerate plastids and chloroplasts, osmophilic inclusions, and crystalline arrays of spherical particles smaller than the actual virus particles (Rubio-Huertos et al., 1967). These X-bodies are separated from the central vacuole by tonoplasm and are connected directly with the normal cytoplasm.

Tobacco mosaic virus crystalline inclusions, as stated earlier, are composed of virus particles. The crystalline inclusions of other rod-shaped viruses, while showing striate patterns as in the case of intranuclear crystals induced by severe etch virus, are composed of rod-shaped particles of similar size as virus particles. There is some doubt about their composition. Infectivity of the crystals has not been proved and some authors think that they are probably formed only by protein. The rods could be only capsids.

Petunia ringspot virus particles may be icosahedral and may form "tubular crystals" like catalase (Kieselev et al., 1967). The big rhomboid crystalline inclusions may be true crystals formed by virus particles.

Virus particles within X-bodies or in the cytoplasm can be arranged in different patterns depending on the infective virus (Edwardson, 1966).

V. THE ROLE OF INCLUSION BODIES IN VIRUS IDENTIFICATION

While many viruses in their host cells form distinct cytological entities, which can be used for a rapid identification of the virus, their utility is rather limited in the case of viruses where the inclusion bodies are not very characteristic. Many viruses do not form inclusion bodies.

Etch viruses and tobacco mosaic virus are easily recognized from other tobacco viruses by their inclusion bodies. Observation of epidermal strips even without fixation or staining is sufficient for a quick identification. Cytological examination appeared to be enough for diagnosis of dahlia mosaic-infected plants. In the field, sometimes crops of *Brassica oleracea* or *B. pekinensis* infected with cauliflower mosaic virus and cabbage black ring virus show different types of mosaic and vein-banding patterns which are not distinctive. The best way for a quick identification in these cases is the examination of epidermal strips of different samples with phloxine without fixation. The inclusion bodies induced by the two viruses are easily distinguished.

Among legume viruses, identification by inclusion bodies is possible in only a few cases, although this criterion is very useful in distinguishing the groups. This method has been successfully used in a survey of legume viruses in Spain (Rubio-Huertos, et al., 1967).

Petunia ringspot virus is easy to identify by the inclusion bodies, which are specially useful in case of symptomless, masked, and necrotic strains.

Inclusion bodies characteristic of lily viruses have been described in Japan by several investigators and can be utilized for their identification (see McWhorter, 1965).

Examples of viruses which produce inclusion bodies that are distinct enough for rapid identification are listed below:

Bean yellow mosaic virus (when forming crystalline inclusions as well as X-bodies)
Cabbage black ring virus
Cactus virus
Cauliflower mosaic virus
Cotton leaf crumple virus
Dahlia mosaic virus
Lily viruses
Petunia ringspot virus
Petunia virus 1
Pisum virus 2
Red clover vein mosaic virus
Tobacco etch viruses

Tobacco mosaic virus
Watermelon mosaic virus

Inclusion bodies due to some viruses are not very specific but their study, combined with other techniques such as electron microscopy, especially the "dip" method and negative staining, can provide reliable identification. A good example of this is inclusion bodies due to broad bean mottle virus which are similar to those formed by phaseolus virus 2 in an early infection. The former are composed of spherical virus particles which are easily detected in the "dip" preparations in contrast to the latter which consist of elongated particles.

VI. EXAMINATION OF TISSUES FOR INCLUSION BODIES

Most of the classical histological methods used for both animal and plant tissues have been employed. Sheffield (1941) tested many fixatives, stains, and combined stains on inclusion bodies produced by tobacco etch virus. Several other authors applied numerous staining combinations to fixed tissues and paraffin sections. Bald (1949*a*, 1949*b*) realized the importance of fixation and developed new fixation methods for plant tissues. The use of semivital stains, without previous fixation, by McWhorter was a great step towards simplicity and accuracy (McWhorter, 1941*b*, 1957).

Methods that indicate the chemical nature of the inclusion bodies or react with RNA or some known portion of the elemental particles were tested. Ultraviolet and fluorescent-microscopy techniques have been usefully applied in this field, mostly in detecting centers of virus formation and in studying early stages in the formation of inclusion bodies (Hirai and Hirai, 1964). Fluorescent antibody technique has been used mostly for tobacco mosaic virus. Lately phase contrast microscopy has been used for the observation of inclusion bodies in vivo.

A. Sampling of Tissues

Inclusions are usually abundant in leaves showing distinct symptoms and are located mostly in the epidermis of the leaves. They are found even in the epidermis of stems, and may be present in yellow areas of the leaves. They are found also in the integuments of flowers. Usually inclusion bodies are present in cells corresponding to areas showing macroscopic symptoms. Epidermal strips should be taken from these areas for preliminary survey of inclusions. Epidermal strips can be examined with or without fixation.

Small portions of areas showing macroscopic symptoms should be used. Sometimes a part of green tissue is included to study differences.

Sections should be used in cases where the nature of the leaf makes it difficult to take epidermal strips or when the surface of the epidermis has some secretion (wax or gum) making observations very confusing or impossible. Sections are employed when inclusion bodies are located in specific tissues and are absent from the epidermis.

Trichomes from stems or hairy leaves of infected plants are especially good material even without previous fixation and staining. Detached recently infected tissues bearing the trichomes are kept alive by floating them in nutrient solutions. In some cases, hair cells can be inoculated directly near the tip or in the basal cell. This has been used mostly for studies on tobacco mosaic virus multiplication (Zech, 1952; Hirai and Hirai, 1964).

B. Methods

1. RAPID METHODS WITHOUT FIXATION

Direct observation of living unstained tissues: Microscopic examination of epidermal strips mounted in tap water is sufficient to reveal inclusions. Cytoplasmic streaming and movements and formation of inclusions are easily observed in large trichomes. Phase contrast optics bring out better contrast and resolution.

Rapid staining methods: The trypan blue method (McWhorter, 1941*b*) employs 0.5% trypan blue in 0.9% NaCl. The dye is dissolved in hot 0.9% aqueous NaCl. The stock solution is diluted in 0.9% NaCl to 1/2000-1/50,000.

Epidermal strips are mounted in this dilute trypan blue solution. Nuclei turn blue within 30 sec after the strips are immersed, and as portions of amorphous inclusions stain strongly, the inclusion masses are readily distinguishable from the nuclei. The tissues of some plants, particularly trichomes, resist penetration by trypan blue. Rich (1948) devised an ether-alcohol mixture to facilitate penetration of the stain.

While some crystalline inclusions stain deeply, trypan blue does not stain others such as tobacco mosaic virus crystals.

The phloxine method (McWhorter 1941*b*) was first developed by using phloxine in combination with trypan blue. Rubio-Huertos (1950) used 1% aqueous phloxine alone. After 30 sec in phloxine epidermal strips are washed in tap water and mounted in a drop of water. Nuclei stain pink, nucleoli and inclusion bodies bright red. Crystalline inclusions of tobacco mosaic virus remain colorless. Plastids, chloroplast, and cytoplasm also remain colorless or are very pale pink.

Crystalline inclusions associated with most viruses are stained bright red by phloxine and even crystals or acicular virus formations within X-bodies can be easily distinguished from the matrix of the amorphous inclusion by their deeper staining. The crystalline inclusions formed by tobacco mosaic virus which do not stain with phloxine without fixation can be stained easily after fixation with Carney's or other fixatives.

The pyronin-methyl green staining method uses 0.5% methyl green and 0.2% pyronin in 0.1 M acetate buffer, pH 5.3. Pure methyl green should be used. It is customary to extract the dye with chloroform to remove methyl violet.

Fresh epidermal strips are immersed in the above solution for 5-10 min, rinsed, and mounted in water. Nuclei stain blue and inclusions red. This method is rapid and gives some information about the chemical nature of the material. The blue color indicates the presence of deoxyribonucleic acid (DNA) and the red ribonucleic acid (RNA). This reaction has given good results only with certain viral inclusions. The method is not completely specific in distinguishing the nucleic acids.

Recently Christie (1967) has developed some rapid staining procedures for viral inclusions using a mixture of methyl cellosolve (ethylene glycol monomethyl ether), 95% ethyl alcohol, and distilled water (2:1:1). The stains used in the above solvent are: calcomine orange 2 Rs., congo rubin, "Luxol" brilliant green Bl. methyl green, methylene blue chloride, and phloxine B. All of them are used as 1% solutions.

With these stains, three combinations are used: (a) calcomine orange-"Luxol" brilliant green (1:4); (b) congo rubin-methyl green (1:10); (c) phloxine-methylene blue. Combination (a) was found best for differentiating tobacco mosaic virus. It stained crystalline inclusions brilliant green, X-bodies an olive brown, plastids light yellow, and nuclei orange. The three methods were found suitable for differentiating inclusion bodies formed by several viruses. Results with different stain combinations and tobacco mosaic virus, tobacco etch virus, and clover yellow mosaic virus have been reported by Christie (1967).

All the rapid methods without previous fixation of the tissues described herein are not suitable for permanent microscopic preparations. Only quick dehydration in alcohol and mounting in Europal can preserve some of the stains.

2. *METHODS USING PREVIOUS TISSUE FIXATION*

Most of the stains used in classical cytological studies are unable to stain the cell contents directly without disturbing their structure. They are not even able to penetrate the cells without a previous treatment of the tissues. The purpose of fixation is to kill the cells, preserving the structure of the whole cell as nearly as possible to the living state, and facilitate the entrance and deposit of the stains in specific sites. A good fixative will rapidly coagulate or precipitate the cell contents while maintaining the morphology and structure of cell organelles preventing any further damage by staining, washing, dehydration, and embedding of the tissues.

Most classical fixatives of different composition have been used for inclusion bodies and all of them provide a morphological preservation of inclusions and nuclei. Fixatives containing acetic acid or formalin are not able to preserve cytological details and certain cellular organelles.

The following fixatives have been widely used for plant inclusions: Champy's fluid, Bouin's fixative and Allen's modification (Sheffield, 1941), Zenker's fluid and Carnoy's fluid, chromic acid, acetic acid, and formalin saturated water (Esau, 1965). Dalton's fixative consisting of mercuric chloride and a small quantity of acetic acid followed by washing with iodine in alcohol has been excellent. Zenker's fluid contains 5 g mercuric chloride, 2.5 g potassium dichromate, 1.0 g sodium sulphate, 5 ml glacial acetic acid, and 100.0 ml distilled water. This fixative does not keep well. This problem is avoided by preparing a stock solution without the acetic acid; the acid is added just before use.

The tissues are fixed for 1-12 hr, washed overnight in running tap water, and then transferred to 50% and subsequently to 70% ethanol containing iodine, several hours in each. Sufficient iodine is added to the ethanol to make it dark brown.

Carnoy's fluid contains 60 ml absolute ethanol, 30 ml chloroform, 10 ml glacial acetic acid. The tissues are fixed for ten minutes to an hour and washed in 95% or absolute alcohol. This is one of the most penetrating fixatives known.

Flemming's fluid contains 15 ml 1% chromic acid, 14 ml 2% aqueous osmium tetroxide, 1 ml or less glacial acetic acid. The solutions are kept separately and mixed just before use.

Bald (1948, 1949, 1964) devised a method which retains accurately both viral and cytological structures. He found that iodine dissolved in potassium iodide acts as a fixative that gives exceptional staining properties to the tissues. Solberg and Bald (1964) improved this formula and used the following solution: 0.90 g potassium iodide, 0.64 g calcium iodide, 0.90 g iodine in 100 ml 50% ethanol.

3. *PERMANENT PREPARATIONS*

Permanent preparations are made in two different ways. One uses epidermal strips which after fixation are stained, washed, dehydrated, and mounted on a slide with Canada balsam using the following schedule:

(1) Fixation (any of the methods described above)
(2) Washing (according to the method described)
(3) Staining (following any of the methods described in the section on staining)
(4) Washing (according to the stain used)
(5) 50% ethanol, 2 min
(6) 70% ethanol, 2 min
(7) 80% ethanol, 5 min
(8) 96% ethanol, 10 min
(9) Absolute ethanol, 5 min
(10) Absolute ethanol, 10 min
(11) Clove oil, 2 min (changing the oil two times)
(12) Washing with xylene until all the clove oil has been eliminated
(13) Mounted in Canada balsam on a glass slide and covered with a cover slide

The second method uses small pieces of tissues, embedding them in paraffin or other embedding materials, sectioning, and after removal of the embedding material, staining, dehydrating, and mounting in Canada balsam.

The following paraffin embedding schedule is used:

(1) *Fixation*—Depending on the kind of fixative used, that is, with Carnoy's fluid the tissue can be washed in 95% ethanol.

(2) *Dehydration*
   (a) 70% ethanol, 30 min
   (b) 80% ethanol, 15 min
   (c) 96% ethanol, 15 min
   (d) Absolute ethanol (1), 20 min
   (e) Absolute ethanol (2), 20 min
   (f) Absolute ethanol (3), 20 min

(3) *Clearing*
   (a) Xylene (1), 10 min
   (b) Xylene (2 changes), 10 min

(4) *Impregnation*
   (a) Paraffin in the oven at 54-60° C, 1 hr
   (b) Paraffin 2nd change at 54-60° C, 1½ hr
   (c) Paraffin 3rd change at 54-60° C, 30 min

(5) *Casting*

After casting the blocks and sectioning in an appropriate paraffin microtome, the sections are always compressed and usually pleated due to the process of cutting. They need to be flattened out and mounted on slides before they can be stained and an adhesive solution is needed.

Egg albumen is mixed with an equal volume of glycerol (known as Mayer's albumen solution); it is then filtered through a Buchner funnel by vacuum filtration and a crystal of thymol is added to the filtrate as a preservative. The solution will keep for months in a well-stoppered bottle.

For mounting sections, dilute the glycerol-albumen solution by adding 3-4 drops of the stock mixture to 15 ml of distilled water. Place a small drop of Mayer's albumen solution on a slide. Spread the solution uniformly into a very thin film. Place a few drops of distilled water on the albumen-coated slide and then float the sections on the drop avoiding air bubbles. Place the slide in an oven at 38-40° C. The sections will flatten in a few minutes. Hold the slide edgeways to eliminate the surplus fluid and let it air dry. The following procedure may be used for removing the paraffin and hydration for staining:

(1) Xylene, 5 min
(2) Xylene, 5 min
(3) Xylene, plus absolute ethanol, 2 min
(4) Absolute ethanol, 2 min
(5) 96% ethanol, 2 min
(6) 70% ethanol, 2 min
(7) 50% ethanol, 2 min
(8) 30% ethanol, 2 min
(9) Aqueous stain

After staining and washing, the sections are dehydrated again, cleared, and mounted in Canada balsam as follows:

(1) 50% ethanol, 2 min
(2) 70% ethanol, 2 min
(3) 95% ethanol, 2 min
(4) Absolute ethanol, 2 min
(5) Absolute ethanol, 5 min
(6) Clove oil, 2 min
(7) Xylene, 2 min
(8) Xylene, 2 min
(9) Canada balsam

4. STAINING METHODS AFTER FIXATION

Most of the histological methods used for animal tissues could also be used for inclusion bodies (Sheffield, 1941). Useful tabulations were made of the staining reactions of tobacco mosaic virus and severe etch virus infected cells against several well-known staining methods and dyes.

Differentiation of inclusion bodies from normal elements of the cell can be better achieved by some combination of dyes or methods such as hematoxylin-eosine, Mallory's triple stain, or Flemming's or Giemsa's stain. The last has been widely used for inclusion bodies.

Giemsa's Method. Giemsa's stain consists of definite proportions of methylene blue, methylene azure, and the eosinates of both. It produces consistently good results, provided it is correctly made and is applied in solutions of the proper pH (6.5-6.8). The stain is commercially available.

Zenker's fluid is the best fixative for Giemsa's method, but Rawlins (1957) recommended Bald's fixative, and the following procedure:

(1) After washing away the fixative, cover epidermal strips or sections with several drops of undiluted Giemsa's stain (without orange G) for about 7 min.

(2) Wash in several changes of distilled water for 20 min (the sections or strips will appear purple after this treatment).

(3) Pour into a large dish of water and again pick up on a microscope slide.

(4) Absorb the water from around the strip or section on filter paper.

(5) Add several drops of absolute ethanol and leave for only a few seconds. Place the slide on white paper so that the color change in the strip may be seen. As soon as the strip changes from purple to blue or green, quickly pour off the ethanol and add several drops of clove oil to the strip. If left in the alcohol too long the cell inclusions will become blue to green, like the host chromatin and cell walls.

(6) After a few minutes, pour off the clove oil and replace with fresh clove oil; continue this process until the oil shows no green and the cell inclusions appear well differentiated when viewed under the low power of the microscope. It may require 10 min or more to obtain the best differentiation.

(7) Remove excess clove oil with filter paper.

(8) Flood the slide with xylene to remove the clove oil and change xylene at intervals of 10 min. Gently move the strip on the slide during this treatment so that the xylene will wash the clove oil from both surfaces of the strip. Do not allow the strip to become dry. Mount in Canada balsam.

VII. CHEMICAL NATURE OF INCLUSION BODIES

General staining methods rarely give information on the chemical nature of inclusion bodies. More specific cytological methods or cytochemical reactions are employed. The best known and most reliable method employs the Feulgen's reagents. It is widely used also for chromatin and eventually other DNA-containing structures. Feulgen's or Schiff's reagent is a solution of basic fuchsin which has been decolorized by SO_2 or by sulfurous acid as described below. It returns to its original red color in the presence of free aldehyde groups such as the 2′, 3′ moeties of ribose.

Basic fuchsin, 1 g
Sodium metabisulphate (anhydrous), 1.9 g
Normal hydrochloric acid, 15 ml
Distilled water, 85 ml

Shake at intervals for 2 hr and then add 0.5 g of activated charcoal; shake for 2 min, filter. Store the solution in a dark bottle.

The Feulgen's method is as follows:

(1) Ready the fixed sections (Zenker's or Helley's fixative) for aqueous staining. Treat them with iodine solution followed by sodium thiosulphate solution.

(2) Wash for 3 min in running water.

(3) Place in 1 N hydrochloric acid maintained at 60° C and leave for 10 min.

(4) Rinse in distilled water.

(5) Place for 1½ hr in Schiff's reagent. Do not rinse.

(6) Wash for 5 min in each of three lots of sulphurous acid solution (6 ml of 10% potassium metabisulphite and 5 ml NH_4CL per 100 ml).

(7) Wash for 3 min in running water.

(8) Counterstain for 15 sec in the light green solution (0.25% in 70% ethanol).

(9) Alcohols, xylene, balsam, or Xam.

(10) Treat another slide in the same way omitting step 3.

A purple or bluish-red color which appears when hydrolysis is performed but does not appear without hydrolysis shows the presence of deoxyribonucleic acid. Nucleic acid will be purple or red and crystalline and amorphous inclusion bodies light green.

Pyronin-methyl green with previous fixation can also be used for differentiating DNA but it is not very specific.

Cytochemical methods such as Sakaguchi reaction, Millory's, and others have been used for a long time. The advances in the knowledge of chemical composition of isolated cellular components have made the task of identifying the chemical composition of inclusion bodies easier and more accurate by electron microscopy. Examination of ultrathin sections of an amorphous inclusion body may reveal that it is composed by cytoplasmic material in which virus particles are embedded. Ribonucleoprotein from ribosomal material, ribonucleoprotein from the virus particles, and lipo- and phospholipoproteins from the membranes of endoplasmic reticulum and mitochondria can be distinguished. Lipid droplets are also easily identified by electron microscopy. In case of crystalline inclusions it is not easy to distinguish their chemical composition. In some cases, as with tobacco mosaic virus, inclusions are composed of infective virus particles. On the other hand, most authors think that intranuclear crystalline inclusions induced by severe etch virus are just capsid material without nucleic acid although they are formed by rod-shaped particles similar to severe etch virus particles. Here, the knowledge of the morphology alone does not help and it would be necessary to resort to other electron microscopic techniques such as autoradiography.

Ultraviolet light microscopy has also been used to obtain information about the RNA or virus contents of inclusion bodies by direct staining of epidermal strips or sections in acridine orange (0.1% or 0.01% solutions in 0.15 M phosphate buffer at pH 6.3-6.8) or indirectly by the use of immunofluorescent antibody techniques. These techniques have been used mostly for studies on virus multiplication.

4

Electron Microscopy of Viruses

ROBERT G. MILNE*

Plant Pathology Department
Rothamsted Experimental Station
Harpenden, England

I. INTRODUCTION

As a spot-check on infected material and virus preparations, in revealing virus structure, and in showing the distribution and development of viruses in infected tissue and the effects they have on the host, electron microscopy has become indispensable. Perhaps, though, the uses of electron microscopy are obvious, so I shall touch briefly on the problems.

For example, when looking for viruses in sectioned cells it is well to think of a tobacco mosaic particle in a cell as being roughly in proportion to a needle in a haystack, and in purified preparations, it proves difficult to find virus particles in suspensions more dilute than 10^8 particles per ml (equivalent to about 0.1 μg per ml of virus).

Although electron microscopes can resolve objects separated by as little as 5 Å (sometimes less), biological preparations seldom offer better than 20 Å resolution because fine structures get degraded during preparation and examination or obscured by the coarser grain of staining materials used to enhance contrast. With or without the use of stains, fine detail may in principle be resolvable, but may not be seen because contrast is too low.

Near the practical limit of resolution, weak "signals" may be obscured by random variations in contrast or "noise," though indirect methods of recovery can be used if the structure to be "rescued" is periodic. These methods include the superimposition of images and optical diffraction analysis. It is also possible, but only in favorable cases, to distinguish the contributions of each

*Present address: Clinical Research Center, Northwick Park, Hassaw, Middlesex, England.

side to images produced when both the top and bottom surfaces of virus particles are represented.

Sections of cells used in electron microscopy sample about a thousandth of the cell volume; they also represent events only at one point in time. Since plant virus infections in individual cells cannot yet be synchronized routinely, great care is needed to reconstruct a process such as virus maturation. (But see two important papers by Takebe et al. [1968] and Takebe and Otsuki [1969] on isolated protoplasts; also see Sakai and Takebe, 1970.)

Electron microscopy has until recently been unsatisfactory in one more respect: it has provided almost no biochemical information. Now, techniques are becoming available for digesting preparations with a variety of specific enzymes, for detecting at least some enzyme activities in tissues, for detecting antigens with labeled antibodies, for following autoradiographically the fate of precursors, and for sequencing nucleic acids. So the versatility of the electron microscope is growing, to complement its ability to reveal fine physical detail.

Perhaps here a word of apologia and caution: Although electron microscopy may be part of science, its pursuit involves a good deal of both art and prejudice. Certain techniques work well on one side of the Atlantic and not on the other; certain opinions are likewise held strongly in some laboratories, are anathema in others. I have picked my way as I may, since I have not space or skill to be comprehensive.

II. INSTRUMENT AND LABORATORY

Good descriptions of electron microscopes exist (Sjöstrand, 1967; Siegel, 1964; Kay, 1965) and will not be duplicated here. Users need not understand all the electronics and mechanics involved so long as they can correct astigmatism and tell whether the machine is in good order, and provided that one experienced person can maintain the instrument.

For biological work, 60-80 kv is used since at lower voltages, resolution and penetration of the specimen fall off and specimen damage increases, whereas at higher voltages contrast decreases. Very high and very low voltage electron microscopes are at present of little use in virology.

A. Calibration

The magnification of the electron microscope is usually read from the meter indicating the intermediate lens current, with the other lenses at fixed current. However, this reading may be in error by ±5-10% for two main reasons. First, the magnification produced by electron lenses is influenced by the residual effect of recent settings (hysteresis). This can be normalized somewhat by switching the lenses rapidly off and on three or four times in a fixed sequence. Second, the current in the objective lens must be varied to focus the specimen, and this produces further magnification changes.

A magnification error of 10% is acceptable for many purposes but it is sometimes important to make independent calibrations as outlined below. For further discussion, see Bahr and Zeitler (1965) and Elbers and Pieters (1964).

1. *BUILT-IN CALIBRATION*

In the Siemens Elmiskop I and IA microscopes, for example, there are apertures of known diameter whose images can be projected onto the intermediate and final screens. Adjustments are made until the images coincide with scribed circles on the screens. This should ensure that magnifications are within 2% of those stated, though Hitchborn and Hills (1968) have found that errors may be larger.

2. *POLYSTYRENE LATEX*

Polystyrene latex spheres (see Fig. 4-7) have in the past been used as calibration standards; one early batch was remarkably uniform in diameter. However, more recent batches are less uniform and errors can arise from shrinkage or swelling during storage, preparation, or irradiation in the beam. Polystyrene spheres are not therefore recommended as precise standards.

3. *DIFFRACTION GRATING REPLICAS*

Shadowed replicas of square-ruled diffraction gratings [*1*]* ruled at 54,800 lines per inch (=2160 lines per mm) make excellent standards (see Bahr and Zeitler, 1965) but have some disadvantages. At high magnification (say 75,000 ×) the lines are too far apart to be measured accurately; therefore the microscope must be calibrated at lower magnification, including in the field of view a piece of dust, the dimensions of which are measured. The microscope is then set at the high magnification and the particle of dust remeasured for the final calibration. Further, the grating cannot be imaged with the specimen or subjected to identical conditions of preparation and irradiation. These sources of inconvenience and error are avoided if the preparation includes internal magnification standards as described below.

4. *PLATINUM PHTHALOCYANIN CRYSTALS*

Markham et al. (1964) used these crystals as an internal standard, the crystals being negatively stained on the same grid as the specimen to be calibrated. The only lattice spacing resolved is stated to be 11.94 Å. The substance makes a good standard but is not available commercially, is rather tedious to prepare, and being sensitive to irradiation, may collapse in the beam. Luftig (1967), who discussed crystal lattices as magnification standards, used copper phthalocyanin (spacing 12.4 Å).

5. *INDANTHRENE OLIVE CRYSTALS*

(Labaw, 1964, and personal communication). This material is easier to prepare than copper or platinum phthalocyanin and its lattice spacing is said to

---

*Italicized numbers in brackets refer to sources of materials listed in the appendix.

be 24.9 Å. To prepare it, proceed as follows:

(1) Mix one part (by volume) of the powdered dye indanthrene olive [2] with two parts quinoline (b.p. 235-237° C) and boil while stirring for 30 min. Small crystals are formed; cool the mixture.

(2) Concentrate and wash the crystals in quinoline by alternate high and low speed centrifugation.

(3) Dip a glass rod into a dense suspension of crystals in quinoline and touch it to a water surface.

(4) Pick up some crystals from the surface with a filmed grid.

6. CATALASE CRYSTALS

(Wrigley, 1968; Cox and Horne, 1968; Luftig, 1967, 1968; Longley, 1967). Beef liver catalase [3] crystal-lattice standards are easily obtained; the crystals are stable in the electron beam and their lattice spacing is consistent within 3% from batch to batch and in different negative stains. Cox and Horne (1968) recommended using the crystals straight from the bottle (probably the best procedure) though Wrigley (1968) recommended recrystallizing the catalase and fixing it in glutaraldehyde before use.

Extensive and careful measurements have been made on the catalase lattice spacing (the distance between the lines that run parallel to the long axis of the crystal, that is, actually half the true lattice spacing) by X-ray and electron diffraction and by using accurately calibrated electron microscopes. The most reliable figures range between 83.5 Å and 89 Å and Wrigley has suggested a working figure of 86 Å with a probable error of $\pm 2.5\%$ for absolute length measurements. Relative measurements made within one laboratory may have an accuracy of $\pm 1.5\%$.

7. TOBACCO MOSAIC VIRUS PARTICLES

The normal length of tobacco mosaic virus is within 3% of 3000 Å. When the dimensions of viruses other than tobacco mosaic virus are being measured, tobacco mosaic virus makes a good standard since it can be mixed with the preparation and subjected to the same influences from staining, shadowing, or stretching of the support film. But each particle of this virus, even in a purified and fractionated preparation, is not of a completely standard length, so a length distribution histogram must be made and compared with that of the unknown virus.

Finally, measurements should always be made directly from near the middle of the negative. Distortion in the image may produce differences in magnification of 5% between the middle and the edge of a negative, and measurements from prints are inaccurate due to enlarger errors and paper shrinkage.

B. Photography

To record the image and to resolve finer detail than is visible on the phosphor screen, the screen is displaced and photographic material exposed to the electrons, the emulsion reacting in much the same way as to light. For maximum contrast a rather dense negative should be made but for maximum resolution a thinner negative is better. A compromise must be made between slow, fine-grain emulsions and faster, grainier ones, and a relatively fine-grain emulsion should be chosen allowing ten times enlargement and requiring exposures of 2-3 sec. Longer exposures often blur because of instability in the microscope or specimen drift.

Whether plates or cut films are used is a matter of personal choice and microscope design. Plates are probably easier to process individually and to handle and store but cut film allows more exposures before breaking the vacuum to change cassettes. Manufacturers are increasingly reluctant to make plates.

III. MOUNTING OF SPECIMENS

Specimens for the electron microscope must be very thin (usually between 50 Å and 1000 Å thick) because image formation depends on transmitted electrons and because resolution does not exceed about one-tenth the thickness of the specimen. Such specimens are fragile and need rigid support, usually from a copper grid. Grids of steel, nickel, gold, and other metals are available, and alternative supports having single holes or slots can be obtained [*4*].

For examining cell fractions or purified virus preparations, a support film is placed on the grid as described below and the material is mounted on that. Sections can be supported on slotted or wide-mesh grids previously covered with a film, especially if large areas are to be scanned at low magnification. If higher resolution is required, the thinnest possible specimen must be made and the support film is therefore abandoned. Very thin sections (silver or gray interference colors, 200-500 Å thick) are laid directly on a fine-mesh grid (say 400 bars to the inch). If the sections are fragile they may be stabilized on the grid by coating lightly with carbon, or the sections can be mounted on a wide-mesh grid covered with a support film itself containing large holes.

A. Adhesion of Sections and Support Films to Grids

Sections and support films usually stick well to grids though some trouble may be experienced with Araldite sections. Difficulties may be overcome as follows:

(1) Before films or sections are picked up, bend the grids very slightly and collect on the convex side.

(2) Films and sections adhere more strongly if the grids are first dipped in concentrated HCl or in carbon tetrachloride, then washed.

(3) Grids can be made sticky by dipping them in a dilute solution of cellulose tape adhesive in chloroform.

B. Storing Grids

LKB grid boxes [*5*] provide for easy reference and compact storage.

Preparations in negative stain may deteriorate in time, but shadowed specimens and stained sections that have been in the beam keep well.

C. Cleaning Grids

Old grids can be cleaned and used again. Formvar or Collodion films that have not been in the beam can be removed in the appropriate solvent but grids covered with other materials must be sonicated, then washed and dried.

IV. SUPPORT FILMS

The films used in biological electron microscopy are made of nitrocellulose (Collodion, Parlodion, Pyroxylin, Necoloidine), polyvinyl formaldehyde (Formvar), evaporated carbon, or combinations of these. Bradley's comprehensive account in Kay (1965) should be consulted for details of other kinds of film.

A. Characteristics of Collodion, Formvar, and Carbon Films

Collodion films are rather weak and unstable in the beam, but are easy to make and are hydrophilic so that preparations such as virus suspensions in negative stain spread well. Formvar films are also easily prepared and are somewhat stronger than Collodion but are more hydrophobic. Carbon films are the strongest and most stable in the electron beam but can be brittle unless backed by Formvar or Collodion. They are usually 50-100 Å thick but can be used when only 20 Å thick, whereas Collodion and Formvar films need to be at least 200 Å thick. Because they are thin and stable, carbon films are best for high-resolution work but they take longer to make and are rather hydrophobic.

All kinds of film tend to become more hydrophobic with time so they are best used fresh. Carbon films are hydrophobic partly because oil vapor from the rotary pump contaminates the vacuum chamber during carbon evaporation, and the remedy is to fit a liquid nitrogen oil trap. Alternatively carbon films, while still under vacuum, may be subjected to a high-voltage discharge for a few seconds. Carbon films may also be made more hydrophilic by washing in chloroform.

Bovine serum albumin (0.01%) can be added to a negative stain preparation as a spreader to improve the distribution of stain and virus particles.

Virus particles tend to flatten when dried down on hydrophilic films. This can be overcome by fixing the particles chemically, mixing the preparation with different amounts of negative stain, or freeze-drying, alternatives discussed under Section VI.

B. Preparation of Films

1. FORMVAR [6]

The film is cast on a glass surface, then stripped on water, as follows:

(1) Make up a solution of 0.2%, w/v, Formvar (0.05-0.3% have been used) in

either chloroform or 1,2-dichloroethane (ethylene dichloride) and store it at room temperature in a dark bottle with a wide mouth. It keeps well.

(2) "Clean" a microscope slide with a chamois leather. If the slide is too clean the film may not strip (see also the preparation of carbon films); films may also fail to strip when prepared from a fresh solution.

(3) Hold the slide by forceps, dip it in the Formvar, withdraw it vertically, and allow it to dry.

(4) Scrape the edges of the slide to help release the film from the glass, and lower the slide at a shallow angle into glass-distilled water the surface of which is free of dust. (The surface may be cleaned by overfilling the container and sweeping the surface several times with the glass rod.) The film should part from the glass and float; breathing on the film before immersion may help to release it though this is not usually necessary if the film is stripped at once.

(5) The film can be mounted in various ways.

(i) Press grids onto the floating film, then touch a round bottle to the edge of the film and rotate it so that film and grids are drawn up. Alternatively, hold a piece of newspaper horizontally above the floating grids, touch it to the surface and withdraw it immediately with the filmed grids adhering.

(ii) Put a fine wire gauze on the bottom of the water vessel and load it with grids before the film is stripped. After stripping, lift the gauze so that the film is stranded across the grids.

(iii) If the film has been cut into squares with a sharp point while still on the slide, individual floating squares can be captured from above or below on grids held in forceps.

2. *COLLODION* [*7*]

The film is cast on a clean water surface by letting fall a single drop of 2% Collodion in amyl acetate. The solution spreads and the solvent evaporates leaving a film which is skimmed off. A second similar film, stronger and cleaner than the first, is now cast and collected by methods (i) or (ii) above. Stronger films result when the water is saturated with amyl acetate. Collodion films can also be made in the same way as those of Formvar, using a 0.5% Collodion solution. The solution is best kept in the refrigerator.

3. *CARBON*

An arc is struck across a small area of contact between two carbon electrodes in a vacuum preferably better than 10^{-4} torr and the carbon vapor condenses on a flat surface to form the required film, the thickness of which is judged (crudely but fairly effectively) by the depth of color built up on a piece of white glazed porcelain.

A vacuum unit is set up as in Fig. 4-1 and the ends of the carbon rods [*8*] are shaped with a pencil sharpener and razor blade. A spring is adjusted to push the rods against each other lightly. A cleaned microscope slide is put on the floor of the vacuum chamber 15-20 cm below the arc, and beside it, the porcelain on which stands a drop of vacuum oil.

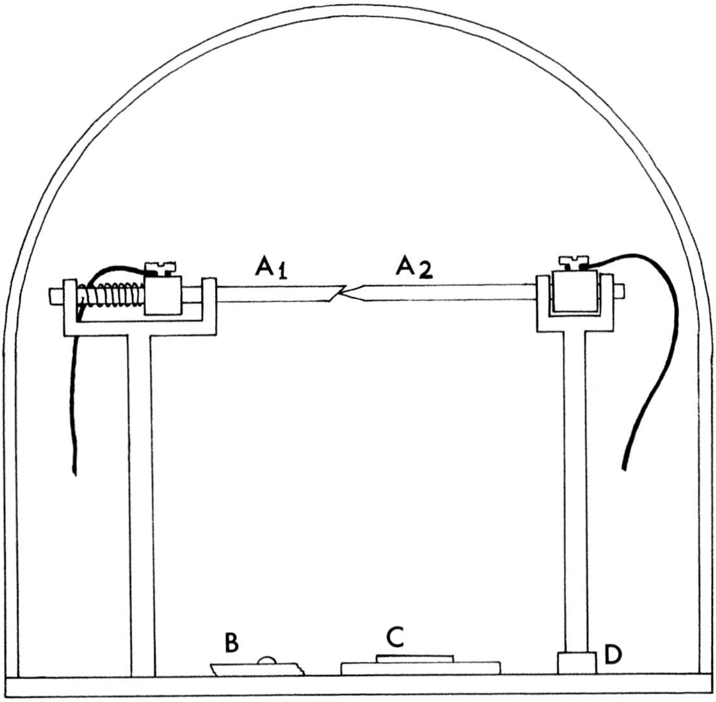

Fig. 4-1. A vacuum unit set up for carbon film deposition. A_1 and A_2, shaped carbon rods, A_1 spring-loaded and A_2 fixed. B, white porcelain with drop of vacuum oil. C, microscope slide. D, insulator.

The chamber is evacuated and the arc struck for about half a second, either by switching directly to a 25 v, 50 amp supply or by increasing the current with a rheostat until arcing begins. After arcing, the area covered by the oil drop stays white but the surrounding porcelain will have darkened; when a difference is just visible the carbon thickness is some 50 Å but the film should usually be just a little thicker. Experience must guide the compromise between thinness and stability. Finally the carbon film is transferred from the glass slide to grids as in the Formvar method.

Fragile carbon films less than 100 Å thick may be prepared as follows: A very thin carbon layer is deposited on Formvar-covered grids and the Formvar is then dissolved by placing the grids on a fine wire gauze and lowering this gently into a chloroform surface for a few seconds.

C. Slide "Cleaning"

Both carbon and Formvar films part reluctantly from clean glass and it is usual to prepare the glass surface in one of the following ways:

(1) Spread a drop of detergent on the slide and polish it off with a cloth.

(2) Smear the slide with grease from the fingers, then polish with a chamois leather.

(3) Wipe the slide with a siliconized spectacle tissue.

(4) Difficulty in stripping carbon films from glass or mica surfaces may be experienced when the evaporating chamber is pumped out to a very good vacuum (say 5×10^{-6} torr) or when backstreaming of oil vapor from the diffusion pump is almost totally prevented by a good cold-trap. Siddall (1960) suggested that some oil contamination on the glass or mica surface is necessary to prevent strong adhesion of the carbon and recommended that a small watchglass of Apiezon, a vacuum oil (vapor pressure 10^{-5} torr at 20° C), should be placed in the work chamber during evaporation.

Some think a glass surface unsatisfactory as a substrate for carbon films because it may be dirty or irregular. Freshly split mica [9] is the best alternative as it is very clean and smooth. The surface is used without further preparation.

D. Films with Holes

A microscope's performance is frequently checked by examining the Fresnel fringes at the edges of round holes in carbon films, and holey or fenestrated films are also useful in making negative stain preparations or mounting sections. The following (Harris, 1962) is a good way to make films containing holes of the required size:

(1) Make a 0.25% solution of Formvar in ethylene dichloride and emulsify it with 1 part in 200 of glycerol.

(2) Cast the emulsion on a microscope slide as usual.

(3) Let the slide dry for 10 min, then play a steam jet over it for 1 min.

(4) Strip the film on water and collect it on grids, then quickly dip them in acetone to perforate any "pseudoholes" (thin unperforated craters).

(5) Finally, increase the contrast and stability of the holey film with a carbon coat. Holes about 2 μ in diameter result.

(6) Larger holes can be produced by increasing the ratio of glycerol to Formvar. For instance, 1/30 glycerol/Formvar gives approximately 7 μ holes and 1/8 gives 25 μ holes.

Formvar or carbon nets for supporting sections can be made as follows (Sjöstrand 1956):

(1) Prepare 2% Formvar in ethylene dichloride.

(2) Line a small beaker with filter paper and saturate its volume with ethylene dichloride vapor.

(3) Coat a microscope slide with the Formvar and at once transfer it to the beaker. Cover the top.

(4) After 20 min blow warm wet air into the beaker via a flask of water at 50-80° C.

(5) After a minute remove the slide and strip the film on water.

(6) Discard the upper third of the film and mount the rest on grids, coating it with carbon if desired.

V. PREPARATIONS FROM FRESH PLANTS

In this and the following section, ways of increasing contrast in preparations are mentioned, but Section VIII describes them in more detail.

A. Dip Preparations

The simplest way to check whether a plant contains recognizable virus particles is to put a drop of water on a clean glass slide and draw a freshly cut plant surface through the drop (Brandes, 1957, 1964; Hitchborn and Hills, 1965). Part of the contents of the ruptured cells flows out into the drop and also forms a film on its surface (see below). A filmed grid is touched to the surface of the drop and adhering material can then be given contrast. Small regions of tissue can be tested quickly in this way and virus particles are easily seen against a reasonably clean background (Fig. 4-2).

As an extension of the dip method, tissue can be gently macerated in negative stain (see Almeida and Howatson, 1963; Purcifull and Edwardson, 1967) and touched to a filmed grid which is then drained with filter paper. The result is not as clean as a dip preparation but organelles and virus particles may retain something of their original relationships.

B. Leaf Exudate Method

Johnson (1951) found that when water pressure is applied to cut stems, droplets exude from the attached leaves or from their cut ends. With virus-infected plants, virus particles can be detected by touching the drops to grids and examining these in the electron microscope. The technique has been useful but the dip method is easier and is generally preferred.

C. Spreading

As long ago as 1948, Fernandez-Moran pushed needles into animal tissue, withdrew them, and slowly lowered them onto water so that the material on the needle was spread in a very thin surface film suitable for examination in the electron microscope. Parsons (1963) lowered such needles into negative stain. The Kleinschmidt and Zahn spreading technique (for a recent version see Lang et al., 1967, and below) for preparing single nucleic acid molecules has been adapted for plant material as follows (Kleinschmidt and Milne, unpublished):

(1) Make a trough of Perspex (Plexiglass) about 25 cm long, 12 cm wide, and 1 cm deep. Paint it black outside and wash the inside briefly with a solution of parrafin wax in chloroform, to make it hydrophobic.

(2) Slightly overfill the trough with double-distilled water and sweep the surface clean with a glass rod.

(3) Sprinkle the surface lightly with talc.

(4) Cut a 1 mm square piece of leaf (say) from the plant, crush it on a slide with a flat-ended glass rod, and immediately touch the rod to the water surface.

(5) The cell contents spread on the water surface, displacing the talc to a diameter of some 3 cm. Collect a sample by gently touching a filmed grid to the

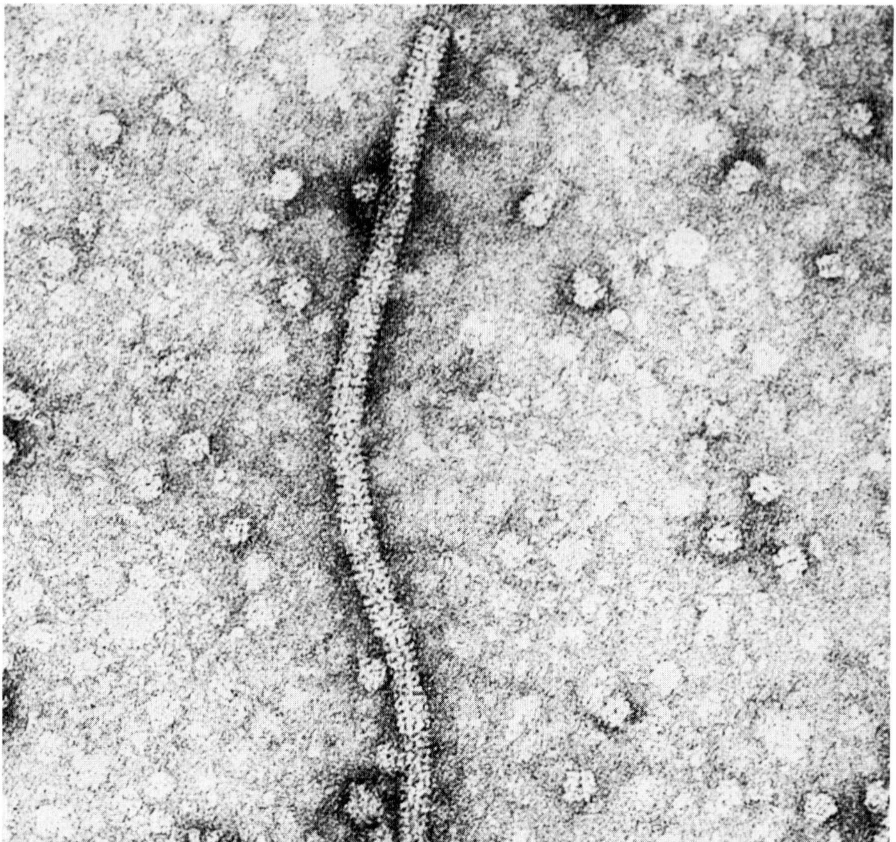

Fig. 4-2. A particle of potato virus X negatively stained with uranyl formate, from a dip preparation of potato virus X-infected tobacco leaf, about ×300,000. The particles in the background are mostly "fraction one protein," a normal cell constituent. (Preparation and photo by Roy Woods.)

edge of the spread material. Failure to spread usually means that a contaminating film is already present on the surface due to dirty water.

The resulting preparation will contain virus particles, polyribosomes and single ribosomes, membranes, nucleic acid filaments, chloroplast fragments, and proteins but will be free of salts, sugars, and other low molecular weight compounds (see Figs. 4-3 and 4-4). Stripped leaf epidermis spreads less violently than whole-leaf material. Surface activity can be further moderated by putting 0.01-0.001% sodium dodecyl sulphate (detergent) in the trough.

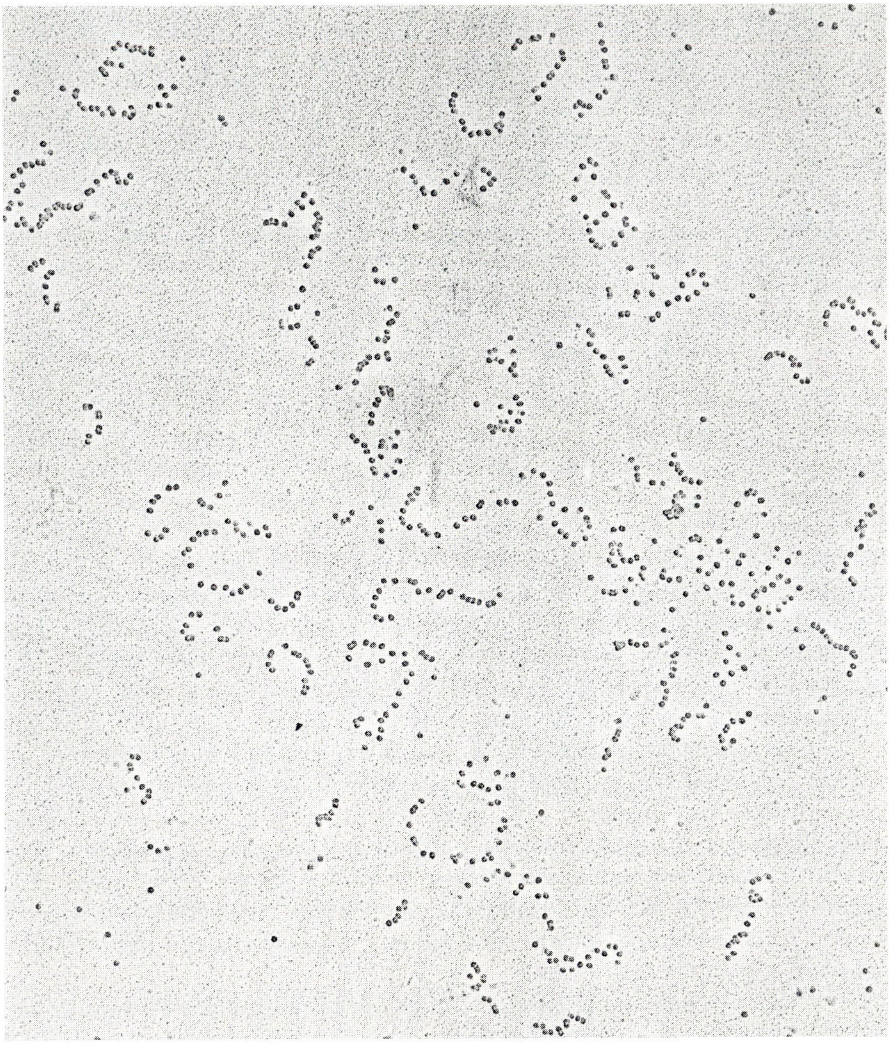

Fig. 4-3. A field of polyribosomes, the result of spreading cells from healthy tobacco leaf epidermis on water. Rotary shadow with uranium. ×40,000.

(4) Dust the surface with talc.

(5) Mix 0.01% cytochrome c (the spreader protein) with the nucleic acid and deliver a small volume down the ramp. Allow spreading to occur.

(6) Slide the free barrier so that the film is slightly compressed. Touch a carbon-coated grid to the film and then to a drop of absolute alcohol. Drain the grid with the edge of a filter paper.

(7) Give the preparation contrast by shadowing at 5° from the horizontal, using either a rotary shadow (see below) or two separate shadowings at right angles.

Further refinements are often made, for instance, Mitra et al. (1967) cited two methods for spreading single-stranded DNA:

Method A. DNA plus cytochrome c in 2 M ammonium acetate is spread onto the surface of 0.1 M ammonium acetate.

Method B. DNA plus cytochrome c in a solution of 0.1 M NaCl, 0.0067 M KH_2PO_4, 0.0034 M Na_2EDTA, and 10% formaldehyde is spread onto 0.5% formaldehyde in water.

Mayor and Jordon (1968) have described a micro-method for which only 0.01 μg of nucleic acid is required. It goes as follows:

(1) Prepare a clean surface of Teflon (polytetrafluoroethylene or PTFE). A polythene surface might do as well.

(2) Add a drop of 0.2 M ammonium acetate containing the nucleic acid.

(3) Dip a needle into a cytochrome c solution (concentration experimentally determined) and touch it to the drop. Cover to prevent evaporation.

(4) After 20 min pick up the film on a grid.

For further references, see Lang et al. (1967), Freifelder and Kleinschmidt (1965).

VII. PURIFIED AND PARTLY PURIFIED VIRUS PREPARATIONS

A. Direct Addition of Material to the Grid

Preparations in solution can be spread and picked up as described above, can simply be touched to a filmed grid, or can be sprayed onto the grid with an atomizer. The easiest method is the second. The preparation is touched to a filmed grid held in forceps and before the grid dries it is washed with a few drops of negative stain from a dropper held just above. The grid is then drained with the edge of a filter paper. This technique has the advantage of washing out small molecules such as salts or sucrose from a gradient, which might otherwise mar the quality of the image.

As an alternative to washing the grid with negative stain, a mixture of equal volumes of stain and virus material can be put on the grid, which is then drained and dried. If the result is poor the proportions of stain and virus material can be adjusted.

B. Spraying Methods

A useful low-velocity spray gun is the glass Vaponefrin nebulizer [*10*] (Fig. 4-6). The horizontal delivery tube should be fitted with a downward right-angle extension and the grid should be sprayed from a distance of about 3 cm while held in forceps or perched on a small metal pedestal. Most of the droplets are deflected if the grid lies on an extended flat surface or if the electrostatically charged droplets are allowed to build up a like charge on the grid.

A good alternative gun is the high-velocity Aerograph airbrush [*11*], which must be driven by an air pressure of some 12 psi but which can be adjusted to give a majority of droplets of a desired size. The grid is sprayed from a distance of 20 cm.

Spraying is a little more tedious than adding material directly to the grid and may damage elongated virus particles. However, the resulting droplet patterns may give a better disposition of stain around the particles and is also useful for counting particles as outlined below.

C. Particle Counting Using Spraying Methods

Probably the best way to estimate virus particle concentration is that proposed by Williams and Backus (1949) and Backus and Williams (1950), modified by

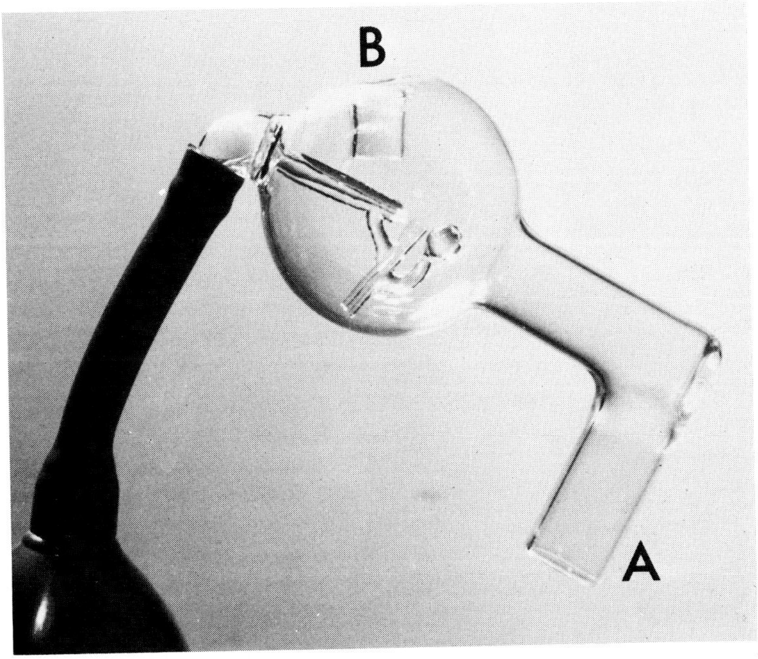

Fig. 4-6. The Vaponefrin nebulizer, fitted with a downward extension, A. A few drops of the material to be sprayed are inserted through the top at B.

Nixon and Fisher (1958) and used, for instance, by Frost et al. (1967), in which a polystyrene latex standard [*12*] is used. A latex suspension of known particle concentration is made up, mixed with the virus particles, and sprayed on a grid. The ratio of latex to virus particles is then determined for whole droplet patterns (Fig. 4-7). The steps are as follows:

(1) Determine the average volume, V, of the latex spheres from their stated or measured diameter.

(2) Take the density of dry latex as 1.05 g/ml. The weight of one sphere is $V \times 1.05$ g.

(3) Determine the dry weight, D, of latex in a given volume of suspension. The number of particles in this volume is then $D/(V \times 1.05)$.

(4) Mix the standardized latex in known proportions by volume with the unknown virus suspension.

(5) Spray the mixture onto filmed grids with a high-velocity spray gun. (With a low-velocity gun many of the droplets evaporate before they reach the grid and clumping of particles results).

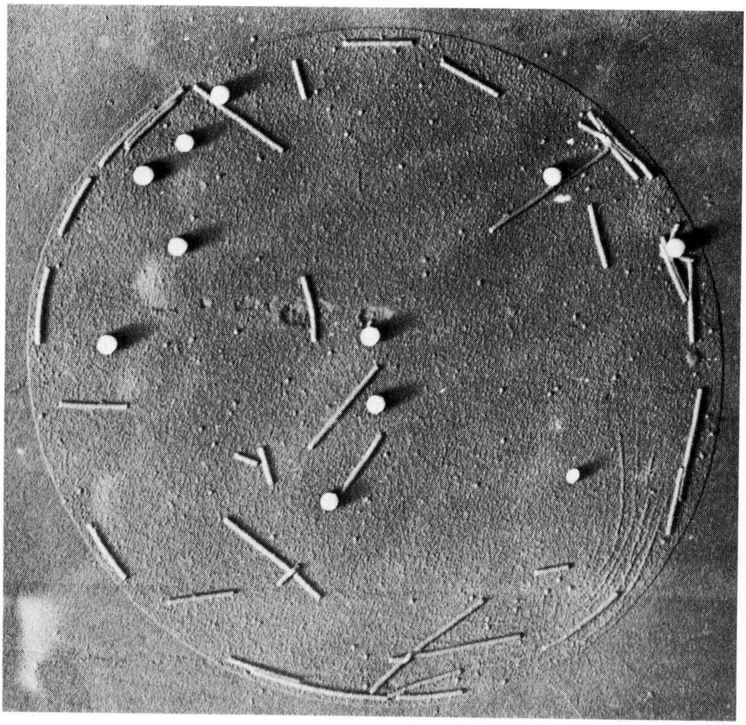

Fig. 4-7. A spray droplet from a mixed suspension of latex spheres and tobacco mosaic virus particles, shadowed with palladium. From such droplets, virus concentrations can be determined as described in the text. (Preparation and photo by Roy Woods.)

(6) Determine the ratio of latex to virus particles within the profiles of entire single drops. This should represent the ratio in the original mixture.

A cascade impactor [*13*] was used by Nixon and Fisher (1958) to select drops of a suitable size and this device is described by them and Horne, in Kay (1965).

The main sources of error are probably the estimation of the volume and density of the latex. Nixon and Fisher suggested the error may be as large as 30%.

The presence of any solid material in the dried droplet other than latex and virus particles interferes with the electron microscope image, so water or volatile buffers are the best suspending media. Backus and Williams (1950) suggested using 2% ammonium carbonate or ammonium acetate. It should be noted that, as these buffers dry, the pH of ammonium acetate drops to about 4 and that of ammonium carbonate rises to 10, although these extremes may be no greater than those occurring when nonvolatile-buffered or unbuffered solutions are dried. A modification that alleviates this problem was suggested by Kellenberger and Arber (1957). The virus suspension is sprayed onto a Collodion film deposited on agar and after low molecular weight substances have diffused into the agar the Collodion film is remounted on grids.

For a fuller account of spraying techniques, see Horne's Chapter in Kay (1965).

D. Fixation of Preparations

Virus particles may collapse or be altered in various ways under the influence of surface tension, salt concentrations, pH extremes, enzyme attack, or degradation by negative stain. This can be prevented by fixing the material for 10 min with an equal volume of 2% neutralized formaldehyde or glutaraldehyde.

E. Freeze Drying

Freeze drying can be a useful way to preserve biological structures for electron microscopy (Williams, 1953; Horne, in Kay 1965; Sjöstrand, 1967) but is seldom used by plant virologists. This may be because negative staining helps to support and preserve structures during drying better than was possible when preparations were only shadowed. Also, chemical fixation as described above is quick and convenient whereas the results of freeze drying may be quickly undone by readsorption of moisture before the specimen can be put in the microscope.

VIII. SHADOWING, NEGATIVE STAINING, AND POSITIVE STAINING

Valentine (1961), Zobel and Beer (1965), and Sjöstrand (1967) discussed contrast and image formation in detail; here only a short introduction is necessary.

In the electron microscope image, contrast is obtained largely because the denser parts of the specimen scatter electrons out of their normal paths; the objective aperture then excludes these deflected electrons and prevents them from reaching the final plane of focus. At high resolution, phase contrast effects are also important but these will not be considered.

Different regions of, for example, virus particles or thin sections of plants vary little in density and are similar in density to their supporting film or embedding medium, so their images in the electron microscope are faint. Biological specimens are therefore given added contrast by attaching to them atoms of a heavy element whose distribution, it is hoped, parallels that of the structures investigated. There are three principal ways of providing this contrast: by shadowing, by negative staining, and by positive staining.

A. Shadowing

Shadowing was very popular before the advent of negative staining but it more severely limits resolution, obscures internal details, and is more laborious. It is still useful for specimens such as spray droplets, nucleic acid preparations, and replicas of freeze-etched material, (see Figs. 4-3, 4-4, 4-5, 4-7, and 4-8) and also provides a check on the results of negative staining.

The specimen to be shadowed is placed face up and horizontal in a vacuum and is coated with a thin layer of electron-dense material by evaporating from a source above and to one side. The evaporated atoms travel in approximately straight lines so they coat the presented surfaces of projections but not areas "unseen" by the source; such areas in the image resemble shadows and the resemblance is often enhanced by reversing the normal contrast in the final print, so that these shadows appear dark. After the evaporated particles land they do in fact creep short distances, and for this reason or because of later bombardment in the beam, the contrasting materials may become granular, decreasing resolution. Bradley, in Kay (1965), suggests that in any event hydrocarbons from gaskets and pump oil may coat the specimen, in the vacuum chamber, with a structured film which limits resolution to about 30 Å.

Shadowing is usually done with heavy metals used individually or alloyed in an attempt to reduce the size of the final crystalline grains. Examples are uranium, palladium, gold, and platinum, and gold-palladium, platinum-palladium, and platinum-iridium alloys. In general, metals with the highest melting points (which are therefore more difficult to evaporate) give the least granular result and platinum is the best of these. For very fine shadowing, a platinum-carbon mixture may be evaporated by arcing between carbon rods one of which has a platinum core [*14*], but for many purposes a uranium shadow is satisfactory, especially since negative staining provides a high-resolution alternative.

In practice, tungsten wire about 7 cm long and 0.5-1.0 mm thick is bent into a V and clamped between terminals in the vacuum chamber with the V pointing at the specimen. Tungsten wire is brittle when cold and is easier to bend under hot water or in a flame. Five or ten milligrams of the metal to be evaporated, as foil or wire, is looped over the filament and the chamber is evacuated, preferably to 10^{-4} mm Hg or better since a good vacuum reduces graininess. The filament is slowly heated by passing through it a large current at low voltage until the metal charge melts and evaporation occurs. The process is conveniently watched through a fogged electron microscope plate. The optimal angle of shadowing (about 30°), the amount of metal evaporated, and the distance between filament and specimen (5-15 cm) are found by trial. See Kay (1965) for further details.

If very sharp shadows are desired, an aperture can be interposed between filament and specimen. A simple aperture consists of a round 2-10 mm diameter hole in a metal foil placed 1-2 cm in front of the source. It is also advisable to clean the inside of the bell jar to reduce metal "bounce" off the walls.

B. Rotary Shadowing

Rotary shadowing is used for contrasting nucleic acid molecules or other strands of very thin material (see Fig. 4-3, 4-4, and 4-5). Double-stranded DNA or RNA (about 20 Å thick) cannot easily be contrasted with either positive or negative staining and single-stranded molecules are even thinner. A metal shadow originating from one side is also unsuitable since parts of the molecule may lie parallel to the shadowing direction and remain uncontrasted.

The specimens are placed in the vacuum chamber on a horizontal surface which can be rotated uniformly at about 60 rpm, either by a small motor within the chamber (this will increase pumping time) or preferably from outside via a gasket or a magnetic drive through the base plate. The external drive can be turned by a motor or by hand. Shadowing proceeds normally except that the metal is evaporated slowly at an angle of 5° from the horizontal.

An alternative to rotary shadowing is to shadow the specimen from two directions at right angles.

C. Negative Staining

A mixture of the negative stain and the material to be contrasted is placed on a support film and dried. The essence of the process (and the derivation of its name) is that the stain does not react with the particle but merely surrounds it, outlining it is electron-dense material. There is some objection to calling the contrasting substance a "stain" when that is exactly what it is not, but the meaning seems clear and a better word is hard to find. The main reasons why the technique succeeds are that it is quick and simple, the stain often interpenetrates the particle substructure outlining morphological subunits, and it dries to a nearly structureless "glass" which, in pools of the right depth, supports the particles against the severe distorting forces of drying. Resolution is better than that obtainable with shadowing (see Fig. 4-2).

A drawback is that negative stains, particularly phosphotungstate, may alter or disrupt the specimen. Various negative stains should therefore be tested at different pH values and concentrations on fixed and unfixed material and the results compared with those of shadowing.

Many substances have been tested as negative stains and some successful ones are sodium or potassium tungstate and phosphotungstate, ammonium molybdate, sodium phosphomolybdate, lanthanum nitrate, uranyl acetate, uranyl nitrate, and uranyl formate. For most purposes the virologist can confine himself to four: potassium phosphotungstate, ammonium molybdate, and uranyl acetate and formate.

1. *PHOSPHOTUNGSTATE*

Potassium or sodium phosphotungstate is made up as a 1% or 2% aqueous solution which is stable in the light at room temperature, sometimes buffered with 0.1 M ammonium acetate or carbonate buffer. It is a good general negative stain but destroys some kinds of specimen. For instance, ribosomes and alfalfa mosaic virus particles must first be fixed in formaldehyde to survive phosphotungstate treatment though they can both be negatively stained in uranyl acetate without prior fixation. Phosphotungstate may also empty a proportion of initially "full" particles of certain spherical viruses. In one instance (Agrawal, Bruening, and Milne, unpublished) pure "bottom component" (as judged in the analytical ultracentrifuge) of cowpea mosaic virus was treated on the grid with 1% aqueous uranyl acetate (pH 4.2) or with 2% sodium phosphotungstate (pH 7) and then examined in the electron microscope. Of 2000 particles counted in each case, 2.5% were "empty" in uranyl acetate and 19.2% in phosphotungstate.

2. *AMMONIUM MOLYBDATE*

A 2% or saturated solution (about 4%) is made up in water and may be adjusted to pH 6-8 with ammonia. It keeps in the light at room temperature and is a good alternative stain to phosphotungstate or uranyl acetate.

3. *URANYL ACETATE*

A 1% aqueous solution is made up and kept in a dark bottle since it precipitates in strong light. The pH is near 4.2 but this does not damage the majority of specimens. The stain gives better resolution than phosphotungstate. Uranyl acetate precipitates above pH 5.5 but van Bruggen et al. (1962) added EDTA (ethylenediamine tetraacetic acid) to make a complex soluble at neutrality. One percent uranyl acetate was mixed with equal volumes of 0.5% disodium EDTA dihydrate and 0.1% ammonium acetate and the pH was adjusted with ammonium hydroxide or acetic acid. Haemocyanin was successfully stained with the preparation and Bradley (1962, 1965) used it on bacteriophages but results have not been reported with plant viruses.

Some materials, especially nucleic acids, may become positively stained when kept for some time in uranyl acetate, so positive and negative staining may both occur and complicate interpretation.

4. URANYL FORMATE

The solid salt is slightly unstable in the light and the solution more so. The substance is available commercially but may also be made as follows (Leberman, 1965):

(1) Dissolve uranium trioxide in a fivefold excess of hot dilute formic acid.

(2) Filter, then evaporate the solution by boiling until crystallization begins.

(3) Cool the solution, collect the crystals by filtration, wash them with cold ethanol, and dry them on the filter paper.

All operations are done in dim light. The solid can be stored at 20° C in the dark for at least a year but 1% working solutions (pH about 5) are best prepared at least weekly and kept dark. Bradley (1965) has an alternative recipe for the aqueous stain starting with uranyl nitrate.

With spherical viruses such as turnip yellow mosaic virus (Finch and Klug, 1966) and broad bean mottle virus (Finch and Klug, 1967), uranyl acetate and uranyl formate perform about equally; with rod-shaped viruses so far tried (see Fig. 4-2), uranyl formate resolves fine structure better than any other negative stain (see Finch, 1964; Varma et al., 1968).

5. INTERPRETATION OF NEGATIVE STAIN IMAGES

Negative stain images are more difficult to interpret than images from sections or shadowed preparations, because the image of a particle in negative stain may derive contrast from the side nearer the support film or, especially in stain suspended over holes, from both sides. Sometimes the two sides make unequal contributions to the image (see Finch and Klug, 1965) and the resulting pattern varies with the orientation of the particle. This variation is at first confusing, but with help from models and from micrographs obtained with a tilting stage, it may lead to a more accurate interpretation of the structure. Optical diffraction can also help to distinguish and separate the contributions of each side of the particle to the image.

A second problem is that images of single virus particles or ordered aggregates of them often contain the same, only partially resolved, information repeated many times against a background of "noise." Several ways of extracting this information have been developed, including optical diffraction, rotation, and linear integration.

a. Optical Diffraction. The technique is discussed elsewhere in this book and also by Rosier and Klug (1968), Finch and Klug (1966, 1967), Caspar (1966), and Klug and Finch (1965). Polaron Instruments Ltd. [*4*] market an electron micrograph optical diffractometer designed in conjunction with Markham and Horne, and Gall (1967) described how a light microscope can be modified for optical diffraction more cheaply but at some sacrifice of performance.

b. Rotation and Linear Integration. Markham et al. (1963, 1964) used optical and photographic methods to clarify symmetry elements whose number or repeat distance is known, or to find these quantities when they are not clear from the original image. The technique is simple, and useful provided the result is interpreted with caution (see Agrawal et al., 1965 and Frisch, 1969). Rotation is applicable to spherical viruses and linear integration to rods. The rotation procedure of Markham et al. (1963) is as follows:

(1) Enlarge the image (in more than one step if necessary) until it is some 8 cm across when projected in the enlarger.

(2) Put a pin through the center of a large white card and mark the card's periphery into a number, n, of equal angular divisions. Prepare a set of such cards where n has values of, say, 5, 6, and 7 or 10 to 20 according to the symmetry of the image, known or suspected.

(3) Determine the exposure time, t seconds, for a normal print.

(4) Center the symmetry axis of the image on the card.

(5) Center a photographic paper on the pin.

(6) Make an exposure of t/n seconds and rotate the paper by $360/n$ degrees.

(7) Repeat step (6) n times and develop the paper.

(8) If the symmetry of the particle is n-fold, reinforcement will occur, but if not, the superimposed images will be blurred; therefore repeat steps 4-7 using different values for n until the best reinforcement is found.

(9) Check whether real symmetry elements in the original object have been reinforced by comparing the superimposed print with the original and repeating all steps with several, preferably at least ten, original images.

C. Linear Integration

The apparatus and its working cannot here be fully described but see Markham et al. (1964). The method was designed for use with tobacco mosaic virus and can be employed only with linearly repeating structures such as straight sections of rod-shaped viruses or uniformly crystalline arrays of spherical viruses. A composite photographic image is built up, as with the rotation method, by photographing an enlarged print of the virus, which is then moved a specific distance along its long axis and rephotographed on the same negative. When the distance by which the print is moved corresponds with the repeat distance of some structure in the print, reinforcement occurs, but if not, the composite negative is blurred.

The same cautions apply as to the rotation method, with one more: If the print used for the master image is distorted during enlarging or drying (as is likely), calculation of repeat distances will be inaccurate and there may be no reinforcement. Distortion-free enlargements can be made on aluminum-backed paper.

D. Positive Staining

Positive staining of extracted particles offers few advantages over negative staining and shadowing and is used less often.

One percent aqueous uranyl acetate is most commonly employed and has some specificity for nucleic acids. The stain is left in contact with the preparation for half an hour or more and then rinsed off. It may be warmed, and ethanolic or methanolic solutions are often used.

IX. REPLICAS

Transmission electron microscopy of thin replicas is still the only way to obtain high-resolution pictures of the surfaces of thick objects. Many replication techniques exist and Bradley's chapter in Kay (1965) should be consulted for a detailed account. The plant virologist may find replication methods useful for examining fresh plant surfaces such as inoculated leaf surfaces (see Gerola et al., 1969) and freeze-etch preparations.

A. Replication of Leaf Surfaces

The method of Williams and Juniper (1968) is recommended.

(1) Cut the leaf into 1 cm squares and attach them to glass slides with cellulose tape.

(2) Evaporate onto the leaf surface about 150 Å of carbon from a source 15 cm distant, with a vacuum better than 10^{-3} mm Hg.

(3) Dip 200-mesh grids in 2% Formvar in chloroform and lay them gently, matt side down, on the carbon-coated leaf surface.

(4) Let the grids dry for a few seconds, then flood the whole leaf surface with 2% Formvar and let it drain.

(5) Let the preparation dry, then press cellulose adhesive tape down onto it rather firmly.

(6) Strip the tape from the leaf. The Formvar, the grids, and the carbon replicas adhere to the tape.

(7) Cut around the grids with a sharp needle and place the resulting discs tape-side down in chloroform. The grids are released from the tape and the Formvar dissolves.

(8) Lift out the grids bearing the carbon replicas and wash them in a second bath of chloroform. Let them dry.

(9) With the replica on the underside (matt side down), place the grid on a clean glass slide in the vacuum chamber and shadow it.

As little time as possible should elapse during stages (1)-(6) because the leaf deteriorates rapidly in the vacuum and under the action of chloroform. If these stages are prolonged it becomes impossible to strip the tape from the specimen.

In the scheme outlined above, the shadowing metal will have been deposited on the back side of the replica, that is, that side away from the leaf surface. The advantage of this is that a positive image of the leaf surface results, in other words, bumps on the leaf are represented by bumps on the replica. The back of the replica follows rather precisely the contours of the front (and of the leaf surface) but some resolution is sacrificed. Tobacco mosaic virus particles on leaf surfaces can be imaged but the resolution is no better than about 50 Å. The alternative is to shadow the other side of the replica and this results in a higher-resolution image that, being a negative of the leaf surface, is more difficult to interpret.

B. Freeze-Etching

Freeze-etching was introduced by Steere (1957) to study virus crystals, including those of tobacco mosaic virus, and was modified by Moor et al. (1961) and Moor and Muhlethaler (1963). Branton and Park (1967) used it to study chloroplasts and Willison and Cocking (1969) to study virus intracellularly (see Fig. 4-8).

The method is expensive, time-consuming, and difficult. In some ways it is similar to sectioning but it has two special characteristics. First, freeze-etching can present views of a cell that has not been fixed irreversibly or subjected to solvents, but has only been frozen very rapidly in the presence of about 20% glycerol. Many kinds of cell survive such treatment alive and the image

Fig. 4-8. A carbon-platinum replica of a tobacco mosaic virus-infected tobacco leaf mesophyll cell, glycerol-impregnated, glutaraldehyde-fixed, and freeze-etched. ×50,000. CW, cell wall; PL, plasmalemma in surface view; C, cytoplasm; T, tonoplast; V, vacuole; TMV, tobacco mosaic virus crystal fractured transversely; P, plastid surface. Inset, a tobacco mosaic virus particle fractured longitudinally showing the axial canal. ×130,000. (Preparation and photos by Dr. E. C. Cocking.)

obtained may therefore be claimed to represent the living state fairly closely. Second, the fracturing process that is part of freeze-etching reveals surfaces not seen in thin sections or by any other methods. The need to use glycerol is unfortunate but without it ice crystals form during freezing and disrupt the specimen.

Freeze-etching is complex and cannot be described here in detail but in principle one proceeds as follows:

(1) Equilibrate the fresh material with 20% glycerol in water. If roots are used, this may be done by growing the roots for some days in the solution.

(2) Rapidly freeze a small piece of the material by plunging it into liquid Freon at −150° C.

(3) Transfer the frozen tissue to the cold table of a freeze-etching unit [*15*] consisting of a fracturing device mounted inside a cold vacuum chamber equipped for shadowing.

(4) Pump out the chamber and then fracture the specimen, discarding the upper part.

(5) Hold the specimen at −100° C for some minutes. Water in the fractured surface sublimes to "etch" the specimen.

(6) Evaporate a platinum-carbon mixture onto the etched surface, making a shadowed replica of it.

(7) Dissolve the specimen in, for example, hot saturated sodium hydroxide solution and mount the replica on a grid for viewing.

X. FIXATION, EMBEDDING, SECTIONING, AND STAINING

Matsui and Yamaguchi (1966) reviewed electron microscope work on embedded and sectioned virus-infected plants but their survey is now rather out of date and did not refer to techniques. Bald (1966) reviewed the light microscope cytology of plant viruses and Esau discussed the anatomy of plant virus infections as seen in both the light and electron microscope. Reviews of methods by Pease (1964), Glauert, in Kay (1965), Morgan and Rose (1967), Wischnitzer (1967), and Sjöstrand (1967) made little reference to plants. Plant virus workers may of course wish to follow their virus into its arthropod or other vector, fixing and embedding animal tissue as did, for instance, Paliwal and Slykhuis (1967) (mites), Shikata and Maramorosch (1967) (leafhoppers), and O'Loughlin and Chambers (1967) (aphids).

A. Fixation

Good fixation requires that in the light microscope the fixed, embedded specimen should resemble its fresh, unfixed counterpart and that where comparisons exist, dimensions of structures should agree with those obtained by independent means such as X-ray diffraction; in the electron microscope organelles should not be broken, displaced, or distorted and a variety of different fixations, including "physical" methods such as freeze-substitution and freeze-etching, should give largely consistent results (for descriptions of freeze-substitution see Bullivant [1965], Zalokar [1965], and Bernhard and Leduc [1967*a*, *b*]). Perhaps as important as achieving a lifelike appearance in the specimen is a realization in the operator that the specimen has been considerably altered by the time it reaches the microscope.

1. FIXATIVES

The useful alternatives are osmium tetroxide, double fixation with an aldehyde followed by osmium tetroxide, chrome-osmium, and permanganate. These are discussed in order.

a. Osmium Tetroxide. As a buffered 1-2% solution this fixative has been used successfully by, for example, Rubio-Huertos and Lopez-Abella (1966), Edwardson (1966), and Purcifull and Edwardson (1967), but unless aldehyde fixatives are unsuitable, they should be used before osmium. The osmium alone does not preserve some of the finest structures such as microtubules and destroys the activity of enzymes so that histochemical techniques cannot then be applied.

Osmium tetroxide is first made up as a stock solution in water or buffer by putting, say, a 0.5 g ampoule containing the osmium and 25 ml water into a strong, dark, ground-glass-stoppered bottle, to make a 2% solution. The ampoule is cut with a file or crushed with a stout glass rod, the stopper quickly replaced, and the bottle set aside for 12 hr for the crystals to dissolve. At greater expense, 2% aqueous osmium tetroxide can be purchased. The solution is stable at 4° C but if it darkens due to oxidation, discard it. Care should be taken to protect eyes, skin, and lungs from the poisonous solution and vapor.

b. Double Fixation: An Aldehyde Followed by Osmium Tetroxide. Sabatini et al. (1963, 1964) fixed animal tissues with a range of aldehydes and determined the best conditions for preservation of structure and enzyme activity. Glutaraldehyde at 4-6.5% was best and acrolein (already used by Luft in 1959) preserved structures well but destroyed enzymes. Acrolein penetrated tissue better than glutaraldehyde. Since 1963 many plant tissues, including virus-infected ones, have been fixed in aldehydes followed by osmium tetroxide and results seem comparable. After aldehyde fixation the tissue is washed in buffer or in some cases can merely be drained and blotted. Fixation is then continued in 1% or 2% osmium tetroxide. Harrison and Roberts (1968) have found 0.1% osmium tetroxide satisfactory.

c. Aldehyde Alone. If aldehyde fixation is not followed by osmium, structures such as ribosomes and some virus particles may become hard to distinguish and membranes appear with reversed contrast. Although fixation is sometimes poor, many enzymes survive and substances such as proteins and nucleic acids remain susceptible to enzyme digestion. This characteristic can best be exploited by using a water-soluble resin for embedding.

Hills and Plaskitt (1968) claimed that osmium fixation degrades some types of isometric plant virus and they therefore used glutaraldehyde only. After fixation, dehydration in ethanol and acetone solutions saturated with uranyl acetate, embedding in Epon, and staining of the sections in lead acetate (Millonig, 1961), they found that the protein coats of isometric viruses were stained and that distinguishing between virus particles and ribosomes was no longer difficult (it is well known that after double fixation the protein coats of viruses are hard to stain, so virus particles and ribosomes often then look similar). The penalty of not using double fixation in the scheme of Hills and Plaskitt is that general fixation of tissues is only moderately good and membranes are indistinguishable.

d. Glutaraldehyde. This is a more popular fixative than acrolein because it is less obnoxious and better preserves enzyme activity. It is used as a buffered solution at 2-5%, made from a 25% or 50% commercial stock solution which slowly oxidizes and becomes acidic even at 4° C. If the pH of the stock falls below 3.5, the results of fixation are poor. Taab Laboratories [*6*] offer "glutaraldehyde EM," a purified solution of pH 5-6 claimed to be stable. Stability is enhanced by keeping the solution well stoppered, cold, and in darkness.

Kay (1965) states that the stock solution may be purified by adding a small quantity of barium carbonate. A precipitate forms and after centrifugation the clean supernatant is used. Frigerio and Shaw (1969) describe a volumetric method for determining glutaraldehyde and discuss conditions under which it may polymerize or oxidize.

Examples of fixation of virus-infected plant tissue by glutaraldehyde and then osmium tetroxide are given by Shalla and Amici (1967), Milne (1967), Shikata and Maramorosch (1967), Weintraub and Ragetli (1966), Rubio-Huertos et al. (1967), and Gardner (1967). See also Figs. 4-9 and 4-10.

e. Acrolein (Acrolaldehyde). Acrolein is purchased as a liquid stabilized with quinol and should be kept in a dark bottle at 4° C. It tends to polymerize and should be discarded when it no longer mixes easily with water. A freshly made 5-10% buffered solution is used for fixation.

f. Glutaraldehyde Plus Acrolein. This mixture has been used occasionally. For instance, Israel and Ross (1967) used 1.25% glutaraldehyde plus 1.25% acrolein to fix tobacco mosaic virus-infected tobacco, and Arnott and Smith (1967) used 6% glutaraldehyde plus 6% acrolein to investigate a virus in sunflower leaves.

g. Glutaraldehyde Plus Formaldehyde. Karnovsky (1965) said that 5% glutaraldehyde plus 4% formaldehyde prefixed mammalian tissue better than

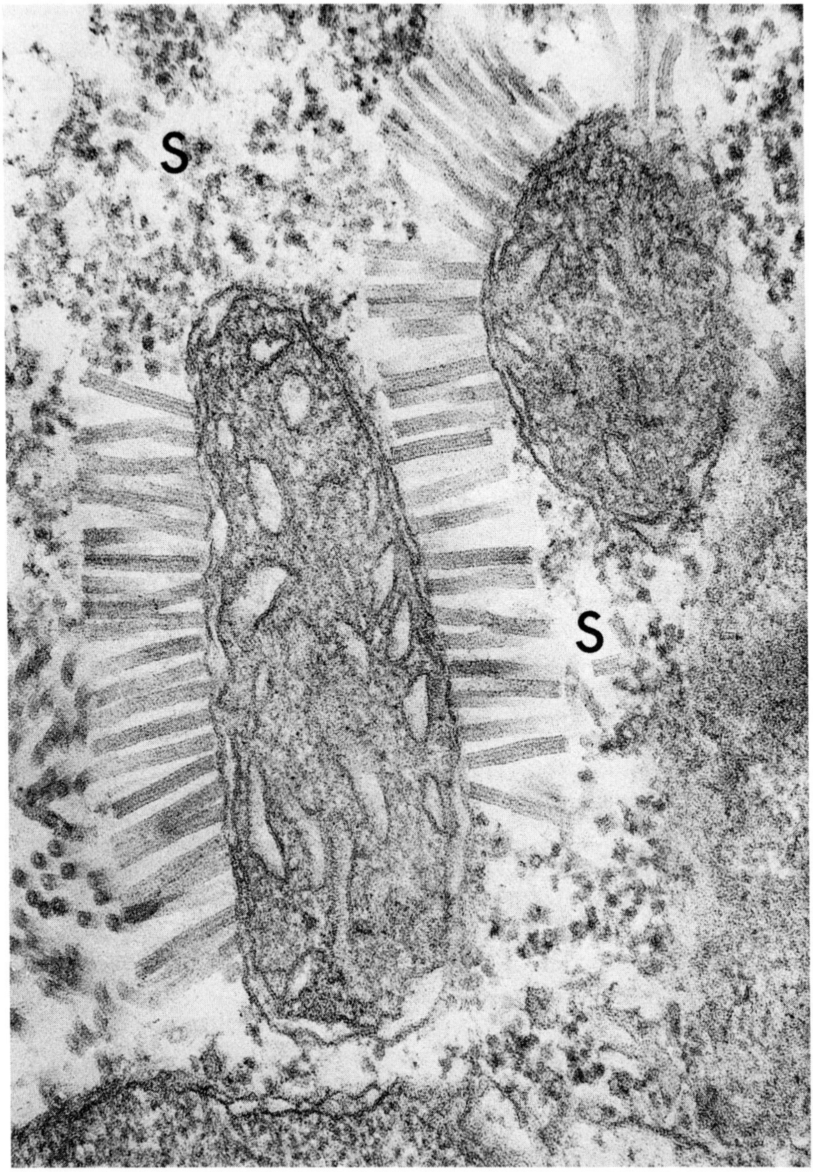

Fig. 4-9. Region of a tobacco leaf cell containing particles of tobacco rattle virus, embedded in styrene-butyl methacrylate and stained in uranyl acetate and lead citrate. ×75,000. The "long" particles are sticking to the mitochondria but the "short" particles, S, are free in the cytoplasm. (Preparation and photo by Ian M. Roberts [Harrison and Roberts, 1968].)

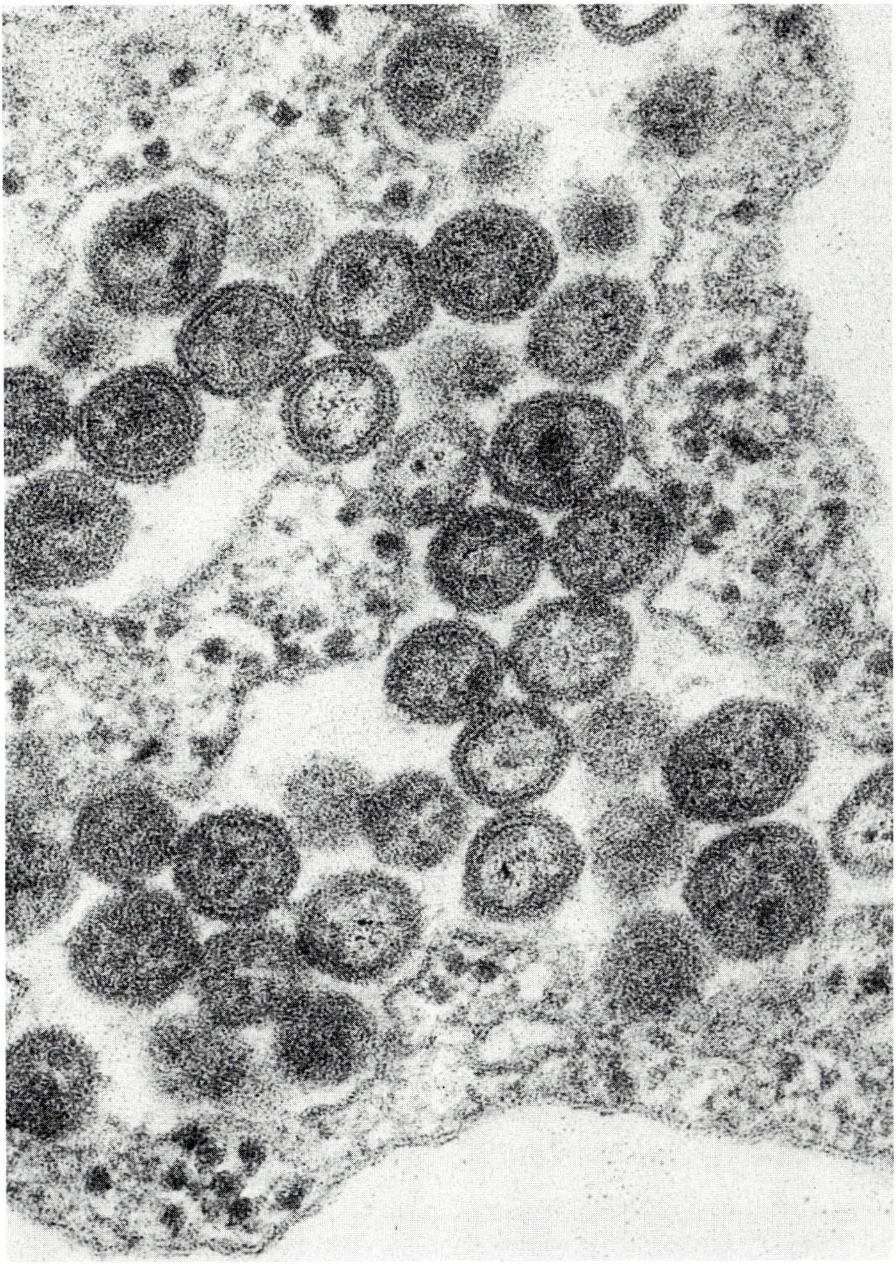

Fig. 4-10. Region of a tomato leaf palisade cell containing particles of tomato spotted wilt virus, embedded in Epon and stained in uranyl acetate and lead citrate. ×200,000.

5% glutaraldehyde alone, probably because the formaldehyde penetrated rapidly. Esau and Cronshaw (1967*a, b*) used the fixative successfully on *Nicotiana* leaves, Karnovsky's directions are as follows:

(1) Dissolve 2 g *para*formaldehyde powder in 25 ml warm water.

(2) Add 1-3 drops 0.1 N NaOH until the solution clears. A slight cloudiness may persist. Cool the solution.

(3) Add 5 ml 50% glutaraldehyde (or 10 ml 25%) and make up to 50 ml with 0.2 M cacodylate or phosphate buffer pH 7.4-7.6. The final pH is 7.2. If cacodylate is used, 25 mg anhydrous $CaCl_2$ is added.

h. Acrolein Plus Formaldehyde. Weintraub and Ragetli (1968) fixed leaves of *Phaseolus* and *Vicia* in aldehydes alone. They found glutaraldehyde unsuitable and obtained better fixation and enzyme digestion by using 5% acrolein plus 5% formaldehyde in 0.1 M phosphate buffer.

i. Chrome-Osmium. Dalton's chrome-osmium fixation (Dalton, 1955) has been used on plant virus-infected tissues by Lee (1967), Kitajima (1965), Kolehmainen et al. (1965), and deZoeten (1966). It can give good results but Milne and deZoeten (1967) found glutaraldehyde plus osmium tetroxide generally better in tests on tobacco mosaic virus-infected tobacco and spotted wilt virus-infected tomato. However, deZoeten and Schlegel (1967*b*), in autoradiographic studies, used chrome-osmium exclusively as serious loss of label was said to occur following glutaraldehyde prefixation. To make chrome-osmium, proceed as follows:

(1) To 80 ml 5% w/v potassium dichromate add sufficient 2.5 N KOH to bring the pH to 7.2. Make the solution up to 100 ml (to give 4% dichromate).

(2) Mix the dichromate solution (one volume) with 3.4% NaCl (one volume) and 2% osmium tetroxide (two volumes). The solution is stable at 4° C.

j. Permanganate. This fixative (Luft, 1956) is usually inferior to aldehyde-osmium. It preserves membranes well but ribosomes and much of the ground structure of organelles are destroyed and enzyme functions are lost. Bradbury and Meek (1960) tested the action of permanganate on animal tissue and gelatin-based models of cytoplasm and concluded that it does not act as a fixative and that any fixation occurring is caused by the subsequent dehydration. This may have been rather an extreme view (see Hake, 1965). Luft's formulation is 0.6% $KMnO_4$ in veronal-acetate buffer adjusted to pH 7.4 with HCl. Permanganate solutions should be kept in the cold in dark bottles filled to the top.

Warmke and Edwardson (1966*a, b*) claimed that very brief permanganate fixation preserved the orientation of intracellular tobacco mosaic virus particles better than glutaraldehyde-osmium. They used unbuffered 1.2% $KMnO_4$ for 2-8 min or 0.6% $KMnO_4$ in acetone for 0.5-2 min at room temperature. Fixation was poor but particle orientation was well preserved. However, Warmke and Christie (1967) had similar or better results with cold dilute osmium tetroxide.

B. Buffers for Fixatives

The concepts of the osmotic pressure and pH of a cell are of uncertain value and the quantities difficult to measure, because so many different membrane-bound systems exist in cells. However, the values of osmotic pressure and pH for plant cells range widely and often differ considerably from those of mammalian cells. This is interesting because the fixatives used successfully on plants are usually those developed for mammalian tissues.

To assess the importance of pH, Milne (1968, unpublished) fixed various healthy and virus-infected leaves in cacodylate-buffered 5% glutaraldehyde, then veronal-acetate-buffered 1% osmium tetroxide at pH values of 5.0, 5.9, 6.8, and 7.4. Good fixation resulted from all treatments and it was concluded that pH could be varied widely without detriment.

Little is known about the effect of the osmotic pressure of fixatives, but with double fixation, control may be less important during the second (osmium) fixation. Buffers are commonly used at a strength of 0.02-0.1 M.

1. TYPES OF BUFFER

Aqueous osmium tetroxide (pH 5.6-6.1 at 1%) and potassium permanganate (pH 5.5 at 1%) have been used unbuffered, though they are more often used with phosphate, veronal-acetate, or cacodylate. Results with all three are good but veronal-acetate buffers poorly in the pH range 5.2-7.5 and reacts adversely with aldehydes, with which it should not be used. Cacodylate is not a very effective buffer above pH 7. Phosphate buffers well over the required range and is compatible with the usual fixatives.

Bennett and Luft (1959) suggested using s-collidine (2,4,6-trimethyl pyridine) but it may cause extraction of proteins (see Glauert in Kay, 1965). Tris (tris [hydroxymethyl] aminomethane), like veronal-acetate, reacts with aldehydes.

Directions for making some buffers or buffered fixatives are given below:

Millonig's phosphate-buffered osmium tetroxide (see Mercer and Birbeck, 1966). Mix 7 ml 2.76% $NaH_2PO_4 \cdot H_2O$ with 43 ml 3.56% Na_2HPO_4 and add 50 ml 2% aqueous OsO_4, to give a pH of 7.3.

Palade's veronal-acetate-buffered osmium tetroxide. Mix together 5.0 ml veronal-acetate buffer (see below), 2.5 ml water, and approximately 5 ml 0.1 N HCl. Adjust the pH to the desired value, then add 12.5 ml 2% OsO_4 and mix.

Phosphate buffer. To make 0.2 M phosphate buffer of any pH between 5.8 and 7.4, use "A" ml 0.5 M KH_2PO_4 plus "B" ml 0.5 M Na_2HPO_4 according to Table 4-1 and make up to 100 ml with water (Long, 1961).

Cacodylate buffer. Mercer and Birbeck (1966) formulated the buffer as follows: mix 100 ml 0.1 M sodium cacodylate with 8.3 ml 0.1 N HCl to obtain pH 7.2 and with 5.4 ml 0.1 N HCl to obtain pH 7.4. Dilute to 200 ml with water.

For a wider range of cacodylate buffers, refer to Table 4-2 (Long, 1961).

Veronal-acetate buffer. Prepare as follows: (1) Dissolve 2.94 g sodium

TABLE 4-1
FORMULATION OF PHOSPHATE BUFFER OF 0.2M IONIC STRENGTH BY MIXING "A" ml 0.5M KH_2PO_4 WITH "B" ml 0.5 M Na_2HPO_4 AND MAKING UP TO 100 ml WITH WATER.

pH at 0°C	*A*	*B*
5.8	32.2	2.60
6.0	28.9	3.70
6.2	24.9	5.04
6.4	20.4	6.54
6.6	15.8	8.06
6.8	11.7	9.44
7.0	8.28	10.6
7.2	5.66	11.4
7.4	3.76	12.1

TABLE 4-2
FORMULATION OF CACODYLATE BUFFER OF 0.2M IONIC STRENGTH BY MIXING "A" ml M CACODYLIC ACID WITH "B" ml M NaOH AND MAKING UP TO 100 ml WITH WATER.

pH at 0°C	*A*	*B*
5.8	63.8	20
6.0	47.6	20
6.2	37.4	20
6.4	31.0	20
6.6	26.9	20
6.8	24.4	20
7.0	22.76	20
7.2	21.74	20

veronal (barbitone sodium, sodium barbital) and 1.94 g hydrated sodium acetate in water. Make up to 100 ml. (2) Use this stock solution at fivefold dilution as, for example, in Palade's fixative (above), adjusting the pH with 0.1 N HCl.

2. *ADDITION OF SUCROSE TO THE FIXATIVE*

Five percent glutaraldehyde in 0.1 M buffer has an osmotic pressure of about 12 atm, probably similar to that of most plant tissue to be fixed. One percent osmium tetroxide in 0.1 M buffer has a pressure of about 3 atm and this is often raised by adding 0.1-0.5 M sucrose. This may be helpful if the osmium is to be used on fresh tissue but is probably not if glutaraldehyde has been used first, because the semipermeability of the membranes will by then have been destroyed.

3. *ADDITION OF SALTS TO THE FIXATIVE*

The inclusion of divalent cations in fixatives has been claimed to improve preservation of nuclei and membranes. Either or both 0.01 M calcium and magnesium chloride have been used, but their effect is problematical.

C. Treatments During Fixation

1. *DURATION OF FIXATION*

Fixation times depend on the tissue and the fixative but should be as short as possible. As a guide, thin leaves are well fixed after infiltration and fixation with glutaraldehyde for one hour, then further fixation of 1 mm wide strips of leaf in osmium tetroxide for three hours, in the cold or at room temperature. Times half as long as these are insufficient, but prolonged fixation, especially in osmium, may leach the specimen. If delay is necessary, for example in sending material to another laboratory or waiting for assay results, aldehyde-fixed material can be kept for several days in the aldehyde or in buffer before fixing in osmium and further processing.

2. *TEMPERATURE OF FIXATION*

Fixation is done either at 0-5° C or at room temperature. Both should be tried as there may be slight differences depending on the kind of fixative and tissue used.

3. *SELECTION AND TRIMMING OF TISSUE*

Where possible, young "soft" plants should be used as they are easier to fix and section. Small roots can be fixed whole, but leaves are too big, as blocks should have two dimensions less than 1 mm to help fixation, dehydration, and embedding. Sometimes blocks of the final size can be cut from fresh tissue, but cells near the edges are often damaged, so it is better to start with larger pieces and trim them after aldehyde fixation, when cutting is no longer injurious.

4. *INFILTRATION OF LEAVES WITH FIXATIVE*

Leaves contain air that prevents liquid fixative from coming immediately into contact with most of the cells, though fixative vapor may do so. The air in the leaf is therefore sometimes removed under vacuum and replaced with the fixation liquid by restoring atmospheric pressure with the leaf submerged. Such infiltration seems a good idea but has not been clearly shown to improve fixation. It can be done as follows:

(1) Put about 2 ml fixative and one or more pieces of leaf about 1 cm square into an 8 × 1.5 cm thick-walled tube. Add a small annular weight to sink the samples.

(1) Image contrast high, sectioning easy, tissue preservation comparable with Epon.

Epon 812	25 ml
Araldite M (CY 212)	15 ml
DDSA	55 ml
dibutyl phthalate	2-4 ml

DMP-30 1.5% or BDMA 3% by volume

(2) Image contrast lower, sectioning easier, tissue preservation better than (1).

Epon 812	62 ml
DDSA	100 ml
Araldite 506	81 ml
dibutyl phthalate	4-7 ml

DMP-30 1.5% or BDMA 3% by volume

(3) Image contrast lower than (1) or (2) and more difficult to section, but tissue preservation excellent, and good for tissues that do not bond well to the resin.

Araldite 506	50 ml
Cardolite NC 513	25 ml
dibutyl phthalate	1-2 ml
BDMA	3% by volume

(If products are not obtainable from the sources mentioned, see Mollenhauer, 1964).

6. IMPREGNATION OF TISSUES WITH EPOXY RESINS

A convenient way to remove acetone from the tissue and substitute Epon or Araldite is described below (see Milne and deZoeten, 1967). Where ethanol has been used for dehydration, it may be necessary to use propylene oxide as an intermediate solvent, especially if penetration of the resin is poor. However, propylene oxide is unpleasant and poisonous and constitutes an extra solvent system that may leach the tissue.

(1) Transfer the acetone-dehydrated tissue to 6 ml of equal parts of acetone and complete epoxy resin in a 5 × 2.5 cm specimen tube. Polyethylene tubes are convenient and are easily cleaned by baking them, then shelling out the hardened resin.

(2) Evaporate the acetone from the open tubes by rotating them in an oven at 40° C, on a motor-driven wheel inclined at 45° (e.g. from Taab Laboratories [*6*]).

(3) When, in a few hours, the smell of acetone has gone, drain the tissue on a paper towel and soak it in fresh resin at the same temperature for a further 4 hr.

(4) Drain the tissue once more and transfer it to fresh resin in capsules [*19*]. Gelatin capsules should be oven-dried before use.

(5) Place the Epon blocks at 40° C until they are fairly hard (about 48 hr) and continue the curing at 60° C for a further 12 hr. With Araldite, cure the

blocks at 60° C for 24-48 hr.

Coulter (1967) suggested three ways to hasten embedding: completing dehydration within only five minutes, vacuum removal of the final solvent (in his case, propylene oxide) in the presence of preevacuated resin warmed to 40° C, and curing the blocks at high temperature (95° C) for 1-12 hr for Araldite and for 3-12 hr for Epon. These accelerated treatments were said to have done no damage and the traditional view that rapid dehydration damages tissues and that high-temperature curing prevents good cutting should be reexamined.*

7. POLYESTER RESINS: VESTOPAL W

Vestopal W (Ryter and Kellenberger, 1958) has been used successfully and Sjöstrand (1967) recommends it highly. It has not become popular, perhaps because batches are said to vary, the components do not keep well, and cutting is not easy.

The resin [*20*] is mixed thoroughly with 1% by volume of the initiator tertiary butyl perbenzoate, and 0.5% of the activator cobalt naphthenate is then added. All components must be stored in the dark. Vestopal W keeps for several months but the initiator and activator must be discarded after some six weeks. The resin is cured in open capsules at 60° C for 12-24 hr. Embedding procedure is as follows:

(1) Dehydrate the tissue in acetone.

(2) Replace the acetone slowly with complete Vestopal, either as described for epoxy resins or stepwise, using 3:1, 1:1, and 1:3 mixtures of acetone to Vestopal, each for one hour.

(3) Drain the tissue on a paper towel and soak it for several hours at room temperature in complete Vestopal.

(4) Drain again and transfer to capsules.

Kurtz (1961) and Sjöstrand and Elfvin (1962) used ethanol dehydration but since ethanol and Vestopal are not miscible, styrene was used as an intermediate. The procedure was said to improve embedding.

8. WATER-SOLUBLE RESINS. GLYCOL METHACRYLATE [*21*]

Water-soluble resins have the great advantage that enzyme digestions can be done on tissues embedded and sectioned in them. Also, tissue can go from the aqueous phase into the resin monomer without passing through intermediate solvents such as acetone. Unfortunately a formulation has not yet been found that gives consistently good preservation of fine structure, though glycol methacrylate (GMA), as used by Leduc and Bernhard (1967) and Granboulan and Leduc (1967), has approached the required standard with animal tissues and bacteria. GMA sections must be laid on a support film and this decreases the resolution possible. The resin was used successfully by Weintraub and

---

*We now cure Epon blocks at 80° C for 4 hr, with no adverse effects.

Ragetli (1968) to study virus-infected beans. The material slowly polymerizes so a fresh preparation must be bought and discarded when not completely miscible with water. All operations, including the UV polymerization, must be done at 0-3° C.

The directions of Leduc and Bernhard (1967) are as follows:

(1) Prepolymerize the resin before embedding, to avoid swelling artifacts. (a) Mix glycol methacrylate with 3% by volume of water; (b) mix butyl methacrylate (still containing stabilizer) with 2% by volume of Luperco accelerator; (c) mix 7 volumes of the glycol methacrylate from (a) with 3 volumes of the butyl methacrylate from (b); (d) put a small amount of this mixture in a large Erlenmeyer flask so that a thin layer not over 1 cm deep (preferably less) covers the bottom; (e) cap the flask with foil and heat it over a flame with very rapid swirling until boiling is about to begin. This should take less than a minute; (f) plunge the flask at once into ice water and continue swirling until the resin has cooled to 0-3° C. The resin should now be a thick syrup, but if it is still too thin, heat it once more. Store the prepolymer in the deep-freeze as long as required and warm it to 0-3° C for use.

(2) Fix the tissue in glutaraldehyde, then rinse it in buffer for 1-12 hr (6-12 hr if nuclease digestion follows). Do not further fix in osmium tetroxide.

(3) Dehydrate the tissue in 80% GMA monomer plus 20% water for 20 min, then in 97% GMA monomer plus 3% water for a further 20 min.

(4) Impregnate the tissue with prepolymer mixture overnight.

(5) Put the tissue into oven-dried gelatin (not polyethylene) capsules filled to the top with prepolymer. Remove all air bubbles, then add the caps, excluding as much air as possible.

(6) Hold the capsules upright with wire or clear plastic supports that permit maximum UV penetration. Polymerize the resin with long-wavelength UV light [*22*]. The tops of the capsules should be 1 cm from the lamp. Polymerization takes 1-3 days.

(7) Mount the sections on Formvar-carbon-coated grids and stain them with saturated aqueous uranyl acetate for 2 hr at 60° C.

Sectioning of glutaraldehyde-fixed material embedded in GMA is routine, but after osmium tetroxide fixation, GMA-embedded material is extremely brittle.

9. *OTHER WATER-SOLUBLE EMBEDDING MEDIA*

Two other water-soluble embedding media of some potential are the epoxy resins Aquon (derived from Epon) and Durcupan. Details of these are in Glauert's review in Kay (1965). Like other epoxy compounds, these are more stable than the methacrylates and allow pepsin and trypsin (but not nucleases) to act on sections made from them.

F. Sectioning

Plant virus materials demand no special techniques for cutting thin sections except that plant tissues in general and vascular bundles in particular are somewhat more difficult to cut than most animal tissues. Good descriptions of sectioning exist (Glauert and Phillips, in Kay [1965], Porter, in Siegel [1964], Morgan and Rose [1967], and Sjöstrand [1967]).

1. *KNIVES* [23]

A good diamond knife, in my experience, cuts thinner and more unblemished sections than any glass knife, and is more convenient to use. It may be set up permanently, with its water bath; it has a straight edge and constant knife angle; and the same part of the knife can be used for trimming and for cutting both thick sections for the light microscope and thin ones for electron microscopy. These merits add up to valuable savings in time and frustration. Knives should remain sharp for several years and can be resharpened for about half price.

However, diamond knives are expensive and easily damaged by carelessness. Delivery times are very long and the quality of individual knives is unpredictable, though makers usually offer free replacement of unsatisfactory ones. For these reasons, many laboratories make and use glass knives. These are cheap, easy to make, disposable, and can produce results not far behind those of a good diamond. The procedure for making a glass knife is briefly as follows.

(1) Cut plate glass about 6 mm thick into rectangles measuring about 25 $\times$ 30 mm. If LKB glass strips [5] are used, they need only be cut transversely at 25 mm intervals.

(2) Lightly score the glass on one side along a diagonal so that the score falls about 1 mm short of each corner.

(3) Apply gentle symmetrical pressure downward on either side of the score mark together with upward pressure beneath it, so that the rectangle breaks into two triangles, one or both possessing a sharp, approximately 40° edge. Because the plane of the break does not continue right into the corner but curves, the actual angle of the cutting edge is difficult to determine, but may be about 50°. LKB [5], or Reichert [24] provide directions for glass knife evaluation with their microtomes.

A glass cutter and a specially beveled pair of pliers supplied by Reichert [25] make knife-making relatively easy, especially if glass strips from LKB are used. The LKB knifemaker is a good investment as it allows high-quality knives to be made routinely by unskilled operators. It should be borne in mind, however, that several diamond knives can be had for the same price. Batches of glass may behave very differently, and if trouble is experienced, different sources should be tried.

Glass knives are best used new but can be stored or transported inserted in Plasticine in a dust-free box.

2. *THE WATER BATH*

As the sections are cut they float onto a water bath in contact with the knife edge; normally this bath is made of adhesive tape or metal foil. A useful alternative is to drape a strand of rubber solution or Durofix glue horizontally across the front of the knife some 2-3 mm below the edge. A small drop of water supported on this hydrophobic barrier makes the water bath, and the height of its surface is adjusted with a finely drawn mouth-operated pipette. This kind of bath is quickly and easily set up, is clean, rapidly replaceable, and does not leak.

3. *SECTION THICKNESS*

Section thickness is judged by interference color, which is not a very accurate guide but sufficient for most purposes. Gray sections are probably thinner than 500 Å, silver sections 500-800 Å in thickness, and gold sections up to 1500 Å in thickness. For high resolution, gray sections should be used whenever possible.

4. *EXPANSION OF SECTIONS ON THE WATER BATH*

Epon sections may be picked up without further treatment but sections of methacrylate and Araldite remain compressed in the direction of cutting after they arrive on the water bath and must be expanded. A filter paper or paintbrush dipped in xylene is brought near the sections and xylene vapor is allowed to flow over them briefly until they flatten out.

5. *PICKING UP SECTIONS FROM THE WATERBATH*

The best method is to hold the grid in fine forceps and lower it on to the sections, which adhere. For sections that do not adhere, remedies are offered in Section III. Sections may be picked up by stranding them across a grid from below, but the sections cannot be so accurately placed, and a large water bath is required.

6. *STAINING OF SECTIONS*

Various stains have been tried in an attempt to give contrast to sections but only two are very successful and routinely used. They are uranyl acetate and lead citrate, normally used one after the other with a rinse in between.

a. Uranyl Acetate. Prepare 5% aqueous uranyl acetate and store it in a dark bottle. Immerse the grids for five minutes at room temperature, then rinse. Staining thus achieved should be adequate, but if it is not, saturated aqueous uranyl acetate at 60° C or 15% methanolic uranyl acetate at room temperature may be applied for one hour. Note that uranyl acetate staining is often more easily done, with similar results, by staining the bulk tissue as described under dehydration.

b. Lead Citrate. Various lead stains have been used but the formulation of Reynolds (1963) seems best. A clean result free from lead carbonate deposits depends on a careful exclusion of carbon dioxide when the stain is made up and during storage and staining. The grids should be immersed in the stain rather than floated on the surface because dirty (though invisible) surface films easily form. The preparation of lead citrate stain is set out below.

(1) Put 1.3 g lead nitrate from a fresh bottle plus 1.75 g hydrated sodium citrate in a 50 ml volumetric flask and add about 30 ml distilled water recently boiled or fresh from the still.

(2) Close the flask and shake it vigorously for 1 min, then occasionally for 30 min.

(3) Add 8.0 ml carbonate-free N NaOH (best purchased as a prepared volumetric solution) and make up to the mark with carbonate-free distilled water.

(4) As the solution is mixed, all milkiness should disappear. If not, discard the stain and seek the source of the contaminating CO_2. Store the stain in the cold in the original flask or in a flexible polyethylene bottle from which the air space can be excluded. The stain keeps for a month or more.

To stain sections on grids, proceed as follows:

(1) Cover the bottom of a 5 cm petri dish with pellets of sodium hydroxide and place among them a small polyethylene saucer (e.g. the lid of a polyethylene tube or a radiation-counting planchet). Put a filter paper in the petri dish lid and wet it with strong sodium hydroxide solution. Replace the lid.

(2) Prepare four lidless 5 cm petri dishes containing distilled water and a small beaker of distilled water for rinsing forceps.

(3) Draw up about 1 ml of lead citrate, discard the terminal drop, and put the stain in the saucer.

(4) Slightly bend the edge of the grid to be stained so that it is easily gripped. Immerse the grid face up in the stain, close the dish, and stain for 1-5 min.

(5) Rinse the grid rapidly in each of the four water baths, each time releasing the grid and rinsing the forceps.

(6) Put the grid on filter paper and at once withdraw the forceps while the water drop sticks the grid to the paper.

Some prefer to rinse the sections in 0.1 N sodium hydroxide after staining and before washing in water. The exposed stain should be discarded after about 15 min.

Sections stained in lead citrate should initially be "warmed up" in the microscope, using a low-beam current and a defocused condenser. If sections are strongly irradiated at once, they develop a granular deposit.

XI. CYTOCHEMISTRY, AUTORADIOGRAPHY, AND IMMUNOCHEMISTRY

These techniques have not been extensively applied to plants infected with viruses but their potential seems great. Their successful development and application is the logical step forward from the static and, in some ways, uninformative electron micrographs obtained by standard methods.

A. Cytochemistry

Cytochemical methods at present fall into two main groups. Either an enzyme is made to reveal its activity and location by forming an electron-dense reaction product in the tissue, or particular substrates are identified and localized by digesting them with specific enzymes and comparing digested material with undigested controls. The simple approach of identifying materials in sections or on support films by using specific electron stains fails because the available stains are not at present specific enough.

1. *IDENTIFICATION AND LOCALIZATION OF ENZYMES*

The principle of the method is to introduce into the tissue a substrate and a staining agent. Where the enzyme is present, the substrate is split and the product of this reaction combines with the staining agent to form an electron-dense precipitate. For instance, in the Gomori reaction for detecting phosphatases, the tissue is incubated with the phosphate substrate and lead ions. Phosphate anions released by the enzyme are precipitated as electron-dense lead phosphate which reveals the site of activity.

The reaction must be modified for each enzyme studied and very careful controls are needed because nonspecific precipitates can easily form. For detailed methods, original papers should be consulted, and many of these are tabulated by Glauert, in Kay (1965) or summarized in the Leiden Symposium on Electron Microscopy and Cytochemistry (1966) and by Roodyn (1967).

To be detected, enzyme activities must survive fixation. Aldehyde fixatives largely permit this though fixation should be as brief as possible. Both substrate and the staining reagent must penetrate into the tissue, so tissues are usually washed thoroughly after fixation in aldehyde, frozen, then sliced at a thickness of some 50 μ on a cryostat or freezing microtome. After incubation, the tissue is washed, osmium-fixed, dehydrated, and embedded. If the reaction product has low contrast, osmium fixation and subsequent staining with uranium and lead may have to be omitted.

Unfortunately, one compromise must always be made. If the reagents are to penetrate effectively, the cells and even organelles must be opened, so much disorganization may result and enzymes may be displaced or lost. These restrictions also apply to the techniques discussed below.

2. *IDENTIFICATION AND LOCALIZATION OF SUBSTRATES BY ENZYME DIGESTION*

The tissue can be exposed to enzyme digestion before or after embedding. Each alternative has its advantages and problems, as illustrated in an interesting paper (Weintraub and Ragetli, 1968) on the chemical nature of bean yellow mosaic virus inclusions.

3. *DIGESTION BEFORE EMBEDDING*

The tissue is fixed in glutaraldehyde, rinsed, frozen, cut into slices 50-100 μ thick, and immersed in the enzyme or control solution. After incubation for 5-

60 min at 37° C the slices are rinsed, fixed in osmium tetroxide, and embedded for thin-sectioning as usual. The enzyme solutions used by Weintraub and Ragetli were as follows:

(1) Pepsin, crystallized three times (Nutritional Biochemicals Corporation, Cleveland, Ohio, USA) 0.25% in 0.1 N HCl.

(2) Papain, (British Drug Houses Ltd., Poole, Dorset, U.K.) 1% in 0.01 M phosphate buffer, pH 5.5.

(3) Trypsin, twice crystallized, salt-free (Nutritional Biochemicals) 0.3% in 0.05 M Tris (hydroxymethyl) aminomethane-maleate buffer, pH 8.0.

(4) Ribonuclease, five times crystallized (beef pancreas, Mann Research Laboratories Inc., New York) 0.1% in distilled water adjusted to pH 6.8 with 0.01 N NaOH.

(5) Deoxyribonuclease, five times crystallized (beef pancreas, Mann Research Laboratories and Worthington Biochemical Corporation, Freehold, New Jersey, USA) 0.05% in distilled water or distilled water with 0.003 M $MgSO_4$ adjusted to pH 6.8 with N NaOH.

4. DIGESTION AFTER EMBEDDING

To avoid damage by slicing, freezing, and thawing, the material can be embedded and thin-sectioned before enzyme treatment. Alternate serial sections can be used as controls or treated with different enzymes, and selected regions of the cell can be investigated more precisely than is possible with disrupted cells. However, the presence of the resin may hinder enzyme penetration or block reaction sites, so the method is complementary to that of digestion before embedding, rather than alternative to it. Platinum supports or gold grids are best for supporting the sections.

Monneron (1966*a, b*) and Monneron and Bernhard (1966) outlined methods for doing enzyme digestions on sections of tissue fixed in glutaraldehyde and osmium tetroxide and embedded in Epon. After oxidation in 3% hydrogen peroxide for 30 min at room temperature, the sections are rinsed and are then susceptible to digestion by various enzymes applied as 0.5% solutions at 40° C for several hours. Monneron used pronase (a general protease) at pH 7, pepsin in 0.1 N HCl, trypsin at pH 8, and amylase at pH 5-7.

Shepard (1968) avoided osmium fixation and subsequent treatment with peroxide. He used the following technique for subtilisin digestion of tobacco etch virus inclusions in tobacco leaves.

(1) Fix 1 mm wide strips of leaf for 3 hr at room temperature in 6.5% glutaraldehyde in 0.1 M phosphate buffer pH 7.

(2) Rinse twice in buffer, dehydrate in acetone, embed in Araldite.

(3) Float mounted sections on subtilisin (0.5% in 0.04 M Tris-HCl buffer pH 7.8) a broad-spectrum proteolytic enzyme (Sigma Chemical Co., St. Louis, Missouri, USA). Incubate for 1-20 hr at 37° C.

(4) Rinse the sections and stain them in uranyl acetate and lead citrate.

Nucleases do not work on epoxy resin-embedded sections, and for nuclease digestion glycol methacrylate embedding should be used and osmium fixation omitted. Glycol methacrylate is also useful for digestion with other enzymes, but as observed before, it is not a good routine medium for preserving structure. Weintraub and Ragetli (1968) said the best way to treat *Vicia* and *Phaseolus* leaves was:

(1) Fix for 30 min at room temperature in a mixture of 5% acrolein and 5% formaldehyde in 0.1 M phosphate buffer (pH not given but probably 7.2).

(2) Rinse in buffer for 1-15 hr, the longer period being used for DNase digestion.

(3) Embed in glycol methacrylate.

(4) Mount the sections on filmed copper grids and float them face down on the enzyme or control solution. The enzyme solutions are those listed under "digestion before embedding," with the exception of RNase, formulated as 0.5% in disodium ethylenediamine tetraacetic acid (EDTA) and KCN, both at 2×10^{-4} M pH 5.4.

(5) Incubate at 37° C for 15-90 min, then rinse in water.

(6) Stain with uranyl acetate for 1 hr at 60° C, followed by 5 min in lead citrate. (These times are longer than those needed for epoxy sections).

5. *DIFFERENTIAL STAINING OF DNA AND RNA IN SECTIONS*

Bernhard (1969) found that when sections stained with uranyl acetate are subjected to chelation by EDTA, the uranyl acetate is lost from DNA structures but retained by RNA structures, giving a difference in electron density. The conditions are as follows:

Stain Epon sections for 1 min in 0.5-5% aqueous uranyl acetate. Differentiate in 0.2 M EDTA in water for 30-60 min. Rinse and stain with lead citrate.

Stain GMA sections for 5 sec-1 min in 0.05-0.5% uranyl acetate. Differentiate in 0.004-0.02 M EDTA in water for 2 min. Rinse and stain with lead citrate.

EDTA itself dissolves with difficulty; the disodium salt is more soluble and works equally well. The procedure also works if uranyl acetate has been incorporated into the bulk tissue before embedding. In practice, the time for which the sections are incubated with EDTA is critical: With too brief a treatment, DNA structures remain dense, and with prolonged treatment, both DNA and RNA structures become electron-transparent. With a given EDTA treatment, differentiation between DNA and RNA structures varies with section thickness, so this also must be standardized. For reliable results, known embedded DNA and RNA standards should be constantly checked.

B. Autoradiography

The principles of autoradiography are the same for light and electron microscopy. A soluble radioactively labeled substance is introduced into a living system so that the process being studied incorporates it into an insoluble product. The tissue is then washed, fixed, embedded, and sectioned and the sections are coated in photographic emulsion and stored in the dark. During this time some of the electrons emitted by the radioactive decay of the label strike silver halide crystals in the emulsion and convert them into a latent image, sensitive to reduction to silver by the developer.

After sufficient exposure, the emulsion is developed and fixed with the sections still in place. Both the sections and the superimposed emulsion are now viewed in the microscope, and silver grains indicate the sites in the section where the label has been incorporated.

Autoradiography has been used by de Zoeten and Schlegel (1967*b*), Langenberg and Schlegel (1967), and Hayashi and Matsui (1967) to study syntheses of plant viruses. To study entry of virus and the processes of uncoating or inactivation, it is possible to label the virus directly, then follow its fate (Brants, 1965; Mayo and Cocking, 1968). Autoradiography has been discussed generally by Granboulan (1967), Sjöstrand (1967), Bachman and Salpeter (1965), and Glauert, in Kay (1965).

1. EMULSIONS

Current favorites are Ilford L-4 [*25*] and Gevaert (Agfa-Gevaert) [*26*] NUC 307 emulsions. Ilford L-4 has a somewhat large grain size (1400 Å) but is easy to use. It is insensitive to yellow-green light, has a high sensitivity to electrons, develops a low background, and can be stored in the cold for 2-3 months before use. Gevaert NUC 307 has a grain diameter of 700 Å and is therefore capable of better resolution than Ilford L-4 but is more sensitive to safelights, develops a slightly higher background, is less sensitive to electrons, and must be used within 2-3 weeks of receipt.

2. DEVELOPERS

The emulsion can be developed in Kodak D19 developer for 2-4 min at 20° C, then fixed for about 30 sec in a rapid fixer, as recommended by Caro and van Tubergen (1962). These workers added Kodak Anti-Fog No. 1 to the D19 to keep the background grain low during longer developments of up to 8 min.

Caro and van Tubergen also used a "physical" developer consisting of 1 M sodium sulphite and 0.01 M p-phenylenediamine. The sulphite was dissolved in water at 50° C and the p-phenylenediamine was then added. The solution is unstable so it is made up, filtered, and used at once. Development takes place for one minute at 20° C and results in very small comma-shaped silver grains which improve resolution considerably. However, the results are hard to control.

3. RESOLUTION

With ordinary photographic development there is no means of knowing where the emitted electron has struck the halide crystal that develops into the ribbon-like silver grain. Autoradiographic resolution must therefore be taken as the horizontal distance between the source of the radiation and the furthest edge of the silver grain that results. With "physical" development there is some evidence that the pointed end of the comma-shaped silver grain lies nearest to the radiation source, so resolution is much better. With skill and by using a fine-grain emulsion and a "physical" developer, a resolution better than 0.1 μ can be achieved, but at present both the emulsions and developers giving the highest resolution are the more difficult to use.

Before a description of the preparation of specimens for electron microscope autoradiography, the following points may be helpful. Roots generally take up and incorporate considerably more label than leaves although virus multiplication may be more controllable in leaves. DeZoeten and Schlegel (1967) used chrome-osmium fixation because, after glutaraldehyde-plus-osmium, they found that label was leached out during dehydration.

4. PREPARATION OF MATERIAL

One of the best procedures, due largely to P. Granboulan and to Glauert, is described below. It is assumed that labeled material has been prepared and embedded.

(1) Coat a microscope slide with 1% Collodion in amyl acetate by dipping the slide into the solution, withdrawing it, and letting it drain vertically.

(2) Dip the lower 5 mm of the slide again. This will facilitate later stripping of the Collodion film.

(3) With a diamond or Carborundum stylus, mark crosses on one side of the glass some 2 cm from the bottom.

(4) Prepare gold (or perhaps silver) sections of the labeled material. Very thin sections are unsuitable because they provide poor contrast and because they contain little radioactivity so exposure times are long. The high resolution available in very thin sections is not useful because the resolution limit of the label is only about 0.1 μ.

(5) With a small black plastic loop remove sections from the water bath and place them over the crosses but on the opposite side of the slide (the crosses act as markers). Remove the water drop with filter paper and let the sections dry.

(6) Stain the sections on the slide with uranyl acetate and lead citrate.

(7) Evaporate over the sections a thin layer of carbon. This prevents interaction between the sections and the emulsion and also provides a uniform base for the emulsion so that the silver halide grains become evenly spread.

(8) Under a safelight, dip the slide into the melted and diluted emulsion, withdraw it, and let it drain at an angle of 30° from the vertical with the sections upward.

(9) Store the slide in the dark over a drying reagent at room temperature for 2-12 weeks. Estimate the correct exposure time by developing test slides at intervals. Alternatively, examine similarly treated and developed 1 μ sections in the light microscope; the correct exposure for electron microscopy will be about five times as long.

(10) Develop, fix, wash, and dry the emulsion on the slide. Fixation for very short periods may prevent loss of contrast in the section.

(11) Float the Collodion film, bearing the sections, carbon film, and

emulsion, onto water and collect the areas over the crosses on 200-mesh grids, as described in Section IVB.

(12) Before examining the grids the Collodion may be removed by standing the grids on filter paper soaked in amyl acetate.

Various alternative procedures are possible. For instance, Martinez-Palomo and Granboulan (1967) did not stain the sections at step (6) or use carbon at step (7), but after step (11) they removed the gelatin of the emulsion by treating the grids with 0.025 M acetic acid for 30 min. They then stained the sections with uranyl acetate and lead citrate.

It is also possible to mount the sections first on grids and attach these to glass slides before the emulsion is applied (de Zoeten and Schlegel, 1967; Hayashi and Matsui, 1967; Mayo and Cocking, 1968), but the emulsion, or the developing and fixing process, may interact with the metal of the grid, and the emulsion may be unevenly spread.

C. Immunochemistry

The antigen-antibody reaction is very specific and provides an excellent means of identifying antigens except that the reaction product is not itself visible in the light or electron microscope, so the antibody must be combined with a marker. Further, the antibody-plus-marker complex is large and will not penetrate into cells or across unbroken membranes.

For electron microscopy the antibody is usually conjugated with ferritin, and electron-dense iron-containing molecule (see Shalla and Amici, 1967; Howe et al., 1967; Oshiro et al., 1967; Higashi et al., 1967). As an alternative to ferritin conjugation, Langenberg and Schlegel (1967) coupled virus antibodies with iodine. The theory and practice of antibody production and conjugation are treated elsewhere; here, attention is given to the penetration of the antibody-marker conjugate into the tissue. Shalla and Amici (1967) described what is yet the best method for plants. It involves disrupting the tissue in three stages.

Stage 1. Cutting open the cells.

(1) Fix the tissue in glutaraldehyde, cut it into small pieces, rinse it in water, and freeze it.

(2) On a freezing microtome shave away some tissue leaving an opened layer of cells.

(3) Thaw the remaining tissue with its opened cell layer and transfer it to the ferritin-antibody preparation.

Stage 2. Preparation of isolated chloroplasts and nuclei. To provide greater access of antibodies to chloroplasts and nuclei, proceed as follows:

(1) Darken the plants for two days to destarch the chloroplasts.

(2) Fix the leaves in glutaraldehyde, homogenize them, press the homogenate through cheese cloth, and centrifuge the juice at 5000 g for 10 min to sediment chloroplasts and nuclei.

(3) Resuspend the pellet in 0.05 M phosphate buffer pH 7 and separate chloroplasts from nuclei on a sucrose density gradient containing 750, 700, 600,

and 500 mg sucrose per ml in 0.05 M phosphate buffer pH 7, by spinning for 30 min at 10,000 rpm in a SW 25 Spinco rotor (or equivalent). Separate zones of chloroplasts and nuclei (directly below the chloroplasts) are formed.

(4) Remove chloroplasts and nuclei separately, dialyze them against cold 0.05 M phosphate buffer pH 7, sediment them with a low-speed spin, and resuspend them in the same buffer.

After this treatment, ferritin-conjugated antibody readily diffuses into the nuclei. In contrast, intact chloroplasts are still impenetrable; therefore proceed to stage (3).

Stage 3. Thick-sectioning of chloroplasts.

(1) Embed chloroplasts in 6% gelatin, freeze them, and section them at 10 μ.

(2) Transfer the sections to phosphate buffer and warm to 45° C to dissolve the gelatin.

Conjugated antibody now has access to the cut surfaces of the chloroplasts, but even now does not penetrate the interior.

All preparations are allowed to react overnight with the ferritin-antibody preparation. They are then washed in water, fixed in osmium tetroxide, dehydrated, and embedded. Sections may be left unstained to allow the ferritin to show up more clearly.

ACKNOWLEDGMENT

I thank Mr. Roy Woods and Dr. Peter Dart for helpful discussions and Dr. Hari Agrawal for his great patience.

XII. APPENDIX

Source of products mentioned in the text:

1. Diffraction grating replicas. Bausch and Lomb Inc., Rochester, N.Y. 14602, U.S.A.; Ernest Fullam Inc., P.O. Box 444, Schenectady, N.Y., U.S.A.
2. Indanthrene olive TP (ex. con. pdr.). General Aniline and Film Corp., P.O. Box 700, Linden, N.J. 07036, U.S.A.
3. Catalase, bovine liver, suspension. C. F. Boehringer, Mannheim, West Germany (Cat. no. KAT-1 15674), or from Boehringer Ltd., Brentford, Middlesex, U.K.
4. Specimen grids. Smethurst High-Light Ltd., Sidcot Heaton, Bolton, Lancs., U.K.; Polaron Instruments Ltd., 4 Shakespeare Rd., Finchley, London N. 3, U.K.; Ernest Fullam Inc., P.O. Box 444, Schenectady, N.Y., U.S.A.; Ladd Research Inc., 159 Wagon Road, Rowlyn Heights,

N.Y., U.S.A.; Veco Zeefplaten Fabrick N.V., Eerbeek, Netherlands.

5. Grid storage boxes. LKB Produkter AB, P.O. Box 76, Stockholm-Bromma 1, Sweden. Also Polaron Instruments Ltd. (see 4).
6. Formvar. Taab Laboratories, 52 Kidmore End Road, Emmer Green, Reading, U.K.; Polaron Instruments Ltd., U.K. (see 4); Shawinigan Ltd., Marlow House, Lloyd's Avenue, London E.C. 3, U.K.
7. Collodion. British Drug Houses Ltd., Poole, Dorset, U.K.; Hopkin and Williams Ltd., Chadwell Heath, Essex, U.K.
8. Carbon rods. Morgan Crucible Co., Ltd., Battersea, London, U.K.; Edwards High Vacuum Ltd., Crawley, Sussex, U.K.
9. Mica. The Essex Wire Corp., 247 Lynnfield Street, Peabody, Mass. 01960, U.S.A.
10. Low velocity spray gun. The Vaponefrin Co., Edison, N.J., U.S.A.
11. Artist's airbrush. Aerograph DeVilbiss Co., Ltd., Holborn Viaduct, London E.C. 1, U.K.
12. Polystyrene latex. Dow Chemical Co., Midland, Mich., U.S.A.
13. Cascade impactor. C. F. Casella and Co., Ltd., Regent House, Britannia Walk, London N.1., U.K.
14. Platinum-cored carbon rods. Edwards High Vacuum Ltd., Crawley, Sussex, U.K.
15. Freeze-etching unit. Balzer's Ltd., Liechtenstein.
16. Epon from the U.S.A. Epon resin 812 (Shell Chemical Co.) repacked by E. V. Roberts and Associates, 9601 W. Jefferson Blvd., Culver City, Calif. Dodecenyl succinic anhydride (DDSA) and Nadic methyl anhydride were obtained from Fischer Scientific Company, Fairlawn, N.J. Trimethyl amino methyl phenol (DMP-30) was obtained from Rohm and Haas Company, Philadelphia, Pa. From U.K.: Taab Laboratories (see 6).
17. Graduated polyethylene beakers. Tri-Por (No. A 3608) Clay-Adams Inc., New York 10, N.Y., U.S.A.
18. Araldite components. Ciba Products Corp., Fairlawn, N.J., U.S.A.; Ciba Ltd. (ARL), Duxford, Cambridge, U.K.: Taab Laboratories (see 6).
19. Gelatin capsules. Size 00 is normally used, obtainable from Parke, Davis and Co. Ltd., Hounslow, Middlesex, U.K. Shaped polyethylene capsules are marketed by B.E.E.M. Inc., P.O. Box 132, Jerome Avenue, Station, Bronx, N.Y. 10468, U.S.A.
20. Vestopal W. Martin Jaeger, Vesenaz, Geneva, Switzerland.
21. Glycol methacrylate. Rohm and Haas, Philadelphia, Pa., U.S.A.; Societe Ugilor, 5, Rue du General Foy, Paris, France. Request a fresh sample.
22. Long wavelength UV light. ($>$ 3150 Å) is obtainable from Type A 405 lamp (P. W. Allen and Co., London) containing a Philips 9 inch 6 watt fluorescent tube having an actinic blue phosphor, color type 05.
23. Diamond knives. Freidrich Dehmer, 8202 Bad Aibling, West Germany; E. I. Dupont de Nemours and Co., Inc., Wilmington, Del., U.S.A.; R. E. Sugg, 508 Foulkestone Road, Wilmington, Del., U.S.A.; Ge-Fe-Ri, Via

Maritima, 03100 Frosinone, Italy; Polaron Instruments Ltd. (see 4).

24. Glass cutter and pliers for glass knives. C. Reichert Optische Werke A. G., A-1171 Vienna, Austria, or Shandon Scientific Co., Ltd., Willesden, London N.W. 10, U.K.
25. Ilford Ltd., Ilford, Essex, U.K.
26. Agfa-Gevaert A.G., Leverkusen, West Germany.

PART TWO

TRANSMISSION OF PLANT VIRUSES

5

Transmission of Plant Viruses by Aphids

MARION A. WATSON

Plant Pathology Department
Rothamsted Experimental Station
Harpenden, England

I. INTRODUCTION

Aphids constitute the largest group of known plant virus vectors and more than 100 viruses are reported to be transmitted by them. They are economically important because they directly damage crops as well as spread viruses from plant to plant. Crop loss by feeding occurs when enormous populations rapidly develop but even greater losses can occur when virus is spread into major growing areas. The biology of the aphid and aphid-borne viruses is discussed in this chapter.

II. BIOLOGY OF APHIDS

Aphids have a complex life history involving alternation of asexual generations with a sexual generation. They have different forms (morphs) each adapted for a particular vital function (polymorphism). Plants on which sexual forms of aphids mate and lay eggs are called primary hosts; secondary hosts are those where asexual generations reproduce with great efficiency by parthenogenesis (without fertilization of the ova) and young are born viviparously (alive and active). The secondary hosts are often agricultural crops.

A. Primary Hosts

Aphids of most genera lay their eggs on perennial plants, shrubs, or trees. The eggs remain dormant (diapausing) during the cooler seasons of the year.

Migration takes place from the primary to the secondary (usually herbaceous) hosts. In tropical or arid regions eggs may be laid on suitable hosts in mountainous areas or oases. Alternation of hosts from deciduous shrubs or trees (primary hosts) to herbaceous secondary hosts is called heteroecism, and these aphids are called dioecious aphids. The specific name of the primary host is often incorporated into the name of the aphid. Thus *Myzus persicae* (Sulz.) lays its eggs on peach (and other *Prunus* spp.); *Rhopalosiphum padi* (L.) on *Prunus padi* (L.) the "Bird-cherry" tree.

Other aphids (monoecious) spend the whole year on the same plant family; for instance *Brevicoryne brassicae* (L.), the waxy cabbage aphid, lays its eggs in biennial *Brassica* spp., similar to those it infests in the summer. Thus cabbages and Brussels sprouts may serve as their primary and secondary hosts.

All aphids eggs hatch when new leaves start to develop on the primary host, giving rise to wingless females (apterae), whose offspring when immature resemble the parent in all except wing development and reproductive maturity; these are called nymphs. The nymphs develop through five instars, casting their integument (ecdysis) at the end of each instar. After one or two generations, the apterous female produces different nymphs that develop into winged females (alatae). The alatae leave their primary hosts and migrate to secondary hosts.

Annual agricultural crops are particularly vulnerable at this time to aphid attack because the plants are all of the same species and at the same stage of growth, and are cared for by man. During migration, aphids feed on weeds, on volunteer crops, or biennials in their seeding year, and may be virus-infected. Polyphagous aphid species feed on various plant species whereas monophagous and oligophagous aphids can alight and feed briefly on disfavored hosts but usually prefer one type of plant in a given plant family.

Primary hosts of dioecious aphids rarely harbor viruses that can infect secondary hosts, but viruses of the intervening hosts do. One or other of the aphid-borne viruses can be spread whatever the aphid's feeding time, movement, and permanent or temporary shelter on plants. Aphid-transmitted viruses seem to have evolved so as to exploit fully all the different kinds of aphid behavior.

B. Secondary Hosts

Winged female aphids colonize an intermediate host or a crop, and become the founders (fundatrigenae) of viviparous generations called alienicolae. Each female produces 40-80 nymphs, depending on the aphid species, host, and environment. The nymphs are born in an advanced stage of development and may themselves reproduce after one week. In the summer aphids multiply for several weeks until they become too populated, or the host-plant ages, or weather causes them to move (as apterae), or start producing more winged generations that migrate to other plants and eventually to other areas. Most plant viruses are spread during these movements within and between crops.

Aphids may still remain alienicolous although they have moved from the crops. *Myzus persicae,* for instance, a notorious vector of many plant viruses, almost disappears in England during August from potato and sugar beet crops that it infests during June and July, but it continues to breed on other herbaceous hosts until the autumn, or even right through the winter when weather conditions permit. Aphids that overwinter as alienicolae are able to multiply on crops and weeds earlier in the spring than if they had to hatch from eggs, and overwintering hosts of the same plant family as the crops may carry viruses. In tropical areas where day length and temperature rarely stimulate the production of sexual forms and where alternative hosts are missing, many aphid species can maintain populations on secondary hosts. Likewise aphids maintain themselves on plants in commercial glasshouses.

C. Host Selection and Adoption

Aphids begin probing plants after migration by inserting the tips of their stylets into a suitable area of host epidermis, and will search out preferred plants. Monophagous aphids preferentially select plants of related species, whereas polyphagous aphids are not fastidious and feed on a wide range of hosts. They usually probe oftener than the monophagous aphids. The aphids then: (1) may become dissatisfied with the plant and leave; (2) may be satisfied, and remain, perhaps for most of their lives; or (3) may be satisfied, reproduce, and leave after a few hours, but continually repeat this behavior, producing small isolated groups of nymphs about the crop. *Myzus persicae* is known for this last behavior pattern.

These three behavior patterns are reflected in three distinctive types of virus transmission. (1) Viruses transmitted by probing aphids are acquired and transmitted within a few minutes; they are quickly eliminated from the vectors and thus usually called *nonpersistent* or, by some, stylet-borne viruses (Kennedy et al., 1962). (2) Aphids such as *M. persicae,* which move frequently while colonizing a crop, may transmit *semipersistent* viruses after feeding for as little as 15 min, and these viruses can then be transmitted for up to two days to a succession of plants. (3) More sedentary aphids often transmit *persistent* viruses which may need many hours to be acquired, and (because of the latent period, see p. 142) even more to be inoculated into plants: Vectors can retain these viruses for life. Because of their brief transmissibility, nonpersistent viruses need to be within or near their host, which may be in gardens and fields where older plants grow beside newer plantings. Semipersistent and persistent viruses spread better from nearby than from distant sources. Persistent viruses may be carried great distances, even from crops grown in other regions where climate and cultivation are different. In the northern hemisphere, alatae migrating from mature crops in a southern area may infect young crops further north. Bruehl (1961) observed this for barley yellow dwarf virus in North America.

III. INTERNAL ANATOMY

A. Alimentary System and Hemocoel

Figure 5-1 shows a median longitudinal section of the head and thorax of *Myzus persicae* feeding on a tobacco leaf. The stylet bundle (sb) projects from the tip of the beak and the inset shows a portion from the next section, where the stylet sheath (S) has two branches into phloem cells. The head contains the upper and lower cerebral ganglia (B and SB) which form the brain. Between is the pharynx, a strongly chitinized region of the mouth, having very strong muscles attached to a projection from the roof of the cranium. These muscles operate the pharynx as a suction pump (ep) contracting in waves so as to keep liquid flowing. The pharynx may close, by elasticity against pressure from without, for aphids acquire most of their food by merely allowing the internal pressure of plants to inject the phloem contents down their gullets. Even though they eliminate excess sugary solutions very rapidly by excretion, some control of input is probably necessary for the stylets to remain in the cell. The esophageal valve (an invagination of the foregut) is usually said to prevent regurgitation from the stomach (S) but may also help to control ingestion. It is often found in dissections in vivo filled with fluid and blocking the fore end of the stomach.

Below the pharynx is a strong bundle of muscles that operate the salivary pump. This is shaped, and operates, something like a hypodermic syringe. The muscle retracts a plunger which returns to its cup by elasticity, forcing salivary secretions into the exhalent stylet canal. The secretions contain, besides some enzymes and unknown substance, a lot of gelatinous material (seen in the dark, swollen, salivary reservoir below the salivary muscle bundle) which hardens around the stylet bundle into a stylet sheath forming a tube in which the stylets can move up and down.

Surrounding the stomach and all the other organs is the hemocoel, which contains nutritive and excretory fluid bathing the stomach, salivary glands, and ovaries. Anything that manages to penetrate the gut wall is therefore in direct contact with all the vital organs.

B. Integument

The integument of an aphid is a rigid chitinous structure with both protective and skeletal functions. Nymphs ecdyze five times during their development into an imago, and the exuvium includes all tissues derived from the embryonic ectoderm, including the skin, mouthparts (stylets), pharynx, and linings of fore and hind guts. If an aphid is still infective after ecdysis, the virus must have been in some tissue derived from mesoderm or endoderm, that is, the hemocoel or the organs in it. If the virus is not passed through the molt, it is assumed to have been held on tissues that are ecdyzed, but this is not certain because the physiological disturbance of ecdysis could get rid of virus. Also the aphid

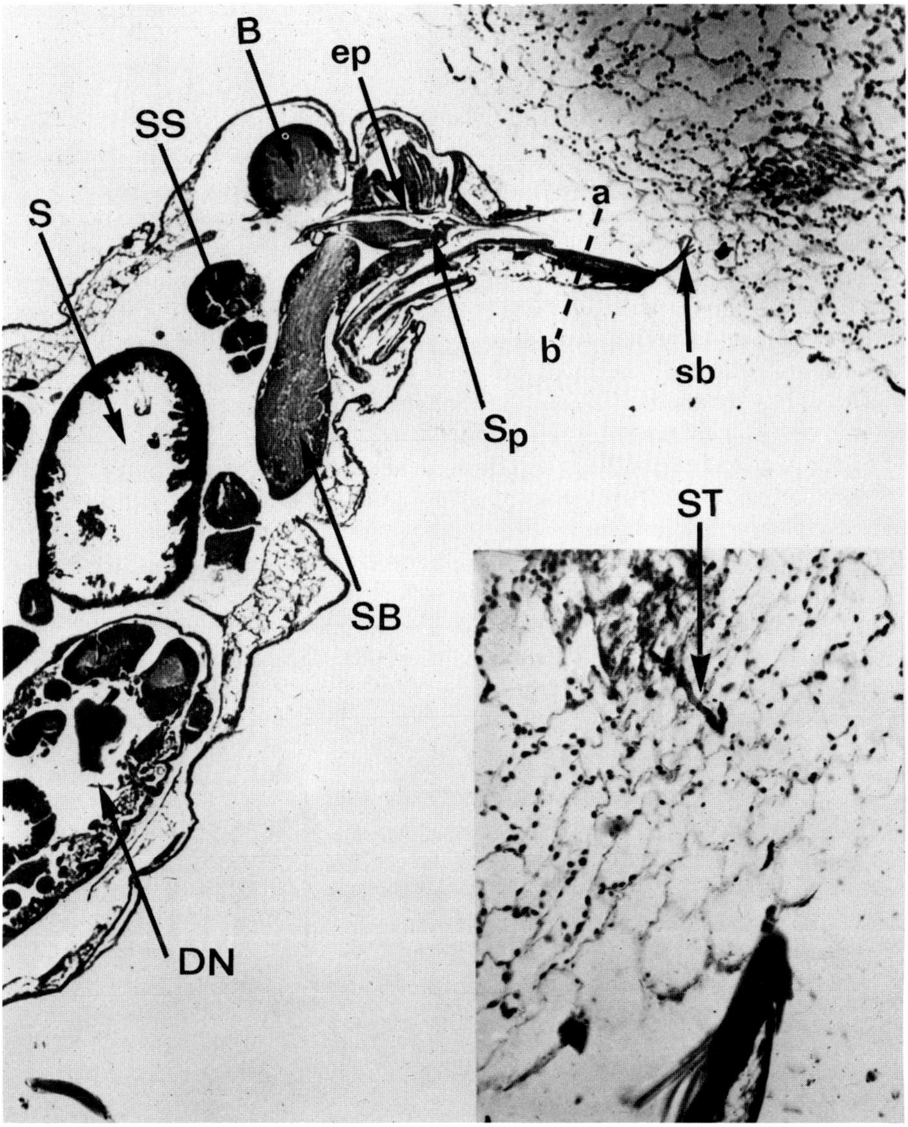

Fig. 5-1. Longitudinal section of an aphid feeding on the phloem region of a tobacco leaf. Anatomical parts of the aphid: sb, stylet bundle; a---b, cross-section of rostrum shown in Fig. 5-2; ep, epipharyngeal (cibarial) suction pump; Sp, salivary pump behind which is located the salivary pump muscle seen as a spindle shaped object; B, brain, SB subesophageal ganglion; SS, salivary gland; S, stomach; DN, developing nymph. Inset: magnified section showing ends of stylet tube, ST, in phloem cells.

cannot feed after ecdysis until the new mouthparts are hardened so virus may be lost even though the tissue that contained it remains.

C. Mouthparts

Figure 5-2 is a transverse section through the rostrum about halfway down its length. The two inner stylets (maxillae) are shown locked together to form inhalent and exhalent feeding tubes, clasped by the outer pair of stylets, the mandibles, which support and help to guide them in the plant. Nerves, lying in the holes down the centers of the mandibles, may influence feeding behavior, especially probing (Bradley, 1962*b*; Forbes, 1966). When even one mandible is removed the aphid will not probe or feed at all, and other treatments, such as the application of formalin or electrostatic charges to the tips of the stylets, also inhibit probing (Bradley and Ganong, 1955; Bradley, 1966) and prevent transmission of nonpersistent viruses.

The stylets are about 500 μ long, the inhalent duct having a diameter of about 1 μ and the exhalent (salivary) duct, about 0.4 μ (Fig. 5-2).

McLean and Weight (1967) described an interesting technique that may help to solve some of the problems of virus uptake, especially of nonpersistent viruses. They measured changes in electrical conductivity between aphids and the leaf by amplifying a small charge between an electrode attached to the aphid and one attached to the leaf. The voltage increased and decreased as the aphid's stylets filled and emptied so the actual moment of ingestion of fluid could be determined. They observed that the ends of the salivary tubes are open

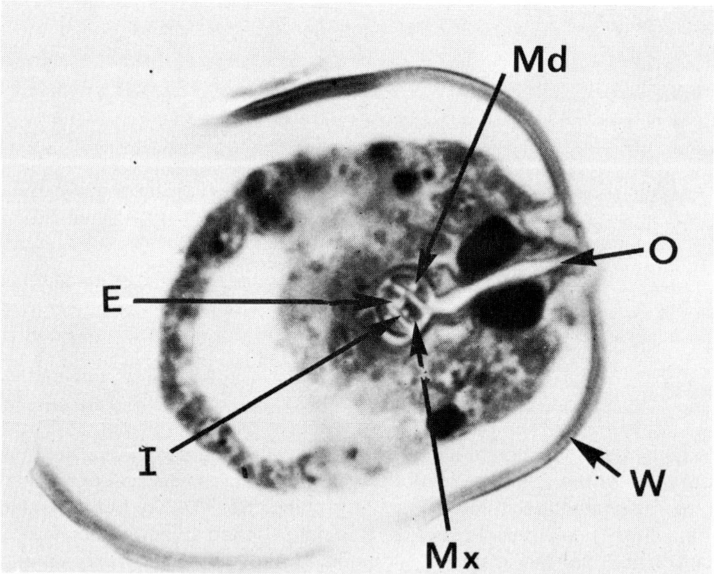

Fig. 5-2. Cross section a---b of rostrum shown in Fig. 5-1. Symbols: W, wall of rostrum; O, opening of rostrum; Mx, maxillae locked together to form the small exhalent, E (salivary) and larger inhalent, I (food) canals; Md, Mandibles clasping the maxillae. Their central canals are now known to be innervated.

when liquid is being imbibed, and that "short ingestion or sampling" occurs when aphids are probing plant tissue (McLean and Kinsey, 1967).

IV. TRANSFER OF VIRUS

A. Definition of Terms

The following terms are used to describe the various stages of transmission:

Term	*Definition*
Infected (source) plant	Plant from which virus is acquired
Test plant	Plant to which virus is inoculated
Acquisition time	Time of feeding on source plants
Test feeding time	Time of feeding on test plants
Infective	Able to infect plants
Inoculation time	Time needed to initiate infection
Transmission time	The sum of acquisition and inoculation time
Latent period	Time between acquisition and inoculation
Persistence	Time that aphids retain infectivity after leaving source plants

B. The Probing Position

Figure 5-3A represents, diagrammatically, a half-section of an aphid in the probing position in which it acquires nonpersistent viruses. The stylet tips just enter an epidermal cell where there are elongated particles of a virus such as potato virus Y. There are several hypotheses concerning the sites where such viruses are carried during transmission. Bradley (1966) concluded that they were carried only on the distal 15 μ of the stylet tips. Sylvester and Richardson (1964) suggest, among other ideas, that they adhere to fragments of saliva.

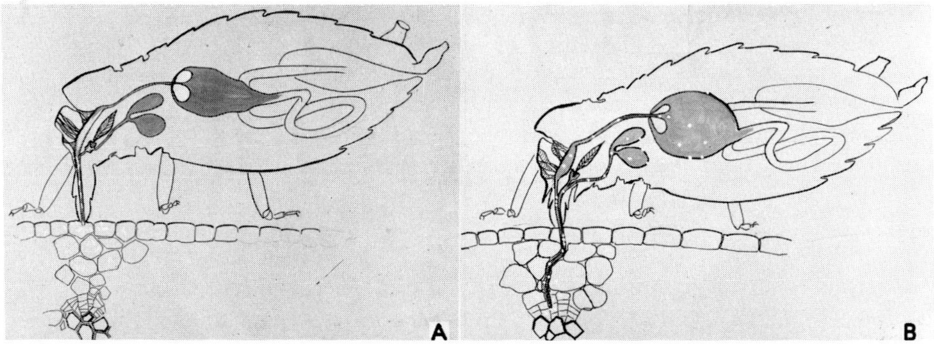

Fig. 5-3. Illustration of aphids in probing position. (A) Aphids in position to probe virus-infected tissues of the epidermis. (B) Aphids in feeding position as in Fig. 5-1, acquiring virus from phloem cells. (Republished by permission of Outlook on Agriculture 5, 1967.)

There seems no a priori reason why they should not be sucked up into the pharyngeal pump, and apparently good reasons why they should, because sap is probably taken up as far as the gustatory organs situated at the end of the pharynx, as part of the probing performance concerned with sampling plant tissues. They would most likely be ejected by compression of the pharynx during subsequent probing and so infect fresh plants within a minute or two of acquisition.

Apterous aphids do not usually move far, so neighboring plants are most likely to be infected by them. Spring alatae migrating from plant to plant in search of suitable food hosts probably cause most outbreaks of infection.

When probing, the beak (rostrum) is extended at right angles to the leaf, the head and body lifted well above the surface and antennae laid back.

C. The Feeding Position

Figure 5-3B shows the position of an aphid whose stylets have penetrated to the phloem. The segments of the rostrum are telescoped and it is pushed back into the head to allow maximum prolongation of the stylets. The body is close to the leaf, the antennae laid back. *Myzus persicae* takes, on the average, about 20 min to reach the phloem (Roberts, 1940). Twenty to thirty minutes is also about the minimum time needed for aphids to acquire persistent circulative viruses, for example, potato leaf roll virus (Kassanis, 1952*b*) and barley yellow dwarf virus (Watson and Mulligan, 1960). Acquisition from and inoculation to the phloem are widely assumed to be essential for transmission.

V. CULTURING OF APHIDS

Aphids grow well at temperatures between 15° C and 20° C and under continuous illumination by a single fluorescent lighting tube. These conditions also tend to suppress the development of alatae, which are not always good vectors and are difficult to handle and control. Plants will not grow for long with such meager illumination, and should be changed every week. The containers should be washed and sprayed with fungicide to control parasitic and saprophytic fungi, and great care must be taken to prevent attacks by hymenopterous parasites. Many aphid species will multiply at widely differing temperatures, and will survive even great extremes, but for experimental use cultures are best kept cool.

A. Aphid Cages

Containers for aphid culture are usually of two types. So-called "larvae cages" (Fig. 5-4) (Watkins and Doncaster, London) consist of an enameled metal base large enough to take a potted plant in a saucer, and a cylinder of transparent plastic with a lid that also fits the base. The lid has a perforated disc let into it and the plastic walls can be ventilated by windows of silk or synthetic gauze or muslin. The meshes should not be wider apart than 0.3 mm. The transparent

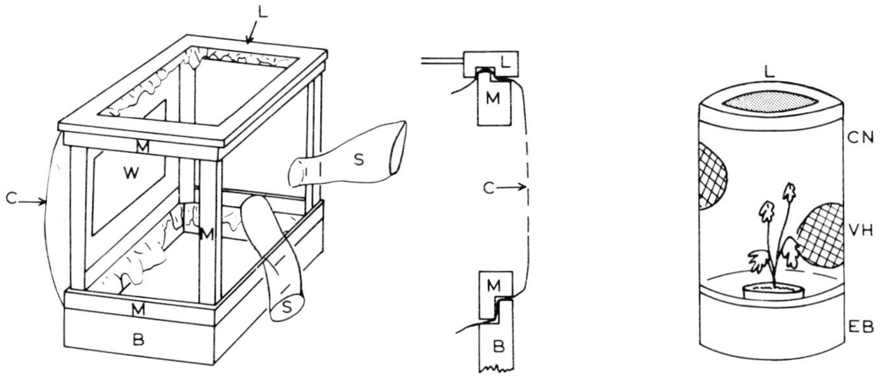

Fig. 5-4. Examples of cages for housing plants with aphids. Left and center: The muslin cover (C) is held between the roof lid (L) and floor base (B) of the wooden cage, so that the cage can easily be taken apart for cleaning. A glass window is part of lid (L). The middle section (M) has a window (W) of cellulose nitrate for observation; the other three sides are covered with the muslin cover (C). The total assembly rests on base (B). Sleeves (S) are for access into the cage. On the right is the "cake-tin" cage with perforated lid L), cellulose nitrate walls (CN) with windows of gauze (VH), and a solid base (EB).

plastic should be cellulose nitrate, not cellulose acetate which is toxic to aphids, and it should be about 6 weeks old, not fresh. Sealing of the gauze windows is done by moistening window edges with acetone and immediately pressing on the gauze.

The other kind of cage is a wooden or wire frame covered with cotton or nylon material. Some of these have doors that open, but this allows winged aphids to fly out when the cage is opened for watering. We use wooden frames made in three portions with a loose muslin cover, kept in place by the weight of the other portions, and with a sleeve for watering or withdrawing samples of aphids (Fig. 5-4). Cotton muslin covers should have a cellulose nitrate window, and the lid can be glass. Very translucent nylon or terylene fabrics such as nylon monofilament, about 20 mesh, will do without windows. The covers should be washed and dried every time cultures are changed, and cages washed and sprayed with fungicide.

Whenever possible, aphids should be cultured on plants that are not hosts of the viruses to be tested. For example, turnip, *Brassica rapa,* and Chinese cabbage, *B. pekinensis,* are good hosts for *Myzus persicae* but they harbor not only easily diagnosable viruses such as turnip mosaic and cauliflower mosaic, but also symptomless ones such as turnip latent virus (McKinnon, 1965) and turnip mild yellowing virus (Watson, 1967), which may be confused in other hosts with potato leaf roll and pea leaf roll viruses. *Hyoscyamus niger* (Solanaceae) is a useful alternative host as it is not susceptible to these *Brassica* viruses.

Barley yellow dwarf virus infects most species of *Graminae,* and when cereals are used as hosts for the vectors, a "nuclear stock" can be maintained on *Arrhenatherum elatius:* not a particularly good host for the aphids, but one of the few that is immune to the virus.

B. Environment

Constant environment is invaluable where conditions cannot be controlled adequately in glasshouses, but is usually too expensive for maintaining aphid cultures. However, reasonably constant temperature and light conditions can be improvised in insulated containers kept in cold rooms at about 5° C. Fluorescent lighting tubes with their transformers provide both light and heat, and the insulation of the boxes can be adjusted to give the desired temperatures (15-18° C). Aphid cultures remain in good condition away from experimental plants, and this secludes them from virus contamination, parasites, and predators. The boxes (wood or fiberglass) can hold about a dozen plants and the walls are insulated with polystyrene covered with metal foil. Insulation of the box is adjusted to obtain the desired temperature. The plants need little watering, but need care to prevent condensation. Aphids will not voluntarily fly out into the dark when cages are opened.

I. J. Wyatt and N. E. A. Scopes of the Glasshouse Crops Research Institute, England, demonstrated to the Association of Applied Biologists, in 1968, aphid-rearing in transparent plastic containers, airtight, and therefore aphid-proof, but aerated by means of small electric blowers. They had artifical lighting and irrigation. This method would have obvious advantages where insect-proof rearing space is not available in glasshouses or cold rooms.

VI. BIOLOGY OF APHID-BORNE VIRUSES

Aphid-transmitted viruses differ in a number of properties, of which at least five can be listed: (1) persistence in the aphids; (2) the latent period; (3) passage through the molt; (4) mechanical inoculation; and (5) virus-particle shape (Table 5-1).

A. Persistence

Aphids usually lose the ability to transmit nonpersistent viruses within an hour after acquisition feeding, but these viruses are retained much longer during fasting (e.g. fasted aphids can transmit potato virus Y or henbane mosaic virus after 24 hr). The loss of infectivity (or transmissibility) is logarithmic whether aphids are fasting or feeding.

Persistent viruses are viruses whose vectors continue to be infective for long periods. Semipersistent viruses are retained by the vector longer than nonpersistent viruses and much less than persistent viruses. They are excluded as nonpersistent viruses because they are not affected by stylet treatments that inhibit transmission of stylet-borne viruses, for example washing the stylets with water or formalin, or irradiating with ultraviolet light (Sylvester and

EXAMPLES OF APHID-TRANSMITTED VIRUSES GROUPED ACCORDING TO THEIR PROPERTIES AND SHAPES

	Behavior	*Noncirculative*		*Circulative*	
	Group	*Nonpersistent (stylet borne)*	*Semipersistent*	*Persistent Mosaics*	*Persistent Yellowing*
	Transmission	*Immediate*	*Delayed*	*Delayed*	*Delayed*
PROPERTIES	Persistence—usually	Less than 1 hr	More than 1 hr		
	Latent period—usually	None	Less than 3 hr	More than 3 hr	More than 3 hr
	Passage through molt	—	—	+	+
	Manual inoculation	++[a]	+[a]	+[a]	—
SHAPES OF VIRUS PARTICLES	Long flexible rods	Henbane mosaic Potato virus Y Beet mosaic Turnip mosaic	Beet yellows Citrus Tristeza		
	Spheres	Cucumber mosaic Cauliflower mosaic		Pea enation mosaic Carrot mottle virus Parsnip mottle Groundnut rosette	Potato leaf roll Beet mild yellowing and yellow-net[b] Turnip mild yellows[b] Pea leaf roll[b] Groundnut assistor[b] Barley yellow dwarf Carrot red leaf Beet western yellows
	Bacilliform (blunt ovoids)	Lucerne mosaic		Lettuce necrotic yellows	Sowthistle yellow vein
	Stiff rods	Red clover vein mosaic Potato virus S Potato virus M Carnation latent			

[a] ++ easily; + with difficulty.

[b] Not finally identified; probable shape of particles, or not yet seen.

Bradley, 1962), and as persistent ones because they do not pass through the molt.

Semipersistent viruses are acquired by aphids after an acquisition feeding time of ten or more minutes. The delayed acquisition suggests that aphids may need to feed on the phloem to acquire viruses of this group. Tristeza virus seems to be confined to the phloem (Price, 1968).

B. The Latent Period

The time between acquiring virus and being able to transmit it is called the latent or the incubation period. The term incubation period is confusing because it is also used for the time plants need to develop virus symptoms. The term latent period is also used in a clinical sense, but it is less exacting, it has been adopted by most workers.

Nonpersistent viruses have no latent period as they can be transmitted immediately after acquisition feeding. It is doubtful whether particles of nonpersistent viruses that pass beyond the region of the mouth of an aphid are ever injected into plants.

Also, semipersistent viruses are usually not thought of as having a latent period, although they are not transmitted immediately.

A latent period could merely be the time for aphids to feed on the phloem, and for viruses then to pass through the gut wall of the aphid and become concentrated enough in the hemocoel to invade the salivary glands and be ejected with the saliva. Its length could depend greatly on the ease with which the aphid acquires virus, and the permeability of its gut wall, but no doubt there are other factors.

There is good evidence that some viruses multiply in aphids, as do most hopper-transmitted viruses. Potato leaf roll virus has been maintained by aphid-to-aphid injection through 15 generations of *Myzus persicae* without apparent loss of probability of infection (Stegwee and Ponsen, 1958). Other viruses have been detected in fat bodies and organs of their vectors in semicrystalline forms, or aggregates resembling those occurring in plants, for example pea enation mosaic in the body of *Acyrthosiphon pisum* (Shikata and Maramorosch, 1965). Lettuce necrotic yellows virus particles are abundant in all parts of the vector *Hyperomyzus lactucae*, and this probably indicates multiplication (O'Loughlin and Chambers, 1967).

C. Passage through the Molt (Ecdysis)

Passage through the molt is an important criterion for distinguishing between noncirculative (nonpersistent and semipersistent) and circulative (persistent) viruses. Only persistent viruses survive through ecdysis and are transmitted by adults that as nymphs acquired the virus. Most of these viruses are better transmitted by nymphs that acquired them as nymphs, than by adults that acquired them when adult. This may be because immature tissues are more easily invaded by the viruses, or because the period of immaturity is too short for long latent periods to be completed.

D. Mechanical Inoculation

Nonpersistent viruses can easily be transmitted by mechanical inoculation (see Chapter 1). Some semipersistent viruses, and persistent viruses such as groundnut rosette, are also transmitted by mechanical inoculation, but with considerable difficulty. Among the useful hosts for some viruses difficult to transmit are *Chenopodium amaranticolor* and *Nicotiana clevelandii*, both found by C. W. Bennett (unpublished) in California to be hosts for sugar beet viruses, but used for many others (e.g. Hollings, 1959).

Some nonmanually transmitted persistent viruses, especially those that cause yellowing symptoms in their hosts, such as potato leaf roll and barley yellow dwarf viruses, also have useful alternative hosts. For example, I have referred to potato leaf roll as a yellowing virus although it does not usually cause yellowing symptoms in potato. It does so in *Physalis floridana*, which is more susceptible to leaf roll virus and a better source of the virus for aphids. Other yellowing viruses such as turnip latent virus (McKinnon, 1965) and turnip mild yellows also infect *P. floridana*.

E. Virus-Particle Size and Shape

Although nonpersistent viruses are alike in their transmission by aphids, they differ in many of their physical properties, such as thermal inactivation, and particle size and shape (Fig. 5-5).

The group containing potato virus Y, one of the first aphid-transmitted viruses to be described, is the largest and best known. Viruses of this group, for example henbane mosaic and "mosaic" viruses of turnip, pea, bean, lettuce, and many other hosts, are also easily transmitted by many different species of aphids. They have long, flexible particles about 13 × 725 nm (Brandes, 1964). Although they cause distinct diseases, some of the viruses are related serologically. The long, slender particles are easily fractured or damaged during extraction, and partially purified preparations are often less infective than freshly extracted saps.

A second group contains nonpersistent viruses with shorter, stiffer rods, about 13 × 600-650 nm; they include potato viruses S, M, and A, carnation latent virus, red clover vein mosaic and pea streak virus, and others. Some are more difficult to transmit by aphids than those of the potato virus Y group but they are usually transmitted in a similar manner. Many seem to be symptomless in some of their hosts.

Cucumber mosaic virus has rounded particles 27-30 nm diameter. Lucerne (alfalfa) mosaic virus is shaped as "bacilliform" particles, short, fat rods with rounded ends 18 × 50 nm.

It seems that several different kinds of plant viruses have become adapted to transmission on or within the mouthparts of probing aphids.

yellow vein viruses occur profusely in tissues of their aphid vectors, and it seems obvious that they multiply there. The sites of aggregation or multiplication of viruses in insects used to be difficult to find but some are now being identified by electron microscopy of ultrathin sections.

VII. HANDLING APHIDS FOR VIRUS TRANSMISSION

The methods used in most transmission experiments are simple. Aphids are gently disturbed on their food leaf by lightly touching their antennae, or breathing on them, which makes them withdraw their stylets. Aphids are picked up and transferred with the moistened tip of a No. 1 or 2 watercolor paintbrush.

Aphids transmit nonpersistent viruses best after fasting, so they are withheld from food before use, usually in glass or plastic containers. We use petri dishes covered with nonwaterproofed tightly stretched cellophane (300 substance). The cellophane has a 1.5 cm hole in the middle to admit the aphids. The hole is sealed temporarily with a square of moistened cellophane (self-adhesive when wet). The dishes are covered with their lids and kept in a cool place (10-15° C) for 1-18 hr, until used. Then the aphids are placed (on their feet) on a detached young infected leaf. This is held in an inverted small, clean, glass vessel (a crystallizing dish or beaker of 50-100 ml capacity) to keep the microclimate of the aphid cool and moist, and also to prevent its being breathed on, because increased CO_2 concentration inhibits attempts to feed.

The aphids are watched with a hand lens, in profile, through the glass wall of the vessel, and when they assume the probing positions (see Fig. 5-3) the duration of the probe is timed with a stopwatch. Probes of 20-60 sec duration are the most likely to lead to infection.

For less critical work, two or three aphids can be placed on the infected leaf and each removed when it has finished a probe of more than 15 sec, or, when many are needed, larger groups can be left for 2-3 min, when those that have probed or are probing can be used.

When the probing is observed with a hand lens, the actual moment of penetration of the leaf can often be detected because the aphid grasps the surface of the leaf and moves with a sharp thrust. Aphids can be made to penetrate artificial membranes or pieces of peeled epidermis. They then can be watched under a low-power stereoscopic microscope (about × 150) in reflected light. Viewed through the membrane from above, the rostrum of an aphid feeding beneath can be seen pressed against it, as a black circle on the underside. A flickering point of light appears suddenly; this is the tip of the stylet bundle.

For bulk inoculation or for transmitting persistent or semipersistent viruses, aphids can be shaken from food plants, after breathing on them, onto deep enameled or smooth plastic trays, about 12 in. × 8 in. × 3 in. deep. Tapping the back of the tray collects them into one corner where they can be shaken onto the infected plant or leaf. Cut leaves are put into labeled glass jars about 5 in. high by 3 in. diameter. In our experiments these are usually 1 lb glass jam jars with about 1 in. of moist sand in the bottom; surplus water at the bottom of the jar can be removed with a pipette. The cut ends of leaves or petioles are inserted not too tightly into clefts in the sand, the aphids dropped into the chamber, and the top covered by a moistened cellophane disc secured by an elastic band. They are kept cool and in the shade.

After feeding, the aphids are collected carefully, and when necessary they are selected according to the stage of development required (all stages of development will transmit most viruses, but many persistent viruses are more efficiently transmitted when acquired by nymphs). They are placed on seedling plants in groups the size of which may differ for different viruses. For example, in optimum conditions beet yellows virus needs about 5 aphids per plant for 80% transmission, and groundnut rosette virus requires about 10 aphids per plant.

The test plants to which aphids are transferred should usually be as young as possible. Some viruses, for instance, cocoa swollen shoot virus (Posnette and Strickland, 1948) and groundnut rosette virus (Storey and Ryland, 1955), are transmitted to the cotyledons, or plumules of newly sprouted seeds *before* these are planted. Young plants are usually the most susceptible and it is easy to provide uniform seedlings and to keep them in optimum conditions of storage. After culturing, the test seedlings should be covered to prevent aphids from escaping. The covers can be plastic or glass beakers, or tubes tightly covered at one end with cellophane or muslin. Cylindrical cages made from cellulose nitrate sheeting are very useful and can be made cheaply in any size.

VIII. STYLET-TIP TREATMENTS

To determine where nonpersistent virus is held by aphids before inoculation Bradley and Ganong (1955) anesthetized the aphids one at a time using a small funnel, or merely the end of a rubber tube, held over it for about a minute with CO_2 gently flowing from a cylinder. They mechanically exposed the stylet bundle using a thick-walled capillary glass tube pressed against the tip of the extended rostrum so that only the rostrum was held back by the tube (Fig. 5-6). Treating these stylet tips, after acquisition-feeding, with ultraviolet irradiation or dipping them in formalin solution, rendered the aphids noninfective. Later Bradley (1962) found that the formalin treatment inhibited probing, but there is so far no simple explanation for the effect of irradiation in preventing transmission, except that the nonpersistent viruses are carried by simple contamination of the extreme tips of the stylets, hence the term "stylet-borne."

There are, however, difficulties about the complete acceptance of this interpretation. Very contagious viruses such as tobacco mosaic are not readily transmitted from systemically infected plants, or by artificial feeding (Pirone, 1967), nor when the stylets are dipped or painted with concentrated virus. In

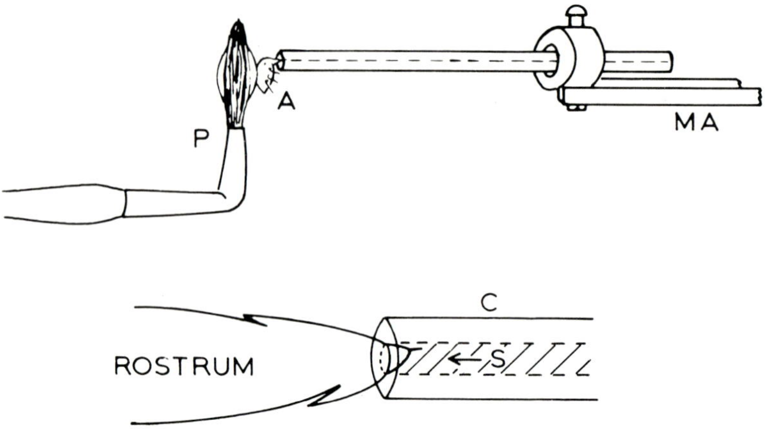

Fig. 5-6. Bradley's method of exposing the stylets of aphids for inactivating acquired virus for inhibiting future probing by treatment. Paintbrush (P) bent to form a vertical support for the aphid (A) held in droplet on the bristles. Capillary tube (C) held in a micromanipulator (MA). In the enlarged diagram below, the rostrum is pushed back by capillary tube (C) and the stylet bundle (S) is extruded into the tube.

contrast, *Myzus persicae* has recently been shown to acquire tobacco mosaic virus from leaves sprayed with concentrated virus and then to inoculate it into healthy leaves (Lojek and Orlob, 1969). No stylet-borne virus has been transmitted by deliberate contaminations of stylets. Also Gamez and Watson (1964) found that when stylets of aphids that have acquired virus by probing are pricked into cells while the aphids are anesthetized, the virus they are carrying is not transmitted, but when they recover consciousness they can inoculate plants. A fine needle pricked into an infected leaf and then into a healthy one will often transmit virus, and there seems no reason why the stylets of unconscious aphids should not do the same if they carry the viruses similarly. Nevertheless, nonpersistent viruses are transmitted so rapidly that they are probably held somewhere within or upon the pharyngeal region or the stylets.

IX. TRANSMISSION OF NONPERSISTENT VIRUSES

A. Effect of Fasting on Vector Efficiency

Watson and Roberts (1939) showed that henbane mosaic virus, potato virus Y, and cucumber mosaic virus were transmitted best when aphids were allowed access to infected leaves for only about 2 min after fasting. With longer times, transmission decreased although aphids were still taken from the same infected leaves.

The efficiency of transmitting henbane mosaic virus increased when aphids were made to fast before acquisition feeding (Table 5-2). Without this preliminary fasting only 10% of the aphids infected, but those given 1 hr of

TABLE 5-2
TRANSMISSION OF HENBANE MOSAIC VIRUS BY *MYZUS PERSICAE* WITH VARYING PERIODS OF FASTING AND ACQUISITION FEEDING[a]

Times of Preliminary Fasting	*TIMES OF ACQUISITION FEEDING*			
	2 min	*1 hr*	*24 hr*	*Mean*
0	5	3	5	4
15 min	21	5	7	11
1 hr	29	4	6	13
6 hr	34	7	10	17 s.e. 2.06
Mean	22	5	7 s.e. 2.34	

[a]Experiments consisted of 5 plants per treatment, repeated on 10 occasions; 1 aphid per plant.

fasting and then allowed 2 min access to infected leaves infected nearly 70% of test plants. The length of fasting is not critical, and 15 or even 5 min increases transmission over unfasted aphids. When acquisition feeding increased, transmission by prefasted aphids decreased, and there was no effect of fasting after 1 hr.

B. Consecutive Infection

After acquiring nonpersistent viruses, aphids have often been thought to infect one plant only but they can actually transmit viruses repeatedly during the first hour after leaving the infected plants. When aphids were fed for 2 min on potato virus Y-infected plants and then moved to fresh plants every 5 min, several aphids infected four or five plants and all except two infected more than once (Table 5-3). In a similar experiment with henbane mosaic virus one aphid infected consecutively seven out of ten plants, and ten of the thirty-six aphids tested infected consecutively four or five plants each (Watson and Roberts, 1940).

C. Factorial Design of Experiments

The factorial experimental design shown in Table 5-2 is useful for many purposes including field experiments that compare, for example, times of infection with different seed-sowing dates, or different insecticidal treatments. It tests the effect of varying several factors in all combinations. The statistical analysis is given by Yates (1937). Three preliminary fasting and four acquisition feeding times were tested (giving 3×4 treatments). The treatments were given to groups of 5 plants on each of 10 separate occasions giving a total of 120 observations with 99 degrees of freedom for error. The analysis of variance is shown in Table 5-4.

The small values for the probability p that the observed differences could

TABLE 5-3
CONSECUTIVE INFECTIONS BY *M. PERSICAE* TRANSMITTING POTATO VIRUS Y[a]

Aphid	A	B	C	D	E	F	G	H	Totals
	Consecutive Healthy Plants								
1	+	+	0	0	0	0	0	0	2
2	+	+	+	+	0	0	+	0	5
3	+	+	0	0	0	0	0	0	2
4	+	+	0	0	0	0	0	0	2
5	+	0	0	+	0	0	0	0	2
6	+	0	0	0	+	0	0	0	2
7	+	0	+	0	+	0	0	0	3
8	+	+	+	0	+	0	0	0	4
9	0	0	0	+	0	0	0	+	2
10	+	0	0	+	0	0	0	0	2
11	0	+	0	0	0	0	0	0	1
12	0	+	+	+	0	+	0	0	4
13	0	+	0	0	0	0	0	0	1
14	0	+	+	0	0	0	+	+	4
15	+	+	0	+	0	0	0	0	3
16	0	+	+	0	0	0	0	0	2
Totals	10	11	6	6	3	1	2	2	

[a] Aphids probed for 1 min, then stayed 5 min on each successive test plant (+ = infected plant; 0 = uninfected plant).

TABLE 5-4
ANALYSIS OF VARIANCE OF DATA FROM A FACTORIAL DESIGN EXPERIMENT[a]

	Degrees of Freedom	*Mean Square*	*Variance Ratio*	
Replications	9	4.67	2.83	
Between Fasting	3	17.9	10.85	$p<0.01$
Between Feeding	2	10.23	6.85	$p<0.01$
Interaction	6	3.9	2.36	$p<0.05$
Error	99	1.65		

[a] Tested were 3 preliminary fasting periods and 4 acquisition feeding times.

have occurred by accident show that the main treatments were well tested. The interaction between feeding and fasting reflects the lack of effect of different feeding times on unfasted aphids (Table 5-2).

D. Effect of Fasting on Brief Probing

Nonpersistent viruses are acquired and transmitted mainly during the first minute of penetration of the leaf. Aphids that have fasted usually end their first probes naturally within one or two minutes, often within a few seconds. Figure 5-7 shows the frequency distribution of the numbers of infecting aphids as a function of the duration of natural probes made by fasted and unfasted *Myzus persicae* in tobacco plants. After fasting, probes lasting about 30 sec were the most frequent *and* the most successful in transmitting virus. The proportion of infective aphids was as great between one and two minutes as it was for times shorter than a minute, but few aphids made first punctures lasting longer than a minute.

Without fasting, many aphids probed (or seemed to probe) briefly soon after being placed on the infective leaf, but few of these brief probes led to infections. A larger proportion of aphids penetrated deeply at their first attempt, and no probes that lasted longer than 5 min caused infection. On the average, aphids took longer to make a feeding attempt, but this did not affect probing time or infectivity.

E. Probability of Infection

Using data such as in Table 5-2 where one aphid was used per plant, the probability of infection can be calculated. The probability of infection (p) by any aphid is assumed constant and independent of whether or not other aphids are feeding on the plant. The probability q of any aphid not transmitting virus is $q = (1 - p)$. The probability of n aphids all feeding at once without transmitting would be q^n. In Table 5-2, 34 of 50 aphids (68%) caused infection when starved for 6 hr and given 2 min access to infected plants; q therefore equals 0.32. If $n = 3$, then $q = [1 - (\text{antilog } 0.32 \times 3)] = 0.0515$. Groups of three aphids should therefore have infected about 95% of the plants. This simple calculation is useful for roughly comparing experiments done with different numbers of insects, but because of various parameters not included, the estimate is likely to be only approximate.

F. Are Inoculations by Aphids Local and Independent?

If infections by aphids are not local and independent, the assumption that $p = (1 - q^n)$ would not be true because the aphids would not act independently, and groups would perform relatively better instead of worse than single individuals.

Table 5-5 shows the results of an experiment made with groups of 1, 5, 10, and 20 aphids to test whether the observed p values have a wider or narrower range than expected. The observed p, based on the performance of one aphid per plant, was 0.117, but the expected number of plants infected by groups of 20 aphids would be 220 ($p = 1 - 0.883^{20}$) which is greater than the observed value. A value for p estimated by the method of maximum likelihood (Fisher, 1928) was 0.10 ± 0.0005. This value fits the whole range of figures better, although the

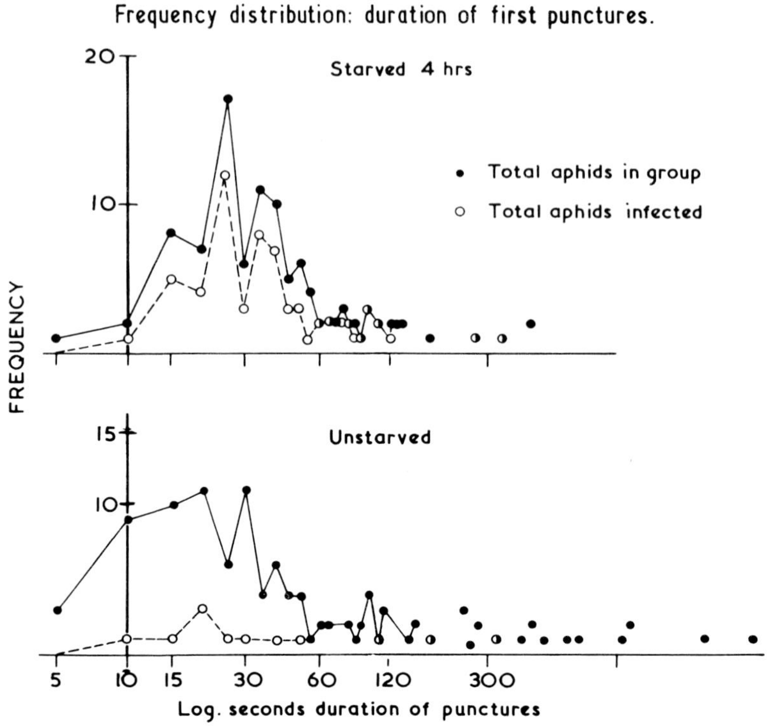

Fig. 5-7. Transmission of henbane mosaic virus. Distribution of durations of first probes made by *Myzus persicae* on tobacco leaves as a function of time. After fasting for 4 hr most probes occurred between 15 and 45 sec and most virus was acquired during these probes. Probes by unstarved aphids still seem to be more frequent around 15-30 sec than during any other interval of time, but many aphids continue to penetrate for 5 or 10 min.

TABLE 5-5
DISTRIBUTION OF INFECTION WITH A NONPERSISTENT VIRUS BY APHID GROUPS OF VARYING NUMBER

Number of Plants Infected	*Number of Aphids per Plant* [a]			
	1	5	10	20
Actual Number Infected	28	127	163	190
Expect Number (if $p = 0.117$) [b]	28	111	170	220
Expected Number (if $p = 0.1005$) [c]	24	99	157	211

[a] Acquisition time: 18 hr, overnight, 240 plants per group.
[b] Observed value for 1 aphid per plant is $\frac{28}{240} = 0.117$.
[c] 0.1005 ± 0.0005 = maximum likelihood value for best fit.

expected value of 24 for one aphid per plant is too small and that of 211 for 20 aphids per plant is still too large.

Observed values for the performance of large groups of aphids are therefore less than those expected even on the hypothesis that infections are local and independent. As the accumulation of subinfective doses inoculated by large groups of aphids would give a greater range of values than expected, this seems to confirm that each infection by aphids is essentially local, and independent of other infections.

Storey (1938) used the maximum likelihood method to show that infections by the leafhopper, *Cicadulina mbila,* transmitting maize streak virus, were also local and independent, and this is usually assumed in interpreting results of transmission experiments. However, it is not necessarily true with all plant viruses, for there seems no reason why virus introduced directly into the phloem should not accumulate at junctions in the food stream before it is carried to the site where it multiples, unless, of course, the first multiplication site is the phloem cell to which it is introduced.

X. TRANSMISSION OF SEMIPERSISTENT VIRUSES

A. Effect of Feeding Times

Beet yellows virus, transmitted by several aphid species, can be transmitted to sugar beet with difficulty by mechanical inoculation if the plants are kept for two days in the dark and receive special care to keep the inoculated leaves alive until symptoms develop (Kassanis, 1949).

M. persicae can acquire and inoculate the virus within 30 min, but not during brief probes, and preliminary fasting has no effect on transmission. The aphid's infectivity increases with feeding times on both infected and healthy plants up to a total of about 12 hr, and aphids can continue to infect test plants exposed to them successively up to 2 days, although they lose much of their infectivity in one day. Table 5-6 shows an experiment illustrating much of this behavior. The aphids were fed for different periods on infected plants, then on test plants for 1, 4, or 20 hr, and then on second test plants, where they were left for 24 hr. Infectivity increased rapidly with increasing acquisition feeding time up to 6 hr, and with test feeding up to about 4 hr, then increasing slowly with longer feeding periods (Watson, 1940).

The increases are nearly proportional to the logarithms of the times of feeding and are therefore exponential.

The proportion of groups causing second infections was greatest after 1 hr on first plants and least after 20 hr, showing that persistence of virus in the aphids was relatively short, although much longer than with nonpersistent viruses. About half the aphids transmitted for the first time after feeding between 1 and 4 hr on test plants, which suggest that there is some delay in development of infectivity, but it is short compared with the latent periods of persistent (circulative) viruses.

TABLE 5-6

TRANSMISSION OF BEET YELLOWS VIRUS BY *M. PERSICAE* FED FOR VARYING TIMES ON INFECTED AND ON TWO SUCCESSIVE TEST PLANTS [a]

Feeding times on First Test Plants (24 hr on second)			*On Infected Plants (hours)* 1/2	2	6	18	24	*Total*
1 hr	Distribution	(+ 0)	3	6	7	5	4	29
		(+ +)	0	12	16	17	21	66
		(0 +)	1	9	14	19	11	54
	Total	1st plant (1 hr)	3	18	23	22	25[b1]	91[c1]
		2nd plant (24 hr)	1	21	30	36	32[b2]	120[c2]
4 hr	Distribution	(+ 0)	7	10	15	12	16	60
		(+ +)	0	14	18	22	18	72
		(0 +)	1	1	2	6	3	13
	Total	1st plant (4 hr)	7	24	33	34	34	132
		2nd plant (24 hr)	1	15	20	28	21	85
20 hr	Distribution	(+ 0)	7	21	27	33	33	121
		(+ +)	1	1	9	8	8	27
		(0 +)	0	1	0	1	0	2
	Total	1st plant (20 hr)	8	32	36	42	41	159
		2nd plant (24 hr)	1	2	9	9	8	29
All Times on First Plants	Distribution	(+ 0)	17	37	49	50	53	206
		(+ +)	1	27	43	47	47	165
		(0 +)	2	11	16	26	14	69
	Total	1st plant	19	75	92	98	100[d1]	384[e1]
		2nd plant	3	38	59	73	61[d2]	234[e2]
	Total		14	113	151	171	161	618

[a] The aphids were fed for different periods on infected plants, then for 1, 4, or 20 hr on the first and 24 hr on the second test plants. Five aphids per plant; 50 plants per treatment.

[b-e] Standard errors:

	1	2
b	3.25	3.45
c	7.25	7.77
d	5.	—
e	12.55	13.46

The results of factorial experiments such as this can be plotted on isometric graph paper as shown in Fig. 5-8.

Figure 5-8 shows the proportions of first test plants infected with beet yellows virus as a function of the logarithms of time. The small figure shows a similar diagram made with the original data, uncoverted. The larger figure shows more clearly what happens at the shorter feeding times; the points are better spaced for statistical treatment and a straight line passing through the origins would probably give a fair idea of the slopes of the lines, even though curves would fit the points better. There are not enough points here for separate regression lines to be fitted properly.

B. Consecutive Infections

In the experiments shown in Table 5-6 it would have been better to use one aphid per plant to ensure that all consecutive infections are caused by the same aphid. However, we calculated previously that even in optimum conditions (about 24 hr on infected and healthy plants) the probability of infection by a single aphid is less than 30%. Hence, the number of plants needed for statistical accuracy would be too great. The results indicate the duration of the latent period and the persistence of infectivity in the aphids, but provide no direct information on the frequency with which individual aphids infect plants.

In another experiment single aphids were removed hourly to eight consecutive test plants after 24 hr feeding on infected ones. Nearly all

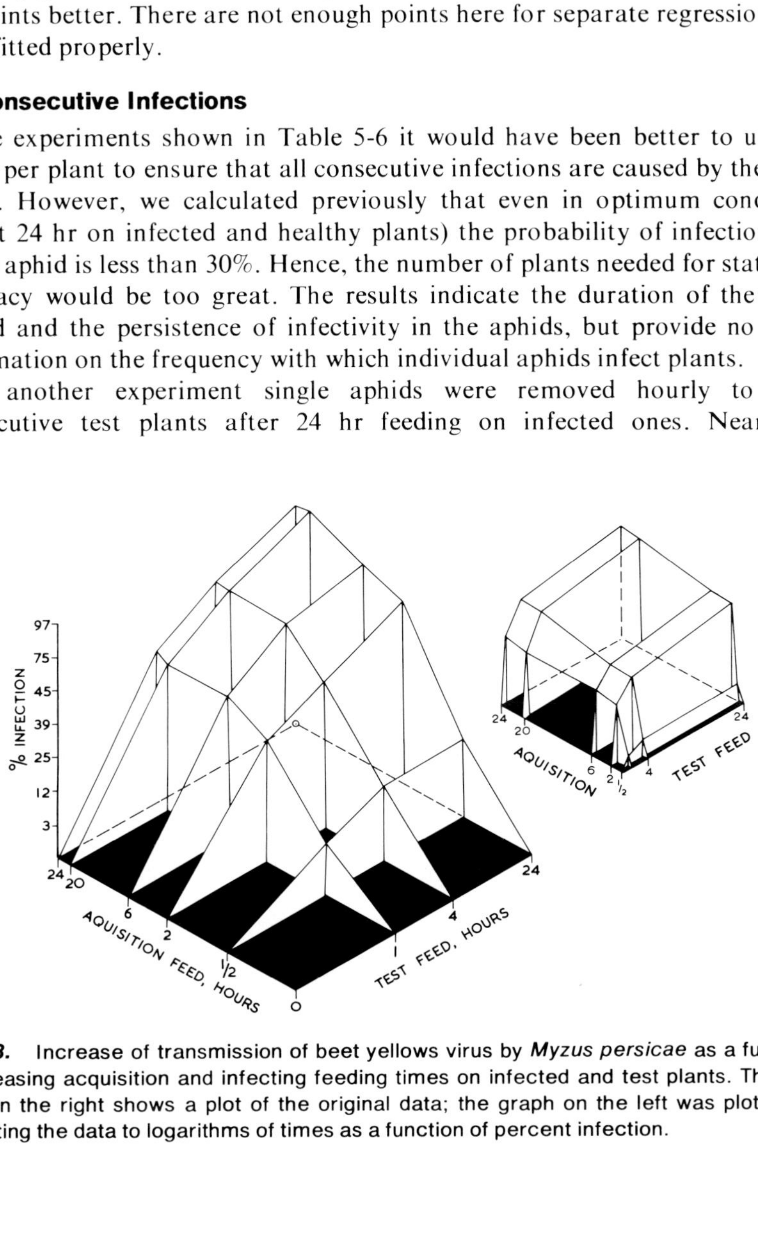

Fig. 5-8. Increase of transmission of beet yellows virus by *Myzus persicae* as a function of increasing acquisition and infecting feeding times on infected and test plants. The diagram on the right shows a plot of the original data; the graph on the left was plotted by converting the data to logarithms of times as a function of percent infection.

transmitted within 3 hr of leaving the infected plant (Table 5-7), and about 50% infected half or more than half of the plants tested; 20% of those that infected within 6 hr also infected next day and none infected on the second day only.

TABLE 5-7
NUMBER AND DISTRIBUTION OF SUCCESSIVE INFECTIONS CAUSED BY *MYZUS PERSICAE* IN TRANSMITTING BEET YELLOWS VIRUS [a]

	Performance at Successive 1 hr Test-Feeding Periods							
Plants	*1*	*2*	*3*	*4*	*5*	*6*	*7*	*8*
1	+	0	0	0	0	0	0	0
2	+	+	0	0	0	0	0	0
3	+	+	0	0	+	0	0	—
4	+	+	+	0	0	+	+	0
5	+	0	+	0	+	0	0	+
6	+	0	0	+	0	+	—	—
7	0	+	0	0	0	0	0	0
8	0	+	0	+	0	+	0	0
9	0	+	0	+	0	+	0	—
10	0	+	+	0	0	+	0	+
11	0	+	+	+	0	0	0	0
12	0	0	+	0	+	0	+	0
13	0	0	+	0	0	+	0	0
14	0	0	+	0	0	+	0	0
15	0	0	+	+	+	+	+	0
16	0	0	+	+	0	0	0	0
17	0	0	0	+	0	+	—	—
18	0	0	0	0	0	+	+	+
Total	6	8	9	7	4	10	4	3

[a] + = aphid infected plant; — = aphid died; 0 = no plants infected.
Fifty-two aphids failed to cause infection on any plant.

This experiment fills in the gaps of information left by the previous one. Eighteen aphids out of 70 tested singly caused infection which compares reasonably with the 29% probability of infection by a single aphid estimated previously.

C. Separation of Viruses by Aphid Feeding

Viruses can be isolated from plants infected with more than one virus by the method of consecutive feeding. The method requires acquisition feeding on the infected plant and then successive transfers of the aphid to healthy plants. It was used to separate strawberry viruses (Prentice and Harris, 1946), and for beet yellows and beet mosaic viruses (Table 5-8). Aphids were given 16 hr

TABLE 5-8
SEPARATION OF BEET MOSAIC VIRUS AND BEET YELLOWS VIRUS FROM A MIXED INFECTION BY DIFFERENT FEEDING TREATMENTS OF *MYZUS PERSICAE*

Time on Infected Plants	*Virus*	*Times Feeding on Successive Healthy Plants*				
		10 min	*2 hr*	*24 hr*	*24 hr*	*Total*
2 min	BMV[a]	15[b]	2[b]	0	0	17
	BYV	0	0	0	0	0
1 hr	BMV	2[b]	1	0	0	3
	BYV	0	2	1[c]	0	3
24 hr	BMV	6	2	0	0	8
	BYV	4	18	22[c]	6[c]	50

[a] BMV = beet mosaic virus; BYV = beet yellows virus.
[b] Plants infected by BMV only.
[c] Plants infected by BYV only.

preliminary fasting then fed for 2 min or 24 hr on plants infected with both viruses, and finally, successive feeds of 10 min, 2 hr, 24 hr, and 48 hr on healthy test plants. Beet mosaic virus was transmitted after 2 min acquisition feeding, but beet yellows virus was not. However, beet mosaic virus was not transmitted to the second and third plants fed on after 24 hr acquisition feeding, and conversely, beet yellows virus was transmitted. Each virus was thus isolated free from the other.

In exploratory tests on unknown viruses, preliminary fasting followed by either 2 min or 24 hr access to infected plants should indicate whether the virus is persistent or nonpersistent. If 2 min is clearly more successful than 24 hr the virus is nonpersistent and vice versa.

XI. TRANSMISSION OF PERSISTENT VIRUSES

A. Persistence and the Latent Period

Although some persistent viruses may be acquired by aphids in a minimum of 20 min acquisition feeding, all are very much more efficiently transmitted with longer times. The average for different viruses ranges from about 6 to 24 hr. The latent period is measured by consecutive feeding. With circulative viruses we are often more interested in the varying performance of individual aphids than of groups, because each aphid behaves as a reservoir of the virus, whose concentrations and viability largely determine the efficiency of the whole populations in spreading viruses.

Figure 5-9 compares the fluctuations in infectivity of individual aphids removed daily to fresh test plants after feeding for one of two periods on infected plants. The results are expressed as the proportion of plants infected each day by those aphids that infected at least one plant. Beet yellows virus has been included to show that the distribution of infections is similar although the times taken are much shorter than with beet yellow net virus and groundnut rosette viruses.

Nearly all the aphids transmitted beet yellows virus during the first day, even when the acquisition feeding period was only an hour. After longer acquisition times, 30% of aphids also infected on the second day and two on the third, but after one hour on infected plants only 2% infected during day 2 and none on day 3. Thus fewer plants were infected by aphids given short acquisition times, and they lost infectivity sooner, so we must presume that they initially acquired less virus and the supply became exhausted.

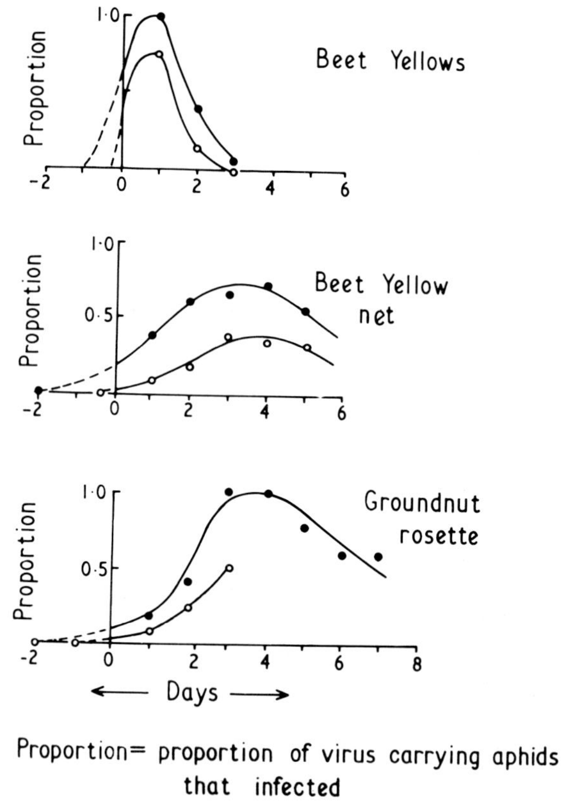

Fig. 5-9. Comparison between beet yellows virus and two persistent viruses, beet yellow net and groundnut rosette, for the effect of acquisition feeding time (negative values on abscissa) on persistence of virus in the aphid. The ordinate shows the proportion of virus-carrying aphids that infected on each day of successive feeding on test plants.

Groundnut rosette virus, transmitted by *Aphis craccivora* (Watson and Okusanya, 1967), and beet yellow net virus, transmitted by *M. persicae* (Sylvester, 1958), resembled beet yellows virus in the way they gained and lost infectivity after different feeding treatments, but the changes were slower. Many aphids started to infect two or three days after leaving the virus source, that is, they exhibited much longer latent periods (average) than did vectors of beet yellows virus.

Aphids given the shorter acquisition feeding times infected daily for several days once the latent periods had ended, but then their infectivity declined. Those aphids given long acquisition feeding periods infected more frequently and lost infectivity later than those given short feeding times. The infectivity of all aphids declined after five or six days, and this too might be attributed to simple exhaustion of the supply of virus, but Sylvester and Richardson (1966) have another explanation. They compared the transmission of pea enation mosaic virus by aphids fed as nymphs for a 3-6 hr period on infected plants, and as eight-day-old adults, fed for 24 hr. The nymphs were able to infect sooner (had shorter latent periods) and retained the virus much longer than the adults (Fig. 5-10). Nymphal infectivity started to decline 12 days, and that of the adults 6-7 days, after virus acquisition. With both stages the decline was accompanied by aging of the populations and deaths of individual aphids. Most aphids ceased infecting one or two days before death, so the decline could result from poor physiological condition of the vectors, rather than exhaustion of the supply of virus.

Some aphids given their acquisition feeds (for groundnut rosette virus) as adults, died soon after the end of the consecutive feeding tests. Therefore, some, or all, of the decline of infectivity could have been caused by aging or possibly by handling. However, those given short acquisition feeds retained infectivity

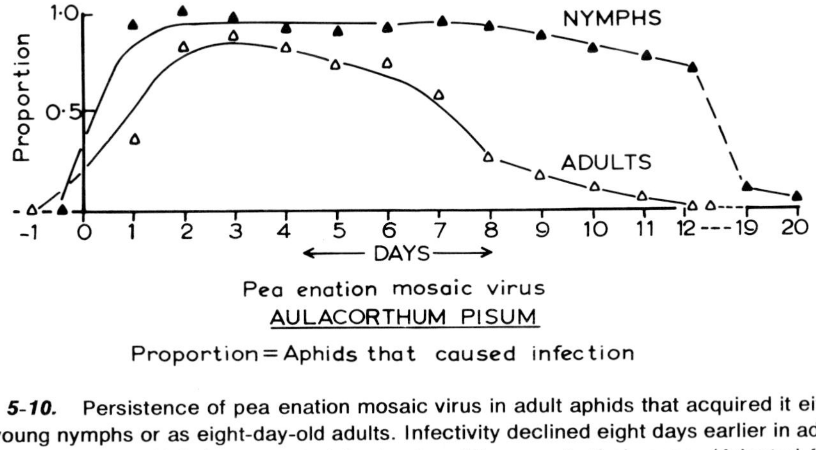

Fig. 5-10. Persistence of pea enation mosaic virus in adult aphids that acquired it either as young nymphs or as eight-day-old adults. Infectivity declined eight days earlier in adults than in nymphs, which is accounted for by the difference in their ages. (Adapted from Sylvester and Richardson, 1966.)

more briefly than those given long ones, although they were a day or two younger (having all started to feed at the same time). Thus performances seems to be affected by the amount of virus acquired as well as by the age of the aphid when it is tested.

B. Effect of Temperature

Persistence and latency are affected not only by the age and state of development of the vectors, but by the conditions in which experiments are done. Figure 5-11 (adapted from Duffus, 1963) summarizes some of the results of consecutive feeding experiments with *Amphorophora lactucae* transmitting sowthistle yellow vein virus, done in controlled environment held at temperatures of 25° C and 15° C. This virus is unusual among aphid-transmitted plant viruses in having a very long latent period. When kept constantly at 25° C, aphids fed for 2 days on infected plants could not transmit until 7 days after leaving them. Then their infectivity increased rapidly to a point where each aphid transmitted at least once every day. After about 16 days this infectivity declined but so did the biological activity of the colony. When the same clone of aphids was maintained at 15° C the latent period of the virus was much longer, so aphids remained noninfective for 20 days after leaving infected plants. Their lives were much prolonged and so was the persistence of the virus.

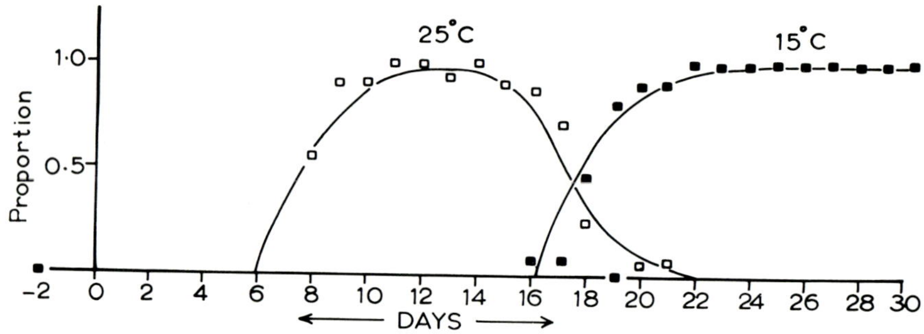

Fig. 5-11. Effect of temperature on the latent period of sowthistle yellow vein virus in its vector, *Amphorophora lactucae*, and on the persistence of the virus. Aphids kept at 15° C took longer to become infective, transmitted more regularly and longer than those kept at 25° C. (Adapted from Duffus, 1963.)

XII. BIOASSAYS

A. Injection Method

Starting with the work of Stegwee and Ponsen (1958) on potato leaf roll virus, which, in turn, was inspired by American experiments with leafhopper-transmitted viruses, a number of aphid-transmitted persistent viruses, including pea enation mosaic (Schmutterer and Ehrhardt, 1964) and barley yellow dwarf virus (Mueller and Rochow, 1961), have been transmitted to aphids by injection of virus preparations from plants, or hemocoelic fluid from infective aphids.

Injections are done with fine glass needles, the points drawn to a diameter of 15-25 μ and obliquely broken off so that a sharp cutting edge is presented to the point of insertion. Starting with 5 mm soft glass pulled evenly out to a diameter of about 1 mm fairly near to the shoulder, a second pull is made over a small flame by hand or using a micro-forge technique. The new tips are examined with a microscope, using a 100 μ grid to select those with diameter about 20 μ. They are carefully broken under a binocular microscope to give the correct section.

Aphids are anesthetized by being held in a gentle flow of CO_2 for about 30 sec or whatever time is necessary to immobilize them during the operation, but no more. Of the various methods devised for holding them during injection, many involve some form of suction. The simplest is a 5 mm tube attached to a suction pump, having the end covered with fine silk, against which the aphid is held. Various forms of perspex slides can be devised, with crevices made to take the head end of the aphid and help to keep it in place; sometimes thick perspex slides are perforated so that suction can be used as well.

The needle is attached to the tip of a micro-syringe, which is first filled with silicone oil (Diamethicone 20) to control the flow of the test fluid. The receptacle and the needle can be attached to a micromanipulator or to the mechanical stage and focusing ratchet of an old microscope which can be modified to serve. The focusing screw moves the plunger of the micro-syringe up and down and the mechanical stage adjusts the position of the aphid, which is injected under a binocular microscope (about $\times$ 50). The needle is inserted beneath one of the tergites (the back) into the thorax, or into a joint at the base of a rear leg. The joint nearest the body is called the coxa. If the coxa is removed and the body slightly depressed a droplet of fluid oozes out, which can be collected in the syringe and be injected straight into another aphid. In this way body fluid free from the detritus of crushed organs can be injected directly from one aphid to another. However, the most usual method is to crush a large number of bodies of viruliferous aphids in buffer (2000 yield about 0.5 ml of fluid) with suitable antioxidants or virus stabilizers, and partially purify the preparation by high-speed centrifugation. The supernatant fluid from the resuspended pellet is used as inoculum. Plant material treated similarly can be used.

Mueller and Rochow (1961) used a long connection from the syringe to a rubber syringe bulb that could be compressed by the foot, leaving the hands free for manipulation.

B. Artifical Feeding Method

Aphids can acquire some viruses by feeding through thin parafilm membranes. The virus is prepared as for injection and diluted in strong sugar solution (about 20% sucrose), which stimulates aphid feeding. Parafilm M (Marathon;

American Can Company, Menasha, Wis.) used for the membrane can be stretched both ways into a very thin waterproof film and is self-adhesive to glass. Cells made of ½ in. tubing cut into ¼ in. rings are fitted, membrane downwards, into a slightly larger and longer section of glass tube, into which aphids have been dropped. The drop of fluid to be tested is placed on the membrane (Fig. 5-12). The holder should be darkened and light coming through the membrane is best filtered through a yellow glass screen, which also encourages feeding. To time the feeding the aphids can be observed in profile between the darkened area of the holder and its membrane.

A more elaborate cage is described by Mittler and Dadd (1963), and in a later modification (Dadd and Mittler, 1964) they use a "sachet" made by stretching the membrane over the mouth of the vessel containing the aphids to be fed, placing a droplet on it, and then covering with another piece of membrane, forming a waterproof bubble of liquid.

Virus preparations from plants and from aphids have been fed to other aphids through such membranes to help to determine in vitro properties of viruses that are not manually transmissible. Gold and Duffus (1967) incubated concentrated beet western yellows virus, prepared by density gradient separation, with antiserum prepared against the same virus and also with normal rabbit serum and healthy plant protein antisera. Aphids recovered virus by artificial feeding only from the test fluids that did not contain homologous antisera. Rochow and Ball (1967) using the same technique demonstrated that transmission of *R. padi*-transmitted isolates of barley yellow dwarf virus is not blocked by antiserum prepared against *S. avenae*-transmitted isolates, and these are not closely related strains of the same virus.

Some nonpersistent viruses can be transmitted if aphids probe briefly into

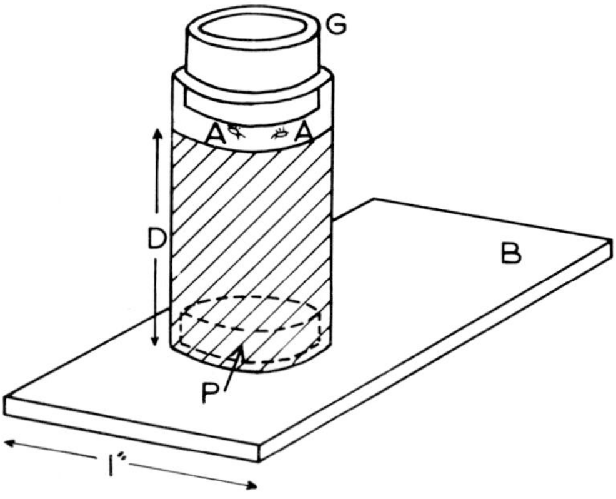

Fig. 5-12. Aphids on parafilm membrane feeding on an artificial diet. Aphids (A) being observed between the darkened region of the tube (D) and the membrane attached to the glass ring (G) inserted into its end. B = base (microscope slide); P = plug (¼ in. ring of plastic tubing) sealed to base.

concentrated solutions presented to them on thin membranes. Some strains of cucumber mosaic virus transmit from partially purified preparations almost as easily as from infected leaves. Lucerne mosaic virus has been transmitted in this way (Pirone and Megahed, 1966). Unfortunately, no long-particle viruses (potato virus Y, henbane mosaic virus, etc.) have been transmitted, although the technique has been tried repeatedly and would be invaluable for investigating the problems concerned with their transmission.* Purified preparations of red clover vein mosaic can be highly concentrated and are stable, but aphids fed on these again have failed to transmit. Tobacco mosaic virus can be acquired by aphids from purified preparations and can be returned to fluids subsequently probed during artifical feeding (Pirone, 1967) but the same aphids do not infect plants. There is no explanation why this highly infectious virus is not transmitted by aphids during brief probes that transmit the nonpersistent viruses, and artifical feeding may eventually solve the problem.

XIII. VECTOR SPECIFICITY

Nonpersistent viruses can be transmitted by many species of aphids, although not necessarily by all of them; often several different species must be tried, to find the vector of an unknown virus. Potato virus Y and henbane mosaic viruses, for instance, are easily transmitted by several *Myzus* and *Macrosiphum* species (see Kennedy et al., 1962), but not *Myzus cymbellaria* (the "viola" aphid), and only with difficulty, if at all, by some *Aphis* species and *Brevicoryne brassicae* (L.). The reasons are not fully understood, but *B. brassicae* does not seem to respond to periods of fasting by frequent probing as does, for example, *M. persicae.*

Figure 5-13 shows the transmission of turnip mosaic virus from infected turnip plants by *M. persicae* and *B. brassicae* that were reared on turnip, and the lengths of first probes made by the aphids were observed. *M. persicae* made many short probes and acquired turnip mosaic virus. *B. brassicae* made few short probes and often seemed to feed for several hours; only 4 of 27 aphids tested transmitted turnip mosaic virus (compare with Fig. 5-7).

Infrequency of probing, however, does not account for the failure of *B. brassicae* to transmit henbane mosaic virus because if *M. persicae* and *B. brassicae* are allowed to probe *Nicotiana glutinosa* plants mixedly infected with henbane mosaic virus and turnip mosaic virus, *M. persicae* easily transmits both viruses, but *B. brassicae* successfully transmits only turnip mosaic virus (Table 5-9).

The results shown in Table 5-9 also demonstrate the advantage of using local-

---

*Kassanis and Govier (1971) have transmitted potato aucuba and potato C viruses using artificial feeding, but only when assisted (see Section XIV A) by potato virus Y that has been previously acquired from plants. This opens up a whole new area of speculation.

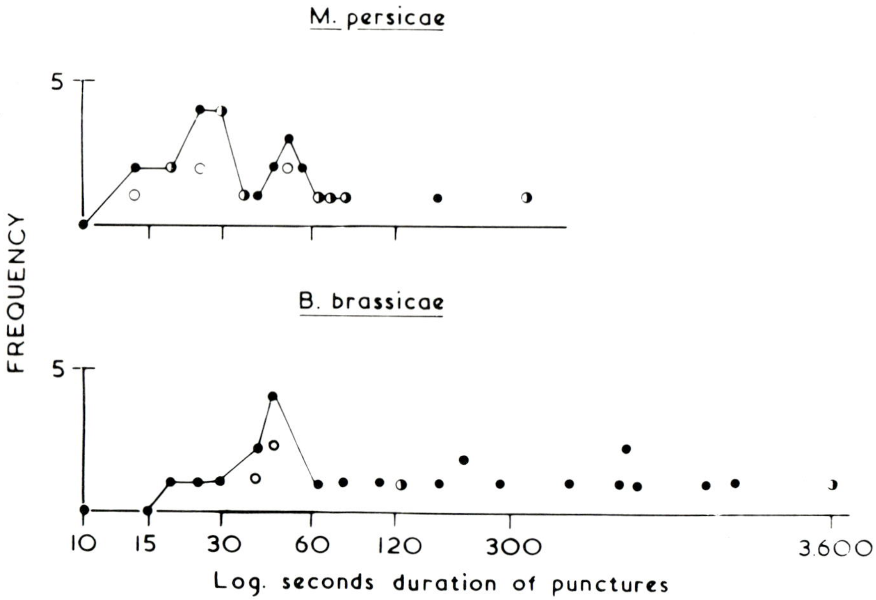

Fig. 5-13. Transmission of turnip mosaic virus (cabbage black ringspot strain) as a function of probing time. Many probes made by *Myzus persicae* after fasting lasted between 20 and 60 sec, and these most frequently transmitted virus. Few *Brevicoryne brassicae* finish their probes in less than a minute and their success is little greater at this time than with the more frequent longer probes of the 25 aphids tested, *M. persicae* transmitted to 15 plants and *B. brassicae* to 5 plants.

TABLE 5-9
TRANSMISSION OF HENBANE MOSAIC VIRUS AND TURNIP MOSAIC VIRUS TO TOBACCO SEEDLINGS

	Vector B. brassicae		*Vector M. persicae*	
Virus	*TrMV* [a]	*HMV* [a]	*TrMV*	*HMV*
Plants Infected Out of 60 Tested	39	1	60	60
Necrotic Local Lesions [b]	99	—	467	—

[a] TrMV = turnip mosaic virus (cabbage black ringspot strain); HMV = henbane mosaic virus; the viruses were acquired from *Nicotiana glutinosa* plants systemically infected with both 10 aphids per plant.
[b] Henbane mosaic virus does not cause local lesions.

lesion hosts when comparing the effects of multiple infections by large groups of aphids. *M. persicae* seems less than twice as successful as *B. brassicae* when transmitting turnip mosaic to whole plants, but it caused five times more lesions. However, the 467 local lesions did not necessarily reflect the number of aphids that became infective, because some caused three or four lesions each, and others none.

Some persistent circulative viruses have many vector species; potato leaf roll virus, for instance, is said to have eleven aphid vectors, but many viruses have only one or two species.

Barley yellow dwarf virus is of great interest because, although 10 species of aphids are vectors, they do not all transmit the same isolates of the virus. There are two main groups of isolates and each is transmitted mainly, sometimes exclusively, by a different species, or genus, of aphid. In Britain, aphids of the genus *Rhopalosiphum* (*R. padi, R. maidis*) transmit one group; *Sitobium* (*Macrosiphum*) species and similar aphids (e.g. *Metopolophium dirhodum*) transmit the other group (Table 5-10). In the United States, *Toxoptera graminum* from different localities transmits different isolates of barley yellow dwarf virus. As Bruehl (1961) says: "It is evident that considerable caution is needed in discussing, in general terms, the relative merits, of species of aphids as vectors."

TABLE 5-10
AVERAGE PERCENT INFECTION WITH BARLEY YELLOW DWARF VIRUS ISOLATES FROM DIFFERENT AREAS IN BRITAIN IN 1959

Isolates \ *Vectors*	*Rhopalosiphum padi*	*Sitobium fragariae*	*Sitobium avenae*	*Metapolophium dirhodum*
Herts	100	0	0	0
N. Wales	100	100	33	not tested
Kent	97	5	51	25
Bristol	98	4	45	22
Bedfordshire	0	8	16	75

Strains of groundnut rosette virus are transmitted by different "biotypes" of the same aphid species, *Aphis craccivora* (Koch.). Aphids native to East Africa will transmit the East African strains of the virus and those from West Africa transmit mainly the West African strains. The specific vectors, although of the same species, differ from each other in host preferences and to some extent in measurements, although these overlap. Figure 5-14 is a photograph of aphids of the two types feeding on a leaf. The larger one is the West African type, and the smaller, the East African one (Watson and Okusanya, 1967).

Probably this vector specificity has arisen as a result of geographic isolation. Groundnut rosette virus introduced on different occasions to East and West Africa, or barley yellow dwarf virus to the east and west of the United States, may have become adapted to transmission by locally dominant aphids or aphid

Fig. 5-14. Biotypes of *Aphis craccivora.* The aphid from Nigeria (in the center) transmits virus from Nigeria and Kenya, but the smaller one on the right transmits only the Kenya virus. The nymph on the left of the group shows wing buds and will become an alate adult. Its waxy covering gives it the gray or lighter color.

biotypes, which may in turn have been dictated by the species of weed hosts of aphids and viruses that are common in a region.

XIV. SYNERGISTIC TRANSMISSION

A. Nonpersistent Viruses

Potato virus Y, which causes leaf drop streak disease in potatoes, is transmitted as a nonpersistent virus by many aphid species. A closely related strain, potato virus C, has lost its ability to be transmitted by aphids and now exists only in old potato varieties where it is propagated vegetatively (Bawden and Kassanis, 1947*b*). Potato virus Y and potato virus C cause similar symptoms in tobacco and other *Nicotiana* species; they react with the antisera of each other and they can multiply together in the same plant. When aphids feed on leaves infected with both viruses, some virus isolates that they transmit are typical of potato virus Y, and causes leaf drop streak symptoms in potatoes, but other isolates resemble potato virus C in causing only local lesions, and they remain transmissible by aphids during several subcultures even when apparently isolated free from potato virus Y.

So the mixing of aphid-transmitted potato virus Y with vectorless potato virus C causes the latter virus to behave as an aphid-transmissible virus, and also confers other properties of potato virus Y (Watson, 1960). Other vectorless potato viruses, potato aucuba mosaic and potato virus A, also have their transmission assisted by potato virus Y (Kassanis, 1961) but apparently without any change in their heritable properties.

B. Persistent Viruses

Some persistent viruses are manually transmitted only when assisted by another virus. Tobacco rosette disease (Smith, 1946) is caused by two viruses, tobacco vein distorting virus which is transmitted by aphid, but not by mechanical inoculation, and tobacco mottle virus which is mechanically inoculable but is not transmitted by aphids, except from plants that have rosette, that is, that are also infected with tobacco rosette virus which acts as a "carrier" or assisting virus. Therefore, aphids do not transmit virus from plants that have been infected by manual inoculation of sap from rosette plants.

Carrot red leaf virus and carrot mottle virus behave similarly. Carrot red leaf virus, the "assistor," is always transmissible by the aphid *Cavariella aegopodii*. Carrot mottle virus is aphid-transmitted only from plants that were infected with carrot red leaf virus also; carrot mottle virus then behaves as a persistent-circulative virus but not when it is alone. Carrot mottle virus can be transmitted by mechanical inoculation (Watson, Sergeant, and Lennon, 1964) but carrot red leaf virus cannot.

Similarly, groundnut rosette virus can be transmitted manually to groundnuts, but the vector *A. craccivora* no longer transmits it from plants infected in this way. Recently Hull and Adams (1968) described an "assistor virus" that does not cause symptoms in groundnut, but seems to perform the same functions as carrot red leaf virus and tobacco vein distorting virus, in assisting the aphid transmission of the manually transmissible partner in the virus complex.

The carrier, or assistor, relationship between viruses is not understood; it may differ with different viruses. With some of the nonpersistent viruses there may be alteration of plant cells by one virus, making them susceptible to another, or there may be complex interaction, or even some form of genetic recombination between the viruses.

6

Transmission of Viruses and Mycoplasma by the Auchenorrhynchous Homoptera

ROBERT F. WHITCOMB

Entomology Research Division
Agricultural Research Service
United States Department of Agriculture
Beltsville, Maryland

I. INTRODUCTION

The disease agents transmitted by leafhoppers have been the subject of many reviews. Early reviewers attempted a broad coverage of the entire field, but in the last decade reviewers have tended to consider only the more recent advances. In this chapter I have attempted an overall survey of the transmission of plant disease agents by the auchenorrhynchous Homoptera. Within recent years a large body of information has accumulated on transmission by insects other than cicadellids; evidence has been presented that many leafhopper-borne agents previously thought to be viruses are microorganisms resembling mycoplasma or bacterial L-forms; several phytarboviruses (viruses which multiply in plants and which are transmitted in the circulatory manner by insects) can now be grown in insect cells cultured in vitro; a number of plant disease agents whose nature remained unknown for many years have now been partially purified and shown to be viruses. Stressing such advances here requires that heavy reliance be placed on the coverage provided by previous reviews (Black, 1959; Carter, 1962; Maramorosch, 1963, 1969; Sinha, 1968*a*).

In a recent review (Whitcomb and Davis, 1970*b*) we described the different types of plant disease agents borne persistently by insects, with special emphasis on the differentiation of mycoplasma-like organisms and phytarboviruses. Future reviews are apt to be restricted to one or the other type of disease agent. Furthermore, the time is rapidly approaching when the phytarboviruses will be classified not by their hosts but by the intrinsic nature of the virion, with emphasis on such properties as morphology, serology, nucleic acid composition, and mode of replication. Such classifications have already been generally well received in other fields of virology (Melnick, 1970). Many of the plant viruses discussed in this chapter have structural similarities to viruses affecting other hosts, such as the small RNA viruses (Kaper 1968), the rhabdoviruses (Melnick and McCombs, 1966), and the diplornaviruses (Verwoerd, 1969). The recently discovered structure of carrot mottle virus (Murant et al., 1969) is reminiscent of some of the ungrouped arboviruses. Some groups of viruses common in vertebrates, such as the poxviruses (Joklik, 1968) or their structural counterparts in insects (Weiser, 1965, Bergoin et al., 1969), have not been found to affect plant hosts. Some viral plant pathogens may be confined to the plant kingdom (Bradfute and Nault, 1969; Nault and Styer, 1970). At present there is no general agreement on an integrated classification of viruses affecting such diverse hosts as animals and plants. It is likely that many of the principles and methods outlined in this chapter will contribute to the development of such an integrated classification.

II. BIOLOGY OF THE VECTORS

A. Taxonomic and Phylogenetic Position

The suborder Auchenorrhyncha has classically been divided into the cicadas (Cicadidae), planthoppers (Fulgoridae), treehoppers (Membracidae), leafhoppers (Cicadellidae), and spittlebugs or froghoppers (Cercopidae). Newer classifications tend to break up larger families and to elevate previous subfamilies to the family rank (Brues et al., 1954). The erection of such families as the Delphacidae, Flatidae, and Cixidae within a superfamily Fulgoroideae has found wide acceptance. Acceptance of a concept of Cicadelloideae as a superfamily has been more hesitant. At present the type genus of the Cicadellidae is located in a large and distinct group of xylem-feeding insects. Although this group is widely separated from other leafhoppers, the most recent revision (Young, 1968) treats it as a subfamily, Cicadellinae.

No plant disease agent is at present known to be borne by cicadas, but if one were found it would surely be an agent of great biological interest. One plant disease agent is known to be borne by treehoppers (Simons and Coe, 1958; Simons, 1962), and two agents by spittlebugs (Delong and Severin, 1950; Severin, 1950; Harpaz, 1962). Leafhoppers are the auchenorrhynchous insects implicated most often in transmission of plant disease (Nielson, 1968; Ishihara, 1969). Among the Fulgoroideae are to be found a number of vectors of plant disease (Ishihara, 1969) most of which colonize gramineous hosts. This reflects two facts: (1) that homopterous insects in general, but leafhoppers and planthoppers especially, have become well adapted to the Gramineae, and a large proportion of the species of these two families colonize gramineous hosts; and (2) that grasses and grains are of particular agricultural interest and application, so gramineous plants constitute a sizable fraction of green plants of any kind.

Recent concepts of worldwide distributions of the Auchenorrhyncha are based upon fragmentary knowledge of the tropical fauna. Indeed, descriptive work continues on the Nearctic fauna, even in taxa where intensive work has already been done, and the next century will probably not see completion of the inventory of world species.

B. Biology and Ecology

DeLong (1971) reviewed the bionomics of leafhoppers. Most members of the subfamily Cicadellinae are xylem-feeding insects. The "honeydew" produced in large amounts as a result of their continuous feeding in woody tissues is sometimes very conspicuous. Certain leafhoppers, particularly members of the Typhlocybinae, feed in the parenchyma of leaves, where their feeding punctures leave conspicuous spots. Most leafhoppers and other Auchenorrhyncha, however, feed in the phloem, as demonstrated by careful sectioning of insect feeding punctures (Smith and Poos, 1931; Putnam, 1941; Houston et al., 1947; Day et al., 1952). Certain species apparently are root feeders, but most feed on green plant parts. Eggs are laid singly or in groups, and may be inserted into soft or woody tissue, or on exterior leaf or stem surfaces. Individuals of most species, especially nymphal forms, feed for long periods on a single plant. Usually there are five nymphal instars. Maturation commonly requires 3-5 weeks. Males of most leafhopper species mature more rapidly than the females. The ability to produce sounds seems to be widespread in the Auchenorrhyncha (Ossiannilsson, 1949) and is apparently prefatory to mating. Upon maturation, adults frequently disperse from their nymphal host plant or stand. This is observed most dramatically in the long-winged forms of species, most of whose individuals are brachypterous. Such forms are often taken at considerable distances from their host plants. Brachyptery is especially common in the Fulgoroideae (Kisimoto, 1965; Nasu, 1969*b*). In certain cases, migration of adults must encompass many miles (Lawson et al., 1951; Medler, 1957; Chiykowski and Chapman, 1965; Drake and Chapman, 1965; Cook, 1967).

III. BIOLOGY OF THE DISEASE AGENTS

A. Viruses with Double-Stranded RNA (Diplornaviruses)

In this group are the plant viruses discussed below; the reoviruses (Rosen, 1968; Shatkin, 1968; Joklik, 1970), which infect a wide array of mammalian species; the cytoplasmic polyhedrosis viruses of insects (Arnott et al., 1968;

Hosaka and Aizawa, 1964; Stairs, 1968); and certain arboviruses (Murphy et al., 1968; Verwoerd, 1969). Double-strandedness of the RNA has been demonstrated by electron microscopy (Kleinschmidt et al., 1964; Miura et al., 1968; Nishimura and Hosaka, 1969), base-pairing studies (Gomatos and Tamm, 1963; Black and Markham, 1963; Hayashi and Kawase, 1964; Miura et al., 1966, 1968; Verwoerd, 1969), X-ray diffraction studies (Tomita and Rich, 1964; Sato et al., 1966), and several other lines of evidence. The RNA can be extracted by conventional means, but only fragments of the RNA are recovered even after careful extraction. The fragments may represent functional segments of the genome (Bellamy and Joklik, 1967; Bellamy et al., 1967; Prevec et al., 1968; Watanabe et al., 1967, 1968; Joklik, 1970). In L cells, reovirus virions are phagocytized and viral coat protein is stripped after uptake into lysosomes (Silverstein and Dales, 1968). Mature particles differ in their infective ability, and proteolytic enzymes may enhance specific infectivity (Spendlove et al., 1966). Synthesis of maize rough dwarf virus (Vidano, 1970) is accompanied by the formation of microtubules, within which the virions mature.

Although the similar sizes of the particles of reoviruses (Vasquez and Tournier, 1964) and wound tumor virus, the arrangement of their 92 capsomeres, and similarities between base ratios of their RNA suggested a possible relationship, serological evidence was negative (Gamez et al., 1967), and the host ranges do not overlap (Streissle et al., 1968). Another divergence between diplornaviruses was revealed when bluetongue virus (Els and Verwoerd, 1969) and rice dwarf virus (Kimura and Shikata, 1968) were reported to have 32 capsomeres instead of the 92 reported for other viruses of the group.

This group may include several plant viruses whose nucleic acids have not been characterized, among them Fiji disease of sugarcane (Teakle and Steindl, 1969) and rice black streaked dwarf (Kitagawa and Shikata, 1969*a*, *b*). A third plant virus, maize rough dwarf mosaic, is discussed in more detail below.

1. WOUND TUMOR VIRUS

This virus (Black, 1965) is transmitted by leafhoppers of the subfamily Agalliinae. Its 60 nm diameter capsid, which encloses double-stranded RNA that accounts for 30% of the whole virus weight, has 92 capsomeres, each 7.5 nm in diameter (Bils and Hall, 1962) with smaller subunits measuring 2.5 nm RNA isolated by the phenol method was fragmented (Kleinschmidt et al., 1964; Dunnebacke and Kleinschmidt, 1967).

The virions require divalent cations and amino acids for optimum stability (Black and Markham, 1963). Symptoms of the disease in plants include enations, veinal hyperplasia, and tumors at wound sites on stems and roots. The agent is mechanically transmissible to crimson clover by pin-pricking preparations of the virus into the crown. The virus is passed transovarially, and has a minimum latent period of 8 days in the vector. Many organs of the vector are affected. Cytopathic changes visible by electron microscopy occur in the nerve cells (Hirumi et al., 1967). Multiplication of the virus in plant and insect is accompanied by the appearance of viroplasmic loci (Shikata and Maramorosch, 1967).

2. RICE DWARF VIRUS

This virus (Fukushi, 1933, 1934, 1937, 1939, 1969) is borne by two deltocephaline leafhoppers. The membranous coat (Shikata, 1966) of the 70 nm particles (Shikata, 1962) can be digested by snake venom phospholipase without destroying infectivity (Toyoda et al., 1965). The temperature optimum is lower than that of wound tumor virus (18° C vs. 27° C). Rice dwarf virus affects the mycetome and fat body of the leafhopper cytopathologically (Nasu, 1963), as viewed under the light microscope, but is not oncogenic. Like wound tumor virus, the particles contain double-stranded RNA, and the latent periods in the vectors are similar.

3. MAIZE ROUGH DWARF VIRUS

Maize rough dwarf virus is the same size as rice dwarf virus and, like rice dwarf virus, induces veinal hyperplasia in plant hosts (Gerola and Bassi, 1966). It is transmitted by a delphacid planthopper (Harpaz, 1966). Maize rough dwarf virus can be detected by thin section electron microscope in many organs of its vector. Its multiplication is also associated with the appearance of viroplasmic matrices and microtubules (Vidano, 1970). The nature of its RNA has not been determined, but its similarities to rice dwarf virus and wound tumor virus indicate that it is likely to be double-stranded.

B. Viruses with Bacilliform Shape (Rhabdoviruses)

A number of plant viruses—as well as viruses of animals, fish, and insects—resemble rabies or vesicular stomatitis viruses in having a bacilliform shape (Simpson and Hauser, 1966; Melnick and McCombs, 1966). The sigma virus of *Drosophila* (Seecof, 1968) was first thought to be a cytoplasmically transmitted genetic factor. Vesicular stomatitis virus is an insect virus in a sense, having been shown to multiply in mosquitoes (Mussgay and Suarez, 1962; Bergold et al., 1968). Among the plant viruses can be included Hitchborn's plantain virus (Hitchborn et al., 1966), mottled dwarf of eggplant (Martelli, 1969), and aphid-borne viruses such as sowthistle yellow vein (Hackett et al., 1968) and lettuce necrotic yellows (Harrison and Crowley, 1965). Leafhopper-borne viruses, in addition to those discussed below, include rice transitory yellowing (Chiu et al., 1965, 1968; Shikata and Chen, 1969) and northern cereal mosaic virus (Lu et al., 1968), which is planthopper-borne.

1. POTATO YELLOW DWARF VIRUS

Potato yellow dwarf virus is borne by several species of agalliian leafhopper. There are two strains in eastern North America, one of them adapted primarily to *Agallia constricta* Van Duzee and the other to *Aceratagallia sanguinolenta*

Provancher (Black, 1944). The virus is bacilliform in shape, with dimensions of 380 × 75 nm (Macleod et al., 1966). Negatively stained particles show prominent outer projections (Black et al., 1965). Upon extraction in conventional buffers, however, this characteristic morphology is apparently lost. The virions have at least 20% essential lipid (Ahmed et al., 1964). Divalent cations and amino acids stabilize the virions (Brakke, 1956). The virus can be mechanically inoculated to *Nicotiana rustica*, followed by the appearance of local lesions, systemic invasion of the vascular tissues, and in later stages, invasion of the parenchyma tissues. Potato yellow dwarf virus is passed transovarially in its vectors (Black, 1953*c*). The minimum latent period in the vector is about 4 or 5 days.

2. WHEAT STRIATE MOSAIC VIRUS

This virus has a minimum latent period of 3-5 days in its deltocephaline leafhopper vector (Lee, 1968). Lee (1968) partially purified the virus by using a buffer containing mannitol, EDTA, $MgCl_2$, and bovine serum albumin. The particles, which measure 260× 80 nm often break into two pieces, 170 × 80 nm and 90 × 80 nm. Wheat striate mosaic virus has not been transmitted mechanically to plant hosts.

3. MAIZE MOSAIC VIRUS

The latent period in the insect of this delphacid-borne virus may be as short as 4 days (Carter, 1943) but is often longer (McEwen and Kawanishi, 1967). The purified virions (Herold et al., 1960) measure 242 ± 10 nm × 48 ± 10 nm and similar particles have been envisioned in sectioned plant and insect hosts (Herold and Munz, 1965).

C. Small Viruses

The small viruses transmitted by species of auchenorrhynchous Homoptera resemble not only the small viruses of animals and bacteria (Kaper, 1968) but also other plant viruses transmitted by such diverse vectors as aphids, beetles, fungi, and nematodes. The viruses range in size from 22 to 33 nm across and have single-stranded RNA. Transovarial passage of such viruses has been claimed in only one instance (Miyamoto and Miyamoto, 1966).

1. OAT BLUE DWARF VIRUS

Oat blue dwarf virus (Banttari and Moore, 1962) is transmitted by *Macrosteles fascifrons* (Stål), with a minimum latent period in the vector of 7 days after acquisition feeding or 1 day after injection (Banttari and Zeyen, 1969). The purification of virions from plant hosts was achieved (Banttari and Zeyen, 1969) by sequential steps of polyethylene glycol chromatography and density-gradient centrifugation. The virus, 28-30 nm in diameter, is relatively stable compared with other phytarboviruses.

2. RICE TUNGRO VIRUS

Virions of rice tungro virus measure 30-33 nm, and are stable for 1 hr at 40° or 1 day at room temperature (Galvez, 1968). The virus is not persistent in its deltocephaline leafhopper vector (Ling, 1966). This is the only known case of nonpersistent transmission by a species of auchenorrhynchous Homoptera. Rice tungro virus is not transmitted mechanically to plants or insects, and its bioassay must be accomplished by allowing insects to feed on infectious solutions (Galvez, 1968).

3. BEET CURLY TOP VIRUS

On the basis of its short latent period, filterability (Severin and Swezy, 1928; Severin and Freitag, 1933), and stability to ethanol (Bennett and Wallace, 1938), as well as preliminary data on its purification (J. Duffus and A. H. Gold, personal communication), curly top virus is tentatively placed in this group. Argentine curly top (Bennett et al., 1946), Brazilian curly top (Bennett and Costa, 1949), and pseudo curly top (Simons and Coe, 1958; Simons, 1962) may also be small stable viruses.

D. Mycoplasma-Like Organisms

This large group of plant disease agents (Whitcomb and Davis, 1970*a*; Maramorosch et al., 1970) is worldwide in distribution and contains many agents which cause diseases of economic importance. Some of the agents have broad host ranges. Symptoms of yellows diseases include general chlorosis, dwarfing, proliferation of axillary shoots, witches'-broom, and virescence of floral parts. The agents have not been transmitted mechanically to plants but can be transmitted by injection of infectious extracts into abdomens of noninfected insects. Transmission from plant to plant can be achieved by grafting or by dodder.

The causal agents have not been isolated. Until 1967 it was commonly assumed that the agents were viruses. Clues contradicting a virus hypothesis were scarce. The details of transmission of many viruses were similar to those of the yellows agents. The search for microorganisms in yellowed plants had failed (Dobroscky, 1931*a, b*), although Hartzell (1937) had observed possible microorganisms in yellowed peach tissue. Black (1943*b*) interpreted his data from ultrafiltration and sedimentation as possibly indicating that the aster yellows agent was larger than any known virus. Yet, viruses commonly adsorb to components in extracts (Tomlinson et al., 1959). Although antibiotic therapy of leafhopper-borne plant disease was attempted several times, in each case the workers made unfortunate choices of either antibiotics or disease agents (Beale and Jones, 1951; Maramorosch, 1958*b*; Jensen et al., 1967).

A new insight into the cause of yellows diseases was provided by the electron microscopy of Doi et al. (1967), who observed numerous bodies in the phloem of plants infected with mulberry dwarf disease. These bodies they interpreted as mycoplasmalike or PLT-like (Psittacosis-Lymphogranuloma-Trachoma-like) organisms. Similar bodies in plants infected with Japanese aster yellows, potato witches'-broom, and Paulownia witches'-broom (all leafhopper-borne diseases with yellows-type symptoms) provided further evidence that the envisioned organisms might, in fact, be the causal agents. Ishiie et al. (1967), in a companion paper, reported that dwarfed mulberries showed a tendency to recover when treated with tetracycline antibiotics.

The report of Doi et al. (1967) has now been confirmed for a number of other yellows diseases. More than 50 diseases are currently thought to belong in the yellows group, by criteria of electron microscopy, response to tetracycline antibiotic treatment, or symptomatic resemblance plus transmissibility by leafhoppers (Maramorosch et al., 1970).

Electron microscopy of phloem containing the suspected particles of yellows diseases has revealed bodies bounded by unit membranes, with large numbers of ribosomes, and filamentous strands which are presumably nucleic acid (Doi et al., 1967; and many subsequent works reviewed by Whitcomb and Davis, 1970*b*; and Maramorosch et al., 1970). The diameters of the bodies varied from 70 to 800 nm or in some cases, 2 μ. Stages were seen which resemble those observed in avian or human mycoplasma (Hayflick, 1967, 1969) including filamentous forms, chains, and buds (Worley, 1969, 1970).

Bodies observed in salivary glands of the leafhopper were somewhat different (Hirumi and Maramorosch, 1969). Bodies were found resembling those in certain stages of multiplication of *Mycoplasma laidlawii,* including evidence of elementary body formation within larger bodies. These and other suggested methods of multiplication based on electron microscope evidence alone must be considered purely conjectural.

Antibiotic data of Ishiie et al. (1967) have been confirmed and extended for American aster yellows (Davis et al., 1968; Davis and Whitcomb, 1970; Whitcomb and Davis, 1970*a*). Supplying antibiotics continuously to plants in hydroponic cultures enabled them to produce normal flowers and resulted in the reappearance of chlorophyll in yellowed leaves. Insects were able to acquire the aster yellows agent from treated plants, but with greatly reduced efficiency. Application of the antibiotic to plants before inoculation by insects delayed or blocked infection. Injection of an efficaceous antibiotic into inoculative insects temporarily impaired their ability to transmit, but within 1-2 weeks the insects had regained their full inoculative potential. Antibiotic acquired through membranes also suppressed transmission (Freitag and Smith, 1969). Insects acquiring antibiotic from plants to which antibiotic had been supplied hydroponically were not protected from infection, although the latent periods of infection in the vectors after acquisition feeding were prolonged. Antibiotic applied to plants after inoculation by leafhoppers, but before symptom expression, greatly delayed the appearance of symptoms. All tetracycline antibiotics tested (chlorotetracycline, oxytetracycline, methacycline, and tetracycline) were effective. Tylosin tartrate, a macrolide antibiotic effective mental transmission may occur in other rare instances (Bennett and Costa, 1961). In all other cases the disease agents are transmitted by a persistent mechanism. After the insects feed on diseased plants, transmission is preceded by a latent period varying from less than one day to as many as 50 days. Although transmission usually continues for most of the life of a vector, inoculativity usually declines with age. This persistent mode of transmission is common in other insect groups which are vectors of plant viruses (Sylvester, 1969; Pirone, 1969) and in vectors of animal viruses (Day, 1955*a*; Chamberlain, 1968).

Persistent transmission usually seems to be supported by multiplication of the disease agent in the vector (propagative agents). In the case of beet curly top, intensive and careful investigations (Bennett, 1962, 1967*a*) have, as yet, failed to demonstrate multiplication in the leafhopper vector, *Circulifer tenellus* (Baker). Bennett hypothesized that virus passes through the gut wall into the hemolymph and from the hemolymph into the saliva as a result of normal transport mechanisms of the insect (circulatory, nonpropagative transmission). Yet the case for failure of curly top virus to multiply is far from closed (Whitcomb and Davis, 1970), and the possibility of limited multiplication in certain tissues, such as intestinal epithelia or salivary glands, cannot be excluded. Even the most careful workers (Harrison, 1958) have experienced difficulty in demonstrating multiplication of similar viruses.

There are at least two widely different types of persistent transmission of agents which multiply in homopterous insects. The first occurs (Day, 1955*b*) with smaller phytarboviruses, such as potato leafroll virus, which multiplies in aphids (Stegwee and Ponsen, 1958), and two other phytarboviruses which may multiply in their aphid vectors but have not yet been shown to do so: barley yellow dwarf virus (Rochow, 1969; Paliwal and Sinha, 1970) and pea enation mosaic virus (Simons, 1954). Certain plant viruses transmitted by white flies may also be of this type (Cohen and Nitzany, 1966). Transmission begins after a very short latent period—no more than 2 days. Transovarial passage has been claimed in only one instance (Miyamoto and Miyamoto, 1966). Frequently ingested or injected virus dosage markedly affects subsequent transmission efficiency (Simons, 1954; Sylvester, 1967; Paliwal and Sinha, 1970). In many cases it is difficult to transmit virus by injection techniques (Maramorosch, 1955; Richardson and Sylvester, 1965). Virions of such viruses prove to be small (25-35 nm) and are relatively stable to butanol and chloroform (Rochow and Brakke, 1964). Examples of transmission by auchenorrhynchous insects by this mechanism include transmission of curly top virus by *C. tenellus* (Baker) (Freitag, 1936), transmission of maize streak mosaic (Storey, 1939), transmission of pseudo curly top virus by a membracid (Simons and Coe, 1958), and transmission of Argentine sugar beet curly top virus by an agalliian leafhopper (Bennett et al., 1946; Bennett and Costa, 1949). It is not known whether any of these viruses are propagative in their vectors.

Transmission of the smaller phytarboviruses is in sharp contrast to that of agents which are transmitted only after a long latent period—usually 8 or more days, but as short as 4 days in the case of wheat striate mosaic virus and potato yellow dwarf virus. There is no experimental evidence that agents with such latent periods are nonpropagative, and as a starting point, multiplication should be tentatively assumed. Although viruses of this type are frequently passed by females to their progeny (Nasu, 1969*a*), the yellows agents are rarely transmitted transovarially, exceptions being cited by Posnette and Ellenberger (1963) and Frazier and Posnette (1957). None of the agents with long latent periods in their vectors have the simple structures of the smaller, more stable viruses; none are particularly stable to organic solvents. In some cases this is especially understandable in view of the content of essential lipid (Ahmed et al., 1964). Many of these agents are optimally stable in media containing divalent cations (especially Mg^{2+}) and amino acids (Brakke, 1956) and other substances (Lee, 1967). A singular fact is the similarity of the transmission mechanisms of plant disease agents as diverse as rhabdoviruses, diplornaviruses, and the mycoplasma-like yellows agents. Only elucidation of their structures has permitted recognition of their profound dissimilarities.

The rickettsial pathogens of louse-borne typhus and trench fever are transmitted by the fecal route by their anopluran vectors (Day, 1955*a*). Mites may transmit plant viruses in this way (Paliwal and Slykhuis, 1967; Slykhuis, 1969). Although homopterous insects seem ill adapted to this mode of transmission, it must remain as a hypothetical possibility.

Transmission from one individual vector to another (as in mating) is unknown. In many cases, however, a propagative virus harbored by a female vector can be passed to the progeny (transovarial passage) (Fukushi, 1933; Black, 1953*c*). The frequency of transovarial passage may vary widely with the insect clone being investigated (Nagaraj and Black, 1962). Vegatative propagation of the plant host, of course, may result in "transmission" of many disease agents, including those discussed in this review (Bennett, 1967*b*).

As yet, no true "insect viruses" (viruses which propagate in, and are usually lethal to insects which are not propagative in an intermediate plant host) have been reported for homopterous insects (Ignoffo, 1968). One insect virus, the Chilo iridescent virus, has been taken experimentally to a leafhopper host (Mitsuhashi, 1967). Virus-like particles observed by some workers in leafhoppers (Lee, 1965), planthoppers (Herold and Munz, 1967), and aphids (Peters, 1965) are of uncertain significance. The absence of true insect viruses in homopterous insects is usually ascribed to the vectors' mode of feeding; whereas "insect viruses" are acquired by the oral route by chewing insects, a virus acquired by the oral route by a homopterous insect (with piercing, sucking mouth parts) should theoretically reside in the plant tissues where the vector probes or feeds. On the other hand, leafhoppers can acquire bacteria from the surface of leaves (Whitcomb et al., 1966*c*) and could doubtless acquire virus also, at least experimentally. Hence, a search for a virus pathogenic to homopterous insects might eventually be successful.

V. EVOLUTION OF DISEASE TRANSMISSION

Although it may be premature to speculate upon broad aspects of evolution (Andrewes, 1965; Bawden, 1964), it seems likely that most taxa of vertebrate pathogens will be found to be associated in some way with invertebrate hosts. Unfortunately, the conservative taxonomic features of viruses, other than gross structure, are located in the nucleic acid genome (Lanni, 1964), so retracing evolutionary steps will be a difficult and laborious task. As yet, no nucleic acid hybridization experiments (McCarthy and Bolton, 1963) have been attempted with any of the plant disease agents discussed in this chapter.

Exhaustive attempts (Gamez et al., 1967) to demonstrate serological affinities between viruses of invertebrates and vertebrates have met with failure in the end, despite promising early reports (Streissle and Maramorosch, 1963). Viruses such as potato yellow dwarf virus and wound tumor virus may be distantly related to animal viruses, since vertebrate hosts have certainly been amply exposed to propagative plant-insect viruses in the course of evolution.

It seems certain that the classification of disease by plant or insect hosts is an artificial one and that a concept such as "leafhopper virus" is no more specific or useful than "cereal fungus" or "pineapple bacterium." This would be entirely true if no differences existed in the biologies of homopterous families. The longer life spans of leafhoppers, for example, are especially favorable for propagative transmission. Vectors whose life span includes periods of nomadic wandering or migratory behavior are particularly apt to serve as reservoirs of propagative viruses. Many plant disease agents are transmitted inefficiently by their insect vectors to all known hosts (Jensen, 1957). Leafhoppers are apparently not as well suited as aphids for the persistent transmission of small phytarboviruses, and are also poorly suited for stylet transmission of viruses. The yellows disease agents (mycoplasma-like organisms) are best adapted to leafhoppers, and several such agents are transmitted by a wide spectrum of leafhopper vectors (Severin, 1945; Gilmer et al., 1966).

Evolution of viruses can be observed in the laboratory (Black et al., 1958). Evolution of virus diseases can be observed in the field (Iida and Shinkai, 1969). Outbreaks of "new diseases" will often be found to result from changes in the ecological status quo (Shepherd, 1965). First recognition of mild diseases, such as barley yellow dwarf, no doubt present for many years, may come with a particularly noticeable epidemic. Some severe diseases, such as elm phloem necrosis, have a tendency to occur in sporadic outbreaks (Swingle, 1942). On the other hand, there are definite cases where disease introduction from one continent to another is suspected (Oman, 1969). There are also many documented cases of introduction of homopterous insects from one continent to another. When an introduced insect becomes a disease vector there is an opportunity to observe the evolution of disease transmission from its inception (Oman, 1969).

An additional factor which can alter the ecological balance (Ossiannilsson,

1966) is a change in cultural practices, such as a widespread change to a new crop variety. Also, shifts in the ecological balance of the plant community have occurred as man has colonized natural areas, and have upset faunal balances greatly. It is likely, for example, that populations of nomadic polyphagous leafhoppers such as *Agallia constricta* (Van Duzee), *Aceratagallia sanguinolenta* (Provancher), *Scaphytopius acutus* (Say), and *Macrosteles fascifrons* (Stål) have been greatly increased by the conversion of vast acreages from natural flora to cultivated crops. There are doubtless many unknown viruses in wild hosts. Pierce's disease has been recovered from grasses at high elevations in the Sierra Mountains (Freitag and Frazier, 1954). Also, collection of vectors from a location between Washington, D.C. and Annapolis, Maryland, unearthed two new disease agents—the clover club leaf agent and wound tumor virus (Black, 1944). Both of these agents are now classical biological study objects which have not been observed to produce natural plant disease. Potato yellow dwarf virus, which is probably mainly a virus of leguminous plants (Black, 1937) would be in this category if it did not cause sporadic damage to potatoes. The method of searching for virus diseases by testing selected vectors is an established technique in animal virology (Chamberlain and Sudia, 1967).

VI. PRIMARY RESEARCH

Certain initial phases of research are essential to any disease problem. A worker confronted with a new plant disease would do well to reread the fundamentals of plant pathology, beginning with Koch's postulates (Koch, 1882). There is, for example, always the danger of falsely assuming that a condition is caused by an infectious agent, as in the numerous cases of insect and mite toxins (Medler, 1941; Severin et al., 1945; Flemion et al., 1954; Carter, 1962; Bos and Grancini, 1968; Miles, 1968). The reverse error is equally easy to make. A number of "physiological diseases" have, in fact, turned out to be viruses (Oswald and Houston, 1951, 1953). Thus, the worker will want first to ascertain whether a condition is due to an infectious agent or perhaps even a complex of an infectious agent plus an insect toxin (Carter, 1952; Nuorteva, 1962).

A. Control

At some point early in the investigation a decision must be made whether to attempt empirical control (Broadbent, 1969) of a disease. Particularly if each host plant is of high value (e.g. orchard trees), control efforts may be indicated as a first step in research efforts. The most obvious control possibility is "roguing" (removal) of known diseased trees. This measure alone succeeded in controlling the phony peach disease (Hutchins, 1933). It is of interest that even graft transmission of this disease was accompanied by special problems (Hutchins, 1939), and that a lengthy large-scale vector search, while eventually identifying vectors (Turner, 1949), was not especially contributory to the control solution (Persons, 1952). Roguing diseased plants will be of little value in cases where most transmission is from wild plants to economic host plants rather than between the economic host plants (Palmiter and Hildebrand, 1943; Carter and Suah, 1964). Thus, before a roguing program is attempted, it may be wise to study the pattern of spread of a disease. Also, in some cases where infected trees continue to bear, a roguing program may be difficult to implement if it causes short-range economic hardship on growers. Where the principal reservoir of virus is a wild plant, roguing of that plant may be a desirable step. For example, chokecherry was found to be the principal source of the X-disease agent transmitted to peaches in the northeastern United States (Palmiter and Hildebrand, 1943). In the western United States, control of X-disease has been achieved (Rawlins and Parker, 1934) in three different ways: roguing of diseased trees selection of resistant root stocks and control of sporadic outbreaks by insecticidal control of vectors. Where herbaceous crop plants are attacked, control measures are often not economically feasible. Growers may be forced to abandon the culture of crops for a period during a disease outbreak. Often, cyclical fluctuations will provide temporary relief (Winkler, 1949)—until the next outbreak. The eventual outcome of control considerations will frequently be a search for tolerance or resistance among available genetic stock (Sakurai, 1969).

B. Vector Search

The search for vectors ranges from an immediate *fait accompli,* with initial discovery of the virus in the insect, to lengthy, laborious, and exhaustive field studies. Possible vectors include any organism in the ecosystem, including man. If insects of any kind are under serious consideration as vectors, it will benefit the worker to gain some knowledge of the status of various insects associated with the diseased plants. The ultimate choice of insects for exhaustive tests will depend in part on intuition, and in part on factual evidence obtained by studying the ecosystem. Success will also entail a measure of good fortune.

Most vectors can be classified either as general feeders, whose life cycles include periods of nomadic wanderings to many plant species, or as host-specific feeders. Host-specific leafhoppers whose occurrence on the crop plant is accidental should at first be excluded from serious contention. For example, grass-inhabiting leafhoppers are frequent "accidentals" on woody or herbaceous crop plants.

If experiments are suitably controlled, insects collected in the field can be used in transmission trials (Freitag, 1951). Again, a general knowledge of the biology of the suspected vector will assist in selection of nymphs or young adults (usually more suitable for transmission tests) and will make it easier to collect enough test insects. When an insect is strongly suspected of being a vector, the worker will usually wish to attempt rearing.

C. Rearing of Leafhoppers

Many leafhopper species can be easily reared on their known host plant or on related plants. If colonies are easily established, the only remaining problem is demonstration that the colony is free of transovarially passed virus or of a disease agent whose symptoms in the host plant may be masked.

If difficulties are encountered in rearing, the workers will wish to consider the following possible causes: (1) Many leafhoppers are host-specific (Oman, 1949; DeLong, 1948; Beirne, 1956), breeding on only one genus or, in rare cases, on one species of plant. (2) In some leafhoppers there are photoperiodic responses, especially to short photoperiods, which may induce diapause. Supplemental illumination to give 16 hr of light per day should be sufficient to prevent diapause. (3) Leafhoppers from northern areas may have a diapause (Raine, 1967; Swenson, 1967; Kisimoto, 1969; Nasu, 1969*b*). Such species may require cold treatment to permit continuation of their life cycle. (4) Temperature may not be adjusted correctly. Rearing is often attempted at too low a temperature. We have found that 28° C is suitable for many leafhoppers. (5) In colonies derived from only a few parents, inbreeding can be a serious problem (Kisimoto and Watson, 1965). (6) The life cycle of a vector may be long, as with certain leafhoppers which have only one generation per year. (7) If the vector is adapted to woody plants, herbaceous plants may be inadequate for its maintenance. (8) It is possible that certain species of leafhoppers which inhabit dry uplands have a different water absorption mechanism and may be intolerant of high humidities. Other types of humidity tolerance may also be important. (9) Because it is necessary to provide an actively photosynthesizing plant for adequate nutrition, the cage must be designed to assure adequate light. (10) Especially when species are of low fecundity, many individuals may be lost by failure to transfer the laying adults at proper intervals. When the host plant collapses, the eggs laid on it will be lost. (11) Insect colonies are frequently beset by microbiological problems (Whitcomb et al., 1966*c*). (12) Many leafhoppers may have complex life cycles. For example, *Colladonus clitellarius* (Say) oviposits in woody tissue but the nymphs feed on herbaceous plants (George and Davidson, 1959). In the event that a complex life cycle is suspected, there is no alternative to careful field observations on the life cycle of the insect. In certain cases, for example, successful rearing of leafhoppers would require ants (Kirkaldy, 1906).

D. Microbial Problems in Insect Colonies and in Nature

Until the recent report that agents of yellows diseases may be mycoplasmas or chlamydiae (Doi et al., 1967), little microbiological work had been done with homopterous insects.

Circulifer tenellus (Baker) was found to harbor "rickettsia-like" microorganisms in the gut and in other tissues (Swezy and Severin, 1930).

Similar organisms were found in beet plants, but it now seems certain that these agents were not etiologically related to curly top virus.

Colladonus montanus (Van Duzee) colonies were found to carry a large number of bacterial and fungal species (Whitcomb et al., 1966*c*). Most of these species were external contaminants which could be inactivated by surface sterilization and which were not found to cause difficulty in rearing or injection work. However, an *Erwinia*-like organism, which multiplied in the gut of the insect, was found to cause premature death in many individuals. Leafhoppers were able to acquire bacteria, presumably through their stylets, from contaminated surfaces of leaves. From 10 to 100 bacteria were lethal when injected into healthy leafhoppers. The bacterial titer varied from 10^7 to 10^8 per insect. Bacteria can be excluded from extracts by filtration (Jensen et al., 1967; Cohen et al., 1969) or repeated centrifugation, or controlled by antibiotics.

Mortality from bacterial infection can occur as late as 28 days after injection if low numbers of bacteria are injected. This observation has importance in attempts to pass virus serially by injection of extract, for when injected insects are harvested and used for subsequent passage, bacteria are harvested as well as virus. A serial-passage experiment with the western X agent met such a fate (Jensen et al., 1967), despite routine addition of high concentrations of penicillin, streptomycin, and chloromycetin to the extract prior to centrifugal clarification. This technique permitted passage of a *Pseudomonas* species which was resistant to many antibiotics in subsequent laboratory tests.

Many homopterous insects have well-developed mycetomes which contain various organisms (Buchner, 1953). These organisms have never been cultured nor have they caused difficulty after injection of extracts. Some leafhoppers, *Agallia constricta* Van Duzee among them, may harbor specific symbionts in the alimentary tract (Gil-Fernandez and Black, 1965). Such symbionts may be difficult or impossible to culture on simple media.

Enterobacteriaceae are present not only in laboratory cultures of leafhoppers but also in nature (Poos and Elliott, 1936). More recently, three isolates obtained from field specimens were identified as an *Escherichia* species (not *E. coli*), a *Proteus* species, and a species of the *Aerobacter C* group (Whitcomb, unpublished data). Bergey's Manual of Determinative Bacteriology (Breed et al., 1957) states that "It has long been known that the *coli*-form organisms that occur in dairy products are frequently derived from grain, and that they occur on panicles of the grass family in open fields where there is no reason to think that they originated from fecal contamination." The role of homopterous insects feeding on gramineous hosts as vectors of Enterobacteriaceae has perhaps been overlooked.

For certain purposes, including establishment of tissue cultures, it may prove worthwhile to rear insects under aseptic conditions (Mitsuhashi and Maramorosch, 1963) by sterilization of eggs and use of plants grown under aseptic conditions as food. Insects grown in this way would be free of organisms except for mycetome symbionts, or other transovarially passed organisms.

E. Experimental Proof of Vector Relationship

1. *ACQUISITION EXPOSURE*

Nonviruliferous insects, preferably reared in a virus-free colony, are confined upon the diseased plant. Several factors may cause poor acquisition. Acquisition from many hosts is a low-probability event, so prolonged feeding may be necessary for successful acquisition by a sizable fraction of vectors (Jensen, 1959). Older adults may be less susceptible, or may die before completion of the latent period. Survival of insects should always be recorded since transmission failures of dead insects are of no significance. Also, there is frequently an adverse effect when leafhoppers are transferred from one host to another (Kunkel, 1954; Whitcomb et al., 1970*b*). If possible, it is suggested that nymphs of the third and fourth instar would be ideal for most transmission trials.

2. *LATENT PERIOD*

One of the first parameters the worker will wish to determine is the latent period of the agent in its vector. Unfortunately, this will be unknown in initial tests and may possibly vary from 0 to 50 days. If the latent period is unknown, vectors should be transferred to healthy test plants immediately after acquisition exposure. If the latent period is known, it is desirable to maintain the exposed insects on an immune host during that period. Search for a suitable plant and demonstration of its immunity may, in itself, pose a challenge (Black and Brakke, 1952).

3. *INOCULATION EXPOSURE AND INCUBATION PERIOD IN PLANT*

Although some vectors may transmit a disease agent daily (Kunkel, 1954) other vectors may be very inefficient. Thus, if prior data are lacking, insects should be held on the test plant as long as possible. It may be important to test source plant species and test plants. *Aphrodes bicincta* (Schrank) will rarely acquire clover phyllody from or transmit to aster, although it survives well on that host (Chiykowski, 1967). It is, however, the major vector from clover to strawberry in the field. Incubation periods in plants may vary from three days (Bennett, 1962) to several years (Carter and Suah, 1964). It should be remembered that this incubation period may vary greatly with the age of the test plant and the temperature at which it is held.

4. *PRODUCTION OF DISEASE IN THE FIELD*

In many cases it will be desirable to produce a given disease syndrome under field conditions (Jensen et al., 1964). Glasshouse symptoms, especially in young plants, may be quite different from those observed in the field. It is always possible that a previously unsuspected virus, unrelated to the disease under study, may be present (Tomlinson et al., 1968).

F. Host Range Studies

One of the important initial investigations should be a study of the experimental host range of the agent (Kunkel, 1931; Freitag and Severin, 1936; Hill, 1943; Frazier and Severin, 1945; Freitag, 1951). If infectious vectors are available, this is a relatively simple task. Only after such a study should selection be made of experimental plants—for virus source, anatomical studies, test plants, etc. Symptoms of a particular agent are usually similar in different hosts. However, examination of roots of sweet clover revealed the root tumors which have since become a standard source of wound tumor virus (Black, 1965).

G. Virus Storage and Vector Repositories

1. VIRUS STORAGE

Recent advances in techniques for storage of biological materials make it seem likely that any infectious agent can be stored by one method or another. This being so, arrangements are being made for a storage bank of plant disease agents which have been used in published studies or are of special interest for some other reason. Wound tumor virus has been stored in frozen root tumor tissue for periods up to 5 years (Black, 1965). Potato yellow dwarf virus has been stored in infected leaves of *Nicotiana rustica* dried under vacuum, followed by storage in ampoules with a desiccating agent (Whitcomb and L. M. Black, unpublished data). The agent of western X-disease could be recovered 3 weeks after freezing vectors or extracts from them (Whitcomb et al., 1968*b*). Communications about possible storage of viruses should be directed to the American Type Culture Collection, Rockville, Maryland, U.S.A.

2. VECTOR REPOSITORY

At this writing definite arrangements are being made for storage of voucher specimens from studies on the transmission of disease. Responsibility for these arrangements rests with Dr. L. R. Nault, Ohio Agricultural Research and Development Center, Wooster, Ohio, U.S.A., and Dr. Reece I. Sailer, Leader, Insect Introduction and Identification Branch, Plant Industry Station, Beltsville, Maryland, U.S.A. Arrangements for storage of specimens should be made with them.

3. TAXONOMIC IDENTIFICATION

The taxonomic burden in a prolonged vector search can be immense. However, the number of required identifications can be reduced by adequate study of the ecosystem so that identifications of the many possible "accidentals" can be deferred, at least during preliminary work. Reference to available taxonomic monographs will enable the worker to make tentative identification of many species. These identifications will help him to define the

1. ELECTRON MICROSCOPY

Rewards of careful thin-section electron microscopy may include determination of the morphology of the disease agent, its distribution in its plant or insect host, or its relationship to the infected cell. The reader who wishes to attempt the technique should study available works on rice dwarf virus (Shikata, 1966, 1969), wound tumor virus (Shikata and Maramorosch, 1965, 1967, 1969), potato yellow dwarf virus (Macleod et al., 1966), maize rough dwarf virus (Vidano, 1970), wheat striate mosaic virus (Lee, 1967), and maize mosaic virus (Herold and Munz, 1965), and the many reports on the presumed agents of yellows diseases (Whitcomb and Davis, 1970; Maramorosch et al., 1970).

Electron microscopy of purified virus preparations may be routine with small polyhedral viruses but more difficult with the larger viruses. An instructive case is that of potato yellow dwarf virus, which is clearly ellipsoidal in vivo (Macleod et al., 1966) but which upon extraction becomes spherical (Macleod, 1968). Specialized techniques applicable to larger viruses were developed for potato yellow dwarf virus, wound tumor virus, and tomato spotted wilt virus (Black et al., 1963).

2. STUDIES OF THE HEMOLYMPH OF VECTORS

By dissection of adults of *Agallia constricta* Van Duzee, Gil-Fernandez and Black (1965) were able to observe blood cells being pumped through the aorta. Study of these blood cells is a recent concern (Granados et al., 1968), but we do not yet know how the distribution of cell types relates to the much better studied blood cells of holometabolous insects (Jones, 1962). Hemolymph can be easily obtained by first removing a leg from an immobilized vector, then taking it up in a micro-needle by capillary action. The infectivity of hemolymph can then be assessed by inoculating it into a healthy vector (Whitcomb et al., 1966*b*).

3. MICROSURGERY

Several surgical techniques other than injection have been devised for "operating" on vectors without killing them. The first was gut puncture (Storey, 1933) with a fine insect needle, an operation which transformed nonvectors into transmitters. Collection of saliva was achieved by carbon dioxide anesthesia (Braun and Maramorosch, 1951). Removal of one leg permitted "bleeding" of insects suspected of carrying virus (Sinha et al., 1964) and permitted confirmatory diagnosis of wound tumor virus infection by fluorescent antibody staining.

4. DISSECTION OF VECTORS

Dissection of insects is relatively easy to accomplish (Sinha, 1965*b*; Whitcomb et al., 1970*b*) by first embedding the insect in wax or upon an adhesive surface and slitting the cuticle. The internal organs can then be removed. Certain systems, particularly the alimentary tract and the salivary gland, will be very different in the various homopterous families. It will prove easiest to dissect the alimentary tract and the male and female reproductive systems. Salivary glands may or may not prove difficult to obtain from dissected material. Mycetomes and cephalothoracic nerve masses or supraesophagal ganglia will be relatively difficult to isolate.

E. Artificial Feeding Techniques

Several workers have attempted to transmit propagative viruses by feeding vectors on an infectious artifical medium. The technique is applicable to certain stable viruses such as beet curly top virus (Carter, 1927, 1928), barley yellow dwarf virus (Rochow, 1960), and beet western yellows virus (Duffus and Gold, 1965). Actually, most viruses are relatively stable under proper ionic conditions, and it is possible that such viruses as wound tumor virus, potato yellow dwarf virus, and rice dwarf virus might be acquired by their vectors if protective factors were present in the medium. Standardization of the uptake dose would be quite difficult, however. Duffus and Gold (1967) measured, rather than controlled, uptake volumes by P^{32} labeling.

Recent work on aphid nutrition has been directed to some of the variables in membrane feeding techniques (Dadd and Mittler, 1965; Auclair, 1963; Auclair and Cartier, 1963). Among them are microbial contamination of extracts, sucrose concentration, color of the medium, attractive substances, etc. Most artificial feeding has been done through membranes, but *Macrosteles fascifrons* (Stål) has been maintained successfully on a wick supplied with a complex feeding medium (Mitsuhashi and Maramorosch, 1963). Nymphs survived for two weeks but eventually died. Adults could be maintained for at least 30 days. Adult spittlebugs have recently been maintained on diets (Hagley, 1967).

F. Bioassay

An adequate bioassay must be developed for purification, study of distribution and synthesis of infectious components, or any property related to numbers of infective virions or infectious units (Whitcomb, 1969).

1. MECHANICAL INOCULATION

Most plant disease agents propagative in their vectors cannot be transmitted by mechanical inoculation from plant to plant by any methods yet employed (see Chapter 1). In certain cases, as with wound tumor virus and potato yellow dwarf virus, virus can be inoculated into the phloem tissue in the crown of susceptible plants with a fine pin puncture (Brakke et al., 1954). With certain viruses, standard techniques of rubbing leaves can be used to produce local lesions (Black, 1938). Interpretation of local lesion numbers requires, as a minimum, that a standard curve of virus concentration vs. lesion number be determined experimentally. In some cases, a statistical analysis may be desirable (Kleczkowski, 1950; Furumoto and Mickey, 1967*a, b;* Whitcomb,

1967). Even if results are not analyzed statistically, a balanced experimental design should be used (Youden and Beale, 1934).

2. INSECT INJECTION

Last-instar nymphs collected shortly before molting to the adult stage are usually most desirable as test insects for injection. Such nymphs are of known age, always seem to be highly susceptible to infection, are easily pierced by sharp needles, and are more likely to survive the injection operation than adults. However, if insects are to be maintained for long periods on test plants, it may be necessary to select young males since test plants may be killed by progeny insects. If young male adults are used, they will often be found to be equal in susceptibility to nymphs. After collection, the insects are stored at temperatures slightly above freezing, in a moist chamber to prevent desiccation. They are injected under a continuous flow of moist CO_2 at room temperature under a dissecting microscope, after which they should be returned as soon as possible to a healthy, preferably immune plant.

Although much of the classical work was done with needles pulled in gas flames (Storey, 1933; Black and Brakke, 1952), many workers (Whitcomb et al., 1966*a*) now use machine-pulled needles.

Factors known to produce mortality (Jensen et al., 1967) of leafhoppers injected with western X include wounding, aging, the western X agent, or bacterial contaminants (Whitcomb et al., 1966*c*; Jensen et al., 1967). Premature death occasionally occurs in young adults which have not been injected. These deaths could possibly be ascribed to microbial infection or to poor nymphal nutrition.

3. ANALYSIS

Results of insect injection can be analyzed as quantal responses or graded responses (Whitcomb, 1969; 1971). In each case, analysis depends upon experimental determination of a standard curve from dilutions of an inoculum. Theoretically, plotting $\log_{10}\{\log_{10}[1/(1-P)]\}$ vs. $\log_{10} c$ (where P is the observed fraction of infected insects, and c the relative virus concentration) should yield a standard curve that is linear (Plus, 1954; Brakke, 1970). Analysis of the dilution curves of leafhopper-borne viruses has not borne out this expectation. Possible factors responsible for nonconformity include aggregation, variation in susceptibility of the insects, or more likely, a combination of various factors. Graded-response assays have the advantage that estimates of virus concentration can be made with a limited number of observations. Such an assay initially requires selection of some parameter of the transmission curve such as the T-50 (time after injection when half of the insects eventually transmitting have done so), LP_{av} (average latent period), and LP-50 (the time after injection when 50% of the insects have completed their latent period). These values are then plotted as a function of $\log c$ on a standard curve, and the values observed can then be read into the standard curve. The most meticulous analysis of latent-period data has been done with pea enation mosaic virus (Sylvester, 1965). The computerized program used in those studies should be adaptable to other viruses and may be preferred if latent periods in insects are utilized. Two variations of graded-response assays are of special interest. Reddy and Black (1966) used the serological titer of insect groups on a fixed date after injection as the chosen parameter. In our work with aster yellows (Whitcomb, 1969, 1971) we have used as the chosen parameter the total latent period (the time between injection of the insect and the final appearance of symptoms in plants inoculated by the insects).

G. Culture of Agent in Defined Systems

1. ARTIFICIAL MEDIA

With recent suggestions that agents of yellows diseases may be mycoplasma, there is a possibility that the agents may be cultured in cell-free media. Several groups have recently claimed partial success (Chen and Granados, 1970; Lin et al., 1970; Faivre-Amiot et al., 1970; Giannotti et al., 1971). In recent years, a number of mycoplasmas, including the agent of primary atypical pneumonia, have been cultured in complex media (Chanock et al., 1962; Hayflick, 1967).

2. TISSUE CULTURE

Recent advances (Black, 1969; Mitsuhashi, 1969) in the growth of leafhopper cells and their infection with virus (Chiu et al., 1966; Chiu and Black, 1967, 1969) offer the prospect of rapid advances in our understanding of virus-cell interactions.

What might be attempted is briefly outlined here. Maintenance of infected tissue in complex media may be accompanied by increases in the titer of the infecting agent (Maramorosch, 1956). Dissected organs may also be preserved in media, with demonstration of retention of virus or virus antigens (Sinha, 1965*b*). Primary cultures of cells may be obtained (Vago and Flandre, 1963), with embryonic tissue or hemocytes representing favorable starting tissue (Hirumi and Maramorosch, 1964*a, b*). When continuous cultures can be induced to grow in monolayers (Chiu and Black, 1967; Gamez and Chiu, 1968), the system can be utilized for routine assay (Chiu and Black, 1969). The first question will be whether or not the agent multiplies in the cells, as determined by conventional bioassay. If the investigator is especially fortunate, there will be a cytopathic effect, and plaques will be formed on the monolayers. Such a cytopathic effect has not yet been found, but the fluorescent antibody cell count may provide an even more rapid and convenient assay (Chiu et al., 1966; Chiu and Black, 1969).

H. Isolation of Causative Agent

1. CLASSIFICATION

The first step in isolation (Brakke, 1969; Suzuki, 1969; Suzuki and Kimura, 1969) of a disease agent should be a preliminary survey of the properties which might be of taxonomic importance. Special interest will focus on the size of the agent. The two simplest techniques for size estimation are filtration (Hsiung, 1965; Casals, 1968; Cohen et al., 1969) and centrifugation (Brakke, 1958). Provided that the agent is not aggregated, its approximate size can be estimated by either technique.

The relative stability of an agent to such organic solvents as butanol, chloroform, and ether will give an important indication of its nature. Athough agents which do not contain essential lipid may be destroyed by these agents, one can be certain that survival after standard treatment (Andrewes and Horstmann, 1949) with these agents precludes the presence of essential surface lipid.

In the light of recent work, it is clear that other properties are of distinct interest. Phenol or detergent treatment may detect an agent present only in the form of free nucleic acid (Diener and Raymer, 1967, 1969; Raymer and Diener, 1969). Response to antibiotic treatment, coupled with large size, may indicate a mycoplasma- or chlamydia-like agent (Whitcomb and Davis, 1970). Large, complex, and distinctive disease agents may be first envisioned and identified by electron microscopy (Bradfute and Nault, 1969). Some agents may apparently be stable even after treatment with 0.35% formalin (Stamp, 1967; Adams, 1970).

2. *IONIC ENVIRONMENT*

A second step essential to isolation is the discovery of an ionic environment in which the agent is soluble and stable. Some factors which may enhance stability are 0.1-0.3 M glycine, 0.005-0.03 M $MgCl_2$, cysteine, 0.01 M sodium sulfite or other reducing agents, and bovine serum albumin. Rapid gel filtration through a 7% agar column will remove most enzymes from virus suspensions and may enhance stability of many disease agents; it also provides a rapid means of transferring an agent to a desired medium (Steere, 1967). The pH range most likely to enhance stability is 6.0-8.0, but solubility and stability may prove to be independent or even antagonistic properties. Molarity of buffers may be more critical than pH in the pH range 6.0-8.0 (Whitcomb et al., 1968*b*). Although predictions based on past experience may be misleading, one should consult the case histories of tomato spotted wilt virus (Black et al., 1963), wound tumor virus (Brakke et al., 1954), western X agent (Whitcomb et al., 1968*b*), aster yellows agent (Whitcomb et al., 1969, 1970*a*), and potato leaf roll virus (Peters, 1967). If ionic conditions are studied in detail, a specific adsorption-elution procedure of use in purification (Black et al., 1963; Suzuki, 1969; Suzuki and Kimura, 1969) might be developed.

3. *SOURCE*

From his studies of availability, host range, and recovery from vectors, the worker will be able to choose a source material which it is hoped: (1) will contain a high virus concentration; (2) will contain a minimum of materials properties could be studied. Separation of viruses from normal subcellular components should be possible by employing lower percentages of agar.

12. ADSORPTION TECHNIQUES

Normal host components of crude extracts may themselves serve as adsorbent, as with tomato spotted wilt virus, where virus adsorbed at 0.1 M could be eluted at 0.01 M phosphate buffer (Black et al., 1963). Column phosphate gels have been used for lettuce necrotic yellows virus (McLean and Francki, 1967) and could probably be adapted to a number of less stable viruses. DEAE has been used for purification of rice dwarf virus (Suzuki, 1969). It was found that rice dwarf virus adsorbed to columns at 0.1 M buffer but eluted when buffer concentration reached 0.2-0.25 M. Polyethylene glycol columns have been used for isolation of the oat blue dwarf virus (Banttari and Zeyen, 1969).

13. SEROLOGICAL METHODS

Serology is a useful technique of great potential which has not yet been applied to purification of less stable agents. The basic principles for ultracentrifugal analyses of soluble complexes (Bradish and Crawford, 1960), and pretreatment of antibodies (Gold, 1961), have been worked out for other viruses, and some details of virus binding followed by density-gradient centrifugation have been worked out (Whitcomb and Spendlove, 1966; Ball and Brakke, 1969). The principles of binding and removal of antigenic host components from preparations should be readily derived from these sources. Avrameas et al. (1969) have recently used virus-specific antibodies for purification.

The advantage of a serological technique for purification is its utilization of a parameter of separation which is unique and independent of other methods. Stable host nucleoproteins, which might be principal contaminants in a final virus preparation, should be good antigens. It should be possible to remove such materials by antibodies directed specifically against them. Attempts in the past to utilize serological parameters have often failed, partially because antibodies were not fractionated from nonspecific antiviral substances present in crude serum, and/or because too little attention was paid to simple but essential principles of the antigen-antibody reaction. For example, systems have been overloaded with antibody, or serological purification was attempted too early in the overall procedure. Serological methods will prove to be a fine scalpel for removal of specific irritating components, but not a bludgeon for wholesale removal of host material.

I. Serology

1. INDICATIONS

With the exception of wound tumor virus, serological study of phytarboviruses is fragmentary. In its most highly sophisticated form, serology can be a powerful tool, useful for investigation of the distribution of viral protein in insects (Sinha, 1965*b*) and plants (Nagaraj and Black, 1961), study of the sequence of events following infection (Sinha, 1965*b*), a rapid technique for detection of virus in plant or vector (Whitcomb and Black, 1961; Nagaraj et al., 1961; Whitcomb, 1964), detection of infected cells in tissue culture (Chiu et al., 1966), and analysis of relationships between strains.

2. LIMITATIONS

Serological work has been limited by three principal obstacles: (1) requirement for highly purified virus for production of antiserum; (2) low antigen titers in plants or vectors; and (3) low antibody titers.

With viruses multiplying in both plants and insects, there is, theoretically, an unusual opportunity to circumvent purification by preparing serum against crude or partially purified extracts from insects for later use against plant extracts, and vice versa. There should be few common antigens. This has recently been done with wheat striate mosaic virus (Sinha, 1968).

An interesting approach to antiserum production was the precipitation of all nonsedimentable protein from plant tissue infected with tomato spotted wilt virus with ammonium sulfate; the resuspended protein precipitate was used as antigen, and antiserum reacting to virus was obtained (Alden, 1956). Feldman and Boninsegna (1968) used heated antigen for preparation of tomato spotted wilt virus antiserum.

3. TECHNIQUES

Since the text of Matthews (1957) covered serological work intensively, a lengthy treatment is not indicated here. Techniques so far employed to advantage have been the precipitin ring test (Whitcomb and Black, 1961), fluorescent antibody (Nagaraj and Black, 1961; Sinha, 1965*b*), hemagglutination (Saito, 1969), hemagglutination-inhibition (Gamez et al., 1967), and density-gradient analysis of soluble complexes (Whitcomb and Spendlove, 1966; Ball and Brakke, 1969). The Ouchterlony test (Ouchterlony, 1958; Wilson and Pringle, 1955) should be useful for systems in which the antigen titer is sufficiently high. It should also be pointed out that the infectivity neutralization test has been applied only sparingly to the phytarboviruses infecting the Auchenorrhyncha, compared with its elegant use in the study of certain aphid-borne viruses (Gold and Duffus, 1967; Rochow and Ball, 1967) and in other fields of virology (Fazekas de St. Groth, 1962). For complete details on serological methods refer to Chapter 17.

VIII. PRINCIPAL RESEARCH RESULTS

A. Dynamics of Multiplication

After acquisition or inoculation, the titer of an infectious agent in a plant or insect host increases according to a typical growth curve with a steep logarithmic phase (Reddy and Black, 1966). This logarthmic phase is followed by the appearance of primary symptoms in the plant, or transmission of the agent by the insect (Whitcomb and Black, 1961; Whitcomb et al., 1966*a, b*). In some cases the transmission curve may not parallel the growth curve (Gamez and Black, 1968; Paliwal, 1968). After acquisition of a disease agent by its vector, the insect's tissues have been shown, by the fluorescent antibody technique (Sinha, 1965*b*), infectivity studies (Sinha and Chiykowski, 1967*b*), and histopathological methods (Whitcomb et al., 1968*c*), to become infected sequentially. The rate of multiplication lessens after maximum concentration of agent is attained. This phase is accompanied by a decline in the inoculativity of the insect (Reddy and Black, 1966; Whitcomb et al., 1966*a*; Slykhuis, 1963).

B. Biogenesis of Virions

Although many electron microscopic observations are available on virion biogenesis (Shikata and Maramorosch, 1967; Vidano, 1970), biochemical studies have lagged, awaiting the development of a tissue-culture system. The growth curve for synthesis of protein subunits (soluble antigen) can be determined (Whitcomb and Black, 1961). Proteins absent from the virion, but coded by viral nucleic acid, have not yet been discovered. Analysis of external and internal antigens has not been attempted, although the presence of an internal protein in wound tumor virus is suggested by electron microscopic evidence (Streissle and Granados, 1968). The details of lipid incorporation into virions or yellows organisms in the respective plant and animal systems is also obscure.

C. Influence of Environment on Course of Infection

At temperatures significantly higher or lower than optimal, remission or complete cure can be achieved in plant or vector. Growth curves for synthesis of agent in the vector have decreased slopes at lower temperatures. Suboptimal temperatures delay events associated with the attainment of maximum titer, such as transmission, symptom expression, or lethality of an agent to its vector.

The probability of transmission per unit time changes throughout the diurnal cycle (Maramorosch, 1964). Feeding and other behavior of vectors may be affected by intensity or duration of light. A number of factors affect the susceptibility of plants to disease agents. The incubation period before disease symptoms appear in glasshouse plants may vary with the season, and the nature and sequence of appearance of symptoms in the plant may be quite different in different seasons.

D. Effect of Plant and Vector Age

Plants are usually most susceptible (Swenson, 1969) to infection when young; in certain cases (Maramorosch, 1950) susceptibility decreases greatly with aging. This effect is perhaps least noticeable with the yellows disease agents. The interactions between the growth curves of the host and the disease agent in the host have not been studied in detail. Vectors also tend to be maximally susceptible when young, in rare cases transmitting only after nymphal acquisition (Sukhov and Sukhova, 1940; Zazhurilo and Sitnikova, 1941).

E. Distribution of Agents in Their Hosts

Most phytarboviruses are restricted to the cytoplasm. Potato yellow dwarf virus is apparently localized, in part, in cytoplasmic invaginations into the nucleus (MacLeod et al., 1966), or within the nucleus of cells grown in tissue culture (Black, 1969). Wheat striate mosaic virus also associates closely with the nuclear membrane, apparently both within the nucleus and in invaginations (Lee, 1967). Maize mosaic virus was found in association with the endoplasmic reticulum (Herold and Munz, 1965). Actual sites of virus synthesis in the cells have been tentatively identified (Shikata and Maramorosch, 1967; Vidano, 1970), and shown to be centers of the production of mature virions. Such centers have been identified in fat body, hemocytes, brain, mycetome, and alimentary tract.

Most agents not mechanically transmissible to plants are localized in the phloem or closely associated tissues. In the vector (Sinha, 1969), rice dwarf virus (Fukushi and Shikata, 1963), wound tumor virus (Shikata and Maramorosch, 1965), wheat striate mosaic virus (Sinha and Chiykowski, 1969), and maize rough dwarf virus (Vidano, 1970) occur in many tissues. Certain yellows agents may be less invasive (Sinha and Chiykowski, 1967*b*, 1968). Cytopathic effects of the western X agent in many tissues indicated widespread invasion of that agent (Whitcomb et al., 1968*c*). Nasu et al. (1970) recently confirmed this indication electron-microscopically. Recovery of virus infectivity from various organs of vectors (Sinha, 1965*a*, 1969) also indicated widespread distribution. Inefficient transmission of wound tumor virus is paralleled by low concentrations of particles in salivary glands (Shikata and Maramorosch, 1965).

F. Dose Relationships

Latent or incubation periods in vector or plant are dependent on dose (Maramorosch, 1953; Whitcomb, 1969). Because transmission usually follows a rapid logarithmic increase of titer in the vector, the concentration of agent in saliva quickly passes through the range of concentrations in which the dose effect is detectable. For example, the first transmission by an insect may appear only after a lengthened incubation period in the plant (Kunkel, 1937). A dose phenomenon is demonstrated most easily in systems where the test plant is poorly susceptible (Whitcomb, 1969, 1971). The dose of the agent injected into insects directly influences the latent periods (Maramorosch, 1953; Whitcomb et al., 1966*a*, *b*, 1969). It is not known whether high multiplicities of infection produce such effects as abortive or deficient infections (Fraser, 1967), or genetic interaction between genomes in the infected cells.

G. Modification of Vector Host Specificity

Acquisition of certain of the yellows agents enables certain vectors to survive on hosts otherwise unacceptable to them (Severin, 1946; Maramorosch, 1958*a*). It is possible that success of such a disease agent under field conditions depends to some extent upon the infected plants being acceptable or attractive to potential vectors.

H. Deleterious Effects of Plant Disease Agents on Their Vectors

Effects of plant disease agents on their vectors (Jensen, 1969) are more common than previously supposed. These may range from minor alterations in the form of the fat body (Littau and Maramorosch, 1960) to intense cytopathology (Whitcomb, et al., 1968*c*) resulting in premature death (Jensen, 1959). Vectors affected so severely may not be important in the field spread of such agents (Jensen, 1957), but in the case of hoja blanca (Jennings and Pineda, 1971), selection against transmitting races may occur in the field. Death of embryos (Sinha, 1960) ascribed to European wheat striate mosaic virus may have been confounded with an inbreeding effect (Kisimoto and Watson, 1965).

I. Lack of Immune Reaction

There is no evidence of defense mechanisms in the plant capable of combating phloem infections. Although there is ample evidence for the ability of genetic and/or physiological factors to alter susceptibility of insects to disease agents, there is no evidence of an extracellular defense mechanism induced by invasion of disease agents. Infection of whiteflies by tomato yellow leafcurl virus may induce an intracellular defense reaction (Cohen, 1969), and other cases merit careful reinvestigation.

J. Interactions Between Strains

Several reviews (Kunkel, 1955; Kassanis, 1963; Freitag, 1969) of interactions between strains make detailed treatment unnecessary here. Briefly, distinctly different disease agents do not interefere with the normal course of transmission of one another (Nagaich, 1960; Frederiksen, 1964). In one case, that of curly top virus, there is no interference between strains in either plant or vector (Giddings, 1950). Many challenging strains of the aster yellows agent are excluded by other strains in the plant or insect (Kunkel, 1955); but such relationships are complex, and replacement of one strain by another occurred with some combinations (Freitag, 1964; Valenta, 1959*a, b*). Thus, although significant interactions between strains probably indicate relationship or similarity, lack of a protective mechanism is not necessarily significant.

K. Genetics

Sampling from different geographical localities (Kanervo et al., 1957; Shinkai, 1956; Iida and Shinkai (1969) or simple selection from laboratory stock (Kisimoto, 1967; Storey, 1932) has often revealed clones of vectors which differ widely in ability to transmit disease agents to healthy plants or to their progeny. The inheritance of such characters, on the other hand, usually proves to be complex (Black, 1943*a;* Nagaraj and Black, 1962).

Disease agents have also been found to vary, once again by chance geographical sampling or by selective pressure applied in the laboratory. Continued passage of a virus in one host often alters its virulence to another host. Continual passage of virus through plants may render it totally unable to infect its insect vector. Such vectorless strains have been observed with wound tumor virus (Black et al., 1958; Black, 1969) and potato yellow dwarf (Black, 1953*b*).

Gut punctures of nontransmitting *Cicadulina mbila* (Naude) rendered them infective, suggesting that the maize streak virus did not penetrate the cells of the alimentary tract (Storey, 1933). The aster yellows agent undergoes limited multiplication in the gut of a nonvector agalliian species (Sinha and Chiykowski, 1967*a*). Fluorescent antibody studies indicated that the decrease in ability to transmit wound tumor virus that accompanies aging of its vector could be explained by a failure of the infection in the intestinal epithelia to proceed further (Sinha, 1967). It was noted (Whitcomb et al., 1967) that failure of a vector to transmit the western X agent was associated with a lack of cytopathology, raising the possibility that the agent failed to multiply in the anterior lobe of the salivary glands.

L. Hyperplasia

Instances of hyperplasia in invertebrates (Harshbarger and Taylor, 1968) are relatively rare and poorly studied. Proliferative symptoms observed occasionally in vectors carrying the western X agent were not noted in healthy insects, and probably were related to infection by that agent (Whitcomb and Jensen, 1968).

Hyperplasia in plants as a result of virus infection is more common (Black, 1965). Girolami (1955) and Rasa and Esau (1961) studied the hyperplasia resulting from infection with the agent of aster yellows disease. Hyperplasia resulting from infection of plants by wound tumor virus has been well studied (Lee, 1955*a, b*), yet we do not know at present the sequence of events leading from viral infection to transformation of cells.

ACKNOWLEDGMENTS

The author is deeply indebted to Drs. K. Swenson, E. Shikata, L. R. Nault, J. Kramer, P. W. Oman, L. Chiykowski, and R. C. Sinha, whose suggestions greatly improved the manuscript. Any errors remaining, however, are the author's responsibility.

7

Transmission of Plant Viruses by Eriophyid Mites*

J. T. SLYKHUIS

Ottawa Research Station
Canada Department of Agriculture
Ottawa, Canada

I. INTRODUCTION

Plant pathologists have become accustomed to labeling as a virus any infectious agent that has not been observed with the light microscope. The recent realization that some plant pathogens that were assumed for many years to be viruses may instead be organisms resembling mycoplasma or the PLT group (Doi et al., 1967) indicates that the above simple criterion is deficient. However, except where evidence for a different identity has been obtained, we shall continue to presume that each of the plant-infecting agents carried by eriophyid mites is a virus.

The essential principles involved in the investigation of plant virus transmission are no different if the vector is an eriophyid mite than if it is an insect, but the techniques of investigation must be adapted to overcome the difficulties created by the microscopic size and intimate host associations of the eriophyids.

The first principle in vector investigation is to prove that the disease in question is caused by an infectious agent and not by the suspected vector. This can be difficult when eriophyids are involved, especially if the virus is not sap transmissible. Some eriophyids cause toxicogenic effects similar to symptoms caused by some viruses, and the mites are so small and often so well hidden

*Contribution No. 635, Cell Biology Research Institute.

within buds or leaf whorls that the presence of small numbers can be difficult to determine.

Another essential principle is to prove that a suspect associated with a virus is actually a vector. This, too, can be difficult if the suspect transmits only rarely or under special circumstances, and if other possible vectors are also present.

After the above principles have been established, other relationships between the virus and vector can be investigated. Evidence has been reported that nine plant diseases are caused by viruses transmitted by eriophyid mites (Table 7-1), but for most of these the difficulties encountered in establishing the above two principles have been so great that little else has been discovered about vector-virus relationships.

Proof that the cause of the disease is a virus that is transmitted by an eriophyid mite can be verified more easily for wheat streak mosaic virus than for any of the other viruses so far investigated. The presence of the virus can be determined by manual transmission to healthy wheat and other susceptible plants, by serological tests, and by observing the particles with the electron microscope. The vector mites, *Aceria tulipae,* which are normally present on naturally diseased wheat, can be transferred readily to healthy test plants. They transmit the virus efficiently and symptoms appear in 6-10 days (Fig. 7-1). Nonviruliferous mites can be obtained readily by hatching eggs on healthy plants and allowing the mites to multiply. These noninfective mites can become infective if they feed on diseased plants (Slykhuis, 1955).

Evidence that three other diseases of grasses now associated with mites were caused by infectious agents can also be verified easily. One of them, wheat spot mosaic (Fig. 7-2), occurs in association with wheat streak mosaic in Alberta, Canada (Slykhuis, 1956). It is transmitted by the same mite *A. tulipae.* Unlike wheat streak mosaic virus it has not been transmitted by any means other than by the vector mite, and no particles have been recognized with the electron microscope. Evidence that it is an infectious agent is based primarily on repeated tests that show that the symptoms are induced on test plants by mites transferred directly from diseased plants, but not by mites from eggs transferred from diseased plants and hatched on healthy wheat. However, such noninfective mites become able to induce the symptoms after they have fed on diseased plants. Furthermore, the disease persists on plants that have been freed from mites by treatment with miticides.

The other two mite-transmitted grass viruses, ryegrass mosaic virus and agropyron mosaic virus, are more comparable with wheat streak mosaic virus. They were known to be viruses before their association with mites was detected. They are readily transmitted manually by the leaf rub method, the particles are flexuous rods visible with the electron microscope, and specific antisera have been prepared for each. Ryegrass mosaic virus was transmitted by the eriophyid mite *Abacarus hystrix* in England (Mulligan, 1960), and agropyron mosaic virus by the same mite species in Canada (Slykhuis, 1969*a*). It was difficult to prove that this virus was transmitted by *A. hystrix* because few test

TABLE 7-1.
PLANT DISEASES CAUSED BY VIRUSES TRANSMITTED BY ERIOPHYID MITES

Disease	*Vector Mite*	*Percent Mites Transmitting*	*Retention by Mites*	*Transmission Through Eggs*	*Other Transmission*	*Incubation in Host*
Grasses						
Wheat streak mosaic	*Aceria tulipae*	34	9 days	0	leaf rub	5-10 days
Wheat spot mosaic	*Aceria tulipae*	65	13 days	0	—	3-10 days
Ryegrass mosaic	*Abacarus hystrix*	30	12 hr	0	leaf rub	10-14 days
Agropyron mosaic	*Abacarus hystrix*	< 1	—	0	leaf rub	5-10 days
Woody perennials						
Currant reversion	*Phytoptus ribis*	< 1	—	—	patch graft	1-3 years
Fig mosaic	*Aceria ficus*	70	—	0	patch graft	10-90 days
Peach mosaic	*Eriophyes insidiosus*	2.5	2 days	0	patch graft	14-100 days
Pigeon pea sterility	*Aceria cajanus*	<10	—	—	wedge graft	3-5 weeks
Rose rosette	*Phyllocoptes fructiphilus*	<10	—	—	graft	30-146 days

Fig. 7-1. Wheat leaves with wheat streak mosaic symptoms. The leaf on the left is normal, the others have progressively more severe chlorotic streaks. The virus is readily transmitted manually by the leaf rub method as well as by the mite *Aceria tulipae.*

Fig. 7-2. Wheat spot mosaic on maize leaves. The leaf on the right is normal. The cause of this disease has not been transmitted manually but it is an infectious agent that can be acquired and transmitted by *A. tulipae,* but does not pass to successive generations through eggs.

plants developed symptoms even when large numbers of mites from diseased plants were transferred to them.

Five diseases of woody perennials have been reported to be caused by viruses transmitted by eriophyids but none of the viruses have been transmitted artificially except by grafting. One of these, currant reversion, was the first plant disease suspected to be caused by a mite-transmitted virus (Amos et al. 1927; Massee, 1952). The mites were known to be present when the disease was first recognized. Transmission of currant reversion by grafting was difficult to prove since mites, being minute, could be in the buds. In addition, the mites were difficult to manage experimentally. Many mites seemed to be needed to induce symptoms. The symptoms were slow and uncertain in their development. There has been no success in rearing virus-free cultures of mites from eggs hatched on healthy plants, hence the effects of known nonviruliferous mites on plants could not be compared with the effects of related mites carrying the virus. The best evidence that the disease was caused by a virus was obtained by infesting small currant seedlings with mites taken from diseased bushes, then, after 4 days, dipping the plants in 0.05% miticidal endrin solution to destroy the mites. Some of the seedlings developed reversion symptoms and the presence of the virus was confirmed by patch bark grafting to healthy bushes (Thresh, 1964*b*).

As with currant reversion, evidence that pigeon pea sterility and rose rosette are caused by viruses rests on transmission of the disease by grafting and the continued presence of disease symptoms after the plants were treated with miticides (Seth, 1962, 1965; Allington et al., 1968). The causes of fig mosaic and peach mosaic have been transmitted by patch bark grafting. In addition, noninfective colonies of the vector mites have been obtained by hatching eggs on healthy host seedlings (Flock and Wallace, 1955; Wilson et al., 1955).

Further investigations of all the above diseases are desirable, and there are undoubtedly other mite-plant virus relationships awaiting discovery. It may therefore be beneficial to discuss some of the techniques that have been used in investigations of transmission of plant viruses by eriophyid mites.

II. DETECTING AND ESTIMATING POPULATIONS OF ERIOPHYIDS

The first requirement for investigating the transmission of a plant virus by an eriophyid mite is to find the mite. This may sound facetious, but the significance of the statement may be appreciated when it is realized that wheat streak mosaic virus was known for 20 years before anyone noticed that the mite, *A. tulipae,* was always associated with naturally diseased wheat. All eriophyids are so small that they are easily overlooked even when present in large numbers. Techniques adapted to cope with their microscopic size and secluded habitats are required for detecting and estimating populations of eriophyids in the field.

A. Detection

Most eriophyid mites are 0.25 mm or less in length. A 10× lens or a wide-field microscope equipped for 20-40× magnification is needed to see them. A higher-power microscope with oil-immersion lens is required to study morphological features used for taxonomy.

The eriophyid mites have intimate and usually highly specific host relations (Keifer, 1952). Species that transmit viruses can usually be found on the common hosts of the viruses. They feed and survive only on living tissues, preferring younger plant cells. Those mites infecting woody plants occur primarily in the buds. *Phytoptus ribis* vector of currant reversion virus resides deep within buds and causes them to swell abnormally (Massee, 1928). The mites leave as the buds flare during spring and summer and may be found on outer bud scales, leaves, and flowers until new buds are sufficiently large for them to enter (Smith, 1960). *Eriophyes insidiosus* vector of peach mosaic virus is also found principally in the buds of its hosts, peach and plum, but the mites become dispersed when bud scales flare (Keifer and Wilson, 1956; Wilson et al., 1955). *Aceria ficus* vector of fig mosaic virus may occur on leaves and in the fruit as well as in buds of the fig (Flock and Wallace, 1955). The grass-infesting mites, *Aceria tulipae* vector of wheat streak mosaic virus and *Abacarus hystrix* vector of ryegrass mosaic in England and of agropyron mosaic virus in Canada, feed in grooves between the veins on the leaf blades, and in the leaf whorls (Fig. 7-3). *A. tulipae* may also occur between bracts of the inflorescence of graminaceous hosts including the husks of maize (Connin, 1956; Mulligan, 1960; Nault and Styer, 1968; Slykhuis, 1955, 1961).

The presence of some of the eriophyids may be indicated by certain abnormalities of the infested plant parts. *P. ribis* is called "black currant gall mite" because it causes swelling of buds, but it may also cause leaf deformities. *A. ficus* may cause russeting, browning, chlorosis, mosaic, and distortion of leaves of the fig tree. *A. tulipae* is called "wheat curl mite" because it causes a tight inrolling of wheat leaves (Fig. 7-4). The tip of the youngest leaf is often trapped in the rolled portion of the preceding leaf resulting in a looping or curling of the leaf. Leaves of some grasses with hairy surfaces and deep depressions between the veins may not become rolled or show any abnormality as a result of mite infestations.

Fig. 7-3. Eggs and various stages of *A. tulipae* on the upper surface of a wheat leaf. The adult mites are 173-285μ long.

Fig. 7-4. Diseased wheat plants heavily infested with *A. tulipae* (left) and healthy, non-infested wheat (right). Note the tight inrolling of the infested leaves.

B. Populations on Infested Plants

Mites may be observed and counted directly on infested plant parts, but special techniques may be needed to observe and count mites in buds if large numbers of buds are to be considered. To remove mites from black currant buds for counting, Smith (1961) macerated specific quantities of buds in water in a homogenizer operated at 7000 rpm for one minute, which separated the maximum number of mites from the buds with minimum damage to the mites. The macerate was poured on a brass sieve (900 holes per square inch) set in a filter funnel. The residue was washed on the sieve with 200 ml of water previously used for rinsing the macerating vessel. The washings were collected in a beaker. The suspension was stirred as drop samples were removed and placed on a black filter paper on which the mites could be observed with a low-power microscope and counted. This procedure is useful even for estimating low populations of mites in buds, but mites killed by the maceration cannot be distinguished from mites that were dead in the buds.

Another apparatus was used for counting live mites emerging from black currant buds (Smith, 1961). A glass suction tube of 5 mm bore was fitted to a test tube containing one inch of water and connected by a long rubber hose to a small suction pump. Mites emerging from buds were sucked into the tube and carried below the surface of the water. For counting the mites, the contents of the collecting tube were filtered with black filter paper on which the mites could be seen with a low-power microscope.

C. Detecting Dispersal by Exposure of Susceptible Plants

Highly susceptible seedlings in pots or flats can be exposed for specific periods to indicate the presence of mites and the rate of natural spread of a virus at particular locations in the field (Slykhuis, 1955, 1961). To prevent infestation by insects, the pots or flats of plants can be placed in cages covered with cloth or plastic screen with openings of 0.17-0.24 mm (see Section IVA). After exposure in the field for the desired period, the plants may be examined immediately or placed in mite-proof cages in the greenhouse to allow the mites to multiply or the virus to incubate for a suitable time before final observations are made.

The presence and times of mite and virus spread may also be indicated by planting a susceptible host at a series of dates. When winter wheat is sown at different dates from spring to late fall, some of the plantings become infested if wheat-infesting mites occur in the area, indicating times of mite dispersal.

D. Estimating Dispersal with Sticky Traps

Glass microscope slides coated with petroleum jelly and clipped to stakes were used to trap *A. tulipae* wind-borne in wheat fields and dispersed by fans in the greenhouse (Slykhuis, 1955). Pady (1955), in trapping rust spores, found eriophyid mites on glass slides that had been coated with silicone grease (Dow Corning 4 compound) and exposed for 24 and 48 hr periods. Each slide was exposed at an angle of 45° in a holder mounted on a wind vane to keep it facing the wind. Staples and Allington (1956) also used slides coated with a high-vacuum grease (Dow Corning), which resists rain, does not change its physical characteristics noticeably with fluctuating temperatures, is optically desirable for microscopic examination of entrapped mites, and does not strip from the slides when they are treated with mite-clearing and-staining media. The mites were cleared and stained in situ on the slides with Keifer's "A" solution (Keifer, 1952), then rinsed in water. Keifer's "A" solution is made up as follows:

(1) Gum arabic powder	1.0 g
(2) Resorcinol crystals	3.0 g
(3) Potassium iodide crystals	0.2 g
(4) Iodine crystals	0.2-0.35 g

Grind in mortar, transfer powder to screw-cap vial and add

(5) Lactic acid, commercial solution	10.0 ml
(6) Hydrochloric acid (conc.)	8 drops

Screw the cap down tightly and heat in an oven at 45° C for 4-5 hr or until the gum arabic is dissolved. This medium clears and expands the mites when heated.

Nault and Styer (1968) found that mites trapped on glass slides coated with silicone grease and exposed in the field were too severely desiccated for accurate species identification by direct observation, but if they were processed in a suitable way they could be identified. The mites were removed from the grease with a minute pin and placed in concentrated KOH in a depression slide and warmed to dissolve the grease and to expand and clear the mites. They were then transferred to Keifer's "A" solution and heated for a few minutes then transferred to Keifer's "C" medium on a glass slide. A cover slip was placed over them to make a permanent mount which can be examined microscopically.

Keifer's "C" medium for permanent mounts (Keifer, 1952) is prepared as follows:

(1) Gum arabic	0.5 g
(2) Sucrose	0.5 g
(3) Chloral hydrate crystals	7.0 g
(4) Potassium iodide crystals	0.2 g
(5) Iodine crystals	0.2-0.35 g

Grind in mortar, transfer to screw-cap vial.

(6) Formaldehyde, ¼ to ½ strength 18 drops (or more)

Since gum arabic will not dissolve readily in the above mixture, put 2 g of chloral hydrate in the original mix, heat at 45° under a tightly screwed cap until the gum arabic is in solution, then add the rest of the chloral hydrate.

III. COLLECTING AND TRANSFERRING

When investigating any vector of a plant virus it is important to collect and preserve specimens for accurate identification, but preservation and identification will be useless unless live individuals can be successfully transferred to suitable experimental plants for vector tests.

Eriophyids collected for taxonomic purposes must be preserved in ways that will not destroy critical features. Most liquid preservatives including alcohol or formaldehyde are too harsh. On the other hand, eriophyids can be preserved simply by wrapping heavily infested plant parts in a soft tissue, placing in an envelope, and drying (Keifer, 1952). A suitable preserving medium used by H. H. Keifer (California Department of Agriculture, Sacramento, California) can be prepared as follows: To a nearly saturated solution of sucrose, add ¼ volume of 95% ethyl alcohol. If separation occurs add 25% ethyl alcohol until the sugar dissolves. Add just sufficient iodine crystals to barely tint the solution. Dispense in screw-cap vials. Mite-infested plant material may be immersed directly in this medium.

Live mites for vector tests are collected with live host plant material. Small plants can be transplanted directly from the field into pots. Polyethylene bags are most valuable for keeping plants or plant parts from wilting until plants can be potted or mites transferred to test plants. Some mites, such as *A. tulipae* on wheat, may be transferred to new plants at any stage and any time of the year (Slykhuis, 1955), but others, for example *P. ribis* on black currant, can be transferred successfully only during the natural period of dispersal in late spring and early summer (Smith, 1962).

The methods for transferring live mites may vary according to the purpose of the experiment, the type of plant material, the numbers of mites available or to be transferred, and the presence of unwanted arthropods.

A. Transfer with Infested Plant Material

Plants may be infested with eriophyids by placing infested plant material on or above the plants to be infested, but this method does not preclude contamination with other organisms that may be present on the source material. The procedure may vary from scattering infested plants or leaves over the test plants (Slykhuis et al., 1957), to placing small portions of infested material in the axils of leaves or in buds of the test plants. Migration of mites from infested material placed on or clipped to the test plants can be hastened by the drying effects of warm, dry air. Del Rosario and Sill (1958) described the use of a beam of light to induce mites to move from pieces of infested leaves to test plants. If contact of source plant material with the test plants is to be avoided, the source material may be suspended above the test plants or enclosed in a screen or in a small cloth bag with a weave sufficiently open to allow the mites to escape.

B. Transfer of Individual Mites

To manipulate eriophyids individually a low-power microscope (20×-40×) is needed to observe and select the mites. Since some mites, such as *A. tulipae,* are hidden in rolled leaves of some hosts, a means of holding the leaves flat is required to expose the mites. Rolled leaves of infested wheat plants can be held flat with hair clips or a strip of double-coated adhesive tape on a microscope slide. Fellows (1956) described other methods to open rolled wheat leaves to facilitate examination for mites. Although a small brush can be used to transfer mites from the surface of a leaf, a single hair cemented to a handle about the size of a pencil is more convenient for manipulating individual mites of any stage (Slykhuis, 1955). Hairs with fine, tapered tips, such as those from an artist's camel's hair brush, or eyebrow hairs enable gentle handling of even the youngest nymphs, which will cling to the hair but can be deposited where required without injury. Eggs and molting mites can be pried from a leaf more readily if the fine tip is clipped off and the stiffer region of the hair is used. Some workers prefer coarser hairs including "medium-coarse human hair" for transferring eggs (Del Rosario and Sill, 1958). Others have used steel needles.

C. Dispersal by Wind

Since wind is a major factor in the dispersal of mites in nature, electric fans can be used effectively for dispersing mites from suitable sources (Slykhuis, 1955). Under warm conditions when mites are active on the surface of leaves, only a gentle breeze is required to dislodge them and carry them away like flecks of dust. A small fan in a small room or growth chamber will disperse mites from a few heavily infested source plants to all test plants in the room.

Wind can also be utilized to infest plants in the field. If infested plants are transplanted around plots of plants to be infested the wind will do the rest. Conversely, plants to be infested may be placed or planted near naturally diseased plants.

D. Aspirators

Aspirators used for insects are useless for the tiny eriophyids, but specially adapted aspirators have been used for collecting eriophyids (Slykhuis, 1967). One of these, operated by mouth (Fig. 7-5), was useful for depositing large numbers of *Aceria tulipae* from wheat leaves onto a small filter paper disc inserted behind the glass tip of the aspirator. Mites deposited on the filter paper disc could be transferred to a plant or into a medium for crushing or mounting the mites (Paliwal and Slykhuis, 1967; Slykhuis, 1967). Another aspirator used suction from a vacuum cleaner to suck mites from heavily infested plants and deposited them directly onto potted plants.

IV. CONFINING ERIOPHYIDS

For most experiments on the transmission of viruses by eriophyids, cages and other devices are required to prevent the mites from spreading to other plants, and to prevent contamination from other mites, predators of the mites, or pests of the host plant. There has been little success in attempts to force the mites to remain on a plant or on a particular plant part. These mites are so small, and their means of locomotion so limited that they cannot independently move more than a few centimeters over the soil to another plant (Del Rosario and Sill, 1964). However, they are readily carried on air currents or by clinging to insects or other moving bodies; hence caging procedures are designed primarily to prevent spread by such agencies.

A. Cages for Potted Plants

Glass lamp chimneys with a close-weave cloth fastened over the top have been used to cover mite-infested plants, but more durable, better ventilated cages can be made of clear sheets of various types of plastic 0.25 or 0.50 mm thick. There should be an easily used solvent such as acetone for cementing the plastic for cage construction. The plastic should be tested to ascertain that it is not toxic to plants or mites. Although cellulose acetate plastics are generally satisfactory, some preparations are toxic. Some clear butyrate plastics have been very satisfactory for small cages. Cages made of such plastics require ventilation by openings covered with screens, but screens with sufficiently small pore size to prevent the passage of mites allow very limited passage of air. Cloths of cotton or other natural fibers have not been satisfactory as screens for mites because of uneven thread thicknesses resulting in great variations in openings between threads. Cloth made of synthetic fibers that are uniform in thickness, such as a close-weave white Dacron cloth, resistant to the effects of light and moisture,

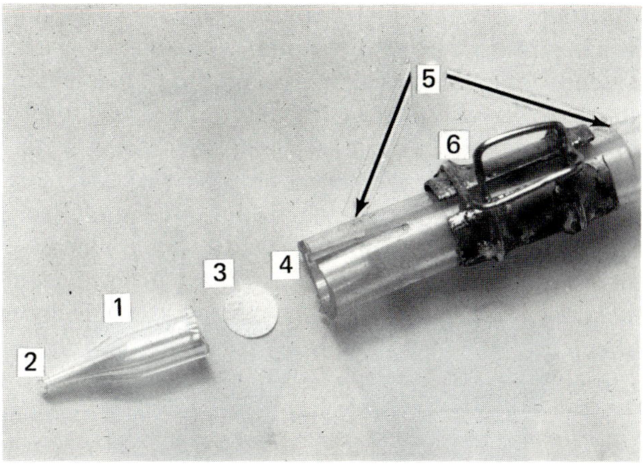

Fig. 7-5. Tip end of a mouth-operated aspirator for collecting large numbers of live eriophyid mites. The short glass tip (1) with a beveled point (2) is backed by a 6 mm disc of open-pore filter paper (3). Both are inserted in the slit end of a length (40-70 cm) of flexible Tygon tubing with 6 mm inside diameter and pressed firmly against a backstop ring of Tygon tubing (5), 6 mm outside diameter, 2-3 cm long, fitted within the larger tubing. The spring clamp (6) is then slipped up to hold the assembly firmly in place.

has served well as screening for mite cages. Any cloth to be used as a screen should be tested for its ability to retain mites by placing several mites on a piece of the cloth while observing under a 20× -40× stereoscopic microscope.

Cylindrical cages constructed of clear plastics, and with ventilator holes in the sides as well as the top covered with a close-weave white Dacron cloth, have been constructed in several sizes depending on the numbers and sizes of plants to be covered (Fig. 7-6). Large cages may require metal reinforcement rings on the ends of the cylinders unless the plastic is thick and rigid enough to be self supporting and resistant to warping.

Cages of varied designs and dimensions can be made with plastic or glass to admit light, and close-weave cloth for ventilation (Del Rosario and Sill, 1964). Plastic cages may be joined with cloth sleeves for tests on the movement of mites from one plant to another in the absence of air currents (Slykhuis, 1955).

To reduce contamination of cultures of eriophyid mites by predaceous mites, cylindrical plastic cages were fitted to screw-cap glass jars used as containers for soil for the culture plants. The plants were watered through an S-shaped copper tube soldered into the metal base of the cage; thus the cage could be kept sealed to exclude the predators until the eriophyids were ready for use (Slykhuis, 1956).

B. Test Tubes as Cages

Large test tubes (2.5 × 25 cm) have been used to contain small wheat plants for starting cultures from single mites or eggs and for testing the infectivity of single mites (Slykhuis, 1955). A partially sterilized soil mixture is placed in

B. Initiating Nonviruliferous Colonies

There has been no evidence that any of the mite-transmitted plant viruses is transovarially transmitted, so eggs provide the most reliable source of nonviruliferous mites. Noninfective colonies of the vectors of fig mosaic and the mite vectors of the known grass viruses were initiated by transferring eggs to healthy seedlings of the host species. High humidity is essential for the eggs to hatch. Eggs of *A. tulipae* hatch well and a good stock of virus-free mites soon develops on wheat seedlings grown in soil in large test tubes (2.5 × 25 cm). The plants can be kept in the test tubes for several weeks with little attention except to provide suitable temperature, light, and watering while the mites multiply for several weeks. Since *A. tulipae* is parthenogenetic, colonies originating from single eggs have been developed on wheat in test tubes. Mites from the test tube stocks can be transferred to caged plants growing in pots, or used directly for experiments (Slykhuis, 1955).

Del Rosario and Sill (1958) described a multistep procedure for rearing *A. tulipae* from eggs. The eggs were hatched on detached wheat leaves kept in a moist chamber. The nymphs were transferred to 2-week-old wheat plants in test tubes plugged with synthetic sponge rubber and inverted in a nutrient solution. Subsequently the plants were transplanted into soil in pots and covered with cages.

C. Seasonal Factors

A. tulipae collected on wheat in the field in any season can be transferred and established on wheat in the greenhouse (Slykhuis, 1955), but timing is critical for the transfer of mites like *P. ribis* on black currants. The latter mite can be established on new plants only if transferred during the natural period of dispersal in late spring and early summer (Massee, 1952; Smith, 1962). At this time the mites are active, and growing points of new shoots permit entrance.

D. Avoiding Predators in Cultures

A. tulipae is frequently attacked by a predaceous mite, *Typhlodromus cucumeris,* which can quickly deplete a prolific culture (Slykhuis, 1956). These active predators are difficult to exclude. Invaded stock can be saved by carefully transferring eriophyids singly to colonize new host plants using care to avoid transferring any of the fast-moving predators. Special cages have been used to minimize invasion of stock colonies by predators.

E. Controlling Mildew on Host Plants

Cages that will exclude eriophyids will also exclude powdery mildew spores, but mildew may infect plants before they are caged or when cages are removed for inspecting or removing mites. When mildew infection occurs it may be necessary to use a fungicide to eradicate it, but most effective mildew eradicants, such as sulfur or Karathane, also kill eriophyids. However, actidione sprays have been successfully used to eradicate mildew on wheat without damaging colonies of *A. tulipae.*

VI. TESTING FOR VIRUS TRANSMISSION

The eriophyids that transmit plant viruses can usually be found on plants infected with the viruses they transmit; therefore initial vector tests are usually made by transferring mites from naturally diseased plants to healthy, young plants of the same species. Some viruses are so efficiently transmitted by their vectors that transmission can be demonstrated by transferring one or more mites to each test plant (Slykhuis, 1955, 1956). Other viruses are so inefficiently transmitted that an adequate test may require the transfer of large numbers of mites to each test plant (Massee, 1952; Slykhuis, 1969*a*). Some eriophyids produce toxins that cause symptoms that resemble virus diseases on plants. Therefore, particularly when large numbers of mites are needed for tests, various mite-induced symptoms presumed to be caused by viruses may in reality be caused by toxins introduced into the plant by the mites while feeding. Additional tests are therefore necessary to determine if specific symptoms induced by the mites are caused by a virus.

A. Detection of an Infectious Agent by Transmission Without Mites

An initial procedure for determining if disease symptoms are caused by a virus is to test for transmission by techniques known to transmit viruses. Mechanical inoculation with infective plant juice has been a convenient and efficient method of testing for the presence of wheat streak mosaic, ryegrass mosaic, and *Agropyron* mosaic viruses. This method has been unsuccessful with other plant diseases reported to be caused by mite-transmitted viruses, but grafting has been used as a means of transmission when the hosts were woody plants.

Whether the transmission test for presence of an infectious agent is done by the leaf rub method or by grafting, precautions must be taken to assure that mites are not transferred from the diseased plants to the test plants. There is little chance for mites to survive in juice expressed from diseased leaves but the juice for inoculum can be filtered to assure that no mites are transferred with the inoculum (Slykhuis, 1953). It is more difficult to avoid the presence of mites when transmission is tested by grafting because the tiny eriophyids may be present but unseen on the diseased scions. Various miticides including sulfur (Flock and Wallace, 1955) have been used to destroy mites on cuttings before grafting, but mites deep within buds are not killed. Systemic miticides or heat treatments can be useful. However, mites can be avoided in graft transmission tests on woody plants by grafting only small patches of bark without buds from diseased to healthy plants (Cropley et al., 1964; Wilson et al., 1955).

B. Different Effects of Infectious and Noninfectious Mites

None of the known mite-transmitted viruses appears to be transovarially transmitted. Therefore, a simple procedure to determine if disease symptoms induced by mites are caused by an infectious agent or by a toxin produced by the mites is to transfer mites from the diseased plants to one group of test plants, and also transfer eggs from the same source to other test plants. If the eggs hatch and a colony of mites develops without causing the symptoms that are caused by mites from the same source, it can be concluded that the symptoms are caused by an infectious agent that is not carried through the eggs. Since the cause of wheat spot mosaic could not be transmitted by the leaf rub or graft methods, lack of transovarial transmission has provided the principal evidence that the cause is an infectious agent rather than a mite toxin (Slykhuis, 1956). However, there is still not enough evidence to conclude that the infectious agent was specifically a virus.

C. Development of Disease After Elimination of Mites

If mite-induced symptoms on plants are caused by a toxicogenic effect of the mites, the plants should recover after the mites are removed, but if the symptoms are caused by a virus transmitted by the mites, the continued presence of the mites should not be necessary for the disease to develop after inoculation has taken place. In experiments with viruses transmitted by insects it is common practice to feed infective insects on test plants long enough to transmit the virus, then remove the insects so the symptoms caused by the virus will not be complicated by toxicogenic or other damaging effects of the insects. Similar procedures are difficult to perform satisfactorily with viruses transmitted by the eriophyids because when these mites are placed on test plants they generally invade leaf whorls or buds from which they cannot be removed and are difficult to kill with miticides. Therefore, when symptoms develop it may be difficult to ascertain that there were no mites present to cause them.

Sulfur dust was used to kill mites on fig cuttings infected with fig mosaic (Flock and Wallace, 1955) but there was no definite assurance that all mites were eliminated. Lime sulfur sprays have been used for many years to kill big bud mites on black currants (Morgan and Smith, 1961), but these sprays kill mites only after they emerge from infested buds. Ekatox (0.1%) was sprayed on pigeon pea to kill mites associated with pigeon pea sterility (Seth, 1962), but there was no assurance given that all mites were eliminated.

Systemic miticides offer a good possibility of freeing test plants from mites after inoculation feeds for the transmission of a virus. A 10 min emersion of mite-infested black currant shoots in a 0.05% solution of fluoroacetamide killed all mites but was phytotoxic. A 0.02% solution was less phytotoxic but allowed a few mites to survive and was less effective against eggs (Collingwood et al., 1960). Repeated spraying of black currants with a 0.04% solution of endrin, which has systemic activity, was the most effective treatment for the control of the black currant gall mite in the field in England (Collingwood and Decker, 1960). A 0.05% solution of endrin was used as a dip to eliminate mites from black currant seedlings 4-24 hr after they were infested with mites from diseased plants. The mites appeared to be eliminated, so currant reversion symptoms that developed on the seedlings were evidently caused by an infectious agent transmitted by the mites and not by toxicogenic effects of the mites.

Warm water treatments including 40° C for 40 min, 42.5° C for 30 min, 45° C for 15 min or 47.5° C for 5 min have been found effective for eliminating *P.* ribis from dormant buds of infested black currant cuttings.

Mites can be eliminated from wheat plants infected with wheat spot mosaic virus by cutting off all growth above the crown and fumigating repeatedly to eliminate any mites that may have remained and could reinfest the new growth from the crown (Slykhuis, 1956).

Cages have been used to confine *A. tulipae* on a leaf blade without allowing invasion of the leaf whorl. When the mites were successfully confined on the leaf blade they could be removed individually or by cutting off the leaf blade, or they could easily be killed by a miticide. However, this procedure requires considerable care to assure that no mites excape into the leaf whorl where they cannot be eliminated easily.

D. Evidence That the Infectious Agent Is a Virus

The causes of wheat streak mosaic, ryegrass mosaic, and *Agropyron* mosaic are readily identifiable as viruses by the ease with which they are transmitted by mechanical inoculation and by electron microscopic detection of particles which are recognized to be viruses. Confirmation of the viral nature of the other mite-transmitted pathogens will await the revelations of the electron microscopic and other techniques in conjunction with infection experiments.

VII. VECTOR-VIRUS RELATIONS

A. Vector Efficiency

Tests for the percentage of mites transmitting at specific stages of development have been done by transferring one or a specific number of mites to each of a number of very young test plants. For tests with peach mosaic, the test plants were peach seedlings 2.5-8 cm high (Wilson et al., 1955). In such tests with woody plants the mites should be placed on a growing point or in a bud where they can feed on young tissues (Smith, 1962). A single mite is not liable to move to and infect other plants than the one on which it was placed, but if progeny develop some of them may spread to other test plants, or test plants may become infested from other sources. Hence the test plants should be caged separately or treated with a systemic miticide such as endrin (Thresh, 1964*b*) after a suitable period to prevent the spread of mites. In tests for the efficiency of transmission of wheat streak mosaic virus by *A. tulipae,* the mites were tested on young wheat seedlings growing in soil in large test tubes (2.5 × 25 cm) plugged with cotton to exclude contaminants (Slykhuis, 1955, 1956). Since the incubation period for the virus is relatively short (6-12 days), the symptoms developed before the plants were too large for the test tubes.

B. Retention Period

The retention of infectivity by insects that have fed on diseased plants is conveniently tested by giving the infective insects short feeding periods on each of a series of test plants, but this procedure has not been satisfactory with the eriophyids. Many of the mites move to the youngest tissues in secluded places from which they are difficult or impossible to retrieve without injuring the host, and the ratio of recovery from each successive test plant is discouragingly low. Methods used to determine virus retention by mites have avoided serial feeding on susceptible plants. Instead, large numbers of viruliferous mites have been placed either on a virus-immune plant or on an artificial surface, then at specific intervals, samples of these have been transferred to test plants to determine infectivity.

In early experiments with wheat streak mosaic virus, a *Triticum X Agropyron* hybrid was used as a virus-immune host for infective *A. tulipae* (Slykhuis, 1955), but readily available varieties of *Lolium perenne* L. can also be used. Mulligan (1960) used wheat as a holding host for *A. hystrix* carrying ryegrass mosaic virus. A difficulty with the above procedure is that progeny of the original mites mature so quickly that after a week the parents are difficult to distinguish from their progeny and there is no convenient means of knowing how long the original mites survive.

An alternative method of testing for duration of virus retention by mites after removal from diseased plants involves keeping the mites on an inert surface for various periods before testing for infectivity. Wilson and Jones (1957) kept the mites on glass cover slips. We used 6 mm discs of heavy paper mounted edgewise on the tip of a pin inserted horizontally through a cork base. Few mites found the pin, which was their only means of crawling off the paper disc. The latter should be dark in color to facilitate observing the light-colored mites. When mites are retained in inert surfaces such as glass or paper, a moist chamber should be used to prevent rapid desiccation of the mites (Slykhuis, 1955).

Agar gel has several advantages as a substrate for mites for virus retention tests. The mites can be sustained by nutrients and moisture in the agar. Suitable gels include potato dextrose agar containing 20 g of dextrose and 40 g of agar per liter, or a wheat decoction agar made with juice extracted from 500 g of wheat leaves, 20 g of dextrose, and 40 g of agar per liter (Del Rosario and Sill, 1964). To inhibit growth of fungi and bacteria, 1 ml of a 1% solution of polymyxin sulfate B was added to each 10 ml of medium before pouring into petri dishes. On these media adult *A. tulipae* did not lay eggs, and although eggs placed on the media hatched, the nymphs soon died. At 21°C the adults survived 80 days until the agar was nearly dried out. When placed on wheat the surviving mites fed normally and recommenced laying eggs.

C. Acquisition and Inoculation Access Periods

Acquisition access periods can be timed by placing noninfective mites of the desired instar on leaves from virus-infected mite-free plants (Slykhuis, 1955; Mulligan, 1960). The leaves are placed on moist filter paper in a petri dish so the mites can be observed under a stereoscopic microscope. Mites that settle down are assumed to feed and may be transferred singly or in groups to test plants. Orlob (1966) transferred four mites to each test plant then determined infectivity of individual mites by an equation used by Gibbs and Gower (1960) for aphids:

$$p^* = 1 - (1 - R/N)^{1/i}$$

where N is the number of test plants used; i is the number of mites tested on each plant; R is the number of test plants which become infected; p^* is the maximum likelihood estimate of the proportion (p) of infected mites in the population.

Inoculation access periods for *A. tulipae* carrying wheat streak mosaic virus may be timed by placing single mites on small test plants in a petri dish; then, after the intended time, each mite is located under a stereoscopic microscope and removed. If the mite cannot be found the plant must be discarded (Orlob, 1966).

There have been no reports of success in attempts to determine actual feeding times required for the acquisition or transmission of a virus by an eriophyid. Few mites transferred to new leaves settle down to feed in locations where they can be observed. They usually move into leaf whorls or buds where the youngest tissues occur, and many cannot be found. Even when they become quiescent it is difficult to determine if they are feeding. Because of these difficulties in timing successive feeds on source plants and on a series of test plants, no evidence has been reported on the presence of a preinfective period after the acquisition of a virus by any of the eriophyid mites.

D. Detection of Virus in Eriophyids

Although vector mites may be fed on test plants to determine if they are carrying a virus, other procedures including mechanical inoculation, serology, and electron microscopy have been used to test for the presence of virus in non-vector as well as vector species of mites (Orlob, 1966; Paliwal and Slykhuis, 1967). For these methods it has been necessary to collect large groups of mites either for maceration or for embedding in a mounting medium for ultrathin sectioning (for methods see Chapter 4). The mites can be collected from source plants with the single hair transfer method, but the mouth-operated aspirator referred to above may speed the collection of the large numbers needed. For macerating, the mites may be deposited in a drop of distilled water in the concavity of a depression slide, then the water evaporated to a thin film and the mites ground with the Teflon tip of a microgrinder or a glass micropestle and the desired diluent added to the macerate.

For transmission by mechanical inoculation, extracts of wheat streak mosaic virus from mites reared on infected wheat were prepared by macerating groups of 150-250 mites in 0.05-0.2 ml 0.1 M phosphate buffer, pH 7.0. Celite or 600 mesh Carborundum was added as an abrasive and the macerate rubbed on the

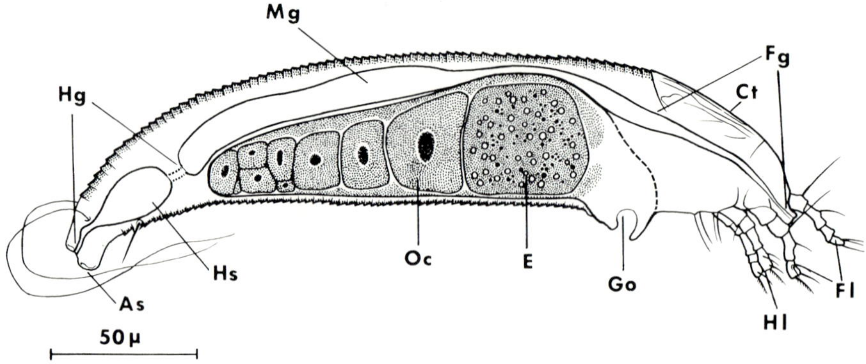

Fig. 7-7. Diagram of young adult *Aceria tulipae* showing the alimentary canal and ovary. Fg, foregut; Mg, midgut; Hg, hindgut; Ct, cephalothorax; Fl, forelegs; Hl, hind legs; Go, genital opening; E, egg; Oc, oocyte; Hs, rectum-like sac of the hindgut; As, anal sucker. Flexuous rod-shaped particles considered to be wheat streak mosaic virus were found in large concentrations in the hindgut and posterior midgut of viruliferous mites.

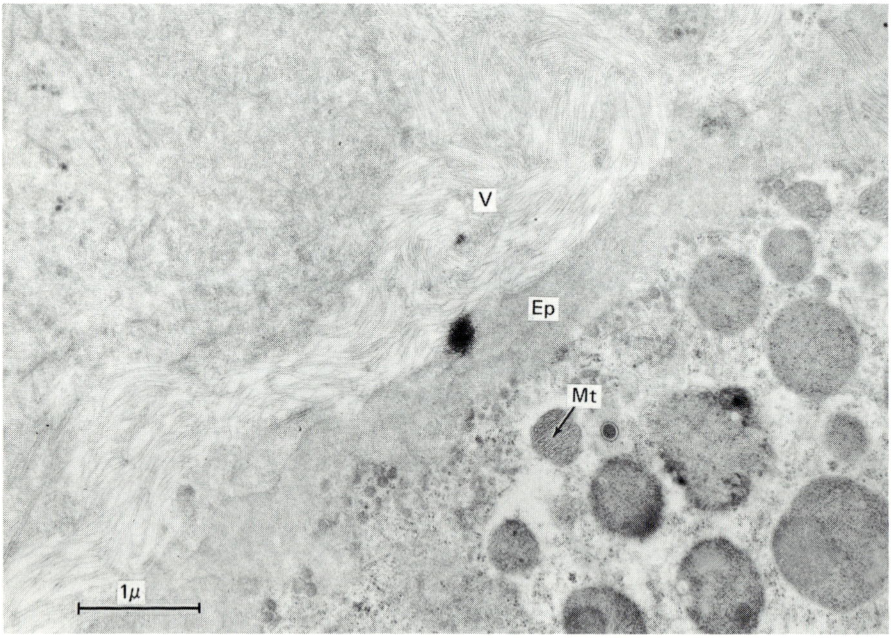

Fig. 7-8. A portion of an ultrathin section through a viruliferous *A. tulipae.* V, an accumulation of wheat streak mosaic virus particles lining the epithelium (Ep) of the wall of the posterior midgut; Mt, mitochondrion.

leaves of wheat test plants with a small glass pestle or with the finger tips. In tests reported, only a few of the inoculated plants (3/46 and 3/39) became infected with this virus from infective *A. tulipae*.

For serological detection of wheat streak mosaic virus in extracts of *A. tulipae*, extracts of 700-800 mites were prepared in 0.2 ml of a solution containing 0.85% NaCl in 0.01 M phosphate buffer, pH 7.0, then centrifuged for 10 min at 6590 *g*. Virus in the supernatant solution was detected by reaction against a specific antiserum using the precipitin ring test (for methods see Chapter 17).

For detection of the rod-shaped particles of wheat streak mosaic virus with the electron microscope, homogenates were prepared from 50-100 mites macerated in a drop of distilled water. A micropipette was used to place droplets of the clear extract on grids for the electron microscope and the preparation was shadowed or negatively stained for examination. The locations of massive concentrations of these virus particles in the midgut and hindgut of *A. tulipae* (Figs. 7-7 and 7-8) were detected by examining ultrathin sections of the mite (Paliwal and Slykhuis, 1967).

8

Transmission of Viruses by Nematodes

C. E. TAYLOR

Scottish Horticultural Research Institute
Dundee, Scotland

I. INTRODUCTION

Interest in nematodes as vectors of plant viruses stems from the convincing demonstration by Hewitt et al. (1958) that *Xiphinema index* transmitted fan-leaf virus in grapevine. Discoveries of nematode vectors of other soil-borne viruses quickly followed: *X. diversicaudatum* was associated with the transmission of arabis mosaic virus (Harrison and Cadman, 1959; Jha and Posnette, 1959), *X. americanum* with tomato and tobacco ringspot viruses (Breece and Hart, 1959; Fulton, 1962), and *Longidorus elongatus* with tomato black ring and raspberry ringspot viruses (Harrison et al., 1961; Taylor, 1962); species of *Trichodorus* were shown to transmit tobacco rattle virus (Harrison, 1961) and pea early browning virus (Gibbs and Harrison, 1964*a*; van Hoof, 1962).

So far, 19 nematode species are known to transmit plant viruses and all are in the Dorylaimid genera *Xiphinema, Longidorus,* and *Trichodorus* (Table 8-1). These genera are fairly cosmopolitan and widely distributed throughout the world. Where the host ranges have been investigated the vector species seem to be polyphagous, although they may multiply at varying rates on different hosts. Most of the species, particularly those concerned with virus transmission, are probably obligate parasites and are migratory ectoparasites feeding mainly at the root tips and to a lesser extent along the sides of the roots.

At least nine serologically distinct viruses are transmitted by nematodes; several of them occur as antigenically distinct strains, each one transmitted by a different nematode species (Table 8-1). All infect a wide range of woody and herbaceous hosts and some are seed- and pollen-borne (Lister and Murant,

1967) (see Chapter 10). They are readily transmitted by mechanical inoculation to herbaceous test plants and are relatively stable in vitro. The viruses can be divided into two groups on the basis of particle shape. Cadman (1963) named those with polyhedral particles "Nepoviruses," that is, *ne*matode-transmitted viruses with *po*lyhedral particles; viruses in this group are transmitted by *Longidorus* or *Xiphinema* species. The second group includes the viruses with rod-shaped or tubular particles for which Harrison (1964*a*) derived the name Netu[a], that is, *ne*matode-transmitted viruses with *tu*bular particles; these are transmitted by *Trichodorus* species.

TABLE 8-1
NEMATODE VECTORS OF PLANT VIRUSES

Vector	Virus	Reference
	A. Nepoviruses	
Xiphinema americanum	Tobacco ringspot—Arkansas isolate	Fulton, 1962
	Tomato ringspot—peach yellow bud isolate	Breece and Hart, 1959
	—grape yellow vein isolate	Teliz et al., 1966
X. coxi	Arabis mosaic—type strain and rhubarb mosaic isolate	Fritzsche, 1964
	Cherry leaf roll—type strain	Fritzsche and Kegler, 1964
X. diversicaudatum	Arabis mosaic—type strain	Harrison and Cadman, 1959
		Jha and Posnette, 1959
	Carnation ringspot—	Fritzsche and Schmelzer, 1967
	Cherry leaf roll—type strain	Fritzsche and Kegler, 1964
	Brome mosaic—type strain	Fritzsche, 1964
	Strawberry latent—type strain ringspot	Lister, 1964
X. index	Grapevine fanleaf virus	Hewitt et al., 1958
X. italiae	Grapevine fanleaf virus	Cohn et al., 1970
Longidorus attenuatus	Tomato black ring—English strain	Harrison et al., 1961
L. elongatus	Raspberry ringspot—type strain	Taylor, 1962
	Tomato black ring—Scottish strain	Harrison et al., 1961
L. macrosoma	Raspberry ringspot—English strain	Harrison, 1962
	B. Netuviruses[a]	
Trichodorus allius	Tobacco rattle—USA (Oregon) isolate	Jensen and Allen, 1964
	—USA (California) isolate	Ayala and Allen, 1966
T. anemones	Pea early browning—English isolate	Harrison, 1967*b*
T. christiei	Tobacco rattle—USA (Wisconsin) isolate	Walkinshaw et al., 1961
	—USA (California) isolate	Ayala and Allen, 1966
T. cylindricus	Tobacco rattle—Dutch isolate	van Hoof, 1968
T. nanus	Tobacco rattle—Dutch isolate	van Hoof, 1968
	—Scottish isolate	Cooper and Thomas, 1970
T. pachydermus	Tobacco rattle—Dutch isolate	Sol and Seinhorst, 1961
	Pea early browning—Dutch isolate	van Hoof, 1962
	—English isolate	Gibbs and Harrison, 1964*b*
T. porosus	Tobacco rattle—USA (California) isolate	Ayala and Allen, 1966
T. primitivus	Tobacco rattle—English isolate	Harrison, 1961
	—German isolate	Sänger, 1961
	—Scottish isolate	Mowat and Taylor, 1962
T. similis	Tobacco rattle—Dutch (*Gladiolus* notch-leaf) isolate	Cremer and Kooistra, 1964
T. teres	Pea early browning—Dutch isolate	van Hoof, 1962
	Tobacco rattle—Dutch isolate	van Hoof, 1968
T. viruliferus	Pea early browning—English isolate	Gibbs and Harrison, 1964*b*

[a] Recently the group name 'Tobravirus' was proposed by the plant virus subcommittee, International Committee on Nomenclature of Viruses, 1971.

II. THE VIRUSES TRANSMITTED BY NEMATODES

Nematode-borne viruses are readily transmitted by mechanical inoculation of sap to a wide range of herbaceous test plants (for methods see Chapter 1).

Although it is possible to identify a virus tentatively from the characteristic symptoms produced on test plants, positive identification is made usually by serological tests using antisera specific to particular viruses (for serological methods see Chapter 17.).

A. Nepoviruses

These are transmitted by *Xiphinema* and *Longidorus* species and include arabis mosaic, strawberry latent ringspot, cherry leaf roll, tobacco ringspot, tomato ringspot, raspberry ringspot, and tomato black ring viruses (Table 8-1). These viruses share no antigenic determinants but they form a group with close natural affinities and all possess polyhedral particles of about 27 nm. Their properties are similar (Table 8-2) and they cause similar symptoms in herbaceous test plants.

Viruses associated with raspberry yellow dwarf and strawberry yellow crinkle diseases were identified as identical with the type strain of arabis mosaic virus (Cadman, 1960) and were shown to be transmitted by *Xiphinema diversicaudatum* (Harrison and Cadman, 1959; Jha and Posnette, 1959). The type strain of arabis mosaic virus has been found only in Europe so far, but it is distantly serologically related to grapevine fanleaf virus, which infects grapevines and probably occurs wherever the crop is cultivated. In the United States the two principal diseases caused by grapevine fanleaf virus are called fanleaf and yellow mosaic; in Europe the same diseases are named respectively arriciamento, roncet, urticado, court-noué or Reisigkrankheit, and panachure or clorose infecciosa (for other names see Hewitt, 1968). Other strains of the virus are associated with veinbanding, wood pitting, and enation (Hewitt et al., 1970). *X. index* was shown to be the vector in California (Hewitt et al., 1958)

TABLE 8-2
PHYSICAL PROPERTIES OF NEMATODE-TRANSMITTED NEPOVIRUSES

Virus	Thermal Inactivation Point (°C)	Dilution End Point (Reciprocal)	Longevity Sap at 18-20°C (Days)	Reference
Arabis mosaic (type strain)	60	10^3-10^5	7-14	Harrison, 1958*b*
Brome mosaic	79	10^3-10^5		Bancroft, 1970
Carnation ringspot	80	10^4-10^5	50-70	Hollings and Stone, 1970
Cherry leaf roll	55	10^3-10^4	5-10	Cropley, 1961
Grapevine fanleaf	65	10^3-10^4	10-28	Cadman et al., 1960
Raspberry ringspot (type strain)	70	10^3-10^4	15-21	Harrison, 1958*b*
Raspberry ringspot (English strain)	74	10^4-10^5	28-35	Debrot, 1964
Strawberry latent ringspot	56	10^3-10^5	50	Lister, 1964
Tobacco ringspot	65	10^3-10^4	10-14	Lister et al., 1963
Tomato black ring (Scottish strain)	60	10^3-10^4	14-21	Harrison, 1954
Tomato ringspot	60	10^3-10^4	3-6	Cadman and Lister, 1961
Tomato ringspot (Grape Yellow vein)	62	10^3-10^4	6-8	Gooding and Hewitt, 1962

and subsequently the vector has been found in Europe in association with infected grapevines (Amici, 1965; Lamberti and Raski, 1964; Martelli and Raski, 1963; Vuittenez and Legin, 1964; Weischer, 1966). In Israel, both *X. index* and *X. italiae* transmit grapevine fanleaf virus (Cohn et al., 1970).

Tobacco ringspot and tomato ringspot viruses are both transmitted by *X. americanum* (Breece and Hart, 1959; Fulton, 1962). Both viruses infect a wide variety of crop and weed hosts, but despite the cosmopolitan distribution of the vector, *X. americanum*, no outbreaks of the virus have been found outside the North American continent. However, Lima (1968) regards *X. americanum* as a complex of seven species and most of the European records of this species should now be referred to one of the new species, *X. mediterraneum*.

Three other viruses in this group are transmitted by *Xiphinema* species. Strawberry latent ringspot virus infects raspberry, cherry, black currant, and several weed hosts as well as strawberry (Lister, 1964; Taylor and Thomas, 1968); it frequently occurs with arabis mosaic virus and is transmitted by *X. diversicaudatum*. Cherry leaf roll, virus of cherry trees in England (Cropley, 1961) has been transmitted by *X. diversicaudatum* and *X. coxi* in eastern Germany (Fritzsche and Kegler, 1964). Brome mosaic virus is transmitted by *X. diversicaudatum* (Fritzsche, 1964). Carnation ringspot virus, which is present in many carnation cultivars (Hollings and Stone, 1965), is transmitted by both *X. diversicaudatum* and *Longidorus macrosoma* (Fritzsche and Schmelzer, 1967).

Raspberry ringspot virus strains cause different disease symptoms. In Scotland the virus causes the economically important "raspberry leaf curl" disease and in the Netherlands an antigenically similar strain is associated with spoonleaf disease of redcurrant. These virus strains are transmitted by *L. elongatus* (Taylor, 1962; van der Meer, 1965). Strains of this virus isolated from blackberry and raspberry in England, and serologically distinguishable from the Scottish strain, are transmitted by *L. macrosoma* (Harrison, 1964*b*). There are also several tomato black ring virus strains that have different *Longidorus* vectors. *L. elongatus* is the vector of the Scottish strain which causes beet ringspot disease; the strain associated with celery yellow vein and lettuce ringspot in England and potato bouquet and potato pseudo-aucuba in Germany is transmitted by *L. attenuatus* (Harrison et al., 1961).

All the Nepoviruses known presently are transmitted through the seed of at least some of their hosts (Lister and Murant, 1967; Taylor and Thomas, 1968) and some have also been shown to be transmitted through pollen (for details see Chapter 10). However, the relevance of this characteristic facility to the ecology of the viruses varies. Tomato black ring and raspberry ringspot viruses, transmitted by *Longidorus elongatus*, are commonly present in seeds of various weed species in soil from outbreak areas, but this is rarely so with other Nepoviruses transmitted by *Xiphinema* spp. Only a single infection of arabis mosaic virus was detected in a total of 444 seedlings that germinated in soil (air-dried to kill all nematodes present) from different outbreaks in England and

Germany, compared with over 10% infection of seedlings in samples of soils from outbreaks of tomato black ring and raspberry ringspot viruses in Scotland (Lister and Murant, 1967). Taylor and Thomas (1968) found arabis mosaic and strawberry latent ringspot viruses transmitted through the seeds of several weed species growing in a patch of diseased raspberries in Scotland but the rate of infection was generally less than 4%. In Indiana, U.S.A., no infections were found among 202 weed seedlings tested in soil from an outbreak of tobacco ringspot virus. It has been shown, fairly conclusively, that infected weed seeds form an important reservoir of infection for raspberry ringspot and tomato black ring viruses transmitted by *L. elongatus* and serve as a means whereby the viruses perennate in soils (Murant and Lister, 1967; Murant and Taylor, 1965; Taylor and Murant, 1968). It is therefore interesting that these viruses do not persist for more than about 8 weeks within their vector. On the other hand, some, if not all, of the viruses transmitted by *Xiphinema* (e.g. arabis mosaic, strawberry latent ringspot and grapevine fanleaf viruses) are rarely or never found in weed seeds from infested soils, but persist in their vectors for 8 months or more (Taylor, 1968*a*; Taylor and Raski, 1964).

B. Netuviruses (Tobraviruses)

Tobacco rattle virus is the cause of a disease of tobacco first described and named "mauche" by Behrens in 1899, but it is nowadays better known as the cause of economically important diseases in bulbous crops in the Netherlands and internal necrosis of potato tubers in Britain, continental Europe, and the United States. The virus is widespread among a great variety of crop and weed hosts in Europe, and in North and South America. Preparations of the virus contain straight tubular particles of two main lengths, about 75 nm and 185 nm, but only the long particles are infective (Harrison and Nixon, 1959; Harrison and Woods, 1966). On their own, the shorter particles are not infective but they nevertheless appear to be essential for the production of stable virus (Frost et al., 1967; Lister, 1966). Both experimentally and in the field, plants showing severe and persistent symptoms may fail to yield virus antigen and few, or no, characteristic virus particles. Such infections are believed to be associated with virus nucleic acid unassociated with virus protein (Cadman, 1962; Sänger and Brandenburg, 1961).

Pea early browning virus occurs naturally in pea and lucerne crops in the Netherlands and in Britain, but it has not been reported from the United States (Bos and van der Want, 1962; Gibbs and Harrison, 1964*b*). This virus also has straight tubular particles of two main lengths, about 210 nm and 105 nm, and, like tobacco rattle virus, it does not readily infect plants systemically.

In Europe, several species of *Trichodorus* have been implicated as vectors of pea early browning and tobacco rattle viruses (Table 8-1). In the United States, tobacco rattle virus has been transmitted by *T. allius, T. christiei,* and *T. porosus* (Ayala and Allen, 1966, 1968; Jensen and Allen, 1964; Walkinshaw et al., 1961).

III. THE NEMATODE VECTORS

A. Collection, Preservation, and Identification

The known nematode vectors of plant viruses are ectoparasitic and free-living in the soil. They may be collected by taking soil samples from an infested area using a spade, trowel, screw-auger, cheese-auger, or some form of soil-corer. *Trichodorus* species are particularly prone to injury and should be collected with as little disturbance of the soil as possible in large soil blocks or cores. Polyethylene bags are ideal for transporting and storing soil samples.

The largest numbers of nematodes are likely to be found in the region where the young roots of the host plant are growing and this should be remembered when taking soil samples, particularly those for quantitative comparisons. Most of the preferred hosts of *Longidorus elongatus* are surface-rooting and consequently the bulk of the populations are found in the top 20 cm of soil (Taylor, 1967). *Xiphinema diversicaudatum,* on the other hand, feeds on plants ranging from surface-rooting herbaceous crops and weeds to deeper-rooting woody crops such as raspberry, and populations may be distributed to a depth of 60 cm or more (Taylor and Thomas, 1968). *X. index* is widely distributed on grapevines in Europe and the United States; in the stony, shallow soils of the Cote d'Or the largest populations were found at 30-60 cm (Vuittenez, 1961) but in the more fertile and deeper soils of the California Napa Valley populations were found to a depth of 3.6 m (Raski et al., 1965).

Both the horizontal and vertical distribution of *Trichodorus* species in the soil may be more variable than that of *Xiphinema* and *Longidorus,* because many of them are capable of increasing rapidly if suitable host plants are present and this may give rise to relatively rapid fluctuations in populations. In 13 out of 16 samples taken from different places in the Netherlands, *T. teres* was found below 20 cm in the spring, but in the summer and autumn large populations were found in the upper layers of the soil (Hijink and Kuiper, 1966). *T. pachydermus* were found mainly between 20 cm and 40 cm deep in one site (Taconis and Kuiper, 1963) but elsewhere they were distributed to a depth of 50 cm (Sol, 1963). *Trichodorus* species feeding on woody hosts may be very localized in their distribution: Pitcher (1967) found that *T. viruliferus* aggregated around the tips of the extending roots of apple and relatively few were found in soil samples taken outside this immediate zone.

Many methods have been described for separating nematodes from soil and the relative efficiency of some of them have been assessed (Townshend, 1967; Whitehead and Hemming, 1965). For *Longidorus* and *Xiphinema* species there seems little doubt that a modification of Cobb's (1918) decanting and sieving technique is the most satisfactory in relation to time expended, efficiency of extraction, and simplicity of equipment. The technique consists essentially of a period of elutriation or flotation, when nematodes and particles of similar size and density are separated from the soil mass in water and removed from the bulk of the fluid by sieving. The sievings with nematodes may then be examined

Trichodorus species can be satisfactorily extracted from soil samples by sieving and Baermann funnel extraction or by using apparatus developed by Oostenbrink (1960) or Seinhorst (1962).

In the Oostenbrink apparatus (Fig. 8-2) the moist soil sample is washed through a 1 mm aperture sieve with a jet of water (700 ml per minute) into the conical flotation funnel. At the same time an upward stream of water enters the bottom of the apparatus and this keeps eelworms and fine particles in suspension but allows heavier particles to settle to the bottom. The samples take 10-15 min to wash into the apparatus and the nematodes are then drawn

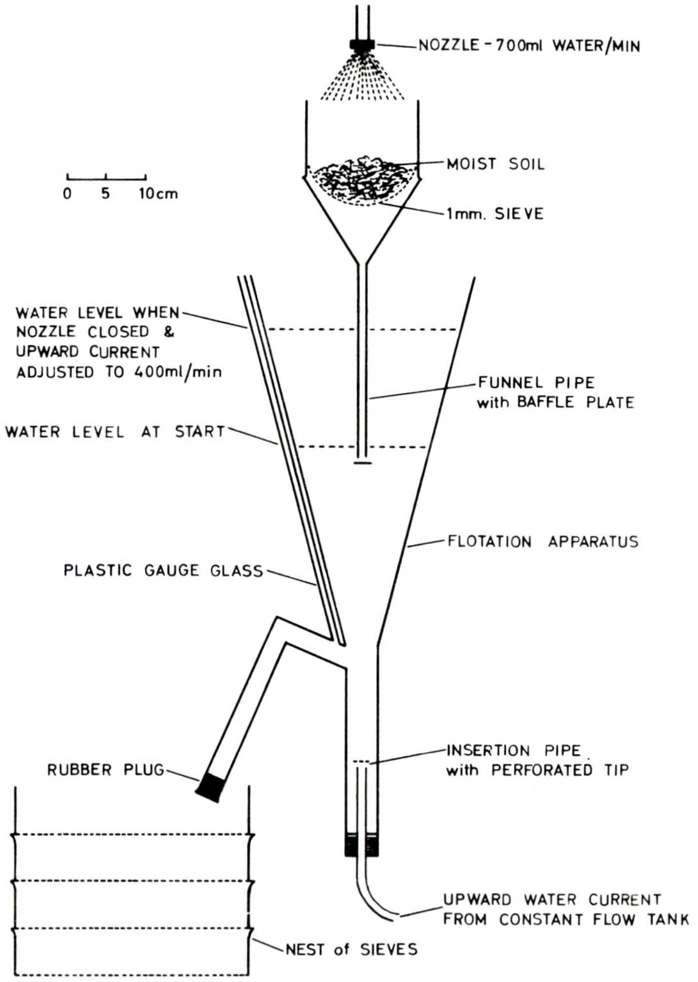

Fig. 8-2. Oostenbrink's (1960) elutriation apparatus for the extraction of nematodes from soil samples (for details and procedure for use see text).

off into a nest of sieves (50 μ) by pulling out the rubber plug of the outlet. The nematodes, plus debris, are washed from the sieves and the nematodes are finally separated in a Baermann funnel.

R. D. Winslow (private communication) has developed a simplified form of Oostenbrink apparatus (Fig. 8-3). A constant flow of water (about 1200 ml per minute for *Xiphinema* species) passes through a specially constructed inlet pipe at the base of a cone 23 in. high, 4¾ in. diameter at the lip, and 1 in. diameter at the inlet end. This extraction process consists of mixing the soil sample (150-400 ml) in 500 ml of water and washing the suspension quickly into the cone. The overflow from the cone is collected for 10 min on the sieves. The debris from the sieves is then washed and nematodes are picked directly from the sievings or separated from debris by a petri dish modification of the Baermann funnel extraction.

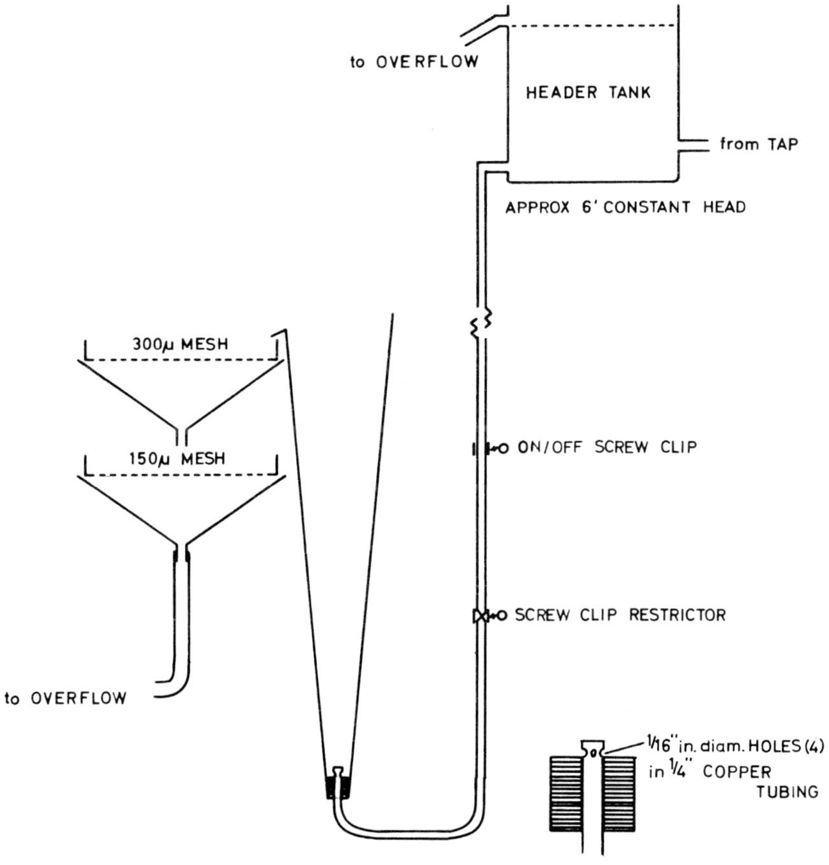

Fig. 8-3. A simplified form of Oostenbrink's elutriator for the extraction of nematodes from soil, developed by R. D. Winslow (for details see text).

The Seinhorst (1962) elutriation apparatus is designed to allow the larger soil particles to settle to the bottom of a column of water while nematodes and small particles remain in suspension. The upward flow of water and the time allowed for elutriation can be varied according to the specific nematode to be extracted. The nematodes are finally extracted by washing the suspension from the column onto a bank of sieves and separating nematodes from debris in a Baermann funnel.

Large samples of nematodes can be killed and fixed in bulk. They should be allowed to settle to the bottom of the tube and the supernatant water then poured off to leave the suspension of nematodes in 5 ml or less. After shaking the tube to distribute the nematodes freely in the suspension the tube is plunged into a water bath at 65° C for about 2 min until the nematodes are dead. An equal volume of double-strength fixative is then added. Southey (1970) lists the formulas for several different fixatives including formalin:acetic acid, 4:1, (%/%) which is widely used:

Formalin (40% formaldehyde)	10 ml
Glacial acetic acid	1 ml
Distilled water	to make up to 100 ml

To make permanent mounts the nematodes must be thoroughly impregnated with glycerol. Seinhorst (1959) has evolved a rapid method of processing in which the nematodes are transferred from the fixative to a small glass dish containing about 0.5 ml of 20% ethanol, with about 1% glycerol to prevent too rapid evaporation of the ethanol. The container is placed in a closed glass vessel or desiccator containing in excess of 96% ethanol and kept at 35-40° C for at least 12 hr. This removes most of the water and leaves the nematodes in a mixture of glycerol and ethanol. Then the container is filled with a solution of 5 parts glycerol and 95 parts of 96% ethanol and placed in a closed petri dish at 40° C until all the ethanol has evaporated (about 3 hr). The container can then be kept in a desiccator with dry silica gel or calcium chloride until the nematodes are required for mounting.

The nematodes are mounted in pure glycerol on clean glass or aluminium slides (see Southey, 1970, for information on the latter). Up to ten nematodes may be transferred to each slide by means of a bamboo splinter, fine steel pin, or hair stuck on a needle, and suitably arranged on the slide. About three pieces of glass-fibre equal in diameter to the nematodes should be placed at the periphery of the drop of mounting fluid on the slide to support the weight of the cover slip. The cover slip is sealed with a suitable compound such as "Zut" or "Glyceel" or substitutes for these such as a quick-drying enamel paint or fingernail polish. Each slide should be labeled as soon as it is made with data on name of species, locality, host, and date of collection.

B. Characteristics of Virus Vectors

Morphologically the genera *Xiphinema* and *Longidorus* appear closely related and are commonly placed within the subfamily Longidorinae. *Xiphinema* and

Longidorus species range in size from about 1.4 mm to 10 mm long in the adult stage, and possess a long, hollow stylet or odontostyle with which they penetrate plant roots and feed on cellular contents. The long, hollow stylet is not unique within these two genera however. In the adult the odontostyle is about 90 μ long, and although it may not be fully exerted in feeding, there is evidence that *Xiphinema* species are capable of reaching the vascular tissues of the fine roots of host plants (Pitcher and Posnette, 1963; Radewald, 1962). Some nematodes which are virus vectors are also plant pathogens. Fisher and Raski (1967) observed the feeding behavior of *X. index* and *X. diversicaudatum* on seedling roots grown in thin agar layers in petri dishes. Juvenile and adult stages fed at the root tips. Galls resulted from the feeding and appeared within 24 hr.

In field soils *X. americanum* infestations were associated with necrosis and destruction of feeder roots of laurel oak (*Quercus laurifolia*) (Christie, 1952), stunting in nursery stocks of ornamental spruce (*Picea pungens, P. glauca densata*) (Griffin and Epstein, 1964), black root rot of strawberry (*Fragaria* x *ananassa*) (Perry, 1958), unthriftiness of red clover (Norton, 1967), and severe decline of periwinkle (*Vinca minor*) (Epstein and Barker, 1966). In the eastern United States, Schindler (1957) and Schindler and Braun (1957) reported that *X. diversicaudatum* caused severe galling on the roots of fig (*Ficus carica*), rose (*Rosa* sp.), strawberry, and groundnut (*Arachis hypogaea*), and in Great Britain Harrison and Winslow (1961) found galls on the roots of strawberry, raspberry (*Rubus idaeus*), celery (*Apium dulce*), and chickweed (*Stellaria media*). *Longidorus africanus* causes swollen tips in roots of lettuce (Radewald et al., 1969). Feeding by *L. elongatus* appears to cause less severe galling than that caused by *Xiphinema* (Thomas, 1969, 1970) but in sandy soils the growth of strawberry may be affected by moderate infestations (Seinhorst, 1966; Sharma, 1965), and in Oregon, *L. elongatus* was associated with severe necrosis or "die-back" of the roots of peppermint (*Mentha piperita*) (Jensen, 1961). In Great Britain *L. attenuatus,* with *Trichodorus* spp., was associated with root damage to sugar beet (*Beta vulgaris*) in soils where "Docking Disorder" occurs (Whitehead, 1965) and in Israel, *L. vineacola,* which is not known to be a virus vector, was associated with stunting of onions (Cohn and Krikun, 1966).

Trichodorus species are relatively short, thick nematodes with the odontostyle in the form of a nonaxial, dorsal tooth from which the functional lumen seen in *Xiphinema* and *Longidorus* is absent. Feeding is probably by repeated, direct thrusts of the odontostyle into the cells with the head appressed to the cell wall; only the root tips are attacked. *T. christiei* feeding causes "stubby root" which results from mild cellular hypertrophy associated with the suppression of mitosis in the root meristem, and this condition is probably more or less the typical reaction to feeding by other *Trichodorus* species. Usually parasitism by *Trichodorus* is not associated with necrosis, but *T. viruliferus* feeding on apple roots caused necrosis, halted growth, and caused the tips to swell, and did not induce proliferation of secondary roots (Pitcher and Flegg, 1965).

C. Distribution of Virus Vectors

The distribution and spread of the viruses ultimately depends on the presence of the specific nematode vectors. Except for *X. index,* which appears to thrive only on grapevine and fig, all other vector species have wide host ranges and host range is thus unlikely to be a factor limiting their distribution, although cultivations associated with annual cropping may adversely affect species having a low rate of reproduction. Hence *X. diversicaudatum* and *X. americanum* are usually associated with woody perennial plants and clover/ grass leys (Griffin and Darling, 1964; Morgan, 1964; Pitcher and Jha, 1961) in conditions where populations may remain undisturbed for long periods; they tend not to be associated with herbaceous crops, although they are able to feed on them (McGuire, 1964; Taylor and Thomas, 1968). Soil type, possibly by its effect on water capacity, limits the distribution of many vector species. *X. diversicaudatum* is often found in heavy clay soils or more particularly in moist, shady sites in the vicinity of water or in medium soils with a high soil moisture level (Fritzsche, 1966). *L. elongatus* is found in light to medium soils, *L. macrosoma* in clays and heavy loams, and *L. attenuatus* in light, sandy soils (Harrison, 1964*a*). *Trichodorus* species are usually found in lighter soils; *T. teres* is found in marine, sandy soils whereas *T. pachydermus* is generally widespread but usually associated with lighter rather than heavier soils (van Hoof, 1962).

D. Spread of Plant Virus Disease

The distribution and use of infected planting material is probably the most important way in which nematode-borne viruses are disseminated. Grapevine fanleaf virus has undoubtedly been distributed from Europe to new areas such as America and Australia in this way; in California the virus is widespread in grapevine plantings where *Vitis rupestris* St. George rootstocks were used in earlier years, indicating that the nematode and virus were carried to these fields as a contaminant of planting material (Raski and Hewitt, 1963). Ornamental bulbs are an effective way of disseminating tobacco rattle virus and doubtless also tomato black ring, raspberry ringspot, and arabis mosaic viruses since, for example, many commercial Narcissus stocks appear to be infected with these viruses (Brunt, 1966). Other kinds of ornamental plants are probably equally efficient.

The dissemination of virus-containing seeds, which germinate more or less normally to produce infected seedlings, would enable nematode viruses to be spread effectively and to become established in areas where their vectors are present. Whether or not this happens in nature is a matter of conjecture, because specific evidence, such as finding virus-infected seeds or seedlings occurring in soil without the vectors, both near and remote from infected sites, is difficult to obtain and no detailed investigations have been made (Murant and Lister, 1967). However, there is little doubt that infected weed seeds provide an effective reservoir of raspberry ringspot and tomato black ring infection in a disease outbreak and are the most important, if not the only, overwintering sources of infection (Murant and Lister, 1967; Murant and Taylor, 1965; Taylor and Murant, 1968).

Infected weed seeds appear to play no comparable role in the survival and dissemination of *Xiphinema*-transmitted viruses. Nevertheless, in common with all the Nepoviruses, infection persists in the soil for long periods. During limited periods without infector plants the viruses may persist in the long-lived adults but over long periods infection may be maintained in buried pieces of roots of woody plants, as happens with grapevine fanleaf virus (Raski et al., 1965).

IV. PROOF OF VIRUS TRANSMISSION BY NEMATODES

A. Association of Nematodes with Virus Disease

The appearance of a patch or patches of diseased plants in a crop is often the first indication of the existence of nematode-borne virus infection. In perennial crops these outward symptoms usually appear two to three years after planting, rarely in the first year of growth, and the spread of infection from the periphery of the patches rarely exceeds a few feet each year. A further characteristic is the persistence of infection in the soil for long periods, even during a fallow or when crops that resist infection are grown.

The problem of associating a vector organism with the causative viruses has been approached in different ways but several procedures and techniques are now routinely and widely used in investigations of soil-borne virus outbreaks.

B. Soil Treatments

Virus infestation in the soil may be abolished by air-drying the soil (Cadman and Harrison, 1960), by grinding the soil between blocks of wood or concrete tiles to create friction between soil particles (Sol et al., 1960), or by treating the soil with chemicals that do not inactivate the virus in vitro (Cadman and Harrison, 1960; Harrison and Cadman, 1959). Loss of infectivity indicates an association with a biological agent, but none of the treatments positively identifies the nature of the agent involved. Many chemicals used as nematicides have a wide biocidal spectrum, and erroneous conclusions may be drawn if this is ignored. Insecticides and fungicides are also not specific in their action when applied to the soil; for example, infectivity of tomato black ring virus in soil was abolished with the fungicide quintozene (pentachloronitrobenzene) (Cadman and Harrison, 1960) by killing *L. elongatus*, the nematode vector of the virus (Taylor and Murant, 1966).

C. Testing Field Soil

The soil-borne nature of many viruses can be demonstrated by growing healthy plants in soil taken from around the roots of infected plants and comparing their appearance with comparable plants grown in infested soil that has been autoclaved or in soil from around healthy plants. Wagnon and Breece(1955) showed that peach seedlings became infected within a year when grown in soil taken from around the roots of trees infected with peach yellow bud mosaic virus but remained uninfected when grown in soil from adjacent healthy trees. This technique was also used to demonstrate the transmission of grapevine fanleaf virus to grapevine (Hewitt et al., 1958), tomato black ring virus to peach (Harrison, 1957), and arabis mosaic virus to strawberry (Jha and Posnette, 1959). However, it is often months, or even years, from the time of exposure to the development of disease symptoms in woody hosts and they are now generally superceded in transmission tests by suitable "bait plants" (Cadman and Harrison, 1960). "Bait plants" are seedlings, preferably of herbaceous hosts of the virus, which, when exposed to infection by planting in soils, detect the presence of virus. Usually bait plants are exposed for 3-6 weeks and are then assayed for virus in the sap by mechanical inoculation of suitable test plants. The viruses in systemically infected test plants can, if necessary, be identified by serological methods (Chapter 17). Because most nematode vectors feed on a wide range of plants the range of bait plants that can be used is equally wide; these are extensively referred to in the literature. *Xiphinema index* is an exception in that it appears to have a limited host range confined to woody plants. However, grape seedlings have proved very successful as bait plants for grapevine fanleaf virus and after about 4 weeks exposure to infection, sap from the roots can be inoculated to herbaceous assay plants to demonstrate the presence of virus (Taylor and Raski, 1964).

D. Identifying the Vector

By taking soil samples from within and without patches of diseased plants, a correlation may be revealed between a specific nematode and the outbreak of the disease. For example, the coincidence of *X. americanum* with patches of blueberry infected with tobacco ringspot virus was shown in this way (Griffin et al., 1963). If this procedure fails to reveal a likely vector species, fractionating infected soil by sieving can help to narrow the search. The usual procedure is to pass a suspension of the soil in water successively through a series of sieves of diminishing mesh size, for example, 240 μ, 150 μ, and 100 μ, and then to test the infectivity of the debris remaining on each grade of sieving by adding it to bait plants growing in sterilized soil. Infectivity may thus be associated with groups of organisms of different sizes (Cadman and Harrison, 1960; Sol et al., 1960). Nematodes in each group can be separated into species and used for transfer to healthy plants, which is the only test that can positively associate the virus with a specific nematode vector.

The techniques used for the demonstration of virus transmission by handpicked nematodes vary. In early experiments with grapevine fanleaf virus, Hewitt et al. (1958) grew healthy grapevines in the same container as diseased ones when *X. index* were added to the soil. This double plant test has now largely been superceded by the technique in which infective nematodes, washed in several changes of distilled water, are added to bait plants growing in sterilized soil and the results assessed by inoculating sap from roots or tops of the plants to suitable assay plants.

For a variety of reasons nematodes obtained directly from field soil may vary considerably in their capacity to transmit—factors such as poor growth of the infected crop plants, low concentration of virus in the roots, inactivity of the nematode vector because of low soil temperatures, and so on—but van Hoof (1966) also suggests that populations of *X. diversicaudatum* and *L. elongatus* from different localities may vary as vectors. Efficiency of transmission can often be improved by feeding the nematodes on suitable virus-infected host plants before exposing them to bait plants. Harrison et al. (1961) used cucumber seedlings and potatoes inoculated with tomato black ring virus as "infector" plants in soil containing *L. elongatus* in order to maintain the nematode populations and to ensure a high level of infectivity. Walkinshaw et al. (1961) found that tobacco rattle virus was transmitted to tobacco bait plants more often when soil containing *Trichodorus christiei* was first cropped with healthy maize than when the tobacco plants were planted directly with field soil; maize was a good host for both nematode and virus but tobacco was a poor host for the nematode.

As indicated earlier, the procedure used for extracting nematodes from soil can affect their ability to transmit. Nematodes used in transmission experiments should be freshly extracted and active. The numbers required for each test are usually suspended in a small volume of water (about 1 ml) and poured around the roots of the bait plant growing in sterilized soil. It is preferable to use small containers for the bait plants in order to confine the nematodes to the roots; many workers use ordinary 4 in. plant pots, but successful results are usually obtainable by using small quantities of soil in watch glasses (Fritzsche, 1967) or small petri dishes and transferring the plants to pots after a few days exposure to infection. With small containers some arrangements should be made to ensure that the atmosphere is kept humid to prevent drying out of the soil.

The number of individuals that transmit the virus in a population may be calculated by the formula

$$P = 1 - n\sqrt{Q}$$

where P is the probability that a single nematode transmits and Q is the ratio of the number of pots of uninfected to infected bait plants when n nematodes were added to each pot.

V. CHARACTERISTICS OF VIRUS TRANSMISSION BY NEMATODES

A. Efficiency of Transmission

Experimental evidence indicates that adult and larval stages of most of the nematode vectors can transmit virus equally efficiently. Grapevine fanleaf virus is transmitted by all stages of *Xiphinema index* (Raski and Hewitt, 1963) and tomato ringspot virus by all stages of *X. americanum* (Teliz et al., 1966). Arabis mosaic and strawberry latent ringspot viruses are transmitted by adults and larvae of *X. diversicaudatum* (Jha and Posnette, 1961; Harrison, 1967*a*) but males appear to transmit strawberry latent ringspot less frequently than females (Harrison, 1967*a*; Taylor and Thomas, 1968). Tobacco rattle virus is transmitted by females, males, and larvae of *T. pachydermus* (Gibbs and Harrison, 1964*a*, *b*; Sol, 1963) and by adults and larvae of *T. teres* (van Hoof, 1964*a*). All these results were obtained by adding varying numbers of vectors to bait plants. Experiments have also been done with single specimens: Ayala and Allen (1968) showed that a single specimen of any stage of *T. allius*, except the first stage larva which was not tested, could transmit rattle virus; Teliz (1967) transmitted tobacco ringspot virus by single *X. americanum*; and Harrison (1967*a*) simultaneously transmitted strawberry latent ringspot and arabis mosaic viruses with single *X. diversicaudatum*.

B. Acquisition and Inoculation

Techniques for determining the actual feeding time of a nematode on a host plant have yet to be satisfactorily developed for experiments on virus transmission. The minimum periods variously recorded for nematodes to acquire viruses from plants and then to transmit them to other plants are essentially access periods and not feeding periods. The term *Acquisition Access Threshold* has been proposed to express the minimum time that a nematode must have access to a virus source in order to become viruliferous and *Inoculation Access Threshold* for the minimum time that a viruliferous nematode must have access to a healthy plant for transmission to occur (Teliz et al., 1966). Neither definition implies that the nematode feeds continuously during this time.

In some early experiments with *X. diversicaudatum* and arabis mosaic virus (Jha and Posnette, 1961) and *X. index* and grapevine fanleaf virus (Hewitt and Raski, 1962), transmission was obtained after an acquisition access period of one day, the minimum time tried. More recently *X. americanum* has been shown to acquire tomato ringspot virus within an hour (Teliz et al., 1966), but with access periods up to and including 48 hr the rate of transmission was low and variable and was not proportional to the length of the access period; in excess of 48 hr high rates of transmission were obtained. Using single *X. americanum*, McGuire (1964) obtained a few transmissions of tobacco ringspot after 24 hr access periods and found that the rate of transmission increased proportionately with time up to 10 days.

The inoculation access threshold of all nematode vectors is probably less than 24 hr. Teliz et al. (1966) obtained a few transmissions of tomato ringspot virus to cucumber seedlings after exposure to *X. americanum* for 1 hr, and 100% transmission after 4 days exposure. Similarly the rate of transmission of tobacco rattle virus increased as the access time was extended from 1 hr up to 48 hr (Ayala and Allen, 1966, 1968). The results of such experiments are all affected by factors such as the numbers of nematodes used per test, the physiological state and kind of infector and bait plants used, temperature, and soil type. Moreover, nematodes feed intermittently, possibly at intervals of several days. This is apparent in the experiments of van Hoof (1964*b*) in which single *T. pachydermus* infected with tobacco rattle virus were serially transferred each day to fresh *Nicotiana rustica* bait seedlings; transmissions with each individual occurred intermittently up to 5 days. Similarly Harrison (1967*a*) transferred single female *X. diversicaudatum* infected with strawberry latent ringspot at 2 day intervals to fresh cucumber seedlings and found that although an individual nematode might transmit to several bait plants the transmissions were rarely in serial succession.

C. Persistence

Nematode vectors kept in fallow soil lose their infectivity slowly, the period of retention much exceeding the in vitro longevity of the virus in plant sap (Taylor, 1968*a*). In *Chenopodium* sap at 18-20° C grapevine fanleaf virus remains infective for 10-28 days (Cadman et al., 1960) but *X. index* in moist, fallow soil retained the virus for up to 8 months. Taylor and Thomas (1968) showed that *X. diversicaudatum* retained strawberry latent ringspot virus for 84 days and arabis mosaic virus for 112 days in the absence of host plants, and *X. americanum* was shown to transmit tobacco ringspot virus after a starvation period of 49 weeks at 10° C (Bergeson and Athow, 1963). *Longidorus* vectors may not retain infectivity for as long as *Xiphinema*. Raspberry ringspot and tomato black ring viruses persist in *L. elongatus* for a maximum of about 8 weeks (Lister and Murant, 1967; Taylor, 1968*a*), and Debrot (1964) found that *L. macrosoma* retained the English strain of raspberry ringspot virus for 34 days but did not experiment further. Less is known about *Trichodorus* vectors but tobacco rattle can persist for at least 36 days in *T. pachydermus* (Sol, 1963), and van Hoof (1964*b*) reported transmission of the virus with *T. pachydermus* in soil stored for 10 months at temperatures fluctuating between 3° C and 20° C. The Californian strain of tobacco rattle virus persisted in *T. allius* for 20 weeks when maintained without a host, but for 27 weeks when feeding on virus-immune sweet pea (Ayala and Allen, 1968).

The procedure for ascertaining the retention of virus in starving nematodes is simply to obtain the nematodes from a highly infective population and add batches of these to pots of sterile soil. At intervals bait seedlings are planted and after 4-5 weeks growth these are assayed for virus infection in the usual way by inoculating sap from roots or hypocotyls to suitable test plants. Alternatively, or as a complementary procedure, samples of naturally infected field soil can be put in pots or small wooden boxes and planted at intervals with bait plants. In this case, however, care must be taken to remove all weeds as soon as possible after they germinate as they may provide sources of infection with virus that has been transmitted through the seed (Lister and Murant, 1967; Murant and Lister, 1967; Taylor and Thomas, 1968) (refer to Chapter 10).

Most of the data on virus retention in nematodes are from experiments using fallow soil. Persistence may differ considerably when the vector is actively feeding. This can be examined by allowing the nematodes to feed on plants that are immune to the virus being studied. For example, fig (*Ficus carica*) is a good host for *X. index* but is immune to grapevine fanleaf virus (Taylor and Raski, 1964), and rye grass (*Lolium perenne*) is a host for *L. elongatus* but is immune to raspberry ringspot virus (Taylor, 1967). Despite the long retention of viruses in their nematode vectors, it cannot be assumed that the biological association is similar to that between "persistent" viruses and aphids. In fact, unlike aphids, the little evidence available indicates that nematodes do not retain the virus through the molt. Using newly molted *X. diversicaudatum*, Harrison and Winslow (1961) failed to transmit arabis mosaic virus, but too few nematodes were used to exclude the possibility that infectivity is occasionally retained. Taylor and Raski (1964) allowed fourth instar *X. index*, obtained from populations raised on fanleaf-infected grapevines, to molt to adults on virus-immune fig; these adults failed to transmit the virus when transferred to grapevine bait seedlings. Experiments which demonstrate that viruses are not retained through the egg stage have been made by maintaining virus-carrying gravid females on virus-immune host plants and using the resulting larvae in transmission tests; grapevine fanleaf virus in *X. index*, arabis mosaic virus in *X. diversicaudatum*, and raspberry ringspot and tomato black ring viruses in *L. elongatus* were not transmitted in such tests (Taylor and Raski, 1964; Taylor, unpublished). Ayala and Allen (1966) also failed to obtain any evidence of transmission of tobacco rattle virus through the molt, or through the egg, of *T. allius*.

D. Specificity of Transmission

Experimental evidence suggests that among Nepoviruses, serologically distinct strains of a particular virus require different, although closely related, species of the same nematode genus for their transmission. Thus although grapevine fanleaf and arabis mosaic viruses are serologically related, they have different species of *Xiphinema* as vectors (Dias and Harrison, 1963). Scottish strains of raspberry ringspot and tomato black ring viruses are transmitted by *Longidorus elongatus*, but the English strains, which share only about half of their antigenic groups with the Scottish strains, have respectively *L. macrosoma* and *L. attenuatus* as vectors (Table 8-1).

On the other hand, variants of viruses that show slight serological differences in gel-diffusion tests but are distinguishable in host range do not necessarily have different vectors. *X. americanum* can transmit tomato ringspot virus and the serologically distinguishable grape yellow vein strain (Teliz et al., 1966) and single specimens of *X. americanum* have been shown to transmit two serologically distinct strains of tobacco ringspot virus (Sauer, 1966). Rush

(1970) transmitted five U.S. isolates of tobacco ringspot by *X. americanum* but failed to transmit the *Eucharis* mottle strain from Peru. Murant et al. (1968) found a variant of raspberry ringspot virus infecting Lloyd George raspberry, which is immune to the normal Scottish strain of the virus, and showed that it was readily transmitted by *L. elongatus*. It was suggested that several strains of a virus may co-exist in soils but that one may become dominant through the interplay of various ecological factors. Contrary to the findings of Harrison (1946*b*) and Debrot (1964), which supported the idea of specificity of transmission, Taylor and Murant (1969) showed that under experimental conditions *L. elongatus* can transmit Scottish and English strains of both raspberry ringspot and tomato black ring viruses, although not all were transmitted equally efficiently. The question of specificity of transmission thus remains an open one, since it is apparent in nature but not under experimental conditions.

Little is known about the specific transmission of Netuviruses.* American and European cultures of tobacco rattle virus are serologically distinguishable (Cadman, 1963; Oswald and Bowman, 1958) and they are transmitted by different *Trichodorus* species. However, each isolate of the virus can be transmitted by more than one vector species, and conversely, a single *Trichodorus* species can transmit more than one isolate (Table 8-1; Ayala and Allen, 1968; van Hoof et al., 1966). Isolates of pea early browning virus are also serologically distinguishable and are transmitted by several species of *Trichodorus* (Gibbs and Harrison, 1964*a, b;* van Hoof, 1962). Thus there is little evidence of specificity of transmission among Netuviruses, but definite conclusions cannot be drawn until further experiments with specific isolates are undertaken.

The experimental procedure of Taylor and Murant (1969) can be adapted to suit most experiments on specificity of transmission. They used field soil containing a virus-free population of *L. elongatus*, which they diluted with a sufficient volume of sterilized soil to provide about 40 nematodes per 400 g. The mixed soil was put into wooden boxes, approximately 7 × 7 × 2 in. and in each were planted eight *Chenopodium quinoa* seedlings which were inoculated, when they had reached the six-leaf stage, with different isolates of raspberry ringspot or tomato black ring viruses. Each of the isolates was cultured in *Nicotiana rustica* prior to use. Controls were established by inoculating some of the boxes of plants with water and by inoculating plants growing in nematode-free soil with the type isolates of the viruses. The infector plants were removed about 4 weeks later, by which time most were dead or severely diseased (except the controls) and the boxes were replanted with *C. quinoa* bait seedlings. The bait plants were tested for virus infection about 5 weeks later by the usual procedure of mechanical inoculation using extracts from hypocotyls of the *C. quinoa* test plants. The identity of the viruses in systemically infected test plants was confirmed in agar-gel serological tests.

*See footnote in Table 8-1.

There may be objections to the use of field soil, because of the presence of other organisms, and it may be necessary to handpick nematodes into pots of sterilized soil into which infector plants are then planted. However, if field soil is used, the vector nematodes can be extracted from the different treatments after the removal of the bait plants and if necessary used in further transmission tests to confirm any results obtained.

E. Mechanism of Transmission

The nature of the association between nematode vectors and the viruses they transmit is virtually unknown, although tentative suggestions have been made that some viruses have a close biological association with their vectors, whereas others are transmitted by mechanical inoculation (Fritzsche, 1970; Taylor, 1968*b*, 1970; Taylor and Raski, 1964). Observations on the persistence of virus in the vector, absence of transmission through the nematode molt, and specificity of transmission are relative to a consideration of the mechanism of transmission, together with the following experimental approaches to the problem.

Some viruses may be detected in their vectors by cutting or grinding infective nematodes in a drop of water or phosphate buffer and using this suspension to inoculate young assay plants. Tobacco rattle virus was detected in *Trichodorus* in this way using *Chenopodium amaranticolor* assay plants (Sänger et al., 1962). Raski and Hewitt (1963) similarly detected grapevine fanleaf virus in *Xiphinema index,* but transmissions were obtained only within the first few days of the nematodes feeding on young, infected roots and not at a later time, although the nematodes remain infective for long periods (Taylor and Raski, 1964). Raspberry ringspot and tomato black ring viruses are readily detected in *Longidorus elongatus* (Taylor 1964), but Debrot (1964) had little success using this technique with *L. macrosoma* and the English strain of raspberry ringspot virus. The virus detected in this way is probably that in the intestine and is unlikely to play any part in the transmission process. If this is so, then it would appear from the results referred to that viruses transmitted by *L. elongatus* pass into the intestine unchanged whereas those in *X. index* and *X. diversicaudatum* appear to lose their infectivity by the time they reach the intestine. In this context it is interesting that arabis mosaic and strawberry latent ringspot viruses may be detected in *L. elongatus* although they are not transmitted (Taylor, 1968*c*).

Electron microscope examination of ultrathin sections of *L. elongatus* has revealed that the stylet has a dorsal, longitudinal slit running along its length which opens to the lumen enclosed by the guiding sheath, and it has led to the suggestion that virus particles imbibed with plant sap pass through this channel into the lumen where they form a reservoir of virus infection (Taylor et al., 1970). Taylor and Robertson (1969) have presented evidence from electron microscopy of thin sections that the inner surface of the stylet guiding sheath of *L. elongatus* is probably the main site of accumulation and retention of raspberry ringspot and tomato black ring viruses. They suggest that the virus particles are selectively and specifically adsorbed onto the cuticular guiding sheath from the flow of sap passing to the intestine.

Comparing viruliferous with virus-free *X. index*, Roggen (1966) observed morphological and physiological differences in the hypodermal chords, but in what way, if any, these differences are associated with virus infection is unclear. Betto and Raski (1966) inoculated *X. index* with suspensions of grapevine fanleaf virus but those nematodes that survived the operation did not transmit virus. Earlier work on electron microscopy of *Xiphinema* species failed to provide any clues to the location of virus retained for transmission (Wright, 1965; Lopez-Abella et al., 1967; Roggen et al., 1967), but more recent studies on *X. index* and *X. diversicaudatum* (Taylor and Robertson, 1970*b*) have revealed virus particles apparently specifically adsorbed onto the cuticular lining of the stylet extension and the esophagus, and it is suggested that transmissible virus is retained at these sites and not within the tissues. This type of transmission mechanism resembles that found in *T. pachydermus* in which particles of tobacco rattle virus were found in close association with the cuticle lining the lumina of the pharynx and the esophagus (Taylor and Robertson, 1970*a*). Thus in the three virus vector genera it appears that viruses are retained in a mechanical but specific manner.

9

Transmission of Plant Viruses by Fungi

D. S. TEAKLE

Department of Microbiology
University of Queensland Medical School
Brisbane, Australia

I. INTRODUCTION

The study of fungal transmission of plant viruses is important for two main reasons; first, plant viruses appear to be biologically unique in having fungal vectors, and second, the diseases they cause are important economically. An indication of the interest in fungal transmission of viruses can be obtained from the number of reviews dealing with the subject, either in connection with the transmission of soil-borne viruses in general (Cadman, 1963; Harrison, 1964*a*; Hewitt and Grogan, 1967) or as a field in its own right (Grogan and Campbell, 1966; Teakle, 1967, 1969).

Since the first claims were made for vector-like associations between *Olpidium brassicae* and the viruses of tobacco stunt (Hidaka, 1960), tobacco necrosis (Teakle, 1960), and lettuce big vein (Campbell et al., 1961; Tomlinson et al., 1962), additional virus-vector combinations have been suggested. To date six fungi and nine viruses have been associated (Table 9-1), while additional reports can be expected.

II. THE VIRUSES

Fungus-transmitted viruses generally comprise a heterogeneous group with respect to many of the usual properties used in grouping viruses (Table 9-1). For instance, four viruses (tobacco necrosis, satellite, cucumber necrosis, and tobacco stunt viruses) have polyhedral particles, whereas three others (potato X, wheat mosaic, and potato mop top viruses) have rod-shaped particles which are either flexuous or rigid; the morphology of the other two viruses (lettuce big vein and pea false leaf roll viruses) has not been clearly established.

TABLE 9-1
SOME PROPERTIES OF VIRUSES TRANSMITTED BY FUNGI

Virus	Shape	Size (nm)	Sedimentation constant (S)	Sap Transmission	Thermal Inactivation Point	Longevity in Sap	Vector	Reference
Tobacco necrosis	Polyhedral	26-30	116-120	Easy	80-95	Weeks	*Olpidium brassicae*	Kassanis, 1964
Satellite	Polyhedral	17	50	Easy	90-95	Many years (3°)	*O. brassicae*	Kassanis, 1962
Cucumber necrosis	Polyhedral	31	133	Easy	75-80	Weeks	*O. cucurbitacearum*	Dias and Doane, 1968
Lettuce big vein				Unknown			*O. brassicae*	Tomlinson and Garrett, 1964
Tobacco stunt	Polyhedral			Difficult	75-80		*O. brassicae*	Hiruki, 1965
Potato X	Flexuous rod	515 × 13		Easy	70	Weeks	*Synchytrium endobioticum*	Smith, 1957
Wheat mosaic	Rigid rod	160 × 25		Difficult	55-60		*Polymyxa graminis*	Brakke et al., 1965
Potato mop top	Rod			Difficult			*Spongospora subterranea*	Calvert and Harrison, 1966
Pea false leaf roll				Difficult	75-80	Few days	*Pythium ultimum*	Thottappilly and Schmutterer, 1968

Infection by mechanical inoculation with these viruses is also variable. At one extreme, potato virus X is so readily transmitted by rubbing that this is believed to be its major natural method of spread, whereas at the other extreme lettuce big vein virus has not yet been transmitted by mechanical inoculation despite many attempts using a wide range of techniques.

Only one of the viruses (pea false leaf roll virus) is known to be transmitted in seed. However, three of the viruses (potato X, potato mop top, and tobacco necrosis viruses) may infect potato tubers, and presumably can be disseminated and/or transmitted in this way.

The only fungus-transmitted viruses with strikingly similar properties are tobacco necrosis virus and cucumber necrosis virus. These two viruses appear to be biologically related in producing similar symptoms in many host plants (McKeen, 1959) and in having vectors belonging to the genus *Olpidium* (Dias, private communication). Also, the physical properties of the viruses are similar, although small but consistent differences in size and sedimentation constant have been reported (Table 9-1).

There are no typical symptoms which indicate that plants are infected by fungus-transmitted viruses rather than by other viruses. A wide range of symptoms has been recorded, including necrosis of leaves (tobacco and cucumber necrosis viruses), oak leaf patterns and yellow chevrons on the leaves (potato mop top virus), mild leaf mosaic (potato virus X), rosette (wheat mosaic virus), and yellow vein banding (lettuce big vein virus). Plants which have localized root infections are usually symptomless but some legumes and other plants show root necrosis when infected by tobacco necrosis virus (Teakle, 1962*a*).

Since the viruses are typically soil-borne and usually enter plants through roots, it is perhaps not unexpected that the viruses are usually obtained earlier and more regularly from the roots than the leaves. For instance, Brakke et al. (1965) reported that incidence of wheat mosaic virus infection of wheat roots was 100% 3-5 weeks after sowing, whereas in leaves incidence was still only 25-50% after 7-11 weeks. Hence indexing roots is often a more reliable test of virus infection than indexing leaves. In some cases, for example with tobacco stunt virus (Hidaka et al., 1956), passage of virus from roots to leaves is inhibited as the plant passes the seedling stage; aboveground symptoms of tobacco stunt develop only if young seedlings with fewer than six leaves including the cotyledons are exposed to infective soil.

In addition to being fungus-transmitted, two of the viruses have other means of transmission. Pea false leaf roll virus is also seed- and aphid-transmitted (Thottappilly and Schmutterer, 1968). Potato virus X under natural conditions is transmitted through mechanical inoculation between leaves, and possibly occasionally by root infection mechanically (Roberts, 1950).

One of the viruses, satellite virus, is unique among plant viruses in its small size (Table 9-1) and in its dependence on an "activator" virus, tobacco necrosis virus, for multiplication. Since satellite virus is serologically unrelated to tobacco necrosis virus, it apparently codes for its own protein coat, and also carries sufficient information in its nucleic acid to code for an additional protein (for details see Chapter 16).

III. THE VECTORS

The fungi involved in virus transmission belong in three different classes, the Chytridiomycetes (*Olpidium brassicae, O. cucurbitacearum*, and *Synchytrium endobioticum*), the Plasmodiophoromycetes (*Polymyxa graminis* and *Spongospora subterranea*), and the Oomycetes (*Pythium ultimum*). Drawings illustrating the three most characteristic stages in the life cycles of the fungi are given in Fig. 9-1. Suggested life cycles of the vectors or of closely related species are shown by Alexopoulos (1962) or Teakle (1969), and a generalized and simplified life cycle is given in Fig. 9-2.

The vector fungi all produce motile zoospores. Differences in flagellation (Fig. 9-1) are considered to be of taxonomic importance (Alexopoulos, 1962). Usually both asexual thin-walled sporangia and sexual resting sporangia may produce zoospores. In the case of *P. ultimum*, however, the thin-walled sporangia and oospores typically germinate by germ tubes (Middleton, 1943), but zoospores may be produced by sporangia originating from oospores at 10° C (Drechsler, 1952). Usually the zoospores are potential virus vectors whether originating from thin-walled sporangia or resting sporangia.

With the exception of *Olpidium cucurbitacearum*, which has been described only recently (Barr, 1968), all of the vectors are known to have a geographic distribution extending over several continents. *O. brassicae* is of worldwide distribution in temperate regions, and since it is present in tropical Queensland (Teakle, unpublished data), it may also occur in other tropical areas. *Pythium ultimum, Synchytrium endobioticum*, and *Spongospora subterranea* are also widely distributed in temperate regions (Middleton, 1943, Commonwealth Mycological Institute Distribution Maps of Plant Diseases Nos. 1 and 34). *Polymyxa graminis* has been recorded only from temperate areas of North America, Japan, Italy, and Denmark. The geographic distribution of this fungus is possibly relatively limited. In support of this suggestion *P. graminis* has not been detected in cereal roots in temperate parts of Queensland, whereas *O. brassicae* has been found commonly following wet weather (Finlay and Teakle, unpublished data).

Because all of the viruses, except potato X and pea false leaf roll viruses, are dependent on their vectors for spread, the presence of the viruses usually indicates the presence of their vectors. For instance, wheat mosaic virus has been reported from the United States, Japan, and Italy, and in each country *Polymyxa graminis* has been found associated with it (Linford and McKinney, 1954; Saito et al., 1964; Canova, 1966). Similarly *Olpidium brassicae* has been found wherever lettuce big vein has occurred. On the other hand, detection of the vectors in the *absence* of their viruses usually depends either on the

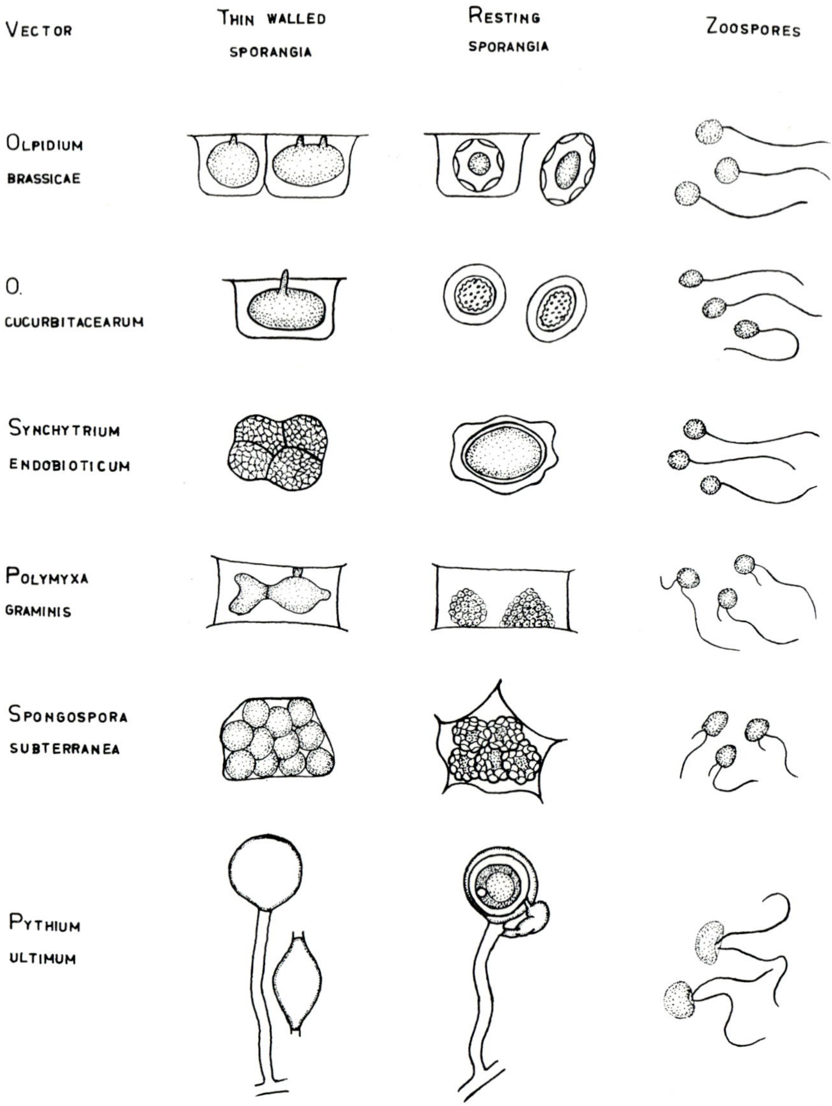

Fig. 9-1. Drawings illustrating the thin-walled sporangia, resting sporangia, and zoospores of the fungal vectors.

production of distinctive disease symptoms by the fungi alone, or on careful microscopic examination of roots or root washings. Thus, the presence of *Synchytrium endobioticum* is readily indicated by potato wart symptoms; *Spongospora subterranea* by potato powdery scab; and *Pythium ultimum* by damping-off symptoms. In contrast, the *Olpidium* spp. and *Polymyxa graminis* cause inconspicuous effects in plant roots, and are detected by their

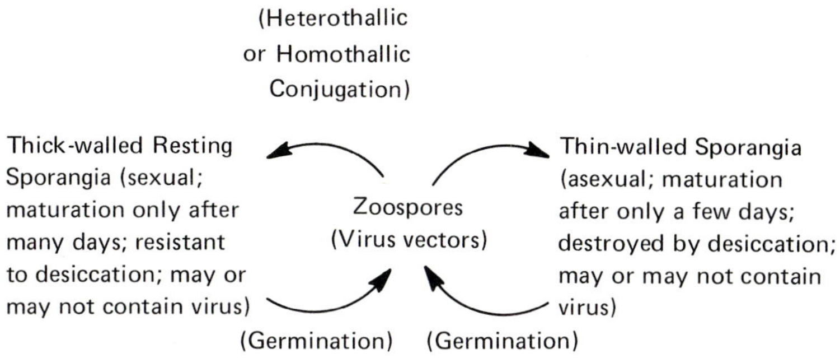

Fig. 9-2. **Generalized and simplified life cycle of the fungal vectors. The thin walled and resting sporangia of *Pythium ultimum* differ from those of the other vectors in typically germinating by germ tubes, except under special conditions (Drechsler, 1952).**

characteristic zoosporangia and resting sporangia in roots or by the typical size, motility, and flagellation of their zoospores in root washings.

A. Culture of the Fungi

With the exception of *Pythium ultimum*, all of the fungal vectors have been satisfactorily cultured only in plant roots. Therefore the experimenter must develop skill in handling the fungi as parasites of plants. Cultural conditions must be carefully controlled, since one adverse factor can prevent efficient virus transmission. *Pythium ultimum*, in contrast, is readily grown in pure culture on laboratory media (Middleton, 1943) and cultures of this fungus have been shown to transmit virus to peas (Thottappilly, 1968).

B. Host Plant

Usually the host plant from which the fungal vector was originally obtained is suitable for the multiplication of the fungus. However, the vector will often multiply satisfactorily in certain other species belonging to the same plant family as the original host. In the case of *Olpidium brassicae*, the fungus may multiply abundantly in hosts belonging to several different plant families, for example, the lettuce strain of the fungus may also multiply abundantly in beans.

Adequate growth of the host plant must be maintained to ensure adequate multiplication of the vectors. The fungi typically infect rapidly growing root or shoot tissues, and if the growth of the host plants is checked, fungal multiplication will also be checked. Thus, appropriate cultural conditions of nutrition, moisture, aeration, temperature, pH, and absence of other pathogens must be provided for the host plant.

Although cultures of the fungi are usually maintained in plants grown in soil, they have also been maintained in the absence of soil. *Olpidium brassicae* has been cultured in the roots of seedlings grown in 20% Knop's solution in the laboratory (Sahtiyanci, 1962), whereas *Synchytrium endobioticum* has been cultured in buds of potato tubers (Nienhaus and Stille, 1965). Presumably vector strains of *Spongospora subterranea* can be cultured in plants grown in nutrient solutions, since this has been done with the form of the fungus which attacks watercress (Tomlinson, 1958).

C. Moisture

The release of zoospores by zoosporangia is highly dependent on the presence of abundant moisture. Further, abundant moisture is necesssary for the zoospores to swim to and enter plant roots. Mature zoosporangia of the chytrids will often germinate soon after infected roots from soil are irrigated, for example, zoosporangia of *Olpidium brassicae* release a "flush" of zoospores 3-10 min after the roots are immersed in water. Such abundant zoospore discharge is not characteristic of plasmodiophoraceous vectors such as *Polymyxa graminis* (Rao, 1968).

Release of zoospores by resting sporangia may sometimes be stimulated by desiccation followed by irrigation. This is probably the reason why potato mop top virus may be best detected in soil infested with *Spongospora subterranea* by air-drying it before moistening and planting it with bait plants (Jones and Harrison, 1969). Efficient transmission of pea false leaf roll virus following drying of infested soil may reflect a similar stimulation in the germination of oospores of *Pythium ultimum* (Thottappilly and Schmutterer, 1968).

D. Temperature

All of the vectors are highly active at comparatively low temperatures, for example 10-20° C. At these temperatures they produce zoospores which maintain their motility and infectivity for a much longer period than at higher temperatures such as 25-30° C. Hence artificial cooling of glasshouses and laboratories is desirable for year-round experimentation in tropical or subtropical areas, or in temperate areas where summer heat prevails.

E. The Suspending Medium

Both tap and distilled waters have often been successfully used as suspending media for zoospores. However, some tap and distilled waters contain toxic materials such as chlorine, zinc, or copper. Therefore it is advisable to test these waters before using them routinely.

Buffer and other solutions may provide a more favorable medium for preserving the infectivity of zoospores. Some of those used successfully are 20% Knop's solution (Sahtiyanci, 1962); 0.01 M phosphate buffer, pH 7.0; dilute sera; or 0.01-0.1 M amino acids (Teakle and Gold, 1964); or 1:20 strength Hoagland's solution (Kassanis and Macfarlane, 1965). A pH near neutrality is usually most suitable.

Retention of motility in zoospore suspensions probably always enhances plant infection. Motile zoospores have the advantage over nonmotile spores in that their motility enables them to collect in areas favorable for infection, thus increasing their inoculum potential (Hickman and Ho, 1966). Infection obtained by inoculating plants in soil with *Olpidium* zoospore suspensions which have lost motility (Campbell and Grogan, 1964) may possibly be attributable either to the zoospores regaining motility (e.g. Teakle, 1962*b*) or to the zoospores producing germ tubes (Barr, 1968); motility appears to be essential for infectivity in aqueous suspensions.

F. Transfer of the Fungi

Since the fungi typically release zoospores when suitably irrigated, a convenient method of transferring the fungi from infected to healthy plants is to transfer root washings. The infected roots may be washed free of soil and placed into solutions in petri dishes or beakers for zoospore discharge. Alternatively, infected plants growing in soil may be watered and the leachings containing zoospores collected for use. Seedlings may be inoculated by placing them with their roots in the zoospore suspension, or the inoculum may be poured around seedlings growing in soil. In the case of *Synchytrium endobioticum*, young shoots on tubers are inoculated using drops of zoospore suspension (Nienhaus and Stille, 1965).

G. Preservation of the Fungi

Air-dried soil infested with the fungi usually remains infested for many years when stored in the laboratory. The fungi can usually be cultured by growing susceptible plants in the soil. Infected roots or tuber lesions can also be air-dried and stored in the laboratory for many years without losing viability. The roots can be homogenized in suspension medium, incorporated in soil, and recovered by bait plants.

IV. TRANSMISSION OF THE VIRUSES BY THE VECTORS

A. Tobacco Necrosis Virus by *Olpidium brassicae*

This subject has been studied independently by several groups of workers and a considerable amount of information has been accumulated. The studies have been facilitated by the fact that tobacco necrosis virus is relatively stable, easily transmitted by mechanical inoculation and readily assayed by the local lesion method, and has been intensively studied since it was first described by Smith and Bald (1935). Further, the vector is relatively easy to handle (Sahtiyanci, 1962; Garrett and Tomlinson, 1967; Teakle, 1967).

The first reports of the association of this virus and *Olpidium brassicae* indicated that the fungus was regularly present in roots shown to be naturally infected by the virus (Teakle, 1960; Teakle and Yarwood, 1962). Furthermore, when zoospore suspensions were taken from lettuce roots naturally infected with tobacco necrosis virus, or when zoospore suspensions from virus-free lettuce roots were mixed with this virus from cowpea leaves, they were more than 50 times as infective to roots as were virus suspensions alone.

Later experiments (Teakle, 1962*b*) established a correlation between the infection of roots by zoospores of *O. brassicae* and by tobacco necrosis virus. For instance, treatments which permanently immobilized and hence inactivated the delicate zoospores (e.g. aging for ½ hr or more, heating to 35° C for 10 min or addition of 0.001% $CuSO_4 \cdot 5H_2O$) prevented transmission of this virus to seedlings placed in the suspensions. Other experiments indicated that both the zoospores and the virus penetrated plant roots at approximately the same time and place, that is, usually two or more hours after encystment of zoospores on the roots and mainly in the zone of root elongation behind the root tip. Some of these correlations, and others also supporting the fungus-virus association, have been observed by Kassanis and Macfarlane (1964) and Campbell and Fry (1966).

Attempts have been made to determine the exact nature of the association of tobacco necrosis virus and *O. brassicae* with only partial success. Teakle and Gold (1963) used an antiserum produced against purified tobacco necrosis virus to study the effect of time of inactivation of the virus by antiserum on the ability of zoospores to transmit the virus. It was found that no virus was transmitted if it and diluted antiserum were mixed *before* the zoospores and the assay seedlings were added. However, some virus was transmitted if the virus and zoospores were mixed for 5 min before adding dilute antiserum and, 5 min later, assay seedlings. This protective action of zoospores against virus inactivation by dilute antiserum was interpreted as indicating acquisition and carriage of the virus either externally or internally by the motile zoospores.

Further work on the effect of antiserum on the ability of zoospores to transmit tobacco necrosis virus has been done by other workers. Kassanis and Macfarlane (1964) found that a concentrated antiserum prevented virus transmission by zoospores which were previously exposed to the virus. They postulated that the virus associated with the zoospore must have been at least partially exposed to the antiserum, and hence was unlikely to be internally borne as suggested by Teakle and Gold (1963). Campbell and Fry (1966) have confirmed Kassanis and Macfarlane, and therefore accepted the concept that the virus is surface-borne, but have emphasized the firmness of the association. They were unable to prevent tobacco necrosis virus transmission by repeatedly pelleting zoospores and resuspending them in water, although this removed all detectable virus from the supernatant.

Kassanis and Macfarlane (1964) and Grogan and Campbell (1966) have criticized the experiments of Teakle and Gold (1963) on the transmission of tobacco necrosis virus by zoospores in the presence of antiserum on the grounds that (1) the antiserum used may have been prepared against a different strain of the virus, (2) the antiserum may have been of low titer, and (3) the serotype of the virus was not reported. However, as stated in Teakle and Gold (1963), (1) the same purified preparation of virus was used both to prepare the antiserum and to carry out the fungus transmission experiments; and (2) the experiments were carried out near the "equivalence point." This term was intended to indicate that the antiserum was diluted to a point where it approximately inactivated the virus present. (3) In view of these points, the question of antiserum titer and of serotype of the virus would seem irrelevant. In fact, it was of the D-serotype (Teakle, unpublished data).

In addition to acquiring tobacco necrosis virus when the zoospores are motile, it is likely that zoospores can acquire the virus by encysting on roots which have been previously exposed to the virus. However, after encystment the acquisition of the virus from solution is inefficient (Teakle and Gold, 1963; Kassanis and Macfarlane, 1964).

The relationship between *Olpidium* and tobacco necrosis virus inside the root has been investigated. Teakle and Gold (1963) found that zoospores released directly into dilute antiserum from roots infected by both *O. brassicae* and tobacco necrosis virus transmitted only 20-50% as efficiently as zoospores released into normal serum. They suggested that some of the zoospores might be carrying tobacco necrosis virus at discharge. However, using similar methods, Campbell and Fry (1966) eliminated transmission, presumably by adding an excess of antiserum. Their conclusion was that the virus and *Olpidium* are released as independent entities. The available evidence still does not show that the virus is integrated with *O. brassicae* inside the root. In fact, the infection of roots by tobacco necrosis virus may considerably inhibit the development of *O. brassicae* in the roots (Teakle, 1964; Fry and Campbell, 1966).

The relationship between the virus and resting sporangia has been studied by Campbell and Fry (1966). They found that resting sporangia from virus-infected roots lost ability to transmit tobacco necrosis virus after treatment with hydrochloric acid, whereas similarly treated resting sporangia from lettuce big vein-infected roots retained ability to transmit lettuce big vein virus. These results were interpreted to indicate that lettuce big vein virus was internally borne in resting sporangia, whereas the tobacco necrosis virus was not.

In summary, the strongest evidence to date indicates that zoospores and tobacco necrosis virus are released independently from roots; the virus is acquired by zoospores either before or during encystment and penetration of the roots; the virus is initially superficial on the zoospore but its later location is not known; and the virus is external to the resting sporangia.

The phenomenon of vector specificity has been observed, and was first reported in detail by Teakle and Hiruki (1964). They found that a wild mustard isolate of *O. brassicae* which multiplied abundantly in cabbage and poorly in lettuce would not transmit tobacco necrosis virus to either host, whereas a lettuce isolate which multiplied abundantly in lettuce and poorly in cabbage readily transmitted the virus to both hosts. Thus, whether the host was favorable or unfavorable for multiplication did not determine vector specificity. Kassanis and Macfarlane (1965) found other fungal isolates with transmission characteristics intermediate between those reported by Teakle and Hiruki. One strain of *O. brassicae* from lettuce transmitted tobacco necrosis virus better than another strain also from lettuce. The relatively inefficient vector strain could transmit one strain of the virus to tobacco callus tissues but not to lettuce or cress seedlings; the more efficient vector strain could transmit the same strain of the virus to all three hosts.

The nature of vector specificity may be complex. Teakle and Hiruki (1964) suggested that zoospores of nonvector strains did not acquire the virus efficiently, since mechanical assay of washed, disrupted nonvector zoospores yielded fewer lesions than did assay of similarly treated vector zoospores. This suggestion has recently been confirmed by Temmink et al. (1970). Using electron microscopic methods they observed numerous particles resembling tobacco necrosis virus attached to washed vector zoospores, but few or none attached to washed nonvector zoospores. In other works Mowat (1968) showed that vector and nonvector strains differed in electrophoretic mobility, and suggested that nonvector strains might acquire but not release infective virus. Kassanis and Macfarlane (1965) concluded that vector specificity was determined, at least in part, by responses of the host cells to the invading olpidia, the response being favorable to virus infection in some cases and unfavorable in others. Infection by tobacco necrosis virus in cress roots, obtained by using a vector strain of *O. brassicae,* was inhibited by inoculating the roots later with a nonvector strain. The inhibition was destroyed by heating the doubly inoculated roots at 50° C for 10 sec, which inactivated the fungus but not the virus.

Obviously further information is required before all the factors involved in tobacco necrosis virus transmission by *Olpidium brassicae* can be adequately explained.

Tobacco necrosis virus may be transmitted by *Olpidium* by the following procedure:

(1) Prepare a virus suspension either by (a) grinding bean or cowpea leaves with coalescing local lesions in 50× the weight of a suitable suspending medium, such as 0.05 M glycine or 0.01 M phosphate buffer, pH 7.0, or (b) purifying the virus by a method such as that of Kassanis (1962) and diluting with a suitable suspending medium.

(2) Prepare a zoospore suspension by placing *Olpidium*-infected lettuce roots in glycine or phosphate buffer for 10-20 min.

(3) Gently mix the virus and zoospore suspensions. Leave for 15 min.

(4) Place young mung bean or lettuce seedlings in the suspension, or pour the suspension over healthy seedlings growing in steamed soil.

(5) Assay the lettuce seedlings by grinding the roots in 1% K_2HPO_4 and inoculating the sap extract into French bean or "Blackeye" cowpea leaves, or count the local lesions on the mung bean seedlings.

NOTE: Suspensions obtained from roots of plants which are infected by both *O. brassicae* and the virus usually contain both zoospores and virus and are infective.

B. Satellite Virus by *Olpidium brassicae*

Tobacco necrosis satellite virus multiplies only in the presence of tobacco necrosis virus (Kassanis, 1962). In naturally infected plants, satellite virus is always found with tobacco necrosis virus, suggesting that the two viruses have the same vector, *Olpidium brassicae* (Kassanis and Macfarlane, 1968). Experimental work has supported this suggestion. In 23 out of 36 tests where plant roots were exposed to both tobacco necrosis and satellite viruses in the presence of *O. brassicae* zoospores, the roots became infected with both viruses; in 9 tests tobacco necrosis virus only was transmitted, whereas in the remaining 4 tests no virus was transmitted. No information on the intimate satellite virus-zoospore relations was reported, except that it was found that different isolates of the fungus seemed to be specific vectors for different strains of satellite virus. For instance, satellite virus-2 was transmitted only by *O. brassicae*-6 and satellite virus-3 only by *O. brassicae*-3. The original strain of satellite virus (SV-1) was not transmitted by any of the four isolates of *O. brassicae* tested, and it was speculated that this strain might have lost the ability to be fungal-transmitted through continued mechanical transfer over more than 20 years.

Satellite virus may be transmitted by *Olpidium* by the following procedure:

(1) Obtain a culture of tobacco necrosis virus which contains satellite virus. (Cultures of tobacco necrosis virus which naturally contain satellite virus when originally isolated are more likely to give good transmission than artificial mixtures of tobacco necrosis and satellite viruses).

(2) Grind leaves heavily infected with the above mixture in a mortar and dilute with 50× weight of 0.05 M glycine or 0.01 M phosphate buffer, pH 7.0.

(3) Prepare a concentrated zoospore suspension of a virus-transmitting strain of *O. brassicae*. (Different strains of *O. brassicae* may transmit different strains of satellite virus.)

(4) Mix equal volumes of the virus and zoospore suspensions. Let the mixture stand for 15 min.

(5) Expose seedling roots in glass containers to the combined suspension, or pour the combined suspension over seedlings growing in soil.

(6) After several days, grind up the inoculated roots and test for satellite virus transmission serologically or by electron microscopy.

C. Cucumber Necrosis Virus by *Olpidium cucurbitacearum*

Cucumber necrosis virus resembles tobacco necrosis virus in pathogenicity to many hosts and in its natural infection of plants by way of roots (McKeen, 1959). Dias (private communication, 1968) has found that cucumber necrosis virus is transmitted by *Olpidium cucurbitacearum* in a manner resembling transmission of tobacco necrosis virus by *O. brassicae*. However, cucumber necrosis virus was not transmitted by *O. brassicae* and tobacco necrosis was not transmitted by *O. cucurbitacearum* (Dias, 1970).

Work with homologous antiserum indicated that cucumber necrosis virus and zoospores of *O. cucurbitacearum* were released independently from roots and formed a firm association only after about 10 min; the addition of antiserum *before* zoospores had been exposed to the virus for 10 min prevented transmission, whereas the addition after longer exposure did not. Dias considered that although the need for an acquisition period did not prove that the virus was contained within the zoospores, it strongly suggested that something other than surface association was involved in transmission.

The methods of transmitting this virus by *Olpidium cucurbitacearum* are identical with those for tobacco necrosis virus and *O. brassicae.*

D. Lettuce Big Vein Virus by *Olpidium brassicae*

Lettuce big vein was first described by Jagger and Chandler (1934), who believed it to be possibly caused by a soil-borne virus. They found that it affected lettuce varieties resistant to the brown blight disease, with which *Asterocystis radicis* (*O. brassicae*) was associated (Jagger, 1928, 1940). Later work showed that big vein disease was favored by low temperature and high soil moisture (Thompson and Doolittle, 1942; Pryor, 1944, 1946), but the constant association with *O. brassicae* was not recognized until the work of Fry (1958) and Grogan et al. (1958). These workers found that suspensions containing zoospores or resting sporangia taken from infected roots induced the big vein disease. Further, treatment of the zoospore suspension with heat, chemicals, or filtration to destroy or remove the zoospores destroyed the ability of the suspensions to induce big vein. They considered the possibility that the *O. brassicae* was acting as a virus vector, but rejected it in favor of the theory that the fungus was producing a toxin which was translocated from the roots to the leaves. However, the theory that a virus rather than a fungal toxin is involved in lettuce big vein has again received general support following the graft transmission of the disease (Campbell et al., 1961; Tomlinson et al., 1962).

The nature of the virus-vector relationship between lettuce big vein virus and *O. brassicae* appears to be distinctly different from that of tobacco necrosis virus and *O. brassicae.* In contrast to tobacco necrosis virus, lettuce big vein virus is not acquired from leaf extracts by zoospores (Campbell and Grogan, 1964). The virus appears to be acquired by virus-free *O. brassicae* during its development within infected roots (Tomlinson and Garrett, 1962). Tobacco necrosis virus is probably acquired only outside the root. Lettuce big vein virus appears to be internally borne in resting sporangia of the fungus, whereas tobacco necrosis virus is not, since resting sporangia treated with concentrated HCl lost tobacco necrosis virus infectivity but not lettuce big vein virus infectivity (Campbell, 1962; Campbell and Fry, 1966). Lettuce big vein virus carried inside the resting sporangium probably explains the survival of the virus for many years in air-dried soil (Pryor, 1946).

The known host range of lettuce big vein virus is confined to species of the Compositae family (Campbell, 1965). However, the present criteria for susceptibility, namely production of symptoms, or maintenance of big vein virus infectivity by *Olpidium* for considerably longer than 12 weeks (Campbell, 1965), appear to be too stringent. Probably a wider host range will be demonstrated in the future. It may be relevant that the wide host range of tobacco stunt virus (Hiruki, 1967) was shown only after sap transmission of the virus had been achieved.

The procedure for transmitting this virus is as follows:

(1) Obtain lettuce plants infected with both lettuce big vein virus and *O. brassicae* by growing seedlings in infested soil.

(2) Wash the soil from the lettuce roots and place the plants for 20 min in the zoospore suspending medium described previously.

(3) Pour the zoospore suspension over healthy young lettuce seedlings growing in steamed soil.

(4) Grow the lettuce seedlings at 15-20° C for 4-6 weeks, to allow symptoms of big vein to develop.

E. Tobacco Stunt Virus by *Olpidium brassicae*

This association was first suggested by Hidaka (1960), because conditions which favored the disease, such as use of contaminated seed-bed materials, low temperatures, and absence of fungicides (Hidaka et al., 1956), also favored root infection by the fungus. This correlation was subsequently confirmed following fractionation of infective soils in water by settling or centrifugation treatments (Hidaka, 1961). Further, transmission was achieved using zoospore suspensions or resting sporangia from diseased plants, but not if the infectivity of the zoospores was destroyed or greatly reduced by Seitz filtration, dilution by more than 10^{-3}, or heating at 45° C for 20 min (Hidaka and Tagawa, 1962).

This correlation between infection of tobacco seedlings by tobacco stunt virus and *Olpidium brassicae* was confirmed and extended by Hiruki (1965). Treatments which destroyed the infectivity of zoospores from virus-infected plants, such as heating 10 min at 55° C, freezing 24 hr at −10° C, or exposure to chelating agents, prevented virus transmission, although the same treatments did not inactivate virus used as inoculum in mechanical transfers. Further, a virus-free strain of *O. brassicae* became viruliferous when allowed to multiply in the roots of tobacco plants mechanically inoculated with tobacco stunt virus to leaves. Conversely, when zoospores of a virus-free strain of *O. brassicae* were released into the drainage water of stunt-affected seedlings, they failed to acquire infectivity. These results indicate that tobacco stunt virus is acquired by *O. brassicae* only inside roots. Thus, the vector relationship appears to resemble that of lettuce big vein virus rather than that of tobacco necrosis virus.

The effect of the nature of the test plant on the fungal transmission of tobacco stunt virus has been studied by Hiruki (1967). When 71 species in 24 plant families were tested for susceptibility to the virus by inoculating with viruliferous zoospores of *O. brassicae*, 35 became infected by the virus. Mechanical inoculations of virus into leaves, however, infected some plants which were resistant when inoculated by means of viruliferous zoospores. Other plants were susceptible when inoculated by zoospores but not mechanically. The 71 plant species could be placed in four groups based on their reactions, namely (1) susceptible to both tobacco stunt virus and *O. brassicae*, (2) susceptible to the fungus but not the virus, (3) susceptible to the virus but not the fungus, or (4) susceptible to neither. Apparently host specificity is an important factor in the fungal transmission of tobacco stunt virus as well as of tobacco necrosis virus.

The procedure for transmitting this virus is:

(1) Obtain tobacco stunt virus-infected plants and *O. brassicae* by growing seedlings of a susceptible variety of tobacco, for instance "Bright Yellow," in infested soil.

(2) Wash the soil from the roots and quickly place the washed roots for 20 min in zoospore suspending medium.

(3) Drench young susceptible tobacco seedlings growing in steamed soil with the suspension.

(4) Grow the tobacco seedlings at 15-20° C for 3-4 weeks until stunt symptoms develop.

(5) Assay the roots and shoots of the plants by grinding in chelate buffer (0.01 M phosphate buffer containing 0.001 M 1-phenylthiosemicarbazide, pH 6.8) and rubbing the sap over the leaves of *Chenopodium amaranticolor* (Hiruki, 1964).

F. Potato Virus X by *Synchytrium endobioticum*

The transmission of potato virus X by *S. endobioticum* was achieved under artificial glasshouse conditions; it has not yet been reported to occur under natural field conditions. Nienhaus and Stille (1965) inoculated young buds on tubers infected with potato virus X with zoospore suspensions of the fungus. After incubation for a month the tumorous tissue which had developed was irrigated and the zoospores released were transferred to buds on healthy potato tubers. Indexing the newly inoculated tubers after 4 weeks revealed the presence of potato virus X. Potato virus Y was not transmitted under similar conditions.

Further work, using zoospores from virus-free tubers mixed with purified preparations of potato viruses X or Y, failed to indicate acquisition of infectivity by zoospores external to the plant. These results indicate that the virus is probably carried internally in the zoospores, resembling the lettuce big vein virus relationship to *O. brassicae*.

Potato virus X can be transmitted by the following procedure:

(1) Obtain potato tubers infected with potato virus X.

(2) Inoculate young buds on the tubers with drops of suspension containing zoospores of *Synchytrium endobioticum*. Incubate at 16-18° C for 4 weeks.

(3) Cut off the tumorous tissue which develops on the shoots and place it in distilled water. After numerous zoospores have been released, inoculate young

buds on healthy potato tubers with drops of the zoospore suspension and incubate as before.

(4) After 4 weeks, cut off small pieces of the developing tumorous tissue and test for potato virus X content by sap-inoculating to tobacco plants.

G. Wheat Mosaic Virus and *Polymyxa graminis*

The first association of *P. graminis* with wheat plants naturally infected by wheat mosaic virus was reported by Linford and McKinney (1954). Although a correlation was established between the occurrence of the disease and of the fungus, Linford and McKinney concluded that there was no evidence that the fungus could function as a vector or reservoir host of the virus. In further unpublished experiments after 1954 (reported in Brakke et al., 1965), Linford and McKinney showed that the virus could move through water in a petri dish from an infected root to a healthy wheat seedling. Also, roots infected with *P. graminis* tended to be better sources of infection than other roots.

The work by Linford and McKinney has since been extended and confirmed. Only roots containing resting spores of *P. graminis* were found to serve as a source of inoculum of wheat mosaic virus. Further, usually only wheat seedlings which became infected with *P. graminis* also became infected with the virus (Estes and Brakke, 1966; Brakke and Estes, 1967). Isolates of *P. graminis* became viruliferous when cultured in mosaic-affected wheat plants, but became nonviruliferous on clover roots (Canova, 1966).

Wheat mosaic virus may be located internally in the resting spores of *P. graminis* since air-dried infested soil or roots retain virus infectivity for 3 years. However, zoospores produced by zoosporangia developing in wheat mosaic virus-infected roots also may be viruliferous (Rao, 1968).

The following transmission procedure is outlined:

(1) Obtain roots of wheat plants naturally infected with wheat mosaic virus. Wash the roots briefly under running tap water.

(2) Soak the roots in water at 25° C for 20 min.

(3) Place 50 very young "Michigan Amber" variety wheat seedlings in the root washings for one day at 10° C, and then transplant into pots. Alternatively, pour the root washings over similar wheat seedlings growing in sterilized soil.

(4) Grow the wheat seedlings at 15° C for 1-2 months.

(5) Observe the development of mosaic symptoms and examine sap from roots and leaves for the presence of the rod-shaped particles characteristic of wheat mosaic virus.

H. Potato Mop Top Virus by *Spongospora subterranea*

Potato mop top virus was described as a new soil-borne virus by Calvert and Harrison (1966). It had been previously known to be a component of the potato yellow mottling complex (Todd, 1965), and probably frequently confused with the nematode-transmitted soil-borne virus, tobacco rattle virus.

Spongospora subterranea was suggested as the vector of the mop top virus by microscope has been found useful in assaying roots and leaves for wheat mosaic virus (Brakke et al., 1965).

The relationship between the fungal vector and the virus is complex enough, and when the plant is introduced, the complexities are even greater. The host specificity of *O. brassicae* and tobacco necrosis virus (Kassanis and Macfarlane, 1965) and tobacco stunt virus (Hiruki, 1967) might occur with other virus-vector combinations.

The question of what factors control disease development in plants is the least understood. For instance, the occurrence of systemic necrosis in tulips caused by tobacco necrosis virus is sporadic, whereas this virus is often readily isolated from their roots. Possibly physiological conditions of the host are partly involved as shown by van Slogteren and Visscher (1967) and Mowat (1970).

Fungus-transmitted viruses vary in particle morphology and other properties. It is unlikely therefore that all have evolved in recent times from a single source. It seems more likely that the ability to be fungus-transmitted has been acquired by viruses which have previously been transmitted by other means. The transmission of potato virus X by *S. endobioticum* under glasshouse conditions might indicate how a natural relationship could arise by the fortuitous simultaneous infection of potatoes by a virus and a fungus. Once this relationship has been firmly established a new system could evolve independently.

Other virus-vector combinations will doubtlessly be discovered. One possibility is oat mosaic virus with which both *Polymyxa graminis* and *Olpidium brassicae* have been associated (Bruehl and Damsteegt, 1961). Another is tomato bushy stunt virus, which is also soil-borne and has had its natural transmission inhibited by the fungicide Captan (Lovisolo, 1966). Precise determination of virus-vector relationships and development of control measures provide the challenges for the future.

ACKNOWLEDGMENTS

The unpublished work reported in this chapter was supported by Grant 56 from the Australian Research Grants Committee. Mrs. Helen Beaton drew Figure 1.

10

Transmission of Viruses Through Seed and Pollen

R. J. SHEPHERD

Department of Plant Pathology
University of California
Davis, California

I. INTRODUCTION

The most fundamental property of viruses is transmissibility, which in early work was the only way of detecting virus activity. Hence, from the first, pathologists have been interested in transmission as a test for the presence of virus and to determine means of natural spread.

Plant viruses, in general, are "wound invaders" and depend on various external agents for transmission or introduction into susceptible hosts. Man, by manipulating plants (in harvesting, pruning, cultivating operations, etc.) and other animals such as nematodes, mites, and most notably insects, (in the act of feeding or ovipositing on host plants), wound plants in succession and carry virus from one individual to another. A few viruses are disseminated readily by mere contact of an infected plant with another, for example by wind agitation of closely adjacent individuals, thereby causing minor wounds as entry sites. However, several important exceptions to the wound-invasion type of transmission are recognized. Many plant viruses, for example, are transferred from one individual to another by tissue union or by way of the seed or pollen, which cannot be viewed as wound invasion. In seed and pollen transmission, man is often an important accessory agent, contributing to perpetuation and spread of the disease through carelessness or ignorance during propagation or culture of useful plant species.

II. SEED TRANSMISSION

Seed transmission of plant viruses has been long recognized. Reddick and Stewart (1919) found that common bean mosaic was seed-borne in *Phaseolus vulgaris* and that up to 50% of the progeny from seed from infected plants were diseased. Since then many plant viruses have been found to be seed-borne to varying degrees, and seed is now recognized as a primary source in epidemic development of several serious plant virus diseases.

Seed transmission of viruses has been reviewed by Fulton (1964), Baker and Smith (1966), and most thoroughly by Bennett (1969).

A. Frequency and Extent of Seed Transmission

Seed transmission was formerly regarded as rare with plant viruses, a notion of considerable heuristic value for encouraging study of the phenomenon. Although what proportion of known plant viruses are seed-borne in at least one or more of their hosts is not known, it is now apparent that dispersal in this manner occurs with a variety of viruses (Table 10-1) and plays an important role in their ecology. Usually, however, only a small portion of the progeny resulting from seed from diseased plants are infected.

Seed transmission apparently depends on the host-virus combination more than on any other factor. Success and extent of transmission are, for example, frequently determined by the particular virus strain or host variety. The amount of transmission also is affected by other factors such as the stage at which the plant becomes infected, the temperature after infection, seed maturity, and perhaps to other factors not now apparent.

Extent of seed transmission varies with the particular host-virus combination, ranging from zero to 100%. Transmission to all the progeny is rare, although several viruses have been found in a high proportion (60-90%) of the seed. This is true for the tobacco and tomato ringspot viruses in soybean (Desjardins et al., 1954; Kahn, 1956), sowbane mosaic virus in sowbane (*Chenopodium murale*) (Bennett and Costa, 1961), squash mosaic in muskmelon (Rader et al., 1947), bean common mosaic in bean (Smith and Hewitt, 1938), and a number of soil-borne viruses in various hosts (Table 10-1).

In contrast, some viruses are probably not seed-borne at all in their host plants. Complete absence of seed transmission is virtually impossible to demonstrate with certainty, however, because lack of transmission in a small sample does not preclude the possibility of a rare instance of transmission. Consequently one tends to view negative results as tentative or inconclusive. In other cases, seed transmission is very limited but is probably important in the distant dispersal of viruses. Sugarcane mosaic virus, for example, is transmitted through a very low percentage of the seed from infected corn (Shepherd and Holdeman, 1965).

B. Influence of Host and Virus Genotype on Seed Transmission

The amount of seed transmission of a particular virus varies greatly with

TABLE 10-1

SOME VIRUSES REPORTED TO BE TRANSMITTED THROUGH SEED AND POLLEN OF THEIR HOST PLANTS

Virus	Host Species	Transmission in (%) Seed	Transmission through Pollen [a]	Reference
Abutilon mosaic	*Abutilon* spp.	14-24	—	Keur, 1934
Alfalfa mosaic	alfalfa (*Medicago sativa*)	1- 4	—	Frosheiser, 1964;
Andean potato latent	*Nicotiana clevelandii*	— [a]	—	Gibbs et al., 1966
Apple chlorotic leafspot	raspberry (*Rubus* spp.)	30-40	+	Cadman, 1965; Converse, 1967
Arabis mosaic	*Capsella bursa-pastoris*	6-33	—	Lister and Murant, 1967
	Poa annua	4	—	Taylor and Thomas, 1968
	Petunia hybrida	20	—	Lister, 1960
	rhubarb (*Rheum rhaponticum*)	10-24	—	Tomlinson and Walkey, 1967*a*
	Senecio vulgaris	2	—	Taylor and Thomas, 1968
	soybean (*Glycine max*)	12.5	—	Lister, 1960
	(numerous other spp.)			(See Lister and Murant, 1967)
Avocado sunblotch	avocado (*Persea americana*)	—	—	Wallace and Drake, 1962
Barley stripe mosaic	barley (*Hordeum vulgare*)	15-100	+	Gold et al., 1954
	wheat (*Triticum aestivum*)	71	—	Hagborg, 1954
Bean, common mosaic virus	bean (*Phaseolus vulgaris*)	<1-93	+	Reddick and Stewart, 1919; Smith and Hewitt, 1938; and others
Bean yellow mosaic	broadbean (*Vicia faba*)	—	—	Quantz, 1954
	red clover (*Trifolium pratense*)	12-15	—	Hampton, 1967
	cowpea (*Vigna sinensis*)	1-37	—	Snyder, 1942
	Adzaki bean (*Phaseolus vulgaris*)	—	—	Hino, 1962
	Lupinus spp.	6.2	—	Corbett, 1958
	peas (*Pisum sativum*)	10-30	—	Inouye, 1967
Broad bean true mosaic (Echtes Ackerbohnenmosaik virus)	*Vicia faba*	1- 3	—	Quantz, 1953

TABLE 10-1 (Continued)

Virus	*Host Species*	*Transmission in (%) Seed*	*Transmission through Pollen* [a]	*Reference*
Broad bean stain	*V. faba*	1	–	Varma and Gibbs, 1967
Cherry leafroll	bean (*Phaseolus vulgaris*)	6-48	–	Lister and Murant, 1967
	rhubarb (*Rheum rhaponticum*)	72	–	Tomlinson and Walkey, 1967*a*
	Chenopodium amaranticolor	100	–	Lister and Murant, 1967
	soybean (*Glycine max*)			Lister and Murant, 1967
Cherry necrotic rusty mottle	sweet cherry (*Prunus avium*)	–	–	Nyland, 1962
	sour cherry (*P. cerasus*)	–	–	Nyland, 1962
Citrus xylopororis	*Citrus* spp.	65	–	Childs, 1956
Clover yellow mosaic	red clover (*Trifolium pratense*)	2	–	Hampton, 1963
Cowpea mosaic	*Vigna sesquipedalis*	8	–	Dale, 1949
	Vigna sinensis	10	–	Shepherd, 1964
Coffee ringspot	*Coffea excelsa*	–	–	Ryes, 1961
Cucumber mosaic	Cowpea (*Vigna sinensis*)	4-28	–	Anderson, 1957
	Echinocystis lobata	15	–	Doolittle and Gilbert, 1919; Lindberg et al., 1956
Dodder latent mosaic	*Cuscuta campestris*	4.8	–	Bennett, 1944
Dulcamara mottle	*Solanum dulcamara*	2- 3	–	Gibbs et al., 1966
Elm mosaic	elm (*Ulmus americana*)	1- 3	+	Bretz, 1950; Callahan, 1957
Grapevine fanleaf	*Chenopodium quinoa*	3		Cory and Hewitt, 1968
	Chenopodium amaranticolor	0.7	–	Dias, 1963
Grape yellow vein	*Chenopodium amaranticolor*	57	–	Cory and Hewitt, 1968
	soybean (*Glycine max*)	58	–	Cory and Hewitt, 1968
Lettuce mosaic	lettuce (*Lactuca sativa*)	1- 8	+	Newhall, 1923; Grogan et al., 1952; Ryder, 1964
Lychnis ringspot	*Lychnis divaricata*	58	–	Bennett, 1959
	Silene noctiflora	42	+	Bennett, 1959

TABLE 10-1 (Continued)

Virus	*Host Species*	*Transmission in (%) Seed*	*Transmission through Pollen* [a]	*Reference*
Pea early browning	pea (*Pisum sativum*)	37	−	Bos and van der Want, 1962
Pea false leafroll	pea (*Pisum sativum*)	40	−	Thottappilly and Schmutterer, 1968
Pea mosaic	pea (*Pisum sativum*)			Quantz, 1958
Peanut mottle	peanut (*Arachis hypogaea*)	2	−	Kuhn, 1965
Peanut stunt chlorosis	peanut (*A. hypogaea*)	0.008-0.2	−	Troutman et al., 1967
Potato spindle tuber	potato (*Solanum tuberosum*)	87-100	−	Hunter et al., 1969
Prune dwarf	cherry (*Prunus mahaleb*)	9	−	Cation, 1949
	sour cherry (*P. cerasus*)	3-30	+	Gilmer and Way, 1960
	cherry (*Prunus mahaleb*)	10	−	Cation, 1949
	sour cherry (*P. cerasus*)	30	+	Cation, 1949; Gilmer and Way, 1960
	peach (*P. persica*)	?	−	Cochran, 1950
Raspberry latent virus	raspberry (*Rubus* spp.)	10	−	Converse and Lister, 1964
(Black raspberry latent virus)	*Stellaria media*	8-26	−	Murant et al., 1968
Raspberry ringspot	*Capsella bursa-pastoris*	2-10	−	Lister and Murant, 1967
	Stellaria media	30	−	Lister and Murant, 1967
	soybean (*Glycine max*)	20	−	Lister, 1960
	strawberry (*Fragaria* spp.)	50	+	Lister and Murant, 1967
Sincomas mosaic	sincomas (*Pachyrrhizus erosus*)	40-80	−	Fajardo and Moranon, 1932
Southern bean mosaic	cowpea (*Vigna sinensis*)	3-4	−	Shepherd and Fulton, 1962
Sowbane mosaic	*Chenopodium murale*	20-70	−	Bennett and Costa, 1951
	C. album	30		Bennett and Costa, 1951
	Striplex pacifica	21		Bennett and Costa, 1951
	C. amaranticolor	14.3		Kado, 1967
Soybean mosaic	soybean (*Glycine max*)	0-68	−	Gardner and Kendrick, 1924

TABLE 10-1 (Continued)

Virus	*Host Species*	*Transmission in (%) Seed*	*Transmission through Pollen* [a]	*Reference*
Squash mosaic	muskmelon (*Cucumis melo* L.)	1-27	—	Mahoney, 1935
	muskmelon (*Cucumis melo* L.)	12-93	—	Rader et al., 1947
	squash (*Cucurbita maxima*)	0.2-1.5	—	Grogan et al., 1959
	watermelon (*Citrullus vulgaris*)	1.5	—	Nelson and Knuhtsen, 1969
Strawberry latent ringspot	*Mentha arvensis*	6	—	Taylor and Thomas, 1968
Sugarcane mosaic	corn (*Zea mays*)	0.1-0.4	—	Shepherd and Holdeman, 1965
Tobacco mosaic	tomato (*Lycopersicon esculentum*)	6[b]		Doolittle and Beecher, 1937; Taylor et al., 1961
	Malus platycarpa	38	—	Gilmer and Wilks, 1967
	apple (*Malus sylvestris*)	3-37	—	Gilmer and Wilks, 1967
	pear (*Pyrus communis*)	up to 80	—	Gilmer and Wilks, 1967
	grape (*Vitis* spp.)	20	—	Gilmer and Kelts, 1968
	pepper (*Capsicum frutescens*)	22[b]	—	McKinney, 1952
Tobacco rattle	*Capsella bursa-pastoris*	2	—	Lister and Murant, 1967
Tobacco ringspot	*Petunia hybrida*	20		Henderson, 1931
	soybean (*Glycine max*)	78-82	+	Desjardins et al., 1954; Kahn, 1956
	Taraxacum officinale	4-36		Tuite, 1960
	tobacco (*Nicotiana tabacum*)	5		Valleau, 1932
	lettuce (*Lactuca sativa*)	3	—	Grogan and Schnathorst, 1955
Tobacco streak	bean (*Phaseolus vulgaris*)	1-26	—	Thomas and Graham, 1951
Tomato bushy top[c]	*Physalis peruviana*	29		McClean, 1948
	Solanum incarnum	53		McClean, 1948

TABLE 10-1 (Continued)

Virus	*Host Species*	*Transmission in (%) Seed*	*Transmission through Pollen* [a]	*Reference*
Tomato black ring	cowpea (*Vigna sinensis*)	23	—	Lister, 1960
	Capsella bursa-pastoris	90	—	Lister, 1960
	Fusaria spp.	100	—	Lister, 1960
	Poa annua	3	—	Lister and Murant, 1967
	raspberry (*Rubus* spp.)	5-19	+	Lister and Murant, 1967
	Senecio vulgaris	3-40	—	Lister and Murant, 1967
	soybean (*Glycine max*)	91	—	Lister and Murant, 1967
Tomato ringspot	red clover (*Trifolium pratense*)	3-7	—	Hampton, 1967
	soybean (*Glycine max*)	76	—	Kahn, 1956
White clover mosaic	red clover (*Trifolium pratense*)	1-24	—	Hampton and Hansen, 1968

[a] — indicates pollen transmission was not tested or that the percentage of seed transmission was not stated; + indicates that pollen transmission has been shown to occur to the ovules of healthy blossoms and/or the stigma-bearing parent plants.

[b] The level of embryo transmission is extremely low or nil. Seed-borne virus results in surface-contaminated plants. Infection is dependent upon wounding of contaminated seedlings (Taylor et al., 1961; Broadbent, 1965).

[c] As pointed out by O'Brien and Rayner (1964), this virus may be the same as potato spindle tuber virus in the United States.

variety or species of plant and is thus a conditional phenomenon dependent upon both virus and host genotype. Kendrick and Gardner (1924), for example, in a study of seed-borne soybean mosaic, observed that transmission in soybean (*Glycine max*) ranged from zero in some varities to 68% in others. Smith and Hewitt (1938), in a similar study with 51 bean varieties (*Phaseolus vulgaris*) infected with bean common mosaic virus reported that seed transmission ranged from less than 1% to more than 75%. In their experiments the extent of seed transmission was correlated closely with varietal reaction to the virus as judged by visual symptoms. Varieties which were symptomless carriers showed little seed transmission (1.1-6.8% average). Only 2 of 7 varieties on which bean common mosaic induced no symptoms were found to transmit the virus through seed. Plants that produced mild symptoms, moderate symptoms, and severe symptoms, in contrast, respectively showed average ranges of transmission through seed of 9.4-20.8%, 20.3-23.3%, and 30.4-36.1%. A similar correlation of seed transmission with disease severity was observed by Singh et al. (1959) with barley varieties (*Hordeum vulgare*) infected with barley stripe mosaic virus. Four varieties highly tolerant to the virus showed less than 15% seed transmission, whereas severely diseased varieties produced up to 75% infected progeny. Lettuce mosaic virus, which is very commonly seed-borne in lettuce (*Lactuca sativa*), is apparently not transmitted through the seed of certain lettuce varieties (Kassanis, 1947; Couch, 1955). Seed transmission of cucumber mosaic virus in cowpea (*Vigna sinensis*) is similarly dependent on host variety (Anderson, 1957).

The success and amount of seed transmission depends also on the virus strain. Anderson (1957) tested several varieties of cowpea for the amount of seed transmission of two strains of bean yellow mosaic virus. One strain was transmitted to a very limited extent (0-1%) in different varieties, whereas the other strain was transmitted in amounts up to 55%. Strains of barley stripe mosaic virus vary widely in the rate of seed transmission in particular barley cultivars and some strains seemingly are not seed-borne at all (McKinney and Greeley, 1965). Nelson and Knuhtsen (1969) tested two strains of squash mosaic virus in various cucurbit species and found that one strain infecting watermelon was seed-borne to a considerably greater degree than the other. They offered this as the reason for the greater prevalence of the watermelon strain in cucurbit crops grown in the Arizona desert. Strains of southern bean mosaic virus that are closely related serologically may or may not be seed-borne. The bean-infecting strains are not transmitted through mature seed of bean (Cheo, 1955), whereas a strain infecting cowpea is transmitted through seed of that host (Shepherd and Fulton, 1962). As might be expected, virus strains which induce only local infection are not seed-borne. This is true for the tobacco mosaic virus in certain pepper varieties (McKinney, 1952; Crowley, 1957*b*), some strains of which induce local but no systemic infections.

C. Influence of Environmental Factors on Seed Transmission

Temperature has a marked effect on the amount of seed transmission with some host-virus combinations and is probably the most important environmental factor influencing transmission. Crowley (1957*b*), for example, found that the temperature at which bean plants were grown before flowering had a marked effect on the percentage of seed transmission of bean common mosaic virus. Although Medina and Grogan (1961) were unable to confirm this effect of temperature on seed transmission of this virus in bean, they suggested that the difference was likely related to differences in varietal susceptibility of the beans used or to differences in virus strains. Singh et al. (1959) found that transmission of barley stripe mosaic through barley seed was greater in seed grown at 20-24° C than in seed grown at higher or lower temperatures. Owusu et al. (1968) found that suboptimal temperatures markedly decreased seed transmission of tobacco ringspot virus in soybean (*Glycine max*). Crowley (1959) found with southern bean mosaic virus in beans that 95% of the embryos of immature seed from plants grown at 16-20° C were infected but only 55% from plants grown at 28-30° C.

D. Effect of Infection Timing on Seed Transmission

The amount of transmission to progeny through seed is markedly dependent upon the time at which the seed- or pollen-bearing parents become infected. In general, infection must occur and the virus must develop systemically before or shortly after flowering. Usually, transmission is successful only if plants are infected before fertilization of the ovules. Tobacco ringspot virus in soybean may be transmitted through 100% of the seed if the plants are infected for a considerable period before flowering, but transmission is low if the plants are infected just prior to or immediately after flowering (Crowley, 1959). Flowering of soybeans is accompanied by increased resistance to infection and systemic development of the virus (Owusu et al., 1968). Seed-borne lettuce mosaic is also dependent upon infection of lettuce cultivars before or just after flowering (Couch, 1955), and the same applies to bean common mosaic virus in beans. Barley stripe mosaic virus in susceptible barley varieties may be an exception to this general rule, for Eslick and Afanasiev (1955) reported that a low level of seed transmission still occurred from plants inoculated long after flowering. Crowley (1959) confirmed their observation but pointed out that infection before flowering was a prerequisite for high levels of transmission and that transmission resulting from late infection may result from late tillering. Experiments of Timian (1967) lend support to this; he concluded that transmission of barley stripe mosaic virus results only if plants are infected before anthesis. Southern bean mosaic virus occurs in bean embryos only when mother plants are infected before or immediately after flowering (Crowley, 1959). Seed transmission thus appears to require that embryos be infected at an early stage of development.

III. MECHANISM OF SEED TRANSMISSION

Table 10-1 lists a number of different plant viruses which are seed-borne. Although this list is far from complete, it is obvious that many plant viruses are not known to be transmitted through seed. Several of the most common and widely distributed plant viruses (tomato spotted wilt, for example) are not known to be transmitted through seed. This, in addition to innumerable reports on the apparent lack of seed transmission with a particular virus in its host plants, suggests that seed transmission is not a common phenomenon in plants. It appears that there are many more virus-host combinations in which the virus is not seed-transmitted than otherwise. In addition, with a few exceptions, even when a virus is seed-borne in a particular host, it is present in only a very small percentage of the seeds. It has been suggested that a protective mechanism has evolved in most plants to restrict the perpetuation of viruses through the seed (Bennett, 1956). Thus propagation of plants through seed has, in general, a cleansing action, allowing propagation of a species free of the viruses which plagued the parent. Such a process is perhaps an evolutionary necessity to insure survival; otherwise, certain species might have perished from rapidly spreading and virulent viruses. Thus sexual reproduction in plants has therapeutic advantages in addition to its genetic significance.

The process by which most viruses are prevented from being seed-borne is obscure but is seemingly related to the ability of the virus to enter and persist in the developing embryo. The ability to invade embryos, and hence be transmitted through seed, is not determined solely by the virus or by the host but is clearly dependent upon the genotypic constitution of both, as pointed out above. In some exceptional cases, the failure of a virus to be seed-borne is not related to its inability to invade embryos.

Pertinent topics in any consideration of seed transmissibility are the relationship of various viruses to the gametophytic tissues, their distribution in seed parts, and hypotheses put forward to account for lack of seed transmission.

Most seed-borne viruses are found throughout the seed tissues, but the critical factor enabling transmission to the progeny usually is their ability to invade and survive in the embryo. Others which are not seed-borne commonly occur in the seed coats. All viruses transmitted in a high proportion of seed seem to possess a unique ability to infect the gametophytic tissues during early stages of development and to persist throughout subsequent phases of growth and maturation. Bennett (1969) and others have termed this "embryonic transmission."

A. Distribution of Viruses Within Seeds

The distribution of viruses within seeds has been studied by dissecting seeds and using the parts in preparing extracts for assays on suitable indicator plants, or detecting characteristic virus particles with the electron microscope. Gold et al. (1954) studied the distribution of barley stripe mosaic, a common seed-borne virus in barley. The virus was found in the embryo and endosperm tissues of seed in about the same concentration as in leaves of infected plants. Under favorable conditions this virus is transmitted to almost all the progeny of some barley varieties. Cheo (1955) found southern bean mosaic virus in a high percentage of embryos of immature seed from infected bean plants and in the seed coats of mature seed. In a similar study, Crowley (1957*b*) found that four non-seed-transmitted viruses also infected seed tissues of their hosts, particularly the seed coats, though none were present in the seed embryos. In contrast, bean common mosaic virus, the only seed-borne virus used in his investigations, was found infecting 83% of embryos of bean seed taken from infected plants. In a later investigation Crowley (1959) confirmed Cheo's results on the high level of embryo infection of immature bean seed from plants infected with southern bean mosaic virus but found that the same virus in two varieties of soybeans did not invade the embryos although infection was 100% in the seed coat and perisperm. Tobacco ringspot virus, which is seed-borne in soybean, was found in a high proportion of embryos (Crowley, 1959). Bennett and Esau (1937) found curly top virus to be distributed throughout the tissues of seed from diseased plants with the exception of the embryo. The virus is not seed-borne. Several other investigations dealing with the distribution of seed-borne or non-seed-transmitted viruses in seeds have reaffirmed that seed transmissibility is usually a consequence of embryo infection, not infection of other seed parts.

Tobacco mosaic virus, which has been demonstrated to be seed-borne in tomato and pepper, may be an exceptional case in that it is confined mainly to the seed coats. Infection probably arises through wounds in seedlings contaminated with virus from the seed coats. This sort of transmission would be expected only with viruses which are easily transmitted mechanically. Genuine seed transmission of this virus has been reported in apple and pear, in which case the virus is present in seed embryos in addition to the seed coat (Gilmer and Wilks, 1967). This report needs confirmation, however, because of the common perils involved in the surface contamination of plants with the tobacco mosaic virus. The virus also may be seed-borne in grape (Gilmer and Kelts, 1968).

The seed coat and nucellus, if the latter is present, are commonly infected by viruses regardless of whether the virus is seed-borne. This fact does not seem unusual since those layers of the seed develop from the outermost tissues of the ovule, which are maternal in origin and are probably infected through cytoplasmic connections to the parent plant. However, infection of these tissues per se does not result in seed transmission, for the critical factor is whether the embryo sac is invaded by virus from the surrounding maternal tissues.

Nucellar tissue may be retained as a conspicuous storage organ of some seed and may be infected, as in sugar beet with curly top virus. This is not surprising in view of its origin from the mother plant.

B. Embryonic Transmission

With a few exceptions, passage of viruses to the embryo is limited, and only a small proportion of embryos from infected plants carry virus. This restriction on the amount of seed transmission has been the subject of much speculation in the past. One of the earliest was Duggar's (1930) suggestion that some inactivating substance might be present in seeds. He reasoned that agents which seemed to be totally systemic in their host plants, such as tobacco mosaic virus in tobacco, were destroyed in the seed by elaboration of some specific material. Crowley (1955) found that seeds, including the embryo, contain potent inhibitors of infection, but he found no evidence that these acted to destroy virus. In further experiments Crowley (1957*a*) grew excised embryos in culture in the presence of virus but found no indication of inactivation. He concluded that inactivation cannot account for the lack of seed transmission. Cheo (1955), however, revived this hypothesis to account for the disappearance of southern bean mosaic virus in maturing bean seed.

Caldwell (1934) put forth one of the most plausible suggestions bearing on the lack of seed transmissibility of plant viruses. He hypothesized that the marked differential between the growth rates of the embryonic and endosperm tissues within developing ovules would lead to rupture of the protoplasmic bridges between these tissues and the nucellus. This view is based on the assumption that protoplasmic continuity is necessary to bring about infection of the embryo by translocating virus from the surrounding maternal tissues to the developing ovules. With this hypothesis, viruses which are borne in a high percentage of the seed must be considered as having a unique ability to reach the isolated embryo and infect it in the absence of protoplasmic junctions with the maternal tissues from which virus orginates, or alternatively and more plausibly, to infect the megaspore mother cell and persist in the developing embryo from its inception. Other evidences hint that seed transmissibility is determined at a very early stage in the development of ovules or pollen. More evidence on this point might be obtained through cytological investigations of incipient embryos in developing ovules with the electron microscope.

Some plant viruses appear to be restricted to certain tissues within infected plants and hence are not truly systemic in their hosts. Beet curly top is a notable example of a virus that is restricted largely or wholly to the phloem tissue of infected plants. It is not seed-borne. The lack of seed transmission of curly top virus is probably due to its inability to invade the embryo because the virus moves exclusively in the vascular tissues of the plant and hence cannot reach the embryo, since the latter has no vascular connection with the maternal tissues in the ovule (Bennett, 1936).

As initially suggested by Valleau (1935) when he found less tobacco mosaic virus in meristematic tissue of tobacco, lack of seed transmission for a virus may rest upon its inability to infect embryos of a particular host, or to do so with difficulty. Embryos may be viewed as a particular type of meristematic tissue, which in general does not appear to be readily invaded by plant viruses. Thus the failure of plant viruses to infect embryos may be more a reflection of inability to multiply in meristematic tissues than some other unique feature of the virus. The apical meristematic regions of many plants, for example, escape infection with viruses that otherwise appear to be systemic. This has enabled Morel and Martin (1948, 1952), Holmes (1956*a, b*), and others to excise and propagate apical meristematic tissues free of viruses.

In this regard, the effect of high temperatures on virus activity in infected plants should be considered since heat contributes to the failure of some viruses to invade meristematic tissues. Hollings (1962) observed, for example, that pelargonium leaf curl virus in geranium (*Pelargonium* spp.) became masked at high summer temperatures and that propagated tip cuttings were free of virus. Many other viruses can be similarly eliminated from apical meristematic regions by subjecting plants to artificially high temperatures (Hollings, 1965*b*). The unfavorable effect of high temperatures on seed transmission of several viruses has already been noted.

Caldwell (1962) has suggested that many plant viruses lack the ability to invade meristematic tissue because pathways for virus synthesis cannot compete successfully with cellular systems for high-energy phosphorylated compounds, and that this phenomenon may account for lack of seed transmission. Meristematic tissues, and doubtlessly the included developing embryos, are sites of very intense metabolic activity. They exhibit a remarkably higher level of ribosomes (Lund et al., 1958), an indication of intensive protein synthesis. This observation suggests that meristematic tissues must require large amounts of high-energy phosphorylated compounds and similar materials to support the economy of the cell. Elevated temperatures have a profound effect on respiration and growth and might be expected to accelerate the demand for these compounds to which most cellular syntheses are coupled, thus contributing to depletion or diversion of these materials required for virus replication. Only a few plant viruses may be fit to compete with cellular pathways for the ribosomes and essential substrates so that they can successfully invade meristematic cells.

Caldwell's hypothesis may be an oversimplification since virus RNA, with the exception of its replication, may function like any other messenger RNA in the cell and it seems unlikely that cellular translation mechanisms distinguish one RNA messenger from another. It does seem clear, however, that most plant viruses lack the ability to invade embryonic tissues successfully.

Even embryos which have a somatic origin are free of virus. Weathers and Calavan (1954) reported that citrus seedlings arising from nucellar embryos of virus-infected clones may be selected for propagation of virus-free clones.

These embryos are adventitious in origin and arise from the somatic tissues of the seed.

The nature of the relationship between seed-borne viruses and their ability to become established in meristematic tissue became a matter of speculation by Lister and Murant (1967) for the nepoviruses, which are transmitted so efficiently through seed. They suggested that the barrier which prevents many viruses from invading and surviving in seed is probably the same as that imposing a similar restraint on invasion of meristems, and that the nepoviruses are less affected by it. Hence, those viruses would have an increased capacity to invade meristematic tissues. As indicative of the latter they pointed out the general tendency of plants to recover from the intial "shock" symptoms following infection by nepoviruses, and association of this phenomenon with entry of the virus into meristems. Support for those observations came in part from previous experiments by Lister (1963) with the strawberry latent ringspot virus. In those experiments he propagated plants from crown buds of individuals infected naturally or infected from seed. All crown buds of plants infected from seed which had "recovered" from the disease contained virus, whereas only a small proportion of similar buds from naturally infected plants contained virus. Plants infected from seed may be assumed to be in a permanently recovered state as a result of invasion of their meristems from their genesis.

Recently Roberts et al. (1970) found that another nepovirus, the tobacco ringspot virus, invaded tobacco meristem cells. The virus was found in the apical initials of the meristems of tobacco, a host in which it is seed-borne. In contrast, three viruses which are not seed-borne in tobacco did not invade the meristem.

Some viruses, though not seed-borne in a particular host, are able to invade the embryo successfully. Cheo (1955) found that the majority of immature seeds from bean plants infected with southern bean mosaic virus contained the virus in high concentration. As the seed matured and dehydrated, however, the amount of virus dropped abruptly and disappeared. Completely ripened and dried seed contained no detectable virus in the embryos, and no seed transmission occurred in the progeny of infected plants. Under the same conditions of drying, the virus was readily recovered from seed coats or leaves, showing that dehydration alone was not responsible for failure of the virus to survive in the embryos. Cheo hypothesized that a virus-inactivating system became operative as the seed dried at maturity. A similar situation applies to pea seed from plants infected with pea streak virus which is seed-borne in immature seed but not in mature seed (Ford, 1966). Crowley (1959) confirmed Cheo's results with southern bean mosaic virus and found that virus appeared in embryos from plants infected prior to or soon after flowering. With the same virus in soybean the virus was never detectable in seed embryos although 100% of the seed coats and perisperm contained virus.

The failure of some viruses to be seed-borne may depend upon the profound effect of infection on the gametophytic tissues of the host. Caldwell (1952), for example, named a virus common in chrysanthemums and tomato in England "aspermy" because of its striking effect on the reproductive tissues of the host. The virus has the effect of stopping the division of mother cells that leads to the formation of microspores and megaspores. Seed production of infected plants was markedly reduced, although the few seeds produced gave rise to healthy plants. Obviously a virus cannot be seed-borne if infection prevents the production of seed. The abortion of microspores as a result of infection is another common effect of viruses that may have a bearing on lack of seed transmission (Valleau, 1939).

Some plant viruses are known to invade the nuclei of their host cells which may enhance its invasiveness of embryonic tissues and hence transmission through seed. Barley stripe mosaic virus, for example, has been observed in the nuclei of leaf cells (Gardner, 1967), but not in the nuclei of mature embryos (Carroll, 1969). Southern bean mosaic virus invades the nuclei of bean plants in which it also infects the embryos (de Zoeten and Gaard, 1969). Pea enation mosaic virus, which is apparently not seed-borne in its leguminous hosts, was observed in nuclei of pea (Shikata and Maramorosch, 1966). These invaded nuclei, however, were affected adversely; the residing nucleoli became filled with virus particles and lost their integrity. Other small spherical viruses such as broad bean mottle, cowpea chlorotic mottle (Hills and Plaskitt, 1968), and pelargonium leaf curl (Martelli and Castellano, 1969) have also been found in nuclei of host cells but none of these are known to be seed-borne. Some strains of tobacco mosaic virus have been observed in nuclei of tobacco parenchyma (Esau and Cronshaw, 1967*b*). The intranuclear occurrence of some strains of tobacco mosaic virus may depend on the plants also being infected with cucumber mosaic virus (Honda and Matsui, 1969). Such information as there is on the occurrence of plant viruses in the nuclei of their host cell has come mostly as the result of casual observations, and as yet does not seem to have any bearing on seed transmissibility.

C. Transmission of Nonembryonic-Borne Virus

Another example of seed transmission with considerable epidemiological significance is the transmission of tobacco mosaic virus with seed of tomato and pepper. These plants are usually propagated by growing seedlings in a greenhouse or hot bed and transplanting them in the open when they are of suitable size. Apparently no embryonic transmission per se occurs with this virus in either host, but the seed from infected plants carries virus both externally and internally. A high proportion of seeds carry virus beneath or within the seed coat, and some contain virus in the endosperm (Crowley, 1957*b*; Taylor et al., 1961; Broadbent, 1965). Seedlings are usually contaminated as they emerge from the soil, and infection results from wounding during transplanting. In pear and apple, however, embryonic transmission through seed may occur with this virus (Gilmer and Wilks, 1967) and perhaps in grape also (Gilmer and Kelts, 1968).

IV. POLLEN TRANSMISSION

Many seed-borne viruses are transmitted to ovules of healthy plants by pollen originating on infected plants. Therefore, it is not unusual for viruses to be encountered in seed arising from otherwise healthy plants. With seed-borne viruses, both parental sources, male and female, usually have the capacity to contribute to virus passed to the progeny. Nelson and Down (1933) found that common bean mosaic virus was transmitted to about one-fourth of the progeny of healthy beans when they were fertilized with pollen from infected individuals. Approximately an equal number of infected progeny resulted from infection of the female parent alone. Shippers (1963) reported essentially the same result. In similar tests with the same virus in beans, Medina and Grogan (1961) found 18-21% infection in progeny of one variety of bean when the virus originated in the male gametophyte, but only 8% when the female parent alone was infected. In a second variety, infected pollen gave 60-76% infection of the progeny; infected females gave 67%. When both parents were infected 86% of the progeny were infected. Hence, infection of either parent may give rise to a high percentage of infected progeny with this virus in highly susceptible varieties. With most other viruses for which pollen transmission to ovules have been studied, infection of seeds is less when pollen is the only source of virus. Results of various studies comparing pollen versus ovule transmission have been summarized by Bennett (1969).

Entrance of pollen-borne viruses into ovules may be assumed to occur along with the male gametes, which move through the pollen tube as it grows into the embryo sac. Virus may accompany the male gametes, one of which unites with the egg cell during fertilization. The other male gamete unites with the polar nuclei, which gives rise to the endosperm. This may be the origin of endosperm infection, or alternatively, as pointed out by Baker and Smith (1966), endosperm haustoria may establish cytoplasmic connections with the nucellus, which leads to invasion of the former tissue.

Some seed-borne viruses cause abortion of the male gametophytes. As a consequence, little if any cross-fertilization (or, hence, seed transmission) occurs unless the seed-bearing parent is infected. Certain strains of tobacco ringspot virus cause complete sterility of pollen in tobacco, and affected plants fail to set seed unless cross-pollinated (Valleau, 1939).

Available evidence suggests that pollen transmission to ovules may be of little importance with some viruses in certain crops because of the poor competitive ability of infected pollen in fertilizing ovules. Lister and Murant (1967) found that both tomato black ring and raspberry ringspot viruses were transmitted through both male and female gametes of raspberry. In tests for the amount of passage through male gametes alone, amount of transmission was suppressed markedly when healthy pollen was mixed in approximately equal proportions with infected pollen before application to stigmas. Hence, these experiments suggest that transmission through pollen to seed may be of little importance in nature because of an inability of infected pollen to compete with virus-free pollen in fertilization.

A. Plant-to-Plant Dispersal of Pollen-Borne Viruses

Virus transmission through pollen to the female gametophyte is a fairly common and well-documented phenomenon and until recently was the only known consequence of pollen transmission. It now appears that infection may not in all cases be restricted to the gametophytic tissues of the pollen recipient. Apple chlorotic leafspot, black raspberry latent, cherry necrotic ringspot, and prune dwarf viruses are readily transmitted by infected pollen to healthy ovules and in addition are able to invade and systemically infect the ovule-bearing mother plants. The first evidence for transmission of this sort to the maternal plants came from experiments with the cherry necrotic ringspot virus in squash (*Cucurbita maxima*). Das and Milbrath (1961) found that application of pollen from infected plants to flowers of healthy individuals resulted in about 10% of the healthy individuals becoming systemically infected, thus suggesting a natural mode of transmission for this important stone fruit virus. In later field experiments George and Davidson (1963) obtained good evidence for transmission of necrotic ringspot in Montmorency cherry from one tree to another by pollen moved by bees or perhaps other insects. In their experiments, in which young trees were caged singly at various times of the year to exclude bees and other insects, infections were most frequent in trees exposed during blossoming (7 infections from 52 exposures). Only occasional infections occurred in trees exposed at other times of the year (2 in 148 exposures), and none occurred in trees that were caged continuously. In similar tests in screened compartments the virus spread only in trees exposed to honey bees (*Apis mellifera* L.) during blossoming and never in trees exposed to numerous other insects and mites. In yet other tests with a bearing orchard the virus spread naturally to infect all but one of 93 trees in a 3-year period. In contrast, no infections occurred in 11 trees whose blossoms were removed during the period of the tests. In addition, the necrotic ringspot virus was transmitted to 4 of 14 trees which were emasculated and the blossoms fertilized with pollen from infected trees. Swenson and Milbrath (1964) and others have made extensive trials with a great variety of insects and mites without obtaining evidence for a conventional sort of aerial vector for this virus. Those results strongly indicate that pollen transmission is the natural means of dispersal for this virus.

Another common virus disease of stone fruits, prune dwarf, may also be pollen-borne from tree to tree. Gilmer and Way (1963) applied pollen from infected trees to the flowers of blossoms on young trees in an orchard free of the disease and were able to obtain five positive transmissions. The virus induced no symptoms and was not recovered from trees until the third season after infected pollen was applied.

Results indicate that two diseases of raspberry (*Rubus* spp.) are carried via pollen to blossom-bearing plants. Cadman (1965) found that apple chlorotic leafspot virus was transmissible in the Lloyd George variety when flowers, which had been emasculated, were hand-pollinated with pollen from infected plants. Numerous attempts to find a vector either in soil or among insects which feed on raspberry have been unsuccessful even though the disease spreads rapidly following its initial appearance in raspberry plantings. Therefore, the virus may be pollen-borne in nature. Converse and Lister (1969) found the black raspberry latent virus transmissible via pollen to the flower-bearing parent plants. Thus, present results indicate that at least four viruses may spread from one plant to another by means of infected pollen.

V. EPIDEMIOLOGICAL SIGNIFICANCE

Natural dissemination within seeds of their host plants has several distinct advantages for the spread and survival of viruses. This mode of survival is particularly significant with certain viruses of annual herbaceous plants that have no perennial hosts, and for viruses that have sedentary types of animal vectors such as nematodes.

Seed transmission may be the single most important source of virus for carryover from one season to another and for initiation of spread in the following season for some crops. Even viruses transmitted in seed to a very limited extent may be dependent upon this as a primary source of inoculum to initiate an annual disease cycle. Doolittle and Gilbert (1919), for example, emphasized the role of wild cucumber (*Echinocystis lobata*) as a source for distribution of cucumber mosaic virus when they found the virus to be seed-borne in this host.

The dependency of seed source for virus carryover will vary with the virus, its host range, transmissibility by animal vectors, and related factors, in addition to other plant reservoirs of virus for dispersal. Many seed-borne viruses also infect perennial plants, which maintain virus for dispersal by animal vectors.

A notable example of the significance of seed-borne virus in the epidemiology of an economically important disease is provided by lettuce mosaic. The role of seed-borne inoculum has been thoroughly documented in this case (Grogan et al., 1952; Zink et al., 1956). Lettuce mosaic virus, though transmitted through only a low percentage of seed from infected plants, may be rapidly dispersed by aphids to a high proportion of plants within a relatively short period. Seed-borne virus has been implicated as the most important primary source of disease in certain California growing areas where levels of aphid activity are high throughout the spring and summer. Aphids transmit the virus in a stylet-borne manner, and spread is thus mainly a local phenomenon. Use of disease-free seed is essential for effective control of the disease in areas with a high level of aphid activity, such as the Salinas Valley of California (Zink et al., 1956).

One of the most obvious functions of seed transmission is assurance of the survival of viruses during periods unfavorable for growth of the host or for activity of a natural vector. The great longevity of some seeds insures survival of the virus for prolonged periods in a potentially transmissible state and hence increases the chances for subsequent spread to new host plants. Viruses may retain activity for many years in dormant seeds which carry the virus through hazardous periods.

Bennett (1969) summarized data on the longevity of viruses in seeds. Though some viruses lose activity in seeds in a relatively short time (e.g. in a few months), most remain active for at least one or two years, long enough to insure carryover through the usual cyclic changes in environment. Several viruses, however, are known to survive in seed for more than five years. This applies to barley stripe mosaic virus in barley (Scott, 1961), squash mosaic virus in pumpkin (Middleton and Bohn, 1953), cherry necrotic ringspot in *Prunus pensylvanica* (Fulton, 1964), and raspberry ringspot and tomato black ring viruses in *Capsella bursa-pastoris* and *Stellaria media* (Lister and Murant, 1967). In some notable examples, viruses have been found to remain active in seeds for more than 10 years. Bennett (1969) found Lychnis ringspot to survive in *Lychnis divaricata* for more than 14 years. In the last case, percentage of seed transmission was not reduced by storage although seed viability dropped markedly. Thus it seems probable that some viruses remain active throughout the life span of seed.

Seed transmission provides a highly effective means of survival and establishment of viruses in crops in successive seasons, and is a more effective source of disease than external sources. Seed-borne inoculum in a crop insures optimum distribution of disease within the crop for subsequent spread since viruses which are seed-borne are randomly distributed throughout the crop and provide numerous foci of disease for further dissemination by vectors. Furthermore, disease is present in the crop earlier since infected and healthy plants emerge from the soil simultaneously. In contrast, disease introduced from outside sources by insect vectors is frequently clustered in the crop, especially around the margins, and enters the crop after emergence.

Another obvious consequence of the great persistence of some viruses in seed is establishment of the virus in new and distant localities. Seed transmission is undoubtedly responsible for much of the long-distance dispersal of viruses and provides much better chances for this than dissemination in vegetative plant parts which are more vulnerable to unfavorable conditions. Fajardo (1930) was one of the first to stress the role of seed transmission in the widespread distribution of bean mosaic carried in commercial seed lots.

Dispersal by insect vectors, even of the circulative type, is limited at any given time. Hence, seed transmission, with perhaps a few exceptions, is probably responsible for the establishment of viruses in isolated geographical areas. Recently, the rapid and almost worldwide distribution of certain strains of sugarcane mosaic virus as a disease of corn and perennial Johnson grass (*Sorghum halepense*) has probably occurred in this manner even though seed transmission occurs very rarely (Shepherd and Holdeman, 1965). Seed is a common item in commerce between countries, and seed movement undoubtedly contributes to transport of viruses to new areas. Unfortunately, the virus content of seed is generally ignored in plant quarantine regulations. Birds and other animals which ingest seeds probably contribute also to the spread of seed-borne viruses.

Seed transmission seems to be extremely common with soil-borne viruses that have nematode vectors. Otherwise, seed transmission seems not to be associated with other virus properties or features, with the possible added exception that none of the leafhopper-transmitted viruses are known to be seed-borne. The proclivity and extent of seed passage of nematode-transmitted viruses in their host plants have been well documented by Lister (1960), Lister and Murant (1967), and Murant and Lister (1967). They have established that dissemination of these viruses in seeds seems to be the rule rather than the exception, and have concluded that this phenomenon must have evolutionary significance to insure spread and survival of these viruses. More than a dozen of the 50 or so viruses listed in Table 10-1 are known to have nematode vectors (see also Chapter 8).

Lister (1960) and Lister and Murant (1967) found several of the nematode-borne viruses to be transmitted in the seeds of several herbaceous annual species that are common weeds in northern temperate zones. Notable examples are the high rates of transmission of arabis mosaic, raspberry ringspot, and tomato black ring viruses in seeds of *Capsella bursa-pastoris, Senecio vulgaris,* and *Stellaria media*. These three weed species are ubiquitous in arable lands of the temperate zones. Moreover, it was shown that these viruses are commonly present in seed from naturally infected plants and that nematode vectors of the three viruses readily acquired virus from progeny infected through seed. This demonstrated rather conclusively the epidemiological significance in nature for seed transport of these nematode-borne viruses.

In additional experiments, Lister and Murant (1967) found that a nematode vector of tomato black ring and raspberry ringspot viruses, *Longidorus elongatus,* retained these viruses for only 9 weeks or so when kept in fallow soil without fresh access to virus source plants. When seeds from infected plants were allowed to germinate in these nematode-containing soils, the vectors again became infective. They concluded that the persistence of these viruses in soils is dependent upon a continuing or recurring source of virus other than that in the nematodes themselves. Otherwise the viruses would not survive in soils through periods of fallowing. These observations suggest that both dispersal and survival of these viruses are largely dependent upon their seed-borne nature. Similar viruses, such as arabis mosaic and grapevine fanleaf, which are seed-borne in some of their host plants, are retained by their nematode vectors (*Xiphinema* spp.) for 8 months or more and are probably less dependent upon transmission through seed. These viruses are found only occasionally in weed seeds in infested soils, and the nematode vectors appear to be associated more with perennial plants, which function as carryover sources of these viruses (Lister and Murant, 1967; Murant and Lister, 1967). The role of seed transmission in the ecology of the latter viruses is obviously less significant than with certain other nematode-borne viruses.

The function of pollen transmission in the epidemiology of viruses has a significance similar to that of transmission through ovules since, in general, infected pollen serves simply to dispense virus more widely to seed during the course of fertilization and leads to a higher proportion of infected seed. As mentioned previously, pollen transmission alone, though highly variable, may lead to high percentages of infected seed.

Transmission by pollen would normally occur in fairly localized areas around infected plants because of its short period of viability. In plants that flower continually for prolonged periods during a growing season, pollen transmission could result in a great increase in the amount of seed transmission.

With some seed-borne viruses, pollen transmission probably has little or no significance in nature because the host is more or less wholly self-pollinating, as in barley and other small grains for example, or because infected pollen does not compete successfully with healthy pollen in fertilizing ovules under natural conditions.

The possible participation of infected pollen in the spread of viruses by foliage contact has not been assessed. Barley stripe mosaic virus in barley, for example, may not be disseminated to any great extent to seed by infected pollen since barley is almost exclusively self-pollinated. Some varieties, however, release a relatively large amount of pollen, some of which must settle on the surface of healthy plants. This virus in barley is readily spread among closely adjacent individuals by abrasion from foliage contact, and one cannot discount the possibility that pollen could provide a source of virus for infection of injured leaves.

VI. CONTROL OF SEED-BORNE AND POLLEN-BORNE VIRUSES

Seed and pollen transmission is an added dimension in any consideration of the epidemiology of a particular virus disease. It may pose a serious additional factor in the control of a particular disease if the virus is seed-borne in plants other than the species of interest. If seed transmission in the crop itself is the only reservoir of a virus for maintenance from one season to the next in a particular growing area, control may be greatly simplified since the production and use of disease-free seed should provide satisfactory control of disease. Frequently, however, other factors that maintain and distribute virus into a virus in tobacco (Valleau, 1932). Hosts of these viruses typically show acute symptoms soon after infection but soon recover and produce few or no symptoms thereafter. Seedlings infected by seed transmission are usually symptomless. The black raspberry latent virus, seed-borne in *Rubus* spp., causes no symptoms in raspberry (Converse and Lister, 1969). Obviously, extensive indexing of seedlings for intended use as clonally propagated material is essential to secure virus-free stocks.

The control of barley stripe mosaic virus in barley is based largely on detection and elimination of infected seed lots. This, unfortunately, is sometimes difficult or unsatisfactory because under certain environmental conditions seedlings from infected seeds fail to exhibit distinct symptoms and some seed lots contain latent strains of the virus which escape detection. Infectivity assays on indicator hosts are reliable, but require considerable time and space, and are impractical for screening large lots of commercial seed for infection. Scott (1961) showed that virus could be detected in single seeds by serological means.

Later, Hamilton (1965) developed an ingenious method for separation and serological testing of barley embryos to assess the health of barley seed lots. The method was found more reliable than a seedling inspection method and was adopted for large-scale screening of experimental and commercial seed lots of barley. Similar procedures may prove useful in evaluating the virus content of other crop seeds which carry economically important viruses.

Some investigators have attempted to rid seed of virus by treating with heat and chemicals. Reddick and Stewart (1919) applied elevated temperatures to bean seed carrying common bean mosaic. Infected seed treated with dry heat at 80° C for 10 min, or 75° C for periods up to 40 min, showed poor survival but still exhibited 50% transmission of virus. Owusu et al. (1968) could not inactivate tobacco ringspot virus in seed of soybean without destroying seed viability. They also tested application of nucleic acid base analogues (2-thiouracil, 8-azaguanine) as leaf sprays on infected plants to determine whether those materials would influence the resulting proportion of seed-borne bean mosaic. These trials were similarly ineffective in influencing the degree of seed transmission. Recently Megahed and Moore (1969) have reported that freshly extracted sour cherry seed carrying cherry necrotic ringspot and prune dwarf viruses can be freed of infection by exposing soaked seed to 60° C or dry seed to 90° C for 1 hr without destroying seed viability. This seems to be an exceptional case, however, as it appears that viruses transmitted through seed in an embryonic manner normally cannot be inactivated by external agents without also destroying seed viability.

Ionizing irradiation of seed has been tested for inactivation of seed-borne viruses without success (Halliwell and Langston, 1965; Megahed and Moore, 1969).

The intrinsic resistance of plants to certain viruses can be utilized to prevent seed transmission even when the virus originates from infected pollen. Medina and Grogan (1961) found with common bean mosaic, for example, that progeny resulting from fertilization of resistant bean varieties with pollen from susceptible plants contained large numbers of infected plants when resistance was inherited as a recessive factor but no infected plants when resistance was dominant.

Long-term storage may diminish the virus content of some seed. Unfortunately, however, viruses rarely disappear entirely before infected seed also loses viability. Middleton and Bohn (1947), and Rader et al. (1947) found that a seed-borne cucurbit virus in muskmelon seed (probably squash mosaic virus) decreased markedly during a three-year period of storage, and Valleau (1939) reported that fewer tobacco seeds transmitted tobacco ringspot virus after storage for 5½ years. Fulton (1964) observed that *Prunus pensylvanica* seed carrying prunus necrotic ringspot virus produced up to 70% infected progeny during the first several years of storage but exhibited less than 5% after 6 years. Megahed and Moore (1969) studied the rate of seed transmission of cherry necrotic ringspot and prune dwarf viruses in sour cherry (*Prunus cerasus*) and sweet cherry (*P. avium*) in seed stored over a 5 year period at room temperature. Transmission decreased more rapidly than seed viability. Seed which initially transmitted 60-75% decreased to 0.5% after 5 years. The examples just mentioned seem to be exceptions, however, for most seeds exhibit no decrease in level of seed transmission after storage (Bennett, 1969).

Viruses carried in or on seed in an essentially external manner are more susceptible to inactivation by suitable treatment than are embryonically transmitted viruses. As noted previously, tobacco mosaic virus is present on and within the seed coat and in a small proportion of the endosperms of seed from infected tomato plants. With this virus in tomato, Chamberlain and Fry (1950) found no seed transmission from fermented or acid-treated seed. Virus concentration in the seed was greatly reduced by fermentation or exposure to acid, suggesting that the virus effective in seed transmission was carried as an external contaminant on the seed surface. Taylor et al. (1961) repeated some of their experiments with acid and found that virus on the seed coat was eliminated by soaking seed in hydrochloric acid or exposing seed to trisodium phosphate. The small percentage of virus in the endosperm, however, was not eliminated by these treatments, though it was inactivated slowly by storage. In additional experiments, Taylor et al. (1961) determined that contaminated seed produces seedlings with virus on their surface, particularly on the roots and cotyledons, but that the seedlings do not become infected unless they receive mechanical abrasions. Their results suggested that transplanting of seedlings was responsible for abrasions, and hence infection. In their study about 2500 seedlings were grown by direct seeding of freshly extracted seed from tobacco mosaic virus-infected plants; only two seedlings became infected. They concluded that embryonic transmission of the virus occurred rarely if ever in tomato although seed-borne virus was an important source of inoculum for the initiation of disease. Broadbent (1965) confirmed some of the results of Taylor et al. with seed-borne tobacco mosaic virus on tomato, and showed additionally that heating of seed at 70° C for 3 days would frequently free seed of internally borne virus. Direct seeding of tomato and pepper, rather than transplanting, may be useful in controlling tobacco mosaic virus.

Some of the nematode-transmissible viruses of economically important crops might be controlled effectively by thorough control of weeds in which the viruses are seed-borne. Results of Lister and Murant (1967) with respect to the limited retention of tomato black ring and raspberry ringspot viruses by their nematode vectors suggest that a brief period of fallowing of soils before planting, coupled with conscientious control of weeds thereafter, might exclude these viruses. As discussed previously, these viruses appear to be retained by their nematode vectors for only 9 weeks or less, so their perpetuation is dependent upon fresh access to virus-infected source plants. Thus elimination of the weeds in which these viruses are seed-borne might eradicate them from vector-infested soils.

The control of viruses transmitted from one plant to another via pollen presents a unique problem. The only solution would appear to be absolute exclusion of pollen-dispersed viruses from planting stock in addition to spatial isolation between infected plantings and young orchards.

PART THREE

ISOLATION AND CHARACTERIZATION OF VIRUSES

11

Purification of Viruses

R. I. B. FRANCKI

Department of Plant Pathology
Waite Agricultural Research Institute
University of Adelaide
South Australia

I. INTRODUCTION

It is no exaggeration to say that only limited knowledge about a virus can be acquired until it is purified. The best illustration of this is that although numerous viruses were recognized, and extensively studied for many years, their nature remained a complete mystery until tobacco mosaic virus was purified by Stanley (1935*b*). It cannot be disputed that very pure virus preparations are essential for chemical, physical, and certain biochemical studies, and that many biological investigations are dependent on the availability of at least partially purified preparations. Hence, during the past three decades and more, much effort has been spent in the development of techniques for the purification of plant, animal, and bacterial viruses. Some of the more spectacular results in virus purification have been achieved with plant viruses, as many of these reach high concentrations in the tissues of their hosts and are remarkably stable. Viruses such as those of tobacco mosaic and turnip yellow mosaic belong to this category and have been extensively studied; it can be said that our present understanding of the molecular architecture of viruses is based to a large degree on past researches with these two viruses (Kaper, 1968).

It must be stressed here that no two viruses are exactly alike and consequently there are about as many purification procedures as there are viruses which have been purified. However, we can recognize several different types of purification systems according to the type of virus:

(i) *Single-component nucleoproteins*. Virus particles consist of a centrally located nucleic acid molecule carrying the genetic information, surrounded by a protective protein coat. The nucleic acid may be arranged in a helix resulting in a rod-shaped particle such as that of tobacco mosaic; or the particle maybe polyhedral, like that of cucumber mosaic, in which case the protein subunits are arranged with icosahedral symmetry surrounding the nucleic acid, whose exact configuration is not well understood (Kaper, 1968). Many of the plant viruses so far investigated belong to the single-component nucleoprotein category, and hence most of this chapter will be devoted to the principles of purifying this type of virus.

(ii) *Multicomponent nucleoproteins.* Many viruses have, in addition to a characteristic nucleoprotein, particles which in some cases may be devoid of nucleic acid or in others may also have one or several particles partially deficient in nucleic acid. Recently it has been shown with one virus of this type, cowpea mosaic virus, that two different RNA-containing particles are necessary for biological activity (van Kammen, 1968). Methods used in the purification of these viruses are those usually applied to nucleoproteins, and the components are then separated by making use of differences in their sedimentation coefficients or densities.

(iii) *"Satellite" viruses.* Some viruses are incapable of multiplication unless the host is simultaneously infected with another virus; such is the case with the "satellite" virus of tobacco necrosis (Kassanis, 1962). In such a system the two viruses must be propagated together and then separated during or after purification.

(iv) *Enveloped viruses.* In addition to nucleic acid and protein, some viruses have an outer membrane containing lipid. It appears that the envelope of these viruses is easily damaged, resulting in loss of infectivity. Many of the methods used for the purification of nucleoproteins are unsuitable for this type of virus as they involve the use of procedures designed to damage biological membranes. Very gentle methods have to be used for handling these viruses, and at present very few have been purified to any degree.

(v) *"Incomplete" viruses.* Some viruses apparently lack the ability to synthesize a protein coat, or else a protein is synthesized which cannot combine with the nucleic acid. These "incomplete" viruses can consist of single-stranded RNA as found in some mutants of tobacco mosaic virus (Siegel et al., 1962); but some appear to be double-stranded, as has been suggested for the infectious entity of potato spindle tuber virus (Diener and Raymer, 1967). Methods for the purification of these viruses must be based on those used for the isolation of nucleic acids and will not be discussed in this chapter.

In contrast to the bewildering variety of methods used for their purification, the isolation of most viruses has certain basic steps in common. The virus must first be propagated in a suitable host in which it will reach high concentrations. The virus must be extracted into a liquid medium with minimum loss of activity, and this extract is usually clarified to remove as much host-plant material as possible with minimum loss of virus. The clarified extract is usually concentrated into a volume that is convenient to handle, and further purification may be required according to the intended use for which the virus is being prepared.

The principles involved in carrying out these steps are discussed in the early sections of this chapter; the latter part of the chapter is concerned with methods used in following the progress of purification, and techniques used to ascertain the homogeneity and purity of the final preparations. The scope of this chapter does not allow a detailed description of the methods used in virus purification, nor the details of procedures used in purifying any particular virus. The reader will find some of the techniques discussed here treated in more detail in other chapters of this volume: in particular those concerning ultracentrifugation, electrophoresis, and chromatography. The reader is also referred to the recent publication "Methods in Virology," edited by Maramorosch and Koprowski (1967), in which over half of Volume 2 describes and discusses in detail, techniques and instruments relevant to the purification of viruses; specific references to many parts of that volume are made throughout this work. Other relevant and useful general references include articles by Steere (1959, 1964), Markham (1959, 1966), Cramer (1964), and Brakke (1969*b*).

II. PROPAGATION AND ASSAY OF VIRUS

A. Biological Purity of the Virus

Whenever possible, it is very desirable to work with a virus culture which is known to be biologically pure, that is, the culture is free of other viruses and is not a mixture of strains (see Chapter 20). Nevertheless, it must be realized that viruses mutate and hence complete freedom of a culture from mutant strains is difficult to maintain. The mutation rate of the virus can be minimized by avoiding conditions which are known to increase it, such as elevated temperatures, irradiation, and exposure to mutagenic chemicals (Siegel, 1965).

To eliminate possible contaminating viruses from a culture, the isolate can be passed through a range of different hosts and then back to the host in which the culture is maintained. In this way, possible contaminating viruses may be detected and then eliminated, or may be eliminated by their host specificity. If the virus required for purification has a host plant in which it produces local lesions, then it can be inoculated to this host at a dilution which will produce only a few lesions per leaf, and single lesions can be excised and propagated separately. Several single-lesion transfers will reduce the chances of the isolate being contaminated by other viruses or by strains of the same virus. Viruses that are transmissible by specific vectors can be purified biologically by vector transmission, which will eliminate any contaminating viruses that are not transmitted by the vector. Once pure, every precaution should be taken to ensure that it does not become contaminated.

B. Selection of Host

1. VIRUS SOURCE PLANT

It is important to choose a suitable plant from which large amounts of virus can be obtained since the virus concentration can vary considerably in different host plants. Systemically infected plants are always preferred to those in which infection is localized. The ease and speed of propagating and growing the plants to the stage at which they can be inoculated are also considered. The ease with which host-plant materials can be separated from the virus can vary between plant species and this factor should also be taken into account. For example, cucumber mosaic virus was purified from infected tobacco giving good yields of virus (Scott, 1963), whereas Francki (1964) used essentially the same procedure for the purification of another strain of the virus from cucumber seedlings. On comparing the relative merits of these two host plants it was found that both gave similar yields of virus; however, there were advantages in using cucumber seedlings, as host plant materials were easier to eliminate from virus preparations, and seedlings were quicker to raise, resulting in saving time and glasshouse space (Francki, unpublished results).

It is very desirable that plants selected for virus propagation not contain large amounts of tannins, gums, latex, or phenolic compounds since these interfere with virus purification. These compounds are usually associated with woody plants, which are generally avoided in virus purification work.

Some circulative insect-transmitted viruses are difficult to purify, probably because they occur in plant tissues in low concentrations and may be confined to phloem tissues. Peters (1967) purified potato leaf roll virus from its insect vector, *Myzus persicae,* as previous attempts to isolate the virus from plant tissues were unsuccessful. This indicates that insect vectors should be seriously considered as a source of those plant viruses which multiply in them and are difficult to purify from their plant hosts.

2. VIRUS ASSAY PLANT

During the purification of a virus it is essential to demonstrate that infectivity is associated with specific fractions and to be able to determine relative infectivity of various fractions quantitatively. For this it is necessary to select an assay plant. If possible, a local-lesion host is selected and if more than one is available then a choice of the most suitable will have to be made. Important factors which affect the final choice of assay plants will be their relative sensitivity to infection, reproducibility of lesion numbers when inoculated with suspensions of similar virus content at different times, visibility of the lesions so they can be counted easily, and the ease and speed with which the plants can be grown. When the choice is made it is desirable to establish standard conditions of growing the plants so that assays are reliable within and between experiments. Variations in temperature, light intensity, photoperiod, and available nutrients can cause considerable variation in the local-lesion response of plants, regarding both the numbers of lesions produced and their appearance; the latter may affect the ease with which they can be counted accurately (Diener and Jenifer, 1964). Local-lesion hosts are unknown for some viruses, in which case suitable plants which produce clear symptoms of some other type must be used. Assays with such plants are far less accurate, but in many cases have given reasonably accurate quantitative data. For example Diener and Weaver (1959*b*) obtained good infectivity-dilution curves for necrotic ringspot virus when they plotted the percentage of infected cucumber seedlings against dilution of the inoculum. Some viruses which cause mosaic symptoms actually produce local lesions which are, however, difficult or impossible to see. These can be made visible for counting by staining with iodine (Holmes, 1931).

The assay during purification of circulative insect-transmitted viruses, the majority of which are not sap-transmissible, has posed a problem (see Chapter 6). This has been overcome in some instances by infecting the insects and letting them feed on plants in which the symptoms are allowed to develop. The insects are infected either by allowing them to feed on virus preparations through parafilm membranes (Rochow, 1960) or by injecting virus preparations into the insect abdomens (Stegwee and Ponsen, 1958). With these methods it is sometimes difficult to know if a significant amount of the test solution has been either taken up by the insect or successfully injected into it. This can be tested, and the volume of inoculum determined accurately, by mixing a small amount of radioactive phosphorus into the virus preparation and then determining the amount of the radioactivity in the insect after feeding or injection. The insects can be caged in small, thin gelatin capsules and the radioactivity determined with a Geiger-Müller end-window detector (Garrett and Francki, unpublished results).

C. Methods of Inoculation

Yields of virus are usually higher from systemically infected plants that were initially heavily inoculated. If the virus is localized to lesions the virus yield will be more or less proportional to the number of lesions produced. The efficiency of infection depends both on the susceptibility of the plants and on the strength of the inoculum. In general, young, vigorously growing plants are highly susceptible and capable of supporting rapid virus multiplication.

Perhaps the most convenient method of inoculating plants is to apply the liquid inoculum to Carborundum-dusted leaves with a finger, but many variations of this method are used and every worker has his personal preferences. Methods of inoculation and conditions for infection are described in Chapter 1.

D. Conditions Affecting Virus Multiplication

Virus yield is very dependent on the conditions under which the inoculated plants are grown. Maximum yields are achieved in plants capable of vigorous growth. Environmental conditions such as temperature and light intensity should not be limiting. At low or high temperatures virus will multiply slowly and may fail to become systemic; the temperature most favorable for virus multiplication will depend largely on the host and the virus but generally favorable temperatures between 20 and 25° C are suitable. Higher temperatures may increase the mutation rate of the virus and should be avoided. Low light intensities usually favor rapid virus synthesis, and during summer in most climates, glasshouses need considerable shading. In England, Bawden and Roberts (1947) found that shading of their glasshouse, to reduce the light intensity to one-third of the normal illumination, considerably increased yields of several viruses; in the case of tomato bushy stunt virus increases of up to tenfold were observed. During the Australian summer it has been our practice to reduce the light intensity in our glasshouses to 20% or less of the normal illumination, to ensure suitable conditions for the multiplication of most viruses. Virus-infected plants grown under warm and shaded conditions will have higher virus concentrations and will often contain fewer substances which are difficult to eliminate from virus preparations during purification. The importance of light and temperature in virus propagation well justifies the use of controlled growth facilities. The plant nutrition and day length can also affect virus multiplication, but these factors are probably not as important as temperature and light. In fact, with most viruses satisfactory yields will be obtained if the plants obtain adequate supplies of essential nutrients which can be assured by application of a balanced mineral nutrient solution if deficiencies of any kind are suspected.

E. Harvesting of Infected Plant Material

Some viruses continue to multiply in leaves for long periods of time whereas others quickly reach a peak and then the virus concentration appears to decrease. For example, Francki and Matthews (1962) found that turnip yellow mosaic virus continued to increase in the leaves of Chinese cabbage for periods of 2-3 weeks without signs of a reduction in the rate of multiplication, in spite of the leaves showing signs of senescence. In contrast, lucerne mosaic virus (Ross, 1941) and tomato spotted wilt virus (Black et al., 1963) have been shown to reach a maximum in tobacco and then decrease. It is therefore important to harvest the infected leaves when the virus concentration is near its maximum. Standardized conditions are necessary since any changes in the environment in which the plants are grown will change the rate of virus multiplication, and thus the time at which virus will reach its maximum may change.

In addition to maximizing infection, concentrations of virus may vary in different plant parts and hence much unnecessary work can be avoided if only parts rich in virus are processed. This is of particular importance when dealing with viruses which occur in plants in small amounts, making it necessary to handle large quantities of plant material. It is also important for obtaining inoculum (see Chapter 1). Generally speaking, virus occurs in high concentration in leaves showing severe symptoms; and these are easily selected.

Harvested material may often be stored by freezing and used for virus purification when required. Although commonly practiced for many viruses, some viruses are readily inactivated by freezing. For example, over 50% of lettuce necrotic yellows virus material from *Nicotiana glutinosa* is lost when leaves are frozen for only 1-2 hr (Francki, unpublished results). On the other hand, viruses such as tobacco mosaic and turnip yellow mosaic remain unaffected by freezing, and in fact freezing is an advantage for purification as the cells are more readily broken during grinding. In addition, many host plant proteins are denatured and hence can easily be removed from the extract. However, most viruses may be stored for several days at 4° C in plastic bags without deleterious effects.

III. EXTRACTION OF VIRUS

A. Preliminary Considerations

Before developing a purification scheme for an uncharacterized virus, it is well to consider all the available information about it. By doing so, it is often possible to reject methods of extraction, clarification, concentration, and further purification which are obviously unlikely to be helpful because of particular properties of the virus. For example, it would be deleterious to try lipid solvents such as ether or chloroform for a virus suspected of having an envelope.

There are many easily determined characteristics of a virus, such as its thermal inactivation point and longevity in vitro in crude extracts, which can help expedite its purification. Also knowledge of the particle morphology of the virus is helpful and can be obtained by the method of Hitchborn and Hills (1965) using crude leaf extracts stained with phosphotungstic acid (for details, see Chapter 4). Characteristic particles present in preparations from infected, but not from healthy leaves are assumed to be virus particles. Recently a disease of frangipani brought to our notice was found to be caused by a virus easily transmitted to several host plants, but its identity could not be established by the symptoms it produced. Purified preparations of the virus were required for further studies, and by electron microscopic examination of crude extracts it was observed that the virus particles (Fig. 11-1) were very similar to those of tobacco mosaic; the virus was purified without any further preliminary experiments by a standard method used for the purification of tobacco mosaic virus.

B. Extracting Media

Viruses that are unaffected by plant substances or extremes of pH may be extracted from tissues without the use of buffers or other substances, thus minimizing the volume of the extract. For instance, complex extracting media are unnecessary for stable viruses such as tobacco mosaic or turnip yellow mosaic. However, many viruses have specific requirements of pH and ionic

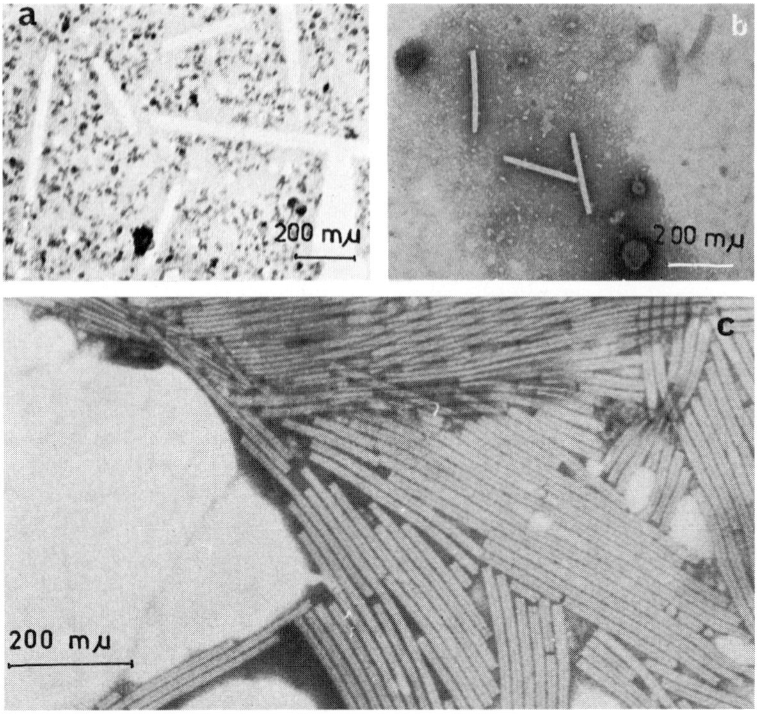

Fig. 11-1. Electron micrographs of a virus isolated from diseased frangipani: a and b preparations made by the phosphotungstic acid staining method of Hitchborn and Hills (1965) from frangipani and *Datura stramonium* leaves, respectively; c, purified virus preparation stained with phosphotungstic acid. The granular background in (a) is due to the latex in sap from frangipani leaves.

strength for stability and may also require the addition of specific substances to protect them from inactivation. It is therefore important to know what conditions are optimum for the stability of a virus if its purification is to be achieved.

1. BUFFERS AND IONIC ENVIRONMENT

The hydrogen ion concentration is a very important factor governing the solubility of proteins and hence of viruses. The charge on virus particles varies with the pH of the solution; the net charge will be zero at the isoelectric point and the virus will precipitate. The majority of viruses have isoelectric points on the acid side of neutrality and hence require neutral or slightly alkaline conditions to remain in solution (Brakke, 1967*b*). However, the pH must not be too high because at low hydrogen ion concentrations the bonding between viral protein and nucleic acid becomes weaker and the two moieties will separate. In fact, alkali treatment is often used as a means of preparing viral proteins free of nucleic acid (Knight, 1964). It has been shown that by raising the pH of virus preparations, virus particles can swell, as reported for cowpea chlorotic mottle virus (Bancroft et al., 1967). This virus has an isoelectric point of 3.65 (Bancroft et al., 1968) and is quite stable at pH 5, but at pH 7 swells and the nucleic acid becomes susceptible to attack by ribonucleases. The exact pH conditions most favorable for retaining the integrity of a virus cannot easily be predicted and must be determined experimentally. A knowledge of these conditions may be essential to the successful purification of the virus.

Viruses also differ in their stability with changes in the ionic environment of solutions in which they are suspended. For example, Brakke (1962) found that barley stripe mosaic virus was more stable in Tris buffer, pH 7, between 0.1 and 0.2 M than at either lower or higher ionic strengths. He also found that the virus was more stable in Tris than in phosphate buffer of the same pH. Tomlinson (1963) has shown with several rod-shaped viruses that preparations contained more infectious virus when purified using borate than phosphate buffer; however, the reverse was true with two polyhedral viruses tested.

Crude extracts may have stabilizing effects on certain viruses, such as barley stripe mosaic virus which becomes less stable on purification (Brakke, 1962). As a virus is purified, stabilizing substances are removed and hence the environment for stability will change. Thus, the solvent system used to extract a virus from plant tissues may have to be modified or changed at various stages of the purification schedule. Some stabilizing substances may be replaced by the addition of specific ions, such as magnesium and calcium in the case of brome mosaic virus (Brakke, 1963), or a variety of additives as with potato yellow-dwarf virus (Brakke, 1956). The changes in the solvent required to maintain virus stability during different stages of purification are well illustrated in the work with cucumber mosaic virus (Scott, 1963). It was found that infected plant material needed to be extracted with 0.5 M citrate buffer, pH 6.5, to avoid loss of virus, whereas later in the purification schedule, 0.005 M borate buffer, pH 9, was found to be necessary.

2. *ADDITIVES TO PREVENT VIRUS INACTIVATION*

When tissues are homogenized to release virus, plant constituents may inactivate viruses by forming insoluble precipitates, degrading them, or by inactivating them in other ways. Compounds most commonly encountered are tannins, polyphenols, and enzymes (see Chapter 1).

Phenolic compounds inactivate virus by forming complexes with them. Formation of these complexes can be reversed or prevented by raising the pH or by dilution (Cadman, 1959), by caffeine (Diener and Weaver, 1959*a*) and nicotine sulphate (Thung and van der Want, 1951), and by proteins such as egg albumin, blood albumin, casein, and hide powder (Brunt and Kenten, 1963). The added alkaloids or proteins probably form complexes with the polyphenols and hence prevent the complexing of the virus (see Kosuge, 1965). Phenolases which form quinones in the presence of copper can be inactivated by chelating agents such as sodium diethyldithiocarbamate or ethylxanthate (Pierpoint and Harrison, 1963) or by reducing agents such as ascorbic acid, cysteine, sodium sulphite, or thioglycollic acid. Many workers routinely use reducing agents and diethyldithiocarbamate in extracting buffers but this is not always wise since the inactivation of some viruses is hastened by some of these compounds, for example lettuce necrotic yellows virus (Crowley, 1967; Atchison et al., 1969). If there is a need to add reducing agents to virus extracts it is advisable to test several agents and find the most efficient one for the particular virus. The concentration of the reducing or chelating agent required to protect the particular virus should also be tested.

Some synthetic polymers such as polyvinylpyrrolidone and polyethylene glycol are known to complex with polyphenols and thereby minimize the substrates for phenolase activity (Kosuge, 1965). Gelatin has also been used to remove polyphenols from virus preparations (Thornberry, 1935).

Although given little attention in the past, it is now known that some viruses are rapidly inactivated by enzymes in crude plant extracts. Viruses consisting of unprotected nucleic acid are known and are sensitive to ribonucleases (Siegel and Zaitlin, 1965). Some, though, are apparently normal nucleoproteins and are also sensitive to ribonucleases (Diener and Weaver, 1959*b*; Francki, 1968), and others may swell under certain conditions and hence become sensitive to the enzyme (Bancroft et al., 1967). If a virus is suspected of being susceptible to ribonuclease the activity of the enzyme can be inhibited by the addition of bentonite, but care must be taken that the virus does not become adsorbed to the clay (Dunn and Hitchborn, 1965). It may also be possible to use other inhibitors of the enzyme. Recently data have been obtained by Atchison et al. (1969) suggesting that some lipid-containing plant viruses may be inactivated by lipolytic enzymes present in plant extracts and that this may account for their extreme instability. If methods can be found of inhibiting the activity of these enzymes in plant extracts, then the lipid-containing viruses may well prove to be very much easier to purify than experience has shown so far.

C. Release of Virus from Plant Tissues

Viruses can be efficiently extracted from plant tissues by grinding with a pestle in a mortar, but the method is laborious if large quantities of tissue are processed. In such cases several types of readily available blenders, juice extractors, or food grinders can be employed (Steere, 1959; Brakke, 1967*b*). For the extraction of viruses which do not require buffers, a very convenient method is to use a domestic meat grinder; however, if buffers are to be used then a blender is more suitable. When grinding methods are used, whether with pestle and mortar or a meat grinder, the tissue can be homogenized more efficiently if it is frozen solid. Some plant tissues are more difficult to homogenize than others. Leaves with long bundles of fibers, such as those of orchids, are particularly difficult to handle, whereas soft leaves, such as those of tobacco, are readily macerated by a variety of different methods.

The homogenate can be freed of large plant debris by straining through cheesecloth. Very often the fibrous material retained by the cheesecloth will still contain considerable amounts of virus and it may be economical to re-extract the material (Steere, 1959). However, in many instances the amount of virus gained is not worth the work involved in handling the increased volume of liquid.

The low yields of some viruses obtained after purification from large quantities of plant material are sometimes due to inefficient release of virus from plant cells rather than to losses during purification. In some cases the virus is released inefficiently because it is enclosed in fairly stable inclusions within plant cells (Markham, 1966), and in others it seems likely that the virus actually becomes adsorbed to particulate fractions of plant cells after they are disrupted. The release of some viruses of mushrooms has been considerably improved by ultransonic treatment of macerated tissues (Hollings et al., 1965). This approach appears to be useful when dealing with polyhedral viruses but not with rod-shaped ones. Tomlinson and Walkey (1967) showed that the infectivity of polyhedral viruses such as cucumber mosaic virus or cherry leaf roll virus was unaffected by 1 hr of sonication; however, in the same tests all infectivity of turnip mosaic virus and potato virus X was lost after only 160 sec of treatment, and electron microscopic examination of the sonicated preparations showed that the rods of these two viruses were broken into short fragments. The poor release of virus from cell seems to be the main reason for using organic solvents such as butanol, chloroform, and fluorinated hydrocarbons as clarifying agents (Steere, 1956). Particular difficulties arise with viruses such as that of lettuce necrotic yellows which is susceptible to organic solvents and sonication. The method of extraction in this case requires to be sufficiently harsh to break the membranes surrounding bundles of virus particles (Chambers et al., 1965) (Fig. 11-2) but gentle enough for the delicate virus particles to remain undamaged.

IV. CLARIFICATION METHODS

On obtaining an extract from infected tissues it is desirable to remove as much plant material as possible before the virus is concentrated. The ideal clarification procedure is one by which no virus is lost, but in practice a compromise is usually reached between the amount of virus retained and the amount of contaminating materials removed. Some of the commonly used clarification methods are described below and their relative merits discussed.

A. Centrifugation

Centrifugation at speeds and periods of time insufficient to sediment the virus may be one of the simplest methods of removing large particulate contaminants. However, with the possible exception of small polyhedral viruses, it is not possible to apply sufficiently high centrifugal force to remove

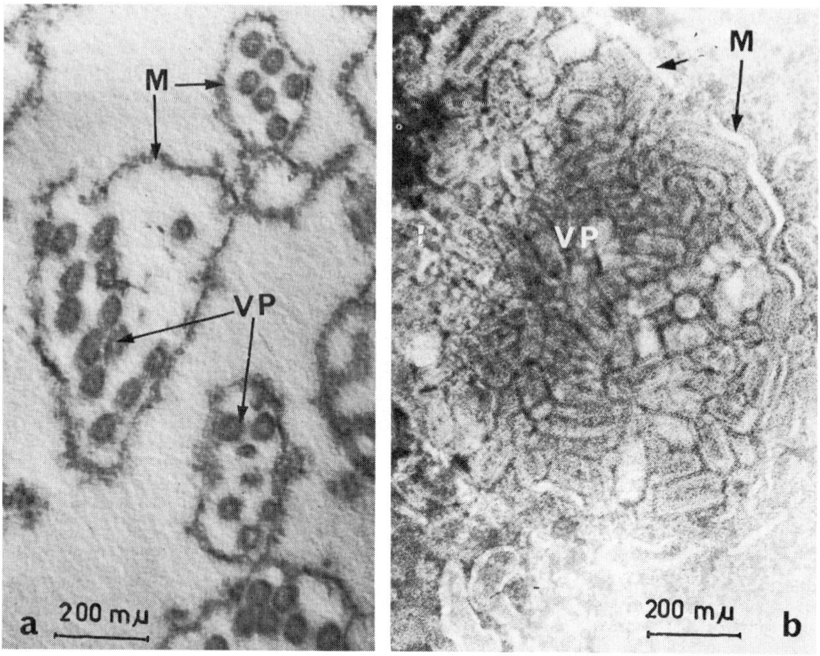

Fig. 11-2. Particles of lettuce necrotic yellows virus as seen in a thin section of a *Nicotiana glutinosa* leaf cell (a) and (b) in a phosphotungstic acid-stained preparation made by the method of Hitchborn and Hills (1965). VP, virus particles; M, membranes enclosing bundles of virus particles. (Courtesy of T. C. Chambers and B. S. Wolanski.)

very large amounts of nonviral materials without serious loss of virus. Thus, the extract is often first treated in some way to coagulate selectively contaminating materials before centrifugation. The speed and time of centrifugation used in clarification will depend on the sedimentation properties of the virus being purified, but will usually be in the range 3,000-17,000 *g* for 5-30 min.

B. Filtration

Filtration through filter paper, or through filter paper supporting a pad of Celite (diatomaceous earth), is sometimes used to replace slow-speed centrifugation in clarifying extracts. If Celite is used it will also act as an adsorbant and under certain conditions is very effective in removing green materials (Steere, 1963; McLean and Francki, 1967). The use of sintered-glass and millipore filters is usually avoided as the pores of these are very easily and rapidly blocked.

C. Heating

Many plant proteins may be coagulated by heating between 50° C and 60° C for about 10 min or at slightly lower temperatures for longer periods. Thus heating may be used as a clarification procedure for those viruses which are unaffected by it. During heat clarification, plant proteins are probably coagulated by enzymatic action in addition to direct heat denaturation. It has often been observed that during heat clarification of extracts of plants infected with viruses, such as those of tobacco mosaic, potato virus X, or cymbidium mosaic, the extracts become dark brown, presumably due to enzymatic activity. These pigments may become adsorbed to the virus and are then extremely difficult to remove later. The ability of a virus to withstand heating is greatly affected by the purity of the preparation and the composition of the suspending medium. For example, crude juice from plants infected with turnip yellow mosaic virus can withstand heating at 55° C without loss of virus (Fig. 11-3). With purified preparations in buffers above pH 6, however, heating at 45° C will inactivate the virus by releasing RNA from apparently undenatured virus protein shells (Lyttleton and Matthews, 1958). This RNA release can be prevented by addition of sucrose to the virus.

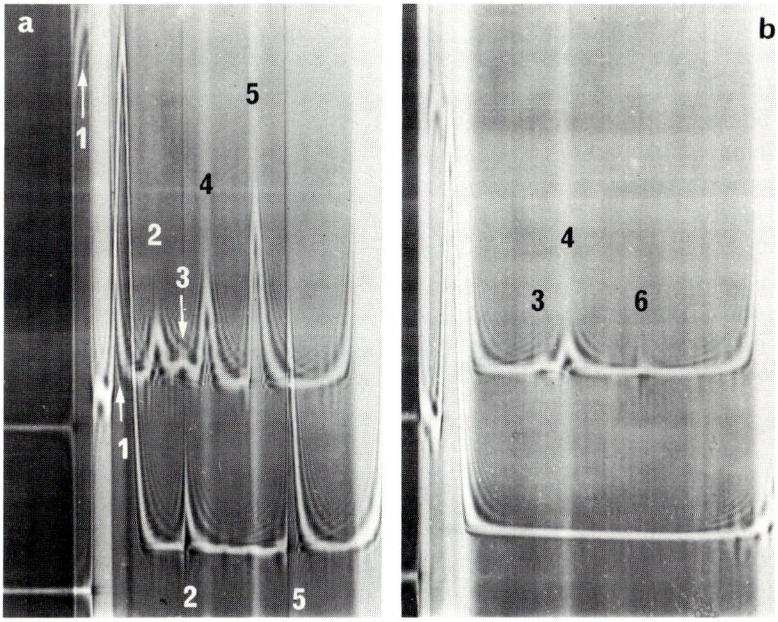

Fig. 11-3. Effect of heat on leaf extracts from virus-infected and healthy plants. Schlieren diagrams of sap from Chinese cabbage leaves infected with turnip yellow mosaic virus (a) and from healthy leaves (b) examined in an analytical ultracentrifuge. Top patterns, freshly expressed sap; bottom patterns, sap heated at 55° C for 10 min. Peaks: 1, Fraction I protein (18*S*); 2, turnip yellow mosaic virus top component (54*S*); 3, chloroplast ribosomes (70*S*); 4, cytoplasmic ribosomes (80*S*); 5, turnip yellow mosaic virus (117*S*); 6, diamers of cytoplasmic ribosomes (125*S*).

clarification; they denature some proteins, disrupt ribosomes, and dissolve lipids without having observable deleterious effects on many viruses. Either virus-infected plant material can be homogenized with the appropriate buffer and solvent to form an emulsion, or else extracted sap can be emulsified with the solvent. The preparation is then centrifuged at low speed to break the emulsion into an aqueous (buffer) layer containing virus and a solvent layer; usually there is an interphase layer containing solid materials, which allows the aqueous layer to be easily removed. The most commonly used organic solvent is probably chloroform, but others, such as ether, carbon tetrachloride, and fluorocarbons have also been used. Some workers have used a combination of solvents such as chloroform with butanol (Steere, 1956).

H. Adsorbants

Plant materials can often be removed by adding adsorbants under conditions in which virus remains in solution. The amount of nonviral material adsorbed will depend very largely on the conditions under which the adsorbant is added, the pH and salt concentration of the extract being particularly important. Usually adsorbants, together with adsorbed materials, are removed by slow-speed centrifugation or filtration. Several different adsorbants have been used successfully in the purification of plant viruses including calcium phosphate gel (Fulton, 1959), charcoal (Corbett, 1961), bentonite (Dunn and Hitchborn, 1965), and DEAE cellulose (Whitcomb, 1965). Adsorbants have been particularly useful in the purification of unstable viruses which are easily inactivated by harsher methods.

Calcium phosphate gel can absorb host-plant materials from leaf extracts, and the efficiency of adsorption will depend largely on the molarity of buffer used in making the extract. Adsorption will be less efficient with higher-molarity buffer; however, at very low ionic strengths both host-plant materials and virus will be adsorbed, so it is very important to determine a buffer system which will allow the maximum adsorption of host-plant materials without serious losses of virus from solution. Figure 11-5 illustrates the effect of buffer molarity on the adsorption of tobacco mosaic virus by calcium phosphate gel from tobacco leaf juice. The pH of the extract will also determine the adsorption properties of the gel. This technique has been used to advantage in the purification of several viruses, in particular several small polyhedral ones (Fulton, 1959, 1967*a*, *b;* deSequeira, 1967; Bock, 1967). The use of calcium phosphate has the disadvantage that the adsorption properties of these gels can vary considerably after storage and according to the way in which they were prepared (Brakke, 1967*b*). This is apparently due to changes in crystal form of the calcium phosphate and can be avoided to a certain extent by always using freshly prepared gel made by adding to a 0.1 M solution of Na_2HPO_4, slightly less than an equal volume of 0.1 M $CaCl_2$ while stirring the mixture thoroughly with a mechanical stirrer (Fulton, 1959). The gel is then washed in several changes of distilled water before use. After the final wash, the gel may be contained much contaminating material there would be a danger of clogging the column. The reader should consult Chapter 13 for further details on column chromatography of viruses.

J. Crystallization

Some viruses crystallize readily in appropriate buffers or salt solutions. For example, purified turnip yellow mosaic virus is easily crystallized, and this method was used to concentrate and purify it (Markham and Smith, 1949). Some viruses will form crystals rather than amorphous precipitates on addition of ammonium sulphate or following acidification. Some viruses such as southern bean mosaic virus will crystallize on dialysis against distilled water or acetate buffer, pH 5.5 (Miller and Price, 1946).

K. Removal of Solvent from the Virus Suspension

As mentioned previously, it is usually better to concentrate virus by removing it from the suspending liquid than to remove the liquid from the virus. If the extracting medium is removed there is less chance of achieving partial purification although the virus is concentrated. Some methods only remove water, causing increases in the concentration of salts which may become deleterious to the virus. Several methods have been suggested by Steere (1964). These include absorption of the liquid by dry gels; dialysis against concentrated sucrose, proteins, or other substances which will pass through dialysis membranes much more slowly than water; pervaporation; freezing out; and surface evaporation.

L. Summary

Ultracentrifugation is used most commonly because the virus retains its structure and biological activity throughout the procedure. Methods of concentrating virus with salts, organic solvents, or pH are avoided by many workers because these treatments are considered unduly harsh on the virus. Although justified with many less stable viruses, many viruses can be concentrated by precipitation very satisfactorily. The precipitation method minimizes time and expense, particularly when large volumes of plant extracts are to be processed. For example, when tobacco mosaic virus is purified by repeated ammonium sulphate precipitation or by acidification to pH 3.5, preparations similar in specific infectivity to centrifuge-prepared virus are obtained (Francki, 1966, and unpublished results). Baudet et al. (1951) reported that tobacco mosaic virus particles aggregated when ammonium sulfate was used, but this observation was not confirmed by Francki (1966). In fact, the ammonium sulphate or acidification methods produced tobacco mosaic virus particles of length distributions similar to those in extracts prior to purification, and yielded significantly higher proportions of 300 nm particles than preparations purified by ultracentrifugation (Francki, 1962).

Little attention has been given to the use of hydrophilic colloids for concentrating viruses. This method prevents exposure of the virus to very high salt concentrations or to extremes of pH, and the precipitation of virus appears to be relatively selective. Dudman (1966) found that tobacco mosaic virus purified by a single precipitation with gelatin was less contaminated by plant materials than was virus isolated by one sedimentation in the ultracentrifuge.

VI. METHODS OF FURTHER PURIFICATION OF VIRUS

Virus may be purified further by a repetition of the procedures used for initial clarification and concentration. For example, many viruses are purified quite satisfactorily by differential centrifugation or by repeated precipitation as outlined in Fig. 11-6. The number of steps required for clarification and concentration depends largely on the virus, the methods used, and the degree of purity required. Contaminating plant material may have properties similar to those of the virus, and repeated clarification and concentration may insufficiently remove them; thus other more selective separation techniques may be necessary.

A. Density-Gradient Centrifugation

The rate-zonal density-gradient centrifugation method of separating macromolecules is one of the most useful methods of separating viruses from contaminating materials. The details of the method, as well as some of its complications, are discussed in more detail in Chapter 12 and in papers by Brakke (1951, 1960, 1964, 1967*b*).

In practice, linear or exponential density gradients are made from sucrose solutions by manually or mechanically layering sucrose solutions of decreasing density on top of each other and allowing diffusion to take place for a few hours, usually overnight. For reproducibility it is preferable to make gradients with mechanical devices, such as that illustrated in Fig. 11-7, which can be used immediately. Cesium chloride, potassium tartrate, sodium bromide and iodide also have been used (see Chapter 12). Unlike sucrose, salts form gradients of high density and low viscosity but may inactivate less stable viruses.

Concentrated virus preparations are usually layered over preformed gradients and then centrifuged. However, with some viruses which reach high concentrations in their host plants, clarified extracts may be layered on gradients directly. This was found to be a very rapid and convenient method of obtaining small samples of relatively pure tobacco mosaic virus (Francki, 1962) which was clearly visible in the centrifuge tube as a light-scattering band (Fig. 11-8). The virus may be recovered from the gradient either with a hypodermic needle pierced through the side of the plastic centrifuge tube in the region of the virus band, or by a bent hypodermic needle lowered into the tube from above. If all the fractions from the density gradient are required, they may be collected by dripping out the contents of the tube by means of a needle inserted into the

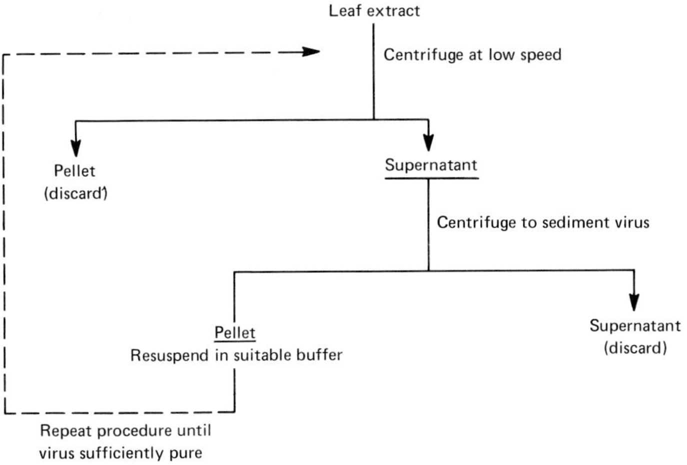

(a)

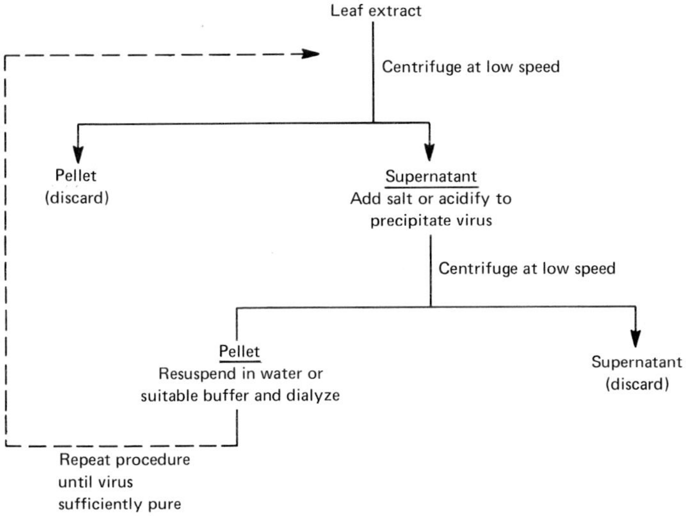

(b)

Fig. 11-6. **Flow diagram outlining the steps carried out during purification of a virus by differential centrifugation (a) and repeated isoelectric or salt precipitation (b).**

bottom; a simple apparatus such as that illustrated in Fig. 11-9(a) may be constructed to collect the fractions. Fractions may also be collected from the top of the gradient with an instrument such as that described by Brakke (1963); this device can be coupled to a fraction collector and recording photometric scanner as illustrated in Fig. 11-9(b).

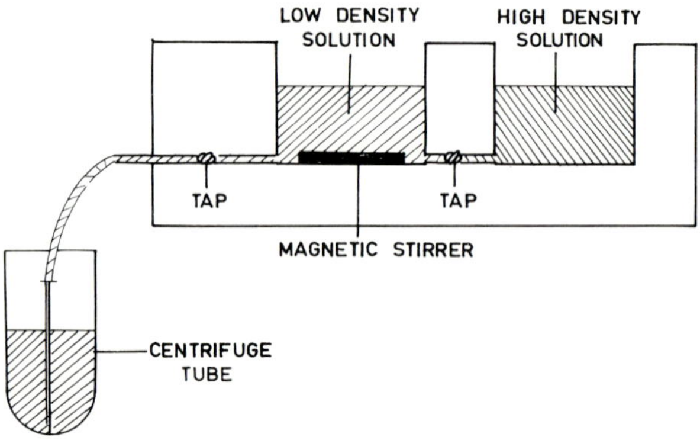

Fig. 11-7. A simple device for preparing linear density gradients. After both chambers are filled with the required solutions both taps are opened and the solution in the chamber containing the solution of lower density is stirred continuously. The solution draining into the bottom of the centrifuge tube increases in density with time, forming a linear density gradient.

Plant virus purification by rate-zonal density-gradient centrifugation is commonly used since this technique separates macromolecules of differing sedimentation velocities. Brakke (1964) showed that nonideal sedimentation occurs when the gradient is overloaded with the virus preparation. Nonideal sedimentation results in an extra wide and smeared virus zone (Fig. 11-10). For maximum resolution it is necessary to minimize the volume of the preparation loaded on the gradient and to ensure that the concentration of virus is appropriate to produce a narrow band. Undue spreading of the zones by diffusion is prevented by minimizing centrifugation times and by recovering the virus immediately after centrifugation. For maximum resolution of closely sedimenting bands, the centrifuge should be allowed to come to rest without braking.

Although ideal for analytical work, the main disadvantage of rate-zonal density-gradient centrifugation is that only relatively small amounts of virus can be handled for preparative work. However, for preparative work zonal centrifuge rotors can be used for the purification of plant viruses (see Chapter 12).

Besides velocity sedimentation, viruses can be separated according to their densities in a density gradient where the virus reaches a position of equal density in the gradient. This procedure, known as isopycnic or density-equilibrium centrifugation has certain inherent disadvantages in purification work. The subject is discussed in Chapter 12. The disadvantages are that only small amounts of materials can be handled, and long centrifugation times are required. Again, the problem of capacity of gradients is overcome by employing zonal rotors (Anderson and Cline, 1967). To overcome the problem of long centrifugation periods, some workers have used procedures which are really a compromise between rate-zonal and isopycnic density-gradient centrifugation. Such procedures have been referred to as quasi-equilibrium-zonal density-gradient centrifugation (Black et al., 1965) and involve centrifugation in a preformed density gradient for a time insufficient for equilibrium to be reached but in excess of that required for separation of virus by rate-zonal sedimentation. This method also avoids the formation of narrow bands of virus which may result in its precipitation.

B. Electrophoretic Techniques

Viruses can be separated from contaminants by boundary electrophoresis in classical Tiselius apparatus, but for preparative purposes it is advantageous to use zone electrophoresis in an immobile support to minimize convection and diffusion. The support can be a gel or a porous solid of any kind, but for virus

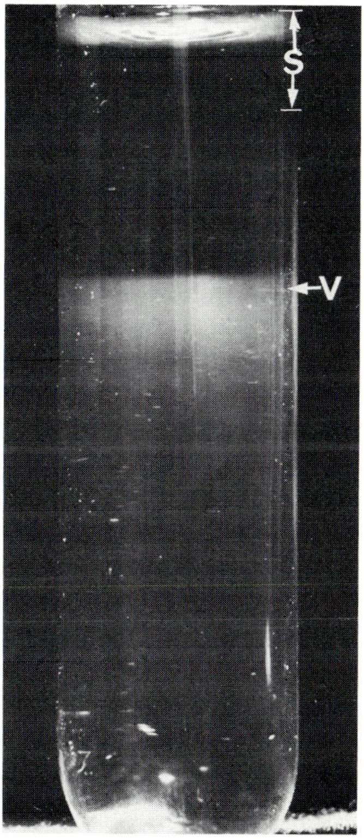

Fig. 11-8. Separation of tobacco mosaic virus from heat-clarified tobacco leaf sap by centrifugation in a sucrose density gradient. Half a milliliter of sap (S) was layered over a 10-40% sucrose density gradient in a 5 ml plastic tube and centrifuged in a swing-out bucket rotor for 30 min at 50,000 *g*. The virus was recovered from the visible light-scattering zone (V).

purification the most popular has been a sucrose density gradient. This has the advantages that the progress of migration may be followed visually, problems of adsorption are eliminated, and there is no electro-osmotic flow (Brakke, 1955). However, some advantages of using certain porous media can be envisaged in that virus separation could be achieved by, say, molecular sieving in addition to electrophoresis.

The electrophoretic method used currently in virus purification is the zonal density-gradient procedure (Brakke, 1955; van Regenmortel, 1964). The method is discussed fully in Chapter 14. Other potentially useful electrophoretic methods include electrophoresis in pH-gradients, gels, and paper curtains (Steere, 1964). Zaitlin (1958) purified tobacco mosaic virus by curtain electrophoresis on filter paper but concluded that the product was poor when compared to virus purified either centrifugally or chemically. Also, the technique of isoelectric focusing of proteins (Haglund, 1967) may well be adapted to virus purification.

The main disadvantages of existing electrophoretic techniques used for virus purification are that the time involved for satisfactory separation to be achieved is usually long and that only relatively small amounts of virus can be handled satisfactorily. The latter problem could possibly be overcome by the development of continuous-flow systems, which would be possible if a curtain electrophoretic procedure could be adopted; possibly "curtains" of materials such as Sephadex or agar-gel granules between glass plates could be

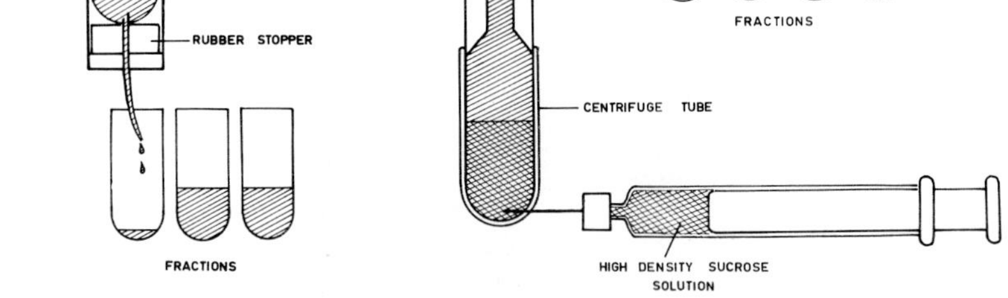

Fig. 11-9. Devices commonly used for sampling density-gradient tubes. (a) Tube contents are sampled from the bottom of the tube by inserting a hypodermic needle and collecting fractions. (b) Tube contents are sampled from the top of the tube by introducing a dense solution into the bottom of the tube which pushes the liquid of the density gradient through an appropriate adaptor from which fractions may be collected. If the density gradient is pushed through a flow cell of a spectrophotometer, a continuous trace of the ultraviolet absorbing materials in the gradient can be obtained on a recorder.

constructed and thus separations by both electrophoresis and molecular sieving could be achieved simultaneously. In addition to the problems already mentioned, a reluctance to use electrophoretic techniques for virus purification is possibly due to the relatively complex nature of the apparatus and manipulations involved.

C. Column Chromatography

The separation of materials by chromatography is based on their selective adsorption and solubility; its uses in clarification of crude virus extracts and in concentrating virus have already been mentioned. The reader should consult Chapter 13 for the principles and methods of column chromatography.

D. Summary

Rate-zonal density-gradient centrifugation is undoubtedly the most widely used method of eliminating host-plant materials from concentrated virus preparations. Usually, it is only when a virus cannot be adequately purified by rate-zonal density-gradient centrifugation that other methods such as isopycnic density-gradient centrifugation, electrophoresis, or column chromatography are attempted. However, a few workers have relied heavily on methods such as electrophoresis; van Regenmortel (1966) lists 18 different plant viruses which he successfully purified by using zone electrophoresis in a sucrose density gradient as a final preparative step, and appears enthusiastic about the results. All these viruses have been purified by alternative and possibly simpler methods by other workers; but the purities of preparations prepared by the various methods have not been compared. Chromatographic methods have also been used for the purification of several viruses by Venekamp and Mosch (1964), but

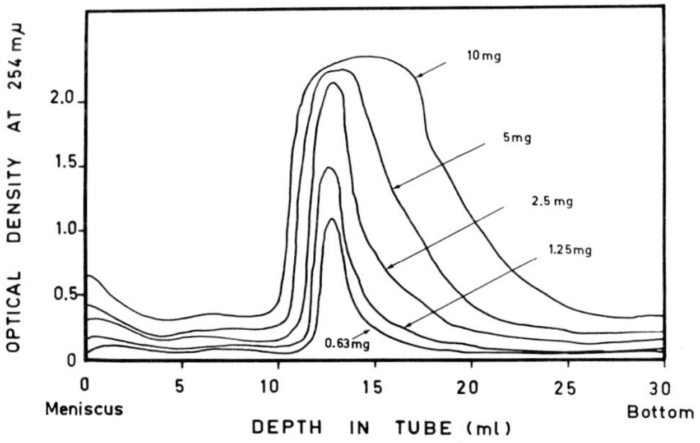

Fig. 11-10. Effect of layering increasing amounts of cucumber mosaic virus (Q strain) on a sucrose density gradient on the width of the band from which the virus is recovered after centrifugation. Each virus sample in 1 ml of buffer was layered on a 5-25% sucrose density gradient (in 0.02 M phosphate buffer, pH 7.2) in a 30 ml plastic tube and centrifuged for 3 hr at 52,000 *g*.

again, for the viruses they purified, alternative and simpler methods are available. It has already been pointed out by Markham (1966) that they probably did not achieve true chromatography but that the viruses were precipitated by the polyethylene glycol, while cellulose columns merely acted as simple filters on which the viruses redissolved when the concentration of polyethylene glycol was reduced. However, Venekamp believes that the mechanism of partition chromatography takes place (see Chapter 13).

VII. FOLLOWING THE PROGRESS OF PURIFICATION

While purifying a virus, particularly a previously uncharacterized one, it is advisable to assess the efficiency of each step taken in the purification schedule; firstly to determine loss of virus, secondly to monitor where the virus goes, and thirdly to assess the efficiency with which contaminating materials are removed. The classical approach has been to follow infectivity and visually inspect the preparation. Under some circumstances, assessment of virus losses by infectivity tests can be misleading (see Chapter 1) and, of course, cannot be used when purifying noninfectious virus-like particles deficient or devoid of nucleic acid or "satellite" viruses which require the presence of "parent" viruses for infectivity. An accurate quantitative and qualitative assessment of materials contaminating a virus preparation can be very helpful in deciding what should be done to further purify the virus. Some of the techniques found useful in following the course of purification of viruses are described below.

A. Infectivity

In spite of its limitations, assessment of virus infectivity during purification is widely used. However, this method introduces considerable errors, particularly during the initial stages of purification. For example, it is possible to lose virus without any apparent loss of infectivity, simply through the removal of a virus inhibitor which may be present in the plant extract. Corbett (1961) actually found that on clarifying an extract from plants infected with potato virus X, by adsorption with charcoal, the infectivity of the extract increased by more than threefold, presumably through the removal of virus inhibitors. Therefore, it is imperative to determine specific infectivity whenever possible.

B. Analytical Ultracentrifugation

The progress of virus purification can be followed with the analytical ultracentrifuge (Markham, 1962) by comparing Schlieren patterns of preparations from healthy and virus-infected plants. It is often possible to identify the virus peak from which a quantitative estimation of virus can be made. Sometimes it is even possible to identify and measure virus peaks in patterns of very crude plant extracts prior to a concentration step. For example, the concentration of turnip yellow mosaic virus in leaves of Chinese cabbage can be determined quantitatively in samples of about 1 g fresh weight containing as little as 100 μg of the virus (Francki and Matthews, 1962). Schlieren patterns also provide a qualitative and quantitative assessment of the contaminating materials in the extract (Fig. 11-3), and the presence of "satellite" viruses or virus-related particles deficient in or devoid of nucleic acid are revealed. Analyses of virus preparations in the analytical ultracentrifuge provide information about the sedimentation properties of the virus from which can be calculated the conditions required for sedimentation in a preparative ultracentrifuge.

In addition to the information obtained from analytical ultracentrifugation, each run rarely takes more than an hour and there is no delay in obtaining the results. Thus it is possible to test the effectiveness of a whole range of, say, clarifying procedures on a virus in a matter of hours; an immediate assessment can be made of virus loss and of their effectiveness in removing contaminating materials. The most promising methods of clarification can then be tested for their effect on the infectivity of the virus.

C. Electron Microscopy

Examination of virus preparations in the electron microscope at various stages of purification can be used to assess the degree of purity and the nature of contaminating materials. It is also possible to determine relative numbers of virus particles in preparations (Backus and Williams, 1950), but this is laborious and the method has not been popular for routine purification. Virus preparations at various stages of purification can be examined in the electron microscope after shadowing with heavy metals or staining with electron-dense compounds such as phosphotungstic acid or uranyl acetate (Horne, 1967). However, there may be certain difficulties in interpretation of electron micrographs due to preparative artifacts, some of which are discussed in Section VIII.

D. Serological Methods

If the virus being purified is one to which an antiserum is already available, then approximate virus losses during the various steps of the purification schedule can be determined by serological assay such as the precipitation reaction by tube, ring, agglutination, or gel-diffusion tests (refer to Chapter 17). These tests are quick and simple to perform and do not require the use of expensive equipment.

Serological tests can also be used for assessing the purity of virus preparations. Antisera to extracts of healthy plants are easily prepared and can then be used for testing virus preparations at various stages of a purification schedule.

VIII. CRITERIA OF PURITY

Although it is often stated "you cannot do clean work with a dirty virus," a high degree of purification is rarely a goal in itself; usually a specific use is intended for the purified virus, and that will determine the degree of purity required. If the virus is intended for the production of antiserum then the main concern will be that it is free of all host-plant antigens. On the other hand, if it is for studying the properties of the viral protein(s) or nucleic acid, then care must be taken to ensure freedom from analogous host-plant derivatives. Tests used for assessing the purity of the virus will be chosen accordingly. It must be stressed that claims of absolute purity of a virus preparation can never be substantiated, as all tests for the presence of impurities have definite limits of detection. Ideally a pure, homogeneous virus preparation would contain a population of identical particles completely free of impurities. In practice this can never be achieved, and even if it were, no tests available at present could substantiate such a claim.

To establish biological homogeneity of a virus preparation it would be necessary to demonstrate that every virus particle in the preparation was potentially infective and was carrying identical genetical information. Considering the very poor efficiency of infectivity assays (see Chapter 1) and the knowledge that viruses are continually mutating, it is probably safe to assume that the purification of a biologically homogeneous virus has never been achieved. Physical homogeneity is easier to test for.

No virus can be claimed to be purified unless infectivity has been correlated with a molecular species. Testing the purity and homogeneity of virus preparations, and correlation of characteristic particles with biological activity, involve the use of similar techniques and hence are discussed together in this section.

A. Ultracentrifugal and Electrophoretic Analyses

The number of molecular species present in a preparation can be determined by several different methods, of which ultracentrifugation and electrophoresis are the most commonly used. Analytical centrifugation is probably the most convenient technique to apply, since molecular species with different sedimentation coefficients are readily observed using either Schlieren, interference, or absorption optics. Electrophoresis, using a Tiselius apparatus, will similarly distinguish between molecular species with different electrophoretic mobilities. Density-gradient centrifugation and electrophoresis in density gradients followed by photometric scanning of the columns may also be used; they utilize the same principles as those using the analytical ultracentrifuge (Fig. 11-11) or the boundary-electrophoresis apparatus. Density-gradient methods have the added advantage that concurrently with the tests for homogeneity and purity, fractions from the columns can be collected to test for the correlation of infectivity with a particular fraction (Fig. 11-12).

The presence of a single sharp peak or component on ultracentrifugal or electrophoretic analysis is sometimes considered sufficient to establish that the virus preparation is pure, in particular if this component can be correlated with specific infectivity. However, the optical systems of the analytical centrifuge and electrophoresis apparatus are efficient in detecting homogeneous

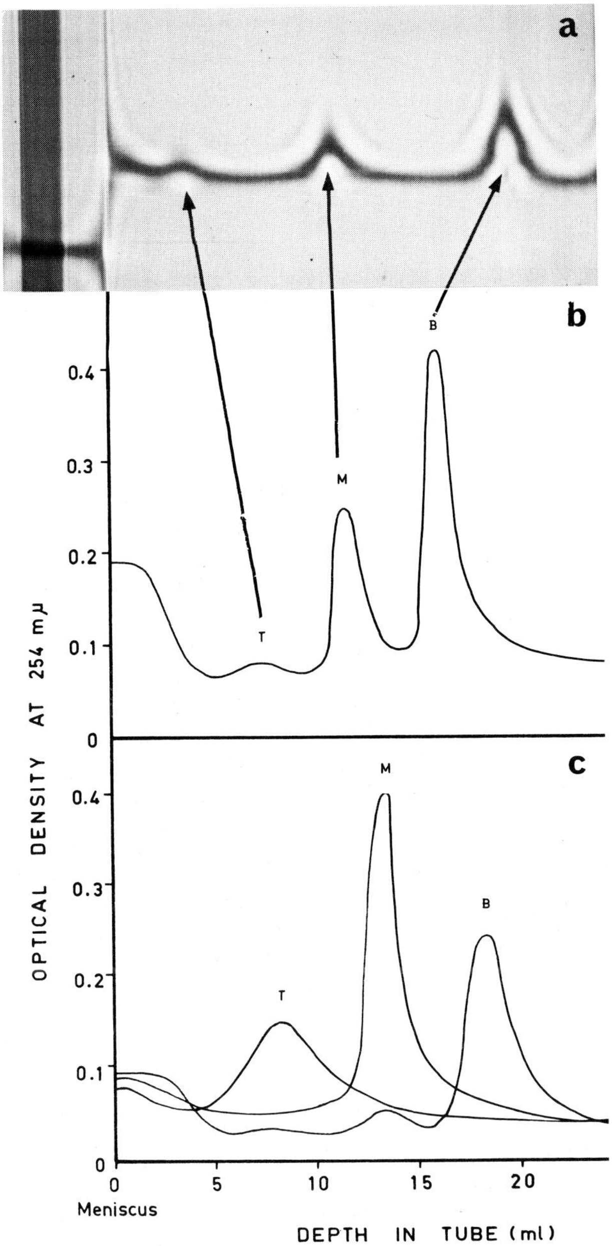

Fig. 11-11. Separation of tobacco ringspot virus (B) from its RNA-deficient components M and T. Component T (57*S*) contains no RNA, Component M (99*S*) 21% and Component B (136*S*) 35% RNA. a, Schlieren diagram of a virus preparation containing all three components examined in an analytical ultracentrifuge. b, Ultraviolet scanning pattern of a similar virus preparation separated by sucrose density-gradient centrifugation. c, Similar ultraviolet scanning patterns of the three components isolated from density gradients and rerun in similar gradients. (Details of density-gradient technique same as described in Fig. 11-10.)

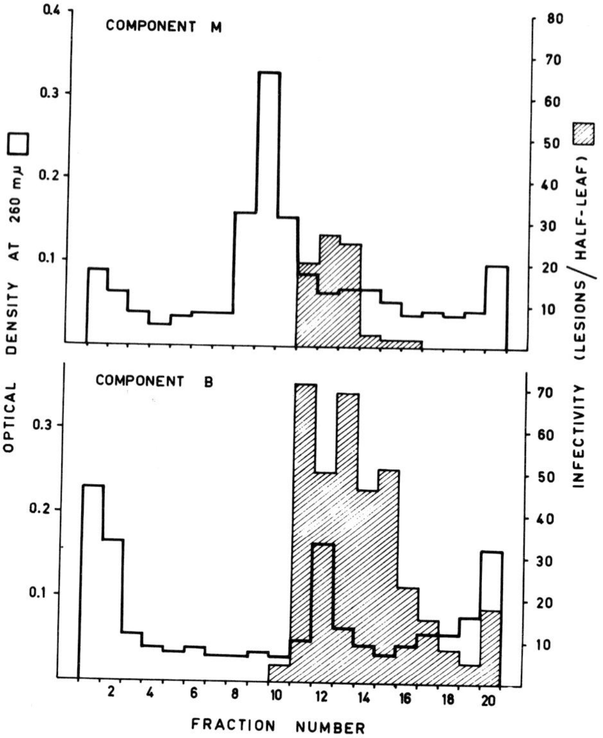

Fig. 11-12. Correlation of infectivity with the B component of tobacco ringspot virus. Components M and B, isolated by sucrose density-gradient centrifugation (as described in Fig. 11-10) were rerun in density gradients and fractions were collected with an apparatus similar to that described in Fig. 11-9(b). The samples were analyzed for optical density at 260 nm and infectivity by the local-lesion assay method. It can be seen from these data that the M component is not infectious and that all infectivity is correlated with the B component.

molecular species but not those which are heterogeneous. Thus, it is possible that a virus preparation contaminated with impurities may sediment or migrate as a single boundary or zone; but if the area under the peak or zone is measured at various times during the experiment and is compared with the expected area, assuming homogeneity, then heterogeneous components may be detected.

Although the presence of a single homogeneous component on centrifugation or electrophoresis may be taken as reasonable evidence that the virus is relatively free of impurities, the presence of more than one component may not necessarily mean that the preparation is contaminated. Although populations of one molecular species, many viruses may under certain conditions aggregate and the aggregates will sediment as distinct peaks in the ultracentrifuge. A good example of this is tobacco necrosis "satellite" virus; the 50*S* particles tend to aggregate in groups of 12 to give an additional peak of 222*S* in the ultracentrifuge. The aggregates can actually be separated by rate-zonal density-gradient centrifugation and observed in an electron microscope (Kassanis, 1966). It is also well known that "multicomponent systems" are found in preparations of some viruses. In such preparations, in addition to the so-called bottom component which has a full RNA complement, there are associated particles consisting of apparently identical protein shells but with an incomplete RNA complement (middle component) and/or particles with no RNA at all (top component). In several viruses of this type, only the bottom component has been demonstrated to be infectious (see Fig. 11-12) but in some cases the middle component has been shown to stimulate the infectivity of the bottom component (Bancroft, 1968). Recently van Kammen (1968) has shown with preparations of cowpea mosaic virus containing top, middle, and bottom components, that neither the middle nor the bottom component alone is infective, but both types of particles are necessary for infection. Thus, it can be said that this virus consists of two macromolecular species. A similar situation has been found with the short and long rod-shaped particles of tobacco rattle virus (Sänger, 1968). In this case, it would appear that the short particles carry the genetical information for synthesis of the viral coat protein whereas the long particles are necessary for synthesis of the RNA replicating enzymes required for virus synthesis (Sänger, 1968).

It should be pointed out here that a virus preparation, demonstrated to contain only one component by one technique, may actually be a multi-component system. For example, when turnip yellow mosaic virus preparations were examined in a Tiselius apparatus by Markham and Smith (1949), they appeared homogeneous with only one component; whereas in the analytical ultracentrifuge, two components were observed. These observations actually led Markham (1951) to suggest that the infective particle consists of RNA surrounded by protein; had the RNA been arranged externally to the protein, the electrophoretic mobilities of top and bottom components would almost certainly have been different.

B. Electron Microscopic Examination

Virus preparations may be examined in the electron microscope for testing homogeneity. The information obtainable is direct, giving details of particle morphology and numbers of particles (Fig. 11-13). Although it is often believed that electron micrographs are easily interpreted, there may be problems arising through preparative artifacts which may mislead the unsuspecting investigator (Fig. 11-13b).

In specimens examined after shadowing, the most common artifact encountered is a flattening or collapse of particles due to surface tension forces accompanying the drying process. This is particularly well illustrated by the work of Cosentino et al. (1956), who examined preparations of top and bottom components of turnip yellow mosaic virus which were air-dried at room temperature and shadowed with uranium. The virus, or bottom component

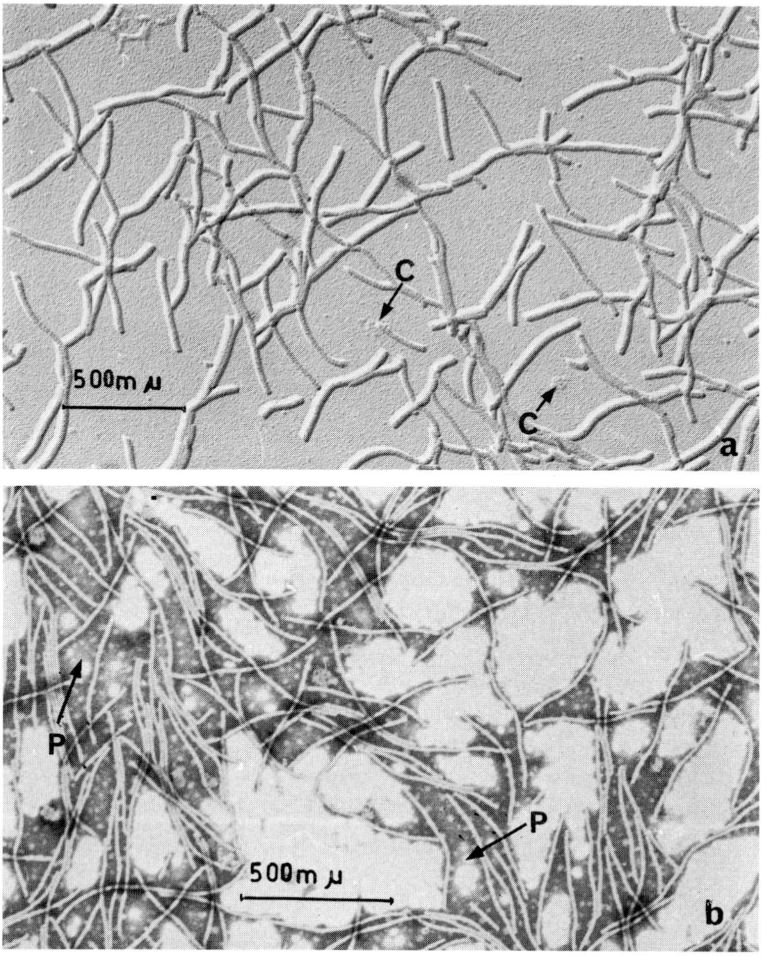

Fig. 11-13. Comparing potato virus X particles in an electron microscope after shadowing with platinum-carbon (a) and after staining with phosphotungstic acid, pH 6 (b). C, contaminating materials in the virus preparation; P, patterns produced by drying of the phosphotungstic acid stain which can easily be confused with materials contaminating the virus.

particles, measured 26 nm in diameter, whereas the top component or empty protein shells were either flattened, measuring 36 nm in diameter, or were clumped together and measured 26 nm in diameter but appeared "dimpled." However, when the specimens were freeze-dried by the method of Williams (1953), particles of both top and bottom components appeared indistinguishable and all measured 26 nm in diameter. A similar problem was encountered with particles of cucumber mosaic virus (Fig. 11-14a), which although containing RNA, measured about 42 nm in diameter when air-dried

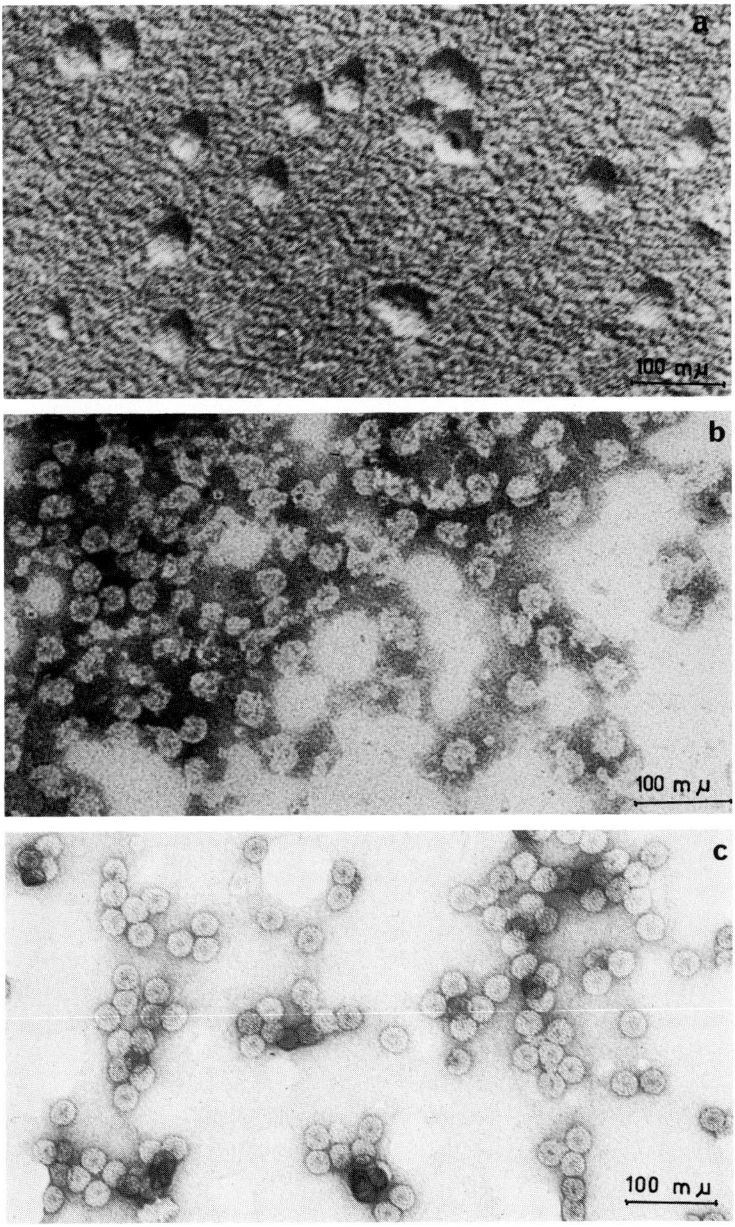

Fig. 11-14. Comparing cucumber mosaic virus (Q strain) particles examined in an electron microscope after shadowing with chromium (a); negative staining with phosphotungstic acid, pH 7, (b); and uranyl acetate, pH 4.5 (c). Notice the large size of the shadowed particles due to flattening during drying and the disrupted particles stained in phosphotungstic acid. (Courtesy of T. C. Chambers.)

and shadowed (Francki et al., 1966) but only 28 nm when freeze-dried and shadowed (Scott, 1963). Procedures for the electron microscope are discussed in Chapter 4.

Negative staining procedures for the electron microscope are discussed in Chapter 4. With this technique several problems can arise which will make the interpretation of micrographs of purified virus preparations difficult. For example, some viruses are unstable in phosphotungstic acid and in this stain the particles disrupt (Fig. 11-14b). In some cases this has been overcome by using the phosphotungstic acid at a lower pH (Finlay and Teakle, 1969) or by using uranyl acetate (Fig. 11-14c) (Francki et al., 1966). It has been shown that some particles of polyhedral viruses may be penetrated by negative stains, and this has sometimes been interpreted as evidence for the presence of particles devoid of nucleic acid (Brenner and Horne, 1959). However, it has been shown by Markham (1962) working with turnip yellow mosaic and by Davison and Francki (1969) with tobacco ringspot virus that in electron micrographs, the presence of particles penetrated by negative stain does not necessarily mean that these particles are devoid of nucleic acid. When virus preparations are examined for the presence of particulate contaminating materials by electron microscopy of negatively stained preparations, difficulties may arise through the patterns produced by drying of the stain on supporting films. Sometimes it is difficult to determine if what appears to be negatively contrasted particulate matter is really that, or just patterns produced by surface tension phenomena during drying of the stain on the supporting membrane (Fig. 11-13b). Some of the most serious problems in interpreting electron micrographs of negatively stained preparations occur with the large enveloped viruses. For example, when Harrison and Crowley (1965) examined preparations of lettuce necrotic yellows virus stained with uranyl acetate (Fig. 11-15a), they observed three different particle types: bacilliform particles with cross-striations, similar shaped particles but without cross-striations, and roughly spheroidal particles. In samples from the same virus preparations, when examined after staining with phosphotungstic acid, pH 7 (Fig. 11-15b), most of the particles appeared to be bullet-shaped (Wolanski et al., 1967). It seems that the true particle shape of lettuce necrotic yellows virus is bacilliform, as observed in sections of infected leaf cells (Chambers et al., 1965), and that the appearance of cross-banding is dependent on whether the stain penetrates the envelope of the particle. The various other shapes of particles observed (Fig. 11-15) are thought to be preparative artifacts due to distortion by osmotic and imbibition forces during staining and drying (Wolanski and Francki, 1969).

C. Serological Tests

Antibodies against a virus preparation can be used to detect the presence or absence of antibodies to host-plant antigens. Another practical method of testing for impurities is to prepare an antiserum against host-plant antigens and

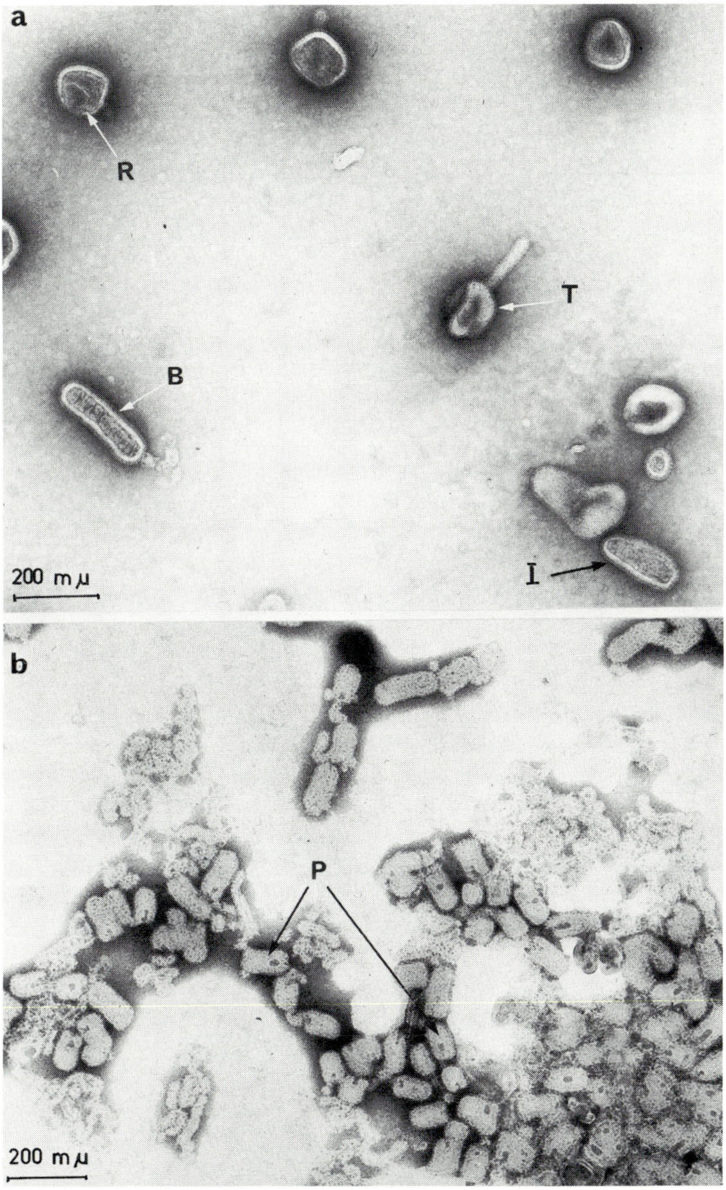

Fig. 11-15. Effect of staining on the morphology of lettuce necrotic yellows virus particles. a, Virus preparation stained with uranyl acetate showing a variety of particle types; B, bacilliform particle whose morphology is little changed to that in cells; I, particle showing swelling but still remaining more or less bacilliform; R, particle deformed into a globular shape; T, particle deformed and having a distinct tail structure. b, Virus preparation stained with phosphotungstic acid, pH 7, showing bullet-shaped particles (P) and also some particles which have disrupted. (Courtesy of T. C. Chambers and B. S. Wolanski.)

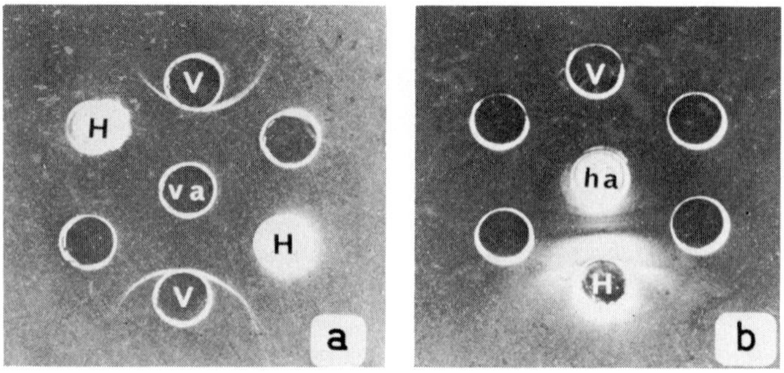

Fig. 11-16. Serological reactions in gel-diffusion tests. a, Reaction between potato virus X antiserum (va) and its homologous antigen (V). Notice that the antiserum does not react with a preparation of host-plant antigens (H), showing that the virus used for immunizing rabbits was relatively free of contaminating antigens. b, Reaction between an antiserum to host-plant materials (ha) against its homologous antigens (H). Notice that the antiserum does not react with the virus preparation (V), indicating that this preparation is relatively free of contaminating antigens. (Courtesy of G. D. McLean.)

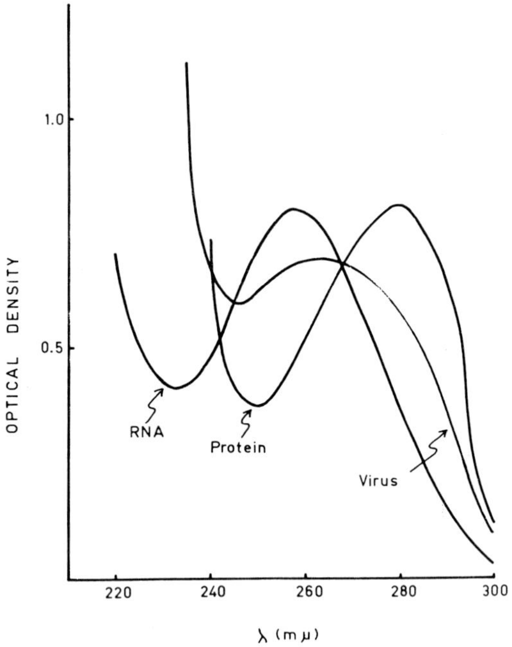

Fig. 11-17. Ultraviolet spectra of potato virus X and its isolated RNA and protein prepared by the method of Francki and McLean (1968).

to test the virus preparation against this antiserum. Figure 11-16 illustrates how antisera to both purified virus and host-plant antigens are used in testing a virus preparation for purity. Other serological tests such as complement fixation or hemagglutination would ensure more sensitivity in the detection of impurities.

D. Ultraviolet Absorption

Viruses, being nucleoproteins, have characteristic absorption spectra (Fig. 11-17). Most viruses show maximum absorption near 260 nm and a minimum at about 240 nm. The 260/230 and 260/280 nm ratios depend on the nucleic acid content, and therefore can be used for the approximation of nucleic acid contents of viruses (Paul, 1959) if appropriate corrections are made for light scattering (Englander and Epstein, 1957). The concentration of virus in a purified prepartion can also be determined from its ultraviolet spectrum by using its specific extinction coefficient ($E^{1\%}_{1CM}$) at 260 nm.

The spectrum of a routinely prepared virus whose ultraviolet absorption is known is often used as a rapid and relatively sensitive test for purity. Changes in the spectrum of a virus being investigated for the first time during successive steps of the purification schedule can also be used as an indication as to whether purification is being achieved. Removal of impurities usually increases the ratio of absorption at the maximum to that at the minimum and decreases the absorption at 280 nm. Decrease in absorption at 320 nm indicates reduced light scattering.

ACKNOWLEDGMENTS

Some of the previously unpublished work reported here was supported by grants from the Australian Research Grants Committee and by the Rural Credits Development Fund of the Reserve Bank of Australia.

12

Preparative Centrifugation in Virus Research*

VERNE SCHUMAKER
and ALLAN REES

The Molecular Biology Institute
and Department of Chemistry
University of California
Los Angeles, California

University of Texas Medical School
San Antonio, Texas

I. INTRODUCTION

Preparative centrifuges are versatile and powerful research tools for studying biological macromolecules. These high-speed centrifuges may be used for isolating, purifying, and physically characterizing viruses.

In this chapter, a brief derivation of the sedimentation equation is presented to avoid directing the reader to more comprehensive treatments (Schachman, 1959). (See Chapter 18 for details.) This is followed by a description of the basic preparative centrifuge equipment desirable for virus research and by a section on the determination of the virus sedimentation coefficient in clarified extracts. Next, density-gradient techniques are described. In the final section, a new and powerful tool, the zonal centrifuge, is discussed.

A. The Virus Window

Figure 12-1 shows an s-ρ plot, that is, the distribution of a variety of biological macromolecules as a function of sedimentation coefficient, s, and density, ρ

---

*Supported in part by United States Public Health Service Grant 13914, Contribution Number 2267 from the Department of Chemistry, University of California, Los Angeles, California 90024.

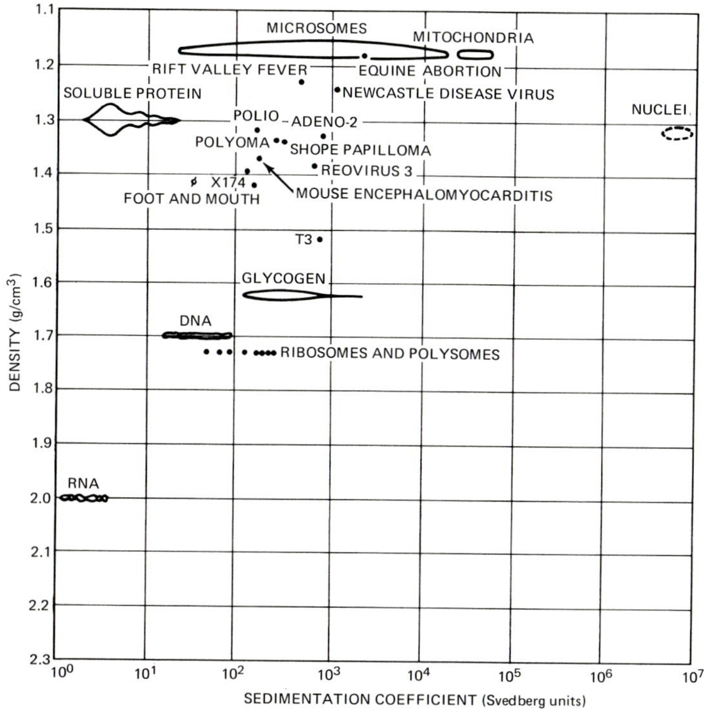

Fig. 12-1. **A variety of biological macromolecules and cell particulates are plotted on this diagram. The horizontal axis is measured in units of sedimentation coefficients, $s_{20,w}$, while the vertical axis is buoyant density. Thus, each type of particle has a unique location on this s-ρ plot. It also may be seen that virus particles mostly are located in a region between 100 and 2000 *S*, and $\rho =$ 1.2-1.6. This region may be called the virus window. (Anderson, 1966.)**

(Anderson, 1966). Each of these macromolecules has its distinct location on the s-ρ plot according to sedimentation coefficient and density, and most plant, animal, and bacterial viruses fall on the s-ρ plot in a region unoccupied for the most part by other molecules and cellular particulates. This region may be called the "*virus window.*" It extends from about 100 to 2000 *S*, and from 1.2 to 1.6 density units. The "virus window" may be used as a guide for roughly selecting the proper conditions for the isolation and purification of virus. The centrifuge may be used to isolate and purify materials according to their sedimentation coefficients and to their buoyant densities. Therefore, most of the unwanted cellular materials can be eliminated by selecting proper conditions of centrifugation.

B. The Sedimentation Equation

The mass m acted upon by a force F will experience an acceleration a according to the equation $F = ma$. If the acceleration is caused by gravity g the last expression may be written $F = mg$. In a centrifugal field, the acceleration is $\omega^2 r$, where r is radial distance and ω the angular velocity. The appropriate force equation becomes $F = m\omega^2 r$. Under the action of this force, a particle may be moved through a viscous medium. But the medium will resist the forward motion of the particle and push back with a frictional force F'. At low velocities, such as those with which we shall be concerned, it turns out, both theoretically and experimentally, that F' is directly proportional to the velocity of the particle, v. Since the frictional force points in a direction opposite to the velocity, however, we must relate the two quantities by a minus sign. Then we can write $F' = -fv$, where f is the coefficient of proportionality. The quantity f is called the translational frictional coefficient, and it is a function of only the size and shape of the particle and the viscosity of the solution.

Imagine that the centrifugal field is applied. Then the following events occur in the first milliseconds: The particle accelerates under the action of the centrifugal force. As it speeds up, the force of frictional resistance increases. This process continues until the force of frictional resistance is equal in magnitude, though opposite in direction, to the acceleratory force. At this point, there is no net force on the molecule. Therefore, it will no longer speed up nor will it slow down, but instead it will continue to move through the solution at a well-defined velocity.

Let us solve for this velocity. The condition of no net force is:

$$F + F' = 0 \tag{1}$$

or

$$m\omega^2 r - fv = 0 \tag{2}$$

Rearranging equation (2) yields:

$$\frac{v}{\omega^2 r} = \frac{m}{f} \tag{3}$$

Equation (3) states that the velocity per unit centrifugal field is equal to the mass of the particle divided by the translational frictional coefficient.

Equation (3), however, is not in final form because we have left out one vital consideration. We have not considered the buoyancy of the medium. Obviously, if the medium had the same density as the particle, then the particle would neither sink nor float under the action of the centrifugal field. Or if the particle were lighter than the medium, it would float to the top instead of sedimenting to the bottom. Equation (3) becomes correct if instead of the mass of the particle, we put in the buoyant mass m_b. The buoyant mass, according to Archimedes, is the mass of the particle minus the mass of an equal volume of the liquid medium. Let $\bar{v}$ be the partial specific volume of the hydrodynamic particle.* Since $\bar{v}$ has the units milliliters of particle per gram of particle, the product $m\bar{v}$ will be the volume of the particle. If this volume is now multiplied

---

*Both m and $\bar{v}$ refer to the hydrodynamic particle, which is the macromolecule plus any bound water, salt, etc., that travels with it under the action of the centrifugal field.

by the density of the embedding liquid medium, ρ, which has units grams of solution per milliliter, the product $m\bar{v}\rho$ will be the weight of an equal volume of solution. Hence the mass of the hydrodynamic particle minus the mass of an equal volume of solution becomes:

$$m_b = m - m\bar{v}\rho = m(1 - \bar{v}\rho)$$

Hence equation (3) becomes:

$$\frac{v}{\omega^2 r} = \frac{m_b}{f} = \frac{m(1 - \bar{v}\rho)}{f} \qquad (4)$$

It is more convenient to speak of molecular weights, M, than masses, m, and these are related through Avogadro's number, N, by $m = M/N$. Thus equation (4) becomes:

$$\frac{v}{\omega^2 r} = \frac{M(1 - \bar{v}\rho)}{Nf} \qquad (5)$$

For a given macromolecule in a particular medium, the right-hand side of equation (5) will be constant, for it is composed mainly of macromolecular parameters such as size, shape, and molecular weight. Thus, the left-hand side, which is the velocity per unit centrifugal field, will be also a constant. It is convenient, then, to give it a special symbol s, called the sedimentation coefficient.

$$s = \frac{v}{\omega^2 r} = \frac{M(1 - \bar{v}\rho)}{Nf} \qquad (6)$$

Using the notation of calculus, the velocity, v, may be written as $\mathrm{d}r/\mathrm{d}t$. Thus,

$$s = \frac{1}{\omega^2 r}\frac{\mathrm{d}r}{\mathrm{d}t} = \frac{1}{\omega^2}\frac{\mathrm{d}\ln r}{\mathrm{d}t} \qquad (7)$$

Equations (6) and (7) are the fundamental equations for centrifugation (refer to Chapter 18 for further details).

C. Basic Centrifugal Equipment

The preparative ultracentrifuge is capable of speeds up to 65,000 rpm. Centrifuges capable of delivering speeds of 40,000-50,000 rpm are generally adequate for sedimenting viruses. Recent centrifuges may be equipped with a "$\omega^2 t$" counter for rapid and accurate determination of the amount of centrifugation experienced by a sample. This electronic device automatically computes the square of the angular velocity and integrates over time to yield $\int\omega^2 \mathrm{d}t$.

Centrifuge rotors for low (10,000-20,000 rpm) and high speed (30,000-50,000 rpm) are essential for purification by differential centrifugation. Large volumes of extracts, for example six 250 ml plastic bottles, are spun at 10,000-20,000 rpm to sediment debris. A series of high-speed angle rotors with top-rated speeds of 30,000, 40,000, and 50,000 rpm is employed for subsequent sedimentation of the virus. For rate zonal centrifugation, swinging-bucket rotors are employed, and a 35,000-40,000 rpm six-bucket rotor probably will be most useful. The three-bucket, 50,000 rpm swinging-bucket rotor is generally used for analytical studies such as for CsCl sedimentation equilibrium studies.

The B-IV, B-XIV, and B-XV zonal rotors are made for large-scale preparative density-gradient centrifugation.

Density gradient centrifugation in virus research has been reviewed by Brakke (1960) and zone centrifugation has been reviewed by Schumaker (1967). A monograph on the zonal ultracentrifuge has been written by Anderson (1966), and the reader can refer to other subjects on centrifugation, for example sedimentation equilibrium in buoyant density gradients (Vinograd, 1961) and density-gradient centrifugation (Moore, 1969; Oster and Yamamoto, 1963; de Duve et al., 1959; and Anderson, 1959; see also Chapter 18 on analytical ultracentrifugation).

II. BOUNDARY SEDIMENTATION

For intelligent design of virus purification by centrifugal methods, it is necessary to have an approximate value for the sedimentation coefficient of the virus. Moreover, it is important to show that the virus is in solution and not bound to cell debris, and that it is not lost or damaged by centrifugation procedures.

One of the more simple and reliable procedures is to design a boundary sedimentation experiment using a partially purified tissue extract. Approximately a dozen centrifuge tubes are filled with the extract and placed in a rotor. The rotor is then spun at a high speed, stopped periodically, and a tube removed for analysis. This process is repeated until all the tubes are removed. Each tube is divided into two fractions by careful removal of the upper seven ninths by a Pasteur pipette. Then both fractions are analyzed for virus, usually by an infectivity assay. From this set of data, a rough estimate of the sedimentation coefficient may be calculated. If this value falls within the "virus window" (see Fig. 12-1), then the virus is not bound to large pieces of cellular debris, although it is possible that small pieces of debris are still attached. If the sum of the infectivities from the upper and lower fractions of each tube is equal to the original infectivity, then most of the virus has not been lost or damaged by centrifugation.

A. Materials

Cellular debris adsorbed to the virus greatly affects its sedimentation properties. The viscosity and density of the medium in turn also affect the sedimentation properties. Therefore, the sedimentation coefficient of the virus is expected to change during purification. To minimize these changes, boundary sedimentation analysis is performed on partially purified extracts rather than the original crude extract.

For determining sedimentation coefficients, infectivity assays are the most valuable, since virus only is being measured, and since low total recoveries would then indicate that virus has been damaged or lost by centrifugation.

A 40,000 rpm, 12 tube, angle-head rotor will be satisfactory for most boundary sedimentation experiments. It will yield the kinds of information desired for the design of differential centrifugation of the virus. The large number of tubes held by the angle-head rotor greatly increases the simplicity and reliability of the experiment, for the sedimentation coefficient values in the angle-head rotor are similar to those obtained in the analytical ultracentrifuge.

B. Method

A dozen plastic, screw-top centrifuge tubes are filled with an accurately measured volume of extract. Enough space should be left so that when the meniscus becomes vertical during centrifugation, it will not leak from the cap in the slanted position in the rotor.

It is valuable to make a full-scale drawing on a sheet of graph paper of a cross section through the center of the rotor (see Fig. 12-2). A drawing is readily constructed from measurements of the distance of the cavity bottom from the axis of rotation, the tube dimensions and thickness, and the angle of inclination. Most manufacturers supply these dimensions or drawings with the rotor, or will do so upon request. A centrifuge tube is filled with a known

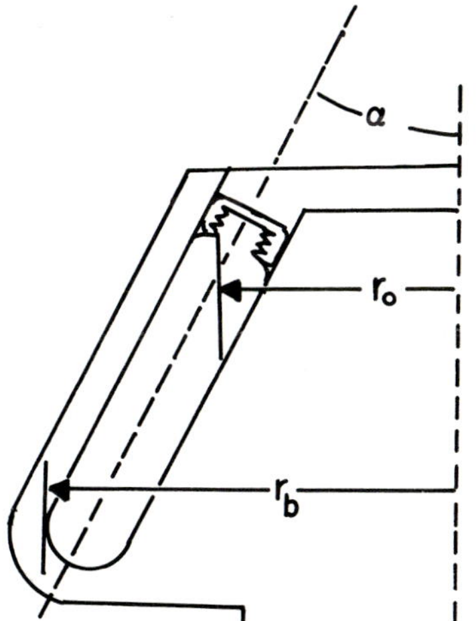

Fig. 12-2. Cross section of an angle-head rotor showing distances r_0, r_b, and the angle α .

volume of liquid, and the height of the meniscus from the tube bottom is accurately measured. When the tube is inclined, the meniscus will rotate around a "hinge-point" at its center, and the distance from the tube bottom to the meniscus along the tube axis will remain almost constant. Hence, the vertical meniscus may be located on the cross-sectional drawing. The dimensions needed for subsequent calculations are readily obtainable from this drawing by direct measurement.

The speeds at which the rotor is to be run and the time interval between removing tubes for analysis are determined by calculations to be described in Section III. The run conditions should be carefully controlled, however, and the rotor not be allowed to warm when it is stopped. The following procedure is suggested: Each time the centrifuge is stopped, the rotor is opened while it is still in the chamber of the centrifuge, a tube is carefully removed and replaced with a balance tube, and the rotor is closed and the centrifuge restarted.

The quantity $\int \omega^2 dt$ should be recorded for each tube removed. If an "$\omega^2 t$ counter" is not available, it will be necessary to measure ω periodically during acceleration and deceleration. Values of ω^2 are then plotted as a function of time, and the integral may be determined by planimetry.

To fractionate the centrifuge tubes, the procedure recommended by Anderson (1968) may be employed, although the volumes mentioned are for a 50,000 rpm rotor: "A 9 ml sample of each particle suspension was placed in each tube. After centrifugation the tubes were rotated very slowly 180° while in place, until the pellet was flat in the bottom of the tube. With a long-stem Pasteur pipette, fluid was carefully removed along the upper wall, keeping the end of the pipette at, or just below the surface. . . . The recovered liquid was placed in a 10 ml graduated centrifuge tube to measure the volume removed. In this manner, 7 ml of the supernatant was recovered, thus avoiding the problem of deciding how much of a loose pellet to remove."

For the determination of the sedimentation coefficient, only the upper fraction of each tube need be assayed. When the sedimenting boundary of virus passes through the upper 7 ml, the assay will abruptly fall to a very low number. From the value obtained by $\int \omega^2 dt$ when this occurs may be computed the sedimentation coefficient of the virus.

C. Convection

For some time it has been felt that convective disturbances which could occur in the angle-head rotor would prevent its use for the measurement of sedimentation coefficients. A theory for this convection was advanced by Pickels (1943). The experimental data presented by Pickels did not support his convective theory very well; however, the theory was so reasonable that it was accepted by most workers in the field. Recently, the theory advanced by Pickels has been challenged on hydrodynamic grounds by Anderson (1968). His theory suggests, and his experiments show, that sedimentation coefficients obtained in the angle-head rotor are close to those obtained in the analytical ultracentrifuge. Anderson used relatively high concentrations of macromolecules in some of his experiments, however, and it might be argued that the density gradients generated by the migration of concentrated macromolecular boundaries stabilized the system against serious convective disturbances. For virus preparations present in great dilution in partially clarified extracts, other soluble macromolecular components might form these stable density gradients during centrifugation, preventing convection. On the other hand, if poor results are obtained indicating that there has occurred some recycling of virus into the upper fraction, then this may be minimized by a superimposed sucrose density gradient. The extract is divided into two equal parts, and solid sucrose added to one half to make an 8-10% solution. The tubes are then half-filled with the sucrose extract, and then the other half is layered on top. A few vertical strokes of a glass rod through the solution will create a density gradient which will stabilize the system against convection.

Finally, Pickels pointed out that convective disturbance during deceleration is likely to occur just before the rotor stops. He reasons that the forces stabilizing the system are proportional to the strength of the centrifugal field, which are small at low speeds. The forces causing convection are proportional to the rate of deceleration. Therefore, he recommends that the rotor be decelerated below 5000 rpm slowly, *without* the use of the brake. Again, a density gradient would minimize such disturbances.

D. Data Analysis

In Fig. 12-3 is shown a plot of the percentage of material remaining in the upper fraction as a function of $\int\omega^2 dt$. A logarithmic interval has been chosen for $\int\omega^2 dt$, and the accuracy of the technique should be assessed with this in mind, that is, only approximate values can be obtained. Below the scale of $\int\omega^2 dt$ is drawn a second scale of $s_{20,w}$. The location of this second scale will change depending upon the dimensions of the rotor used, and the density and viscosity of the medium. If r_0 is the original meniscus position, and r_f the final meniscus position, the integration of equation (7) while holding s constant, and rearranging, yields:

$$s = \frac{1}{\int\omega^2 dt} \ln \frac{r_f}{r_0} \qquad (8)$$

The value of $\int\omega^2 dt$ to be used in equation (8) is the extrapolated intersection point where the material remaining in the upper fraction falls to zero (see Fig. 12-3). The value of s calculated by equation (8) will be the sedimentation coefficient of the virus at the particular viscosity and density of the run.

It is useful to convert the measured value of s to a value of $s_{20,w}$, which is the sedimentation coefficient which would be obtained in a medium having the same density and viscosity as water at 20° C. In order to do this, the formula to be employed is given by Schachman (1959) as:

$$s_{20,w} = \frac{s(1 - \bar{v}\rho_{20,w})}{(1 - \bar{v}\rho)} \frac{\eta}{\eta_{20,w}} \tag{9}$$

where $\rho_{20,w}$ and $\eta_{20,w}$ are equal to 0.09982 g/ml and 0.01002 poise, respectively, and ρ and η are the density and viscosity of the medium at the temperature of the experiment. The best choice for the value used for $\bar{v}$ in equation (9) will probably be an average of measured values for similar kinds of virus, until such time as the partial specific volume is finally measured on a highly purified preparation of the virus being studied.

SELECTION OF AN INITIAL SET OF $\int\omega^2 dt$ VALUES

In order to decide at which values of $\int\omega^2 dt$ the centrifuge should be stopped for the boundary sedimentation experiment, we make use of the "virus window" concept. We wish to look for particles having sedimentation coefficients, $s_{20,w}$, between 100 and 2000 Svedbergs. By measuring the viscosity and density of the purified extract, and adjusting these values to the temperature at which the centrifuge will be run, and assuming a reasonable value for $\bar{v}$, we may use equation (9) in inverted form to compute s from values of $s_{20,w}$ of 100 and 2000 Svedbergs (1 Svedberg $= 10^{-13}$ sec). With these values of s, we may now compute $\int\omega^2 dt$ by rearranging equation (8):

$$\int\omega^2 dt = \frac{1}{s} \ln \frac{r_f}{r_0} \tag{10}$$

Of course, this computation will be for a particular rotor having radii r_0 and r_f before and after removal of seven ninths of the solution.

These two values of $\int\omega_2 dt$ would be located at the end extrapolation points on a plot similar to Fig. 12-3 for virus of 100 and 2000 S. To be certain that the entire curve is obtained, however, it is advisable to remove the first tube at a value of $\int\omega_2 dt$ two logarithm units to the left of the value calculated for the larger virus. Then equally spaced intervals of log $\int\omega_2 dt$ would be taken between this initial point and the value for smaller virus. The number of points would depend upon the number of centrifuge tubes in the rotor.

III. DIFFERENTIAL CENTRIFUGATION

The technique of differential centrifugation may be applied if the virus being studied sediments into a pellet by centrifugation and remains infectious, and if by sedimentation analysis the virus is reasonably free of occluding cell debris. The principle of this technique relies on differences in sedimentation rate, that is, faster-sedimenting components pellet whereas slower-sedimenting components remain in the supernatant fluid. A centrifuge speed is usually chosen to pellet the faster contaminating components first and the virus second. The virus is ultimately purified by repeated centrifugation at low and high speeds. Such procedures of alternating low and high centrifuging speeds are

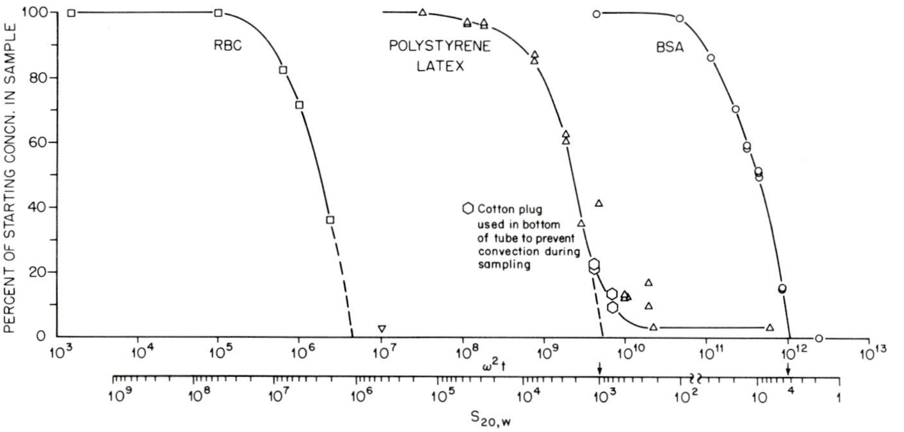

Fig. 12-3. Sedimentation of sheep red blood cells (RBC), polystyrene latex particles, and bovine serum albumin (BSA). Points indicated by squares obtained at 1 *g* using gravity; all other points obtained in Spinco No. 50 angle-head rotor. Points indicated with octagons obtained with small cotton plug in bottom of centrifuge tube. Along the abscissal axis are plotted the integral of $\omega^2 t$ and the sedimentation coefficient in Svedberg units corrected to water at 20° C. The left-hand portion of this scale is for data taken at 5° C, the right-hand segment for data taken at 10° C. Arrows indicate observed sedimentation coefficient for BSA and calculated sedimentation coefficient for polystyrene latex.

called differential centrifugation.

Differential centrifugation was developed as an effective purification procedure for tobacco ringspot virus and tobacco mosaic virus by Stanley and Wyckoff in 1937. It was rapidly adopted by many workers studying plant and animal viruses because the yield, concentration, purity, and infectivity of the recovered virus is good. Some of these procedures are described in review articles by Sharp (1953) and by Steere (1959).

A. Details and Analysis of the Technique

The purification of tomato bushy stunt virus is an example of this technique used by W. M. Stanley in 1940. Five pounds of diseased young tomato plants were cut and stored for 3 days at −12° C and then ground in a meat grinder. The ground material was buffered with dipotassium phosphate and was allowed to thaw. The extract (1400 ml) obtained from the pulp was filtered through celite (Hyflo super-cel). It was then centrifuged at 3000 rpm, the low-speed step which removes large amounts of filterable plant debris. The supernatant fluid was then centrifuged at 60,000 *g* (30,000 rpm) for 1½ hr. The supernatant fluid was discarded, and the pellet resuspended in 200 ml of 0.1 M borate buffer, pH 7. It was recentrifuged for 30 min at 3000 rpm. The supernatant fluid was again centrifuged for 1½ hr at 30,000 rpm. These cycles of low-speed and high-speed centrifugation were repeated two more times. The volume was approximately halved each time so that the final preparation contained 18 ml. The yield was 102 mg of purified virus.

Freezing and thawing of the diseased material reduces nonviral protein by 50% (more or less depending on the material). The extract was then clarified by filtration and low-speed centrifugation which is often necessary to reduce the viscosity and density of the extract to make centrifugation feasible and to remove contaminating substances not easily removed by centrifugation.

The first high-speed centrifugation separates the virus from soluble proteins and low-molecular-weight plant material and also concentrates it. Higher-speed rotors and shorter centrifugation times than previously used are subsequently employed since the volume has been reduced by six or more times.

The first high-speed centrifugation produces a large pellet of material and virus and can be washed to remove most of the virus from the debris.

Repeated cycles of low-speed and high-speed centrifugation improve the purity of the virus, but reduce the yield with each cycle. The mathematics of the recycling process are discussed in detail in the next section. This mathematical treatment is, at best, only a guide, for unexpected behavior is likely to occur. For example, sometimes virus will aggregate when purified of low-molecular-weight contaminating proteins.

Loose pellets, from which it is difficult to remove the supernatant fluid, may cause trouble. This difficulty can be particularly acute during the second and subsequent high-speed centrifugations when pellet mass is greatly decreased. When loose pellets occur, as much fluid as possible is removed with a Pasteur pipette without disturbing the loose pellet. The mathematics of purification are still reasonably good. Two cycles of tenfold purification are equivalent to one cycle of hundredfold purification, and the final yield could be much greater for a loose pellet.

Occasionally, one problem of differential centrifugation is the loss of desirable but noninfectious components such as empty virus capsids. These components contain less mass, and they are progressively lost during differential centrifugation. Capsid components are detected by rate zonal centrifugation and isopycnic banding. Once their presence is suspected and sedimentation characteristics measured, the differential centrifugation procedure may be adjusted for the isolation of these components.

B. The Mathematics of Differential Centrifugation

If r_0 is the original boundary position (usually the meniscus) of particles with a sedimentation coefficient s, then integration of equation (7) and converting it into exponential form will yield the final boundary position, r_f, according to the expression:

$$r_f = r_0 \, e^{s \int \omega^2 \, dt} \tag{11}$$

If the cell bottom is located at r_b, then the fraction q of the original solution still containing particles is given approximately by the expression

$$q = \frac{r_b - r_f}{r_b - r_0} \tag{12}$$

Equation (12) is only approximate because no provision is made for the hemispherical tube bottom. For our purposes, however, this is not a serious defect.

We shall assume, also, that the fraction of the original particles which are pelleted is $p = 1 - q$, and the fraction which remains in solution is q. This assumption ignores the radial dilution effect which occurs in a centrifugal field, but since we are interested in an approximate answer only, again this is not a serious defect.

Equations (9), (11), and (12) may be used to compute p and q as a function of $s_{20,w}$ for a given rotor, medium, and value of $\int \omega^2 dt$. Should q become negative, it should be replaced by zero. This is a simple problem for a programmable desk top computer. These values of p and q may be plotted, and in Fig. 12-4 are shown the results of a plot of p for a 2 hr run at 19,000 rpm, $r_0 = 4.4$ and $r_b = 8.9$ cm, at 5° C in a medium with the viscosity and density of water.

For a high-speed centrifuge run in which the pellet contains the virus, and the supernatant fluid is discarded, a plot of p vs. $s_{20,w}$ gives the fraction of material of each particular sedimentation constant that is retained in the pellet.

For a low-speed run in which the pellet is discarded and the supernatant fluid saved, a plot of q vs. $s_{20,w}$ gives the fraction of material of each particular sedimentation constant that is retained in the supernatant fluid.

Let p be the value of p for a high-speed run, and q' be the value of q during the low-speed run. Then a plot of the product pq' gives the fraction of material of each particular sedimentation constant that is retained during a cycle composed of a high-speed and a low-speed run.

If the differential centrifugation procedure is composed of a number of cycles, then a plot of the product $p_1 p_2 \ldots p_n q_1' q_2' \ldots q_n'$ vs. $s_{20,w}$ gives the fraction of material of each particular sedimentation constant that is

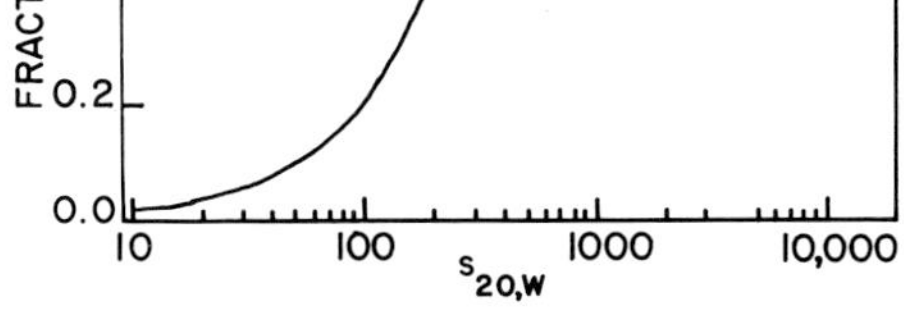

Fig. 12-4. **A plot of the fraction of material at any given sedimentation coefficient, $s_{20,w}$, which would be pelleted according to equation (12). Centrifugation conditions are: $\omega = 19{,}000$ rpm, $t = 2$ hr, $r_0 = 4.4$ cm, $r_b = 8.9$ cm, $T = 5°$ C; and assuming the medium has the viscosity and density of water.**

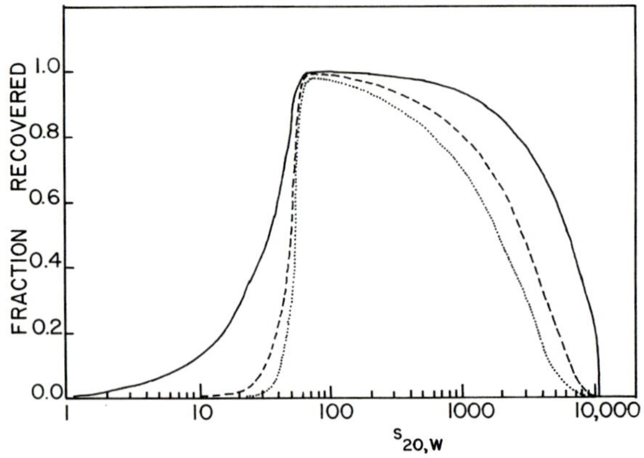

Fig. 12-5. A calculated plot of the fraction of material at any given sedimentation coefficient which would be recovered after 1 (solid line), 3 (dashed line), and 5 (dotted line) cycles of differential centrifugation. Centrifugation conditions for each cycle are: 120 min at 40,000 rpm to form the pellet. The virus is then extracted and clarified at 6000 rpm for 30 min. Rotor dimensions are $r_0 = 4.8$ cm, $r_b = 8.0$ cm, and $T = 5°$ C. The solvent is assumed to have the density and viscosity of water.

retained over the entire experiment. If the conditions for each of the "*n*" cycles are identical, this product may be simplified as $[pq']^n$.

In Figure 12-5 is shown such a plot for 1, 3, and 5 cycles at 5° C in dilute aqueous medium at high and low speeds of, respectively, 40,000 and 6000 rpm in a rotor having values of r_0 and r_b of 4.8 and 8.0. Low-speed runs are for 30 min and high-speed runs are for 2 hr.

From inspection of Fig. 12-5, it may be seen that for 3 and 5 cycles, the "cutoff" between 40 and 60 *S* is sharp. This is a characteristic of the higher-speed runs. But the "low-speed cutoff" is very broad, particularly as the number of cycles is increased.

By changing procedures, however, it is possible to make both the "high-speed cutoff" and the "low-speed cutoff" reasonably sharp. To do so, a much higher speed is chosen for the "low-speed" run. The initial "low-speed" pellet will then contain a sizable fraction of the virus. Next, this pellet is washed with about one-half of the original volume or less of fresh solvent and centrifuged at the same "low speed." The washing procedure is repeated with the second pellet. These two washes are then re-added to the original supernatant. We may calculate that the first "low-speed" pellet contained a fraction of p' of any component. After the first wash, the resulting pellet contained p'^2, and after the second wash, p'^3 of any component. Thus, after re-adding these washes to the original supernatant, the amount of each component conserved in the supernatant fluid is $(1 - p'^3)$. If three high-speed runs are now made to extract the combined supernatant fluids, the fraction of each component recovered in the final pellet will be p^3. Thus the total recovery from both the high-speed and low-speed runs

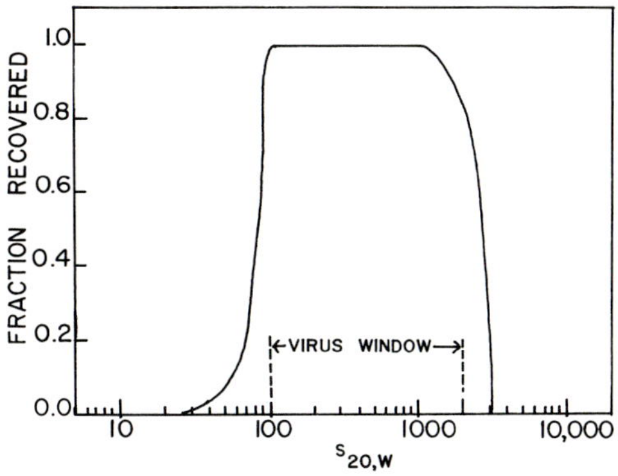

Fig. 12-6. A calculated plot of the fraction of material at any given sedimentation coefficient which would be recovered after 3 low-speed washes followed by 3 high-speed pellets. Centrifugation conditions are chosen to initially form a pellet at 13,500 rpm, saving the supernatant solution. The original pellet is then washed two times at 13,500 rpm, and the wash fluids combined with the original supernatant solution. These combined solutions are then pelleted at 38,000 rpm, and the pellet dissolved and reformed two times more at 38,000 rpm. The fraction recovered in the final pellet is calculated as described in the text. Rotor dimensions are $r_0 = 4.8$ cm, $r_b = 8.0$ cm, and $T = 5°$ C.

will be $p_3(1 - p'^3)$. Such a plot is shown in Fig. 12-6 for three "low-speed" runs at 13,400 rpm, and three "high-speed" runs at 38,000 rpm, in a rotor with dimensions r_0 and r_b of 3.8 and 8.1 cm. High-speed runs are for 120 min, and low-speed runs are for 30 min.

IV. ISOPYCNIC BANDING FOR PURIFICATION

If macromolecules are centrifuged through a solvent containing a steep density gradient of salt, sucrose, or other low-molecular-weight solute, eventually the large molecules will approach that point in the centrifuge tube where the solution density will be equal to the buoyant density of the particles. At this point, the term $(1 - \bar{v}\rho)$ in equation (6) will become zero, and sedimentation will stop. At the conclusion of the run, the macromolecules will be found in a narrow zone centered at this isopycnic (or isodensity) location in the centrifuge tube (Fig. 12-7).

If centrifugation is prolonged until true equilibrium conditions are obtained for both the macromolecules and the low-molecular-weight solute, the technique is called "sedimentation equilibrium in a buoyant density gradient." If equilibrium of all species is not reached, however, the technique is usually called "isopycnic banding." Analyses of purified virus at sedimentation equilibrium in the analytical centrifuge are described in Chapter 18.

As described previously, differential centrifugation is used to purify virus particles from other materials according to differences in sedimentation

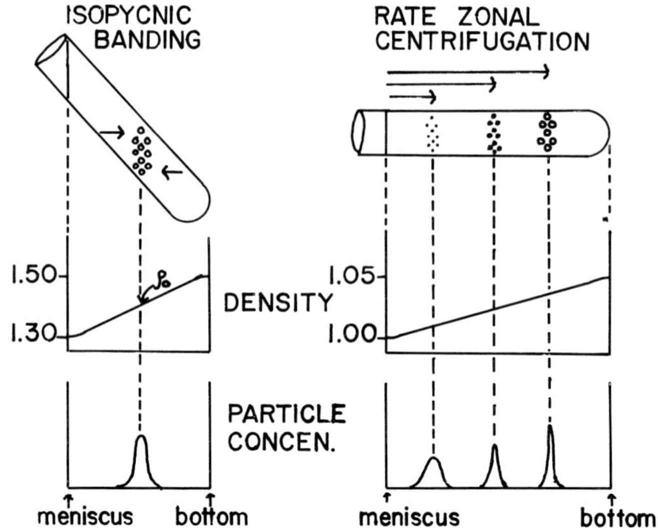

Fig. 12-7. On the left is illustrated the technique of isopycnic banding performed with an angle-head rotor. A density gradient from 1.30 to 1.50 density units is formed in the tube, and the macromolecules, here assumed to have a density of about 1.4, collect and band in the center. Particles are separated according to buoyant density by isopycnic banding. On the right is illustrated the technique of rate zonal centrifugation, using a swinging bucket rotor. A zone, containing macromolecules, is layered on top of a density gradient. The gradient is present in order to prevent convective disturbances, but usually the highest density at the tube bottom is not sufficient for isopycnic banding. Under the action of the centrifugal field, the macromolecules will migrate along the axis of the tube with a velocity equal to $s\omega^2 r$. Particles are separated according to size and shape by rate zonal centrifugation.

properties. But differential centrifugation does not separate virus from those materials possessing different densities but similar sedimentation properties. As can be seen from Fig. 12-1, the region lying between 100 and 2000 Svedbergs, which includes the virus particles, also contains microsomes, polysomes, glycogen, and perhaps a variety of membrane fragments and other debris which could be present in special cases. Sometimes these contaminating materials may be dissolved by detergents such as desoxycholate, or caused to coagulate by salts, acidification, etc., or digested by enzymes such as ribonuclease. Often these procedures will not remove all the contaminating materials, however, or else they will cause excessive loss or damage to the virus. Moreover, particles which have an incomplete content of nucleic acids or are empty altogether may be included with the virus. Fortunately, virus and virus-like particles often are cleanly separated from contaminating materials having similar sedimentation properties, by isopycnic banding.

The use of this purification technique may be described as taking a horizontal slice through the $s-\rho$ plot shown in Fig. 12-1. Differential centrifugation, on the other hand, may be described as taking a vertical slice

Rate zonal sedimentation is an exceptionally gentle technique for purifying fragile biological structures, as the sedimented material is not pelleted, thus need not be redissolved; is not stored or sheared, or exposed to any surfaces or reagents other than the gradient itself. In one variation of the technique, fragile macromolecules which cannot tolerate pelleting and redissolution are freed of slower-sedimenting contaminants and concentrated in one step, by sedimenting them onto a "shelf" of concentrated sucrose solution at the bottom of the tube. The macromolecules can penetrate only a very short distance into this viscous solution, and are stopped without ever touching a solid surface.

A. Materials and Methods

The gradients which are used for rate zonal studies are generally less steep and less concentrated than those employed for the isopycnic banding studies. In fact, it has become almost standard to use 5-20% sucrose gradients for many biological materials including proteins and ribosomes. For plant viruses, however, somewhat steeper gradients have been extensively and successfully used by Brakke (1960).

Because of the more favorable geometry with less sidewall impaction of the sedimenting particles, swinging-bucket rotors have been used in almost all rate zonal studies. Some experiments by Charlwood (1963) have indicated that it might be possible to use angle-head rotors, however.

Rate zonal centrifuge studies are occasionally ruined by serious convective disturbances. Evidently, these occur soon after layering the zone of macromolecules on top of the density gradient and before the tubes are spun in the centrifuge (Nason et al., 1968). The causes of these convective disturbances are not thoroughly understood, but involve the different diffusion rates of the macromolecules and the low-molecular-weight solute. The instability of layered zones to small perturbations is treated mathematically by Sartory (1968).

As a matter of practical experience, it has been found that the magnitude of the convective disturbance is minimized as the concentration of macromolecules in the zone is reduced. The virus should be in a low-density buffer, and centrifugation should be started promptly after layering. On a 5-20% sucrose gradient, the maximum concentration of virus which can be layered without overwhelming convection is about 0.2 mg/ml. Even at this concentration, some convection will ensue. The zones will be considerably sharper if the virus concentration is reduced by an order of magnitude. Another technique which may be used to minimize convection is to layer the initial thin zone of virus in such a manner as to form an inverted virus gradient stabilized by a steeper gradient of sucrose or salt. A small gradient engine is used to generate this kind of initial zone, by placing the virus in the chamber of the gradient engine which also contains the less dense of the two solvents.

These convective disturbances do not seem to occur in the zonal centrifuge, an instrument described in detail in the last section of this chapter. For this reason, as well as because of its large capacity, the zonal centrifuge can handle more material than swinging-bucket rotors in rate zonal centrifugation experiments. With the zonal centrifuge, the technique can be used on a preparatory scale; with swinging-bucket rotors, the technique is analytical, that is, useful primarily for accurate measurement of sedimentation coefficients and number of components, and perhaps to prepare material for infectivity and electron microscope studies.

Since it is desirable to be able to locate with accuracy the radial positions of zones of macromolecules within the centrifuge tubes, an equation must be derived to relate radial distance to fraction number and volume. For swinging-bucket rotors, the volume, V, of solution centrifugal to any radius r is given by:

$$V = 1/2(4/3\,\pi a^3) + (r_b - a - r)\pi a^2 \tag{15}$$

where a is the radius of the centrifuge tube (see Fig. 12-8). Solving for r yields:

$$r = r_b - \frac{a}{3} - \frac{V_0}{\pi a^2} \tag{16}$$

If the volume of solution originally placed in the centrifuge tube is V_0, then the meniscus position, r_0, is calculated by this equation as:

$$r_0 = r_b - \frac{a}{3} - \frac{V_0}{\pi a^2} \tag{17}$$

Subtracting equation (17) from (16) and rearranging yields:

$$r = r_0 + \frac{1}{\pi a^2}\left(\frac{V_0 - V}{V_0}\right)V_0 \tag{18}$$

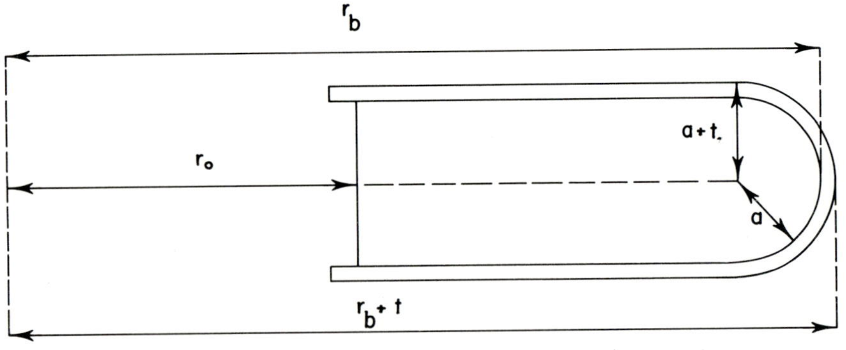

Fig. 12-8. Cross section of a swinging-bucket tube showing the distances measured from the axis of rotation to the meniscus, r_0, and to the tube bottom, r_b. The distance a is the internal radius of the centrifuge tube, while t is the thickness of the plastic forming the tube.

During fractionation, due to material left on the walls, the volume recovered is always less than that which was originally placed in the tubes. Let V_i be the volume recovered up to and including the ith fraction. Let V_0' be the total volume recovered. Then the radial position *to the center of the* i^{th} *fraction*, r_i may be obtained by the following expression:

$$r_i = r_0 + \left[\frac{2V_0' - \left(V_i' + V_{i-1}'\right)}{2V_0'}\right] \frac{V_0}{\pi a^2} \tag{19}$$

(Equation (19) is derived by writing equation (18) for both the ith fraction and the $i - 1$ fraction, and then adding these equations together and dividing by 2.) In equation (19), the primed volumes may be measured in any kind of units, such as drops. The unprimed V_0, however, must be measured in cubic centimeters. The equation is not quite correct for those fractions lying within the hemispherical tube bottom.

Serious errors in the determination of radial distances will occur if the volumes of the consecutive fractions are not equal. In order to obtain volumes of equal size, a small peristaltic pump may be connected to the exit tubing. With each rotation of this pump, a simple noise making device, consisting of a metal strip under tension encountering a hole drilled in the rotating drum, makes a sharp clicking sound. With the use of this device, volumes can be delivered with good reproducibility.

The concentration of macromolecules in each fraction, c_i, is measured and plotted on a histogram against fraction radius, r_i. The radius of the center of mass, $\langle r \rangle$, of the zone of macromolecules is calculated by summation using the defining expression

$$\langle r \rangle = \frac{\Sigma r_i c_i}{\Sigma c_i} \tag{20}$$

The rate of migration of the center of mass is the quantity that is exactly related to the sedimentation velocity of the macromolecules, as shown by Schumaker and Rosenbloom (1965).

B. Mathematics of Rate Zonal Centrifugation

Equation (13) may be rearranged to yield the desired expression for the sedimentation coefficient of the migrating zone of macromolecules:

$$s_{20,w} = \frac{(1 - \bar{v}\rho_{20,w})}{\eta_{20,w} \int \omega^2 dt} \int_{\langle r \rangle_0}^{\langle r \rangle} \frac{\eta}{(1 - \bar{v}\rho)} \, d\ln r \tag{21}$$

where $\langle r \rangle_0$ is the radial distance to the center of mass of the initial zone of macromolecules. The integral on the right-hand side of equation (21) may be evaluated by planimetry, as previously described. Then equation (21) may be used to determine the sedimentation coefficient of the particles, $s_{20,w}$.

A better procedure, both in accuracy and in simplicity, has been devised by Brakke, (1958). A second virus of well-characterized sedimentation properties is used as a marker in the same centrifuge tube or a different tube in the same rotor. A plot is constructed of $\langle r \rangle$ vs. $\int \omega^2 dt$ for both viruses (Fig. 12-9). Smooth curves are drawn through the points. Then values of $\int \omega^2 dt$ are read from the curves for both virus zones at the same $\langle r \rangle$ locations. The ratio of these values $\int \omega^2 dt$ is equal to the inverse ratio of the sedimentation coefficient multiplied by a small term involving the partial specific volumes. To eliminate this small term, the ratio of the values of $\int \omega^2 dt$ is plotted as a function of $(\rho + \rho_{20,w})/2$, and extrapolated to $\rho_{20,w}$. This extrapolated value then will be exactly the ratio of sedimentation coefficients (Fig. 12-10).

VI. SEDIMENTATION EQUILIBRIUM IN BUOYANT DENSITY GRADIENTS

Rate zonal centrifugation, as described in the previous section, is actually a rather precise analytical tool for the measurement of sedimentation coefficients. Differential centrifugation, on the other hand, is a preparative tool which separates particles according to sedimentation coefficient.

Sedimentation equilibrium in a buoyant density gradient, which is described in this section, is actually a rather precise analytical tool for the measurement of buoyant densities. Isopycnic banding, on the other hand, is the preparative tool for the isolation of particles according to density.

Using sedimentation equilibrium methods, it is possible to obtain buoyant densities of macromolecules with impressive precision. Molecular weights also may be measured in these experiments if the macromolecules are homogeneous in density. If they are not homogeneous in density, then it is possible to determine the degree of density heterogeneity once the molecular weight is known.

As previously described, the technique of sedimentation equilibrium in a buoyant density gradient involves centrifuging a solution containing virus and a low-molecular-weight solute such as CsCl until both the macromolecular component and the salt have obtained an equilibrium distribution in the centrifugal field. When equilibrium is obtained, the salt concentration will be lowest at the meniscus, and highest at the bottom, and a gentle gradient of density will be formed throughout the centrifuge tubc. Thc virus will sediment from the meniscus and float from the bottom, and collect in a band of Gaussian shape centered about the isopycnic position. Since the system will approach thermodynamic equilibrium, the macromolecular properties determining the shape and location of the band may be computed with great precision. The resolving power of the technique is very great. For example, two viruses of 10^7 molecular weight present in equal concentrations should be resolvable as a "double-peaked" pattern if their buoyant densities differ by only 0.004 density units. If the virus were 50% DNA, this small density difference would

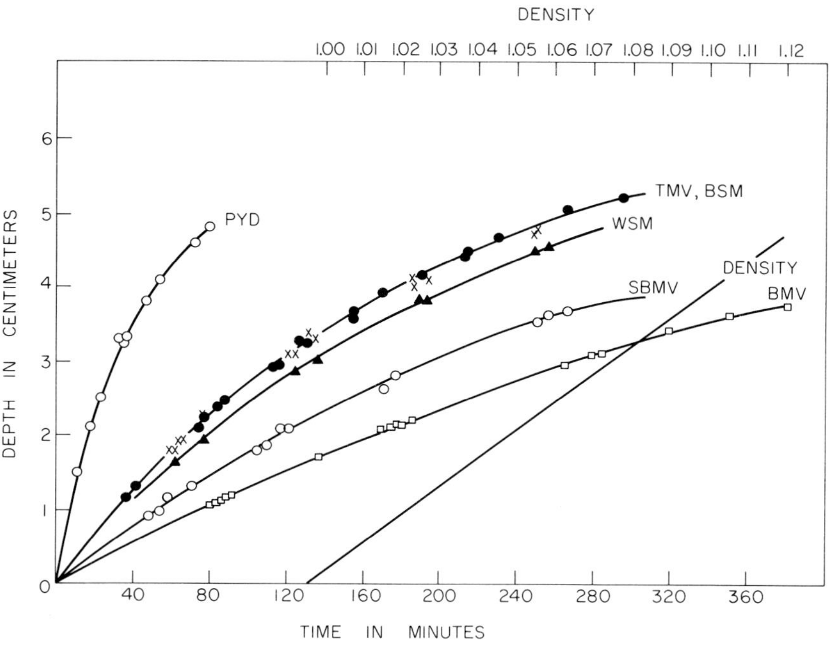

Fig. 12-9. Depth of viruses in a gradient column prepared from 3, 7, 7, 7, and 4 ml of solutions of 0, 100, 200, 300, and 400 g sucrose per liter in a 1×3 inch centrifuge tube after centrifugation at 24,000 rpm for various periods. The sucrose was dissolved in 0.0067 M K_2HPO_4, 0.0033 M KH_2PO_4. Two milliliters of virus solution containing approximately 0.02-0.2 mg virus per milliliter were floated on each column. Temperature after centrifugation was 5-6° C. The density curve is an average curve obtained by measurements on several gradient columns. PYD = potato yellow dwarf virus; TMV (•) = tobacco mosaic virus, BSM (X) = barley stripe mosaic virus, WSM = wheat streak mosaic virus; SBMV = southern bean mosaic virus; BMV = brome grass mosaic virus. (Brakke, 1958.)

correspond to a loss of about 4% of the DNA.

Nucleic acid isolated from the virus may also be analyzed by sedimentation equilibrium in buoyant density gradients. RNA is more dense than DNA and is usually studied in Cs_2SO_4 rather than CsCl solutions. Banding of nucleic acids is probably the best way of removing traces of protein normally associated by electrostatic forces and other types of noncovalent bonds to the DNA. Infectious nucleic acid assays made with material purified in this manner should be free of traces of virus, although suitable nuclease-treated controls must be simultaneously employed.

Circular, intact, double-stranded DNA may be isolated from linear DNA by sedimentation equilibrium in a buoyant density gradient in the presence of 100 μg/ml of ethidium bromide,* as described by Radloff et al. (1967). This

*Ethidium bromide may be obtained from the Boots Pure Drug Co., Ltd., Nottingham, England.

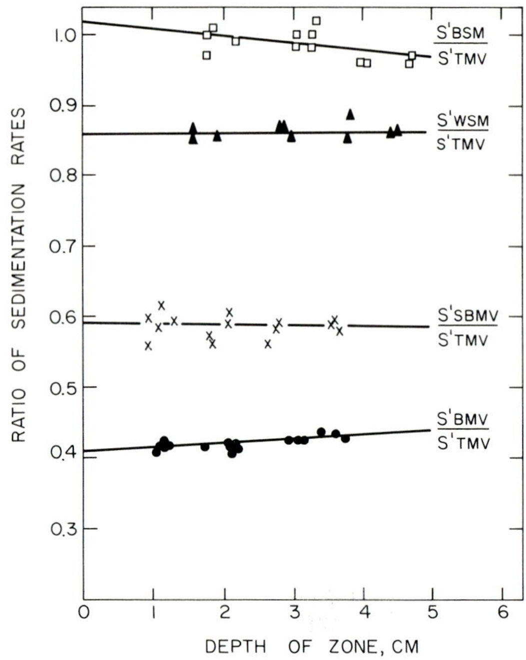

Fig. 12-10. **Ratio of the sedimentation rates (*S'*) of various viruses to the sedimentation rate of tobacco mosaic virus plotted against the depth of the zone of the first virus. Experimental conditions and symbols are the same as in Fig. 12-9. (Brakke, 1958.)**

heterocyclic dye intercalates between the stacked purine and pyrimidine bases in the DNA molecule, and the insertion of each molecule of dye causes a rotation of 12° of the helix (in the direction of unwinding). If the DNA molecule is linear, the ends are free to rotate with the insertion of each dye molecule. But if the DNA molecule is circular and without single-strand interruptions, then it will become a "supertwisted" circle as dye is taken up. Because of the strain imposed by this supertwisting of the DNA molecule, much less dye is bound. Therefore, in a buoyant density gradient, the circular DNA bands at a different position in the tube from the linear DNA. Since the dye is not as dense as the DNA, the linear molecules are found at a less dense location than the circular DNA.

A. Materials and Methods

With CsCl and Cs_2SO_4 solutions, it is useful to have relations for solution density, refractive index, and concentration in weight percent. Some of these relations are given by Hearst and Vinograd (1961), and by Moore (1969).

Concentrated salt solution is usually added to the virus preparation to make a final solution having a buoyant density close to that of the virus. This solution is then placed in the tubes of a swinging-bucket rotor and spun at a high rate of speed for several days. The time required to reach within 0.1% or 1% of the equilibrium distribution may be calculated from the following two formulas:

For the CsCl (Van Holde and Baldwin, 1958):

$$t_{0.1\%} = \frac{(r_b^2 - r_0^2)\, F(\alpha)}{D}$$

For the macromolecule (Meselson and Nazarian, 1963):

$$t_{1\%} = \frac{\sigma^2}{D}\,[\ln((r_b - r_0)/\sigma) + 1.26]$$

where $\alpha = 4D/(s\omega^2(r_0 + r_b))$ and $F(\alpha)$ is a function tabulated by Van Holde and Baldwin (1958). The $\sigma^2 = 2RT/(M\bar{v}d\rho/dr\ \omega^2(r_0 + r_b))$, where $d\rho/dr$ is the equilibrium effective density gradient of the salt (Hearst and Vinograd, 1961).

For example, using a swinging-bucket rotor with $r_0 = 4.7$ cm, $r_b = 9.8$, at 40,000 rpm, it will require 7 days for a cesium chloride solution to approach within 0.1% of the final equilibrium distribution. For a spherical virus particle of 10^7 mol wt ($D = 0.73 \times 10^{-7}$ cm^2/sec, $\bar{v} = 0.6375$) in a CsCl gradient in the same rotor at the same speed, it would require only 9 hr to reach equilibrium. It is obvious, therefore, that several days may be saved by starting with preformed density gradients of the salt.

It is essential that no leakage occur when using this technique for analytical measurements. Even a slow evaporation of liquid will cause a progressive change in concentration of the low-molecular-weight solute, and the system will never reach an equilibrium distribution.

Fractionation techniques and methods used for the detection of the macromolecules in equilibrium studies are similar to those used in rate zonal studies. Since knowledge of the precise shape of the virus band is desired, it is important that the fractions be uniform in size.

Damage or destruction of virus particles with the dissociation and loss of some of the protein subunits is sometimes a problem when using the concentrated salt solutions employed in buoyant gradient studies. Protection of the virus sometimes may result if a low concentration of Mg^{2+} or other divalent cation is added to the solution.

B. Mathematics of Sedimentation Equilibrium in a Buoyant Density Gradient

The rate of mass transport per unit area, J, in units of grams per second per square centimeter, which is due to sedimentation, is given by the expression:

$$J_s = vc \tag{22}$$

where v is the velocity and c the concentration of macromolecules. The rate of mass transport per unit area due to diffusion, J_d, is expressed by Fick's first law (Tanford, 1959):

$$J_d = -D\left(\frac{\partial c}{\partial r}\right)_t \tag{23}$$

where D is the diffusion coefficient, $(\partial c/\partial r)_t$ is the concentration gradient at a particular time, and r is the distance from the axis of rotation. Since $D = RT/Nf$ where R is the gas constant, T is absolute temperature, N is Avogadro's number, and f is the translational frictional coefficient, the last equation may be written as:

$$J_d = -\frac{RT}{Nf}\left(\frac{\partial c}{\partial r}\right)_t \tag{24}$$

At sedimentation equilibrium, the net transport of mass by sedimentation and diffusion must equal zero. Therefore, we may write:

$$J_s + J_d = 0$$

or

$$vc - \frac{RT}{Nf}\left(\frac{\partial c}{\partial r}\right) = 0 \tag{25}$$

Into the last equation may be substituted the defining expression for the sedimentation coefficient, equation (6):

$$s\omega^2 rc - \frac{RT}{Nf}\left(\frac{\partial c}{\partial r}\right) = 0 \tag{26}$$

According to equation (6), $s = M(1 - \bar{v}\rho)/Nf$. Substituting this expression into the previous equation and rearranging yields:

$$\frac{M(1 - \bar{v}\rho)\omega^2}{RT} = \frac{1}{rc}\,\frac{dc}{dr}$$

or

$$\frac{M(1 - \bar{v}\rho)\omega^2}{2\,RT} = \frac{d\ln c}{dr^2} \tag{27}$$

At true equilibrium, virus bands are usually so narrow that it is difficult to obtain sufficient fractions for analysis of the band shape. But if it is possible to do so, then equation (27) may be used for data analysis. The concentration of macromolecules is measured for each fraction and the corresponding radial distance determined by equation (19). Occasional refractive index measurements are also required so that a plot of densities may be constructed. From the values of virus concentration and radius for each fraction, a graph of ln c vs. r^2 is made, and a smoothed curve is carefully drawn through the experimental points. By very careful reading of ordinate values at regular and close intervals along the ln c vs. r^2 curve, the slopes, d ln c/dr^2 may be computed for the midpoints of each interval. These slopes are then plotted as a function of density, ρ. From inspection of equation (27), it may be seen that a straight line should be obtained with slope equal to $M\bar{v}\omega^2/2RT$, while the intercept on the horizontal axis will give the buoyant density of the hydrodynamic unit.

As was discussed in detail in Section IV, the buoyant density of the hydrodynamic unit is a function of the preferential binding of water or solute, Γ'. In addition, the molecular weight which is measured is also affected by this term. The "true" molecular weights and partial specific volumes are given by the expressions:

$$M_{\text{true}} = M/(1 + \Gamma') \qquad (28)$$

$$\bar{v}_{\text{true}} = \bar{v} + \Gamma' (\bar{v} - \bar{v}_w) \qquad (29)$$

where M and $\bar{v}$ are obtained from the slope and intercept of the plot of equation (27).

Equation (28) and (29) are written for systems homogeneous in molecular weights and in densities. If molecular-weight heterogeneity is present without density heterogeneity, equation (27) gives a value usually very close to the weight average molecular weight at band center. But if density heterogeneity is present, the molecular weights will be fictitiously low. An analysis of the degree of density heterogeneity may be performed, but it is beyond the scope of this chapter. Details are given by Sueoka (1959) and by Baldwin (1959).

VII. ZONAL CENTRIFUGATION

The zonal centrifuge rotors constitute a set of powerful new research tools which combine the analytical features of density-gradient centrifugation with the high-capacity feature of differential centrifugation. The quantities of material that may be handled, and the degree of purity that can be obtained, far exceed the capabilities of any other centrifugal technique. Thus, it is not surprising to find that the zonal centrifuge rotors have become a major research tool for the virologist.

The zonal rotor is a large cylindrical vessel capable of withstanding the high pressures generated by great rotational speeds. The liquid being centrifuged nearly fills the whole interior. Thus, these rotors are strikingly different from earlier centrifuge rotors which hold their liquid contents in centrifuge tubes in a number of cavities within the rotor.

The most novel feature of these rotors, however, is the mechanism whereby they are loaded and unloaded. These operations are performed while the rotor is spinning at a few thousand revolutions per minute. First, a density gradient is introduced, filling the rotor. Then, the sample is layered as a zone on top of the gradient. Next, the rotor is disconnected from the feed lines, the armored vacuum chamber closed, and the rotor accelerated to a high speed. After running for an appropriate time, the rotor is decelerated back to the low speed, the chamber opened, and the discharge lines connected. Finally, the gradient is removed from the rotor while it is still spinning, and the solution is fractionated into a large number of samples (Fig. 12-11). Other zonal rotors spin without disconnecting the feed lines since a bearing assembly is attached to the rotor's spindle.

The capacities of different kinds of zonal rotors vary with their dimensions, but the volumes are usually in excess of 650 ml. The size of the sample which may be layered on top of the gradients may be scaled-up accordingly. In addition, it has been found that some of the kinds of convective processes which may disturb the zones layered at one gravity are absent or minimal in the presence of the small centrifugal field generated by the low rotational speed employed during the loading process with the zonal rotor. Therefore, considerably higher concentrations of virus may be layered over the gradients in the zonal centrifuge rotor. This combination of high concentration and large volume means that rate zonal methods may be carried out on a truly preparative scale. And because of the superior resolving power of the density-gradient techniques, the rate zonal method enables separations to be performed which are an order of magnitude better than may be achieved by differential centrifugation.

Since virus may be isolated without pelleting at any stage, zonal centrifugation is the most gentle of the centrifugal isolation techniques. Sterility could be maintained by completely enclosed systems interconnecting different rotors used for rate zonal centrifugation and isopycnic banding. Finally, the purity of vaccines obtained by zonal centrifugation exceeds that from other commonly employed techniques.

A. Materials and Methods

Zonal rotors are constructed of duralumin or titanium, and the solvents employed come directly into contact with the metal. Titanium is quite inert, and is much to be preferred over the anodized duralumin. The cost of titanium rotors is greater by more than 50%. Over a period of years, however, there is no additional expense since the lifetime of the titanium rotor is at least double that of the duralumin rotor. In addition, titanium metal is stronger and the rotors can be used at higher rotational speeds.

A valuable monograph concerning the design, theory, and application of zonal centrifuges can be obtained from the U.S. Government Printing Office (Anderson, 1966). By shape, the rotors can be divided into two classes, those which are broad and squat, and those which are tall and thin. These two shapes were chosen as being dynamically stable when rotated. The broad, squat rotors are made to fit already existing preparative centrifuges, and can be purchased

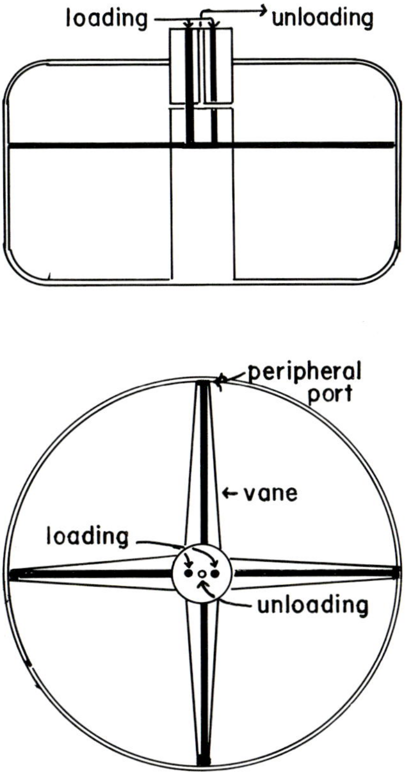

Fig. 12-11. Two views of vertical (top) and horizontal (bottom) cross sections of a typical zonal rotor, such as the B-XIV or B-XV. The gradient is introduced through the loading ports while the rotor is spinning at a few thousand rpm. The gradient enters the rotor through the two loading holes, just off center in the rotor core, and then through passages drilled in the vanes out to the periphery, where it is discharged into the hollow interior. When the rotor is filled, liquid will begin to emerge from the central exit. After this occurs, flow is reversed, and the macromolecule-containing sample is introduced through this central (unloading) exit. This zone of macromolecules is layered on top of the density gradient next to the core. An overlay of even lighter liquid is often introduced to move the initial macromolecule-containing zone farther into the rotor in order to increase field strength at the start of the experiment. Then the feed lines are disconnected, and the rotor is capped and taken to a high speed for the rate zonal or ispycnic banding study. At the end of the run, the rotor is decelerated to a low speed and unloaded. The lines are reconnected, and fractions are removed from the central (unloading) port by the introduction of a very dense solvent at the periphery.

from major manufacturers. Those rotors which are tall and thin include the continuous-flow, very-high-capacity K rotors which require special centrifuge chambers.

For the isolation of very large cell particulates and the fractionation of plasma cells, nuclei, etc., some of the A series rotors, such as the A-XII, which may be purchased commercially, are particularly useful. The A-XII rotor is designed to operate at relatively low speeds. Moreover, it is constructed with a plastic top and bottom and the zones of cell components may often be observed as they migrate in the rotor as well as during fractionation.

For high-speed operation, the all-metal B-XIV and B-XV rotors may be used for the separation of small cell particulates, viruses, and large proteins. These rotors have capacities of 650 ml and 1665 ml respectively. Loading of these rotors is accomplished through a central, detachable, rotating seal assembly. Liquid enters the rotor through the rotating seal close to the axis, and it then flows from the center through holes in the vanes to the periphery of the rotor where it enters the central chamber (Fig. 12-11). The "light end" of the density gradient is introduced into the rotor first. The gradient is continuously displaced toward the center as denser material is pumped into the chamber. When the chamber is completely filled, the light end is forced out through another passage through the central shaft. The sample is applied, flow is reversed, and the sample, now in contact with the light end of the gradient, is pumped into the center. The sample is then followed with a still lighter solution, called the "overlay," which forces the sample some distance away from the center and into a region of high centrifugal field. The external tubing connections may then be removed and rotor speed increased. Centrifugation proceeds as described for rate zonal centrifugation or isopycnic banding. After the run, dense fluid is pumped into the periphery of the rotor and the gradient is displaced out the center for analysis and fractionation.

For loading and unloading the B-XIV and B-XV rotors, a simple gradient engine has proved satisfactory and inexpensive. In Fig. 12-12 is shown a drawing of the apparatus, consisting of a reservoir for the densest solution, a mixing chamber, and a pump, all interconnected to permit either the generation and loading of a gradient or the pumping of dense fluid directly into the rotor in order to displace the gradient from the center.

At the conclusion of the experiment, it is of great value to have a continuous recording of the optical density of the gradient solution while it is being unloaded and fractionated on a fraction collector. The required instrumentation consists of a quartz flow cell and an automatically recording ultraviolet spectrophotometer. Available instrumentation may often be modified for this purpose. Since flow rates during unloading can be 100 ml/min or more, special flow cells are required.

Some new rotors are currently being developed which may be unloaded at the periphery as well as at the rotor center. With the use of these rotors, it should be possible to perform rate zonal separations, and then remove from the rotor both slower and faster contaminants, leaving the virus still inside. Then, a dense gradient may be introduced and isopycnic banding performed with the virus. By combining these two techniques in one run using one rotor, sterility would be much easier to maintain, the virus would be subject to less handling, damage, and loss, and the entire centrifuge procedure would be simplified and reduced in time.

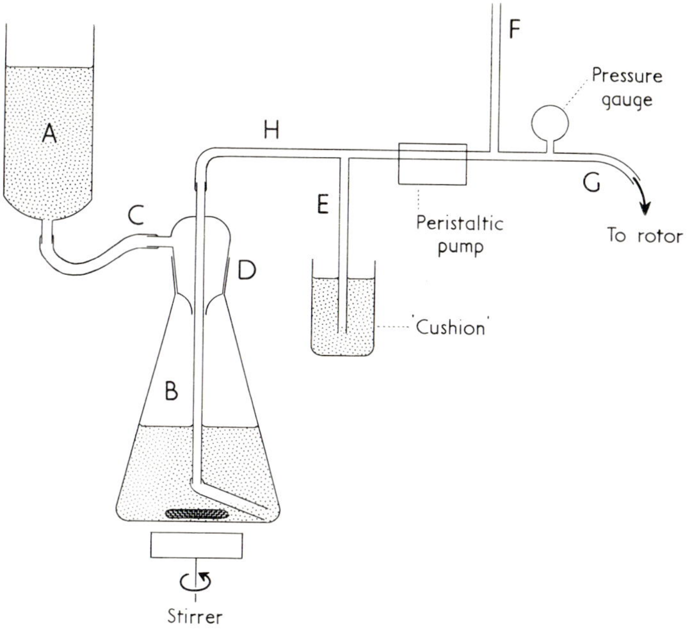

Fig. 12-12. Diagram of density gradient engine for zonal centrifugation. To form a gradient, the volume of heavy sucrose solution required is put in vessel A and the tube from A is filled with the sucrose solution to point C, at which point the tube is clamped. Hemostats are especially handy for clamping the tubes. The volume of light sucrose solution required is put in the mixing vessel (B) and the apparatus is assembled, making sure that the ground-glass joint D is airtight. Tubes E and F are clamped, the delivery tube is connected to the fluid seal leading to the wall of the rotor, and the magnetic stirrer is started. The peristaltic pump is switched on and, simultaneously, the clamp at C is released. To ensure that rapid mixing takes place in vessel B, it is essential that the heavy sucrose solution is introduced into the center, rather than down the walls of vessel B. In view of this, the top of vessel B is so constructed that the heavy sucrose runs down the *outside* of the exit tube from vessel B. Since the volume of gradient solution formed is equal to the volume of heavy sucrose solution introduced into the mixing vessel, the whole gradient has been delivered to the rotor when vessel A is emptied (assuming that the volume of liquid in the tube from vessel B to the rotor is negligible). A "cushion" of heavy sucrose solution can then be loaded into the rotor by clamping the delivery tube to points H and G, unclamping tubes E and F, and continuing to pump until all air has been expelled from E to F. G is then unclamped and F reclamped, and the required volume of cushion solution is pumped into the rotor. The same arrangement can be used to displace the contents of the rotor after centrifugation has been completed. (Birnie and Harvey, 1968.)

The large K rotors, which are used for the commercial production of vaccines, may be operated on the continuous flow principle. After a density gradient is established and the rotor accelerated to a medium speed, tissue extract is permitted to enter the rotor through a bottom seal and to be layered as a thin zone at the center of the rotor on the surface of the density gradient. At the top of the rotor, an exit port allows this same thin zone to be removed. While the extract flows slowly from the bottom to the top of the rotor, virus filtration employs highly pure agar (Agarose, Sepharose) which possesses pore sizes to accommodate viruses, and thus the latter technique is commonly used for virus purification.

The first three types of chromatography have been routinely used to resolve proteins, nucleic acids, lipids, etc., and such methods have now been applied in investigating macromolecules. However, many modifications were necessary particularly with large biological macromolecules such as plant viruses.

II. PREPARATION OF VIRUS EXTRACTS FOR CHROMATOGRAPHY

For chromatographic purification of plant viruses it is desirable to choose a host that yields minimum amounts of virus inhibitors, dark pigments, and undesirable host proteins. Various methods of extracting the virus and keeping various undesirable constituents minimized are described in Chapter 11.

Clarification of cellular extracts of host constituents by various methods is described in Chapter 11, and clarified extracts as described may be employed in chromatography. However, because a number of viruses are unusually labile and difficult to purify by other methods, clarification by adsorption is the method of choice.

Clarification can be performed by batch-wise methods using the same adsorbant as used in column chromatography. In this case the adsorbing material is added to the crude extract, stirred and filtered or centrifuged.

Acid-washed, reagent grade, activated charcoal is commonly used to adsorb plant substances, especially brown pigments (Corbett and Roberts, 1962). Under certain conditions charcoal is known to adsorb virus (Galvez, 1964). A thick layer of charcoal on the filter should be avoided. Centrifugation of the mixture is preferable. The use of chloroform with charcoal seems to improve the clarification of virus extracts (Singh et al., 1966; Timian and Savage, 1966; McLean and Francki, 1967).

Another adsorbant in clarification is bentonite converted to the magnesium form (Dunn and Hitchborn, 1965). Extracts containing turnip yellow mosaic virus were clarified as follows: 100 g of leaf tissue were homogenized with bentonite and $MgSO_4$. The mixture was centrifuged for 15 min at 15,000 rpm and the supernatant was centrifuged for 1 hr at 45,000 rpm. The pellet was resuspended in 5×10^{-3} M $MgSO_4$, 10^{-3} M phosphate buffer, and 0.1 ml of bentonite. After one cycle of low and high speed centrifugation the pellet was resuspended in water. The advantages of using bentonite in plant virus purification are economy of time, reduction or elimination of ribonucleases, and selective adsorption when two viruses or two viral components are present in an extract. According to Hitchborn and Dunn (1965) the adsorption of wild cucumber mosaic virus on magnesium bentonite is dependent on the pH and the magnesium ion concentration. In a 5×10^{-4} M magnesium ion solution the adsorption of each component of the virus (0.5 mg/ml) on bentonite (5 mg/ml) varies from 0% at pH 7.55 to 100% at pH 6.86. Under certain conditions of pH and magnesium ion concentration, the adsorption of the two virus components varies with a maximum at pH 7.15 and a magnesium ion concentration of 5×10^{-4} M; 100% of the top component and only 15% of the bottom component were adsorbed. A solution of sodium ethylenediaminetetra-acetate eluted the virus particles almost quantitatively.

Besides charcoal and bentonite, calcium phosphate gel (hydroxyapatite) has been used for clarification. Fulton (1959) prepared hydrated calcium phosphate by adding to a 0.1 M solution of Na_2HPO_4 slightly less than an equal volume of 0.1 M $CaCl_2$. The flocculent white precipitate settled sufficiently in 4-6 hr to allow one third or more of the supernatant to be decanted. Distilled water was added and the precipitate resuspended. Decantation and resuspension of the precipitate in distilled water was done 15-20 times to remove soluble salts. Centrifugation at low speed yielded a final preparation with a consistency of soft butter and a water content of 97%. It was stored in this form at room temperature. This gel clarified extracts containing sour cherry necrotic ringspot and prune dwarf viruses. Cucumber cotyledons harvested 3-6 days after inoculation were homogenized in 0.02 M phosphate buffer, pH 8.0, containing 0.01 M sodium diethyldithiocarbamate and 0.02 M sodium thioglycolate. Large debris were removed by centrifugation and the calcium phosphate gel was added to the supernatant. The mixture was again centrifuged for 15 min at 2500 rpm. One such treatment removed nearly all green material from the homogenate; repeated treatments with the gel usually removed additional green material, leaving clear amber supernatant. Two cycles of low and high speed centrifugation of the clarified homogenate produced clear colorless pellets of highly infectious virus. Fulton (1967*a*, *b*) applied the same technique to purify tobacco streak, tulare apple mosaic, and rose mosaic viruses. However, Dunez (1966) found that paste of hydrated calcium phosphate irreversibly adsorbed considerable amounts of necrotic ringspot, prune dwarf, and tobacco mosaic viruses.

Diethylaminoethyl cellulose (DEAE-cellulose) has been applied to clarify maize dwarf mosaic virus (Sehgal, 1966, 1968). DEAE-cellulose (0.5 g/10 ml of extract) adsorbed colored contaminants and proteins.

DEAE-cellulose has been used for clarifying lettuce necrotic yellows virus (McLean and Francki, 1967) and for potato yellow dwarf virus (Whitcomb, 1965).

Ion exchange resins have been used to a limited extent. Tremaine (1961) separated plant antigens from carnation virus by using the strongly basic ion exchange resin, Amberlite CG4000. Shainoff and Lauffer (1956) earlier used the basic exchange resin, Amberlite XE67, for southern bean mosaic virus purification.

Clarification methods using adsorption principles do not imply that they should always precede chromatographic virus purification. Such clarification procedures will depend on various parameters as discussed in Chapter 11.

III. ADSORPTION METHODS

A. Cellulose

Paper chromatography used to separate an array of compounds developed rapidly and seemed applicable to viruses by some workers. Cochran (1947) reported a paper chromatographic method to demonstrate tobacco mosaic virus in sap from Turkish tobacco plants. Here the paper acted as an adsorbant and the buffer solution was able to translocate the virus along the cellulose fibers. Since paper chromatography is essentially a special form of partition chromatography, Cochran's technique is not this form of chromatography.

Ragetli and van der Want (1954) used filter paper (57 cm × 6 cm) and applied 90 μl of extract on an area of 3 × 4 cm situated 6 cm from the top edge of the paper strip and 1 cm from both side edges. The top edge was clasped between two (6.5 × 6 cm) glass plates placed in a porcelain vessel. Ten milliliters of distilled water, sodium chloride or sodium acetate, and acetone-water (1/1, v/v) were used. The virus was located on the paper by assaying for infectivity.

Paper chromatography of tobacco mosaic, tobacco rattle, tobacco necrosis, cucumber mosaic, bushy stunt, and potato viruses was reported by Ragetli et al. (1955).

Loebenstein (1960) identified two sweet potato viruses on filter paper. The R_F-value of vein clearing virus was 0.53 and that of ringspot virus was 0.67 when 0.1 M sucrose was used as solvent. This permitted the identification of some viruses even from plants with masked infections.

B. Calcium Phosphate (Hydroxyapatite)

Calcium phosphate gel can be used in two ways. First, it can be used for clarification, especially to remove green plant substances. Second, the gel can be used in column chromatography. The adsorption of proteins on calcium phosphate gel or hydroxyapatite ($Ca_5(OH)(PO_4)_3$) is well known (Tiselius, 1954) and can be prepared in the following way: Add one part 0.5 M Na_2HPO_4 to one part of 0.5 M $CaCl_2$ at room temperature. Wash the precipitate with water to remove the excess of salts. The final gel preparation is preserved under water. Proteins adsorbed at low phosphate concentrations can be eluted at higher concentrations. Tiselius et al. (1956) found that tobacco mosaic virus was eluted from the calcium phosphate column between 0.100 M and 0.120 M phosphate, pH 6.7.

McLean and Francki (1967) purified lettuce necrotic yellows virus by chromatography on calcium phosphate. Considerable amounts of protein material remained adsorbed on the column and a large proportion of infectious lettuce necrotic yellows virus was recovered. In routine experiments 2.5 ml of virus suspensions were applied on a column (1.3 cm × 7 cm) and the initial 15 ml of eluant after discarding the void volume of the columns were taken. The whole procedure took about three hours and therefore the technique is suitable for this unstable virus.

Schito (1966) used magnesium pyrophosphate to purify coliphages T1, T2r$^+$, and N4. Magnesium pyrophosphate was prepared by adding 2 parts 0.1 M sodium pyrophosphate to 10 parts 0.1 M magnesium chloride. The precipitate was collected by centrifugation (5 min, 2000 rpm) and washed. After washing, the pellet was resuspended in water. An extract of *Escherichia coli* infected with coliphages T1, T2r$^+$, or N4 was mixed with magnesium pyrophosphate gel (1/2, v/v). The mixture was stirred during 30 min at room temperature and centrifuged at low speed. After the pellet was washed with water, the virus was recovered from the precipitate.

According to Tiselius et al. (1956), electrostatic interactions may be involved between virus particles and calcium or magnesium ions of the adsorbant. Schito (1966) showed that the quaternary virus structure influences adsorption since phage N4 combined readily with the gel, but its protein subunits were not retained by the same adsorbant.

Chromatography employing phosphate gels can be used in purifying viruses with compact structures. Small viruses are recovered more quantitatively than larger viruses. Moreno (1966) percolated crude alfalfa mosaic virus preparations through hydrated calcium phosphate and recovered all of the infectivity, but Venekamp (1966) recovered no infectivity when crude extracts of apple chlorotic leafspot virus and potato virus Y_N were applied. The calcium phosphate gel adsorbed these viruses irreversibly.

C. Glass

Oxelfelt (1966) found that tobacco mosaic virus suspended in 0.4 M NaCl, 0.05 M phosphate buffer, pH 7.2, was retained by a sintered glass filter G5, and distilled water liberated the virus. Decreasing the NaCl concentration in the virus suspension to 0.3 M resulted in a poor retention of the virus and 0.3 M KCl had the same effect as 0.4 M NaCl. These salts caused the virus to flocculate and be retained by the filter.However, the filters had a limited capacity to adsorb virus particles. Since sedimentation coefficient of the virus was not seriously affected by concentrated salt solutions, it indicated that aggregation was not the main cause of preventing passage of the particles through the filter. Hydration of ions in the suspension appeared to be responsible for increasing the affinity of the virus for glass.

In spite of the above phenomenon, the use of glass filters in the virus purification is very limited since relatively pure virus preparations are required.

D. Chitin

Townsley (1961) described a column chromatographic technique for tobacco mosaic virus using chitin. Ninety grams of finely powdered chitin were suspended in 1 liter of 1 N NaOH and stirred for 30 min. The chitin was washed and resuspended in 1 liter of 2 N HCl. After 30 min the milky liquid was decanted. The remaining material was washed successively with water, 95%

ethanol, and ether. The final chitin preparation was used in the chromatographic procedure.

A column (1.3 cm × 17 cm) was packed with finely divided chitin and equilibrated with 0.01 M tris (hydroxymethyl) amino methane (Tris), pH 6.8, and 0.3-2 mg virus was applied on top of the column. After washing the column with three portions of 1 ml Tris buffer the virus eluted with a linear gradient from water to 0.5 M K_2HPO_4. It appeared in two peaks. The fraction eluting at pH 7.2 was four to ten times as infectious as that eluting at pH 7.7. The ratio of the optical densities at 260 nm and 280 nm of the first fraction was 1.23 and that of the second fraction was 1.29. Townsley did not explain this phenomenon. It is likely that the virus was contaminated with polyphenols and host proteins, thus altering its surface properties, which can account for the differences in adsorption.

E. Methylated Albumin

Methylated albumin coated on kieselguhr (MAK) for fractionating nucleic acids has been described by Fraenkel-Conrat and Olcott (1945) and Mandell and Hershey (1960). Methylated albumin is made by suspending 5 g of albumin in 500 ml of absolute methyl alcohol and then adding 4.2 ml of 12 N HCl; the protein dissolves and eventually precipitates again. The mixture is allowed to stand in the dark for three or more days with occasional stirring. The precipitate is collected and washed twice with methyl alcohol and twice with anhydrous ether. The ether is evaporated in air and in vacuo over KOH. Then the material is ground to a powder and stored over KOH.

For coating the kieselguhr with the above methylated albumin, a suspension of 20 g of kieselguhr in 100 ml of 0.1 M buffered saline (in 0.05 M Na phosphate buffer, pH 6.7) is boiled, then cooled. Five milliliters of 1% esterified albumin are added, the mixture is stirred, and 20 ml of additional saline are added.

In a 2 cm fractionating tube, a pad of paper powder is placed on the sintered glass disc and the column prepared by pouring 5-10 ml portions of MAK. The excess buffered saline is driven down just to the level of packed MAK under air pressure of 2-3 psi. The column has a volume of about 50 ml. The column is washed with 200-300 ml of 0.4 M buffered saline at the designated pressure. The final preparation is suspended in 125 ml of 0.4 M saline and then stored in the refrigerator.

Kubo et al. (1965) used MAK to fractionate tobacco mosaic virus and its components. The viral protein was not adsorbed, but tobacco mosaic virus was specifically eluted. The nucleic acids from infected tobacco leaves were eluted in two peaks while those from healthy leaves produced only one peak.

Ralph et al. (1965) isolated ribonucleic acids from tobacco leaves infected with tobacco mosaic virus, and from Chinese cabbage leaves infected with turnip yellow mosaic virus. The preparations contained a ribonuclease-resistant RNA-fraction (double-stranded RNA) that could be effectively fractionated on MAK. Both viruses yielded such a fraction.

Williams and Whitfeld (1966) developed a method for isolating undamaged tobacco mosaic virus RNA for end-group sequence analyses. The separation was accomplished by using a three-layer MAK column at 2-4° C and a 250 ml or 500 ml linear salt gradient (0.7-1.0 M). Then RNA (0.8 mg in 50 ml of 0.7 M NaCl) was applied and gradient elution started immediately. The first fractions eluted contained degraded ribonucleic acid and little, if any, infectivity. Subsequent fractions showed varying amounts of a 31*S* component and considerable infectivity.

Highly specific adsorbants have been used by binding viral antibodies to diazotized *p*-aminobenzylcellulose (Galvez, 1963). Antiserum to tobacco mosaic virus combined with this cellulose adsorbed tobacco mosaic virus specifically; likewise southern bean mosaic virus antiserum did the same with southern bean mosaic virus. The combination was demonstrated by a batch procedure and by frontal analysis of effluents from a 1 cm diameter column. This method could be used for characterizing the reaction of viruses with antisera, and for the removal of host antigens. If the virus could be eluted (Wakimoto, 1960) it would be a virus purification technique. Such methods were used to remove considerable amounts of bacterial products from a crude phage suspension (Wakimoto, 1960).

IV. ION EXCHANGE CHROMATOGRAPHY

Because extremes in pH occur during ion exchanges, synthetic resins were not widely applied in the separation of proteins. Cellulose ion exchangers, however, offered better conditions to chromatograph macromolecules and have been used with moderate success.

Synthetic ion exchange resins undergo variations in volume during the processes of adsorption and desorption. The exchange resin expands and contracts under these conditions and desorption of material will be slower than adsorption, causing trailing zones during elution.

Synthetic resins also can bind material irreversibly by strong coulombic combinations and entrapment. Such conditions do not occur on ion exchange celluloses, since swelling does not create holes in the macrostructure because the unsubstituted regions prevent excessive expansion and contraction.

Unlike the synthetic resins, the ion exchanging capacity of ion exchange celluloses is limited to about 1 meq/g. This capacity, although low when compared with the conventional ion exchange resins, is nevertheless, on a mass basis, very high for the adsorption of macromolecules. The absolute capacity is of little importance since macromolecules are adsorbed at multiple sites in an open structure of the modified cellulose. Also, modified celluloses have great chemical and physical stability, and this property makes them suitable for separating large molecules, such as viruses or viral components. These celluloses can be regenerated by washing with acid and alkali, the order of washing depending on the type of cellulose exchanger used.

A. Divinylbenzene-Crosslinked Styrene

Although the advantages of celluloses were discussed, there are some practical merits in using synthetic resins. For instance Johnson et al. (1967) used polymers of divinylbenzene crosslinked with styrene for viruses. This material adsorbed tobacco mosaic virus and poliovirus from aqueous suspensions. In this case the virus was considered the contaminant and the method was used to purify water. Analytical applications need to be explored.

B. ECTEOLA-Cellulose

This cellulose contains basic groups derived from triethanolamine bound to cellulose by reaction with epichlorohydrin. It has been used to separate tobacco mosaic virus mixtures into different active fractions (Commoner et al., 1956).

According to Taniguchi (1962*b*) an estimate of the amount of virus adsorbed on an ion exchanger and the dissociation constant of their complex would be very useful in the application of the ion exchanger for the study of viruses. He demonstrated that both values could be determined graphically.

Purified tobacco mosaic virus preparations were separated into several components by stepwise elution of a cation exchanging resin (Amberlite IR-120, Na form) and on ECTEOLA-cellulose columns (Taniguchi, 1964). Determined were the amounts of the components in the preparations from plants infected during different seasons.

C. DEAE-Cellulose

Levin (1958) used diethylaminoethyl-cellulose (DEAE) to purify chromatographically tobacco mosaic virus and potato virus X. One milliliter of the virus (A_{260} of 16 in 0.01 M Tris, pH 7.3) was applied on top of a column containing 1.5 g DEAE. The column was washed with 5 ml of Tris buffer and stepwise elution was carried out with 15 ml of each of the following solutions: 0.10, 0.17, 0.25, 0.32, 0.40, and 0.80 M Tris buffer, pH 7.3. The last three or four stepwise fractions contained virus.

Cochran et al. (1959) used DEAE-cellulose under anaerobic conditions to separate components from virus-infected tobacco leaves. When extracts from healthy Turkish tobacco leaves were chromatographed, four components were obtained. Chromatographic separation of extracts from tobacco mosaic virus-infected leaves showed components 1, 3, and 4, with component 2 missing. Some virus was eluted with component 1. The remaining virus was adsorbed on the cellulose and only eluted following component 4 at low pH and high salt concentration.

Venekamp and Mosch (1963) isolated potato virus X by taking fractions eluted from a column of calcium phosphate and applying these fractions on a column of DEAE-cellulose (diameter 2 cm, height 5 cm, dry weight 5 g). Elution with successive 100 ml lots of 0.2, 0.4, 0.6, and 0.8 M NaCl in 0.02 M phosphate buffer solution, pH 7, yielded four different fractions. More than 90% of the total amount of virus was eluted at 0.4 M NaCl. The nucleoprotein concentration in this fraction was much higher for virus-containing solutions than for solutions prepared from healthy plants. These experiments did not result in a complete purification of the virus.

Toyoda et al. (1965) observed that rice dwarf virus adsorbed strongly on DEAE-cellulose since the virus could not be eluted with NaCl concentrations up to 1.0 M. Elution was achieved with 0.5 M NaOH. The authors suggested that the virus particles were covered with a strong acidic enveloping substance which could easily be removed by treating the preparation with phospholipase. Rice dwarf virus, treated in this way, could be eluted from DEAE-cellulose with 0.2-0.25 M NaCl.

Salama and Lyman (1966) also used DEAE-cellulose to purify internal cork virus of sweet potatoes according to procedures mentioned above.

D. Discussion

From the experiments on the purification of potato virus X, Venekamp and Mosch (1963) concluded that adsorption techniques with calcium phosphate and with DEAE-cellulose did not lead to complete purification of the virus because the same chromatographic results were obtained with healthy plant material. Moreover, each change in solvent in a stepwise procedure elutes more material and thus the virus is partitioned in more than one fraction.

Bradley and Rich (1956) studied the fractionation of ribonucleic acids on ECTEOLA-cellulose and concluded that four or more sites of nucleic acids were active in combining with the column material. The consequence of this would be that elution of nucleic acids will depend on the number of such sites.

The characteristic feature of stepwise elution is a spreading of the protein zone (Boman and Westlund, 1956). At a certain ionic strength the whole column can be covered with protein, and further increase of the salt concentration will result in the appearance of a protein peak in the effluent. Under such conditions it may be very difficult to get one pure component in a single chromatographic run, especially if the component constitutes only a small part of the mixture.

Ginoza et al. (1954) mentioned the effect of a mutual adsorption of different macromolecules on the elution in a chromatographic procedure. The physical properties of a complex of two protein molecules are different from those of separate molecules.

V. PURIFICATION WITH POLYHYDRIC POLYMERS

A. Two-Phase Systems

Partition chromatography has been used extensively for separating low-molecular-weight substances, and usually organic solvents with limited solubility have been used. Although such solvents are used as clarifying agents, they are rarely used for separating high-molecular-weight components because proteins and nucleic acids are denatured and are usually insoluble in them.

Partition chromatography over two liquid phases has been employed to a limited extent in virus purification. Albertsson (1960) described phase systems of aqueous solutions of polyethylene glycol and dextran; polyethylene glycol and dextran sulfate; methylcellulose and dextran; methylcellulose and dextran sulfate; polyvinyl alcohol and dextran; polyvinyl alcohol and dextran sulfate. At certain concentrations polyethylene glycol is not completely miscible with certain concentrations of aqueous dextran. If this mixture is allowed to stand for an hour, an upper phase of polyethylene glycol and a bottom phase of dextran will form. When this system contains 0.01 M phosphate buffer, pH 7, and tobacco mosaic virus, the virus partitions into the upper phase. Addition of NaCl transfers the virus to the interface (0.1 M NaCl transfers more than 90% of the virus). Macromolecules are more soluble in one phase than in the other, whereas low-molecular-weight substances and proteins distribute rather evenly in the system. The ratio of polyethylene glycol to dextran may be adjusted so that the bottom phase is smaller in volume than the upper phase. The virus is made to favor the bottom phase so that it is concentrated as well as partially purified.

Partition of virus in the phases is dependent on certain salts, especially sodium chloride, and on the type of polymers used in the system. In purification the virus may be collected in the bottom phase of a methylcellulose-dextran system. The upper methylcellulose-rich phase is removed and a polyethylene glycol solution is added. It is made to contain 0.01 M phosphate buffer and no NaCl so that the virus will then be found in the newly formed upper phase.

Although polyethylene glycol can precipitate virus, especially in the presence of sodium chloride, liquid-liquid partition chromatography has wide application in virus purification.

Two-phase systems have been applied especially in the purification of animal and bacterial viruses. Frick and Albertsson (1959) concentrated phage T2 from a 2 liter culture of *Escherichia coli* by adding 601 ml of 2% methylcellulose 4000 and 131 ml of 10% dextran 0.68. After the phases were separated the bottom layer of about 200 ml contained the bacteriophage. Frick (1961) added dextran sulfate and methylcellulose to final concentrations of 0.22% and 0.51% respectively to a bacteriophage T2 suspension in 0.15 M NaCl. After standing of the mixture for 36 hr, the bottom layer was isolated, mixed with sodium chloride to a final concentration of 0.5 M, and treated with one part Freon 113 per three parts of the suspension. This yielded a precipitate, which was then centrifuged. The Freon treatment of the supernatant was repeated twice. Addition of 0.058 ml of 1 M $BaCl_2$ or 0.12 ml of 4 M KCl per milliliter of suspension precipitated the dextran sulfate. After centrifugation to remove the precipitate, the supernatant was dialyzed against 0.15 M NaCl + 0.001 M $MgCl_2$. Dextran sulfate and methylcellulose were added as before to the dialyzed material and the purified virus was collected in the bottom phase.

Norrby and Albertsson (1960) concentrated poliovirus in a two-phase system containing 0.2% dextran sulfate, 6.45% polyethylene glycol, and 0.3 M NaCl.

The virus concentrated in the bottom phase. The dextran sulfate was then removed by adding 0.69 ml of 3 M KCl or 0.3 ml of 1 M $BaCl_2$ per gram of bottom phase.

Philipson et al. (1960) used polyhydric polymers to concentrate various viruses. For example, phage T2 was concentrated in the bottom phase (90% of total) using 0.22% dextran sulfate, 0.51% methylcellulose, and 0.15 M NaCl. This phase had a volume of 7/100 of the total volume of the system. ECHO-virus was concentrated using 0.20% dextran sulfate, 6.45% polyethylene glycol, and 0.3 M NaCl. The virus-carrying bottom phase was concentrated further by adding salt and polymers to final concentrations of 1 M NaCl, 17% dextran sulfate, and 0.8% polyethylene glycol, and then collecting the virus in the new top phase. Strains of ECHO-virus were concentrated into the bottom phase using 0.3% dextran sulfate, 3.0% polyvinyl alcohol, and 0.15 M sodium chloride.

Very efficient purification is achieved by selecting the proper polymers, for example, ECHO-virus and cellular debris distributed in the turbid bottom phase of a dextran-polyethylene glycol system. However, when dextran was replaced by dextran sulfate, the lower phase was optically quite clear and contained most of the active virus while the cell particles were collected at the interface.

The concentration of barley stripe mosaic virus in barley sap is relatively low and large amounts of plant material are necessary to obtain enough virus for further studies. Timian et al. (1962) mixed such crude sap with 0.82% methylcellulose and 0.60% dextran and achieved tenfold concentration of the virus.

Bengtsson et al. (1962), employing counter-current distribution, used 7% dextran sulfate, 1.2% polyethylene glycol, and 0.6 M NaCl for the purification of poliovirus. In this system the virus had a partition coefficient of 0.57 and therefore could not be effectively purified since a partition coefficient of about 1 is prerequisite for effective separation in counter-current distribution.

For isolating tobacco mosaic virus from crude extracts of tomato plants Venekamp and Mosch (1964*a*) used the following systems: polyethylene glycol/dextran of 5/10%, 6/8%, 9/2.5% and 5/2%. A mixture of components, mostly green, collected in the dextran-rich bottom phase and the virus distributed into the upper phase when 4.5% glucose was added instead of NaCl.

Leberman (1966) isolated turnip yellow mosaic virus and turnip crinkle virus from extracts of *Brassica chinensis* by using 1.34 g of 20% dextran sulfate, 29 g of 30% polyethylene glycol, and 5 g of 5 M NaCl to each 100 ml of clarified extract. Either virus collects in the bottom phase.

Grapevine fanleaf virus or "virus du Court-noué de la Vigne" (Caudwell, 1966, 1968) was partially purified from *Chenopodium quinoa* extracts by the following procedure. Four grams of leaves were ground in a mixture of 2.5% polyethylene glycol and 1% dextran. The emulsion was broken by centrifugation (10,000 rpm, 10 min) and polyethylene glycol to a final concentration of 8% was added to the supernatant. The mixture was allowed to stand for 30 min and the bottom phase was collected by centrifugation. The virus in this phase as a pellet was resuspended in 8 ml of water.

B. Precipitation of Viruses

Hebert (1963), the first to use polyethylene glycol to purify various viruses, precipitated wheat mosaic virus with 2% polyethylene glycol; tobacco mosaic virus with 4% polyethylene glycol plus 2% NaCl; tobacco ringspot virus with 8% polyethylene glycol and 2% NaCl; and bean pod mottle virus with 8% polyethylene glycol and 1.2% NaCl.

Turnip crinkle virus (in *Brassica chinensis* cv. Petsai) was prepared by Leberman (1966) as follows: Homogenates from the infected plants were centrifuged for 30 min at 12,000 *g* and 5 g celite were added per 100 ml of the crude virus extract, pH 5.8. The centrifugation was then repeated. Thirty-three grams of 30% polyethylene glycol and 0.8 ml of 5 M NaCl were stirred into each 100 ml of the supernatant. The virus was collected by centrifugation (15 min, 9000 rpm) and was extracted with the magnesium phosphate medium of Dunn and Hitchborn (1965) until the extract was no longer opalescent. The virus was further purified by ultracentrifugation.

Gooding and Hebert (1967) modified techniques of Hebert (1963) and of Leberman (1966) to purify tobacco mosaic virus. A homogenate was prepared from infected frozen *Nicotiana tabacum* leaves in 0.5 M phosphate buffer, pH 7.2, containing 1% 2-mercaptoethanol (1 ml buffer/g tissue). The extract was clarified with n-butanol (8 ml per 100 ml of extract) and 4.0 g of polyethylene glycol were added per milliliter of extract to precipitate the virus. The virus was collected by centrifugation and the pellet was resuspended in 0.01 M phosphate buffer, pH 7.2. According to van Kammen (1967), butanol-chloroform treatment causes considerable loss of cowpea mosaic virus and partial breakdown of the viral top component. Precipitation of the virus with 4% polyethylene glycol and 0.2 M NaCl permitted rapid and high recovery of the virus, and resulted in preparations with the same quantitative and qualitative ratios of the three components as those in crude plant sap.

Dudman (1966*a*) studied the precipitation of tobacco mosaic virus by liquefied gelatin and found that 1.1% gelatin was necessary to precipitate the virus irrespective of the viral concentration. A linear relation between turbidity and virus concentration was observed. Precipitation was independent of the virus strain or the type of gelatin used. The presence of neutral salt was required and maximum precipitation occurred outside the interisoelectric pH range, showing that the interaction was *not* a salt-like combination between gelatin and tobacco mosaic virus. It is instead believed to result from a mutual spatial exclusion phenomenon whereby the virus is precipitated by gelatin through a mutual incompatibility and spatial exclusion.

Dudman (1966*b, c*) also found that concentrations of agar (kept liquid at 45°) greater than 0.2% precipitated tobacco mosaic virus. The interaction responsible for this phenomenon did not occur in agar gelled at 20°. When turbid liquid mixtures of agar and tobacco mosaic virus were cooled and gelled, the turbidity decreased to a degree similar to such gels without virus. This type of precipitation of viruses and macromolecules is simple and rapid, and can be used to estimate the effective radii of such macromolecules (Dudman, 1966*b*).

C. Chromatographic Methods

Polyethylene glycol without other polymers may be used in chromatography. Venekamp and Mosch (1963) showed that potato virus X was retained in a cellulose column and a solution containing 6% polyethylene glycol and 4.5% NaCl was used to elute low-molecular-weight substances and plant proteins. The virus was then eluted from the column by reducing the polyethylene glycol and sodium chloride concentration to 1.5% and 1.12% respectively.

Adapting the method used for isolating virus from host constituents with two-phase systems (e.g. polyethylene glycol, dextran, and glucose) Venekamp and Mosch (1964*b*) designed chromatographic procedures employing the system on cellulose columns. The chromatographic column material consists of 10 g of dry cellulose mixed with 5% polyethylene glycol, 0.05% dextran, 4.5% glucose, 2% NaCl, 0.004 M $MgCl_2$, and 0.01 M phosphate buffer, pH 7. This is poured into a column (3 cm × 5 cm bed size) sandwiched between layers of 2 cm of sand. The columns are jacketed and alcohol at 2° C is circulated to keep the column cooled during chromatography.

Virus-containing homogenates are applied to the column, incubated for 30 min, and then allowed to flow into the column. The initial effluent is usually brown and gradually becomes colorless after about 250 ml of the effluent are collected. To elute the virus, 150 ml of the above solvent without polyethylene glycol and NaCl are used. Most of the green pigmented material remains on the column and some debris penetrates into the column but is adsorbed by the cellulose. Virus also adsorbed on the cellulose is selectively eluted.

To the virus-containing effluents, 5% polyethylene glycol and 2% NaCl are added and then passed through a second column. The virus can be eluted selectively by stepwise elutions using 100-ml amounts of elution buffer in which polyethylene glycol, dextran, glucose, $MgCl_2$, and phosphate buffer, pH 7.0, concentrations are held constant while the concentration of NaCl is varied from 2% to 0%. After these percolations the polyethylene glycol is varied from 3% to 0%.

Such cellulose column techniques have been used to purify potato virus X, potato virus Y, potato stem mottle, tobacco mosaic, tobacco rattle, white clover mosaic, carnation ringspot, and carnation mottle viruses. A simplified modification of the method uses 200 ml portions of solvent mixtures to elute the virus. Ultraviolet-absorbing substances were present in effluents from diseased plant extracts but not in healthy plant extracts.

A modification of the procedure was necessary for the purification of tobacco necrosis virus. The first solvent contained 8% polyethylene glycol and

3% NaCl to retain the virus in the column and to elute low-molecular-weight substances. The virus is eluted with the second solvent containing 5% polyethylene glycol and 1.5% NaCl. The virus can be freed of polymers by dialysis (for polymers less than 10,000 mol wt) or by precipitating it with polyethylene glycol and NaCl (final concentration of 5% and 2% respectively). The virus precipitate is collected by centrifugation and resuspended in an appropriate buffer. Also such polymers are removed by passing the virus-carrying effluents through a DEAE-cellulose column. The column is washed with 0.02 M phosphate buffer, pH 7, to remove polyethylene glycol, dextran, and glucose, and the virus is eluted with 0.8 M NaCl in 0.02 M phosphate buffer, pH 7. The salts can be removed by dialysis.

1. POTATO VIRUS X

High-molecular-weight polymers, such as polyethylene glycol, dextran, ficoll, and methylcellulose, cause potato virus X to aggregate. Glucose and glycerin also enhance aggregation. NaCl retards aggregation, and increasing its concentration in polyethylene glycol disperses virus aggregates. Aggregation of potato virus X particles was obviated by using 10% polyethylene glycol, 10% NaCl, 4.5% glucose, 0.01 M ammonium acetate, pH 7, and 0.004 M $MgCl_2$. As the effluent fractions gradually became clear, buffer without polyethylene glycol and NaCl was used to elute the virus. To the virus-containing fraction 10% polyethylene glycol and 10% NaCl were added and then passed through a second cellulose column. The column was washed stepwise by altering only the polyethylene glycol from 10% to 0%. The virus was eluted between 3% and 1% polyethylene glycol.

The potato virus X preparation showed a higher ultraviolet absorbancy than that of the virus isolated by the method using buffer containing 5% polyethylene glycol and 2% NaCl. There apparently was less adsorption of the virus on cell constituents than the 5-2% buffer system. Apparently the presence of excess ions decreases the adsorption. The virus was collected by ultracentrifugation and the preparation appeared homogeneous in the analytical ultracentrifuge.

2. OAT BLUE DWARF VIRUS

The cellulose-polyhydric polymer method was used to purify oat blue dwarf virus (Banttari, 1967; Banttari and Zeyen, 1969). Infected oat leaves were homogenized in 0.02 M phosphate buffer or Tris buffer, pH 7, containing 0.008 M $MgCl_2$. The homogenate was filtered through cheesecloth and filtrate was centrifuged for 10 min at 7000 *g*. The pellets were washed with a mixture of 5% polyethylene glycol, 4.5% glucose, and 2% NaCl and the washings were combined with the supernatant. It was then applied to a column (12 cm × 7 cm), consisting of a layer of 150 g of quartz sand, 30 g of cellulose, and ground filter paper. A slurry of ground filter paper was layered over the cellulose to prevent plugging. The flow rate was adjusted to 8 ml/min.

the inner chamber was then replaced with a solution of virus of unknown diffusion rate. Buffer in the outer chamber at 8° ± 0.5° C was repeatedly withdrawn and tested for virus and replaced by fresh buffer. The concentration of virus and its infectivity were determined spectrophotometrically at 2.60×10^{-7} M and by the local-lesion tests.

The diffusion coefficient (D) is integrated from Fick's first law:

$$\frac{DA}{l} = -\frac{V_i - V_o}{t(V_i V_o)} \ln \frac{(C_i - C_o)_f}{(C_i - C_o)_i}$$

where l and A are the effective length and total cross-sectional areas of pores in the membrane; V_i and V_o are the respective volumes of the inner and outer solutions whose concentrations are C_i and C_o; i and f refer to initial and final conditions. t is the duration of the experiment. The cell constant A/l is estimated by measuring the diffusion of small molecular species of which the diffusion coefficient and size are known. After this calibration the cell is emptied and freed of the calibration standard by prolonged diffusion in any desired solvent. Next the diffusion coefficient for the unknown virus can be estimated.

A. Modified Dextrans

Gels of different porosities may be used for eliminating salts and inhibitory substances which are difficult to remove otherwise. Phenol and soluble plant constituents can be removed immediately after extracting for RNA by gel filtration. According to Fritz et al. (1964) there are some advantages of gel filtration since it yields highly infective preparations. In these experiments 0.1 M phosphate buffer, pH 8, was used to elute the RNA from Sephadex G-25 columns at 5° C. Van Griensven and van Kammen (1969) isolated double-stranded RNA from plants infected with cowpea mosaic virus. RNA preparations were treated with pancreatic ribonuclease A and passed through a Sephadex G-200 column (2.5 cm × 35 cm), and the first fraction, which contained double-stranded RNA, was collected. Schneider et al. (1964) removed low-molecular-weight substances from sugar beet mosaic virus preparations with Sephadex. György and Nemeth (1965) used Sephadex G-200 to eliminate fruit tree virus-inhibiting substances.

Agar gel in concentrations ranging from 2.5% to 15% is prepared in phosphate buffer, pH 7.2. The gel is then chopped up in water into fine particles and sieved according to the gel size desired (e.g. 80 mesh). Agar granules with sizes between 80 and 120 mesh can be collected by using an 80 mesh sieve followed by a 120 mesh sieve. The agar granules have pore widths to accommodate large macromolecules such as viruses. The pore width depends on the agar concentration: Low-percentage agar (1.5-3.0%) possesses larger pore sizes (of the size for plant viruses) than higher-percentage agar (3-8%).

The particles prepared in this manner have angular forms (see Polson, 1961) whereas a method of preparing spherical agar particles giving good flow rates has been described by Bengtsson and Philipson (1964) who adapted a Seitz filter to force melted agar through a fine opening. In this case agar is melted in an autoclave, cooled to 65° C, and then forced through a constricted glass tubing (0.5 mm diameter orifice) by pressure of 2-3 kg/cm^2 of nitrogen into ice-cold ether. The ether is in contact with stirred ice-cold water. The fine stream of liquid agar gels immediately on contact with the cold ether, and the spheres of agar are collected periodically, the water being replenished at those times. The agar spheres can then be sized accordingly.

B. Agar and Agarose

Agar is composed of the polysaccharides agarose and agaropectin of which the latter contains sulfate and carboxyl groups. Agar from different sources varies in sulfate and agaropectin content, and the presence of large numbers of these sulfate groups may affect the efficiency of virus recovery. Therefore, agarose has been used instead of agar (Hjerten, 1962). Agarose is a linear polysaccharide consisting of alternating residues of D-galactose and 3,6 dihydro-L-galactose units.

Agarose is less stable beyond pH 9 and below pH 4, and below O° C and above 40° C.

Steere and Ackers (1962*a*) and Steere (1963) described an agar-gel filtration method to purify tobacco mosaic and southern bean mosaic viruses. Suspensions of 40-60 mesh size, granulated 4% and 8% agar were used. Extracts of virus were applied on an 8% agar column (4.5 cm × 60 cm), previously equilibrated with 0.2 M NaH_2PO_4-0.3 M NaCl, pH 6.8. The virus-containing effluent fractions can be concentrated by ultracentrifugation or by ultrafiltration. A column (60 cm × 2 cm) of 4% was used to eliminate any contaminants not separated from the virus when it was passed through 8% agar.

Van Regenmortel (1962*b*, 1964*b*) used granulated agar columns for the purification of necrotic ringspot virus of stone fruits, turnip yellow mosaic, cucumber mosaic, and watermelon mosaic viruses. He found the technique useful for spherical and elongated plant viruses, and different strains of watermelon mosaic virus prepared in this way were suitable for physical and serological studies.

Čech (1962) employed agar-gel filtration to obtain viruses for electron microscopy. Plant sap was first clarified with chloroform, butanol, or other means to remove chloroplast material and then passed through an agar-gel column.

There are many other applications of agar-gel filtration and only a few examples are cited below. Augier de Montgremier (1963, 1965) observed a yellow fluorescence in ultraviolet light when an extract containing turnip yellow mosaic virus was passed through an agar-gel column stained with acridine orange. This phenomenon was attributed to the spontaneous liberation of nucleic acid from the virus particles. Munro (1967) developed a simple technique to obtain purified potato virus X using agar gel. The virus was then concentrated by dialysis against a polyethylene glycol and used to produce potato virus X-immune rabbit antiserum. Walters and Bryan (1965) purified bean pod mottle virus from Black Valentine bean leaves. An extract from about 80 g of tissue in 0.1 M phosphate buffer, pH 6.5, clarified with charcoal, was placed on a 2 cm × 120 cm column of 40-60 mesh granulated 2% agar gel. The virus was eluted with the same buffer containing 0.005 M $MgCl_2$.

Agrawal and Bruening (1966) used agarose to purify influenza virus labeled with ^{32}P from allantoic fluid. Jacoli (1968*b*) used agarose to isolate turnip mosaic virus from an effluent of a Sephadex G-25 column. After concentrating the effluent, the residue was put on a column 2.5 cm × 40 cm of 4% agarose gel. A 0.05 M phosphate solution, pH 7.5, eluted the virus using a flow rate of 25 ml/hr. The elution diagram showed that turnip mosaic virus came off soon after the void volume separated from two yellowish components. Some retention of virus occurred and it was assumed that the same virus particles penetrated into the gel. Aggregation did not occur as it does with virus prepared by centrifugation.

Fridborg et al. (1965) used a column of 4% agarose gel (diameter of gel particles were 0.04 cm) to purify satellite tobacco necrosis virus. The virus (tobacco necrosis and its satellite) was precipitated with NH_4SO_4 from a crude homogenate of 100 g of infected leaves of French beans or *Nicotiana tabacum* cv. Samsun X White Burley. The concentrated sample (6 ml) was layered on a column (2 cm × 102 cm). The virus was eluted with 0.05 M Na phosphate buffer, pH 7.2, containing .001 M $MgSO_4$. When about 210 ml had passed through the column (flow rate of 6 ml/hr) the satellite virus was collected and further purified by electrophoresis.

ACKNOWLEDGMENT

The author is indebted to Miss J. M. Krijthe and Mr. J. P. W. Noordink for discussing the manuscript.

14

Electrophoresis

MARC H. V. VAN REGENMORTEL

Department of Microbiology and Virology
University of Stellenbosch
Stellenbosch, South Africa

I. INTRODUCTION

Electrophoresis is the migration of charged particles under the influence of an electric field. The charges of virus particles responsible for the migration reside on the outer protein surface of the virions and have their origin in the basic and acidic groups of certain amino acids. At strongly acidic pH, the free acidic groups of aspartic and glutamic acid are uncharged and the free basic groups of lysine, histidine, and arginine are combined with protons, so the particle has a positive charge. At strongly alkaline pH, the acidic groups are dissociated and the basic groups are uncharged, so the particle is negatively charged.

Since the charge on the particle varies with pH, it is necessary in electrophoretic experiments to maintain a constant pH by means of a buffer. The choice of a particular buffer depends on the purpose of the experiment and on the stability of the virus under study. For purification purposes, for example, it is often advantageous to work at a pH as alkaline as is compatible with the stability characteristics of the virus, since greater mobility differences are then usually observed between different components in a mixture.

The isoelectric point (pI) of a protein is defined operationally as the pH at which the particle does not move in the electric field. This occurs when the number of charged acidic groups exactly balances the number of charged basic groups, resulting in an average net charge of zero. The proteins of most plant viruses are fairly acidic with a pI in the range of 3.5-6.0. Isoelectric points have often been used for characterizing viruses and for distinguishing between different strains; since they correspond to the point of minimum solubility (Oster, 1951) they also have a practical significance in the design of purification methods that rely on the differential precipitation of virus particles.

In this chapter the discussion will center around two main applications of electrophoresis: the characterization of viruses by their electrophoretic mobilities as determined with the moving boundary method and the purification of plant viruses by zone electrophoresis in a sugar density gradient. Some theoretical notions useful for an understanding of electrophoretic phenomena will also be presented. For a more complete discussion of electrokinetics, the reader is referred to Overbeek (1952), Overbeek and Lÿklema (1959), and Myzels (1959).

II. ELEMENTARY THEORY OF ELECTROPHORETIC MIGRATION

The force F exerted on a suspended particle by an applied electrical field E is equal to the product of the field strength and the net charge, Q

$$F = QE \tag{1}$$

The particle reaches its terminal velocity v almost instantaneously; the resisting frictional force experienced by a spherical particle can be calculated from Stokes' law and is equal to the applied force:

$$QE = 6\pi \eta a v \tag{2}$$

$$v = \frac{QE}{6\pi \eta a} \tag{3}$$

where v is the terminal velocity, η the viscosity of the medium, and a the radius of the particle.

The electrophoretic mobility μ of the particle is defined as its electrophoretic velocity per unit field strength:

$$\mu = \frac{v}{E} = \frac{Q}{6\pi \eta a} \tag{4}$$

In the literature, electrophoretic phenomena are usually interpreted not in terms of the charge Q carried by the particle, but in terms of the potential effective in moving it. This electrokinetic or zeta potential corresponds to the potential difference between the surface of the hydrated particle (at the slipping plane and thus including the Stern layer, see below) and parts of the medium far away from the surface.

The electrokinetic potential ζ of a charged spherical particle of radius a is given by:

$$\zeta = \frac{Q}{Da} \tag{5}$$

where D is the dielectric constant of the medium. Combining equations (4) and

(5) one obtains

$$\mu = \frac{\zeta D}{6\pi\eta} \tag{6}$$

From equation (6) it may be seen that virus particles of different sizes will have the same mobility, provided that they have the same electrokinetic potential. Thus particles of an elongated virus that differ in length because of aggregation or breakage may still be homogeneous electrophoretically. The relative independence of electrophoretic phenomena from size and molecular-weight parameters is one of the reasons why electrophoresis is often valuable as a purification step following a preliminary partial purification by differential ultracentrifugation.

A number of limitations to the validity of equations (4) and (6) have to be taken into account, for buffer salts are always present in appreciable concentration during electrophoretic measurements and electrical interactions between charged ions and the particles cannot be ignored.

Ions are attracted to the charged surface of particles by electrostatic forces, and it is customary to distinguish a layer of about one ion in thickness (the Stern layer) which is so strongly bound to the surface that it is in fact an integral part of the hydrated moving particle. Further away from the surface the ordered arrangement of ions is increasingly opposed by thermal agitation and the particle is surrounded by a gradually more diffuse cloud of ions. In practice, the effective charge on the particle is lowered by the opposite charge of the Stern layer, and the motion of the particle is further retarded by the ions in the diffuse cloud tending to migrate in a direction opposite to that of the particle. This trailing behind of the diffuse ion layer is known as the relaxation effect. The magnitude of these interactions depends on the electrolyte concentration; the diffuse ionic atmosphere surrounding a particle contracts with increasing ion concentration and increasing ion valency as indicated by equation (7)

$$\frac{1}{\kappa} = \frac{3 \times 10^{-8}}{\sqrt{\Gamma/2}} \text{ cm} \tag{7}$$

where $1/\kappa$ is the thickness of the double layer and $\Gamma/2$ is the ionic strength of the electrolyte solution. Ionic strength is defined as

$$\frac{\Gamma}{2} = 1/2 \Sigma c_i z_i^2 \tag{8}$$

where c_i is the molar concentration of each type of ion and z_i the corresponding valency. (The ionic strength of a NaCl solution is thus one-third that of a Na_2HPO_4 solution of the same molarity).

As the ionic strength is increased, the diffuse ionic atmosphere condenses, more of the ions travel with the particle, and the zeta potential is reduced; this in turn decreases the electrophoretic mobility. This effect is illustrated for tobacco mosaic virus in Fig. 14-1. It is thus essential to use buffers of the same ionic strength when determining the mobility versus pH curve of a virus.

The validity of equations (4) and (6) is also dependent on the magnitude of κa, which is the ratio of the particle radius a to the thickness of the double layer $1/\kappa$. When this ratio is larger than 300, the numerical coefficient in equations (4) and (6) decreases to a limiting value of 4. From equation (7) it can be calculated that the thickness of the double layer in a 0.1 M solution of monovalent ions is approximately 10 Å (see Myzels, 1959). For a virus particle with a radius of 150 Å, the ratio κa under these conditions would be approximately 15 and the coefficient in equation (4) would be close to 5.

When these various effects are taken into account, equation (4) becomes

$$\mu = \frac{Q}{6\pi\eta a(1+\kappa a)} f(\kappa a) \tag{9}$$

where $f(\kappa a)$ varies between 1 and 3/2 depending on the size of the particle. The factor $(1 + \kappa a)$ arises from the decrease in zeta potential brought about by the double layer; this changes equation (5) into equation (10) (see Magdoff, 1960):

$$\zeta = \frac{Q}{Da(1+\kappa a)} \tag{10}$$

Substitution of equation (10) into equation (6) leads to equation (9) above.

Magdoff-Fairchild (1967) made use of equation (9) to calculate the number of charges, or valency, of five southern bean mosaic virus variants. (The valency is given by $z = Q/e$ where $e = 4.08 \times 10^{-10}$e.s.u.).

She found that such variants with widely different pI and mobility-pH curves had approximately the same number of charges (about 3000 per virus particle or 16 per subunit, assuming 180 protein subunits per capsid). This unexpected result was shown to be due to differences in ability of the variants to bind acetate or chloride ions. Magdoff-Fairchild's (1967) results, in fact, provide a remarkable illustration of the extent to which specific binding of ions can influence electrokinetic phenomena. It sometimes happens that certain ions have a strong specific affinity for a particle surface and this binding may vary with the ionic strength; ions may also influence the ionization constant of dissociable groups on the protein surface. Longsworth and Jacobsen (1949) have shown, for instance, that the pI of β-lactoglobulin and bovine serum albumin could vary as much as 0.5 pH unit in different buffers of 0.1 ionic strength.

Some of Magdoff-Fairchild's (1967) data are summarized in Fig. 14-2. The five variants were detected and separated by moving boundary electrophoresis. The isoelectric points of four of the variants were shifted to a lower pH when the buffer system contained chloride ions; in the case of variant I, the pI was lowered by as much as 2.18 pH units, suggesting that the virus bound a large

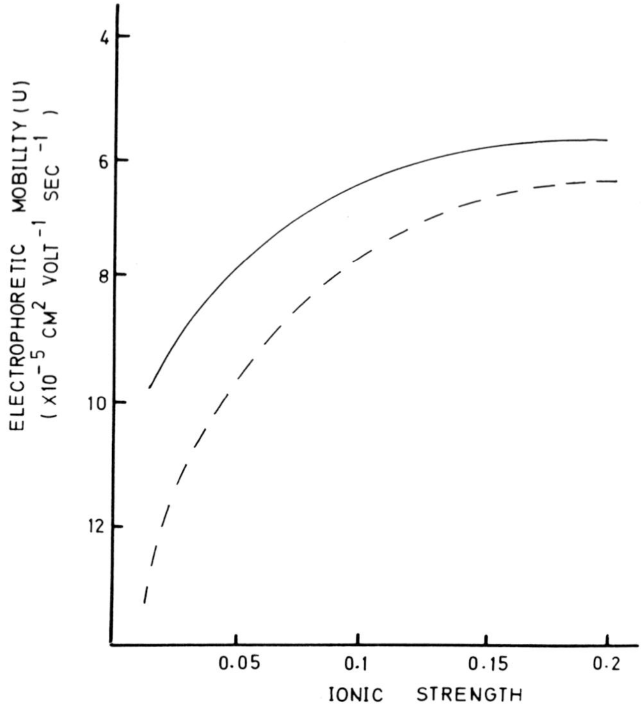

Fig. 14-1. **Dependence of the electrophoretic mobility of tobacco mosaic virus on ionic strength at pH 5.0 (———) and pH 7.0 (— — —). See Kramer (1957).**

amount of chloride ions. From experimental values of the change in pI upon addition of NaCl to the buffer, the fraction of total amphoteric charges of the virus that binds chloride ions was calculated. At the isoelectric point, 50% of the charges are positive and in the case of variant I as much as 38% of the total charges were bound by chloride so that only 12% remained ionizable as free amino groups (see Fig. 14-2). This accounts for the large drop in pI observed with this variant. The large differences in pI of the variants in 0.1 M sodium acetate (see Fig. 14-2) could not be attributed to differences in amino acid composition or charge densities, for the variants appeared to be very similar in this respect. The differences in pI were explained by assuming that variants that have a low pI are binding more acetate ions than those that have a high pI. From the small shift in pI observed with variant IV, for example, it was calculated that this variant did not bind many chloride ions; this would actually be expected if variant IV did bind a large number of acetate ions.

These results show that it is essential to report the composition of the buffers that are used in electrophoresis experiments and stress the importance of making allowance for specific binding effects in the interpretation of electrophoretic data.

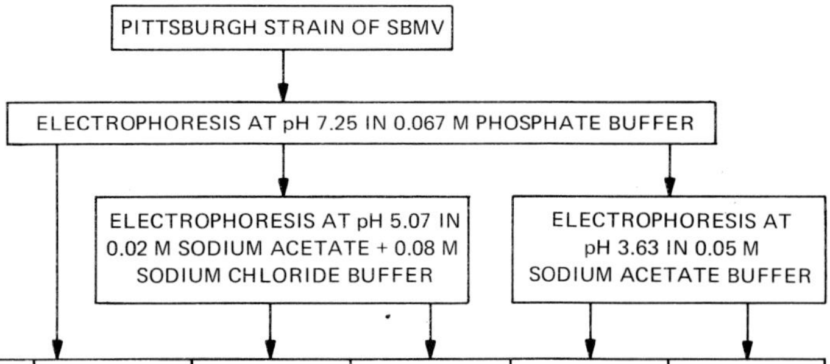

VARIANTS	I	II	III	IV	V
pI IN 0.1 M ACETATE	6.03	5.57	5.00	4.03	3.95
pI IN 0.02 M ACETATE + 0.08 M NACL	3.85	4.27	4.20	3.62	-
% OF CHARGES BOUND BY CL^- AT pI	38	25	15	7	-
% OF CHARGES NOT BOUND BY CL^- AT pI	12	25	35	43	-

Fig. 14-2. **Schematic representation of some results of Magdoff-Fairchild (1967) obtained with electrophoretic variants of the Pittsburgh strain of southern bean mosaic virus (SBMV). The variants were isolated by moving boundary electrophoresis in three different buffer systems.**

III. THE MOVING BOUNDARY METHOD

A. Description

Quantitative measurements of electrophoretic mobility are usually performed by the moving boundary method using the Tiselius electrophoresis cell. A number of commercially available instruments have been listed by Svensson and Thompson (1961) and one of them, the Spinco model H electrophoresis-diffusion apparatus, manufactured by Beckman Instruments, Palo Alto, California, is shown in Fig. 14-3. This instrument allows three electrophoresis experiments to be performed simultaneously in the same water bath and makes provision for the automatic photographic recording of Schlieren and Rayleigh fringe patterns.

The formation of sharp boundaries between a solution and its dialysate by lateral displacement of cell components is illustrated in Fig. 14-4. It is the downward displacement of the boundary in the left-hand channel of the cell and the upward displacement of the boundary in the right-hand channel that are followed optically, by virtue of the difference in refractive indices between solution and solvent. Figure 14-5 illustrates the development of multiple

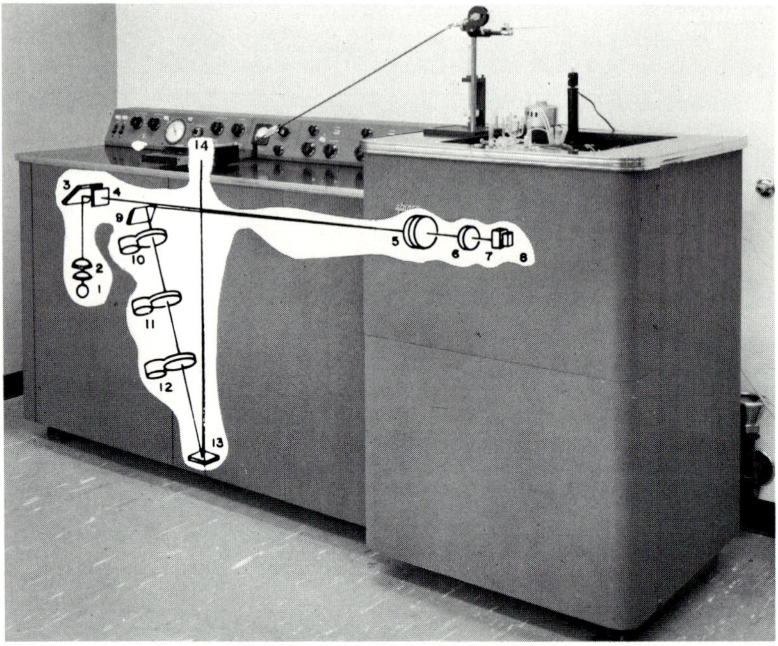

Fig. 14-3. The Spinco model H electrophoresis-diffusion apparatus. The optical path is shown schematically: (1) lamp; (2) light source condensing lens; (3) light source mirror; (4) filter and slit assembly; (5) schlieren lens; (6) water bath window; (7) electrophoresis cell; (8) cell mirror; (9) first camera mirror; (10) interchangeable schlieren diaphragm or phase plate; (11) camera lens; (12) cylindrical lens; (13) mirror; (14) viewing screen and photographic plate.

boundaries in the case of a mixture of different solute components and brings out clearly the limited value of this type of electrophoresis for purification purposes. The method is perfectly satisfactory, however, if only a small quantity of say the fastest component in a mixture of viruses is required for inoculating plants and producing a pure culture (see e.g. Magdoff-Fairchild, 1967). The Spinco model H electrophoresis apparatus is equipped with a sample needle which can be seen and controlled by the operator while looking at the boundaries in the optical system; this makes it possible to control precisely the level at which material is removed from the cell.

The moving boundary method has been used occasionally for purifying plant viruses (e.g. Kassanis, 1962) but it is mainly of value for accurately determining electrophoretic mobilities. The electrophoretic mobility of a charged particle is defined as its velocity per unit field strength [see also equations (4) and (6)]:

$$\mu = \frac{dx/dt}{E} \tag{11}$$

dx/dt being determined by measuring the distance that the boundary moves per

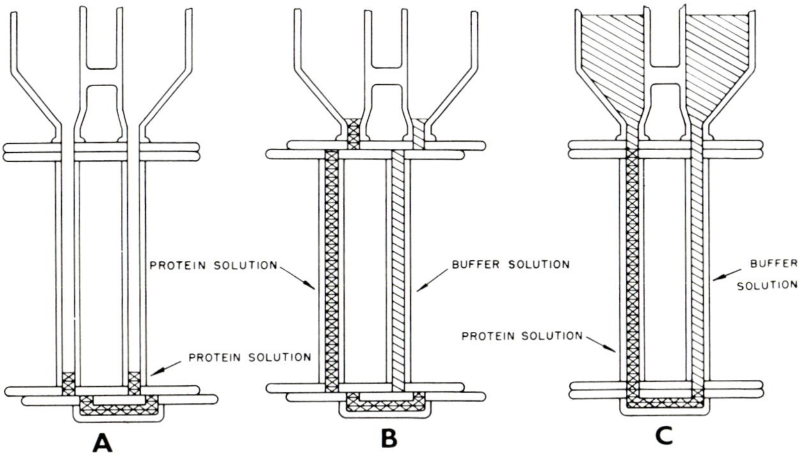

Fig. 14-4. Filling process and formation of boundaries in the electrophoresis cell. Step A: the bottom section of the cell is filled with the protein solution and slid to the right side. Step B: the center section is filled with protein solution and buffer as indicated. Step C: the top section is filled with buffer. By aligning the three sections, boundaries are formed at the ground glass plates. (From Spinco model H electrophoresis apparatus instruction manual.)

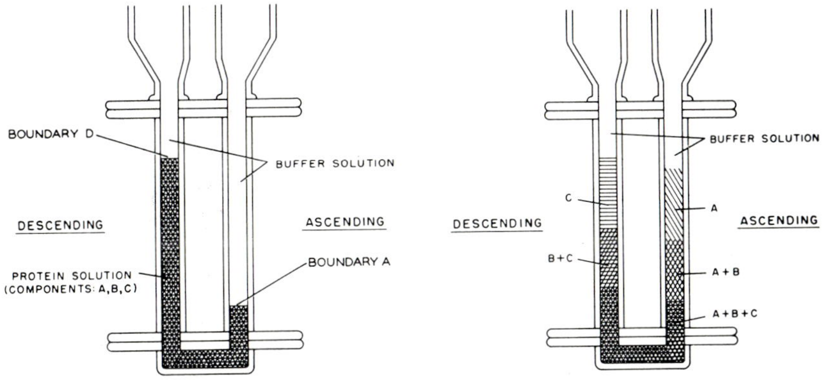

Fig. 14-5. Electrophoretic separation of three components, A, B and C in the electrophoresis cell. (left, before separation; right, after separation). Only a fraction of the fastest and slowest components (A and C) can be obtained in a pure state.

unit time. In the Beckman model H electrophoresis apparatus, a number of photographs of the ascending and descending channels can be taken on the same plate at regular time intervals, using the schlieren-cylindrical lens system with a schlieren diaphragm. A typical composite photograph obtained in this way is illustrated in Fig. 14-6. The electric field strength E in equation (11) varies at different locations in the electrophoresis apparatus and is the greatest in the rectangular channels, since these have the smallest cross-sectional area and offer the greatest resistance to the flow of current. The value of E (in volts per centimeter) to be used in equation (11) is calculated from equation (12)

$$E = \frac{I}{KA} \tag{12}$$

by measuring the current I (in amperes) flowing through the channels of cross-sectional area A (about 0.75 cm^2) when the cell is filled with the solution under study of specific conductivity K (ohm^{-1} cm^{-1}). The specific conductivity is obtained by determining the resistance of the solution with a Wheatstone bridge. (see Alberty, 1948; Longsworth, 1952, 1959). The value of the mobility obtained in this manner applies to the temperature at which the specific conductivity has been measured (usually 0-4° C) and this temperature should therefore be recorded with the results. The sign of μ is that of the net charge of the particle under the conditions of measurement; at the basic side of the pI, proteins will thus have negative μ values.

Under ideal conditions, particle mobilities measured in the ascending and descending channels should be the same. Ideal conditions for electrophoresis

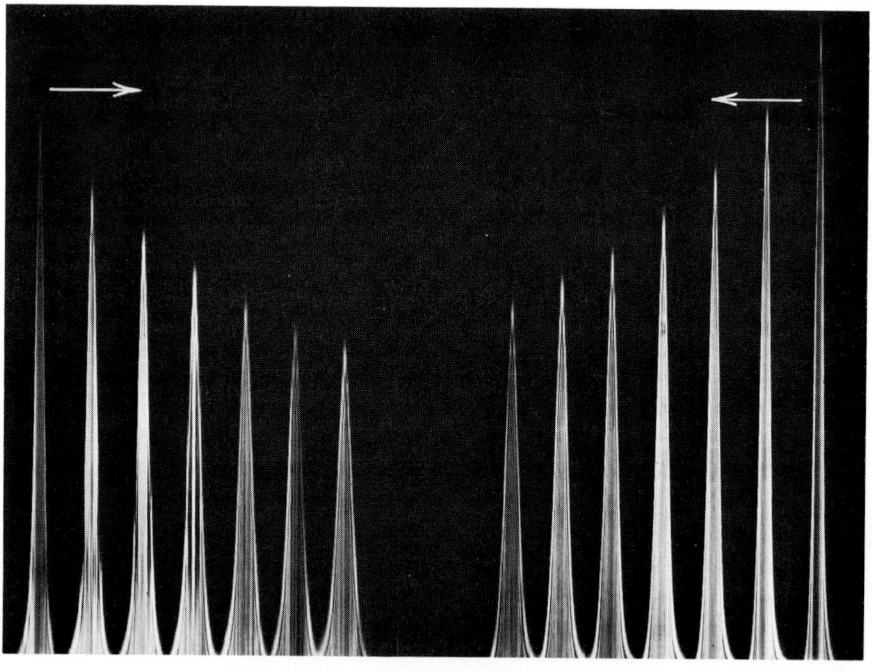

Fig. 14-6. Determination of the electrophoretic mobility of bromegrass mosaic virus at pH 4.0 in 0.1 ionic strength acetate/HCl buffer (von Wechmar and van Regenmortel, 1966). Successive schlieren peaks in each channel of the Tiselius cell were photographed on the same plate, at hourly intervals, using a schlieren diaphragm. Arrows indicate direction of migration in the two channels.

apply when the contribution of the particles to the conductance is negligible, when the same salt concentration is found in the buffer and in the solution (no Donnan effect), and when a uniform pH and electrical field occurs throughout the cell.

Under nonideal conditions and in the case of small protein molecules it is usually found that the boundary in the ascending channel migrates faster and is sharper than the one in the descending channel; in this case the mobility of the descending boundary is used for calculating μ, since this boundary moves into the solution of conductivity K used in equation (12). In addition, so-called stationary boundaries are often observed under nonideal conditions at the positions of the initial protein-buffer boundaries (see Longsworth, 1959). With particles of the size of viruses, however, these anomalies do not occur, for the equivalent concentration of the particles is too small. This is simply a reflection of the fact that only charges on the outer surface of the virus particles are electrophoretically effective. The net surface charge of a virus particle is always small compared with its total number of ionizable groups, and the contribution of suspended virus particles to the conductivity (in a 1-2% solution for example) is negligible.

B. Applications

Results obtained up to 1957 have been reviewed by Brinton and Lauffer (1959). Some pH-mobility curves of plant viruses obtained in recent years are presented in Figs. 14-7 and 14-8. Electrophoretic mobilities of tobacco mosaic virus strains and mutants have been investigated by several authors (Ginoza and Atkinson, 1955; Kramer, 1957; Sarkar, 1960; Sengbusch, 1965) and all the curves that have been published show much resemblance. The common strain of tobacco mosaic virus (Fig. 14-7) and the Dahlemense strain (Fig. 14-7), for example, have pH-mobility curves that are closely related, although the two strains differ by about 30 changes in their amino acid sequences and by a total of 8 net amino acid exchanges (for amino acid composition, see Chapter 20). The close resemblance between the two curves is even more striking in view of the six differences in location of acidic and basic amino acids between the two strains (see Table 14-1).

In his work on tobacco mosaic virus mutants, Sengbusch (1965) showed that changes in the virus capsid protein sequence involving charged groups are not always electrophoretically distinguishable. Intact particles of mutant Ni 109 (Table 14-1), which has a glutamic acid residue in place of a glycine residue in position 97 of the amino acid sequence, cannot be differentiated from type tobacco mosaic virus by electrophoresis. Charges originating at position 97 are presumably "too far" from the surface of the virus rod to be able to exert an electrophoretic effect. On the other hand, the mobility of disaggregated protein subunits of this mutant is affected by the exchange at position 97 (Sengbusch, 1965) which seems to indicate that residue 97 is located on or near the inner surface of the protein subunit that is masked by the formation of the quaternary

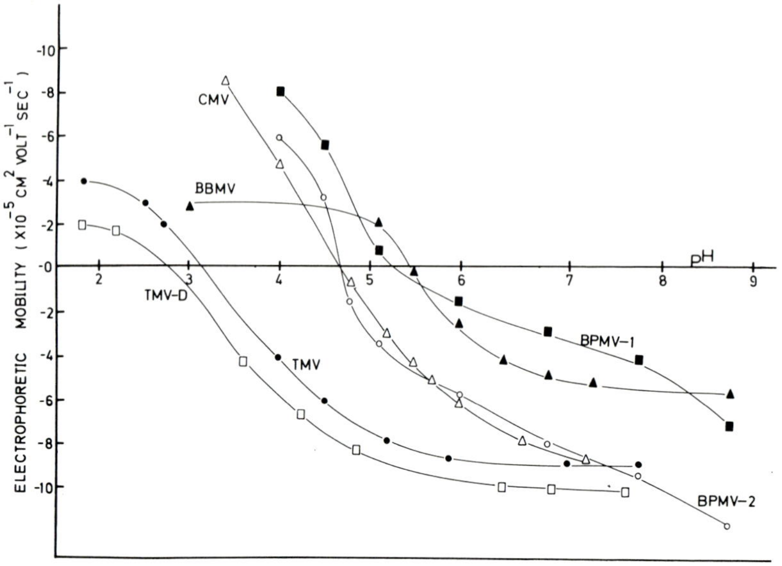

Fig. 14-7. Electrophoretic mobility versus pH curves of plant viruses. TMV: tobacco mosaic virus, common strain (Kramer, 1957); TMV-D: tobacco mosaic virus, Dahlemense strain (Kramer, 1957); BBMV: broad bean mottle virus (Semancik, 1966*b*); CMV: cucumber mosaic virus (van Regenmortel, 1967*a*); BPMV-1 and -2: two electrophoretic components of bean pod mottle virus. (Semancik and Bancroft, 1965.)

structure of the virus rod (see also van Regenmortel, 1966, 1967*b*).

Two other mutants studied by Sengbusch (1965) which have amino acid exchanges at positions 66 and 140 (Table 14-1 and Fig. 14-8) show the same decrease in μ above pH 4, compared to the type virus. On the acidic side of the pI, mutant CP 415, which gained a positive charge at position 140, now has a higher *u* than the type virus, whereas mutant Ni 102, which lost a negative charge at position 66, is indistinguishable from tobacco mosaic virus (Fig. 14-8). These results agree with theoretical expectations based on the pK values of aspartyl and lysyl groups in proteins (see e.g. Fox and Foster, 1957, p. 176).

The close similarity between the curves of the common and Dahlemense strain of tobacco mosaic virus (Fig. 14-7) thus seems to indicate that only one or two amino acid exchanges contribute to the change in mobility (Table 14-1). Since exchanges occurring at amino acid sequence positions 66 and 140 are both electrophoretically and serologically recognizable (Sengbusch, 1965) whereas an exchange at position 46 is not distinguishable serologically (van Regenmortel, 1967*b*), it seems likely that the loss of the guanidyl group at position 141 is the main factor responsible for the position of the curve of the Dahlemense strain in Fig. 14-7 slightly below that of common tobacco mosaic virus.

Difficulties in the interpretation of mobility differences are also apparent in

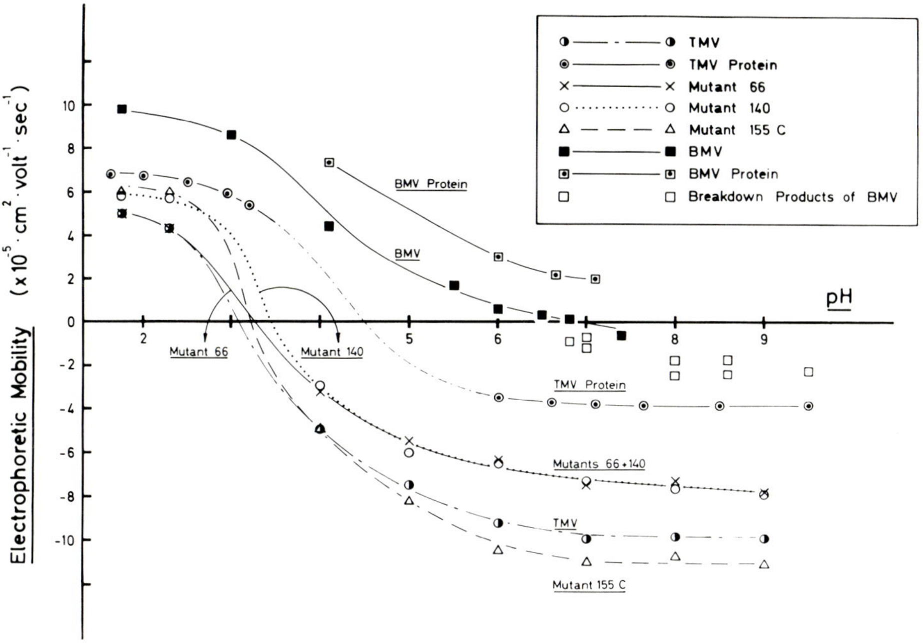

Fig. 14-8. Electrophoretic mobility versus pH curves of plant viruses. Mutants 66, 140, and 155 C are mutants of tobacco mosaic virus (TMV) with exchanges at positions 66, 140, and 156 of the amino acid sequence of tobacco mosaic virus protein (see Table 14-1 and text). TMV and mutants: Sengbusch (1965); TMV-protein: Kramer and Wittmann (1958); BMV (bromegrass mosaic virus) and BMV-protein: von Wechmar (1967).

TABLE 14-1
ACIDIC AND BASIC AMINO ACID SUBSTITUTION OF TOBACCO MOSAIC VIRUS STRAINS AND MUTANTS [a]

Strain or Mutant	*Position in Sequence*	*Nature of Change*	*Type of Charge Gained*	*Type of Charge Lost*
Ni 109 [b]	97	GLU → GLY		−
Ni 102 [b]	66	ASP → GLY		−
CP 415 [b]	140	ASP-N → LYS	+	
Dahlemense [c]	46	ARG → GLU-N		+
Dahlemense	50	GLU-N → GLU	−	
Dahlemense	64	ASP → GLY		−
Dahlemense	65	SER → ASP	−	
Dahlemense	66	ASP → VAL		−
Dahlemense	141	ARG → GLU-N		+

[a] For details of amino acid sequences of tobacco mosaic virus mutants and strains, see Chapter 20.
[b] Sengbusch (1965).
[c] Dahlemense strain (Wittmann-Liebold and Wittmann, 1963: Anderer et al., 1965).

the case of mutant 155 C. This mutant in position 156 has a proline replaced by a leucine, with the result that the three terminal amino acids can be split off by carboxypeptidase. Tobacco mosaic virus rods treated with this enzyme have a higher mobility than untreated virus at most pH values (Mutant 155 C in Fig. 14-8). Harris and Knight (1955) had observed a smaller difference in μ when only the terminal threonine of tobacco mosaic virus protein was removed by carboxypeptidase treatment. Since only neutral amino acids are involved, it is necessary to assume that the removal of the terminal amino acid residues either allows underlying charged residues to exert an electrophoretic effect or alters the configuration of the polypeptide with concomitant changes in the position of ionizing groups.

From the observation that on the alkaline side of the pI disaggregated tobacco mosaic virus protein subunits have a lower mobility than the intact virus rods (see Fig. 14-8) it has been concluded that, as a result of the exposure of previously concealed positively ionizing groups, the net negative charge is less in the depolymerized protein than in the polymerized rod (Kramer and Wittmann, 1958; Kleczkowski, 1959). However, Ansevin et al. (1964) showed by acid-base titrations and by measuring pH shifts during polymerization that when the protein is depolymerized from a rod at pH 6.5, there is an increase of five negative charges per molecular weight unit of 100,000. This seemingly paradoxical result may be reconciled with the lower anodic mobility of tobacco mosaic virus protein by recalling that electrophoretic mobility is proportional to average net charge per unit surface area (Ansevin et al., 1964). Since there is a considerable increase in the specific surface area when the virus depolymerizes and only a small increase in net charge per unit weight, it is perfectly natural that the mobility of the subunit protein should be less than that of the intact virus.

When disaggregated tobacco mosaic virus subunits are repolymerized in the absence of nucleic acid, protein rods of various lengths are obtained which have a mobility similar to that of the original virus particles (Kramer and Wittmann, 1958). From the helical configuration of the nucleic acid inside the virus rod, it can be calculated that the phosphate groups of the RNA could theoretically contribute an additional 3 negative charges to the 14 negative charges present in each protein subunit (10 aspartyl, 3 glutamyl and 1 terminal carboxyl group). It seems likely therefore that the failure of the RNA to exert an electrophoretic effect is due to its location about 35 Å from the outer surface of the virus rod.

In the case of turnip yellow mosaic virus and wild cucumber mosaic virus the presence of RNA inside the protein shell has little or no influence on the electrophoretic mobility of the capsids (Markham, 1959). From the work of Klug et al. (1966) on the location of the RNA inside the protein shell of turnip yellow mosaic virus, it would thus appear that charges buried 20 Å below the outer surface of a virus particle need have no influence on the electrophoretic mobility. The top and bottom components of wild cucumber mosaic virus also have the same mobility in 0.1 M phosphate buffer, but under salt conditions which lead to differential adsorption of the two components by bentonite, the nucleic acid inside the bottom component makes itself felt (Hitchborn and Dunn, 1965). When the two components of wild cucumber mosaic virus were submitted to electrophoresis in 0.01 M pH 7.3 phosphate buffer containing 0.001 M $MgSO_4$, the bottom component had a higher negative charge and a greater mobility than the top component. No such difference could be shown between the two components of turnip yellow mosaic virus.

Particles of bean pod mottle virus and cowpea yellow mosaic virus, on the other hand, display an electrophoretic heterogeneity which is not related to differences in the nucleic acid content of the various centrifugal components found with these viruses (Bancroft, 1962; Agrawal, 1964). This is clearly shown, for example, by the electrophoretic heterogeneity of a preparation of top component of cowpea yellow mosaic virus (Semancik, 1966*b*). Since the proteins of the different electrophoretic components appear to have the same amino acid composition, the differences in mobility probably reflect two distinct protein conformations in the capsids.

IV. ZONE ELECTROPHORESIS IN A DENSITY GRADIENT

It is a commonplace notion that for preparative purposes, zonal separations are preferable to moving boundary procedures. This applies to centrifugal and chromatographic methods as well as to electrophoresis. It has already been indicated (see Fig. 14-5) that when the Tiselius electrophoresis cell is used for purification, only a small portion of the fastest and slowest migrating components of a mixture can be obtained in a purified state. Another disadvantage of the moving boundary method is that hydrodynamic stability in the Tiselius cell is derived from the density increments caused by the different components that are being separated. This means that when very dilute virus solutions are used, the stabilizing influence of the density increment may be too small to prevent convection. In density-gradient electrophoresis, on the other hand, there is no lower limit to the amount of virus material that can be applied to a column.

Brakke (1955) was the first to use a density gradient in the electrophoresis of plant viruses and he designed a simple apparatus made of a U-tube 8 cm long and of 1.8 cm internal diameter. Density gradients were obtained by layering solutions of decreasing sucrose concentration above each other and waiting for diffusion to smooth out the density differences. This apparatus has been used by Brakke and his collaborators in studies of potato yellow dwarf and wheat streak mosaic viruses and for obtaining approximate electrophoretic mobility rates of different plant viruses (see Ball, 1966).

An excellent review covering both theoretical and practical aspects of density-gradient electrophoresis is available (Svensson, 1960) and reviews dealing specifically with zone electrophoresis of viruses have also appeared (Cramer and Svensson, 1961; Polson and Russell, 1968). This section will be limited to a description of an apparatus that has proved to be well adapted to work with plant viruses and will discuss some of the results obtained with this technique.

A. Description of the Apparatus

The apparatus used in the author's laboratory is similar to the one used by Polson and Cramer (1958) and is a simplified version of the original design of Svensson and Valmet (1955). The apparatus is shown schematically in Fig. 14-9 and a photograph of it is shown in Fig. 14-10. It can be easily constructed by a glassblower of moderate skill. It consists of a U-tube of 23 mm internal diameter provided with a wide-bore stopcock (A) in the right arm and connected to two electrode vessels of 5 cm internal diameter, ending in two capillaries (B1 and B2) of 2 mm internal diameter. The bottom of the straight arm of the U-tube is closed by a glass stopper through which pass two capillaries, C1 and C2, of 1 mm and 2 mm internal diameter respectively.

Reversible electrodes consisting of a folded piece of silver mesh, 20 cm × 5 cm, attached to a silver wire are used alternatively as the positive and negative poles. The silver wire is isolated from the solution by a glass tube sealed with wax. Before use, new electrodes must be coated with silver chloride by

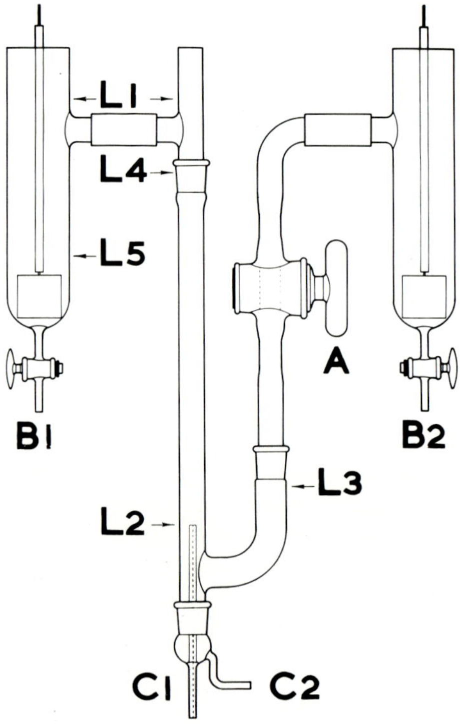

Fig. 14-9. Zone electrophoresis apparatus. For significance of symbols, see text. The sugar density gradient is located between levels L2 and L4, over a distance of 32 cm.

electrolysis in 1 N HCl, using a piece of platinum wire as the negative electrode (see also Longsworth, 1959). When current is passed through the electrophoresis column the following reaction takes place at the anode:

$$Ag + Cl \longrightarrow AgCl + e$$

At the cathode the reaction is reversed. The electrodes have a limited capacity dependent on the amount of available silver chloride at the cathode. In the event of a run extending over 24 hr, hydrogen and chlorine may start to be generated, at the cathode and anode respectively, and the electrodes must then be interchanged. This can be done without stirring the contents of the U-tube if stopcock A is closed temporarily during the change of electrodes.

Reversible Ag-AgCl electrodes when used properly will last almost indefinitely. Platinum electrodes are less suitable for this apparatus; the acid

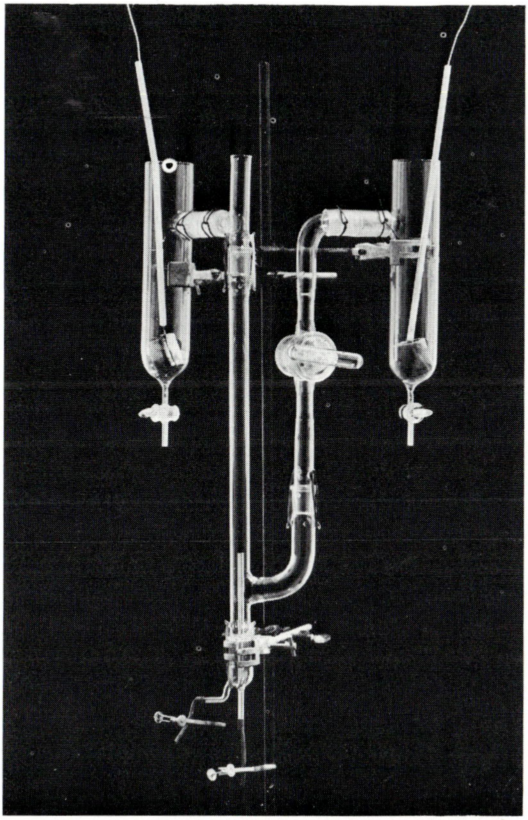

Fig. 14-10. Photograph of zone electrophoresis apparatus drawn schematically in Fig. 14-9.

generated at the anode was found to lower the pH in the region of the density gradient within 16 hr of passing a current of 22 m amp.

B. Buffer and Sucrose Solutions

The following solutions have been used successfully in work with plant viruses. For additional buffer systems, see Miller and Golder (1950) and Polson and Russell (1968).

1. 0.1 M, pH 8.6 BORATE BUFFER

This buffer has the following composition: 0.035 N H_3BO_3; 0.0175 N NaOH; 0.0075 N HCl; 0.037 N NaCl; and is prepared by mixing 350 ml of a solution of 62 g of H_3BO_3 and 20 g of NaOH in 5 liters of distilled water, 150 ml of 0.1 N HCl, 500 ml of normal saline, and 1 liter of distilled water. With this buffer a potential difference of 220 v applied across the electrodes generates a current of about 20-22 m amp. Under these conditions no thermal convection occurs when the experiment is done in a room kept at 15° C. Air-cooling with an electric fan may help to prevent convection in a warmer room or when somewhat higher voltages are used.

2. 0.08M, pH 8.6 BORATE BUFFER

This buffer is prepared in the same way as buffer (1) but using only 250 ml of normal saline and bringing the total volume to 2000 ml with 1250 ml distilled water. The migration of particles may be somewhat accelerated with this buffer, for the lower conductivity allows a potential difference of about 280-300 v (16 m amp) to be used without concomitant heating.

3. 0.03M, pH 7.5 PHOSPHATE BUFFER

With this buffer a potential difference of 270 v generating a current of about 16 m amp can be used without heating.

4. 0.05M, pH 4.0 ACETATE BUFFER

A potential difference of 200 v and current of 18-20 m amp are suitable with this buffer.

5. 50% SUCROSE SOLUTION

This is prepared by dissolving 300 g of reagent enzyme grade sucrose in 300 ml of double-strength buffer (e.g. 0.2 M borate buffer in the case of buffer 1) and bringing the resulting volume of about 500 ml up to exactly 600 ml by adding distilled water. With borate buffers the formation of borate-sugar complexes necessitates the addition of a few milliliters of 1 N NaOH to bring the pH back to a value of 8.6.

C. Operation

The apparatus is first filled through C2 (see Fig. 14-9) with a buffer solution up to level L1; a 50% sucrose solution in the same buffer is then also introduced through C2 until it reaches levels L2 and L3. Level L3 is about 3 cm higher than

L2; this is achieved by letting some buffer flow out through B2. Stopcock A is then closed and a sucrose gradient is introduced through C1. Svensson (1960) has reviewed various devices for obtaining density gradients. A suitable gradient may be produced with two 250 ml Erlenmeyer flasks connected in series by a capillary tube (see Polson and Russell, 1968). One flask containing 150 ml of 50% sucrose solution is left open and the other one containing 150 ml of buffer solution is closed and placed on a magnetic stirrer. The sugar solution is allowed to flow slowly into the closed flask where it is mixed with buffer. The resulting dilute sugar solution is allowed to enter the apparatus through C1 and is layered above the 50% sucrose solution at a rate of about 4 ml/min (stopcock A being closed). The initial dilute solution is slowly displaced upward by the following more concentrated portion of the sugar solution flowing from the Erlenmeyer flask. The sucrose gradient obtained in this way between levels L2 and L4 is illustrated in Fig. 14-11.

About 150 ml of buffer are then let out through B1 and the electrodes are covered up to level L5 with a saturated NaCl solution introduced through B1 and B2. Level L5 should be adjusted in the two electrode vessels to ensure that

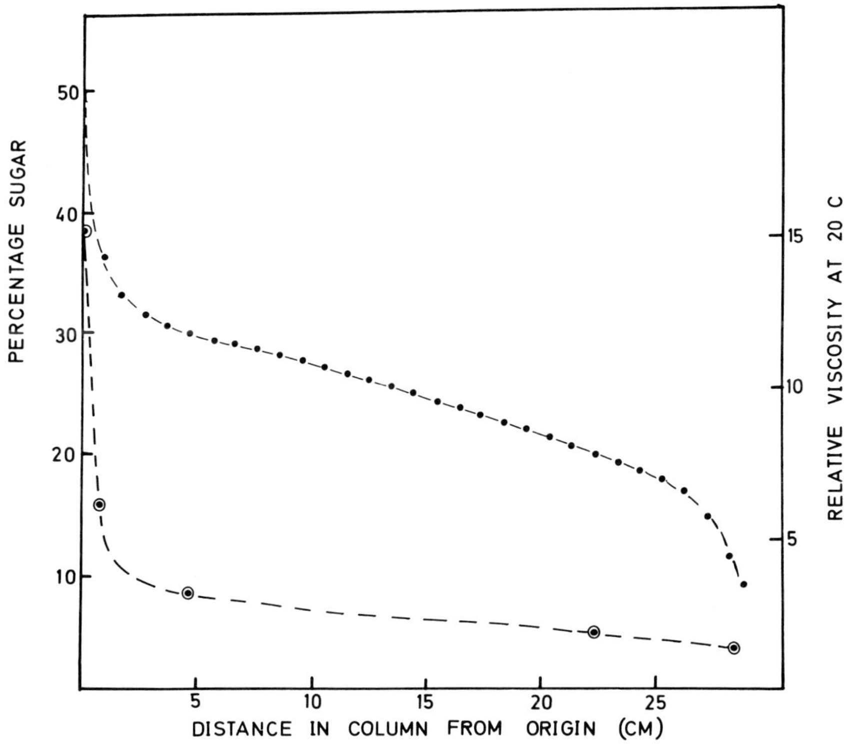

Fig. 14-11. Sugar concentration gradient (•——•) and viscosity gradient (•— — —•) in the zone electrophoresis apparatus.

when stopcock A is opened, level L2 remains stationary. The interface at L2 is then sharpened by letting about 5 ml run out through C1 with stopcock A open. This produces a density shelf (see Svensson, 1960) which will prevent droplet sedimentation of the initial zone of material after it is introduced through C1. A 2 ml sample of the material to be analyzed is adjusted to a sugar concentration of about 45% so that it forms a stable zone between the density gradient and the 50% sucrose. After some phenol red has been added to the sample to serve as a marker, the material is introduced through C1 with a syringe. The residual solution in the C1 capillary is removed by adding about 15 ml of buffer to the right electrode vessel, opening stopcock A so that level L2 is raised, and letting about 10 ml of the 50% sugar solution escape through C1. A sample of 2 ml corresponds to about 0.5 cm column height; the level occupied by the preparation at the beginning of the experiment is known as the origin and the distances that different components migrate are measured from this point.

Within a few minutes of introducing the sample the current may be switched on, care being taken to use a potential difference and current that will not cause heating in the column and a concomitant breakdown of the density gradient (see preceding section). As a practical rule of thumb, the product milliamperes × volts should not exceed a value of 5000, if thermal convection is to be avoided when the experiment is performed in a room kept at 15-20° C.

Electrophoresis is allowed to proceed, usually overnight, until a sufficient separation of the different components has been achieved by their upward migration. Separations are best performed in a room that can be darkened. By illuminating the column with a vertical fluorescent lamp enclosed in a metal cylinder that is provided with a slit, bands of opalescence can usually be detected in the case of most plant virus preparations. The presence of as little as 0.5-1.0 mg of virus can easily be recorded photographically, using Ilford N5-31 film for example.

After completion of the run, stopcock A is closed, the positions of visible zones and the reference substance phenol red are measured, and the contents of the column are collected through C1 in different fractions according to the desired separation. Sampling should be performed very slowly and the capillary may be rinsed between fractions, by letting some 50% sucrose solution from the right side of the U-tube run out through C1 (by opening stopcock A briefly). If no zones are visible, fractions may simply be taken at 1 cm intervals and assayed by infectivity or serological tests (see Polson and Russell, 1968).

D. Purification and Separation of Plant Viruses

Because of the limited capacity of density gradients and the small volumes that can be handled, zone electrophoresis is of value only in the final stages of a purification procedure, after clarified plant sap has been concentrated and most impurities removed. The method has been used successfully for separating 20 different plant viruses from contaminating antigenic plant proteins (van

Regenmortel, 1964*b*, 1966; see Table 14-2) and it is thus of particular value for obtaining virus preparations to be used for immunizing rabbits. When drastic sap clarification methods are used, such as alcohol precipitation, the removal of plant proteins presents, of course, no problems. With less stable viruses, however, methods of sap clarification in common use combined with differential ultracentrifugation often yield virus preparations that are heavily contaminated with plant antigens. Because of the tendency of such antigens to aggregate, it is often difficult to remove them by density-gradient centrifugation (see van Regenmortel, 1966), and in such cases electrophoretic methods, being mainly independent of the size and shape of particles, offer a distinct advantage.

Zone electrophoresis has also been used successfully for separating different viruses from a mixture (see van Regenmortel et al. 1964; and Fig. 14-12), and in view of the wide range of electrophoretic mobilities of plant viruses (Table 14-2) this technique can often reveal the presence of unsuspected contaminating viruses. The presence of sowbane mosaic virus (which is seed-borne in *Chenopodium quinoa* Willd.) in preparations of tobacco necrosis virus propagated in *C. quinoa* is a case in point (Fig. 14-12).

E. Zone Electrophoretic R_ϕ Values of Plant Viruses

After analyzing the results of zone electrophoresis experiments obtained in a 30 cm-long column over a number of years, it was found that the ratio of the electrophoretic mobility of a virus to that of phenol red is remarkably constant and independent of such factors as the precise shape and concentration of the sugar gradient, the potential difference applied across the electrodes, or the duration of the experiment. Brakke (1955), Aach (1963), and Ball (1966) also found that the rate of migration during zone electrophoresis was virtually independent of the sucrose concentration, presumably because the effect of the viscosity gradient (caused by the sugar gradient; see Fig. 14-11) is cancelled out by the increase in field strength that accompanies increases in viscosity. Quacquarelli et al. (1967), using a similar apparatus to the one used by the author, have determined relative electrophoretic mobilities of plant viruses by comparing the distances moved by phenol red and different viruses respectively. They used the midpoints of the phenol red and virus bands as basis for their calculations. Since the width of a virus band supported in a sugar gradient is greatly dependent on the amount of virus material present (because of partial sedimentation when the density gradient has been overloaded), it appears preferable to use the upper boundary position of the virus zone for characterizing the mobility. The diffusion coefficients of viruses are sufficiently small to introduce only insignificant errors when such a procedure is followed. Because of the high diffusion coefficient of phenol red, the midpoint of this band is best chosen as representing the position of the reference substance. The ratio of the position of the upper limit of a virus band to the middle of the phenol red band has been termed the zone electrophoretic R_ϕ value of a virus.

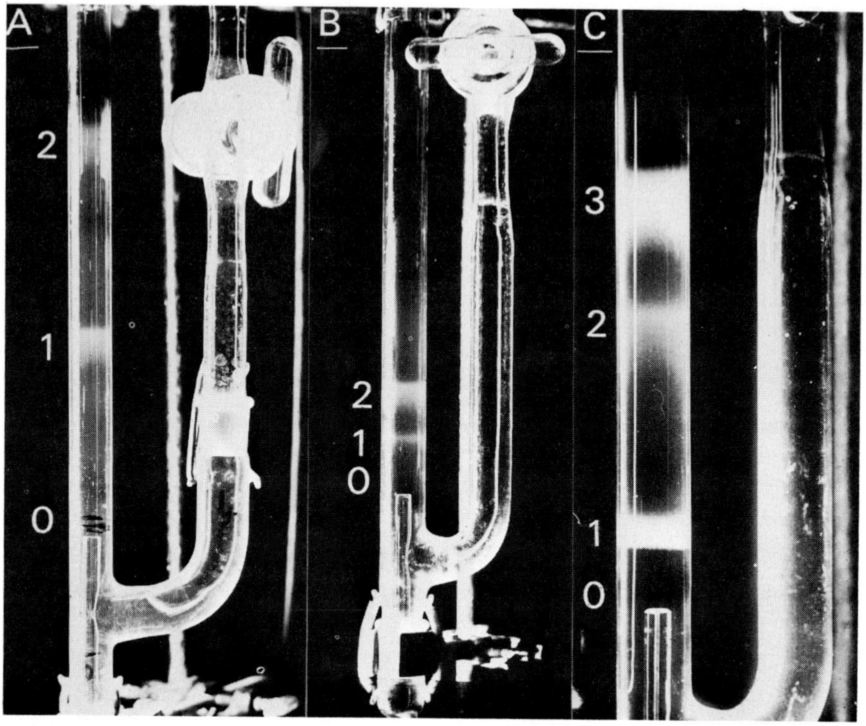

Fig. 14-12. Separation of plant viruses by zone electrophoresis. A: Separation of U1 and U2 strains of tobacco mosaic virus. 0: initial position of mixture; 1: U2 zone, located 10 cm from the origin; 2: U1 zone located 20 cm from the origin. B: Separation of sowbane mosaic virus (SOMV) and tobacco necrosis virus (TNV). 0: initial position of mixture; 1: TNV zone 1.5 cm from the origin; 2: SOMV zone 4 cm from the origin. C: Separation of potato virus X (PVX), alfalfa mosaic virus (AMV), and tobacco mosaic virus (TMV). 0: initial position of mixture; 1: PVX zone 1.5 cm from the origin; 2: AMV zone 8 cm from the origin; 3: TMV zone 12 cm from the origin. (van Regenmortel et al., 1964.)

R_{ϕ} values of different plant viruses are summarized in Table 14-2. Values obtained in individual experiments were always within 2-4% of the reported values. The accuracy and reproducibility of these determinations seem to warrant the use of R_{ϕ} values for characterizing and identifying virus strains. In addition, a comparison of the R_{ϕ} values of viruses with those of plant proteins of different hosts should help to predict the outcome of purification experiments by means of zone electrophoresis (see van Regenmortel, 1966).

The reproducibility of the R_{ϕ} values may be illustrated in the case of cucumber mosaic virus, strain S. Values calculated from 20 experiments varied between .51 and .55 with a mean of $.533 \pm .028$.

TABLE 14-2
ZONE ELECTROPHORETIC $R\phi$ VALUES OF PLANT VIRUSES

Virus	*Rϕ*
Potato virus X	.09 [a]
Tobacco necrosis virus	.14
Wild cucumber mosaic virus	.14
Sugarcane mosaic virus	.21
Tomato bushy stunt virus[b] (Pelargonium leafcurl strain)	.21
Watermelon mosaic virus	.25
Potato virus Y	.29
Tomato bushy stunt virus[b] (Pelargonium leafcurl strain)	.30
White clover mosaic virus	.31
Sowbane mosaic virus	.32
Arabis mosaic virus	.36
Tomato bushy stunt virus[b] (type strain)	.37
Tobacco mosaic virus (U2 strain)	.38
Bean yellow mosaic virus	.46
Alfalfa mosaic virus	.49
Cucumber mosaic virus (S strain)	.53
Cucumber mosaic virus (Y strain)	.53
Turnip yellow mosaic virus	.63
Prunus necrotic ringspot virus	.66
Tobacco mosaic virus (type strain)	.76
Tobacco mosaic virus[b]	.79
Barley stripe mosaic virus	1.37

[a] Values represent the ratio of the position of the upper edge of a virus band to the middle of the tracking dye (phenol red) band; designated as $R\phi$.
[b] Recalculated from Quacquarelli et al., 1967.

V. ADDITIONAL TECHNIQUES

In addition to zone electrophoresis in a density gradient and the moving boundary method, a number of other electrophoretic techniques have also been applied to various facets of plant virus research. The following examples should suffice to illustrate the wide range of electrophoretic separations that can be performed.

Continuous filter paper electrophoresis has been used by Zaitlin (1958) to separate the U1 and U2 strains of tobacco mosaic virus. In this technique, electrophoresis is performed on a paper curtain with solvent flowing down at right angles to the electrical field. A common difficulty encountered with paper electrophoresis, namely the adsorption of virus particles to the paper, is eliminated in the free-flow type of electrophoresis; this technique was used by

Streeter and Gordon (1966) to separate the same U1 and U2 strains (see also Fig. 14-12). Electrophoresis is performed between two glass plates (50 cm × 50 cm) that are separated by 0.5 mm; buffer is applied at the top and flows down between the plates at a constant rate. The principle of continuous sample input and collection inherent to this technique seems to provide a practical method for the large-scale separation and purification of viruses.

Agar gel electrophoresis has been used for the purification of tobacco mosaic virus and potato virus X (Townsley, 1959*a*) and has proved successful for the removal of the ribonuclease associated with tobacco mosaic virus preparations obtained by ultracentrifugation (Townsley, 1959*b*). An adaptation of the same technique was used for fractionating crude plant sap from virus-infected plants and obtaining samples suitable for virus identification by electron microscopy (Weintraub et al., 1962).

Electrophoresis of viruses in agarose layers of increasing concentration (Polson and Russell, 1968) as well as in polyacrylamide gels (Tiselius et al., 1965) has been employed as a method of particle sieving (see Chapter 16 for further details.) For applications of immunoelectrophoresis the reader is referred to a review (van Regenmortel, 1966).

15

Isolation and Properties of Virus Proteins

AKIRA TSUGITA AND AKIKAZU HIRASHIMA

Laboratory of Molecular Genetics
Medical School
University of Osaka
Osaka, Japan

I. INTRODUCTION

It is now common knowledge that plant viruses are composed mainly of protein and nucleic acid, but in the past the roles of these constituents were unclear. Elegant reconstitution experiments with tobacco mosaic virus clearly showed that the virus protein was not responsible for infectivity but played the role of protecting the nucleic acid moiety and stabilizing the virus (Fraenkel-Conrat, 1956). Since then, there has been a steady increase of knowledge in molecular virology and most of the interest in viruses has been directed towards their nucleic acids. Accumulation of information on nucleic acids, in turn, has now resulted in the necessity to further understand proteins whose precise synthesis resides in the nucleotide sequence of nucleic acids. Understandably, therefore, the deciphering of the genetic code was aided by studies on the primary structure of viral protein (see Chapter 20). Thus, studies of proteins are always interrelated with those of nucleic acids, especially when the recent developments have been on the biology of the virus. Proteins are composed of one or more polypeptides consisting of 20 amino acids, and their complicated structure can vary considerably. Because of the diversity of proteins, this chapter will limit its scope to plant virus proteins. Viral nucleic acids are covered in the following chapter.

II. GENERAL CHARACTERISTICS OF PROTEINS

Proteins are essentially condensation products of many amino acids linked by peptide bonds into linear macromolecules (primary structure). The repeating condensation unit can be expressed by

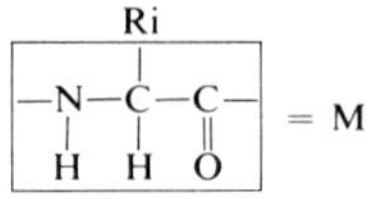

and protein can accordingly be expressed as

$$H-\begin{bmatrix}Ri\\M\end{bmatrix}_n-OH$$

when n is an arbitrary number and Ri is a side chain. The protein's terminal amino group is more reactive than the terminal carboxyl group and sometimes the amino terminus is in a "protected" form, for example, plant virus proteins are usually acetylated. The M unit of the polypeptide chain is repeated in all proteins forming combinations of helical (usually α-helix) and random structures. The tertiary structure is more or less controlled by Ri in the unit in collaboration with water molecules and/or other environmental molecules. Ri can have up to 20 variations (for each amino acid), and some of them possess anionic or cationic groups which contribute to the random structure of the protein whereas others possess hydrophobic groups which favor formation of the helical structure. An exceptional case is proline, which creates a bend in the polypeptide backbone because it possesses the imino group

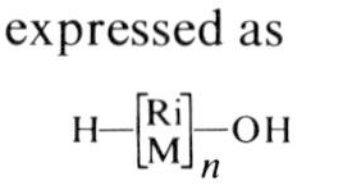

instead of the usual amino group.

Cysteine is important for the formation of intra- and interpeptide bridges between polypeptide chains. These bridges or S—S bonds are formed by autooxidation and are prerequisites for the protein's secondary structure. Amino acids in the polypeptide chain interact through ionic, hydrophobic, and hydrogen bonds (inter- or intramolecular) and with other molecules in the environment, and eventually form into tertiary protein structures. These may be substrates, allosteric effectors, subunits of larger biological molecules, such as virus proteins, etc.

The amino acid sequence of the polypeptide chain as determined by the nucleotide sequence encoded in the genetic message is synthesized in vivo on polyribosomes. The amino acids are joined sequentially beginning with the amino terminus and synthesis proceeds toward the carboxyl terminus. As the nascent polypeptide is formed, the formation of secondary structure follows spontaneously. Then through intra- and intermolecular linkages, tertiary structures evolve to provide the most stable structure in the environment. The first amino acid is blocked in situ (formylmethionine in the case of bacterial systems) since the genetic message specifies initiation of protein synthesis with such amino acids. This blocked amino terminus of the peptide chain can be subsequently digested by specific enzymes, followed by digestion with amino peptidase until the most stable structure is attained. The free amino terminus of a number of viruses are acetylated to form a more stable structure (Marchis-Mouren and Lipmann, 1965). The acetylated amino terminus ovalbumin is proposed to be synthesized de novo like formylated-methionine in bacteria (Narita et al., 1969). With the formation of these stabilized structures the protein is completed, resulting in a biologically active form.

A method for studying the primary structure of protein was first developed by Sanger (1953), who established the amino acid sequence of insulin. Likewise, the coat protein of tobacco mosaic virus was sequenced (Anderer et al., 1960; Tsugita et al., 1960), and in the same year RNase I was sequenced (Hirs et al., 1960).

Various methods have been developed for elucidating the tertiary structure of proteins. Circular dichroism, optical rotatory dispersion, ultraviolet spectral analysis, titration, tritium exchange, viscosity, electrophoresis, etc., are currently employed. The electron microscope and X-ray diffraction methods have been most useful for studies on the shape and construction of viruses. Also, X-ray analysis was developed for *direct analysis* of the tertiary structure of protein molecules (Kendrew, 1961). Plant virus proteins are just beginning to be analyzed by this method.

III. METHODS OF ISOLATING PLANT VIRUS PROTEINS

During infection of the host, more than one type of virus protein is produced, and the number of different virus proteins depends on the number of genes (for RNA viruses see definition by Kado and Knight, 1968) in its nucleic acid. Besides the coat or capsid protein (commonly known as "virus protein"), polymerases (replicases), maturation factors, and other "virus-directed" proteins are being isolated and characterized from infected bacteria, but so far, none have been isolated from virus-infected plants. Therefore this chapter will limit its discussion to the isolation and properties of "virus proteins."

Virus protein is isolated from highly purified virus. Denaturation of the protein component is preferentially avoided during the course of dissociating the viral tertiary structure. The isolation method of course depends on the purpose of the experiment. For instance, for studies on the amino acid composition or the sequence of the protein, an isolation procedure is required which provides relatively large amounts of pure protein (10-100 mg). For reconstitution and serological studies, a method is required which yields protein closely akin to native protein. The isolation method also depends on the type of virus investigated since the tertiary structure, the interacting forces between protein and the other components, and the stability of protein can differ among unrelated viruses and even among strains of a given virus.

Isolation and purification methods can be devised on the following simple principles: (1) breakage of the interacting bonds between protein and other components, usually nucleic acid; (2) denaturation or destruction of the unwanted component(s); (3) avoidance of denaturation and decomposition of the protein component; (4) removal of the unwanted component (e.g. undegraded virus and nucleic acid). In this section, in addition to some popular procedures, the currently reported ones will be cited. The reader should consult the original papers for more details.

A. Tobacco Mosaic Virus

The cold acetic acid method—Fraenkel-Conrat (1957) modified the method of Bawden and Pirie (see Bawden, 1964*b*) to prepare tobacco mosaic virus protein which was then employed for experiments successfully demonstrating the reconstitution of the virus in vitro. This method is widely accepted for preparing proteins from other plant, animal, and bacterial viruses as well as ribosomes. Because acetic acid blocks ionization of the carboxyl groups of proteins and displaces any hydrogen bonds, it dissociates protein from the RNA, and at the same time it precipitates and degrades the nucleic acid while keeping protein in solution.

The procedure is as follows: To an ice-cold, aqueous solution of tobacco mosaic virus (10-30 mg/ml) two volumes of glacial acetic acid are added gradually. The mixture is allowed to stand for 30-60 min with occasional stirring. The solution becomes turbid as the RNA is denatured and precipitated. The insoluble material is removed by centrifugation and the supernatant is dialyzed against distilled water for 2-3 days with occasional changing of water. During dialysis the protein precipitates as the pH of the dialyzate reaches the isoelectric point of the protein. The precipitate is then collected by centrifugation and suspended in distilled water. The pH of the suspension is adjusted to pH 8.0 by adding dilute alkali gradually (overnight) to dissolve the protein. The protein solution is then centrifuged at 100,000 *g* for 1 hr to remove any residual insoluble material (denatured protein and undigested virus). The clear solution contains tobacco mosaic virus protein which may be stored in the frozen or lyophilized state. The yield of the protein is about 95%. The amino acid composition of this virus protein has been studied and the sequence of the protein has been elucidated (Anderer et al., 1960; Tsugita et al., 1960). This method can be applied without modification to spontaneous and chemically induced mutants of tobacco mosaic virus and also to other viruses such as turnip yellow mosaic virus (Harris and Hindley, 1961), polio virus and RNA phages. Tobacco rattle virus protein was prepared in this manner and was used in the reconstitution of the virus in vitro (Semancik and Reynolds, 1969).

B. Broad Bean Mottle Virus

The warm formic acid method—Broad bean mottle virus protein may be prepared by the $CaCl_2$ method (to be discussed later) but this method seemed to adversely denature the protein. Miki and Knight (1965) therefore used warm formic acid for the preparation of this protein. The method is basically that of the acetic acid method since formic acid has properties similar to those of acetic acid and is also a good protein solvent.

The method is as follows: To the purified virus (10-20 mg/ml) are added two volumes of 99% formic acid, mixed and incubated for 18 hr at 37° C. The solution is then dialyzed against water for two days with frequent changes of water to remove the formic acid and RNA fragments. One-half volume of saturated ammonium sulfate is then added and the resulting precipitate (protein) is collected. The ammonium sulfate fractionation is repeated twice to ensure complete removal of nucleic acid. Finally, the solution is dialyzed against water to remove the salt.

This method was modified and used for the isolation of sowbane mosaic virus protein, which was analyzed chemically (Kado, 1967).

C. Potato Virus X

The guanidine hydrochloride method—Urea and guanidine hydrochloride reduce the hydrophobicity of nonpolar groups in proteins (Whitney and Tanford, 1962). The presence of these reagents weakens the intermolecular hydrophobic bonds between proteins and causes dissociation of the virus. Guanidine hydrochloride was used by Miki and Knight (1968) for obtaining potato virus X protein since its efficiency to dissociate the protein structure was better than that of urea. This procedure is summarized as follows: The aqueous solution of the virus is dialyzed overnight against 2 M guanidine hydrochloride at room temperature. The RNA precipitates while the protein remains in solution. The precipitate is removed by centrifugation and the supernatant containing the protein is dialyzed for 1-2 days against water. Any turbid material due to the presence of the undecomposed virus, is removed by centrifugation (105,000 *g* for 2 hr). The recovery of virus protein by this procedure is about 90% and the preparation may be used for determining its amino acid composition and molecular weight.

D. Turnip Yellow Mosaic Virus

The phenol method—Phenol has been used primarily for the isolation of RNA (see Chapter 16) but can be used effectively for the isolation of protein. For instance, it was used to isolate tobacco mosaic virus protein (Anderer, 1959*a*), thus Rappaport et al. (1965) used phenol for the preparation of turnip yellow mosaic virus protein. The procedure is as follows: The virus solution (1-10 mg/ml) in 0.01 M phosphate buffer is shaken for several minutes with an equal volume of water saturated phenol. The virus quickly dissociates into its RNA and protein components, and by centrifugation the protein in the phenol layer is separated from the RNA in the aqueous layer. The phenol containing the protein is extracted with water, and then five volumes of acetone are added. The precipitated protein is washed several times with acetone and then dissolved in water. The solution is then dialyzed against water or buffer.

The phenol method for tobacco mosaic virus described by Anderer (1959*a*) is as follows: To a solution of virus in sodium acetate is added an equal volume of water-saturated phenol. The aqueous phase is removed and five to ten volumes of methanol are added to the phenol containing the protein together with some crystals of sodium acetate. The precipitate is collected and washed with methanol and ether successively. The precipitate is allowed to dry and then solubilized in water by adjusting to pH 7.5 with dilute alkali while the solution is being warmed at 60-80° C. The presence of a reducing agent during the phenol treatment favors the renaturation of the protein (Fraenkel-Conrat, personal communication). The phenol method has the added advantage of isolating RNA during the simultaneous isolation of protein. Proteins obtained by the phenol method, however, are not always renaturable. Fortunately protein in its native state is not required for sequential studies. Anderer et al. (1961) used phenol to prepare protein for amino acid sequence studies. Earlier work used the detergent, sodium dodecyl sulfate, for the preparation of tobacco mosaic virus protein (Anderer, 1963), but it was found to cause modifications of the protein.

E. Bromegrass Mosaic Virus

The cold calcium chloride method—Virus structure is well known to be dissociated by high concentrations of salts. The $CaCl_2$ method was thus developed by Yamazaki and Kaesberg (1963*a*). The divalent Ca^{2+} cation breaks ionic bonds and may form complexes with virus proteins and nucleic acids. The method is applicable to bromegrass mosaic virus. The virus is dialyzed against 1 M $CaCl_2$ (pH 6.3) for 12 hr at 4° C. During dialysis, the solution turns turbid as the RNA comes out of solution and some denaturation of the protein takes place. The precipitate is removed by centrifugation (25,000 *g* for 20 min), and the clear supernatant liquid which contains most of the virus protein is dialyzed against buffer.

The method was improved by Stubbs and Kaesberg (1964), but the same principle applied. This method has been employed for isolating proteins from plant viruses such as broad bean mottle virus (Yamazaki and Kaesberg, 1963*b*), cucumber mosaic virus (van Regenmortel, 1967*a*), barley stripe mosaic virus (Gumpf and Hamilton, 1968; Atabekov et al., 1968) and alfalfa mosaic virus (Hull et al., 1967). A modified $CaCl_2$ method was used for the barley stripe mosaic virus (Atabekov et al., 1968) by reducing the salt concentration to 0.5-0.6 M. Barley stripe mosaic virus protein showed a tendency to polymerize and the polymerization was dependent on the ionic strength of monovalent cations, Na^+ and K^+.

F. Cowpea Chlorotic Mottle Virus

The sodium chloride method—This method, used by Bancroft et al. (1967), is not much different in principle from the above $CaCl_2$ method. NaCl was preferred since it elicits a stronger effect on the decomposition of virus conformation. Purified cowpea chlorotic mottle virus is dialyzed overnight at 4° C, against 0.02 M Tris buffer (pH 7.4) containing 1 M NaCl and Cleland's reagent. The dialyzed material is then centrifuged for 15 hr at 30,000 rpm. Three fourths of the supernatant (from the top) was gently removed and discarded and the remaining liquid was collected avoiding the pellet. This fraction containing the virus protein also contains small amounts of nucleic acid. Such protein preparations were employed for virus reconstitution in vitro. Reconstitution was shown to depend on prepared RNA and not on the contaminating RNA in the protein preparation (Bancroft et al., 1967).

Besides the above two methods, a number of variations in salt method have been developed. Salt was used at elevated temperatures for turnip yellow mosaic virus (Lyttleton and Matthews, 1958) and alfalfa mosaic virus (Kelly and Kaesberg, 1962).

G. White Clover Mosaic Virus

The mild alkali method—This method, previously used by Fraenkel-Conrat and Williams (1955) and Schramm et al. (1955) for the isolation of tobacco mosaic virus protein, was recently modified by Miki and Knight (1967) for the preparation of white clover mosaic virus protein. The virus solution (5-10 mg/ml) is dialyzed against 2 liters of 0.1 M carbonate buffer (pH 10.5) for 2-3 days. The dialyzed material is then centrifuged (105,000 *g* for 2 hr) to remove undecomposed virus. To the supernatant is added an equal volume of saturated $(NH_4)_2SO_4$ solution to precipitate the protein which is collected by centrifugation. The pellet is then dissolved in a small volume of water and digested with pancreatic ribonuclease for 30 min at room temperature. A half volume of saturated $(NH_4)_2SO_4$ solution is then added and the resulting precipitate is redissolved and reprecipitated several times with saturated $(NH_4)_2SO_4$ solution. The pellet is dissolved in water and dialyzed against water to remove the salt. The protein solution can be stored in the lyophilized state or in the frozen state. The recovery was about 50%. The protein can be used for chemical and physico-chemical characterization.

Artificial protein shells free from RNA and small protein fragments can be obtained by raising the pH to pH 11-12 of a solution of turnip yellow mosaic virus (Kaper, 1964). Such protein shells are thought to be the same as the natural top component of this virus observed in the sucrose gradient centrifugation. Besides alkali, other basic reagents such as amino alcohols are also useful for the isolation of virus protein (Knight, 1964).

H. Southern Bean Mosaic Virus

The dilute hydrochloric acid method—Instead of alkali, 1 N HCl has been used to disrupt viruses (e.g. Tremaine, 1966; Ghabrial et al., 1967).

An aqueous solution of southern bean mosaic virus is mixed with an equal volume of 2 N HCl and allowed to sit for 15 hr at room temperature. The precipitate that forms (protein) is collected by centrifugation (15,000 rpm for 30 min) and is washed twice with 1 N HCl to remove RNA. This preparation may be used for amino acid analysis. Proteins of tomato ringspot virus (Tremaine and Stace-Smith, 1968) and alfalfa mosaic virus (Hull et al., 1967) were also isolated by the HCl method. The disadvantage of this method and the dilute alkali method, however, is that they can cause irreversible denaturation of the virus proteins. Therefore, their usefulness, aside from gross chemical analyses is limited.

Various techniques for isolating virus proteins have been reviewed in this chapter. There is no hard and fast rule on the choice of the method for a particular virus, but the essential principle is to choose a method that effectively disrupts the tertiary structure of the virus followed by the depolymerization of the protein into subunits and maintains the protein in its native state. The purity and original configuration of the protein preparation are checked in the usual way by analytical ultracentrifugation, electrophoresis, serology, and chemical analysis such as terminal and amino acid composition analysis. Special precautions should be taken for the physicochemical analysis since the result obtained can sometimes be misleading due to aggregation of pure proteins.

Electrophoresis in polyacrylamide gels containing urea and/or sodium dodecyl sulfate can be used where aggregation is the problem.

Possible contamination of nucleic acid in the protein preparation is routinely checked by taking the ultraviolet absorption spectrum. P^{32} labeling may be used when protein of the highest purity is required.

The original native state of the virus protein is difficult to assess because virus proteins are devoid of biological activity and function only to protect the infectious nucleic acid. In many cases the protein becomes insoluble when denatured and this criterion can serve as an indication of the state of the virus protein. Protein prepared in urea or sodium dodecyl sulfate, even if it is denatured, remain soluble. The best method of determining the nature of the protein is to test its ability to reconstitute with virus nucleic acid.

IV. PROPERTIES OF PLANT VIRUS PROTEINS

A. General Characteristics

Some of the general properties of the plant virus proteins are summarized in Table 15-1. The discrepancies between reported data of individual proteins in the table and those of older references (not cited in Table 15-1) may be due to analyses made on aggregated or denatured proteins.

Nucleic acids of plant viruses are usually single-stranded with the exception

TABLE 15-1
GENERAL PROPERTIES OF PLANT VIRUS PROTEINS

Virus	Preparation Methods	Percent Protein/RNA[1]	Molecular Weight of Subunit	Number of Subunits	Number of Amino Acids[2]	N Terminus	C Terminus
Rod-Shaped Viruses							
Cucumber virus 4[3]	67% acetic acid	95/5	18,000	2100	158[31]	Acetyl-Ala-Tyr[4]	Ala[5]
Potato virus X[6]	2 M guanidine-HCl	94/6	22,300[7]	1475	210[6]	Acetyl[6]	Pro[5]
Tobacco mosaic[8]	67% acetic acid[1]	95/5	17,500	2200	158[3]	Acetyl-Ser-Tyr[9]	Thr[10]
White clover mosaic[4]	Alkali	91/9	14,300	1520	133[11]		Arg[11]
Spherical Viruses							
Alfalfa mosaic[12]	1 N HCl[9]		32,600		297	Acetyl-Ser	Arg
Broad bean mottle	67% formic acid[13]	78/22[14]	20,900	190	194		Ala
	1 M $CaCl_2$[15]		20,500	195	190		Ala[16]
			20,000	182			
Bromegrass mosaic[32]	1 M $CaCl_2$	79/21[33]	40,000	91	189		Arg
Carnation ringspot[17]	1 N HCl	80/20	38,000	148	347		
Cowpea chlorotic mottle[18]	1 M NaCl	76/24	19,600	160-170	183		
Cucumber mosaic[19]	1 M $CaCl_2$	82/18	32,000	176-182	287		
Pea enation mosaic[20]	1 M $CaCl_2$	71/29	21,800	180	199		
Potato virus Y[36]	6 M guanidine-HCl	95/5	22,000		193		
Southern bean mosaic (cowpea type)[21]	1 N HCl[10]	79/21	29,000	180	270		Ser
Sowbane mosaic[22]	10% formic acid	80/20[23]	19,200	180-260[23]	176		Lys[22]
Tobacco etch virus[34]	5.3 M guanidine-HCl	95/5	22,000		194		
Tobacco necrosis[24]	1 N HCl	81/19[24]	33,500	200	312	Ala	Met
Tobacco necrosis (AC36)[25]	1 N HCl	80/20[25]	22,600	180	197		Ile

TABLE 15-1 (Continued)

Virus	Preparation Methods	Percent Protein/RNA[1]	Molecular Weight of Subunit	Number of Subunit	Number of Amino Acid[2]	N Terminus	C Terminus
Tobacco necrosis satellite[26]	1 N HCl	80/20[26]	27,000[27]	48	250	Ala	Ala
Tobacco necrosis satellite SV-C[25]	1 N HCl	79/21[25]	24,200	60	209	(Acetyl)	Leu
Tobacco ringspot[28]	1 N HCl	58/42[1]	24,000[28]	140[28]	217		Tyr
Tomato bushy stunt		83/17[1]	21,500	185	402		Leu[5]
Tomato ringspot[29]	1 N HCl	60/40	24,300	140	217		
Turnip yellow mosaic[30]	67% acetic acid	61/39	21,300	150	195	Acetyl-Met-Glu	Thr
Turnip mosaic virus[35]	guanidine-HCl	95/5	26,000		230		
Wild cucumber mosaic[28]	67% acetic acid	65/35[28]	21,500		190		

[1]The data in this column are given by Knight (1964) with exception of special references.
[2]The data in this column are given by Tremaine and Goldsack (1968).
[3]Tsugita and Fraenkel-Conrat (1960)
[4]Narita (1959)
[5]Niu et al. (1958)
[6]Miki and Knight (1968)
[7]Reichmann (1959)
[8]Tsugita and Fraenkel-Conrat (1963)
[9]Narita (1958)
[10]Harris and Knight (1955)
[11]Miki and Knight (1967)
[12]Hull et al. (1969)
[13]Miki and Knight (1965)
[14]Yamazaki et al. (1961)
[15]Yamazaki and Kaesberg (1963*b*)
[16]Yamazaki and Kaesberg (1963*b*)
[17]Kalmakoff and Tremaine (1967)
[18]Bancroft et al. (1968)
[19]van Regenmortel (1967*a*)
[20]Shepherd et al. (1968)
[21]Ghabrial et al. (1967)
[22]Kado (1967)
[23]Kado and Black (1968)
[24]Lesnaw and Reichmann (1969*a*)
[25]Uyemoto and Grogan (1969)
[26]Reichmann (1964)
[27]Lesnaw and Reichmann (1969*b*)
[28]Yamazaki and Kaesberg (1961)
[29]Tremaine and Stace-Smith (1968)
[30]Harris and Hindley (1961)
[31]van Regenmortel (1967*b*)
[32]Stubbs and Kaesberg (1964)
[33]Bockstahler and Kaesberg (1962)
[34]Damirdagh and Shepherd (1970*a*, *b*)
[35]Hill and Shepherd (1972)
[36]Stace-Smith and Tremaine (1970)

of wound tumor virus, rice dwarf virus (Miura et al., 1966), and probably still other plant viruses (see Chapter 6). Noticeable differences occur in the constitutions of the proteins with respect to the nucleic acid of the viruses listed. The protein content of rod-shaped viruses is nearly 95% with only about 5% RNA, whereas the protein content of spherical viruses is 61-83% with 17-39% RNA. Besides protein and nucleic acid, some spherical viruses (e.g. turnip yellow mosaic virus, broad bean mottle virus, and turnip crinkle virus) contain polyamines, such as 3,3-diaminopropylamine. Polyamines have not been reported in rod-shaped viruses.

The amino acid compositions of various viruses are summarized in Table 15-2. (For the amino acid composition of tobacco mosaic virus strains and mutants see Chapter 20).

1. HOMOGENEITY OF THE SUBUNIT AND MOLECULAR WEIGHT

The coat proteins of plant viruses listed in Table 15-1 are homogeneous, that is, only one type of capsid protein/virus particle. However, Gumpf and Hamilton (1968), reported that the coat protein of barley stripe mosaic virus is composed of three components which could be distinguished by a serological test.

When this protein was prepared by the $CaCl_2$ method and was used as an antigen, three bands were observed in an immunodiffusion pattern (Fig. 15-1). One of them, the major component, behaved as the protein subunit of the virus while the two minor components diffused faster than the major component. These rapidly diffusing antigens correspond to the partially decomposed products of the virus or its protein. It is suggested that the major component consists of the protein subunit and that the aggregation of this subunit caused variations in the serological reaction by the appearance of the minor bands. This obviously implies that structural variations due to aggregation affect antigen-antibody activity. Jeener (1965) found the same kind of variation with tobacco mosaic virus protein. In alfalfa mosaic virus, Moed and Veldstra (1968) also reported that during the process of decomposing the virus into its protein subunits, changes in tertiary structure of the protein took place which resulted in the appearances of new types of antigens.

In general, serological tests alone to show heterogeneity of virus proteins may be misleading. A combination of serological, physico-chemical, and chemical methods should be employed for testing homogeneity of these proteins. Sedimentation analysis has also been used for such tests. Sedimentation constant, diffusion constant, and sedimentation equilibrium of the protein provide excellent criteria for homogeneity and tell us the molecular weight of the subunits. As pointed out previously, aggregation may be a problem. For instance, tobacco mosaic virus protein, obtained by the acetic acid method, shows a molecular weight of about 100,000, which corresponds to an aggregation product of six molecules (Schramn et al., 1955). The same protein when disaggregated by alkali (0.1 M) become trimers (Knight, 1964) or

TABLE 15-2
AMINO ACID COMPOSITION OF PLANT VIRUS PROTEIN

Rod-Shaped Viruses	*Asp*	*Thr*	*Ser*	*Glu*	*Pro*	*Gly*	*Ala*	*Cys*	*Val*	*Met*	*Ile*	*Leu*	*Tyr*	*Phe*	*Lys*	*His*	*Arg*	*Trp*	*Total*
Tobacco etch virus[28]	25	13	9	23	8	13	19	1	12	10	5	13	7	5	10	6	13	2	194
Tobacco mosaic[1]	18	16	16	16	8	6	14	1	14	0	9	12	4	8	2	0	11	3	158
Tobacco mosaic-J14D1[2]	17	16	17	15	8	6	14	1	14	0	9	12	4	8	3	0	11	3	158
Tobacco mosaic-yellow aucuba[2]	19	17	14	16	8	6	14	1	14	0	8	12	4	8	2	0	12	3	158
Tobacco mosaic-green aucuba[2]	19	17	15	16	8	5	14	1	14	0	8	12	4	8	2	0	12	3	158
Tobacco mosaic-OM[3]	19	15	16	16	8	6	14	1	15	0	8	12	4	8	2	0	12	3	158
Tobacco mosaic-Dahlemense[2]	17	17	16	19	8	6	11	1	15	1	7	13	5	8	2	0	9	3	158
Tobacco mosaic Y-TAMV[2]	18	17	15	19	8	6	11	1	15	1	7	13	5	8	2	0	9	3	158
Tobacco mosaic G-TAMV[2]	22	19	10	16	10	4	18	1	12	2	8	11	6	8	1	0	8	2	158
Tobacco mosaic-U2[4]	22	19	10	16	10	5	17	1	12	2	8	11	6	8	1	0	8	2	158
Tobacco mosaic-HR[5]	17	13	13	22	8	4	18	1	10	3	8	11	7	6	2	1	10	2	156
Tobacco rattle (B)[6]	20	10	21	16	13	7	21	—	8	3	3	14	5	11	15	1	10	—	178
Tobacco rattle (C)[6]	20	9	21	16	13	7	22	—	8	3	4	14	4	11	15	1	9	—	178
Turnip mosaic virus[29]	29	16	10	23	9	15	17	1	12	10	11	20	8	9	13	8	17	3	231
Potato virus X[7]	19	24	14	15	14	11	38	2	11	6	10	8	2	10	10	2	8	6	210
Potato virus S[8]	16	8	10	17	11	9	13	1	10	6	10	9	3	4	4	3	11	—	144
Potato virus Y[30]	22	13	10	23	11	13	16	1	13	8	12	10	6	5	13	4	11	2	193
White clover mosaic[9]	12	11	10	9	8	7	19	2	7	2	9	10	3	6	8	2	6	2	133
Spherical Viruses	*Asp*	*Thr*	*Ser*	*Glu*	*Pro*	*Gly*	*Ala*	*Cys*	*Val*	*Met*	*Ile*	*Leu*	*Tyr*	*Phe*	*Lys*	*His*	*Arg*	*Trp*	*Total*
Bean pod mottle[10]	21	14	18	17	13	21	14	—	15	7	12	18	2	11	9	3	6	—	201
Broad bean mottle[11]	14	10	18	17	9	10	23	2	23	2	7	19	4	7	15	2	12	0	194
Bromegrass mosaic[12]	10	11	13	18	7	10	33	1	18	3	8	15	5	5	13	4	13	2	189
Carnation mottle[8]	36	36	28	31	24	29	31	5	35	9	19	28	10	14	25	1	19	2	382
Carnation ringspot[13]	34	37	37	23	20	20	24	3	36	7	16	26	16	12	14	2	16	4	347
Cowpea chlorotic mottle[14, 31]	11	17	16	16	7	10	25	2	19	1	7	16	5	4	12	2	9	4	183
Cucumber green mosaic[15]	20	10	24	10	6	9	21	0	7	0	7	18	4	9	4	1	8	2	160
Cucumber mosaic[16]	30	17	32	20	18	16	17	0	22	8	16	26	11	7	18	4	24	1	287
Cucumber necrosis[8]	46	31	32	24	23	32	41	0	33	1	20	33	12	20	16	3	17	7	391
Pea enation mosaic[17]	21	13	16	14	11	21	17	3	13	3	7	10	5	7	11	4	21	2	199

TABLE 15-2 (Continued)

Spherical Viruses	*Asp*	*Thr*	*Ser*	*Glu*	*Pro*	*Gly*	*Ala*	*Cys*	*Val*	*Met*	*Ile*	*Leu*	*Tyr*	*Phe*	*Lys*	*His*	*Arg*	*Trp*	*Total*
Southern bean mosaic[18] (bean type)	18	32	26	18	14	19	24	3	21	7	12	28	10	4	7	2	20	5	270
Southern bean mosaic[18] (cowpea type)	21	30	17	19	18	16	28	4	23	9	14	23	9	4	12	2	16	5	270
Sowbane mosaic[19]	16	13	14	12	12	16	15	2	13	5	9	12	7	4	12	3	8	3	176
Squash mosaic[20]	21	17	16	14	10	15	19	—	9	4	14	19	3	10	8	3	7	—	189
Tobacco necrosis[21] (satellite)	57	34	24	32	9	28	28	3	25	9	25	29	8	14	15	8	25	—	373
Tobacco necrosis[22]	34	19	21	24	23	25	41	5	18	5	20	21	15	10	10	1	20	—	312
Tobacco necrosis[23] (satellite SVC)	27	25	12	18	4	8	9	2	13	4	13	20	6	11	11	6	24	—	209
Tobacco necrosis-AC36[23]	18	16	14	20	15	8	13	2	14	6	11	10	11	12	12	1	14	—	197
Tobacco ringspot[24]	17	13	14	14	11	15	15	—	11	3	11	14	6	10	9	6	8	—	177
Tomato bushy stunt[25]	44	45	35	21	16	38	37	3	40	3	13	43	10	14	13	5	20	2	402
Tomato ringspot[26]	17	15	16	18	11	18	15	5	10	3	13	24	7	14	10	5	11	5	217
Turnip crinkle[27]	14	14	12	16	9	15	17	1	12	2	5	11	4	6	12	1	9	4	164
Turnip yellow mosaic[27] (typical)	11	26	16	15	20	8	15	4	14	4	15	18	3	5	7	3	3	2	189
Turnip yellow mosaic[27] (cauliflower)	17	20	20	15	20	7	13	5	14	4	13	19	3	3	4	5	6	2	190
Wild cucumber mosaic[27]	15	13	26	11	19	9	16	2	12	1	13	24	3	8	9	3	5	1	190

[1]Tsugita and Fraenkel-Conrat (1960)
[2]Tsugita (1962*a*)
[3]Nozu and Okada (1968)
[4]Rentschler (1967)
[5]Funatsu and Funatsu (1968)
[6]Semancik (1966*a*)
[7]Miki and Knight (1968)
[8]Tremaine et al. (1968)
[9]Miki and Knight (1967)
[10]Semancik (1966*b*)
[11]Yamazaki and Kaesberg (1963*b*)
[12]Stubbs and Kaesberg (1964)
[13]Kalmakoff and Tremaine (1967)
[14]Bancroft et al. (1968)
[15]G. Funatsu, private communication
[16]van Regenmortel (1967*a*)
[17]Shepherd et al. (1968)
[18]Tremaine (1966)
[19]Kado (1967)
[20]Mazzone et al. (1962)
[21]Reichmann (1964)
[22]Lesnaw and Reichmann (1969*a,b*)
[23]Uyemoto and Grogan (1969)
[24]Stace-Smith et al. (1965)
[25]de Fremery and Knight (1955)
[26]Tremaine and Stace-Smith (1968)
[27]Symons et al. (1963)
[28]Damirdagh and Shepherd (1970*b*)
[29]Hill and Shepherd (1972)
[30]Stace-Smith and Tremaine (1970)
[31]Chidlow and Tremaine (1971)

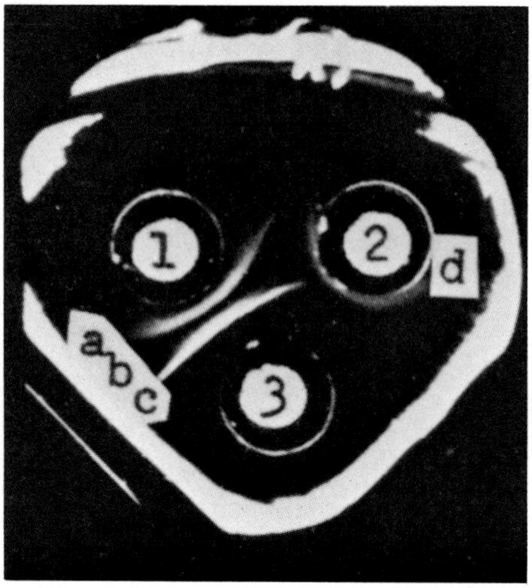

Fig. 15-1. Immunodiffusion patterns of barley stripe mosaic virus and its protein at pH 7.0. Well 1, barley stripe mosaic virus protein; well 2, barley stripe mosaic virus; well 3, undiluted barley stripe mosaic virus antiserum. a: barley stripe mosaic virus-protein major antigen; b and c: barley stripe mosaic virus-protein minor antigen; d: barley stripe mosaic virus major antigen. (From Gumpf and Hamilton, 1968.)

dimers (Newmark and Myers, 1957). The molecular weight of 18,000 (monomer) was obtained by diluting the protein to 0.01%. The same monomer was also shown by sedimentation analysis in 67% acetic acid (Ansevin and Lauffer, 1959; Wittmann, 1959*b*). Adjusting the protein solution to pH 6 causes polymerization of the subunits into rod-shaped structures just like tobacco mosaic virus.

With other viruses, however, it was shown that even 2M guanidine hydrochloride failed to keep potato virus X protein in its dissociated form during sedimentation (Miki and Knight, 1968).

Molecular weights of viral proteins can be estimated from X-ray diffraction analysis (Klug and Casper, 1960). From studies using the methylmercury derivative x the virus, tobacco mosaic virus is believed to be hollow along its axis (40 Å diameter) with the protein units arranged in a gently pitched helical array (49 units in 3 turns, pitch 23 Å) (Franklin and Holmes, 1958). Electron microscopic observations (Nixon and Woods, 1960) and light-scattering data showed the length of the rod to be 3000 Å. Thus simple calculations revealed that 2130 [(49/23×3)×3000] protein subunits form the virus. From particle weight of tobacco mosaic virus (3.9×10^7) and protein content (95%), the total weight of the protein component is 3.7×10^6, and the molecular weight of a protein subunit is therefore 17,400. This unit may be called a "crystallographic subunit."

This value coincides nicely with the value derived from the chemical methods. In fact, the molecular weight of tobacco mosaic virus protein was first established by chemical analysis. Convincing data can be obtained by carboxypeptidase treatment. The carboxylterminal sequence of tobacco mosaic virus protein is -Pro-Ala-Thr; thus, only one mole of threonine per mole of protein is liberated by carboxypeptidase. Further digestion is blocked because of the inhibitory effect of proline. Such quantitative analysis of tobacco mosaic virus protein gave a molecular weight of about 18,000 (Harris and Knight, 1955). Subsequently carboxypeptidase has been used on various virus proteins for the determination of minimum molecular weight and also to establish the protein's purity. However, when more than two amino acids are liberated, these tests cannot be made without referring to the molecular weight values obtained from the other methods mentioned in this section.

Amino terminal analysis methods have been widely used and are considered more precise and conclusive than analysis of the carboxyl termini. However, the amino termini of *Escherichia coli* proteins are found to be limited to a few amino acids.

If this is a general fact for virus proteins, then the amino terminal analysis for checking protein homogeneity becomes invalid. In addition, a number of virus proteins are acetylated at the amino terminus. Thus one may not be able to obtain the molecular weight by way of the amino terminus or use it for checking homogeneity. The acetyl residue itself may be analyzed. Miki and Knight (1968) estimated the acetyl residue in potato virus X by gas chromatography.

Besides end group analysis, amino acid composition together with peptide mapping can be used for the determination of molecular weight. From the data of the amino acid analysis, the molar ratio of the individual amino acid is calculated by taking the amino acid having the least amount as unity.

By this calculation the "minimum molecular weight" of tobacco mosaic virus protein is 18,000, using the value of cysteine as one unit. In the case of potato virus X, if tyrosine is used, the minimum molecular weight is 11,000. The ratios of the other amino acids become arbitrary numbers. The end group method showed that the molecular weight of potato virus X protein was 22,000, which is double that obtained by amino acid analysis. Obviously this difference could be explained by postulating that two moles of tyrosine are present.

2. ULTRAVIOLET ABSORPTION SPECTRA

Proteins absorb ultraviolet light mainly because of the presence of the amino acids tyrosine, tryptophan, and phenylalanine, which possess conjugated double bonds. The measurement of ultraviolet absorption is one of the most efficient and convenient methods for the identification of proteins. The maximum absorbancy of proteins occurs around 280 nm, while the nucleic acid of the virus has an absorbance maximum around 260 nm. The absorbancy coefficient of nucleic acids at the wavelength of maximum absorption is 10 times greater than that of protein on a weight-to-weight basis. When the protein component is isolated from a virus, its UV-absorption spectrum is routinely inspected to check its purity and determine its concentration. In general, pure virus proteins show a value higher than 2.2 in the ratio of absorbancies at 280 nm/250 nm. A typical UV-absorption spectrum is shown in Fig. 15-2. The specimen is bromegrass mosaic virus protein, prepared by the $CaCl_2$ method (Stubbs and Kaesberg, 1964). The ratio of absorbancies (280 nm/250 nm) is within the range of the above criteria.

3. TERMINI OF VIRUS PROTEIN SUBUNITS

Free carboxyl termini have been detected in almost all plant virus proteins, while free amino termini of many plant viruses have not yet been detected, as shown in Table 15-1. As for the amino terminus, an acetyl group has been found in many plant viruses since the discovery of an acetyl group present at the

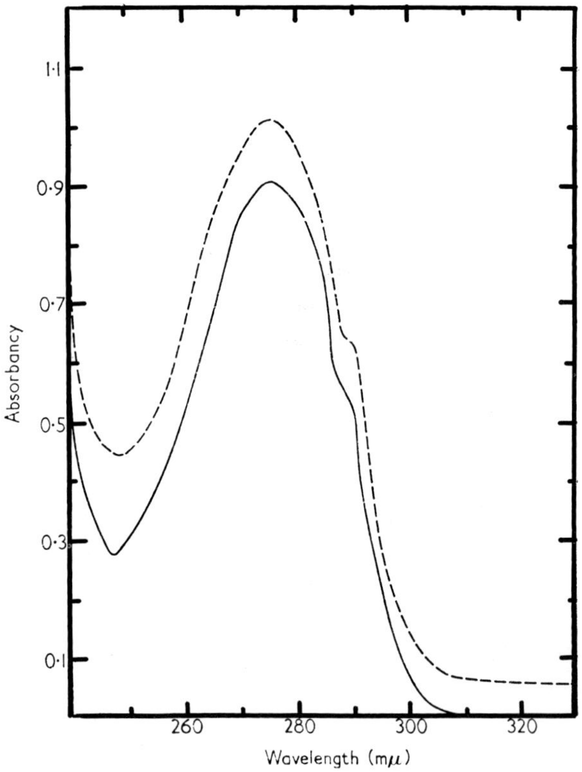

Fig. 15-2. UV absorption curve of purified bromegrass mosaic virus protein at a concentration of 1 mg/ml in 1 M $CaCl_2$-0.05 M sodium cacodylate pH 6.0. The solid curve is corrected for light scattering, and the dashed curve is uncorrected. (From Stubbs and Kaesberg, 1964.)

amino terminus of tobacco mosaic virus protein (Narita, 1958). This finding was the first instance of the presence of acetylated amino terminus in a plant virus protein or in any other proteins. The amino terminal sequences of cucumber virus 4 (Narita, 1959) and turnip yellow mosaic virus (Harris and Hindley, 1961) were found to be as follows:

cucumber virus 4: Acetyl-Ala-Tyr-Asn-Pro-Ile-Thr-Ser- . . .
turnip yellow mosaic virus: Acetyl-Met-Glu-Ile-Asp-Lys-Glu-Leu-Ala-Leu-Glu-Asp-Arg-Thr- . . .

Amino terminal analysis is usually easier to perform than that for the carboxyl terminus, and failure in the recovery and identification of a free amino terminal residue may suggest that it is blocked by an acetyl or some other group that is pyrolidonated. Whether the acetyl amino acid plays any role in the initiation of the protein synthesis is uncertain (see Introduction), and the importance of the absence or blocking of the free amino group of proteins in nature is unknown.

The carboxyl termini of virus proteins have been analyzed by carboxypeptidase as well as by hydrazinolysis. Besides chemical molecular weight determinations, carboxypeptidase can be used for determining the partial sequence of proteins. The following peptide sequences were determined:

turnip yellow mosaic virus: -Asp-Val-Thr-Ser-Thr
broad bean mottle virus: -Ser-Phe-Ala

4. STRUCTURE OF VIRUS PROTEIN SUBUNITS

When the primary structure of tobacco mosaic virus was completely elucidated its structural relation to the nucleic acid became clear. The data obtained from such studies as chemical modifications of the virus, mutant analysis, serological and physico-chemical approaches all aided in explaining this relationship. A summary of the primary structure is illustrated in Fig. 15-3. Such comprehensive studies came from the serological behavior of various mutants with altered amino acid sequences (Sengbusch, 1965; van Regenmortel, 1967*b*). In the figure the amino acids that are underlined are positions where the amino acid was exchanged either by artificial or by spontaneous mutations. The amino acids written above these positions are those that normally occur. It is indicated that these amino acid replacements are not essential for they do not affect the tertiary structure during the formation of the virus rod. These mutants were stable, indicating other structural stabilities.

There are some interesting exceptions. Mutant PM2 is defective in its ability to make infectious progeny particles. The protein of this mutant is present in infected tobacco and can be detected by serological tests. When this protein was isolated and subjected to polymerization in a test tube, the protein polymerized into unusual helical strings. The protein's amino acid replacements were: Thr

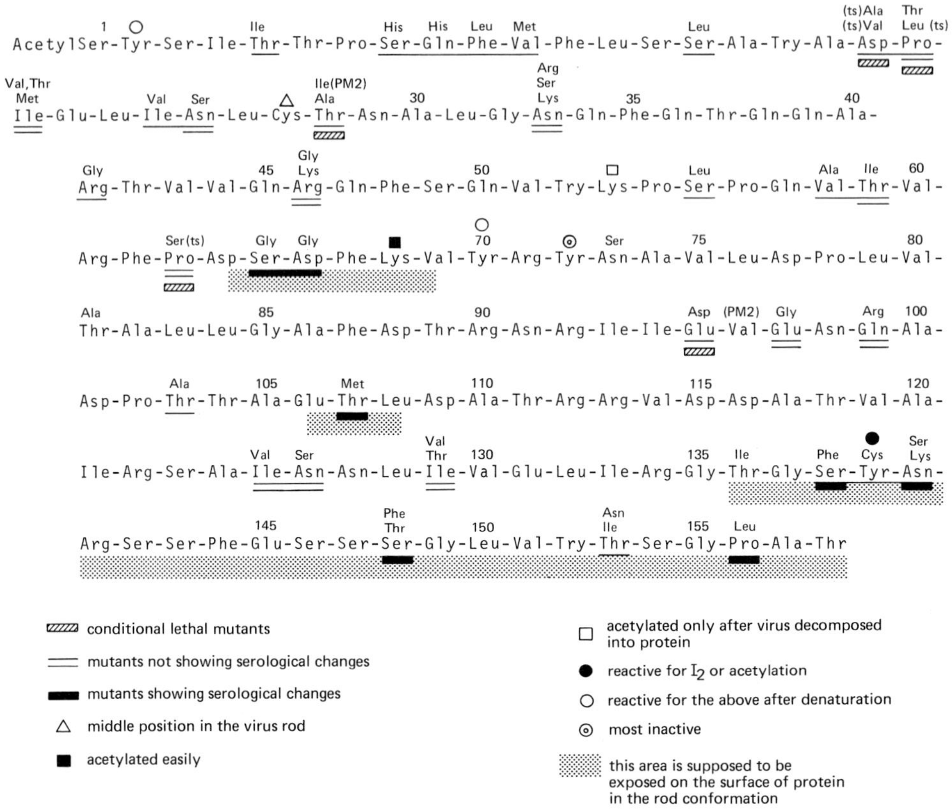

Fig. 15-3. **The amino acid sequence of tobacco mosaic virus protein and the location of amino acid in the tertiary structure.**

→Ile and Glu → Asp, indicating that either (or both) threonine or aspartate are essential for determining the structure of tobacco mosaic virus (Zaitlin and Ferris, 1964). Also, temperature-sensitive (ts) mutants have been isolated (Wittmann, 1966). These mutants can form intact virus particles at 20-25° C but not at 30-35° C. The protein of these mutants also possessed amino acid exchanges (Jockusch, 1966*b*). Amino acid replacements of Pro → Leu at the 20th and 63rd positions in the mutant protein caused temperature sensitivity. The replacement of proline at position 156 did not convert it to a ts mutant (Tsugita et al., 1960). Such studies soon indicated that $Asp,^{19}$ Pro^{20} and Pro^{63} are important for keeping the integrity of the tertiary conformation of tobacco mosaic virus protein (such amino acids are indicated by ▨ in Fig. 15-3). Tobacco mosaic virus mutants differ from the wild type strain by one or a few amino acid exchanges in their primary structures (for details see Chapter 20). The mutants can be distinguished serologically into two types, one of which differs serologically from wild type and the other has the same serological properties in spite of the amino acid alterations (Sengbusch, 1965; Sengbusch and Wittmann, 1965; van Regenmortel, 1967*b*). The serologically active and nascent groups are shown in Fig. 15-3. The serological distinction may indicate that those reactive amino acids are located on the surface of the virus particle, or that they caused some changes on the surface structure of the protein subunit.

Chemical modifications of the virus also provided valuable insight on virus structure (Fraenkel-Conrat, 1965). Such modification was carried out by acetylation whereby one of the two lysines at position 68 was preferentially acetylated and the other (at position 53) was not acetylated until the virus released its protein subunits (marked by ■ and □ respectively in Fig. 15-3) (Fraenkel-Conrat and Colloms, 1967). Also by acetylation (of either virus rods or as the native protein subunit), only tyrosine 139 was affected and the other three at positions 2, 70, and 72 (marked with ○ or ●) remained unaffected. The identical result was obtained by iodination. Tyrosine 139 is most reactive with iodine. Even with denatured protein, tyrosine 2 and 70 reacted with iodine but tyrosine 72 did not react (Fraenkel-Conrat, 1962). Reconstitution of tobacco mosaic virus was possible with proteins without acetylated residues which were acetylatable in the intact virus. The unreactive so-called "masked" tyrosine 72 displays an abnormal UV absorption spectrum of the protein, when one of the four tyrosine residues (possibly tyrosine 72) did not become normalized in alkali (Fraenkel-Conrat and Narita, 1958). These data indicate that tyrosine 72 is located "inside" or in an unaccessible part of the protein conformation and that the tyrosine 139 is located on the surface of the protein subunit.

The single -SH group of tobacco mosaic virus protein combines quantitatively with methyl mercury but not with *p*-chloromercuribenzoate in the intact virus, this difference perhaps being due to the sterically hindered state of cysteine. Iodine also can act with the -SH group in the virus transforming it to an-SI group (Fraenkel-Conrat, 1955). The protein modified by methyl mercury or iodine remains in its native state and forms almost intact virus rods. X-ray diffraction studies were also possible with this mercury derivative (Caspar, 1956). X-ray analysis showed that methylmercuri-cysteine was located 50 Å from the long axis of the virus rod. From the protein radius of 90 Å and the inner RNA radius of 20 Å, the geometrical position of the cysteine is situated in the middle of the virus protein shell. The modification occurring with intact virus rods also clearly indicated that the cysteine is not involved in the bonding of protein subunits and RNA.

With nitrous acid mutants, the amino acid exchange of Pro → Leu at position 156 removes the proline which is inhibitory to carboxypeptidase digestion. Carboxypeptidase treatment of this mutant virus particle caused the liberation of leucine, alanine, and threonine and released more than 16 amino acids. This observation, together with the serological study, suggests that the three amino acids are like a flexible "tail" on the periphery of the protein of the

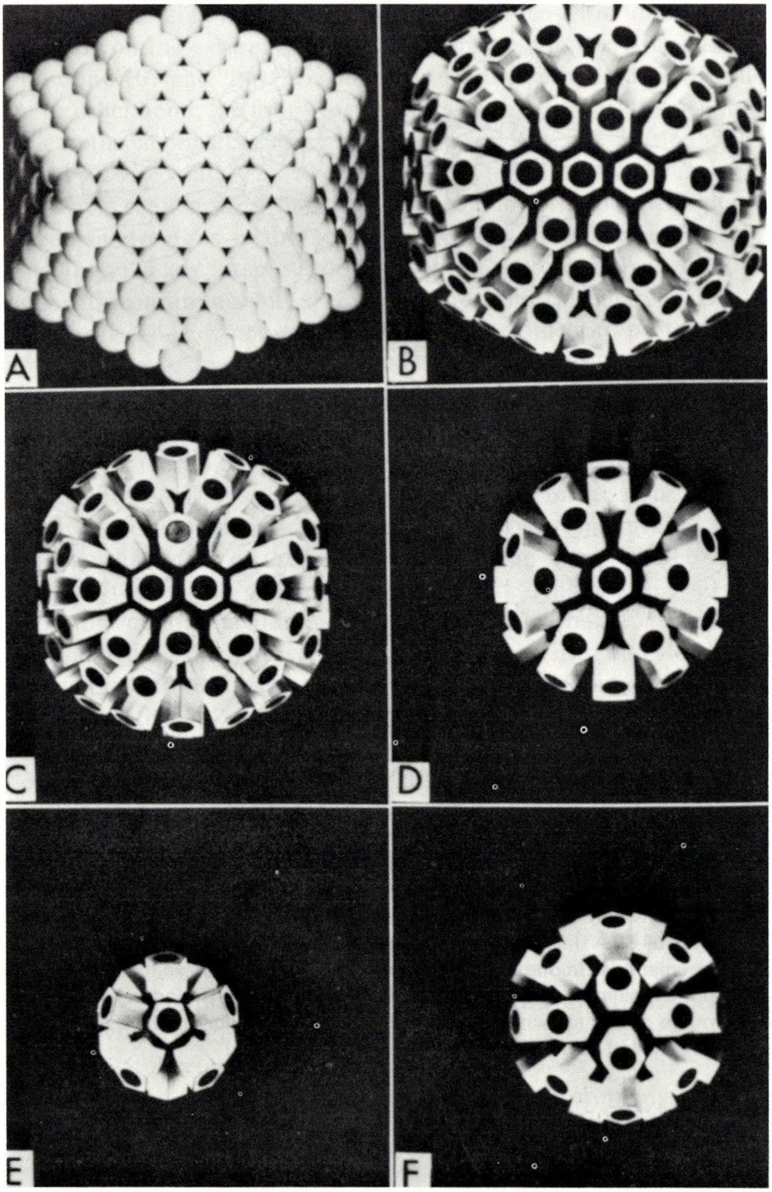

Fig. 15-5. Six models constructed from pentagonal and hexagonal members arranged in accordance with 5:3:2 symmetry. All fall into the series $10X(n-1)^2 + 2$ for models A-E, $X = 1$; for F, $X = 3$. The numbers of members are as follows: A, 252; B, 162; C, 92; D, 42; E, 12; F, 32. (From Horne and Wildy, 1961.)

occurs as a hexamer (Jeener et al., 1954). When the virus is treated with dilute alkali, it gradually degrades into progressively smaller oligomers as shown by the successive reduction of sedimentation constants: 194S → 133S → 94S → 45S → 4S (Harrington and Schachman, 1956).

Lauffer et al. (1967) demonstrated that the reversed process (i.e. polymerization of the protein subunits in vitro follows similar steps as shown in Fig. 15-6). Using X-protein, Commoner et al. (1962) also observed

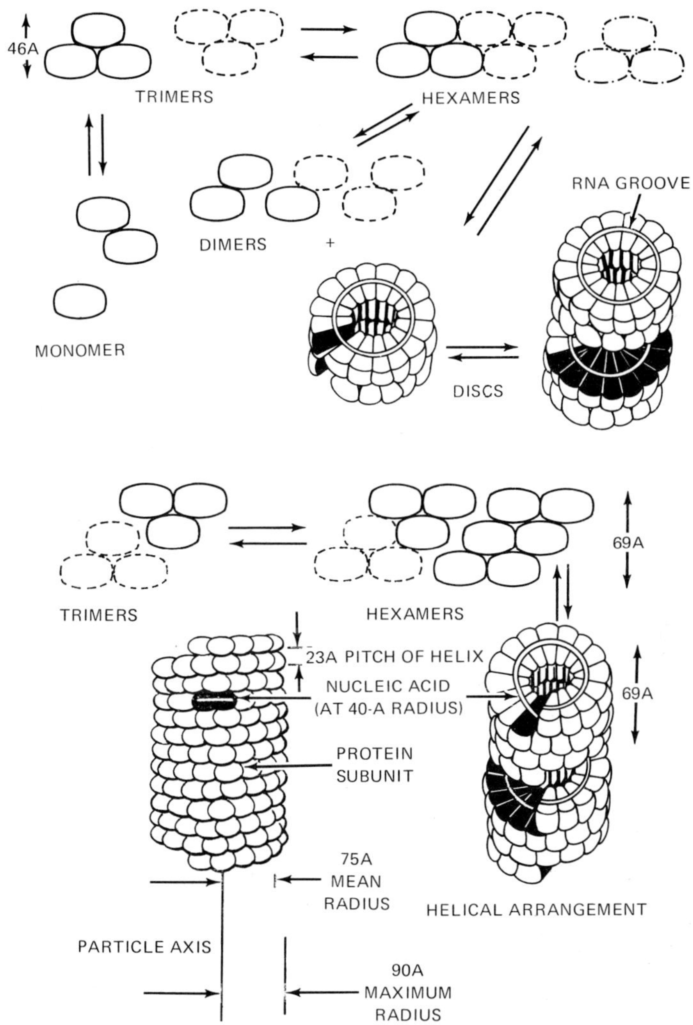

Fig. 15-6. Polymerization process of tobacco mosaic virus protein subunits. Upper: Disc formation during polymerization. Lower: Helical arrangement during the process of polymerization. (From Khalil and Lauffer, 1967.)

polymerization of the protein subunits in vivo. Similar observations were also reported for other viruses (Bancroft et al., 1968; Atabekov et al., 1968).

The question arises whether nucleic acids are involved in the determination of the tertiary structure during polymerization of the protein subunits. Matthews (1966*a, b*) showed that turnip yellow mosaic virus RNA and tobacco mosaic virus protein will reconstitute under certain conditions. Although the reconstitution rate was not sufficient and fluctuated considerably (1-30%, when compared with that of the homologous mixture), the resulting "mixed" reconstituted particles were observed to be rod-shaped like tobacco mosaic virus. Also, Sugiyama (1966) succeeded in "mixed" reconstitution between tobacco mosaic virus protein and coliphage MS2 RNA. Again rods were observed, some with "kinks" and a number with half the length of a tobacco mosaic virus particle. (The size of RNA is considered to determine the length of the rod-shaped particle. The molecular weights of tobacco mosaic virus RNA $= 2.1 \times 10^6$, MS2-RNA $= 1.1 \times 10^6$). These results suggest that the virus protein, not the RNA, directs the morphology of the virus. This concept is also supported by the following "mixed" reconstitution experiments between spherical virus protein and rod virus RNA. Bancroft et al. (1969) mix-reconstituted tobacco mosaic virus-RNA with each of several spherical viral proteins that were isolated by the NaCl method. Proteins of cowpea chlorotic mottle, bromegrass mosaic, or broad bean mottle viruses were used. In these experiments, spherical particles were obtained in every case (Fig. 15-7). Furthermore, among these particles the larger ones proved to be more infectious. It was also observed that the protein component alone forms into spherical virus-like structures (Bancroft et al., 1968*b*).

The formation of particles that are a facsimile to virus particles is well known to occur with tobacco mosaic virus protein. These particles are not uniform and are less stable than with the virus. Although an ambiguity still remains in the requirement of virus RNA for virus particle formation, the role of the virus protein in virus conformation is relatively clear. Since Fraenkel-Conrat (1955) reported his reconstitution experiments with tobacco mosaic virus, many similar experiments with various viruses have been carried out. Among these experiments, the reconstitution of spherical viruses were found to be rather difficult. Sugiyama et al. (1967) succeeded in the reconstitution of a spherical virus, bacteriophage MS2. It was shown that a maturation factor (called the "A" protein) (this "A" protein is different from the A protein of tobacco mosaic virus) was required in addition to the subunits of the coat protein and the RNA. This indicated the existence of heterogeneity in the protein component of the MS2 phage. Using phage R17 (a RNA phage similar to MS2), Robert et al. (1968) isolated its coat protein, an "A" protein and RNA, and demonstrated reconstitution by incubating these constituents in a test tube.

As for plant viruses, Bancroft et al. (1967, 1968*a, b*) studied the auto-assembly process in vitro, using the spherical cowpea chlorotic mottle virus.

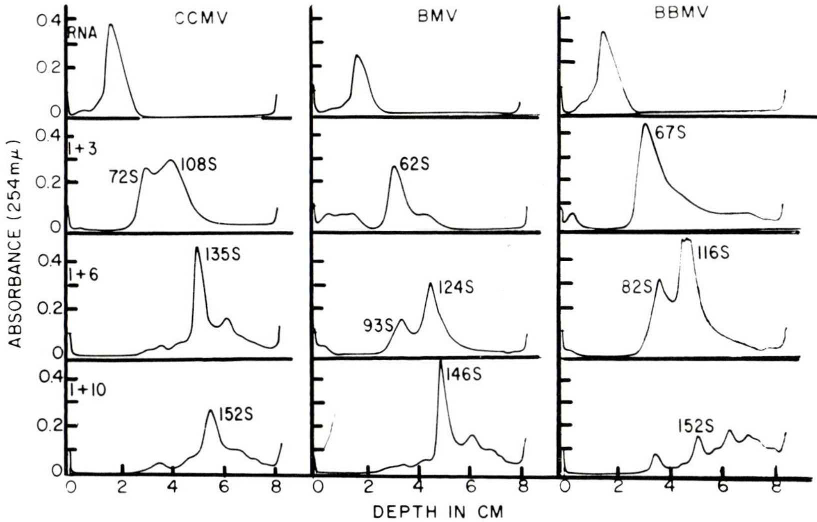

Fig. 15-7. Reconstitution (assembling) process of tobacco mosaic virus RNA with spherical virus proteins. Tracings of the 254-mμ absorbing distributions of tobacco mosaic virus-RNA, and individually assembled 1:3, 1:6, and 1:10 tobacco mosaic virus-RNA; protein of cowpea chlorotic mottle virus (CCMV); brome mosaic virus (BMV), broadbean mottle virus (BBMV) aggregates. Centrifugation was for 2 hr at 39,000 rpm in a SW 41 rotor on a gradient of 100-400 mg of sucrose per ml of 0.1 M , pH 5.7, acetate buffer. Samples in the vertical columns were run in sister tubes. Sedimentation is toward the right and S is for Svedberg units. (From Verduin and Bancroft, 1969.)

The protein of this virus prepared by the NaCl method, was incubated with its RNA and the resulting particle, as revealed by electron microscopy, were identical in shape and size with that of the original virus particle (Fig. 15-8). Infectivity of the reconstituted virus was found to be almost the same as that of the original virus (Table 15-3). Hull (1970) used the protein of alfalfa mosaic virus and the RNA of broad bean mottle virus or turnip yellow mosaic virus and achieved partial reconstitution. Some particles were bacilliform like alfalfa mosaic virus but most were semi-spheres.

Beck and Gordon (1970) were able to assemble a hybrid virus by mixing tobacco mosaic virus protein with the RNA of potato virus X. One of the major reasons for the successful reconstitution was attributed to the similarities between the sizes and base compositions of the nucleic acids of these viruses. The biological activity of the hybrid virus resembled that of potato virus X, but the photochemical behavior of this hybrid closely resembled that of tobacco mosaic virus, indicating that the coat protein (in this case tobacco mosaic virus protein) during ultraviolet irradiation prevents subsequent host photoreactivation.

The biological role of the virus protein is now of particular interest. Host range of hybrid viruses have been tested (Hiebert et al., 1968; Kado and Knight,

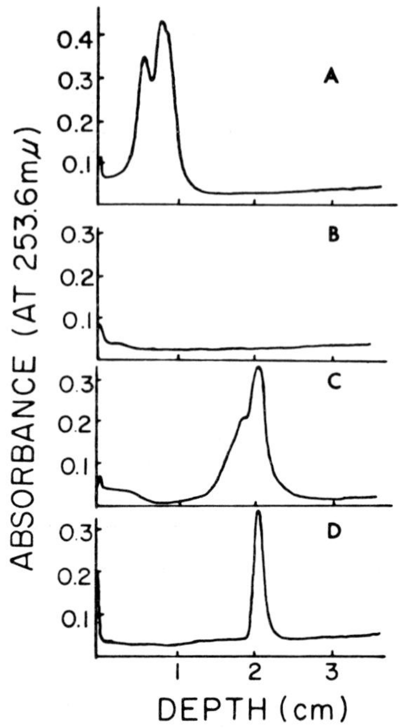

Fig. 15-8. Formation of an infectious nucleoprotein from protein and nucleic acid isolated from cowpea chlorotic mottle virus. Tracings of the 254-mμ absorbing distributions of RNA (A), protein (B), reconstituted nucleoprotein (C), and control virus (D) after density-gradient centrifugation of the cowpea chlorotic mottle virus or its constituents for 105 min at 36,000 rpm in a SW 39L rotor on a gradient of 10-40 mg of sucrose per milliliter of 0.1 M pH 5.0 acetate buffer. Sedimentation is toward the right. (From Bancroft and Hiebert, 1967.)

1970). Hybrid viruses were prepared between RNA of cowpea chlorotic mottle virus and proteins from bromegrass mosaic virus or broad bean mottle virus (Hiebert et al., 1968); also between cucumber virus 4 RNA and tobacco mosaic virus protein (Kado and Knight, 1970). The reconstituted virus then infected each distinct host which was susceptible to each of the above viruses. The results showed that the host range was entirely dependent on the RNA and not on the type of virus protein used. Thus the virus protein does not specify the host range of the virus; its role is limited to protecting the RNA.

C. Components of Plant Viruses

With the advancement of new techniques of purification, multiple components of plant viruses became evident. Viruses which were purified by earlier conventional methods were found occasionally to be composed of several components. The advanced techniques, now taken for granted, such as sucrose density-gradient centrifugation showed that such virus preparations did contain multiple components. Such preparations generally fractionated into three

TABLE 15-3
INFECTIVITY OF THE RECONSTITUTED VIRUS FROM PROTEIN AND RNA ISOLATED FROM COWPEA CHLOROTIC MOTTLE VIRUS

		Days after Preparation		
Material Assayed	*Treatment*	*1*	*2*	*5*
Virus-DG[a]	Enzyme[b]	1496[c]	—	263
	Not treated	1266	365	142
Reassembled-DG	Enzyme	1114	—	248
	Not treated	1109	501	404
RNA	Enzyme	0	—	0
	Not treated	2000	1169	1300
RNA-DG		—	0	—
Protein DG		—	0	—

[a] The designation DG means that the sources of the assayed preparations were the UV-absorbing materials found a little more than halfway down the density-gradient tubes after centrifugation. RNA-DG and protein-DG were removed from the same level as reached by the virus. RNA refers to the starting nucleic acid before density-gradient centrifugation.
[b] Enzyme refers to a final concentration of 20 units of snake venom phosphodiesterase per milliliter of material. Incubation for enzyme-treated and control material was for 20 hr at 22°.
[c] Number of local lesions on 8 half-leaves of soybean at E_{260} = 0.100 with the exception of RNA-DG and protein-DG, which showed no absorption at the level reached by the nucleoproteins. All material was incubated and assayed, and preparations were stored at 4° between treatments requiring storage at 22°. (From Bancroft et al., 1969.)

components (the bottom, the middle, and the top) in a density gradient of sucrose. As an example, the following viruses were found to have three components: turnip yellow mosaic virus (Kaper, 1960, 1964; Longley and Leberman, 1966); tobacco ringspot virus (Stace-Smith et al., 1965); alfalfa mosaic virus (Moed and Veldstra, 1968). Since turnip yellow mosaic virus has been best studied, details on this virus will be discussed. As mentioned, this virus possesses three components. The top component consists of only protein, while the middle and bottom components are nucleoproteins. The infectivity is exclusively observed in the bottom component, which also constitutes the major component which was considered to be the matured virus itself. Kaper (1960) treated the purified virus under alkaline conditions (pH 12 for 30 min) and obtained three components, one of which consisted only of RNA, another of protein shells, and the rest of protein fragments. The protein shell was considered an "artificial top component" since it was identical with the "native top component" obtained by density-gradient sedimentation, when examined by physico-chemical methods. Postulating that the artificial top component is identical with the native top component, and that these components are intermediate products in the process of virus formation, Longley and Leberman (1966) proposed a model for the maturation process of the virus in the host cell.

According to their model, protein subunits are first synthesized, then polymerized to form a protein shell, into which the nucleic acid "enters" and winds up during the last step of maturation. In contrast, from experiments on the polymerization process of tobacco mosaic virus, Lauffer et al. (1967) suggested that the protein subunits polymerized into the rod form by lining up along the nucleic acid molecule (Fig. 15-6).

From these experiments one cannot, at present, conclude which of the two possible models is correct, and it may be conceivable that the mechanism depends on the type of shape of the virus. In any case, the assembly of the protein shell into a rod without RNA discloses the role of protein in the maturation process of the virion.

V. SIGNIFICANCE OF STUDIES ON VIRUS PROTEINS

As mentioned in the preface of this book, tobacco mosaic virus has been playing a leading part in the biological sciences. The crystallization of this virus (Stanley, 1935*b*) was an epoch in establishing our concepts of life. Another finding that followed next was that infectivity of the virus was confined in the RNA and not in the protein component, and proved to be the genetic substance (Fraenkel-Conrat, 1955, 1956; Gierer and Schramm, 1956).

In 1961 two groups, in Tubingen and in Berkeley, elucidated the primary structure of tobacco mosaic virus protein. The work was important not only in the field of protein chemistry, but also gave meaning to the genetic code. When the sequence was almost completed, the Berkeley group (Tsugita and Fraenkel-Conrat, 1961) carried out the coding problem a step further. They isolated infectious RNA from the virus and treated it with nitrous acid to deaminate the bases in the RNA as Schuster and Gierer had shown (Schuster and Schramm, 1958; Gierer and Mundry, 1958). The chemical alteration of the bases can be brought about by nitrous acid (cytosine → uracil, adenine → hypoxanthine (similar to guanine) and guanine → xanthine (similar to guanine) and these workers observed a high frequency of mutations resulting from such treatment (for details see Chapter 20). The chemically altered RNA was reconstituted with the protein component and the reconstituted virus was allowed to infect the leaves of the test plant. High frequency of mutations was observed and the mutants were harvested and isolated. The protein component of these mutants was analyzed for its amino acid composition to note if amino acid exchanges occurred. In one particular strain (No. 171), three amino acids were replaced. Wittmann in Tubingen carried out essentially the same kind of experiments, independently (see Chapter 20). At first, no alteration of the amino acid composition was found (Wittmann, 1959*a*), but subsequently other mutants showed amino acid exchanges. These experiments were published prior to the outstanding experimental system established by Nirenberg (1961). Table 15-4 summarizes the experimental data from the above two groups. The results contributed to the early stage of deciphering the genetic code by the following

TABLE 15-4

AMINO ACID EXCHANGE AND POSSIBLE BASE CHANGE IN MUTANT PROTEIN OF TOBACCO MOSAIC VIRUS

Amino Acid Exchange					
From	*To*	*HNO_2*	*Other Mutagens* [a]	*Possible Base Change* [b]	
Asp	Gly	1,2		GAY → GGY	A → G (2) [c]
	Ala	2	Sp[1]	GAY → GCY	A → C (2)
	Val		NBS[2]	GAY → GUY	A → U (2)
Asn	Ser	2,1,4	NBS[4,2] Sp[1]	AAY → AGY	A → G (2)
	Lys		Sp[1]	AAY → AAZ	U, C → A (3), G
Thr	Asn		Sp[6]	ACY → AAY	C → A (2)
	Ala	2	FU[1]	ACX → GCX	C → G (1)
	Ile	1,7,2		AC^{Y}_{A} → AU^{Y}_{A}	C → U (2)
	Met	1,2		ACG → AUG	C → U (1)
Ser	Thr		NBS[2]	UCX → ACX	U → A (1)
	Gly	1	FU[1] NBS[2]	AGY → GGY	A → G (1)
	Leu	1,2		UCZ → UUZ	C → U (2)
	Phe	4,1,2	NBS, DMS[4] Sp[8]	UCY → UUY	C → U (2)
	His		5	?	
	Arg		Sp[8]	AGY → AGZ	U, C → A (3), G
Glu	Asp	1	4 [d]	GAZ → GAY	A, G → U (3), C
	Gly	1,4,2		GAZ → GGZ	A → G (2)
Gln	His		3 [d]	CAZ → CAY	A, G → U (3), C
	Arg	2		CAZ → CGZ	A → G (2)
	Thr	2		?	
Pro	Ser	1	Sp[3]	CCX → CUX	C → U (2)
	Leu	2,1,9,4	NBS[4,2] DMS[9,2] SP[3]	CCX → UCX	C → U (1)
Ala	Gly		Sp[8]	GCX → GGX	C → G (2)
	Val		Sp[1]	GCX → GUX	C → U (2)
Val	Asp		FU[1]	GUY → GAY	U → A (2)
	Ala		Sp[1]	GUX → GCX	U → C (2)
	Met	2		GUG → AUG	G → A (1)

TABLE 15-4 (Continued)

Amino Acid Exchange						
From	*To*	*HNO2*	*Other Mutagens* [a]	*Possible Base Change* [b]		
Ile	Thr	1	NBS[4,2] Sp[6,7]	AU^{Y}_{A}	→ AC^{Y}_{A}	U → C (2)
	Met	1,2		AU^{Y}_{A}	→ AUG	U, C, A → G (3)
	Val	1,2		AU^{Y}_{A}	→ GU^{Y}_{A}	A → G (1)
Leu	Ile		Sp[8]	$^{U}_{C}C^{Y}_{A}$	→ AC^{Y}_{A}	U, C → A (1); C → U (1)
	Phe	2	Sp[1]	CUY, UUZ	→ UUY	A, G → U, C (3)
Tyr	Cys	1		UAY	→ UGY	A → G (2)
Phe	Leu		Sp[1]	UUY	UUZ, CUY	U, C → A, G (3) or U → C (1)
Arg	Gly	4,2	NBS, DMS[4]	CGX, AGZ	→ GGX	C → G (1); A → G (3)
	Lys	4,2	NBS[4]	AGZ	→ AAZ	G → A (2)

[a] Sp: spontaneous, NBS: N-bromosuccinimide, DMS: dimethylsulfate, FU: fluorouracil.
[b] X stands for U,C,A or G, Y stands for U or C, Z stands for A or G.
[c] Exchange position in codon triplet.
[d] Frequency of mutant found.

[1]Wittmann and Liebold (1966)
[2]Funatsu and Fraenkel-Conrat (1964)
[3]Jockusch (1966*b*)
[4]Tsugita (1962*a*)
[5]Rombauts and Fraenkel-Conrat (1968)
[6]Nozu and Okada (1968)
[7]Wittmann (1966)
[8]Tsugita (1962*b*)
[9]Tsugita and Fraenkel-Conrat (1960)

information: (1) A correlation was established between amino acid exchanges and the possible base alterations; (2) the fact that neighboring amino acids were unaffected by the alterations of individual amino acids proved that the code is nonoverlapping; (3) the variety of exchanges suggested the existence of degeneracy of the code; and (4) the fact that noticeable symptomatic changes were observed, although the coat protein remained unexchanged, indicated the presence of other genes besides the one controlling coat-protein formation in tobacco mosaic virus RNA (polycistronic nature of the RNA).

Recently, the location of the coat protein gene was investigated using tobacco mosaic virus particles partially stripped of its protein and exposing specific pieces of the viral RNA to nitrous acid (Kado and Knight, 1967, 1969). Morphogenesis of plant virus has become an important field of study and its progress would contribute much to our knowledge of the infectious mechanism of virus and simultaneously to the field of developmental biology. The advantage of using plant virus as a model is that it is relatively easy to obtain in large quantities.

VI. ADDENDUM

Since this article was written several reports have appeared that provide some experimental evidence for details of the reconstitution process of tobacco mosaic virus. Using the specificity of phosphomonoesterases and labelled RNA (at its 3′-terminus), Butler and Klug (1971), and Ohno and Okada (1971) clearly indicated that the direction of protein assembly along the RNA chain proceeds from its 5′- to 3′-terminus.

Furthermore, from the evidence of kinetics of the reconstitution and/or mix-reconstitution with cucumber green mottle mosaic virus protein, Durham et al. (1971*a*, *b*) and Okada and Ohno (1971) demonstrated that a core, of 20-30 *S* material corresponding to a double-layered disc of 34 protein subunits, was required for the initial process of reconstitution. The core, of which its formation is rate limiting for the reconstitution reaction, combines with RNA near its 5′-terminus followed by the assembly of protein "building blocks" along the RNA axis. A small discrepancy still exists in the size of building blocks between two research groups; Klug's group indicated the building blocks consist of 20 *S* core discs, but Okada and Ohno suggested that it consisted of a rather small protein oligomer unit.

ACKNOWLEDGMENT

We wish to acknowledge with grateful thanks the help of Dr. Bertil Grahn, and Dr. M. A. Rashid for their discussion throughout this review. I also thank Miss M. Kobayashi for her help in preparing the manuscript.

16

Virus Degradation and Nucleic Acid Isolation

GEORGE E. BRUENING

Department of Biochemistry and Biophysics
University of California
Davis, California

I. INTRODUCTION

Viruses provide a unique resource in biology—a convenient and relatively abundant starting material for the isolation of homogeneous sets of genes.* Viral nucleic acids can be isolated and studied free from the influences of cells, and since the isolated nucleic acids of many viruses can initiate infections (Sanders, 1964), the effects of chemical, enzymatic, or physical modification on biological activity can be assessed. Nucleic acids cannot generally be released from viruses without extensive alteration of the virus structure; complete disruption is frequently required. However, less than complete disruption, either with or without nucleic acid release, also has potential value as a tool for probing the protein, the nucleic acid, and other aspects of virus structure. Three experiments illustrate especially well the diversity of interesting virus structure problems to which partial degradation techniques can be applied. Rueckert et al. (1969) have proposed a model for the capsid structure of an isometric, RNA-containing animal virus which is based on analyses of products released from virus particles under relatively mild conditions. When virions were completely disrupted by treatment with detergent and sulfhydryl reagent, the capsid was dissociated to monomeric polypeptide chains. Electrophoresis of the products on acrylamide gel revealed three major proteins, in approximately equal molar

*See Chapter 1 for definition of genes for RNA viruses.

proportions, and two proteins in minor proportions. However, incomplete disruption with acid or acid plus urea yielded two kinds of particles which sedimented more slowly than complete virions but more rapidly than would be expected for monomeric virus proteins. These (presumed) capsid fragments were found to be composed of equimolar amounts of the three major proteins. Molecular weight estimates indicated that the smaller of the two particles was a simple trimer containing one of each of the three major proteins. The larger particle was composed of fifteen protein molecules, five of each kind. No evidence was obtained for complexes containing only one or two of the three major proteins. The authors concluded that the principal structural elements of the capsid are sixty capsomeres, the capsomere being a complex containing one copy of each of the three major proteins.

A second application of partial virus degradation has been reported by Kado and Knight (1966) in their analysis of the position of a local-lesion gene (and the coat protein gene, Kado and Knight, 1968) on tobacco mosaic virus RNA. When tobacco mosaic virus was suspended in a solution of sodium dodecyl sulfate and was incubated at 37° C, protein subunits were removed in a polar fashion beginning at the end of the virus rod which has the 3′-hydroxyl end of the RNA molecule. Unfortunately, protein subunits were not found to be stripped off synchronously from different virus particles in the same solution. It was possible to fractionate the partially stripped virus by sucrose gradient zone centrifugation into preparations with small ranges in the lengths of exposed RNA. The exposed RNA is more susceptible to mutagenesis by nitrous acid, and to the action of pancreatic ribonuclease, than is RNA protected by protein within the virus rod. The efficiency of production of local-lesion mutants by nitrous acid treatment was low for virus stripped to expose 70% or less of the length of the RNA. However, the efficiency of mutagenesis increased dramatically for partially stripped virus with 74% or more of the RNA exposed, implying that the local-lesion gene is in the last 30% of the RNA molecule to be exposed by sodium dodecyl sulfate stripping. Urea also causes polar stripping of protein subunits from tobacco mosaic virus (Kado and Knight, 1966). The cleavage of turnip yellow mosaic virus RNA, apparently at specific sites, in virions incubated at high pH (Bosch et al., 1967) indicates that it may also be possible to modify the nucleic acid of spherical viruses in a controlled fashion.

In the case of a multicomponent virus it is possible to observe different rates or degrees of partial degradation for the different components. Semancik and Bancroft (1965) have applied chloroform at pH 9.5 as an agent for differentially degrading the middle component in a mixture of the middle and bottom components from bean pod mottle virus. This approach has potential value for defining structural differences between the middle, bottom, and other components of the multicomponent plant viruses.

Virus structure may be studied not only by following controlled disassembly of virions but also by observing the extent and kinetics of the reconstitution of virus structure. Virus reconstitutions have been initiated from the original virus constituents and from mixtures containing some of the original constituents plus foreign materials of either viral or nonviral origin. Only a few plant viruses have been successfully reconstituted. However, the extensive investigations on tobacco mosaic virus (Fraenkel-Conrat, 1969) and other rod-shaped viruses (Semancik and Reynolds, 1969) and on cowpea chlorotic mottle and some other small isometric viruses (e.g. Hiebert et al., 1968) have begun to elucidate the specificity of protein-protein and protein-nucleic acid interactions in these viruses. It is apparent that successful reconstitution requires that the protein be in a certain ("native"?) conformation. The isometric viruses which were the subjects of the latter studies are relatively easy to disrupt, making it possible to isolate viral protein under mild conditions and improving the chances of obtaining protein in a conformation which allows reconstitution. However, tobacco mosaic virus protein can by careful dialysis be returned from a 5 M guanidinium chloride solution (in which the polypeptide chain should assume a random conformation) to a dilute aqueous buffer and be caused to participate in the reconstitution reaction. Thus reconstitution may also be possible for the isometric viruses which require powerful denaturants for disruption, if the isolated viral proteins can be transferred to the proper solvent. The reconstitution or self-assembly of more complex structures has been reviewed by Kushner (1969).

Virions vary widely in stability. If any virion is exposed to solutions of very high or very low pH, to high temperature, or to some other sufficiently extreme condition, it will of course be disrupted. Many such extreme conditions will also degrade (i.e. rupture covalent bonds in) nucleic acids or even proteins. The principal purpose of this chapter is to describe some guidelines for the selection of conditions and reagents which can cause complete or, if it is desired, incomplete virus disruption. Those conditions and reagents which will tend to retain nucleic acid covalent structure are emphasized. The preservation of those nucleic acid or protein structures which are stabilized by noncovalent forces (e.g. the bihelical structure of double-stranded DNA) is also an important consideration. The sensitivities of various types of nucleic acids to chemical, enzymatic, and physical degradation are discussed first. Then classes of disrupting agents (all of which can function under conditions expected to give maximum nucleic acid stability) are compared for their effectiveness and modes of action. The techniques which are used to separate nucleic acid from the other constituents of disrupted virions must also be designed to preserve the nucleic acid. Therefore, separation techniques for nucleic acids are discussed in general, and some specific isolation procedures are reviewed. Finally, methods are described for evaluating the quality of the isolated nucleic acid. Conditions for virus dissociation, when recovery of intact nucleic acid is not a consideration, are given in Chapter 15.

II. CONDITIONS WHICH MAINTAIN NUCLEIC ACID STRUCTURE

One reason for the much slower advance of nucleic acid chemistry in the pre-1950 period, as compared to the advance of protein chemistry, was the lack of information about the extreme lability of nucleic acids under acidic or basic conditions, or when exposed to heat or enzymatic attack. In that period several enzymes had been purified in their active form, but there were very few reports of purified nucleic acids which exhibited biological activity. The early published reports on nucleic acid isolation (reviewed by Spirin, 1963) indicated that when sufficiently mild conditions were employed so that the nucleic acid might not be chemically attacked, nuclease activity caused significant degradation. More severe conditions prevented nuclease attack but chemically degraded the nucleic acid. Other potential difficulties were unknown. The extreme sensitivity of high-molecular-weight DNA to breakage by shear stresses is a physical degradation effect which was not appreciated in the earlier studies. The important routes of nucleic acid degradation are reviewed below to define just what extremes of temperature and other variables one can employ (as they are needed in the disassembly of the more stable viruses) and yet avoid nucleic acid degradation.

The covalent bond in RNA most labile to chemical cleavage is the P-O bond adjacent to the 5′-carbon of ribose. The cleavage of that bond may be catalyzed by either acid or base through a mechanism involving an attack of the 2′-oxygen on the phosphorus and the formation of a 2′-3′-cyclic phosphate intermediate (Michelson, 1963). The reaction is slowest in the pH range 6-6.5 (Ginoza et al., 1964). Cleavage of a single phosphodiester bond is apparently sufficient for inactivation of a viral RNA molecule, though there is some evidence for biological activity being lost to a lesser extent through other reactions as well at high pH (Strauss and Sinsheimer, 1968). The rate of cleavage, at least at elevated temperatures, does not depend on the source of single-stranded RNA (Gordon et al., 1963). RNA hydrolysis is a first-order process, and the rate of hydrolysis under a given set of conditions is conveniently expressed by the half-life of an RNA molecule. The half-life of a nucleic acid molecule may be estimated as the time required for the infectivity of the nucleic acid preparation to decrease to half its original value. For tobacco mosaic virus RNA (or any other single-stranded RNA molecule of similar size) the approximate half-lives at 60° in 0.1 M buffer (free of amines and divalent metal ions) are 10 min, 30 min, 40 min, 25 min, and 5 min at pH values of 4.5, 5.5, 6.5, 7.5, and 8.5, respectively (Ginoza et al., 1964; Gordon et al., 1963; Eigner et al., 1961; Strauss and Sinsheimer, 1968). At pH values higher than 8.5 the half-lives are decreased by a factor of 10 per pH unit, and at pH values lower than 4.5 they are probably decreased by a factor of 10 or more per pH unit. At higher temperatures and neutral pH the half-lives will decrease by a factor of 2.7 or more per 10° of temperature increase. At lower temperatures, especially at high salt concentrations, the magnitudes of the half-lives cannot be predicted because secondary interactions within the RNA molecules (e.g. base stacking) can significantly increase the half-lives of phosphodiester bonds (Bock, 1967). A minimum estimate of the half-lives at lower temperatures can be obtained by extrapolation from data at higher temperatures. The minimum half-life of tobacco mosaic virus at 25° could be expected to be about 50-70 hr in pH 6.5 buffer but less at lower and higher pH values in approximate proportion to the behavior at 60°. The minimum half-life at 0° can be expected to be greater than 100 days at pH 6.5. The half-lives of other RNA molecules of known molecular weight may be easily estimated by comparison with tobacco mosaic virus RNA since the half-life of an RNA molecule and its degree of polymerization (number of nucleotides per molecule) should be inversely related. The inverse relationship has been at least semi-quantitatively confirmed experimentally by Ginoza et al. (1964). First-order kinetics predict that 1% inactivation will be achieved in .015 half-life and 10% inactivation in .152 half-life.

This information is applied in the following example: In a proposed virus disruption procedure the virus solution is to be heated to 60° in a pH 6.5 solution to aid the virus disruption process. However, the RNA, with an estimated degree of polymerization of 4200, is to be inactivated to an extent of no more than 1%. Tobacco mosaic virus RNA has a degree of polymerization of about 6300, so the RNA in this example could be expected to have a minimum half-life of (6300/4200) (40 min) = 60 min at pH 6.5. One percent inactivation would be achieved in (0.015) (60) $\simeq$ 0.9 min. Therefore the solution should be held at 60° for no more than ~50 sec.

The most labile covalent bond in DNA is the N-C bond between the nitrogen at the nine position in the purine bases and the 1′-carbon of deoxyribose. Cleavage of the purine N-glycosidic bond is catalyzed by acid through a mechanism possibly involving protonation of the purine base. However, the details of the mechanism are obscure at this time (Szybalski, 1967). The N-glycoside bonds to pyrimidines are considerably more stable than those of purines. Phosphodiester bonds in DNA are chemically very stable (so long as the N-glycosidic bonds remain intact) because no 2′-oxygen is available to catalyze the hydrolysis internally. Chemical cleavage of the phosphodiester chain usually occurs only after removal of a purine. Loss of a single purine base seems to be sufficient for the inactivation of the DNA, and the biological activity of single-stranded DNA decreases according to a first-order rate law when it is heated in buffer. It is possible to estimate from the inactivation rate of bacteriophage ϕ X-174 DNA (Ginoza et al., 1964) that a single-stranded DNA of the same degree of polymerization as tobacco mosaic virus RNA would have a half-life of about 700 min at 60° and pH 6.5, that is, almost 20 times as long as for RNA under these conditions. The rate of inactivation of DNA would, of course, depend more strongly on base composition than would the rate for RNA. The half-life ratio of [(single-stranded DNA) / (RNA)] can be expected to be even greater at lower temperatures but less at higher temperatures. If the rate of depurination depends on the degree of protonation of the purines, the half-life of inactivation could be expected to decrease by a factor of 10 per unit of pH decrease in the medium. Both the inactivation of ϕX-174 DNA (Ginoza et al., 1964) and the hydrolysis of purine base N-glycoside bonds in mononucleotides (Venner, 1966) depend approximately on the first power of the hydrogen ion activity, the half-life of DNA decreasing by a factor of about 10 for each pH unit below pH 6.5. [Comparisons of DNA and mononucleotide depurinations must be interpreted with caution, however, since different mechanisms may be involved (Szybalski, 1967)]. There are few data on the stability of DNA from pH 6.5 through the alkaline range. Since the purine bases are essentially completely unprotonated at pH 6.5, little further increase in the half-life can be expected at higher pH value. The data of Ginoza et al. (1964) indicate that ϕX-174 DNA is only about three times as stable at pH 8.5 as it is at pH 6.5. In extremely basic solutions (e.g. >0.1 M NaOH) and at elevated temperatures DNA will be degraded since even mononucleotides are slowly destroyed (Hurst and Kuksis, 1958).

Double-stranded DNA will, of course, be considerably more stable than single-stranded DNA because it is completely in the base-stacked configuration. Complexes of DNA with basic polypeptides such as poly-L-arginine are more resistant to depurination at low pH than double-stranded DNA alone (Miano and Ferrini, 1969).

In the isolation of double-stranded DNA or RNA it is frequently as important to preserve the double-helical structure as it is to preserve covalent bonds. Extremes of pH will protonate or deprotonate the base residues of DNA or RNA and reduce their ability to form hydrogen-bonded pairs. The midpoint temperature for the melting transition (i.e. the T_m) of T2 bacteriophage DNA has its maximum value in the pH range 5.5-8.0 (Privalov et al., 1969), indicating maximum stability of the helical structure in this range. The pH 8 boundary of this range should be almost invariant, and the pH 5.5 boundary (Zimmer and Triebel, 1969), only weakly variant, for DNAs of differing percentage contents of guanine plus cytosine. The optimum pH for preserving both the helical structure and the covalent bonds of double-stranded DNA is thus approximately pH 7.5-8.0. It would be especially important to maintain that pH range if elevated temperatures were to be used in the virus disruption procedure or DNA separation steps. Single-stranded DNA should be most stable at a slightly higher pH, but a pH of 6.5 is probably optimal for both single-stranded and double-stranded RNA. Double-stranded RNA has a higher melting transition temperature than DNA of similar base composition (Gomatos and Tamm, 1963*b*; Chamberlin, 1965).

It is well known that the melting transition temperature (T_m) of the double-helical forms of both DNA and RNA tends to increase with increasing ionic strength as expected from considerations of the electrostatic repulsion between the phosphate residues in the two strands (Kotin, 1963). The T_m increases about 18.5° per tenfold increase in concentration of monovalent salt up to a concentration of approximately 0.2 M. DNA cannot be expected to remain in a bihelical configuration at room temperature in solutions free of polyvalent cations and at an ionic strength of less than 0.0001 because the melting transition temperature will be below room temperature. At higher temperatures, higher ionic strength is obviously required to maintain the double-stranded structure.

Large DNA molecules can be easily degraded by the physical process known as shear. Whenever portions of a DNA solution move at different velocities, the ends of the DNA molecules can become oriented in the directions of the solvent flows in which they happen to be located. This exerts a stretching force on the molecules which is maximal at the middle and which can exceed the strengths of the covalent bonds in the phosphodiester chains. DNA molecules can be sheared into half-molecules and even smaller fragments. It is reported by Josse and Eigner (1966) that DNA of bacteriophage T2 (molecular weight estimated to be between 1.1 and 1.3×10^8) can be broken by the velocity gradients produced during the passage of 1 ml of dilute DNA solution through a 0.25 mm orifice in 2 sec. Smaller DNA molecules are, of course, more resistant, and DNA in concentrated solutions is partially protected from shear breakage. DNA isolation procedures which call for vigorous shaking with chloroform-alcohol or phenol solutions to remove proteins are not suited for the isolation of high-molecular-weight DNA because of the large shear forces which are produced.

When conditions for virus disruption have been defined so as to avoid chemical and physical degradation of the nucleic acid, the conditions frequently are also ideal for enzymatic degradation. Several communications describing ribonucleases, and sugar nonspecific nucleases (but not deoxyribonucleases) from higher plants have been reviewed by Barnard (1969) and Privat de Garilhe (1967). These plant enzymes have pH optima between 5 and 7. Proteins will be removed from the nucleic acid in the usual course of a nucleic acid isolation procedure, but nucleases must be separately destroyed, inactivated, or removed very early in the procedure if hydrolysis of phosphodiester bonds is to be avoided. Detergents, such as sodium dodecyl sulfate (SDS), and phenol have been commonly employed to inactivate nucleases, and other protein denaturing agents described in more recently published nucleic acid isolation procedures are apparently similarly able to inactivate nucleases. It is important to distinguish inactivation and destruction. Nucleases are frequently very stable proteins, and it is possible to recover active ribonuclease(s) from water-saturated phenol (~88% phenol in water) by precipitation with acetone or by lyophilization (Rushizky et al., 1963; see also Chapter 1). The enzyme(s) is inactive in phenol-saturated water at room temperature (~12% phenol), but it is obviously not destroyed under those conditions.

The most widely utilized nuclease inhibitors are polyvinyl sulfate, dextran sulfate (Philipson and Kaufman, 1964), bentonite (Fraenkel-Conrat et al., 1961), and macaloid(Mandeles and Bruening, 1968). All three agents appear to act by adsorbing the nucleases. Polyvinyl sulfate is a water-soluble material whereas bentonite and macaloid are anionic clays. An advantage of bentonite and macaloid is that they can easily be removed from the purified nucleic acid solution by low-speed centrifugation. The latter two references cited give details for purifying the clay inhibitors. High concentration of bentonite, and probably also of macaloid, will reduce nucleic acid yields through adsorption (Blanton and Barnett, 1969). Philipson and Kaufman (1964) have investigated the pH dependence for the activities of these inhibitors. More recently diethylpyrocarbonate, an inactivator of nucleases and other enzymes, has been employed in a procedure for isolating plant nucleic acids (Solymozy et al., 1968). The compound was incorporated into the tissue homogenization medium to prevent nuclease action during the first step of the isolation procedure.

Another approach to minimizing the effects of nucleases is to reduce the amount of nuclease activity present in the initial virus suspension. More complete purification of the virions and the use of low-nuclease plants as hosts are means of achieving this end. Since plant nucleases seem to have low pH optima, buffering the initial virus suspension at or slightly above pH 9 has been used to prevent ribonuclease action during the early stages of nucleic acid isolation (Wells and Sisler, 1968). Such a buffer must be used with caution with RNA because of consideration of phosphodiester bond lability. Low temperatures will slow nuclease action, but denaturation or adsorption of nucleases will also be less effective.

When all of the precautions against nuclease degradation have been taken, it is still possible for traces of nuclease activity to remain in a nucleic acid preparation. Therefore, steps in the isolation procedure should be designed so that they can be carried out rapidly. The isolated nucleic acid should be stored as a precipitate or lyophilized powder to minimize nuclease action during storage. Autoclaving of all glassware and utensils or washing with hot detergent or 15% hydrogen peroxide is strongly recommended to avoid the introduction of extraneous nucleases.

III. AGENTS FOR VIRUS DISRUPTION

The labilities of covalent and noncovalent bonds in various types of nucleic acids establish limitations in the selection of agents and conditions for controlled or complete virus disruption, when nucleic acid structure is to be preserved. Since the pH must not be extreme (assuming the disruption is to be carried out in aqueous solution), the agent must be a neutral or nearly neutral ionizable compound or be nonionic. Polypeptide bonds are generally more stable than phosphodiester or purine N-glycoside bonds, and unless proteases are employed, polypeptide bonds can be expected to survive any degradation process from which intact nucleic acid can be recovered. Disruption of virus structure must be accomplished through dissolution of the relatively weak but numerous noncovalent bonds and forces. Reducing agents or other compounds which cleave or prevent the formation of disulfide bonds are also usually added to the medium in which the virions are to be disrupted. The noncovalent bonds and forces presently considered important in stabilizing the structure of viruses and other macromolecular entities include hydrophobic bonds, hydrogen bonds, and electrostatic forces between residues which have a net ionic charge or which are dipolar in nature. An elementary discussion of the roles of noncovalent bonds in maintaining protein structure has been presented by Bernhard (1968).

Viruses vary widely in stability, and superimposed upon this variation are the several degrees of disruption which may be desired in different experiments. Agents which denature proteins differ markedly in their effectiveness. Some, such as detergents, cause alteration in proteins when they are present at millimolar or even lower concentrations, while some salts and nonionic compounds have an effect only at concentrations of several molar. However, denaturants which act at very low concentrations tend to have effects which are difficult to reverse. Agents of several types which act as protein denaturants also lower the T_m of DNA and seem to weaken the noncovalent bonds and forces in other polymers. The effectiveness as a denaturing agent also correlates with the ability to increase the solubility, in water, of certain organic molecules (Hatefi and Hanstein, 1969; von Hippel and Schleich, 1969). This lack of specificity is to be expected if there are only a few kinds of stabilizing bonds and forces in macromolecular structures. Lack of specificity is, of course, disadvantageous to the investigator who wishes to affect differentially one portion of a complex virus structure. However, certain compounds (e.g. the anionic detergents) are very much more effective against some polymers (e.g. proteins) than against others (e.g. nucleic acids), and evidence of other such differential affects has been found.

An understanding of the nature of noncovalent bonds and forces can potentially supply a logical foundation for matching denaturant to the intended measure of virus dissociation or the degree of alteration in a virus dissociation product. The effects of various agents on the secondary and tertiary structures of proteins and nucleic acids have been intensively studied for many years. Recent reviews by Tanford (1968, 1970) and by von Hippel and Schleich (1969) summarize and evaluate the extensive data that have been collected, but they also indicate that completely reliable rules to predict the action of various agents are not available at the present time. No extensive study has been published comparing the effects of denaturing agents on several virions. The discussion below is therefore intended as a guide to the agents likely to be effective in the partial or complete disruption of virus structure.

Hydrophobic bonds appear to be the most important of the noncovalent factors maintaining protein structure and the most significant barrier to the release of nucleic acid from virions. Hydrophobic bonding results from the strong tendency of nonpolar residues in proteins, nucleic acids, lipids, etc. to remain in contact with each other, as a large proportion of them almost certainly are in the native virus structure. In contrast, after virus disruption they can be associated with the aqueous solvent or with polar residues. Such regions of associated hydrophobic residues probably also stabilize hydrogen bonds and electrostatic interactions by providing an environment of low dielectric constant. Caspar (1963) has summarized some of the evidence for anhydrous, nonpolar regions in the interior of tobacco mosaic virus.

Two sources of energy can be invoked to explain the tendency of nonpolar small molecules or nonpolar residues in polymers to be associated with each other, rather than with solvent in an aqueous solution. These sources are (1) the London dispersion (also known as van der Waals) forces and (2) the inability of aqueous solvents to adapt in an energetically favored way to the presence of nonpolar molecules or residues in close contact with water molecules. London dispersion forces arise from electrostatic attractions between the fluctuating (not permanent), mutually induced dipoles which occur upon the very close approach of any pair of atoms (Bernhard, 1968) or even surfaces (Tabor and Winterton, 1968). Since the strength of induced dipoles depends on the polarizabilities of the entities involved, the stability of a London dispersion interaction can be expected to increase with increasing polarizability.

The transition of nonpolar residues in a polymer from a hydrophobically bonded state to a state of association with aqueous solvent is difficult to study directly because of the many simultaneous structural rearrangements which must occur in a polymer during the transition. Evidence concerning the ability of aqueous solvents to accommodate the nonpolar residues has therefore been obtained primarily from model systems. For example, the transfer of a hydrocarbon molecule from a hydrocarbon phase to water at room temperature is usually accompanied by an increase in free energy. The reaction is usually exothermic at room temperature, the change in free energy being positive because of the large negative entropy of the transfer. The analogous transfer of nonpolar (hydrophobic) amino acid residues, for example leucine or phenylalanine side chains, from the presumably nonpolar interior of a capsomere to the aqueous environment outside the virion would therefore also be expected to be energetically not favored. The energetics of the transfer process, of hydrocarbon from hydrocarbon phase to aqueous phase, have been taken as an indication of the unfavored adaptation which the arrangements of water molecules in pure liquid water must undergo in order to accept hydrocarbon molecules into the aqueous phase. It thus seems that hydrophobic bonding is promoted in part by the changes which would be forced on the solvent if the hydrophobic bond were disrupted. This effect is to be distinguished from the positive type of interaction illustrated by the London dispersion forces. A concise summary of the quite limited information thus far accumulated on the contributions of London forces and solvent effects to hydrophobic bonding is given by Jencks (1969). Stevens and Lauffer (1965) have shown that tobacco mosaic virus protein loses water upon polymerization and that the reaction is accompanied by an increase in entropy. The water loss would be expected if water surrounding the unpolymerized protein was returned to the general aqueous solution upon protein polymerization.

A denaturing agent which acts on hydrophobic bonds must decrease the free energy of transfer of a nonpolar residue from a nonpolar phase to an aqueous phase as indicated, for example, by its ability to increase the solubility of a nonpolar substance in an aqueous phase. Evidence for the importance of hydrophobic bonds in stabilizing the structure of biological macromolecules is the correlation, cited above, between the ability of various agents to denature macromolecules and to increase the solubility of nonpolar compounds (von Hippel and Schleich, 1969). For salts composed of a common cation and anions of the series

$$SO_4^{2-} < F^- < CH_3COO^- < Cl^- < Br^- < NO_3^- < ClO_4^- < I^- < NCS^- < Cl_3CCO_2^- < Br_3CCO_2^-$$

Members on the left tend to decrease the solubility of adenine, riboflavin, benzene, benzoic acid and other relatively nonpolar compounds in water. They also tend to stabilize the native conformations of proteins and DNA. However, members from the right-hand side of the series show a weaker ability to "salt out" the low-molecular-weight compounds. The salts of anions on the right actually cause a several-fold increase in the solubility of adenine and riboflavin when they are present at sufficient concentration. They also tend to destabilize DNA and proteins and can solubilize preparations of subcellular membranes (Hatefi and Hanstein, 1969). For example, the T_m of calf thymus DNA is reduced to below 25° in 4 M sodium trichloroacetate (Hamaguchi and Geiduschek, 1962) notwithstanding the expected stabilizing effects of such a high ionic strength solution. Detergents such as sodium dodecyl sulfate are, of course, examples of compounds having anions which are very effective in increasing the solubility of some nonpolar compounds in an aqueous phase, and they are very powerful protein denaturants. Detergents which contain aromatic hydrocarbon structures have been found to be active at even lower concentrations than aliphatic detergents (Decker and Foster, 1966).

For salts composed of a common anion and cations of the series

$$Ca^{2+} < Li^+ < Na^+ < K^+ < NH_4^+ << \text{guanidinium}^+$$

members on the left-hand side tend to decrease the solubility of nonpolar compounds in an aqueous solution. Only the guanidinium cation in this series shows solubilizing power (Tanford, 1970). This order does not in general correspond to the order of effectiveness as denaturants of biological macromolecules for probable reasons given below. The tetrapropyl- and tetrabutylammonium cations are also effective denaturants (von Hippel and Schleich, 1969).

Neutral compounds such as urea and its derivatives, amides, imidazole, phenol, dioxane, and low-molecular-weight aliphatic alcohols also act as denaturants (e.g. Herskovits and Jaillet, 1969; Tanford, 1970). Phenylalanine is more soluble in urea solution than in water, and the solubility of leucine is also increased by urea, but to a lesser extent. The solubility of alanine is decreased by increasing concentrations of urea (Whitney and Tanford, 1962). The calculated free energies of hydrophobic bond formation between two phenylalanine residues is greater than that for bond formation between leucine residues, which is in turn greater than that between two alanine residues (Nemethy and Scheraga, 1962), supporting the notion that urea is a hydrophobic bond breaking agent. The ability of a series of nonionic molecules to increase the solubility of adenine correlated well with their ability to lower the melting transition temperature of DNA (Levine et al., 1963). However, such correlations are more difficult to make with neutral denaturants of proteins because there is considerable variation in effectiveness towards different proteins (Gordon and Jencks, 1963; Herskovits and Jaillet, 1969).

Correlations of denaturing and solubilizing abilities can obviously be of use in choosing reagents for virus degradation, but knowledge of the detailed mechanism(s) by which reagents denature or solubilize could be of even greater value. For example, do the agents form a weak complex with the hydrophobic residues which can then become integrated into an aqueous solution? Micelle formation by detergents is an example of such a complex, but whether similar complexes exist for the hydrophobic bond agents which are effective only at high concentrations is unknown (Tanford, 1970). It may also be necessary to consider not just hydrophobic interactions but also whether aromatic or aliphatic groups are involved. One property which most effective denaturants do seem to have in common and which may provide a clue to their mode of action is polarizability. Hanlon (1966) has shown for a wide variety of compounds (with the exception of some polyhydroxy compounds) that there is a strong inverse correlation between the concentrations of various agents which are required to cause DNA denaturation and the polarizability of the agents.

Hydrogen bonds are a second important stabilizing force for the three-dimensional structures of biological macromolecules. In proteins the most numerous sites for hydrogen bonding are those on the polypeptide backbone, at the carbonyl oxygen and the amide nitrogen. The interaction of potential denaturing agents with peptide bonds has been investigated with model compounds by comparing the solubilities of peptides, for example triglycine, in water and in aqueous solutions of the denaturing agent (e.g. Robinson and Jencks, 1965). Urea and guanidinium chloride both increase the solubility of simple glycine peptides, the effect being more pronounced for the larger peptides with more peptide bonds. Other more direct indications of complex formation with amide functions have been described and reviewed by von Hippel and Schleich (1969). Lithium and calcium cations and the I^-, ClO_4^-, and NCS^- anions all form weak complexes. Since a peptide group complexed to one of these ions presumably could not participate in hydrogen bonding, these observations offer a possible explanation of the effectiveness of Ca^{2+} and Li^+ as denaturants even though they are probably not good hydrophobic bond reagents (i.e., they tend to decrease the solubility of nonpolar compounds in aqueous solutions). Calcium and lithium ions may have a degree of specificity as denaturing agents acting at peptide hydrogen bonds. Urea, guanidinium ions, I^-, ClO_4^-, and NCS^-, on the other hand, apparently act on both peptide hydrogen bonds and hydrophobic bonds.

Electrostatic forces between charged residues of either the same or opposite charge will, like hydrogen bonds, be strongest in regions of low dielectric constant such as are likely to be found in the interior of macromolecular structures. In aqueous solution the addition of a salt will decrease the repulsive or attractive forces between charged residues because of the shielding which results from the presence of large numbers of positively and negatively charged ions. The influence of salt on the T_m of DNA has been discussed above, the T_m being increased with increasing concentrations of salt. High concentrations (e.g. 1 M) of monovalent salts such as NaCl or $NaClO_4$ are employed in many DNA isolation procedures to aid in separating the DNA from electrostatically bound basic proteins. Since high-molecular-weight, single-stranded RNA is poorly soluble in 1 M NaCl, such procedures are not generally applicable to the isolation of single-stranded nucleic acids. However, it is desirable to be able to use certain salts, which potentially dissociate hydrophobic or hydrogen bonds, at concentrations higher than one molar. RNA is soluble in concentrated solutions of guanidinium thiocyanate (Ellem and Rhode, 1969) and perhaps in strontium nitrate (Pirie, 1954). Reichmann and Stace-Smith (1959) found the RNA of potato virus X to be precipitated in 2 M guanidinium chloride. However, Cox (1966, 1968) has shown that ribosomal RNA as well as DNA can be dissolved in 4 M guanidinium chloride. The double-helical structure of the DNA was maintained. RNA showed a tendency to aggregate in this solvent at a concentration greater than 1 mg/ml.

Among the compounds and ions already discussed only a few have been systematically tested for their effects on virus structure, though ability to initiate protein denaturation has been extensively investigated for many of them. The controlled and hopefully reversible disruption of virions will require high concentrations of the denaturing agent in most cases, so solubility is an important consideration. (Though detergents act at very low concentrations, their effects are difficult to reverse. Alcohols such as butanol and propanol can initiate protein denaturation at concentrations only slightly higher than the concentrations required for detergents but lower than the concentration required for most other agents.) It is important to distinguish effectiveness in *initiating* denaturation and the *extent* of denaturation or unfolding which can be achieved with a given agent (Tanford, 1968). Only guanidinium chloride has been shown, at the present time, to convert a wide variety of proteins from compact structures to randomly coiled polypeptide chains, though some other agents will probably be found to be as effective.

The denaturing power of a salt seems to be algebraically predictable from the denaturing abilities of its constituent ions (von Hippel and Wong, 1965). It is possible to suggest the following agents, only a few of which have actually been employed, for virus disruption experiments on the bases of high water solubility and other factors discussed above. Combination of agents could very well show synergistic effects. Where data are available the approximate molal (*m*) or molar (M) solubilities at zero degrees (unless another temperature is indicated) are given in parentheses:

LiBr (17.0 *m*)
$LiClO_4$ (5.6 *m*, 25°)
LiI (5.6*m*)
$LiNO_3$ (2.4 *m*)
LiSCN (8.4 *m*, 25°)
lithium benzoate (2.6 *m*, 25°)
$MgBr_2$ (5.5 *m*, 20°)
$Mg(ClO_4)_2$ (2.2 *m*, 25°)
MgI_2 (3.6 *m*)
$Mg(NO_3)_2$ (2.8 *m*, 25°)
guanidinium chloride
 (>7.5 M, room temperature)
guanidinium nitrate (1.14 M, 25°)
guanidinium thiocyanate
 (>6 M, 30°)
propanol (infinite)
butanol (1 *m*, 20°)
phenol (0.7 *m*, 16°)
$CaBr_2$ (6.2 *m*)
$Ca(ClO_4)_2$ (7.9 *m*, 25°)
CaI_2 (2.2 *m*, 10°)
$Ca(NO_3)_2$ (6.2 *m*)
$Ca(SCN)_2$ (3.8 *m*, 25°)
lithium 2-hydroxy-3,5-diiodobenzoate
$BaBr_2$ (3.3 *m*)
$Ba(ClO_4)_2$ (5.9 *m*, 25°)
BaI_2 (4.3 *m*)
$Ba(SCN)_2$ (2.5 *m*, 25°)
urea (10 M, 25°)
N-propyl urea
1,3-diethyl urea
N,N-dipropyl acetamide
imidazole

The tetrapropyl- or tetrabutylammonium salts of the above anions, and the trichloroacetate and tribromoacetate salts of some of the cations, should also be very soluble. The mild and reversible action of these denaturing agents recommends their use where isolation of both viral proteins and nucleic acids is desired or where some degree of surviving noncovalent structure is of value, such as in virus reconstitution studies. However, precautions must be taken to avoid irreversible denaturation through reaction of impurities or degradation products of the agents with proteins or nucleic acids. Cyanate formation in urea solutions (Marier and Rose, 1964) and air oxidation of iodide ion to iodine are examples of potentially harmful side reactions.

IV. NUCLEIC ACID ISOLATION PROCEDURES

Methods for isolating nucleic acids from viruses, subcellular organelles, and particles have been extensively reviewed in recently published reference works. Many variations of the basic phenol two-phase procedure, including the use of nuclease inhibitors, detergents, and metal ion chelating agents as well as preliminary lipid extractions or protease treatments, have been surveyed by Ralph and Bergquist (1967). They have also reviewed papers on the uses of guanidinium chloride, urea, sodium perchlorate, calcium chloride (agents for which the possible modes of action have been discussed in the previous section), and other agents from publications appearing prior to 1966. Other applications of phenol, detergents, and a few other agents are described in the two volumes of the "Methods in Enzymology" series, edited by Grossman and Moldave (1967, 1968). Scherrer (1969) has given a very complete description of the use of hot phenol for isolating RNA from cells or virus particles, and Massie and Zimm (1965) have described the use of hot phenol for the isolation of DNA. Older but still useful articles on nucleic acid isolation have been published by Schuster (1964) and by Sanders (1964). This array of organized information on nucleic acid isolation makes it unnecessary to repeat here the details of the older methods. Rather, general methods for separating nucleic acids from other materials will be considered first, and then some specific, more recently published procedures will be described.

The chances of isolating "intact" nucleic acid will usually be improved if the nucleic acid can be separated from the other components of disrupted virus particles as rapidly as possible. Sometimes the disruption and separation steps can be combined to further reduce the time for conversion of a preparation of virus particles into free nucleic acid. Several methods are available for separating nucleic acids from other biological polymers: (1) distribution between two liquid phases, (2) centrifugation, (3) precipitation, and (4) adsorption onto a solid.

Phenol, which is probably the most frequently used reagent for nucleic acid isolation, forms two liquid phases when sufficient amounts are added to water. The phase in which water predominates has a lower density, unless sufficient salt is present to make it more dense than the phenol-rich phase. Phenol is a powerful protein denaturant, and it reaches a sufficient concentration even in the aqueous phase to dissociate many viruses into nucleic acid and protein. Denatured proteins tend to be more soluble in the phenol phase than in the aqueous phase, whereas nucleic acids, especially RNAs, are more soluble in the aqueous phase. Thus the addition of phenol phase to an aqueous virus solution, followed by vigorous shaking to bring the two phases into intimate contact, can both disrupt the virus particles and separate proteins and nucleic acids (at least partially) into different phases. In most cases the denatured protein does not completely dissolve in the phenol phase, and some denatured protein forms a solid disc at the interface between the two liquid phases after phase separation by centrifugation. High temperature favors the mutual solubility of phenol and water, causing the two phases to have similar densities. Only one phase will be present above 60-65° if the salt concentration is sufficiently low. Having two phases of similar densities or only one phase makes vigorous shaking unnecessary; yet two phases of unequal density will again be present when the temperature is lowered, allowing effective removal of proteins. Chloroform plus isoamyl alcohol or other higher alcohol can also provide a denaturing second phase, but one with poor ability to dissolve denatured protein and hence less ability to remove large quantities of protein. The denatured protein appears mostly at the interface after centrifugation.

Kirby (1968) has described another kind of two-phase system in which RNA can be extracted into an organic phase as an alkylammonium salt. It is also possible to remove certain carbohydrates from a nucleic acid solution with a two-phase system employing 2-methoxyethanol (Ralph and Bergquist, 1967). Some degree of success has been achieved in the separation and fractionation of nucleic acids in two-phase systems which are formed by dissolving polymers such as polyethylene glycol and dextran, rather than organic solvents, in water (Philipson, 1969). Among these two-phase systems only phenol and chloroform-alcohol have sufficient protein denaturing power and speed of protein removal to be widely useful in the early stages of nucleic acid isolation. Phenol two-phase methods have found wide application, but they are not universally successful. The yields of nucleic acid are seldom greater than ~65% and are often much lower for certain viruses. Under these circumstances it is possible to have significant selective losses of different kinds of nucleic acid and to obtain a distorted picture of the distribution of nucleic acid species in a given material. The addition of cresol to the phenol, as suggested by Kirby (1968) has been shown to increase the yield of RNA from tobacco necrosis virus (Lesnaw and Reichmann, 1969*a*). References to the uses of hot phenol by Scherrer (1969) and by Massie and Zimm (1965) were cited above. Since the loss of nucleic acid in the typical phenol methods is apparently due to transfer of the nucleic acid to the phenol phase and to the solid material at the interface between the phases, an obvious option is to disrupt the virus under single-phase conditions and to isolate the nucleic acid by centrifugation, precipitation, or adsorption.

Among the materials released upon virus disruption, the nucleic acid will almost certainly have the highest sedimentation coefficient. Thus simple high-speed centrifugation should cause the viral nucleic acid to be removed from solution and be concentrated as a "pellet" in the bottom of the centrifuge tube. Unfortunately, such a pellet, while greatly enriched in nucleic acid, would be significantly contaminated with viral proteins. A greater degree of purification can be obtained through zone centrifugation. A solution of virions disrupted by detergent or other agent can be layered on a sucrose gradient (Noll and Stutz, 1968) or on a CsCl (or other dense salt) solution or gradient (Crawford, 1963). In the latter case the proteins cannot sediment to contaminate the nucleic acid because density of the solution exceeds the buoyant density of typical proteins.

The nucleic acid may reach the bottom of the tube or form a zone in the dense salt solution, depending on the type of nucleic acid, the type of salt, and the salt concentration (Szybalski, 1968). In the case of sucrose density-gradient centrifugation, if centrifugation were continued long enough, both nucleic acid and protein would be present at the bottom of the tube. Careful selection of the time of centrifugation results in nucleic acid being found either as a zone in the solution or as a pellet, and the proteins remaining in the upper portion of the tube. It is not even necessary, in preparative work, to construct a gradient. A layer of concentrated (e.g. 25%) sucrose in the bottom one-fourth of the tube may be sufficient to prevent viral proteins and other components from contaminating the high-molecular-weight nucleic acid, which can more rapidly penetrate the sucrose layer. Detergents and chelating agents are frequently incorporated into the density gradients to reduce nuclease action and nucleic acid aggregation, respectively.

A common procedure for precipitating the RNA is the addition to an aqueous solution of RNA of pH 5 potassium acetate to a concentration of 0.05-0.1 M and of two volumes of 95% ethanol at or below 0°. Somewhat less ethanol need be added for effective precipitation of high-molecular-weight DNA, and it is sometimes convenient to wind the DNA onto a glass rod as it is being precipitated since subsequent washing of the precipitate may be carried out while it is still on the rod. Isopropanol (0.5-1.0 volumes) may be substituted for ethanol and is especially useful in precipitating nucleic acids at low concentrations and from sucrose solutions. Nucleic acids can be recovered from guanidinium chloride solutions (Cox, 1968) and polar solvents such as formamide and dimethylsulfoxide (or even from the less polar solvents in which they have been dissolved as their alkylammonium salts) by alcohol precipitation, sometimes after the addition of water. Addition of a salt such as potassium or sodium acetate is even more essential for precipitation from non-aqueous, as opposed to aqueous, solutions. Neutral polysaccharides also tend to be precipitated by alcohols. Cetyltrimethylammonium salts of nucleic acids are insoluble in dilute aqueous salt solutions, and cetyltrimethylammonium bromide causes nucleic acids to be selectively precipitated from solutions containing contaminating neutral polymers (Ralph and Bergquist, 1967). Extremely small quantities of nucleic acids, from solutions of concentration down to 1 ng/ml, can be precipitated with this reagent (Sibatani, 1970). The nucleic acid can be dissolved in aqueous solution after converting it to, for example, the sodium form by washing with cold 70% ethanol containing 0.1 M sodium acetate. Kirby (1968) has used cresol as a nucleic acid precipitating agent.

Adsorption of high-molecular-weight nucleic acid onto traditional anion exchange resins such as DEAE-cellulose has, with rarest exception (Hung, 1969), proved to be irreversible. However, fractionation techniques which rely on adsorption, such as column chromatography, have great potential for bringing nucleic acid preparations to a state of high purity and removing important contaminants such as nucleases. Adsorption and elution have been achieved on column materials composed of methylated protein plus diatomaceous earth (the MAK column, Koch et al., 1969), benzoylated-naphthoylated DEAE-cellulose (Sedat et al., 1969), unsubstituted cellulose (Barber, 1966), and calcium phosphate (hydroxyapatite, Bernardi, 1969). Elution from hydroxyapatite seems to depend principally on the phosphate concentration in the eluting buffer and not on the ionic strength. It should therefore be useful for adsorbing small amounts of nucleic acid from large volumes of high ionic strength solution. The reversed-phase chromatography systems which have been developed at Oak Ridge National Laboratory for transfer RNA fractionation (Novelli, 1969) could also prove useful in the purification of high-molecular-weight RNA.

Investigations of the nucleic acids of a variety of viruses indicate that one or another modification of the two-phase phenol method will usually give at least a small yield of the nucleic acid. As was stated above, phenol is a widely applicable, but not a universal, reagent. Several recently published alternatives to the traditional two-phase phenol procedures are described below. In some of the modified phenol methods, such as in the first example, below, the phenol does not play an important role in disrupting the virus but is used principally to separate nucleic acid from the other viral components. Phenol can also be applied in single-phase procedures as in the second example. High-quality viral nucleic acid can also be isolated without the use of phenol.

A. Proteolysis in Detergent Solution

Cauliflower mosaic virus, the first plant virus which has been shown to contain DNA, appears to have an extremely stable capsid (Shepherd et al., 1970). Incubation with phenol alone, sodium dodecyl sulfate alone, or pronase alone failed to give rapid or complete release of the viral DNA. Other procedures, including heating in NaCl solution, incubation with urea plus sodium dodecyl sulfate, the single-phase phenol procedures described below, and treatment with hot phenol plus sodium dodecyl sulfate also did not effectively disrupt the virus. However, incubation with sodium dodecyl sulfate at a concentration of 2.5 mg/ml, together with pronase (50 μg/mg virus), caused the opalescence, due to undegraded virus particles, to disappear within 10 min at 37°. After 2 hr at 37° the sodium dodecyl sulfate concentration was increased to 10 mg/ml and the solution was incubated for one more hour. The digest was extracted twice with the phenol-rich phase from a mixture of phenol and pH 7 citrate buffer. The recovered aqueous phase was dialyzed against citrate buffer and then was centrifuged with added CsCl until a single zone of DNA was obtained in the equilibrium CsCl gradient. A DNA yield of 90-100% was obtained, and the DNA was infectious.

B. Single-Phase Phenol

Diener and Schneider (1968) have tested the effects of phenol at saturating and subsaturating levels (one phase) on several viruses in aqueous solution. Turnip yellow mosaic virus was incubated at room temperature for 15 min in the presence of 5.2% by weight phenol. The solution was then dialyzed and centrifuged at low speed. Analysis by sucrose density-gradient centrifugation revealed that virus particles did not survive this procedure but that RNA was recovered in almost 90% yield. The RNA had a specific infectivity similar to that observed for RNA isolated by two-phase phenol extraction. Southern bean mosaic virus was not degraded under these conditions, and tobacco mosaic virus was only partially degraded. When tobacco ringspot virus was incubated with 2.6% phenol in 0.02 M phosphate buffer, both free RNA and a ribonucleoprotein fragment were observed. The latter sedimented as a narrow zone and at a rate slower than that of the naturally occurring middle and bottom components of tobacco ringspot virus. Thus it is possible that in this case a controlled degradation with a low concentration of phenol released a specific viral degradation product. Incubation with 2.6% phenol and 0.2 M, instead of 0.02 M, phosphate buffer caused free nucleic acid to be released in good yield. In other experiments phenol was used in conjunction with sodium dodecyl sulfate or ethanol to give more powerful denaturing reagents which allow higher concentrations of phenol in the single phase. The RNA can be separated from the phenol solution by precipitation with ethanol as well as by sucrose density-gradient centrifugation, though the latter method (without a preliminary dialysis step) should be more effective in removing nucleases.

C. Lithium Salts

The known affinity of lithium ions for peptide bonds makes lithium salts excellent potential virus degrading agents. Neurath et al. (1970) found that LiI at concentrations less than 2 M was able to release DNA from adenovirus type 7 particles whereas acetone, formamide, 5 M urea, and heat treatment released only the nucleoprotein core. Virions labeled with C^{14}-amino acids were treated with 1.65 M LiI (containing $Na_2S_2O_3$ to prevent oxidation of iodide ions) for 30 min at 36°. The solution was then layered over 2.5 volumes of saturated CsCl solution and was centrifuged at 109,000 *g* to form a density gradient from which the DNA was recovered at the expected buoyant density. The C^{14} label associated with the DNA amounted to only 4.5% of the amount associated with the original virus particles. DNA prepared by papain digestion plus detergent and phenol treatment similarly retained 4.7% of the C^{14} label. Biological activity of the DNA was not reported, but H^3-thymidine-labeled material sedimenting at the expected rate (31 *S*) was recovered in good yield. Lithium chloride should be a less potent agent than lithium iodide. However it has been successfully applied in nucleic acid isolation procedures. For example, Francki and McLean (1968) degraded potato virus X by freezing a virus solution in 2 M LiCl, pH 7.2 phosphate buffer, at -10° to -15° for a period of three or more hours. An RNA precipitate was recovered by low-speed centrifugation after the solution was thawed. The RNA precipitate was soluble in phosphate buffer.

Lithium chloride-prepared RNA was found to be similar to phenol-prepared RNA in spectral and sedimentation properties and perhaps to be more infectious. The mechanism by which freezing aids the virus disruption reaction has not been elucidated.

D. Sodium Chloride and Other "Inactive" Salts

Sodium chloride and certain other common salts (e.g. KCl, CsCl) have not been shown to have any dissociating effect on hydrophobic bonds or to have any special affinity for peptide bonds. These salts tend to reduce the solubility of organic compounds in water and are considered to be stabilizers of protein structure (von Hippel and Schleich, 1969). However, certain viruses are extensively dissociated by concentrated solutions of "inactive" salts at neutral pH. For example, Hiebert et al. (1968) have isolated the capsid proteins of cowpea chlorotic mottle virus, bromegrass mosaic virus, and broad bean mottle virus by dialyzing the virions at 4° for 24 hr against 1 M NaCl buffered at pH 7.4 and containing 1 mM dithiothreitol (a disulfide reducing agent). The protein remained in the supernatant liquid after high-speed centrifugation of the dialyzed liquid for 12-16 hr. It was free or nearly free of contaminating nucleic acid and was successfully employed together with phenol-isolated viral RNA in virus reconstitution experiments. The mode of action of 1 M NaCl is unknown. but it is possible that electrostatic forces between the viral protein and RNA are involved. The proteins are probably positively charged at neutrality since the isoelectric point of bromegrass mosaic virus is above 7 (Bockstahler and Kaesberg, 1962) and basic peptides predominate in the trypsin digests of both broad bean mottle virus (Yamazaki and Kaesberg, 1963*b*) and bromegrass mosaic virus (Stubbs and Kaesberg, 1964) proteins.

E. Soluble Polyanion

Weller and Morgan (1967) have disrupted *Escherichia coli* ribosomes with a salt form of sulfonated polystyrene. The linear, polyanionic molecules (molecular weight approximately 100,000) apparently formed complexes with the ribosomal proteins during an incubation of one or more hours duration, freeing the ribosomal RNA. The RNA was precipitated from 0.01 M NaCl solution with ethanol added to a final concentration of 52%. The proteins and sulfonated polystyrene remained almost completely soluble under these conditions. This unusual nucleic acid isolation method could have applications in virology, but it must be noted that the residual protein contamination and sedimentation heterogeneity of the ribosomal RNA both appeared to be higher than has been obtained by two-phase phenol procedures.

V. EVALUATION OF THE ISOLATED NUCLEIC ACIDS

The success of a nucleic acid isolation procedure may be assessed by considering the yield and the quality of the product. One of the most important aspects of quality is the degree of covalent integrity, that is, the fraction of the original viral nucleic acid covalent bonds which remain in the isolated nucleic acid. The plant virologist has in many cases a sensitive relative measure of covalent integrity: the infectivity assay. Since it appears that even a single break in the phosphodiester chain or the loss of a single base residue is inactivating, the specific infectivity (e.g. local lesions per unit concentration of nucleic acid in the inoculum) is an excellent property for comparison of nucleic acid preparations when it can be measured with sufficient accuracy. However, since the number of infectious units is always several orders of magnitude smaller than the number of nucleic acid molecules, infectivity cannot be an absolute standard of nucleic acid quality at the present state of technology.

Physical and chemical tests of the size and homogeneity of nucleic acid molecules provide more reliable measures of covalent bond survival. With the exception of shear degradation of double-stranded nucleic acid, most of the damage to nucleic acids during isolation will be random in nature, introducing size heterogeneity into the population of molecules. Sedimentation (see Chapter 18) or gel electrophoretic (Loening, 1967) analyses, perhaps in conjunction with other techniques, can reveal size heterogeneity, and also, either directly or by comparison with nucleic acids of known molecular weight, allow the molecular weight to be estimated. An isolated nucleic acid which appears homogeneous by sedimentation or electrophoretic analysis, and which has an apparent molecular weight equal to the product of the virion particle weight and the virion weight fraction of nucleic acid (assuming one nucleic acid molecule per virion), might appear to warrant the adjective "intact," at least with respect to the phosphodiester chains. However, it has been widely observed (e.g. Bosch et al., 1967) that small numbers of scissions of phosphodiester bonds in a molecule may not be revealed by these analyses. Noncovalent forces can maintain the original configurations of both double-stranded and single-stranded nucleic acids even after the introduction of covalent breaks.

The lack of continuity of the polynucleotide chains can be revealed by placing stress on the molecules. Spirin (1963) has reviewed the early work in which RNA was heated to dissociate polynucleotide chains not held together by covalent bonds. For example, the reduced viscosities (which are qualitatively a measure of polymer molecular dimensions) of tobacco mosaic virus RNA preparations with high specific infectivity were found to increase more than fourfold as the temperature was raised from 20° to 80°. The sigmoid shape of plots of reduced viscosity versus temperature indicates that RNA probably reaches its maximum degree of unfolding and extension at the higher temperatures and that all fragments held together by noncovalent fragments should be dissociated. The correlation of large increases in reduced viscosity upon heating with high specific infectivity supports the notion that the molecular properties observed at 80-90° in 0.1 ionic strength (or less) buffer are those of covalent polynucleotide chains, both the intact molecules and the fragments.

Calculations of the half-lives of RNA molecules at elevated temperatures, as discussed above in Section II, show that phosphodiester bonds are rapidly cleaved at 80-90°, and the reduced viscosity does fall in succeeding measurements at those temperatures as expected. The destructive nature of this test means that it can give only a minimum measure of the covalent integrity. Milder tests have been developed. If a dilute (fraction of 1 mg/ml) solution of single-stranded RNA is heated to 80° for 4-5 min in buffered 2% formaldehyde and is then rapidly cooled, covalent structures will predominate in the resulting preparation. Noncovalently associated fragments were separated at the higher temperature, and formaldehyde reaction with the nucleic acid bases prevented most of the molecular aggregation which would otherwise have occurred. Similar applications of formaldehyde are reviewed by Boedtker (1968).

Tests can also be made by denaturing without heating. Strauss et al. (1968) have sedimented several RNAs in dimethyl sulfoxide solutions (80-99% dimethyl sulfoxide in water, with ethylenediaminetetraacetate and lithium chloride as dissolved electrolytes). Sedimentation coefficients (uncorrected for the viscosity of dimethyl sulfoxide) in 99% dimethyl sulfoxide were found to be related to the molecular weights of the RNAs in a simple exponential fashion over the range 0.5 to 2.0×10^6 in molecular weight (3-5 *S*). Double-stranded RNA was denatured at dimethyl sulfoxide concentrations greater than 70%, so covalent breaks in single-stranded RNA would certainly be revealed unless divalent and trivalent metal ion contamination were significant. Infective RNA could be recovered from this solvent. Density-gradient centrifugation was also carried out in dimethylsulfoxide solution (Strauss et al., 1968). Formamide was found to be a somewhat less effective denaturant than dimethylsulfoxide. Other possible nucleic acid denaturing amide solvents have been evaluated by Herskovits (1963), and Boedtker (1968) has described briefly the use of 7.2 M $NaClO_4$.

The regular nature of bihelical nucleic acids provides an obvious structural feature, the second strand, which allows the original molecular weight to be retained after a scission has occurred in one strand. Such "single-strand breaks" have been extensively studied in DNAs. They can be revealed by heating, by heating in formaldehyde, and by dissolution in a denaturing solvent. However, denaturation in alkali provides a simple means of separating the DNA strands through deprotonation of the thymine and guanine bases. DNA is quite stable at high pH, as indicated in Section II. Alkali-denatured DNA can be examined by sedimentation, either in an analytical ultracentrifuge or on a sucrose gradient. At 1.0 ionic strength (e.g. in 0.9 M NaCl, 0.1 M NaOH) the sedimentation coefficient of the separated, alkali-denatured strands is slightly greater than that of the corresponding double-stranded DNA (if the polynucleotide strands are intact) and thus can be conveniently measured. Simple relationships between the molecular weights and the sedimentation coefficients of alkali-denatured DNAs have been reported (Studier, 1965; Abelson and Thomas, 1966).

17

Plant Virus Serology

R. BERCKS, RENATE KOENIG, AND GERTRUD QUERFURTH

Biologische Bundesanstalt für Land- und Forstwirtschaft,
Institut für Virusserologie
Braunschweig, Germany

I. INTRODUCTION: APPLICATIONS OF SEROLOGY

Injection of an antigen into experimental animals leads to the formation of specific antibodies which are complementary to certain surface structures of the antigen. Under suitable conditions the reaction of antibodies with an antigen can be demonstrated in vitro, for example as precipitation, neutralization of infectivity, or complement fixation.

In practical work antibodies have been most useful for rapid and unequivocal diagnosis of plant virus diseases. Large-scale routine assays are performed in breeding programs with vegetatively propagated plants, especially with potatoes, but more recently also with other plants such as chrysanthemums (Oertel, 1968*) and cherries (Schade, 1967). With rod-shaped viruses, the combined use of electron microscopy and serology is especially valuable for diagnostic purposes. The virus is classified by electron microscopy according to its normal length (Brandes and Bercks, 1965); serological tests are performed accordingly.

Large portions of the surface structures of closely related virus strains are identical or nearly so. Thus, a great number of the antibodies formed in response to one strain will be able to react with other strains. The reaction of an antigen with its own antiserum is called homologous; the reaction with an antiserum against another antigen is called heterologous. A relative measure for the amount of antibodies present in an antiserum is its titer. Homologous titers are usually higher than heterologous titers. Titer comparisons thus serve

*For the sake of brevity literature citations are given for the most recent and comprehensive publication only. References to earlier papers by the same or other authors concerning the same subject will be found in the publication cited.

as a tool for differentiating strains; the differentiation can be refined by the use of cross-absorption techniques.

The replacement or loss of a single amino acid on or near the surface of a virus subunit may be sufficient to change the antigenic structure of a virus, whereas several amino acid exchanges in the center of the virus subunit are not recognized serologically (Harris and Knight, 1955; von Sengbusch, 1965; van Regenmortel, 1967*b*). Subunit aggregation, too, exerts its influence on the surface structure of a virus (Rappaport et al., 1965; van Regenmortel, 1967*b*). The antigenic structure of tobacco mosaic virus has been studied by means of model antigens and tryptic fragments of the polypeptide chain (Anderer and Schlumberger, 1966*a, b;* Benjamini et al., 1968). Serology thus contributes to the elucidation of viral surface structures.

In recent years, serology has developed into a useful tool for the detection of distant relationships between viruses. With distantly related viruses the identical surface structures are relatively small. Thus, only a small portion of their antibodies will be able to react heterologously, that is, the heterologous titers are much lower than the homologous titer. With low-titered antisera a distant relationship may be overlooked frequently (see Brandes and Bercks, 1965).

In all studies, especially those concerning the surface structure of viruses or relationships between viruses, it should be kept in mind that with the use of antisera, a biological method is applied which is subjected to all laws of biological variation. Strong individual differences have been found in the heterologous reactions of antisera from different animals (Bercks, 1963). Moreover, sera taken at different stages of immunization of the same animal show great differences in their content of strain-specific and group-specific antibodies (Koenig and Bercks, 1968). Thus, several antisera may have to be tested before a distant serological relationship is detected.

Serology has served to identify intermediates and by-products of the virus synthesis (Takahashi, 1959; Agrawal, 1964; Jaspers and Moed, 1966).

By means of labeled antibodies the synthesis and distribution of viral antigens in tissues and cells have been followed (Section IV.D).

Serological methods are extremely sensitive for the detection of protein impurities in virus preparations (Oertel, 1961; Gold, 1961; van Regenmortel, 1966). Antisera against plant proteins may be used to remove such impurities from virus preparations (Oertel, 1961).

Plant virus serology has been reviewed by Matthews (1957, 1967, 1970), van der Veken et al. (1962), Ball (1964), Wetter (1965), and van Regenmortel (1966).

II. PRODUCTION AND PROPERTIES OF ANTISERA

Antisera are produced by immunizing animals with partially or highly purified virus preparations. The degree of purity of the preparations should be as high as possible. Impurities may disturb the immunization process and frequently they are antigenic themselves, as is the fraction-l protein which has been found in all higher plants tested (Dorner et al., 1958; van Regenmortel, 1962*a*, 1963). Antibodies against fraction-l protein and other host constituents may cause unwanted serological reactions when crude saps of virus-infected plants or partially purified virus preparations are tested even if the virus used for immunization has been propagated on a host plant which is botanically unrelated to that used as a virus source in serological testing. This is due to the serological relationship of all fraction-l proteins tested so far. It is advisable to use such host plants as antigen sources from which the viruses can be purified fairly well. At the time of harvest, the virus concentration should be at its highest. Certain viruses which tend to degrade during the preparation or the immunization process have been stabilized by using low-percentage formalin solutions (von Wechmar and van Regenmortel, 1968).

Rabbits are widely used for immunization purposes. The antigen is usually introduced intravenously or intramuscularly into the animal. Intravenous injections are given into the marginal vein of one ear. The corresponding vein of the other ear is used for bleedings. Normally, the animals receive several intravenous injections. Repeated injections lead to injuries in the vein, making injections at the same site or above impossible. Therefore, the first injection is started near the tip of the ear. On the other hand, bleedings should begin close to the base of the ear, with further bleedings taken successively from the upper part of the vein.

Thin and short needles are used for intravenous injections. Air bubbles are removed from syringes by holding the tip with the needle straight upright and applying weak pressure. At the place where the needle is to be inserted the ear is shaved and disinfected by means of alcohol. If the vein does not show up well the whole section of the ear may be rubbed with alcohol or the blood may be stowed up temporarily below the site where the injection is to be performed. The needle — bevel up — is pushed gently at a low angle through the skin into the ear vein in the direction of the blood flow, that is, toward the base of the ear.

During injections or bleedings, the animal is placed in a box to avoid movements of its head and body. Experienced persons may work without such an aid.

No general immunization scheme can be recommended. The production of high-titered antisera depends on the amount of virus injected, viral antigenicity, and disposition of the animal. Genetic factors of the animal are important. The animals should be about a year old and in good health when the first injections are given.

With intravenous injections, small doses are administered at first, and then increased gradually. Good results were obtained in our laboratory by using the following procedure. One milliliter of a virus suspension is injected each time on three succeeding days. After four days, three injections of 2 ml each are given on succeeding days. After another four days, the dose is increased to three successive 4 ml injections.

Intramuscular injections are normally given into the thigh muscle. Usually the antigen is emulsified with an equal volume of Freund's complete or incomplete adjuvant. The widely used commercial product of Difco Laboratories consists of 1.5 parts Arlacel A (trade name for mannide monooleate), and 8.5 parts Bayol F (a paraffin oil of the Esso Company). The complete adjuvant contains killed and dried *Mycobacterium butyricum*. The incomplete adjuvant lacks the *Mycobacterium*.

The virus-adjuvant emulsion is prepared in a small homogenizer or by repeatedly (about 50 times) forcing the solution from a syringe through a wide needle. The emulsion is ready for use if a droplet of it does not spread over water. Normally, 2 ml of emulsion are injected. One milliliter portions are injected into the right and left thigh muscles.

Too rapid translocation of the antigen is prevented by the adjuvant. Thus two intramuscular injections given one week apart may initiate a long-term process of immunization which is comparable to that induced by a greater number of intravenous injections. In addition, the adjuvant causes an inflammatory reaction which favorably influences the formation of antibodies.

When serum titer begins to decline, it may frequently be restored by a single injection to its original or to a higher value. This booster phenomenon occurs mainly when an intravenous immunization scheme has been followed. Serum titers obtained after intramuscular injections with sufficient amounts of stable antigens usually remain unchanged for long periods provided bleedings are taken at regular intervals of two to three weeks. Booster shots have little effect under these conditions. Hollings and Stone (1970*b*) recommended one intravenous followed by two intramuscular injections over a period of three weeks.

Bleedings are taken when the antibody level has reached its maximum. This is checked with small test samples. When an intravenous immunization schedule is followed, the maximum will normally be reached one week after the last injection. With intramuscular immunization, three or four weeks are required after the last injection.

The animals should be fasted 12 hr prior to bleeding. The lateral ear vein is severed with a razor blade or a snapper. Very narrow veins which bleed slowly may be widened temporarily by rubbing the ear with xylol. This treatment, however, may lead to hemolysis. As much as 50 ml of blood can be taken at two-week intervals over a period of at least one year without seriously interfering with the health of an animal that weighs about 4 kg (Koenig and Bercks, 1968). Blood samples are also obtained by direct heart puncture.

Blood samples are allowed to coagulate for about two hours at room temperature. The sample is then placed overnight in a refrigerator. The next day the serum is decanted and clarified by centrifuging at low speed. A plastic granulate may be added to achieve better clarification (e.g. SEP-AR-AID of the Serva Co., Heidelberg, Germany).

Preservatives such as Merthiolate, Cialit, sodium azide, or glycerol may be added to the clarified serum, but Merthiolate and Cialit can cause adverse effects (see page 474). Sera are usually stored under refrigeration or in a deep freezer, or as lyophilized powders. Prolonged storage in refrigerators may cause precipitation of serum proteins. The quality of the sera is usually not affected and such sera can be clarified by centrifugation. The quality of antisera depends mainly on the type and the amount of antibodies in the globulin fraction. Three main classes of immunoglobulin are known for which the following nomenclature has been proposed (Ceppellini et al., 1964): IgG (γ, 7Sγ, γ2), IgA (β_2A, γ_1A) and IgM (19Sγ, γ macroglobulin, γ_1M, β_2M). The older terms are set in parentheses. Each immunoglobulin is made up of light and heavy chains which are linked together by disulfide bridges. The heavy and the light chains are different for each class of immunoglobulin. The contiguous N-terminal amino acid sequence is variable and determines antibody specificity. The contiguous C-terminal sequence is constant in each class of immunoglobulins (Lennox and Cohn, 1967). Apparently, IgM antibodies are formed preferably during the early stages of immunization. IgG antibodies appear somewhat later in the blood. Tremaine and Wright (1967) working with antisera against southern bean mosaic virus observed a considerable increase of IgG after secondary immunization. After primary immunization the levels of IgM and IgG were about identical.

Serological tests are usually performed with unfractionated antisera. However, for several reasons fractionation of antisera may be advisable or necessary. Globulin fraction is obtained by using ammonium sulfate fractionation. To avoid losses of antibodies, a final salt concentration of 40% is appropriate for antisera of several plant viruses (Chester, 1936; Bercks, 1960, 1967*a*). For many purposes, however, precipitation with ammonium sulfate is not sufficient. Fractionation of antisera by gel filtration on Sephadex (Pharmacia Fine Chemicals, Uppsala, Sweden) has been performed by Miličić et al. (1966), and Tremaine and Wright (1967). Diethyl diaminoethyl cellulose has been used to fractionate antisera against wound tumor virus (Nagaraj et al., 1961). Fractionation of antisera by electrophoresis has been used by Bercks and Querfurth (1969).

In precipitin tests, clearer results are frequently obtained when purified globulin fraction is used instead of crude antiserum preparations. This may be especially important when the precipitin reactions are very weak. In adsorption tests (Section IV.C) several serum fractions may strongly inhibit precipitation so that reactions are not observed when unfractionated serum is employed (Bercks and Querfurth, 1969).

III. COMMENTS ON SEROLOGICAL REACTIONS

It is usually assumed that precipitates are formed as a result of the reaction between bivalent (IgG) or pentavalent (IgM) antibodies and multivalent virus antigens. A great number of virus antigens become linked together to form a network (lattice hypothesis). After the complexes have reached a certain size they become insoluble. Other hypotheses have been supported in recent years by Kleczkowski (1965) and by Loor (1966). Kleczkowski suggested that hydrophobic complexes are formed as a result of the antigen-antibody reaction. The complexes precipitate in the presence of salt. A prerequisite for the formation of precipitates is that both reactants are present at sufficient concentrations. Low concentrations of viral antigens are frequently found in crude plant saps. For this reason homologous reactions may occasionally not be observed with antisera whereas heterologous precipitations can occur (van Slogteren and de Vos, 1966). Obviously, the ratio of concentration of the reactants is important in the formation of precipitates. No precipitation will occur when there is an extreme excess of one of the reactants. According to the lattice hypothesis this phenomenon is explained by assuming that each molecule of excess reactant is able to react with only one of its valencies. Thus, a network and a precipitate cannot be formed. Serological tests with unknowns should always be performed with a variety of virus and antiserum dilutions since undiluted high-titered antisera or viral antigens occurring at high concentrations in plants may not form precipitates.

At antibody-antigen equivalence none of the reactants are found in the supernatant. With elongated viruses the equivalence zones are broad, whereas with isometric viruses they are narrow.

The respective titer or dilution-endpoint of an antiserum or a virus preparation is defined as the highest dilution at which a reaction is observed. Usually, a series of twofold dilutions is made in saline or buffer. Reactant excess inhibits precipitation; therefore, when the titer of one reactant is to be determined the other reactant should always be used near its dilution endpoint. This has the additional advantage that substances which may produce anomalous reactions are more likely to become diluted out.

Titers vary depending on the method used for determination. Therefore, statements on titers should include the method used for their determination. Also, titers are not absolute measures of antigen or antibody concentrations. More antibodies are necessary to precipitate small isometric viruses than long rod-shaped viruses. Therefore, antisera against isometric viruses usually have lower titers than those against rod-shaped viruses. Adsorption of one of the reactants to larger particles leads to an increase of titer of the other reactant (see Section IV.C).

The specificity of all serological reactions is checked by means of suitable controls. Antisera are tested against the sap from healthy plants and concentrated preparations of normal plant proteins. Antigens are tested against sera of nonimmunized rabbits (normal sera). In some cases antisera against unrelated viruses have to be used. Certain plant saps tend to give rise to nonspecific precipitations. Frequently, this tendency is alleviated by adjusting the pH to 7.0 (e.g. *Antirrhinum*) by heating, or by freezing and thawing the sap. Checks are made to ensure that such treatments do not alter or damage the virus. The presence of normal host proteins in virus preparations is checked by using antiserum against these proteins (Oertel, 1961).

The cross-absorption technique is used for studying serological relationships and viral surface structures. Cross absorption is made by treating antiserum with a heterologous antigen until all antibodies reacting with this antigen are removed. Titer changes due to dilution should be carefully checked. Cross absorption may be performed in liquid or semisolid media (see page 480).

In hapten inhibition tests antisera are saturated with low-molecular-weight monovalent substances which have structures similar to certain surface regions of the antigen. Precipitation of these "haptens" does not occur due to their monovalency. However, the reaction with the homologous antigen is inhibited.

IV. SEROLOGICAL TESTS

A. Precipitin Tests in Liquid Media

Various factors strongly influence precipitin tests such as the electrolyte content, pH, temperature, and concentrations of anomalous substances. Antisera are usually diluted in saline and antigens in weak buffer or buffered saline. All tests except the chloroplast precipitin test are performed with purified virus preparations or clarified plant extracts.

The precipitation reaction is usually examined either macroscopically (e.g. tube and ring-interface precipitin tests) or by dark field microscopy at about 60× (e.g. slide and microprecipitin tests). Minute precipitates are observed by electron microscopy (Anderer et al., 1967). Quantitative determinations of the amount of precipitate are performed by the Kjeldahl or Folin technique. Precipitates are washed thoroughly with ice-cold saline before such tests are performed. Also, turbidimetry and nephelometry can be used (Bradish and Crawford, 1960). Although the tube, slide and microprecipitin tests are employed in studies with all viruses, the double diffusion test in two dimensions (described on page 477) is preferable for isometric viruses. The slide and microprecipitin tests have found wide application, for example in routine assays of potatoes for the presence of potato viruses X, S, M, and Y.

1. *PRECIPITIN TEST IN TUBES*

Small tubes (7 mm ID) are filled with 0.3-0.5 ml of antigen and mixed with an equal volume of antiserum. Usually, the test is performed in a water bath at 37° C. However, at 37° C certain plant saps, such as sugar beet sap, cause nonspecific precipitations. Thus, lower temperatures may be preferable. The results are read against a dark background after 2 hr.

2. *PRECIPITIN TEST ON SLIDES*

Thirty microliters of antiserum and the same amount of virus are mixed on a microscope slide. The mixture is incubated at 25° C under high humidity for 20-45 min depending on the virus and examined by means of dark field microscopy.

3. MICROPRECIPITIN TEST

The microprecipitin test (van Slogteren, 1955) is performed in petri dishes pretreated with Formvar (1% in chloroform). Permanent coating is achieved by treatment with silicon solution (e.g. Serva, Heidelberg, No. 35130). Antigen and antiserum drops are deposited together in an orderly array and then mixed by stirring with a thin glass rod.

The drops are then covered with mineral oil and incubated for 3-6 hr at room temperature. The results are read with a dark field microscope or a stereomicroscope against a dark background. Mineral oil can be omitted as proposed by Vulič and Arenz (1962). In this case, dishes are covered with a lid and incubated for 24 hr at 10-12° C. According to our own experience, the recommended illumination of the dishes with 4000 lux of neon light can be omitted. The long incubation time at low temperature makes this test about four times more sensitive than the tube and the slide-precipitin tests. The microprecipitin test has the added advantage of saving time and space since 40-60 determinations are easily performed in a single petri dish, requiring only small amounts of antigens and antisera.

4. RING INTERFACE PRECIPITIN TEST

Antiserum at various concentrations is mixed with 10% glycerol and 0.1-0.2 ml is pipetted into tubes (2-3 mm, ID). The same amount of antigen is then carefully layered on top. The tubes are incubated for 1-2 hr at room or at some higher temperature in a water bath. The reactants form a ring or disc of precipitation where they meet at optimal concentrations. Inhibition by excess antigen or antibody is widely eliminated by this technique. Readings are taken in a dark room against a strong light source. This method is more sensitive than the tube and microdroplet tests when purified virus preparations are used. False rings can form at antigen excess (more than 0.1 mg/ml) or when impurities are present.

5. CHLOROPLAST PRECIPITIN TEST

This test, frequently described as the chloroplast agglutination test, is performed with crude uncentrifuged plant sap. Chloroplasts and cell particles become enmeshed in the network of the precipitate. The results are read macroscopically or microscopically with a dark field microscope at 60× after 5-30 min incubation at room temperature. The advantage of the test is that centrifugation is unnecessary. However, nonspecific precipitations occur more frequently with this test than with others. This test is also less sensitive than the others. The test has been used in large-scale assays for potato viruses X, Y, and S in potatoes (van Slogteren, 1952).

6. SHEEP RED BLOOD CELL PRECIPITIN TEST

In this test, which has been used by Moorhead and Price (1953), sheep red blood cells are enmeshed in the precipitate. Highly purified virus preparations must be used when the host plants contain hemolysins or hemagglutinins. Equal volumes of antigen, antibody, and a 1% suspension of sheep red blood cells in buffered saline, pH 7.3, are mixed and incubated for 3 hr at 5° C. The test is best performed in plastic agglutination plates. Positive results are marked by uniform distribution of the precipitate on the entire bottom of a well of the plate, whereas in the case of a negative result the red cells will sediment in the center of the well.

B. Precipitin Tests in Semisolid Media

Precipitin tests performed in semisolid media have the deciding advantages since the serological reaction is preceded by one or several fractionating steps such as diffusion and migration in an electrical field. Thus, resolving power is greatly enhanced. Relatively more information is obtained on the purity, homogeneity, and physical properties of the reactants and on the occurrence of nonspecific reactions. A serious limitation of those tests which are based on diffusion is that long rod-shaped viruses and some small viruses with a tendency to aggregate (Hollings, 1965*a;* Hamilton and Ball, 1966) diffuse very poorly in gels. Attempts have been made to overcome this limitation by using degraded virus (Hamilton and Ball, 1966; Purcifull, 1966; Tomlinson and Walkey, 1967*b;* Koenig, 1969*a;* Shepard, 1970). High-grade agar (e.g. Difco Agar noble) is widely used as the stabilizing gel. Saline and a suitable buffer are usually added. Most favorable conditions should be selected for each system, that is, pH and salt concentration (Wetter, 1967*a, b*). A preservative may be added to prevent fungal and bacterial growth, but its effect on the virus should be checked carefully. Mercury-containing preservatives (Merthiolate, Cialit) may destroy or alter the serological activity of a virus or interfere with its diffusion in agar (Bancroft, 1962; Koenig, 1969*b;* 1970). Also Merthiolate can lead to nonspecific precipitations of crude plant saps (Tremaine and Willison, 1962). Azide has not been reported to have adverse effects in agar gel tests so far. Temperature, due to its influence on aggregation, can affect the number and position of precipitin bands (Rappaport et al., 1965). The immunodiffusion techniques have been reviewed elsewhere (Crowle, 1961; Ouchterlony, 1964, 1967; van Regenmortel, 1966).

1. *SINGLE DIFFUSION TECHNIQUES*

This technique has been devised by Oudin (see Crowle, 1961). One of the reactants, the so-called external reactant, is allowed to diffuse into the agar in which the other reactant, the so-called internal reactant, has been incorporated. For bands to form in the agar, the ratio of external to internal reactant concentrations must exceed the ratio at equivalence. Tests may be performed in one or two dimensions (tubes or plates, respectively). Single diffusion techniques rarely have been applied in plant virus work (Jeener et al., 1954; Commoner and Rodenberg, 1955). Wetter (1967*a*) found that in the case of tobacco mosaic virus (the external reactant), the migration of the front zone of the precipitin band was inhibited by the presence of 0.8% sodium chloride in the gel. This was believed to result from an interaction between virus and agar in the presence of electrolyte. With belladonna mottle virus, band formation is strongly influenced by mercury-containing preservatives (Koenig, 1969*b*). The single diffusion method with plates has been used by Shepard (1970) to detect potato viruses X and S in infected plants. The two viruses were degraded by the use of 30% pyridine or 2.5% pyrrolidine, respectively. Antisera against the degraded viruses were incorporated in melted 1% agar containing 0.05 M Tris-saline buffer, pH 7.2.

Procedure for single diffusion in one dimension: Glass tubes (2-7 mm ID and 80 mm long) are precoated by filling the tubes with 0.1% agar, allowing them to drain and dry thoroughly. The internal reactant is incorporated into melted agar at 46° C, bringing the final agar concentration to 0.3%. The tubes are filled 3-4 cm with the agar. After the agar has solidified, a solution containing the external reagent is layered on top of the agar to a height of 2-3 cm. The tubes are sealed with parafilm or wax and kept at constant temperature. At regular intervals over a period of one to two weeks the advancing edge of the precipitation zone is measured.

2. DOUBLE DIFFUSION TECHNIQUES

The essential feature of double diffusion techniques is that both reactants diffuse toward each other into the gel medium originally free of them. A precipitin line is formed at the site where they meet under optimal proportions. The location of the initial precipitin line depends mainly on the diffusibility of the reactants and on their concentration. If the reactants diffuse to the reaction site at equivalent concentrations, the system is called balanced. Balanced conditions are the prerequisite for the formation of sharp stationary precipitin lines. The precipitin line acts as an immunospecific barrier. Diffusion of either reactant does not occur across this line. Under unbalanced conditions, the reactant in excess moves across the initial precipitin line. It may cause a gradual dissolution of the initial precipitate. Reprecipitation occurs on the side of the low-concentration reactant. The result is a migrating band with a unilateral diffuseness on the side of the reactant in excess. Frequently band splitting and the formation of multiple bands (Liesegang bands) will be observed under unbalanced conditions. In balanced systems, the formation of Liesegang bands may be due to variations in temperature. Liesegang bands can be confused with true multiplicity of precipitin lines developing in complex systems. On the other hand, the formation of one line does not necessarily indicate the presence of a single system since one line may be accidentally superimposed on another. Mixtures composed of homologous and heterologous antigens generally form only one line if the homologous antigen is diffusing ahead of the heterologous antigen. All antibodies are removed before they reach the heterologous antigen. If the heterologous antigen diffuses at a higher rate than the homologous antigen, two precipitin lines will form due to the incomplete removal of antibodies by the heterologous antigen. Precipitin lines formed by heterologous antigens can show bilateral diffuseness (Wilson and Pringle, 1956; Crowle, 1961).

a. Double diffusion in one dimension. Double diffusion in one dimension was originally developed by Oakley and Fulthorpe (1953). A micromethod was used by Scott (1961) for detecting barley stripe mosaic virus in seeds.

Approximations (error $\pm 5\%$) of the concentrations and diffusion coefficients of the reactants in double diffusion tests became possible after refinement of the technique by Preer (1956) and Polson (1958*a*, *b*). Preer found a linear relationship between the position of the precipitin bands and the logarithm of the ratio of concentrations of antigen to antibody. Based on the position of the precipitin bands in tubes with variable amounts of antigen and a constant amount of antibody, van Regenmortel and co-workers (1966) have made quantitative estimates of virus and plant proteins. Diffusion coefficients can be calculated from the equation

$$\frac{D_g}{D_b} = \frac{x_g^2}{x_b^2}$$

where x_g and x_b represent the distance of the stationary precipitin line formed under balanced conditions from the antigen and antibody menisci, respectively. D_g and D_b are the respective diffusion coefficients of antigen and antibody. For D_b the value of 4.81×10^{-7} cm^2/sec at 20° C has been found with rabbit, guinea pig, and rat antibodies (Polson and Deeks, 1960). The diffusion coefficients of several plant viruses and fraction-l protein have been calculated using this method (van Regenmortel, 1966). The value obtained for turnip yellow mosaic virus agreed well with that obtained by the usual optical method. The main advantage of this technique is that relatively small amounts of protein are needed. Care should be taken that no errors are introduced into the results due to secondary reactions. Wetter (1967*a*) found that in the case of tobacco mosaic virus the position of the stationary band and the width of all bands were strongly affected by the presence of 0.8% NaCl in the medium. By inserting the diffusion coefficient in the Stokes-Einstein equation, $r = kT/6\pi\eta D$, approximate estimates can be made of the diameter of the hydrated antigen assuming spherical particles.

Procedure: Precoated tubes as described for the single diffusion technique may be used for qualitative investigations. The antiserum is mixed with an equal volume of melted agar (final concentration 0.3-1%) at 50° C. The mixture is carefully placed on the bottom of the tubes, avoiding touching the upper walls of the tube. After the serum-containing agar has solidified, neutral agar of equal concentration is layered on top to a height of 8 mm. The antigen, in liquid or incorporated in agar, is layered on last. The tubes are then sealed and kept at constant temperature. Observations on the precipitin bands are made over a period of one to several weeks. Quantitative measurements can be made using the apparatus designed by Polson (1958*a*) (see also van Regenmortel, 1966).

b. Double diffusion in two dimensions.

Double diffusion in two dimensions (Ouchterlony test) is performed in agar gels in petri dishes or on glass slides. Direct comparisons can be made between several antigens. In this case antigen-containing wells surround a central well of reference antiserum. Under balanced conditions the precipitin lines form characteristic patterns (Fig. 17-1) and conclusions may be drawn on various antigenic relationships. Ouchterlony (1964, 1967) differentiated between the following four basic patterns (reaction types). (1) Complete fusion of the proximal ends of precipitin lines. This fusion indicates that the two antigens have identical antigenic groupings (e.g. group *a* and group *b*). One of the antigens may have additional groupings (e.g. group *c*), but this would not influence the precipitin pattern if the reference serum contains antibodies to *a* and *b* but not against *c*. Antisera against both antigens should be tested to prove serological identity. If one of the antigens has additional antigenic groupings, a pattern of partial fusion or intersection (reaction type 3 or 4) develops when its homologous antiserum is used. (2) Intersection of precipitin lines without mutual interference. This pattern develops when two antigens are tested against a heterologous antiserum. The first antigen may possess antigenic group *a*, the second antigenic group *b*. The precipitin lines would intersect if these antigens were tested against an antiserum containing anti-*a* and anti-*b* antibodies. This is due to the permeability of the precipitin line *a*-anti-*a* for anti-*b* and of *b*-anti-*b* for anti-*a*. Intersection of precipitin lines does not always mean that the two antigens tested are completely unrelated. They may have a common group *c* which remains undetected by an antiserum containing only anti-*a* and anti-*b*. Common antigenic groups can be detected by testing the two antigens against the homologous antisera of each antigen. A pattern of partial fusion or intersection (reaction type 3 or 4) would be observed if the two antigens were related. Reactants at low concentrations can form very short precipitin lines that do not extend beyond the point of intersection; such reactions simulate the type 1 reaction pattern. However, the two precipitin lines usually meet at an angle, whereas in the case of fusion they form a rounded arch. (3) Partial fusion and (4) partial intersection originally have been described by Ouchterlony and other investigators as one type. They are closely related and frequently may be difficult to differentiate. The extension of one precipitin line into a fainter band and the formation of a so-called spur are due to the incomplete trapping of antibodies by the heterologous antigen. Unbound antibodies diffuse across the heterologous precipitin line reacting with the homologous antigen or with another heterologous antigen that shares additional antigenic groupings with the homologous antigen. The degree of serological relationship between the two antigens is indirectly proportional to the length and the intensity of the spur and is directly proportional to the curvature of the spur. The formation of double spurs is observed when two heterologous antigens with antigenic groups *ac* and *bc* are compared using a

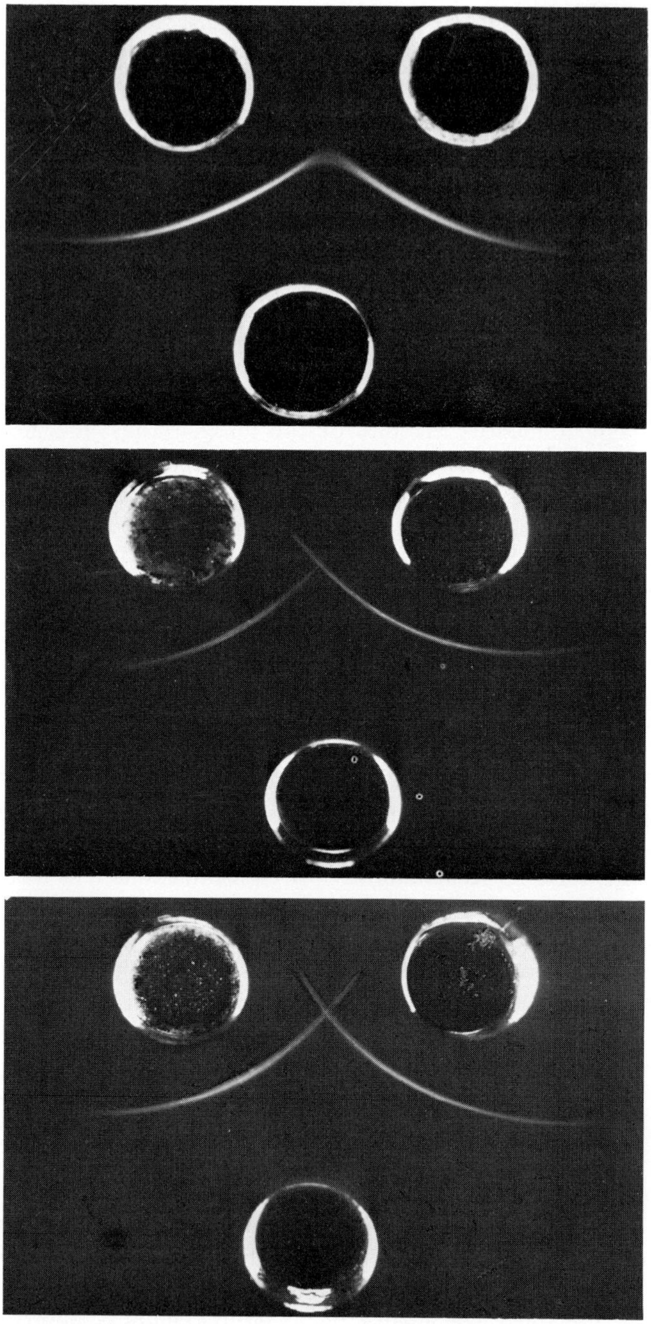

Fig. 17-1. Fusion (upper), spur formation (middle) and intersection (lower) of precipitin lines.

reference serum that reacts with groupings shared by both antigens (e.g. *c*) as well as with groupings specific for each one (*a* and *b*, respectively). Precipitin patterns with double spurs and the closely similar pattern of intersection differ from one another by the presence or absence of fusion between the two intersecting lines, by the abruptly thin spur as compared with the gradual thinning of the intersecting precipitin lines, and by the tendency of spurs to bend as opposed to the straight continuation of the intersecting precipitin lines. Exact information on the presence or absence of common antigenic groupings can be obtained by comparing the two antigens against the homologous antisera of each one.

The precipitin patterns described are a result of the selective permeability of the precipitin lines for all antibodies except those which are leading to its formation. This selective permeability is based on the immunospecific barrier effect which is strictly operative only under balanced conditions as mentioned above. Therefore, the precipitin patterns are conclusive only if the reactants are present at equivalent concentrations. Heterologous antigens have to be used at lower concentrations than homologous antigens because only a portion of the antibodies will be able to react with them. Under unbalanced conditions numerous artifacts may be observed, the most important being the formation of false spurs by identical antigens which are present at different concentrations and the formation of Liesegang bands (page 475). The formation of true spurs is based on the ability of antibodies with different specificities to diffuse independently of each other. No spurs would be formed if a homogeneous virus preparation were placed in a center well and different antisera (e.g. a homologous and a heterologous one) in neighboring outer wells (Grogan et al., 1964). Those antigenic groupings of a virus which do not combine with heterologous antibodies are unable to move independently across the precipitin lines. The formation of a spur under these conditions would indicate that the virus preparation contains other distinct antigens.

Double diffusion in two dimensions has been applied for a variety of purposes. Most work has been done with isometric or shorter rod-shaped viruses. With isometric viruses, a clear interpretation of the results is better obtained by double diffusion tests than by precipitin tests on slides. Attempts to adapt the method to longer rod-shaped viruses by using virus fragments have met with difficulty. When very short fragments are used, the serological properties of nonaggregated virus proteins may partially (van Regenmortel, 1967*b;* Shalla and Shepard, 1970) or completely (Rappaport et al., 1965; van Regenmortel, 1966) differ from those of the complete virus. Fragmentation of rod-shaped viruses causes a decrease of antiserum titers (Koenig, 1969*a*).

The two-dimensional double diffusion test has been employed routinely to detect alfalfa mosaic virus (Mueller, 1963). The technique was used to test the purity of virus preparations by using antisera against normal host constituents (Purcifull, 1966; Hamilton and Ball, 1966). Strain relationships have been studied by others (Ball, 1964; Wetter, 1965). Intragel cross-absorption tests were used in studying antigenic structures of tobacco mosaic virus mutants (van Regenmortel, 1967*b*). A single amino acid exchange in a mutant could lead to its differentiation from the type strain by the formation of a spur. The intragel cross-absorption technique is performed by placing the cross-absorbing antigen into the antiserum wells and allowing it to diffuse into the agar before the actual test is made. Because a gradient of absorbing antigen develops in the agar, cross-reacting antibodies are fully precipitated at the region of optimal proportions. This simple technique can also be used to remove antibodies against normal host constituents from antiviral sera. In this instance, the antiserum wells are filled beforehand with sap or concentrated protein preparations of healthy plants.

Procedure of the double diffusion test in two dimensions: Melted agar (0.7-1.0%) is poured into precoated dishes to a depth of 2-3 mm or on slides to a depth of 1 mm. Wells are cut into the agar with a cork borer or other suitably designed cutters which simultaneously punch several wells. Agar plugs in the wells are removed with a pipette connected to an aspirator. The diameter of the wells and their distance from each other should measure 1-5 mm depending on the diffusion coefficient of the antigen to be tested. Less time is required for the formation of precipitin lines when the distances are small. After the wells are filled with the reactants, the plates are incubated in moist environment for one to several days at constant temperature. Observations are made by looking at the plates, which are illuminated from one side against a dark background.

3. *IMMUNOELECTROPHORESIS*

Immunoelectrophoresis has been developed by Grabar (1964) (see also Ouchterlony, 1967). A mixture of antigens is first fractionated by zone electrophoresis in a supporting gel. Then the separated components are identified by letting them react with specific antisera. The antisera diffuse through the gel at right angles to the direction at which the antigens had been separated electrophoretically. The radially diffusing antigens which have been separated electrophoretically will form arc-shaped precipitin lines (Fig. 17-2). Section B.2 should be consulted for factors governing the formation of precipitin lines and for interpretation of the precipitin patterns.

Besides agar and agarose gels, cellulose acetate membranes and Cellogel strips (gelatinized cellulose acetate, Chemetron Chimica, Ltd., Milan, Italy) and polyacrylamide have been used. After electrophoresis in polyacrylamide the gels are usually implanted in agar into which a parallel trough for the antiserum is cut (Shalla and Shepard, 1970; Wolf and Casper, 1971). Electro-osmosis is an important secondary phenomenon when immunoelectrophoresis is performed with charged carrier media such as agar. The tendency of negatively charged carrier gels to move toward the anode is counterbalanced by the movement of water and all substances dissolved or suspended in it toward the cathode. The electro-osmotic flow changes the relative position of the antigenic components. Weakly charged antigens, which were expected to move

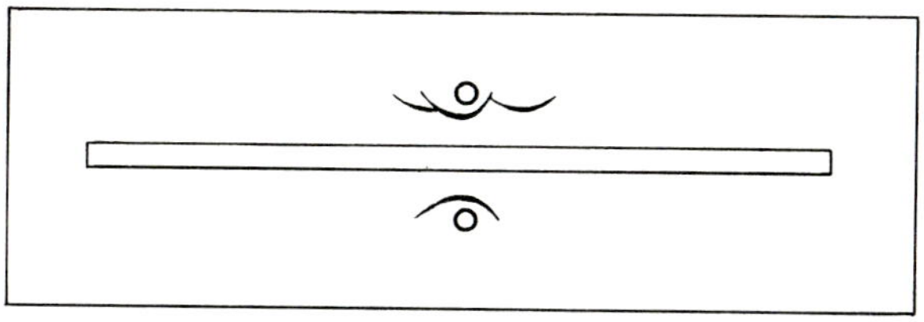

Fig. 17-2. Differentiation of three strains of a virus by immunoelectrophoresis. The upper well contains a mixture of three strains. Only one strain has been placed into the lower antigen well. An antiserum against the latter strain fills the central channel.

toward the anode, may appear on the cathode side after electrophoresis. In tests using agarose or membranes, the electro-osmotic flow has little influence. Immunoelectrophoresis has been reviewed for plant viruses by van Regenmortel (1966). The method has been used to demonstrate a serological relationship between turnip yellow mosaic and wild cucumber mosaic viruses (MacLeod and Markham, 1963), to investigate the electrophoretic heterogeneity of cowpea mosaic virus isolates (Agrawal, 1964), to differentiate three strains of tomato bushy stunt virus (Bercks and Lovisolo, 1965), to characterize soluble antigens and complete virus particles of Prunus necrotic ringspot virus (Allen and Tremaine, 1965), and to demonstrate changes in isometric viruses caused by mercury-containing preservatives (Koenig, 1969*b;* 1970).

Procedure: Immunoelectrophoresis generally should be carried out with buffers between 0.025 and 0.05 ionic strength, at a pH between 5.0 and 8.6. Agar or agarose are used between 1-1.5%. Antigen suspensions are either placed directly on membranes or filled into wells positioned centrally in a gel layer on a glass plate. The voltage gradient of the electric field should be set between 3 and 6 v/cm. The running time depends on the carrier medium, the buffer, the electric potential, the temperature, the type of setup, and the antigen mixture to be separated. After electrophoresis, antisera are filled into lateral channels that have been cut into the gel, or are placed directly on membrane strips parallel to the direction of the electrophoretic migration. The distance between the antigen well and the antiserum channel depends on the diffusion coefficients of the antigens and the concentration of the reactants. Precipitin lines develop when plates or strips are incubated in a moist chamber. Staining of lines may be performed with amido black. The plates are washed beforehand for 1-3 days with several changes of saline to remove proteins which have not participated in the reaction.

4. *IMMUNO-OSMOPHORESIS (ELECTROSYNERESIS)*

Immuno-osmophoresis is based on the migration of negatively charged antigens toward the anode and the simultaneous electro-osmotic flow of the weakly charged antibodies towards the cathode. The reactants are positioned so that they move toward each other. This technique has been used with plant viruses by Ragetli and Weintraub (1964). The micromethod developed by John (1965) requires only 1 μg tobacco mosaic virus in 0.5 μl, which can be detected in 3-5 min.

The original method can be performed only with antigens having a free electrophoretic mobility of at least -6.4×10^{-5} $cm^2/v/sec$ at pH 7.0. Synthetic agar derivatives such as trimethylaminoacetylhydrazone agar (Ragetli and Weintraub, 1965, 1966) with reduced negative liquid flow have been employed with antigens possessing an electrophoretic mobility as low as -1.38×10^{-5} $cm^2/v/sec$ at pH 7.0. Antigens with even lower electrophoretic mobilities (e.g. potato virus X) have been used after enhancing their net negative charge by conjugating them with phenylsulfonate and carboxymethyl groups (Ragetli and Weintraub, 1965).

Weakly charged antigens can also be studied by reversed immuno-osmophoresis. In Bacto agar these antigens are carried towards the cathode due to electro-osmotic flow. Their antibodies are made to move toward the anode by introducing negatively charged phenylsulfonate groups. By combining standard and reversed immuno-osmophoresis techniques (two-way immuno-osmophoresis), all constituents of an antigenically complex preparation can be detected simultaneously. Such a preparation is placed in the center well, native antiserum in the well closest to the anode, and chemically modified antiserum in the well closest to the cathode (Ragetli and Weintraub, 1965).

Procedure of the standard micromethod: One percent Bacto agar, prepared in 0.01 M Tris-succinate buffer pH 6.75, is poured on slides held in close-fitting aluminum foil boats to a depth of 4-5 mm. The foil is removed after the agar has set. Small holes, placed exactly opposite each other in the agar at a distance of 0.5 cm, are filled with 0.5 μl of antigen at the cathode side and with 0.5 μl of a fairly high-titered antiserum at the anode side. The slide is placed on the edges of two open petri dishes each containing about 30 ml of Tris-succinate buffer. Platinum electrodes are placed in each dish. The slide is connected to the buffer using filter paper wicks. A current of 40-50 m amp at 300 v is applied. Visible precipitin lines are formed within a few minutes.

C. Adsorption Tests

These tests are carried out with antigens or antibodies which have been adsorbed to some larger inert particles such as collodium, ion exchange resins, bentonite, acrylic resin (acrylpolymethylmetaacryl), barium sulfate, polystyrene latex, and erythrocytes. The particles become linked together as a result of the precipitin reaction taking place at their surface. Thus, this type of reaction corresponds to agglutination reactions. With an adsorbed reactant it is frequently possible to detect much smaller amounts of the other nonadsorbed reactant than with the previously described precipitin tests. Unfortunately, very little is known about the mechanism of the adsorption process. With many antisera it is necessary to purify the γ-globulin fraction before tests are performed.

1. INDIRECT OR PASSIVE HEMAGGLUTINATION AND HEMAGGLUTINATION INHIBITION

These tests are performed with erythrocytes on which antigen or antibodies have been adsorbed (sensitized erythrocytes). Passive hemagglutination occurs when the nonadsorbed reactant is added to the sensitized erythrocytes. In the inhibition test nonadsorbed antigen and antibody are first allowed to react with each other. The second step is an indicator reaction in which hemagglutination of sensitized erythrocytes determines how much antigen or antibody has been used in the test reaction.

Most work has been done with sheep red blood cells; however, human, rabbit, horse, or guinea pig erythrocytes may be used as well. Most proteins can be adsorbed only on erythrocytes which have been pretreated with tannic acid (Boyden, 1951). Rabbit antiserum is heated to 60° C for 30 min to destroy complement. Certain stabilizers are added to the dilution medium to prevent spontaneous agglutination of the erythrocytes. Usually, normal serum is used; however, bovine serum albumin, bovine γ-globulin, gelatin, or polyvinylpyrrolidione may also serve as stabilizing agents. Fresh erythrocytes of less than 18 hr should be used. However, erythrocytes may be kept at 5° C for several months if treated with formalin (Kabat and Mayer, 1961). Unfortunately, with some proteins the sensitivity of the test is lowered when formalin-treated erythrocytes are used. With erythrocytes of different bleedings the results may not always be reproducible. Some erythrocytes are sensitive to tannic acid. In these cases the concentration of tannic acid is reduced. The antiserum to be tested or the normal serum used as a stabilizer may contain agglutinins which are removed by pretreatment with red blood cells. When the amount of virus added is too high, spontaneous agglutination may occur. The virus preparation must be highly purified if the host plant contains hemagglutinins or hemolysins.

From 1000-fold to 10,000-fold increase in serum titer is obtainable with the passive hemagglutination test as compared with the precipitin test. We have observed about 100-fold increase of the virus dilution endpoints. Saito and Iwata (1964) found the lower limit of the test at 10^{-2} μg of virus. Passive hemagglutination has been used to detect various plant viruses (Saito and Iwata, 1964; Cunningham et al., 1966; Richter, 1967). Gamez et al. (1967) employed the technique to reexamine the purported relationship between wound tumor virus and reovirus. Recently, Abu Salih et al. (1968) obtained good results using formalin-treated erythrocytes for the study of several plant viruses.

Procedure: Sheep red blood cells are washed and suspended in veronal-buffered saline (pH 7.2). One volume of a 2.5-3.5% suspension is mixed with an equal volume of a 0.005% solution of tannic acid in saline. The mixture is incubated in a water bath for 10 min at 37° C and centrifuged. The sediment is first washed and then resuspended in veronal-buffered saline to its original volume. The red blood cells are coated with protein by mixing one volume of tannic acid-treated erythrocytes with four volumes of a suitable buffer and one volume of the protein suspension. The choice of the buffer and the pH depends on the protein to be adsorbed. For different plant viruses the pH may vary between 5.0 and 8.0, for the adsorption of antibodies good results are obtained at pH 7.0. The mixture is centrifuged after it has been kept at room temperature for 15 min. The sediment is washed three times and finally resuspended in veronal-buffered saline (pH 7.2) containing 0.5-1% normal serum or another stabilizer. The final concentration of erythrocytes should be between 2.5 and 3.5%.

In passive hemagglutination tests, the test reactant is diluted with veronal-buffered saline (pH 7.2) and mixed with an equal volume of sensitized erythrocytes. The test is best performed in plastic agglutination plates. The samples are incubated at room temperature for 3-12 hr or at 5° C overnight before the results are read. The entire bottom of a well covered with a thin layer of red blood cells signifies a positive result. Densely packed erythrocytes in the bottom center of the cavity indicate a negative result. In the inhibition test nonadsorbed reactants are incubated at 37° C for 30 min before the sensitized red blood cells are added. The minimum amount of test reagent that is capable of inhibiting hemagglutination completely is determined.

2. *BENTONITE FLOCCULATION TEST*

This technique was devised by Bozicevich et al. (1960). Antigen or antibodies are adsorbed to sodium bentonite particles (Bentonite, Standard Volclay, American Colloid Co., Chicago). Calcium bentonite has not been used successfully. When antibodies are to be adsorbed the sensitivity of the test depends on the use of suitable antisera or purified γ-globulin fractions and on the dilution of the antisera. We found that good results are usually obtained with antisera or serum fractions at dilutions of about half-titer strength, that is, the dilution should be 1:512 for an antiserum with a titer of 1:1024 as determined in the precipitin test. Under these conditions, we found an average of 100-fold increase of the virus dilution endpoint as compared with the precipitin test. Maximally, a 1400-fold increase was observed (Bercks, 1967*a*). Bozicevich et al. (1963) found the lower limit of the test at a virus-protein concentration of 0.05 g/0.1 ml. Using the bentonite technique, Scott et al. (1964) established the presence of potato virus X in potato tubers. We applied the test to several isometric and elongated plant viruses (Bercks, 1967*a*).

Procedure: One volume of the antigen suspension (crude sap or purified preparation) or of the diluted antiserum (crude serum or purified γ-globulin fraction) to be adsorbed is mixed with two volumes of a bentonite suspension (Bozicevich et al., 1963) and incubated overnight at 4° C. One-fifth volume of 0.1% aqueous methylene blue is added to increase the visibility of the agglutinated material. The mixture is centrifuged and washed twice with 0.02 M phosphate-buffered saline, pH 7.2. Finally, the sediment is taken up in an equal volume of distilled water and polyvinylpyrrolidione (final concentration 0.02%) is added as a stabilizer. A preservative is also added. The test is performed by mixing two volumes of the antigen or antiserum with one volume of sensitized bentonite suspension. The antigen is diluted in saline. Some plant saps elicit nonspecific flocculation; they have to be centrifuged and diluted. The reactants are mixed by shaking for 10 min. The test may be performed on plastic agglutination plates or on ringed microscope slides. The results are read macroscopically or microscopically at low magnification after two hours. Flocculation indicates a positive result. The particles remain in suspension for several hours if the results are negative. When antigens are to be tested, the controls should not be performed with normal serum but with adsorbed antibodies against a nonrelated virus. Frequently strong inhibitions of the reaction by excess antigen are observed. Therefore, several antigen dilutions including very high ones should be tested.

3. LATEX TEST

In this test, first used by Singer and Plotz (1956), antigen or antibodies are adsorbed onto latex particles. The sensitivity of the test depends on similar factors as mentioned for the bentonite test.

a. Detection of Small Amounts of Antigens. Using sensitized antisera which were diluted to their precipitin titers before adsorption onto latex we usually found 100-fold increases of virus dilution endpoints compared with the results of the precipitin tests. The maximum increase of a virus dilution endpoint was around 10,000-fold (Bercks, 1967*a;* Bercks and Querfurth, 1969). Using a different technique, Jermoljev and Albrechtová (1965) found an 8-fold increase in the dilution endpoint with potato virus X. The test has been applied for the assay of a variety of isometric and elongated plant viruses; several viruses have been detected directly in crude sap of grapevine leaves (Bercks, 1967*b*).

Procedure: One volume of polystyrene latex suspension (Difco-Bacto Latex, 0.81 μ) is diluted with 14 volumes of saline. The antiserum (crude serum or purified γ -globulin fraction) is diluted in 0.05 M Tris-HCl buffer, pH 7.2, mixed with an equal volume of the latex suspension, incubated for 30 min, and then centrifuged. The sediment is washed with one volume of 0.05 M Tris-HCl buffer, pH 7.2, containing 0.02% polyvinylpyrrolidone as a stabilizer. Finally the sediment is suspended in one volume of the same buffer. A preservative is added.

The test is performed by mixing two volumes of the antigen suspension (crude sap or purified preparation) with one volume of the coated latex suspension in plastic agglutination plates. Antigen dilutions are set up in Tris-HCl buffer, pH 7.2, containing 0.02% polyvinylpyrrolidone. The plates are shaken for 15 min. The results are read during the next two hours. Flocculation indicates a positive result whereas the latex particles will remain in suspension for at least two hours in negative results. The controls are performed using latex coated with antibodies against a nonrelated virus. Similar precautions should be taken as in the bentonite test.

b. Detection of Small Amounts of Antibodies. In this test (Bercks and Querfurth, 1971) the antigen is adsorbed onto the latex. This is most successfully achieved indirectly, that is, highly diluted antisera are first adsorbed onto latex and then reacted with such an excess of antigen that flocculation does not occur. The antigen is thus adsorbed to the latex via adsorbed antibodies. Flocculation occurs after the addition of free antibodies.

This test is especially useful for the detection of distant serological relationships. With heterologous reactions much higher increases of titers (up to 100-fold) are observed when pure γ-globulin fractions are tested instead of unfractionated antisera. With homologous reactions even higher increases in serum titers are observed. With several antisera heterologous reactivities were detected by this method which could not have been demonstrated by the precipitin techniques.

Procedure: Antibodies are adsorbed onto latex as described above. One volume of resulting latex serum is mixed with one volume of antigen suspension. The antigen should be adjusted to a concentration about 10 times as high as its precipitin titer. After incubation at 4° C overnight the mixture is centrifuged for 30 min at 5000 *g*. The sediment is washed three times with one volume 0.05 M Tris-HCl buffer, pH 7.2, containing 0.02% polyvinylpyrrolidone. Finally the sediment is suspended in one volume of the same buffer containing 0.05% sodium azide.

The test is done by mixing 0.025 ml of the latex-sensitized antigen with 0.05 ml antiserum in plastic agglutination plates. The plates are shaken for 15 min and the appearance of flocculations, which indicate a positive reaction, is observed during the next two hours using a magnifier. Between readings the plates are kept in plastic bags to avoid evaporation.

D. Labeled Antibody Techniques

By the introduction of labeling groups into antibody molecules, specific tracers are produced which provide a valuable tool for histochemical and cytochemical studies. Antibodies labeled with ferritin (Singer and Schick, 1961), fluorescent dyes, or radioactive iodine (Langenberg and Schlegel, 1969) are identified by electron microscopy, fluorescent microscopy, or radioautography, respectively. Enzymes have also been used as markers (Wicker and Avrameas, 1969). Most work has been done with fluorescent antibodies. Viral antigens can be located using virus-specific fluorescent antibodies (direct method) in situ. Binding of unlabeled virus-specific rabbit antibodies can be demonstrated by using fluorescent antibodies against rabbit γ-globulin (indirect method). The main difficulties encountered are the occurrence of bright autofluorescence of certain tissue components (e.g. lignified cell walls, chloroplasts) and nonspecific adsorption of fluorescent material other than antibody to certain cell constituents. This material and antibodies against normal plant proteins may be removed by absorbing the antisera with acetone powder preparations from healthy plants (Cremer and van der Veken, 1964). With some tissues, it is difficult to obtain thin sections satisfactorily. This difficulty has been overcome by Nagaraj (1962) by using enzymatically dissociated plant cells. The general aspects of immunofluorescence have been reviewed by Holborow and Johnson (1967).

Fluorescent antibody techniques have been used to study the distribution of virus antigens in plant and insect tissues. The intracellular synthesis of tobacco mosaic virus protein has been followed at different stages of infection. These applications have been reviewed by Ball (1964), Wetter (1965), and van Regenmortel (1966). The production of an insoluble protein by a defective tobacco mosaic virus mutant (PMl) which originally was thought to be "proteinless" was first demonstrated by the use of fluorescent antibodies (Nagaraj, 1965).

Procedure: (1) Preparation of conjugated antisera: All operations are carried out at 4° C. The globulin fraction of the antiserum is precipitated by adding an equal volume of neutralized saturated ammonium sulfate. The precipitate is dissolved and dialyzed against phosphate-buffered saline at pH 7.0. The protein concentration is then adjusted to 1% with 0.025 M sodium carbonate buffer, pH 9.8. The sample is dialyzed against 10 volumes of 0.01% fluorescein isothiocyanate in the same buffer. After 24 hr the conjugated globulin solution is dialyzed against phosphate-buffered saline, pH 7.0, until the dialyzate is free of fluorescence. After low-speed centrifugation the conjugated serum is stored at −40° C. It may be kept for about 4-6 weeks. Thawed samples should not be refrozen since repeated freezing and thawing may increase nonspecific staining. Conjugated antisera against rabbit γ-globulin are available commercially.

(2) Preparation of plant tissue sections: Modification of the method developed by Schramm and Röttger (1959): Leaf pieces measuring about 2 × 5 mm are placed into small molds of aluminum foil containing 20% gelatin in 5% glycerol. The samples are frozen with dry ice and equilibrated to −20° C or less. The aluminum foil is removed and sections of 10-20 μ are cut at the same temperature. The sections may then be lyophilized (Worley and Schneider, 1963) or can be mounted directly on slides smeared with a glycerol gelatin adhesive. The adhesive is prepared by dissolving 10 g gelatin in a hot mixture of 60 ml distilled water, 50 ml glycerol, and 1 g phenol. The slides are kept overnight at 4° C in moist chambers. With some tissues fixation with chloroform or absolute ethanol is advisable. After 20 min most of the fixative is removed with a syringe, and the remainder is allowed to evaporate at room temperature or 37° C. Nagaraj (1965) used a modified paraffin embedding technique which allows thinner sections (5μ) to be cut. Smears of internal organs of leafhoppers can be stained after fixations with acetone for 15 min at 37° C (Nagaraj et al., 1961).

(3) Staining and controls of specificity: When the direct method is to be used, the sections are covered with a few drops of fluorescent antiserum and kept in a moist chamber for 45-60 min in the dark. They are then rinsed for 10-15 min with phosphate-buffered saline at pH 7.2 to remove excess fluorescent material, and mounted in a glycerol-phosphate mixture (1 ml acid free glycerol + 9 ml phosphate buffer, pH 7.2). With the indirect method the sections are first covered with the unlabeled virus-specific antiserum for 40-60 min and then rinsed for 15 min in phosphate-buffered saline at pH 7.2. These steps are followed by treating with fluorescent antiserum against rabbit γ-globulin for 45-60 min. The procedure is then continued as described for the direct method. Observations are performed with a fluorescent microscope.

In each set of experiments, the specificity of the fluorescent staining should carefully be controlled by (1) the use of conjugated preimmunization sera or antisera against a nonrelated virus; (2) staining comparable tissues from healthy plants; and (3) using nonconjugated virus-specific antisera in the case of the direct method. The later staining of these specimens with conjugated virus-specific antiserum should result in very little or no fluorescence.

E. Complement Fixation and Immune Adherence Reactions

Complement, present in all fresh sera, consists of a group of several serum factors. It becomes gradually inactivated with time. Complement is bound by antigen-antibody complexes and then frequently causes secondary reactions. In tests described in this section, two complement-mediated reactions are used: (1) the lysis of red blood cells by a specific antiserum (complement fixation test), and (2) the attachment of antigen-antibody complexes to erythrocytes (immune adherence).

1. *COMPLEMENT FIXATION TEST*

This test involves the participation of five different components: the antigen and antibody to be tested, complement, sheep red blood cells, and a specific antiserum against sheep red blood cells (the hemolytic amboceptor or hemolysin). It comprises a test reaction between the antigen and antiserum, and an indicator reaction between sheep red blood cells and the hemolytic amboceptor. The fixation of complement by antigen-antibody complexes formed as a result of the test reaction is measured by the indicator reaction, that is, by the lysis of red blood cells. If all complement is bound, no lysis will be observed. With the microcomplement fixation test described by Wasserman and Levine (1961) up to 1000-fold increases in titer as compared with the precipitin test have been observed by us. The main disadvantage is that virus preparations must be highly purified if the host plants contain hemolysins or hemagglutinins. Hemolytic sera may inhibit hemolysis.

The test has been applied to several plant viruses (see Wetter, 1965). Tremaine and Wright (1967) have studied temporal changes in the amount of strain-specific and cross-reacting antibodies against two strains of southern bean mosaic virus.

Procedure: Before the test is set up, the "50% hemolytic unit" of the guinea pig serum complement is determined, that is, the amount which is necessary for 50% lysis of a given amount of sheep red blood cells. The test antiserum and the hemolytic amboceptor (rabbit sera) are freed of complement by heating for 30 min at 60° C. The titer of the hemolytic amboceptor in the complement-mediated lysis of red blood cells is also determined. For the actual test the hemolytic amboceptor is used at a concentration four times the titer. The test reaction is performed by mixing the antigen and antibody and two to five hemolytic units of complement. With the higher amounts of complement less interference by impurities in virus preparations is observed. However, the sensitivity of the test decreases, too. The dilution of the reactants, the salt concentration, and the pH are critical. Kolmer's solution or Eagle's buffer are suitable. The pH should be between 6.3 and 7.8. Normally, the mixture is incubated for 60 min at 37° C or for 24 hr at 2-4° C. Incubation time and temperature must be identical in all experiments. A suspension of sheep red blood cells, about 3%, which had been preincubated with hemolytic amboceptor for 30 min at 37° C, is then added. Generally, the whole system is incubated for 30 min at 37° C. The degree of lysis of the red cells which can be determined photometrically will be inversely proportional to the amount of complement utilized in the test reaction. Complete lysis indicates a negative result. No lysis of cells is a positive result, indicated by a sediment at the bottom of the tube. Some antisera may cause nonspecific binding or inactivation of complement. Therefore, excessively high concentrations of the reactants should be avoided. The nonspecific complement fixation should be checked in controls.

2. *IMMUNE ADHERENCE*

The phenomenon of immune adherence has been studied in detail by Nelson (1953). The reaction system consists of four different components: the antigen and antibody, complement, and erythrocytes from humans or other primates. No reaction is observed with red blood cells from nonprimates. Human or animal sera are used as a source of complement. The disadvantage of animal sera is that they contain small amounts of antibodies, causing human erythrocytes to agglutinate. This also applies to the test antisera, which have to be used at a high enough dilution to assure the inefficiency of the hemagglutinins present. The complement is removed from the test antiserum (rabbit serum) by heating at 60° C for 30 min. If the host plants contain hemolysins or hemagglutinins, highly purified virus preparations must be used. This test, which is more sensitive than the precipitin test, has been applied by Nelson and Day (1964) for the assay of cauliflower mosaic virus.

Procedure: Hank's balanced salt solution is used to dilute the reactants. Antigen and antibody are mixed with complement and incubated for 10 min at 37° C. A 1% suspension of primate erythrocytes is added and the mixture is incubated with occasional shaking for another 15 min at 37° C, and then for an additional 60 min at room temperature. When a reaction between antigen and antibody occurs, complement is bound, and the entire complex, attached to the red blood cells, agglutinates. The agglutinated erythrocytes form a diffuse layer covering the whole bottom of a tube. If no reaction has taken place the erythrocytes form a dense sediment in the center of the tube bottom.

F. Neutralization of Infectivity

Treatment of viruses with their respective antisera results in marked reduction of infectivity. Techniques based on neutralization of infectivity have limited application in plant virus work in spite of their high sensitivities. They are much more time consuming than other serological methods. The interpretation of the results is complicated by the fact that sera of nonimmunized rabbits also inhibit infectivity, though to a lesser degree. This nonspecific effect may be partially eliminated by using purified antibody preparations (Chester, 1936) or by lowering the pH to pH 6.0 (Anderer and Schlumberger, 1965*a*). Results can be interpreted without ambiguity when high-titered antisera are employed at dilutions of more than 1:1000 since the effect of normal serum does not extend beyond this dilution (Rappaport and Siegel, 1955). Recently, such tests have been applied to impure preparations of viruses which are transmitted in a persistent manner by insects (Duffus and Gold, 1969). The insects must be able to acquire the virus by feeding through membranes.

The exact details of the test differ with each system and thus should be carefully evaluated in advance. The virus, at a concentration giving reasonable local-lesion numbers, is mixed with antiserum or normal serum. The mixture is allowed to stand at room temperature for at least one hour and is then centrifuged at low speed. The supernatant is applied to plants. Chapter 1 should be consulted for factors which may influence the outcome of the local-lesion assay.

18

Analytical Ultracentrifugation

RODES TRAUTMAN

Plum Island Animal Disease Laboratory
Veterinary Sciences Research Division
Agricultural Research Service, U.S. Dept. of Agriculture
Greenport, New York

and

MARY G. HAMILTON*

Sloan-Kettering Institute
for Cancer Research
New York, New York

I. INTRODUCTION

A. Scope

The history of analytical ultracentrifugation reveals a complete circle. The pioneering work of Svedberg and collaborators in the 1930s involved huge turbine machines with complicated optical systems and even more complicated mathematics (Svedberg and Pedersen, 1940). Then from 1950 to 1965 there was the relatively simple, compact, readily available Model E Beckman/Spinco ultracentrifuge with phaseplate schlieren optics. Now, not only are there several other manufacturers, but the instruments have exceedingly complex optical systems and ancillary devices. The methods are legion, the cells are multichambered, the mathematics are frightening, and secondary and tertiary effects can no longer be neglected. Analytical ultracentrifugation is now swinging back to the domain of specialists.

Both preparative and analytical ultracentrifugation are physical techniques that are uniquely suited to the study of viruses. From a preparative standpoint,

*Supported by U.S. Public Health Service Grant CA 08748.

this is because viruses have size and density properties that are generally different from most other biological materials, and from a practical standpoint, because centrifuges are available to most, if not all, virologists. The preparative ultracentrifuge should be used at the very beginning of an investigation without any misconception that the biological agent has to be "pure" before certain fundamental molecular parameters can be established. Thus, in a *preparative run* the contents of tubes are analyzed by specific biological assays after the centrifugation period. The methods used to determine the sedimentation coefficient and density of the active agent are covered in another chapter of this volume. Here, we are concerned with the *analytical run*, in which the solute distribution along a radius in the cell containing the solution is observed and recorded at selected times during centrifugation. From crude to purified preparations of active agents, the analytical ultracentrifuge offers a means of "seeing" immediately what the solution contains and gives information of help in purifying and characterizing viral components.

Not every optical record has to be processed; in fact, perhaps 90% of all records have never been measured. An attempt is made here, therefore, to single out the *qualitative* information that can be obtained simply by looking at the patterns. For example, the gross purity of a substance can be determined by inspection and from it the progress of purification procedures can be ascertained. Similarly, stability to various chemical and physical agents is easily determined without plate reading.

It can be anticipated that *quantitative* information will require the use of advanced mathematics. For example, it is the redistribution of mass by centrifugation in the presence of diffusion that is measured to determine a molecular weight, and the analysis requires expressions from thermodynamics as well as special transformations of the raw data. Since there are more points measured on an optical record than the number of final quantities to be calculated, it is appropriate to use least squares statistics. Until recently, these manipulations and computations were done only by experts. With the advent of the generally available, inexpensive desk-top computer, any centrifuger can process data by exceedingly complicated procedures with ease. The computer, furthermore, makes obsolete all special devices such as tables, graphs, nomographs, and oversimplified expressions for the mere purpose of easing computation. By taking for granted that at least desk-top computer assistance will be available to the reader, it is feasible to give only those practical equations that need to be programmed for data processing and to omit tedious detailed instructions on their application.* In that respect, this chapter departs from numerous monographs and chapters already published (Williams et al., 1958; Schachman, 1957, 1959; Fujita, 1962; Williams, 1963; Creeth and Pain,

---

*Information on centrifuge programs for the Olivetti Underwood Programma 101 Desk-Top Computer is available from the authors.

1967; Markham, 1967; Chervenka, 1969). (References in this chapter are those that have been helpful to the authors and are not intended to give credit to original sources or to be complete. For a detailed bibliography see Browning, 1969, 1970.)

Three specific data processing protocols are given in the appendix along with a glossary of symbols and important constants to aid in evaluating formulas or in programming.

B. Methods of Centrifugation

The "physical" properties of molecules or particles that have direct bearing on centrifugation procedures are (1) size, (2) shape, (3) mass, and (4) density. The pertinent "chemical" properties are (1) particle interactions with each other at the finite concentrations used, and (2) interactions with the buffer that result in solvation, specific binding, and charge effects. Depending on the purpose of the experiment, any of the effects due to variations in these properties can be either considered as complications or used to measure the appropriate parameter.

Diffusion is transport by Brownian motion and it depends on the first two physical properties and the first chemical property listed above. The movement of any one particle caused by the centrifugal force is superimposed on the much larger Brownian motion. Both diffusion and centrifugal transport occur simultaneously in almost all ultracentrifuge runs. Diffusion is used to advantage in many methods, but it limits resolution in determinations of homogeneity.

There are four major experimental arrangements or methods used in analytical ultracentrifugation to exploit differences in the physical properties of particles or molecules in solution. No general agreement exists on nomenclature and no arguments will be presented for our choice. There are two velocity or *transport* methods—moving boundary and moving zone—that both depend upon size or shape differences. And there are two *equilibrium* methods—classical equilibrium and isodensity equilibrium—that both depend upon mass differences, and the latter, in addition, depends on buoyant density differences. General descriptions, with some practical details of experimental technique and major variations, will be given for three of the four methods: moving boundary, classical equilibrium, and isodensity equilibrium. The fourth method, moving zone, will be described briefly under the current name: band centrifugation.

The range of particle sizes of interest to plant virologists that can be studied by one or all of the methods is truly vast: from whole viruses down to their smallest subunits including the synthetic or degradation products of protein and nucleic acid. Solvents, too, are generally unrestricted and include such extremes as concentrated solutions of urea and inorganic salts, as well as organic liquids. Temperatures as low as 0° C and as high as 40° C are practical in commercial instruments, and the relative centrifugal force can be varied from a thousand to several hundred thousand times gravity.

II. INSTRUMENTATION

A. Optical Systems

Modern instruments (Elias, 1964; Bowen, 1966; Sagaert, 1967) have three major optical systems for observing and recording the redistribution of solutes during centrifugation. Figures 18-1, 18-2, and 18-3, which illustrate methods of centrifugation, also illustrate optical patterns. The details of operation and the underlying optical theory are not simple. Manufacturers' manuals give such information and methods of alignment (see especially Chervenka, 1969). We will comment here only on the comparative properties that are important to the user.

1. SCHLIEREN OPTICS

This system is the simplest, most versatile, and most common. It is based on the interception of light rays that are deflected when passing through an index of refraction gradient. The pattern represents this gradient as a function of radius. The vertical magnification (height of pattern on screen or photographic plate) can be changed during a run by changing the angle of the phaseplate schlieren diaphragm. At a given magnification, the area of the pattern is proportional to the (weight/volume) concentration and is roughly independent of the nature of the solute. However, the boundary region is affected by diffusion: the lower the diffusion coefficient, the sharper the peak. Because of their large sizes and, hence, low diffusion coefficients, viruses give sharp peaks and so can be studied at higher dilution than can proteins, even into the hundred microgram range. Schlieren optics can be used with all except the moving zone method, which requires exceedingly dilute solutions. Multiple cells can be used and there is no restriction on solvents, except that they be transparent at the wavelength of light used and do not react chemically with the centerpiece of the cell. The double-sector cell permits the baseline to be obtained simultaneously with the pattern. This is important when concentrated solutions are used as solvents.

2. RAYLEIGH INTERFERENCE OPTICS

This system is much more complicated to use than schlieren optics. It requires double-sector cells, special optical masks, and critical alignment. It is based on the differential retardation of light that passes through a solvent and a solution at the same distance from the center of rotation. The patterns are interference fringes, representing concentration as a function of radius. The magnification cannot be changed during a run. Rayleigh interference optics depend on (weight/volume) concentration differences with the number of fringes roughly independent of the nature of the solute because of the relative constancy of the specific refractive increment. These optics cannot be used in isodensity equilibrium or moving zone methods, and normally permit the use of only one multichambered cell at a time (except for rarely used "side-wedge" cells). They cannot be used at high concentrations or high speeds, because fringes are not resolved.

It is important to realize that the two *refractometric* methods yield data of a completely different form: The interference system gives the concentration difference whereas the schlieren system gives the concentration derivative with respect to radius. The two are complementary rather than competitive.

3. SCANNER ABSORPTION OPTICS

This system requires more elaborate equipment than either of the others: a monochromator for a variable wavelength light source, a split-beam densitometer, a chart recorder, and a facility for multiplexing several channels or cells. The system is based on the absorption of light passing through a solution, and the registrations represent the concentration as a function of radius. Two optical density ranges and several scanning rates may be available in order to change magnification factors during a run. A slow scan in velocity runs may permit too much change during the scan itself, but it can improve the precision of reading equilibrium patterns. Of the three optical systems, absorption optics allows the use of the lowest concentrations, frequently as low as the microgram range, if the substance has a strong absorbance at an available wavelength. The charts cannot be read as precisely as the photographic plates of the refractometric methods, and the system cannot be used with solvents that absorb too much light (e.g. traces of acetone or phenol make solutions opaque in ultraviolet light). It is the only optical system that provides an absolute scale of concentration, and it can be used for all four methods of centrifugation. Double-sector cells are employed so that absorption of the solvent and imperfections in optical components can be balanced out.

Absorption optics using photographic film for an instantaneous record have not yet been rendered obsolete. With appropriate optical wedges and masks, several single-sector cells can be used simultaneously. There is no possibility of changing the magnification or any simple way to obtain the concentration distribution required for the classical equilibrium method. However, for the moving boundary and isodensity equilibrium methods, where only boundary or band positions are required, the film system is still useful.

4. CALIBRATION CELL

A factory-calibrated prism with scribed lines can be used in all three optical systems. It enables one to determine a precise value for both the radial magnification and the distance from the center of rotation to the rotor or counterbalance reference edges. In the schlieren system, it is also used to give quantities needed for the vertical magnification (the optical lever arm times the cylindrical lens magnification).

5. MOVEMENT OF OPTICAL COMPONENTS

In the schlieren system, the phaseplate can be moved laterally across the optical track to effect centering of the pattern. This adjustment is not usually provided by the manufacturer but can be requested (Trautman, 1956). Such movement is useful during a run when phaseplate angles are changed, or parts of multiple cell patterns are to be photographed. The phaseplate angles should be standardized, to say, 15°, 30°, 45°, 60°, or 75°, so that comparison of plates from different experiments on a light box is simplified. Since the magnification depends on the cotangent of the angle, the effect of an error in setting is not uniform over the range of angles. Care must be taken to approach the desired setting from the same direction to avoid errors from backlash.

6. COMBINED OPTICAL REGISTRATIONS

In the classical equilibrium method, it is sometimes advantageous to have both schlieren and Rayleigh interference patterns on the same frame. The Rayleigh pattern is photographed first, then the optics are changed to schlieren and a narrow object laid perpendicular to the light source to cast a shadow in the region of the plate used for the Rayleigh pattern. Without shifting the plate, a manual "double exposure" is taken (Chervenka, 1966). Such photographs aid in aligning the plate on the comparator and in identifying meniscus positions. In fact, a good schlieren pattern is itself a critical test of the alignment of the Rayleigh system mask.

B. Comparators

All three optical systems provide two-dimensional records, that is, the pattern consists of a curve or trace that can be specified by a series of (x,y) coordinates (see Figs. 18-1, 18-2, and 18-3). The scanner absorption optics provide the trace on ruled paper. Since some commercial systems are not linear, additional measurements of built-in calibration marks are needed.

The simplest comparator for photographic records is a photographic enlarger projecting the pattern onto graph paper. More elaborate equipment such as an "optical comparator" or densitometer is preferred. Measurements on schlieren and interference plates are frequently recorded to the nearest micron, even though most comparators are accurate only to $\pm 2\,\mu$ and measurements of schlieren patterns are difficult to reproduce to better than $\pm 10\,\mu$. Plate reading definitely requires care and skill. By systematizing the plate reading procedure and by using carefully designed equipment, it is possible to facilitate the measurements so that tedium is not really a serious problem.

Completely automated comparators or densitometers have not received general acceptance, both because their cost is exceedingly high and because the human is required to make decisions that have so far defied programming. For example, a human can read a schlieren pattern with streaks due to a dirty light source or lenses, whereas a densitometer cannot. However, data acquisition can be automated at several levels for convenience. Thus, digital display comparators are available, but the numbers must be written down. Printing, digital read-out systems are more useful, but more expensive. Punched paper tape output seems attractive, but it is much more trouble than its worth since elaborate coding, editing, and further computer processing are also involved. For the scanner absorption optics a very expensive digital conversion of the signals for recording on magnetic tape is available from the manufacturer.

No matter which optical system or which comparator is used, the measurements must be converted to meaningful quantities. The abscissa X on an arbitrary scale is converted to the radius r in cm in the rotor by

$$r = r_{ref} + \frac{X - X_{ref}}{M_0 M_e}$$

where X_{ref} corresponds to r_{ref}, the rotor reference edge, M_0 is the camera lens magnification and M_e is the magnification of the enlarger or comparator, including dimension changes.

The comparator ordinate Y on an arbitrary scale is converted differently depending on which optical system is used. For schlieren optics, the concentration gradient dc/dr in the cell is obtained from

$$\frac{dc}{dr} = \frac{(Y - Y_{BL}) \tan \theta}{a'b' M_c M_e R_l}$$

where Y_{BL} is the baseline reading at the same radius, θ is the phaseplate angle with respect to the light source image, a' is the thickness of the cell in the direction the light passes, b' is the optical lever arm, M_c is the magnification of the cylindrical lens, and R_l is the specific refractive increment, which is practically a universal constant.

For Rayleigh interference optics, the plate can be read in two ways. The first, or one-dimensional way, is to record the abscissa for each intersection of a fringe with comparator fiducial mark as the plate is moved horizontally. Let Δj be the increment of fringe number (either half or whole fringe) and n the number of increments, then the difference in concentration between the level called zero and the level called n is

$$\Delta c = \frac{n(\Delta j)\lambda}{a' R_l}$$

where λ is the wavelength of light. In many cases only relative concentrations are needed, so that just the fringe number $J = n\Delta j$ is used. A useful conversion factor is that one fringe corresponds approximately to 0.25 mg/ml in the 1.2 cm cell.

The second, or two-dimensional way, is to record the (X, Y) coordinates along a single fringe from which

$$\Delta c = (Y - Y_{BL}) \left(\frac{\lambda}{a' R_l} \right) k$$

where k is the number of fringes per comparator unit measured vertically in the reference hole or air space.

For the absorption optics, the $Y - Y_{BL}$ can be converted to optical density using the calibration "stair steps," and then the optical density can be converted to concentration by

$$c = \frac{\text{optical density}}{Ea'}$$

where E is the extinction coefficient.

C. Computers

Whereas complete automation from the rotor through the plate reading stage of analytical ultracentrifugation is just beginning to be solved (Spragg and Goodman, 1968) and is not available to most readers, the automation of the computations has been accomplished (Trautman, 1966). The general protocol in data processing is (1) to convert the raw measurements from the comparator to standardized quantities, such as r, c, Δc or dc/dr; (2) to perform linear or quadratic curve fitting by the method of least squares of appropriate transforms of the standardized quantities; and (3) to convert the resulting parameters, with ancillary information, to molecular parameters together with their internal standard errors.

Many laboratories have large computer facilities available and readers may choose to combine the several operations of the general protocol above into one computer pass. Then, the protocol is (1) to keypunch the raw and ancillary data into cards; (2) to submit the resulting data deck with a program deck to the dispatcher for the large computer; (3) to return hours to days later for the output, which has the final parameters with their standard errors, graphs of intermediate quantities, and statistical measures of goodness of fit. The source programs are, for the most part, written in FORTRAN language. The conversion and compilation to machine language for the particular computer used gives some difficulty with complicated programs and has limited their exchange (Trautman, 1966).

For those just beginning to use computer assistance, the desk-top computer is recommended. It costs less than 25% of the price of the centrifuge itself, is available throughout the world, and has the capacity to perform most of the calculations required (Trautman, 1968). At first the manufacturers' programs are used in a "black-box" manner, which means following an operating procedure like a cookbook without question as to what is going on in the machine. The user can, and should, test such programs with literature data or some that have already been computed in his own laboratory. The effect of changes in the data can readily be seen by reprocessing. In effect, this means to "experiment" with the black-box program. By starting at the desk-top computer level, the number of black boxes that must be accepted is fewer. After processing many runs, one is in a good position to instruct the programmers of the local computer facility to synthesize the overall protocol into a single program for larger machines. In this escalation in the hierarchy of computer systems, direct access on a remote terminal is preferable to the use of cards for batch processing.

A very important practical feedback in having computer assistance is that the plate reading can be greatly simplified. Formerly, equal spacing of ordinates on schlieren patterns would be chosen to simplify calculations of area. Now, the points can be read in any convenient way. Similarly, aligning the plate so that the zero of the comparator scale corresponds to the center of rotation is now unnecessary, but still quite convenient (Trautman, 1956).

D. Statistics

The minimum requirement for ultracentrifuge data processing is both linear and quadratic least squares programs that give intercepts as well as standard errors. Most computing centers have such programs, but for those readers who must write their own, the formulas have been collected elsewhere (Trautman et al., 1969).

Application of the least squares principle gives unbiased values for the parameters as well as estimates of their internal standard errors. For example, linear least squares treatment of a table of x,y data pairs determines the best straight line that fits the plot of y against x. The line can be specified by two parameters: the slope and the intercept. With quadratic least squares, which fits a parabola to the data, there are three parameters: the intercept, the first-degree coefficient, and the second-degree coefficient. Choice of origin for the x scale does not affect the magnitude of the second-degree coefficient or its standard error, but does affect the first-degree coefficient since it represents the slope at the origin.

For equally spaced data (in x), the linear least squares slope will have the same numerical value as the tangent to the quadratic least squares parabola at the mean time. It will have a larger standard error if the quadratic coefficient is significantly different from zero. Tests of significance are made with the Student's-t test as follows. The absolute value of the coefficient is divided by its standard error to compute "Student's-t". If this is less than 1.96, the coefficient is not significantly different from zero at the 95% confidence level; if it is greater than 12.7, it is significant; and if it is between these numbers, then it may be significant depending on the number of degrees of freedom, that is, the number of data points less the number of parameters. Tables of the Student's-t distribution are given in most statistical texts.

It should be noted that using *linear* least squares does not test whether the data are "linear" or not. *Quadratic* least squares are used for that. Then, if there is no significant curvature, linear least squares may be used for processing the data. It is important that the intercepts and standard errors be inspected for reasonableness before accepting least squares statistical results. Replicates should be made to determine the reproducibility. The internal standard errors do not provide this information and are usually much smaller than the standard error computed from the list of replicate values.

The standard error, σ, of relations converting one quantity into another is given for some of the centrifuge formulas. For this, the "propagation of error formula" (Deming, 1964) was used.

$$\sigma_z^2 = \left(\frac{\partial f}{\partial x}\right)^2 \sigma_x^2 + \left(\frac{\partial f}{\partial y}\right)^2 \sigma_y^2 + 2r_{xy}\left(\frac{\partial f}{\partial x}\right)\left(\frac{\partial f}{\partial y}\right)\sigma_x\sigma_y$$

where $z = f(x,y)$ and r_{xy} is the correlation coefficient of y on x, which is zero for independent measurements.

E. Ancillary Measurements

1. DENSITY

The weight per unit volume (g/ml) of a solution is called its density, ρ. It is a function of the nature and concentration of solutes and depends slightly on temperature. Extensive tables for common substances can be found in handbooks. The easiest technique for measurement is to discharge a droplet of the solution in an immiscible, organic-fluid density gradient. The droplet will come to rest at a level corresponding to its density, which is calculated by interpolation between the positions of standards (Jacobsen and Linderstrøm-Lang, 1940). Pycnometry with volumes under 1 ml is difficult, but does not require a standard.

2. PARTIAL SPECIFIC VOLUME

The second quantity to be considered is closely related to density but it is not as easily measured. This is the partial specific volume, $\bar{v}$ (ml/g), which can be specified for each constituent as well as for the entire solution. It is approximately the reciprocal of the *dry* density rather than the reciprocal of the density of the *solvated* particle. It is defined thermodynamically as the limit in the change in volume of solution per change in weight of the (dry) constituent as the changes approach zero. Dry weight determinations for macromolecules and small differences in densities are difficult to obtain. For biologicals, the value is often computed from the chemical composition: for proteins, the amino acid composition is used; for viruses, the composition in terms of protein and nucleic acid is used. Typical values are 0.45-0.55 for nucleic acids, 0.71-0.75 for protein, and 0.9-1.1 for lipoproteins. Values of $\bar{v}$ for the individual amino acids and the method for making the weight average calculation are given by Schachman (1957).

3. VISCOSITY

The solution viscosity, η, is measured with a capillary viscometer by comparing the outflow time, t, of the solution with that for water (Schachman, 1959). The density must also be measured. The viscosity is calculated from

$$\eta = \left(\frac{t}{t_w}\right)\left(\frac{\rho}{\rho_w}\right)\eta_w$$

where the subscript w refers to water for which values of its viscosity at various temperatures are tabulated in handbooks and in Svedberg and Pedersen (1940).

III. MOVING BOUNDARY METHOD

A. Description

1. GENERAL

Figure 18-1 shows the schematic distribution in the cell and typical patterns recorded by the three optical systems during centrifugation. The purpose of the method is to separate solutes on the basis of their differences in velocity. As a by-product from ascertaining purity and s-rates, the mass concentration of each component can be determined.

If double-sector cells are used, then any type of syringe that will deliver identical volumes is required (e.g. Chaney, Hamilton). The solution is loaded first into one sector, then the same syringe is rinsed with the solvent and used to deliver solvent into the other sector. Two sizes of cells with optical path lengths of 12 and 30 mm are used routinely (others are available). Long path length cells are used for dilute solutions; they usually have a lower maximum speed, however.

2. SELECTION OF CONCENTRATION, SPEED, AND TIME

It is possible to use as little as 0.5 ml of a 100μg/ml virus solution in the 30 mm cell and get usable schlieren optical patterns even though the cell is less than half full. Normally, 0.7 ml of 1-10 mg/ml are used in the 12 mm cell.

The speed may be selected from as low as 1000 to as high as 60,000 rpm now, and 100,000 rpm in the near future. One should not think that the higher the speed the better, but should observe the pattern during acceleration. If in that period a boundary moves away from the meniscus more than 1/10 of the height of the fluid, then the operating speed is too high. For example, even with small plant and animal viruses it is necessary to operate under 30,000 rpm. When there is a mixture of components with a factor of ten difference in s-rates, after the faster components have moved to the bottom, the speed can be increased so that the time for the slowest component to do the same will not be excessive.

The effective start of centrifugation can be approximated by accelerating the rotor at a constant rate (by using a constant drive current in Beckman equipment) and starting the photographic timer and a running time clock at two-thirds of the final operating speed. Photographs taken at standardized true centrifugation times are much easier to compare when they are laid side by side on a light box for visual interpretation.

3. MAJOR VARIATION

An important variation of the moving boundary method uses a *synthetic boundary cell,* which permits layering dialyzate over solution to form a boundary. Such an initial boundary, located at any desired level in the cell, is in contrast to that formed in a conventional cell where it is at the meniscus between solution and air (for sedimentation) or at the bottom of the cell (for flotation experiments). The synthetic boundary cell procedure permits the

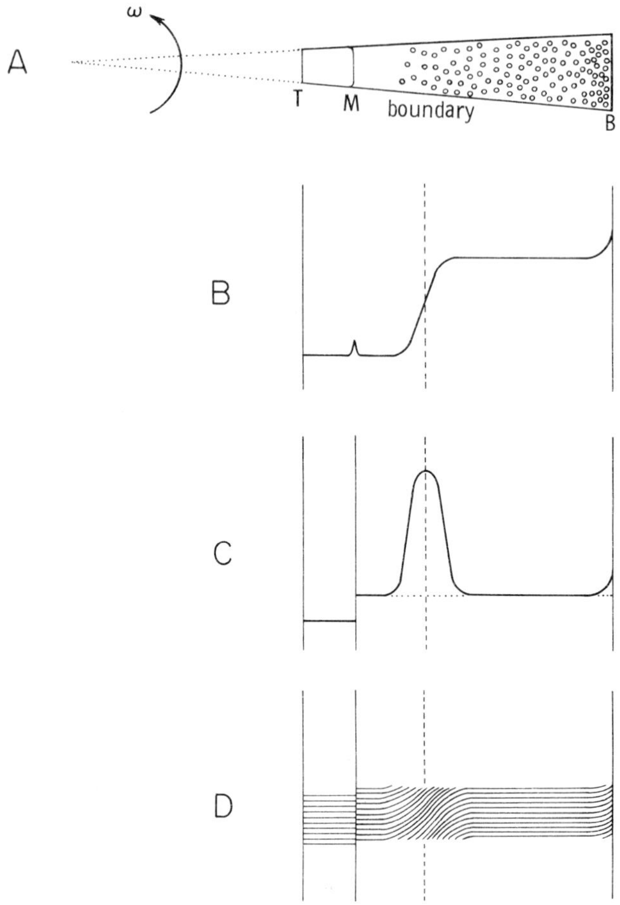

Fig. 18-1. Moving boundary method

A. Schematic sector shaped cell spinning about an axis perpendicular to the plane of the page with angular velocity **ω**. T is the top; B, the bottom; and M, the meniscus between air on the left and solution on the right. Initially, the solute molecules, indicated by circles, were uniformly distributed in the solution. At the time shown, the particles have moved to the right in the direction of the centrifugal force, bringing the initial boundary at the meniscus to the position indicated between the particle-containing solution and the supernatant fluid above it.

B. The absorption optics give concentration as a function of radius (*c* vs. *r*) with the meniscus indicated by a small sharp deflection.

C. The schlieren optics differentiate the distribution such that its derivative is given as a function of the radius (d*c*/d*r* vs. *r*). This reveals a "peak." The baseline that would be present with a double-sector cell is shown dotted.

D. The Rayleigh optics give multiple interference fringes representing concentration against radius. Following any one fringe gives *c* vs. *r*; but following a horizontal line, the intersection with successive fringes *n* gives *r* vs. $n\Delta c$, where Δc is the concentration difference between whole or half fringe intervals.

entire initial solute concentration to be determined, which is essential in purity measurements, especially on labile virus preparations.

With small molecules of s-rate less than unity and a high diffusion coefficient, the duration of the run is so long that diffusion interferes with the separation of the boundary from the meniscus. By using the synthetic boundary cell, the complete boundary can be seen from the beginning and thus can be measured precisely. The sample must be dialyzed, and dialysis must be stopped just prior to the synthetic boundary cell run. Otherwise, even slight differential evaporation can cause either a density inversion or a superimposed buffer boundary. Dialysis is especially important when high concentrations of urea or guanidine hydrochloride are used to dissociate or solubilize protein subunits. To study low concentrations of small molecules that cannot be dialyzed, almost unattainable precision is required in preparation of solutions. If ultrafiltration is used simultaneously with dialysis to concentrate the agent, then the pressure difference must be eliminated several hours before dialysis is stopped.

B. Qualitative Information

1. PURITY

The photographic records reveal immediately the number of components present since each will form a boundary. This number is a minimum, as several components can overlap or be unresolved because of the spreading of the boundaries by diffusion. A system is said to be paucidisperse if only a few distinct peaks are visible.

2. GUIDE TO FURTHER PURIFICATION

The biologically active component is identified by the agreement of analytical and preparative s-rates (Graves et al., 1968). The size of the boundary registration increases with the concentration of the component. By comparing plates taken before and after a purification step, one can readily observe which contaminants were reduced in amount relative to the major component of interest.

3. EFFECT OF CHEMICAL AND PHYSICAL AGENTS

Compare plates taken before and after the treatment. By lining up the menisci of corresponding frames of the two plates side by side on a light box one can see whether the treatment has altered the area of the schlieren peak and whether it has changed its rate of movement.

4. CHEMICALLY INTERACTING SYSTEMS

One must always be on the lookout for evidence of chemical reaction taking place during centrifugation. Thus, it is quite common for proteins to dissociate into subunits as their concentration is lowered, or viruses to aggregate as their concentration is increased. The quantitative aspects are indeed complicated but have been reviewed by Nichol et al. (1964) (see also Adams, 1967). The qualitative observations are given below.

(a) For a dissociating system, comparison of corresponding frames from a dilution series will show that the more concentrated boundaries move farther. This is opposite to the concentration dependence of independent systems where the s-rate increases with *dilution*.

(b) The shapes of boundaries may be peculiar. Thus, the skewness of the boundaries may vary with concentration and also with the solvent at the same concentration. Rapid reversible dimerization produces a unimodal (single peak) boundary, which may be skew. If the dimerization depends on a small molecular constituent (which does not show a peak), then the pattern may be bimodal. With such patterns, the schlieren curve does not reach the baseline between the peaks, and the absorption or Rayleigh pattern is not parallel to the baseline.

With two or more different solutes interacting to form a complex, convective patterns will result unless the complex has a higher sedimentation coefficient than either of the reactants. With these peaks, the pattern does not return to the baseline between the faster two, but may do so ahead of the slowest peak.

(c) Even though the concentration is increased, the slower peak remains approximately constant in size. This constancy is evidence of a monomer-oligomer reaction where the oligomer, which must be a trimer or higher, is favored. With independent components, the relative proportion of slower to faster peak size increases with concentration because of the Johnston-Ogston effect (Trautman, 1964).

(d) The s-rate increases with speed. This is impossible for a noninteracting system, since the speed factor in the velocity has been removed to give the s-rate, but when the equilibrium constant of a monomer-oligomer reaction depends on pressure, the oligomer is favored at higher pressures (higher speeds) and the peak has a higher s-rate (Kegeles et al., 1969). A non-pressure dependent variation of the s-rate with speed is possible with large and heterogeneous nucleic acid systems (Rosenbloom and Schumaker, 1967).

C. Quantitative Information

Some computational procedures are summarized in Protocol 1 of the appendix.

1. *SEDIMENTATION COEFFICIENT*

In order to get a parameter which is independent of the rotor or speed used, the velocity per unit field strength is computed. It is called the sedimentation coefficient (abbreviated as s-rate) and defined from a practical standpoint as follows. Let $\bar{r}$ be the position of the boundary at time t and r_a be the starting level, with ω the angular velocity. Then

$$s = \left(\frac{10^{13}}{\omega^2}\right) \frac{d \ln \bar{r}}{dt}$$

where the factor 10^{13} converts the result to Svedberg units, S, and $\omega = 2\pi$ (rpm)/60 in which (rpm) is the number of revolutions per minute.

The derivative is evaluated as the slope of a tangent to the ln $\bar{r}$ vs. t graph. Now, even though the plot is slightly curved, linear least squares curve fitting gives the (internal) s-rate at the mean time from the slope. Quadratic least squares curve fitting permits a statistical test for the curvature as well as yielding the initial s-rate, which has, however, a much larger standard error than the internal s-rate because of extrapolation to one end of the data. The intercept on the time axis is the zero time correction, that is, the time at which the boundary started from the meniscus. It should not be significantly different from zero if the timer and photographic mechanism were started as indicated above. A positive value indicates convection during acceleration and a negative value indicates that the cell was probably at a higher temperature than the rotor at the start. Significant zero time corrections can be obtained if the data are really curved and only linear least squares curve fitting is used. It should be noted that the s-rate does not depend upon calculation of the zero time correction, and data can be processed no matter when the timer was started. But the zero time correction is of great help in providing an additional point for those components that cannot be measured in every frame. It is important to compute standard errors since they reveal the internal consistency of the data and they permit application of statistical tests.

The boundary position, $\bar{r}$, is physically the location of the step in concentration formed if the mass were redistributed to remove the effect of diffusion in the boundary region. Theoretically, this is the square root of the second moment of the concentration gradient distribution about the center of rotation, if the meniscus concentration is zero, and can be expressed as

$$\bar{r} = \left[\frac{\int_{r_a}^{r_p} r^2(\partial c/\partial r)\, dr}{\int_{r_a}^{r_p} (\partial c/\partial r)\, dr} \right]^{1/2}$$

where r_p is a point ahead of the boundary region where the concentration does not change with distance (called the plateau). The boundary position, $\bar{r}$, is only slightly more radial than the position of the maximum gradient ("peak"), r_{max}. The standard errors are such, in practice, that the s-rate from r_{max} and $\bar{r}$ overlap at the 95% confidence level. Hence, it is better to introduce a little bias and use r_{max} in schlieren optics since many more frames can be read in less plate-reading time. With Rayleigh interference optics r_{max} corresponds to the level of closest fringe spacing—not easily measured—and with absorption optics to the half-concentration level—again, not too precisely located. Hence, with these two optical systems, several ordinates are measured in the boundary region and $\bar{r}$ is computed. As a by-product, a measure of the boundary width is also obtained. This will be considered under "boundary spreading."

Charged macromolecules will have a different s-rate in different solvents, with the lowest value in water if the macromolecule is soluble. An electrolyte at about 0.1 M concentration is needed to dampen the primary charge effect, which lowers the s-rate and which arises from a superimposed electrical force because of redistribution of the counter ion. A secondary effect is less pronounced and can be eliminated by using a solvent such as 0.1 M KCl, which can be weakly buffered near the isoelectric point if the agent is soluble and stable there.

To standardize reporting and to facilitate special calculations, the observed s-rate is converted to "standard conditions" of 20° C and water. It is denoted $s_{20,w}$ and calculated from

$$s_{20,w} = \left(\frac{1 - 0.99823\bar{v}}{1 - \bar{v}\rho}\right)\left(\frac{\eta_T}{\eta_{20,w}}\right)\left(\frac{\eta_{soln}}{\eta_w}\right)_{T'} s$$

where the first viscosity ratio uses handbook values for the viscosity of water at the temperature T and at 20° C, and the second is the ratio of buffer viscosity to that of water at any convenient temperature T′.

2. *MOLECULAR WEIGHT*

There is a great desire on the part of users of the velocity methods to convert an observed s-rate into a molecular weight. The feeling is that as the mass increases, the s-rate should also increase in some known way. Fundamentally, the s-rate depends on the combination of size and shape rather than mass of the moving solvated particle (the "kinetic unit"), and further, it is the conversion to weight that causes the difficulty. Three examples of the complication can be given. (1) Double-stranded RNA and DNA (with double the mass) both actually have a smaller s-rate than does the single-stranded molecule. The reason is a change in shape that drastically affects the frictional resistance. (2) An increase in mass by aggregation of a protein will increase the s-rate (a dimer will move 34% faster than a spherical monomer), but an increase in mass by combining with lipid may have little effect (e.g. the serum β-lipoprotein of about one million molecular weight has one-third the s-rate of the macroglobulin antibody of similar molecular weight). (3) A 1% solution of the rod-shaped tobacco mosaic virus has only about 75% of the s-rate of a highly diluted solution.

To make the conversion from s-rate to molecular weight, one should extrapolate observed s-rates to infinite dilution. Then, the following relations can be programmed for computing any desired parameter from the others, either as known or as assumed values. First, for size and shape

$$s = \frac{(2a_0)^2(\rho' - \rho)}{18\eta\,(f/f_0)}$$

where $2a_0$ is the diameter of a sphere with the same volume (as well as the same density and mass) as the solvated particle, η is the viscosity of the solution, ρ' and ρ are the densities of the solvated particle and solution, respectively, and f/f_0 is the friction factor, which increases from unity as the shape deviates from spherical. Expanded tables of f/f_0 for ellipsoids of revolution have been recomputed (Westley and Cohen, 1966). Second, to convert to mass, let the molecular weight, M, be the weight of Avogadro's number N of anhydrous particles. M_S be the solvated molecular weight, V' be the dry specific volume, and H be the specific solvation consisting of a fluid mantle of density ρ''. Then

$$M_S = (4/3)\pi a_0^3 \rho' N = M(1 + H\rho'')$$

$$\rho' = (1 + H\rho'')/(V' + H)$$

To compute the minimum anhydrous molecular weight, set $f/f_0 = 1$, $V' = \bar{v}$, and $H = 0$.

The solvated particle density changes with the density of the solution, since the solvation mantle changes its density (ρ''). If $\rho'' = \rho$, the additional centrifugal force on the mass of the solvation mantle is canceled by the increased buoyancy, but there is a change in friction because of the increased size that affects both s and the diffusion coefficient, D, to the same degree. For rigid unhydrated spheres, the above relations show s to be a function of $M^{2/3}$, but not for linear and coil molecules, such as the nucleic acids. Common empirical relations over short ranges do not extrapolate to zero and give the exponent as 0.4-0.6. Theoretical relations have been attempted (Gray et al., 1967), which reflect changes in shape and volume as the molecular weight increases.

3. CONCENTRATION

Sector-shaped cells are used so that sedimenting particles will not collide with the walls, and such cells are thus convection-free in the absence of inverse thermal gradients or misalignment. Since both the centrifugal field and the cross-sectional area change with radial distance, there is a dilution of the material as it sediments. The optical systems give the plateau concentration, c_p, but the initial concentration, c^0, is usually desired. This concentration is determined from the square law of radial dilution

$$c^0 = \left(\frac{\bar{r}}{r_a}\right)^2 c_p$$

With schlieren optics, this can be written

$$c^0 = \int_{r_a}^{r_p} \left(\frac{\bar{r}}{r_a}\right)^2 \left(\frac{\partial c}{\partial r}\right) dr$$

which can be evaluated by using the trapezoidal rule for numerical integration without knowing $\bar{r}$ explicitly.

4. BOUNDARY SPREADING

Two main phenomena contribute to boundary width: diffusion and heterogeneity of s-rate. The purpose of boundary spreading analysis is to separate the two effects and to determine whether the material is homogeneous or polydisperse. This latter term means that a continuous distribution of classes is present and a single, but broad, boundary that does not divide into separate peaks is observed. A complication results because the sedimentation coefficient is a function of the concentration, mainly because of a viscous friction effect. The effect on the boundary is to sharpen it, since particles in the trailing edge are more dilute and speed up. This artificial sharpening is by no means negligible. The simplest computational protocol is to assume a Gaussian shape for the boundary and to determine its midpoint and width from a weighted linear least squares fitting of appropriately transformed data (probit analysis). As a result, both the diffusion coefficient (from the slope) and the heterogeneity parameter (from the curvature) are obtained, as well as their standard errors. If the heterogeneity is not pronounced, the curvature will not be significantly different from zero, and a more critical requirement of homogeneity is that the computed diffusion coefficient must not differ significantly from its true value (Trautman et al., 1968).

IV. CLASSICAL EQUILIBRIUM METHOD

A. Description

1. GENERAL

Figure 18-2 shows the schematic distribution of macromolecules in the cell and typical patterns obtained with the three optical systems after equilibrium has been reached. The purpose is to redistribute the solute in such a way that the molecular weight can be determined from the *shape* of the pattern. Creeth and Pain (1967) have written an especially good and comprehensive review of this method. For practical details see both Van Holde (1967) and Richards et al. (1968). No qualitative information can be obtained from the records. Both versions of the method depend upon precise plate reading.

The short-column, low-speed version will be discussed first. A preliminary high-speed velocity run should be made to check that the material is pure enough to warrant the sedimentation equilibrium run. From the s-rate an estimate of the molecular weight can be made; this estimate is needed to select the optimal speed, and to provide the initial concentration (the velocity run is frequently of the synthetic boundary cell type). Samples should be dialyzed because: (1) The definition of components is otherwise ambiguous (Eisenberg, 1968); (2) a reference solvent is required since the long run causes considerable redistribution of the buffer itself; and (3) the synthetic boundary cell run requires it. A new and promising data processing method for homogeneous macromolecules can be used that does not require the initial concentration (DiCamelli et al., 1970).

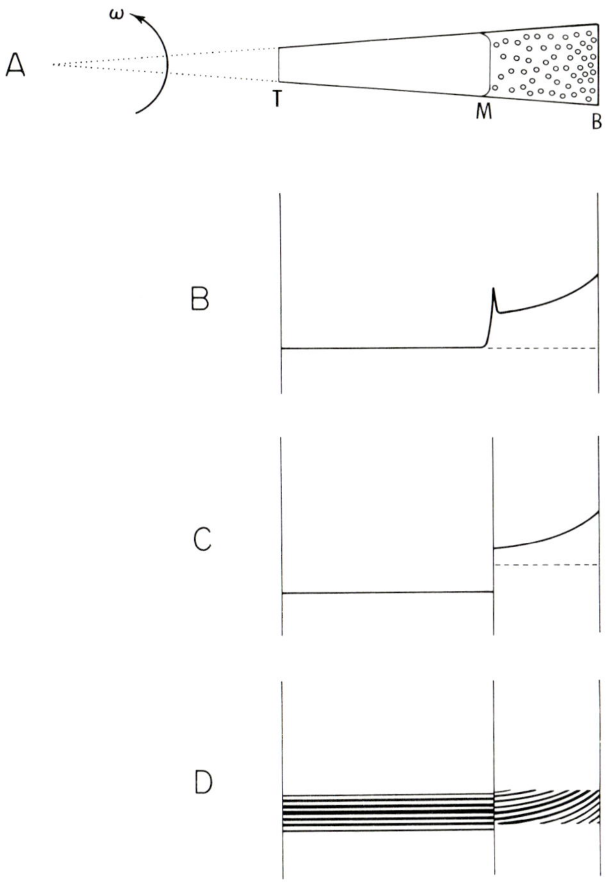

Fig. 18-2. Classical equilibrium method

A. Schematic cell as in Fig. 18-1A, except that the cell is usually about one-third full. Initially, the solute molecules, indicated by circles, were uniformly distributed. At the equilibrium time shown, they have redistributed such that there is no net transport across any level, since sedimentation and diffusion processes are balanced.

B. The absorption optics give concentration as a function of radius (*c* vs. *r*) with the meniscus indicated by a small sharp deflection, and the zero optical density by the dotted line.

C. The schlieren optics give the derivative as a function of radius (d*c*/d*r* vs. *r*). The baseline that would be present with a double-sector cell is shown by the dotted line.

D. The Rayleigh optics give multiple interference fringes representing concentration against radius. Following any one fringe gives *c* vs. *r*; but following a horizontal line, the intersection with successive fringes *n* gives *r* vs. $n\Delta c$, where Δc is the concentration difference between whole or half fringe intervals.

Recent work has shown that short columns of a few millimeters in height are adequate for determining both the molecular weight and the molecular weight heterogeneity (Van Holde and Baldwin, 1958). The reason for the short column is that the time to reach within a given closeness to equilibrium is proportional to the square of the column height. Overnight runs are frequently adequate and over-the-weekend runs are long enough for most protein and virus systems being studied. Because the runs are time-consuming, multicelled rotors and multichambered cells (Yphantis, 1964) are used that permit up to 12 samples to be run simultaneously in the scanning absorption optical system.

2. *MAJOR VARIATION*

An important variation of the classical equilibrium method is the *meniscus depletion*, or high-speed, version (Yphantis, 1964). In this method, the speed is chosen high enough that the concentration at the meniscus will drop essentially to zero, and yet not so high that all the material will be pelleted. Since measurements start at zero concentration, an absolute concentration scale is automatically established for the refractometric methods. A zero meniscus concentration is not necessary for the absorption optics, which always provides an absolute concentration scale. The meniscus depletion method is used for large particles, such as viruses (greater than one million molecular weight) that cannot be studied by the low-speed method because commercial centrifuges cannot be operated precisely at low enough speeds.

3. *SELECTION OF CONCENTRATION, SPEED, AND TIME*

For the Rayleigh optical system the concentration should be about 2 mg/ml for the low-speed and 0.5 mg/ml for the high-speed version; both versions require about 0.1 ml in the 12 mm cell, and double-sector cells must be used. For the absorption optical system solutions with an absorbance of one optical density unit or less are satisfactory. Double-sector cells are used with the scanning system. The method is generally not used with the film system.

A complete study of a particular substance will require many runs at several speeds and concentrations. The following equations can be rearranged to compute the speed for the first run. For the low-speed method the ratio of the final concentration at the bottom to that at the top, c_b/c_a, should be about 4 and so

$$\ln \frac{c_b}{c_a} = \left[\frac{\omega^2 M(1-\bar{v}\rho)}{RT}\right]\left(\frac{r_b^2 - r_a^2}{2}\right) = \left(\frac{\omega^2 s}{D}\right)\left(\frac{r_b^2 - r_a^2}{2}\right) \approx \ln 4$$

For the high-speed method a different relationship seems appropriate:

$$\frac{\omega^2 s}{D} = \frac{\omega^2 M(1-\bar{v}\rho)}{RT} \approx 5$$

Diffusion controls the time to reach equilibrium, which time for the low-speed version is:

$$t = \frac{0.7\,(r_b - r_a)^2}{D}$$

and about ⅓-¼ of that for the high-speed version. Plates taken one hour apart should be measured and processed. If the molecular weights overlap within one standard error, then equilibrium has been reached.

Several techniques involving multicompartmented cells are being explored. The possibility of layering solutions with these during acceleration has led to proposals to shorten the time to reach equilibrium by approximating the final distribution initially (DiCamelli et al., 1970). Alternative ideas include overspeeding followed by underspeeding. When these procedures are programmed for the desk-top computer so that the experimenter can select the proper conditions, they will certainly be much more useful. For the latest references, see Teller et al. (1969).

B. Quantitative Information

Some computational procedures are summarized in Protocol 2 of the appendix.

1. EQUILIBRIUM FORMULATION

The fundamental thermodynamic expression of sedimentation equilibrium for any and all levels is

$$\left(\frac{1}{rc}\right)\frac{dc}{dr} = AM^{\text{app}}$$

where

$$A = (1 - \bar{v}\rho)\omega^2/(RT)$$

$$M^{\text{app}} = M/(1+BMc+\cdots)$$

with $\bar{v}$ the partial specific volume, ρ the solution density, R the gas constant, T the absolute temperature, M the molecular weight, and B (double) the second virial coefficient. A fantastic number of manipulations of this basic relation have been made to provide ways of processing (and exploiting) the experimental data. Only a few will be discussed here.

2. MOLECULAR WEIGHT AND HETEROGENEITY

A measurement of molecular weight has no value without at least an estimate of the heterogeneity. In a mixture, each species follows the fundamental equation, but the different distributions are superimposed. The weight average molecular weight is generally defined as

$$M_w = \frac{\Sigma c_i M_i}{\Sigma c_i} = \frac{\int M\,dc}{C}$$

and the z-average molecular weight as

$$M_z = \frac{\Sigma c_i M_i^2}{\Sigma c_i M_i} = \frac{\int M^2\,dc}{\int M\,dc}$$

where $C = \Sigma c_i = \int dc$, and the limits of summation or integration extend over all the solute species of interest. The standard deviation of this distribution of molecular weights is called the molecular weight heterogeneity parameter, p_m, and is given by

$$p_m{}^2 = \frac{\Sigma(M_i - M_w)^2 c_i}{\Sigma c_i} = \frac{\int (M - M_w)^2 dc}{C} = M_w M_z - M_w{}^2$$

and its standard error by

$$\sigma p_m = \left\{ (M_w \sigma M_z)^2 + [(M_z - 2M_w)\, \sigma M_w]^2 \right\}^{1/2} / (2p_m)$$

Thus, the two averages M_w and M_z, can be combined to characterize the distribution in terms of its width, the heterogeneity parameter p_m. For a homogeneous solute $M_z = M_w$ and $p_m = 0$.

The amazing thing is that the centrifuge can yield both M_w and M_z by proper data processing. By writing the equation as $dc_i/dr = Arc_iM_i$ and summing over all species, it can be shown that

$$\frac{d \ln C}{d(r^2)} = \left(\frac{A}{2}\right) M_w$$

Thus, since a derivative is required, the weight average molecular weight can be evaluated from the slope of the linear least squares line through ln C vs. r^2. It can be further shown that

$$\frac{d \ln \left[\left(\frac{1}{r}\right) \frac{dC}{dr}\right]}{d(r^2)} = \left(\frac{A}{2}\right) M_z$$

and so the z-average molecular weight can be obtained from the slope of the linear least squares line through ln $[(1/r)(dC/dr)]$ vs. r^2 data. If M_w and M_z are not significantly different, then the sample is homogeneous at the midpoint of the r^2 values to which the slopes refer. If they are significantly different, then the solute is heterogeneous.

It may be of interest to compute the weight and z-averages for the original solution, not the solution that has a different composition because of the centrifugation. Such averages are denoted with a bar and called *whole cell* averages. They can be computed from the following formulae, but since least squares are not used the standard error is not simply obtained.

$$M_w = \left(\frac{2}{A}\right) \left[\frac{C_b - C_a}{C^0(r_b{}^2 - r_a{}^2)}\right] = \frac{(M_wC)_b - (M_wC)_a}{C_b - C_a}$$

$$M_z = \left(\frac{1}{A}\right) \left[\frac{\left(\frac{1}{r}\frac{dC}{dr}\right)_b - \left(\frac{1}{r}\frac{dC}{dr}\right)_a}{C_b - C_a}\right]$$

It will also be observed that both C and dC/dr are needed for the complete analysis into Mw, M_z, and p_m. Numerical differentiation of the interface or absorption patterns and numerical integration of the schlieren patterns are thus required. An attractive proposal for the refractometric methods is to take simultaneous schlieren and interference patterns, processing the former for the z-average and the latter for the weight average (Chervenka, 1966).

Since the equilibrium distribution is independent of the cross section, the above relations hold for rectangular, sector-shaped, and even circular cells (Yphantis, 1960). But the total mass does depend on the shape, and in order to use the initial concentration, C^0, there must be an integration over the cell contents. For the sector cell

$$C_a = C^0 - \frac{r_b^2(C_b - C_a) - \int_{C_a}^{C_b} r^2 dC}{r_b^2 - r_a^2} = C^0 - \frac{r_b^2 \int_{r_a}^{r_b} (dC/dr)dr - \int_{r_a}^{r_b} r^2(dC/dr)dr}{r_b^2 - r_a^2}$$

and for the rectangular cell

$$C_a = C^0 - \frac{r_b(C_b - C_a) - \int_{C_a}^{C_b} r dC}{r_b - r_a} = C^0 - \frac{r_b \int_{r_a}^{r_b} (dC/dr)dr - \int_{r_a}^{r_b} r(dC/dr)dr}{r_b - r_a}$$

These integrals can be evaluated using the trapezoidal rule.

An absolute concentration scale implies that the initial concentration, if not known, can be determined from the pattern. So it can be stated, generally, that to determine the whole cell weight average, the initial concentration must be known, whereas it need not be known to determine the z-average. The meniscus depletion version is a special case in that both the weight and z-averages can be obtained without knowing C^0. If the pattern is obscure at the cell bottom or lower meniscus, then this statement applies only to the smaller components that provide the visible pattern. The molecular weight of the smallest component can be obtained only if the next smaller component has sedimented out of the way.

With the heterogeneous model, it can be shown that

$$\frac{d^2 \ln C}{d(r^2)^2} = \left[\frac{\omega^2(1 - \bar{v}\rho)}{2RT}\right]^2 M_w(M_z - M_w)$$

which means that the difference between M_z and M_w is a measure of the curvature (indicated by the second derivative) of the ln C vs. r^2 plot used to compute M_w. This curvature will be positive if there is heterogeneity since $M_z > M_w$.

3. ACTIVITY COEFFICIENTS

One must be constantly on the alert for nonideality, that is, BM_c of the basic equation not being sufficiently close to zero. It can be shown for a single nonideal substance that

$$\frac{d^2 \ln c}{d(r^2)^2} = \left[\frac{\omega^2(1 - \bar{v}\rho)}{2RT}\right]^2 \frac{BM^3c}{(1 + BMc)^3}$$

Since B is usually positive, this equation means that nonideality will give a negative curvature to the ln c vs. r^2 plot. Thus, if the M_W and M_Z are computed from the least squares fitting in the way mentioned above, a simple test evolves: if $M_Z > M_W$, heterogeneity has a greater effect than nonideality; but if $M_Z < M_W$, nonideality has a greater effect than heterogeneity.

The simplest way to remove the nonideality is to perform a series of experiments at different concentrations. Determine either M_W or M_Z, calling either M^{app} for apparent molecular weight. Then plot $1/M^{app}{}_W$ vs. $(C_a + C_b)/2$ or $1/M^{app}{}_Z$ vs. $(C_a + C_b)$. The molecular weight is calculated from the intercept and the second virial coefficient from the limiting slope, which coefficient is related to the variation of the activity coefficient of the macromolecule with concentration.

4. CHEMICALLY REACTING SYSTEMS

Since chemical reactions take place in proportions based on numbers of molecules (molar concentrations) it can be anticipated that the number average molecular weight will be needed in a complete analysis. The number average molecular weight is not as directly measurable as is the weight or z-average (Yphantis, 1964). Also, since volume changes on a molar basis can be truly enormous for macromolecules, the pressure effects are dramatic. Thus, studies must be made at several speeds or with and without overlays of oil to permit the pressure to be varied. A recent conference had a large section devoted to studies of these problems and should be consulted for details (N.Y. Acad. Sci., 1969).

It has been shown that a double equilibrium is possible, that is, both the chemical reaction at each radius as well as the distribution with respect to radius can be at equilibrium. The reaction mechanism must be ascertained from other sources, for example, moving boundary method, then the reaction equilibrium constants and activity coefficients are computed from the sedimentation equilibrium distribution. The problem is that by increasing the total concentration in order to promote the chemical reaction the nonideality effects are also increased. The relationships are both complicated and nonlinear and have not as yet been investigated from the point of view of least squares for data processing.

As an example of the type of calculation required, consider that the reaction is known to be monomer-dimer ($2A \rightleftarrows B$), that the molecular weight, M, of the monomer is known and that there are no nonideality effects. The chemical equilibrium constant, K, is defined as

$$K = \frac{c_B}{c_A{}^2}$$

where concentrations, c, are in weight per unit volume. At each total concentration, C, in the cell, or for several experiments at different total concentrations, the weight average molecular weight, M_W, is determined as indicated above from the ln C vs. r^2 curve fitting. Then K is computed from

$$K = \frac{(M_W - M)M}{C(2M - M_W)^2}$$

If this equation is rearranged to express M_W as a function of C, it will be seen that M_W increases with C, that is, in the same direction as polydispersity, but the reverse of nonideality.

5. DIFFUSION COEFFICIENT

The time required to reach within a given closeness to equilibrium is diffusion-controlled. By processing measurements taken during the transient period, it is possible to determine the diffusion coefficient, D. Neither of two proposed variations (Van Holde and Baldwin, 1958; Sophianopoulos et al., 1962) have come into wide use, perhaps because the diffusion coefficient is not needed now that the molecular weight can be determined directly. The diffusion coefficient can be computed from the measured M and s using the Svedberg equation, which in its usual form is

$$M = \frac{RTs}{D(1 - \bar{v}\rho)}$$

The standard errors of the separate parameters are propagated according to

$$\sigma_M = M\left\{\left(\frac{\sigma_s}{s}\right)^2 + \left(\frac{\sigma_D}{D}\right)^2 + \left(\frac{\rho}{1 - \bar{v}\rho}\right)^2 \sigma_{\bar{v}}^{\,2}\right\}^{1/2}$$

It should be observed that heterogeneity of s, M, or D can occur independently of each other.

V. ISODENSITY EQUILIBRIUM METHOD

A. Description

1. GENERAL

Figure 18-3 shows the schematic distribution in the cell and representative equilibrium patterns for the two applicable optical systems: absorption and schlieren. The method depends on two separate equilibrium processes. First, a small salt molecule, typically CsCl, is used at high concentrations to increase the solvent density to approximately that of the macromolecule. The salt is too small to sediment completely; instead, it redistributes as in the classical equilibrium method. The change in concentration with radius produces a concomitant density gradient, called a "field-formed" gradient. Second, because of the density gradient, the macromolecules, which were uniformly distributed, sediment from the top and float from the bottom of the cell. At equilibrium they form a band located at their buoyant density. The final width of the band results from a balance between diffusion, which tends to broaden it, and the sedimentation-flotation, which tends to narrow it. The purpose of the isodensity method is to separate solutes on the basis of their differences in density, and in special cases to determine the molecular weight from the width

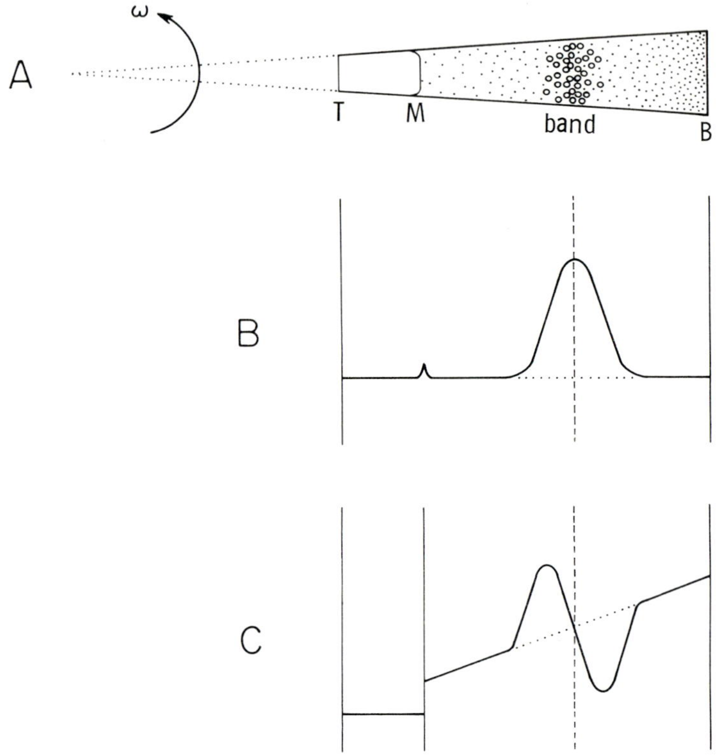

Fig. 18-3. Isodensity equilibrium method

A. Schematic cell as in Fig. 18-1A. Initially, both the large solute molecules, indicated by the open circles, and the small solute molecules, indicated by the dots, were uniformly distributed. At the equilibrium time shown, the small molecules are distributed as in classical equilibrium producing a density gradient in which the large molecules have banded.

B. The absorption optics, if set to a wavelength that does not detect the small molecules, give the concentration against radius distribution of the large molecules (and the sharp deflection at the meniscus). The continuation of the zero optical density line is shown dotted under the zone or band.

C. The schlieren optics give the concentration derivative with respect to the radius of both solutes and give the sloping line for the small molecule and the superimposed biphasic curve for the large molecule. Rayleigh optics cannot be used since there is too great a concentration difference because of the small molecule gradient.

of the band. Because so much is accomplished by the centrifuge from a simple loading of a uniform solution in the cell, this field-formed gradient isodensity equilibrium method is considered one of the most elegant. For a comprehensive review see Vinograd and Hearst (1962). Szybalski (1968) gives additional details for the photographic absorption method.

2. SELECTION OF CONCENTRATION, SPEED, AND TIME

Note that if the material is stable, all the original macromolecules present contribute to the pattern, in sharp contrast to the moving boundary method where only those in the boundary region do so. Thus, the equilibrium isodensity method requires only micrograms of viral materials. If double-sector cells are used, the filling of the solvent side must be done precisely by using the same (Chaney) syringe that was used for the solution.

The time to reach equilibrium for the salt is independent of the speed, and is about 8 hr for a 1 cm CsCl column. The time for the macromolecule depends inversely on the s-rate (in the absence of a gradient) and also inversely on the fourth power of the speed. For viruses, the salt redistribution is the limiting factor, but for proteins and nucleic acids, it is the macromolecule (requiring overnight or over-the-weekend runs). The higher the speed, the steeper the density gradient and the larger the density difference viewed in the run. Resolution is independent of the speed because the bandwidths are also affected, but it varies with different salts because they produce different density gradients at the same nominal density and speed.

3. OPTICAL COMPLICATION

The density gradient is also a refractive index gradient, which deflects the light so much that negative wedge (prismatic) windows may be needed to compensate. This is particularly so for Cs_2SO_4 (Ludlum and Warner, 1965; Szybalski, 1968). In extreme cases, especially at high speed in thick cells with any salt, in addition to the wedge windows, the light source can be moved off axis to bring the deflected light back into the optical system.

B. Qualitative Information

1. PURITY

The number of bands is a direct measure of the number of components of differing buoyant densities. This number is a minimum because there can be overlapping of zones or lack of resolution because of diffusion.

2. GUIDE TO FURTHER PURIFICATION

By inspection, one can see if unwanted components are reduced after each purification step. The size of the band is a measure of the mass of material present.

3. EFFECT OF SALTS

There is sometimes a different number of bands of the same material in different salt gradients. The effect is due to a heterogeneity of binding sites for the various salt ions (Magdoff-Fairchild, 1967). The solubility of some viruses and other macromolecules is lowered in concentrated salts, and any density shift of the precipitate band compared to that of the soluble component can, if it occurs, be detected.

4. EFFECT OF CHEMICAL AND PHYSICAL AGENTS

The effect of pressure can be ascertained by adding a layer of oil and recentrifuging. Comparison of the positions of the band will show whether the added pressure due to the oil shifts the buoyant density. Similarly, the effects of heating, freezing, or chemical treatment can be seen by the shift in band position or by the reduction or elimination of bands.

C. Quantitative Information

There are many corrections and conversions necessary to obtain meaningful parameters from the raw data. Even though the standard error analysis for propagation of errors has not yet been made, the computer programs do provide an orderly way to make the conversions. Some computational procedures are summarized in Protocol 3 of the Appendix.

1. BUOYANT DENSITY

The *composition density gradient* is determined from the known sedimentation equilibrium distribution for the salt in question and which, fortunately, does not depend upon the pressure with less than the 300 atmospheres encountered. For the commonly used salts, numerical solutions have been tabulated as a function of the density in a manner which is independent of the volume or speed used (Trautman, 1960; Ifft et al., 1961; Ludlum and Warner, 1965; Ifft et al., 1970). The effect of temperature has not been adequately studied. In order to use the solutions, the density at atmospheric pressure at one point must be known. Let $1/\beta^0$ be the tabular value for a particular marker of density $\rho_m{}^0$ located at $r=r_m$; then, the composition density gradient is

$$\left(\frac{d\rho}{dr}\right)_{r_m}^{0} = \frac{\omega^2 r_m}{\beta^0}$$

where the superscript zero indicates composition. The notation β^0 is used for historical reasons: actually, $1/\beta^0$ is more useful. The composition density at another level is

$$\rho^0 = \rho_m{}^0 + \int_{r_m}^{r} \left(\frac{d\rho}{dr}\right)^0 dr$$

by representing $1/\beta^0$ as a function of ρ by a polynomial, the composition density and density gradient at any level at any speed can be computed. The solution composition density at band center is the buoyant density of the macromolecule, $\rho_0{}^0$ where the subscript zero means band center.

If no marker is available, then the isoconcentration point, r_e, can be used.

This is the level where the concentration, and hence the density at atmospheric pressure, is equal to that of the initial solution and is commonly called the "hinge point." The equation is

$$r_e^2 = \frac{r_a^2 + r_b^2}{2} + \frac{d(1/\beta^0)}{d\rho^0} \frac{\omega^2(r_b^2 - r_a^2)^2}{48}$$

where the derivative can be evaluated either graphically or from the polynomial approximation of the $1/\beta^0$ vs. ρ^0 function. The square on the second term was inadvertently omitted in the original paper (Ifft et al., 1961) and in a subsequent review (Vinograd and Hearst, 1962). The first term is called the "root-mean-square position" and is an approximation to the true isoconcentration level. A very good practical way for determining buoyant densities, whether or not the $1/\beta^0$ function is known, is to find empirically the solution in which the macromolecule is banded at the isoconcentration level. By using very short columns at low speed the correction to the root-mean-square level is negligible. A less attractive proposal is to use the meniscus as a marker (Ludlum and Warner, 1965).

2. PRESSURE DEPENDENCE

The formula for the pressure, P, at any level takes into account the compressibility as well as the fact that the density changes with composition in the aqueous column (Hearst et al., 1961; Ifft and Vinograd, 1966). With an overlay of oil, this pressure is the sum of that at the aqueous meniscus due to the oil plus that due to the aqueous column

$$P = \left[\frac{(\rho_{oil}{}^0)\omega^2}{1 - \varkappa_{oil}\,(\rho_{oil}{}^0)\,\omega^2(r - r_a^2)/(4\ell)}\right]\left(\frac{r_{oil}{}^2 - r_a^2}{2\ell}\right) + \rho_\alpha{}^0\omega^2\left(\frac{r^2 - r_a^2}{2\ell}\right)$$

where r is the level at which the pressure is computed; r_a, the aqueous-oil interface; r_{oil}, the oil meniscus; $\rho_{oil}{}^0$, the density of the oil at atmospheric pressure; $\rho_\alpha{}^0$, the composition density at the root-mean-square level r_α between r_a and $r (r_\alpha = [(r_a{}^2 + r^2)/2]^{1/2}$, and $\varkappa_{oil}$, the isothermal compressibility of the oil. The conversion factor ℓ is used because atmospheres are practical units for P and so are $atmos^{-1}$ for $\varkappa$.

The approximations involved are such that it is not necessary to take into account the compressibility of the aqueous solution, which increases the density at r_α slightly. However, for purposes of binding and molecular weight calculations, the physical density at band center r_0 is needed

$$\rho_0 = \frac{\rho_0{}^0}{1 - \varkappa P}$$

where $\varkappa$ is the isothermal compressibility of the aqueous salt solution.

The buoyant density will be found to be a linear function of the pressure in the actual run, whether the elevated pressure is caused by an overlay of oil or

VI. OTHER METHODS

A recent technique which may prove useful in virology is *band centrifugation* (Vinograd and Bruner, 1966; Markham, 1967; Studier, 1969). It is a variation of the moving zone method — the fourth fundamental method of centrifugation. It requires the scanner absorption optics so that both the very dilute sample can be detected and the density-gradient material will not contribute to the pattern. The sample is layered on a slightly more dense solution (such as 50% D_2O-H_2O in buffer) by flow through a capillary during acceleration, as in the synthetic boundary cell. The solvent boundary diffuses more rapidly than the band sediments so that there is always a slight density gradient for stabilization. Since the bands broaden rapidly and become skew if there is concentration dependence, it is difficult to assess homogeneity. However, the method may be valuable in detecting viruses in crude extracts, because they will move ahead of the essentially opaque debris.

By recognizing that no material can leave the cell, it is possible to write equations for the balance of sedimentation and diffusion at two special levels: the top and the bottom. These equations can be considered as describing the moving boundary method while the boundary is separating from the meniscus, or as describing the approach to classical equilibrium. Procedures for utilizing these equations to obtain a molecular weight are collectively called the *Archibald method.* It is mentioned only because it has not yet been completely replaced by the short-column, classical equilibrium method. Creeth and Pain (1967) give an excellent review of its applications.

The scanner absorption optics have been used to study preferential interactions of proteins with small molecules (Steinberg and Schachman, 1966) and for measuring the partial specific volume by classical equilibrium in two solvents of different densities (Edelstein and Schachman, 1967).

Even though Brownian motion is involved in all centrifuge experiments, it is usually not measured. However, by arranging for a boundary to spread more by diffusion than by centrifugal effects, it is possible to use the optics of the analytical ultracentrifuge to determine diffusion coefficients of proteins, nucleic acids, and viruses. For a recent application, see Burness and Clothier (1970).

VII. APPENDIX

A. Glossary of Symbols

a	Subscript meaning air-solution interface or meniscus.
d'	Thickness of the cell along the optical path in centimeters (1.2 or 3.0).
a_0	Radius of a sphere of the same volume as a solvated particle in centimeters.
A	$(1-\bar{v}\rho)\omega^2/(RT)$
b	Subscript meaning bottom of cell.

b'	Optical lever arm in centimeters (~60).
B	Twice the second virial coefficient.
c	Concentration in weight per volume units.
c^0	Initial concentration.
c_i	Concentration of species i.
C	Total concentration of all solutes.
D	Diffusion coefficient in square centimeters per second or in Fick units, F.
E	Extinction coefficient in optical density units per concentration units per centimeter for absorption optics.
F	10^{-7} cm^2/sec, one Fick unit.
f/f_0	Friction factor for solvated particle.
H	Specific solvation in milliliters per gram of dry particle.
j, J	Fringe number in Rayleigh optics.
k	Number of fringes per comparator unit measured perpendicular to a radius in the image of the reference hole or air space.
K	Mass action equilibrium constant.
ℓ	Conversion of pressure units, 1.01325×10^6 g cm^{-1} sec^{-2}/atmos.
m	Subscript meaning marker of known buoyant density.
M	Molecular weight.
M^{app}	Apparent molecular weight neglecting corrections for nonideality.
M_c	Cylindrical lens magnification (~3).
M_e	Enlarger or comparator magnification including dimensional change (commonly, 10 mm/cm).
M_0	Camera lens magnification (~2).
M_w	Weight average molecular weight of heterogeneous solute.
M_z	z-Average molecular weight of heterogeneous solute.
M_w, M_z	Whole cell average molecular weights.
n	Number of fringe increments in Rayleigh optics.
N	Avogadro's number, 6.023×10^{23} particles/mole.
^{0}	Superscript meaning "initial" or at infinite dilution; for isodensity method it means "composition" of gradient material.
$_0$	Subscript meaning band center in isodensity method.
p	Subscript meaning plateau region in moving boundary method where concentration is not a function of radius but is a function of time.
p_m, p_s	Heterogeneity parameter, which is the standard deviation of concentration distribution; m for molecular weight and s for s-rate heterogeneity.
P	Pressure above atmospheric in atmospheres.
r	Distance from the center of rotation, i.e., radius in centimeters (5.7-7.3).
$\bar{r}$	Position of moving boundary independent of shape of concentration profile.
r_e	Isoconcentration level in isodensity method where concentration is equal to the initial concentration of gradient solute.
r_{max}	Position of maximum ordinate or peak in schlieren optics, or level of closest fringe spacing in Rayleigh optics, or half-concentration level in absorption optics.
r_{ref}	Radius of rotor reference edge, possibly the average of two reference edges (for Beckman equipment: 7.3 cm for the outer, 5.7 cm for the inner; 6.5 cm for the average of the two edges, or 7.4 cm for the average of the rotor reference hole edges).

R	Gas constant, 8.3143×10^7 ergs/deg/mol.
R_1	Specific refractive increment (0.000186 ml/mg for bovine serum albumin).
(rpm)	Speed in revolutions per minute.
s	Subscript meaning solvated.
s	Sedimentation coefficient or s-rate in seconds or Svedberg units, S.
$s_{20,w}$	s-Rate corrected to standard conditions of 20° C in water.
$s_{20,w}^0$	Corrected s-rate extrapolated to zero concentration of macromolecule.
S	10^{-13} sec, one Svedberg unit.
t	Time in seconds.
T	Temperature in degrees absolute (273.16 + temperature in degrees Celsius).
$\bar{v}$	Partial specific volume in milliliters per gram.
$\bar{v}_1$	Partial specific volume of solvent, usually water (~1).
V'	Dry specific volume in milliliters per gram.
w	Subscript meaning water.
X	Distance on the optical record in the radial direction (abscissa) in comparator units or scale units of the graph paper.
X_{ref}	Abscissa of rotor reference edge on the optical record.
Y	Distance on the optical record perpendicular to radial direction (ordinate) in comparator units or scale units of the graph paper.
Y_{BL}	Ordinate on the optical record of the baseline.
z	Weight fraction of bound salt to macromolecule; as subscript refers to the bound salt.
α	"Nameless parameter" in effective density gradient calculations (~0.2).
α	Subscript meaning root mean square level between two given radii.
$1/\beta^0$	Density gradient function due to composition of gradient material, $(d\rho/dr)/(\omega^2 r)$ ($\sim 8 \times 10^{-10}$ for CsCl at $\rho = 1.7$).
Γ'	Net hydration of salt-free macromolecule in grams H_2O per gram macromolecule to give buoyant density.
Γ_*'	Net hydration of salt-binding macromolecule in grams H_2O per gram macromolecule to give buoyant density.
η	Viscosity in poise (0.010050 for H_2O at 20° C).
θ	Phaseplate angle with respect to the light source image in schlieren optical system.
$\varkappa$	Isothermal compressibility in atmospheres^{-1} (for H_2O at P = O, $\varkappa = 4.52 \times 10^{-5}$ and decreases to 4.30×10^{-5} at P = 200 atmos; for aqueous CsCl at $\rho = 1.3$, $\varkappa = 3.42 \times 10^{-5}$).
λ	Wavelength of light in centimeters (5.461×10^{-5} cm in Rayleigh optics).
ρ	Density of solution in grams per milliliter.
ρ'	Density of solvated particle in grams per milliliter.
ρ''	Density of solvation mantle in grams per milliliter.
σ	Standard error or standard deviation of band in isodensity method in centimeters.
Ψ	Pressure dependence of buoyant density that involves differential compressibilities of macromolecule and solvent in isodensity method in atmospheres^{-1}.
ω	Angular velocity in radians per second.

B. Data Processing Protocols

Typical flow charts for some of the procedures for the three major ultracentrifugation methods discussed in this chapter are illustrated as Protocols 1-3. These indicate the order of computations for the desk-top computer. The PI- numbers in the various blocks refer to programs written for the Olivetti Underwood Programma 101 according to the formulas given in this chapter.

1. MOVING BOUNDARY METHOD

The optical records are converted as indicated in the first computational block of Protocol 1. The linear least squares slope of ln $\bar{r}$ vs. t is computed for the s-rate in PI-1032 and PI-1041. The intercept is used in PI-1032 for the conventional cell to give the time at which the boundary started from the meniscus, but with the synthetic boundary cell it gives the level at which the boundary started. Curvature of the ln $\bar{r}$ vs. t graph because of concentration or pressure dependence is checked with PI-1072, which provides a quadratic least squares fit of ln $\bar{r}$ vs. t. A data editing plot can be made with PI-1019, and conversion of observed s-rates to standard conditions, $s_{20,w}$, with PI-1031. The concentration is determined by trapezoidal integration of schlieren patterns in PI-1021, and plateau levels are converted to initial concentrations for interference or absorption optics with PI-1079.

Additional programs are required for the boundary spreading analysis.

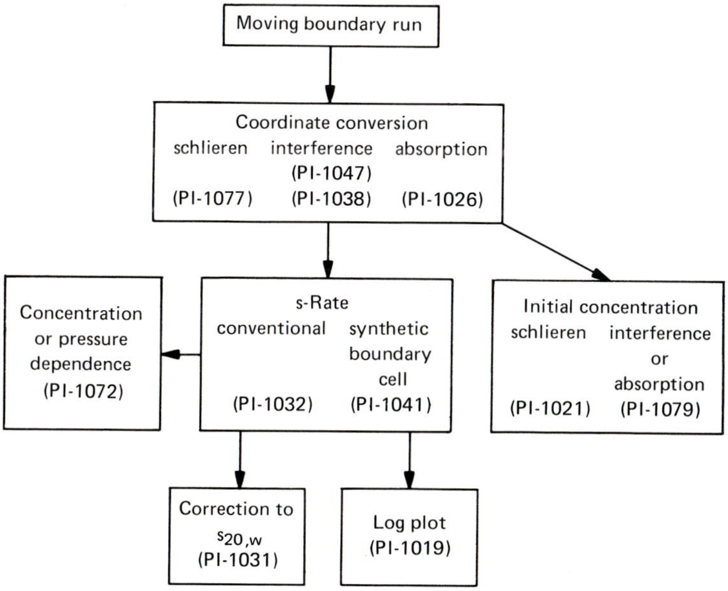

Protocol 1. Moving boundary method.

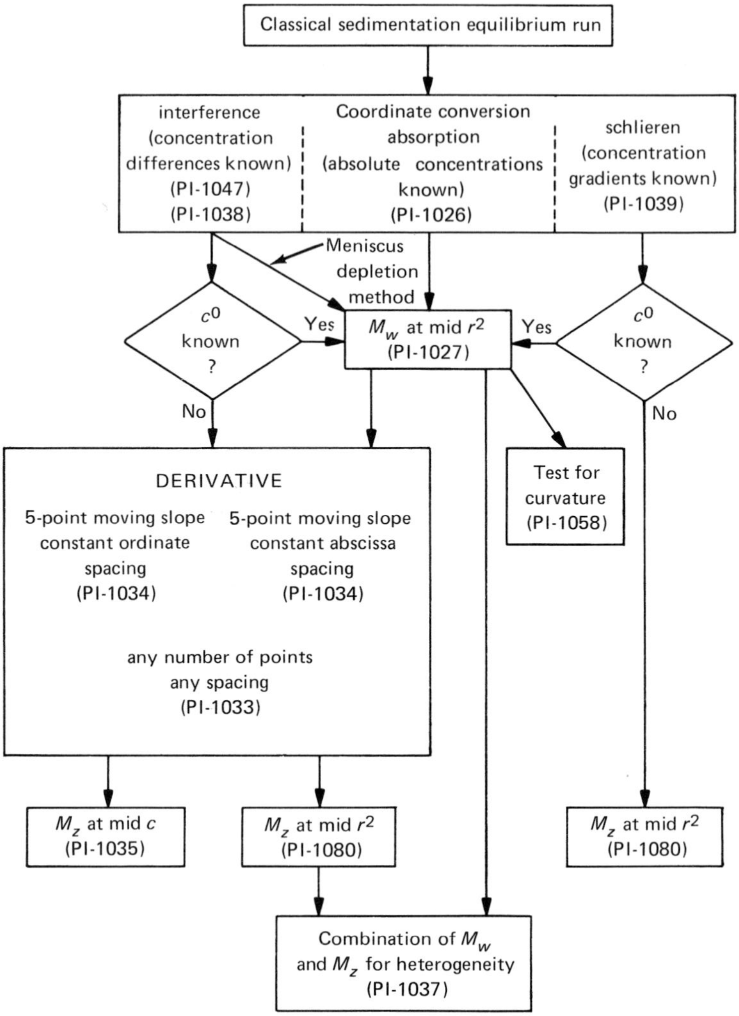

Protocol 2. Classical sedimentation equilibrium method.

2. *CLASSICAL SEDIMENTATION EQUILIBRIUM METHOD*

The center route down Protocol 2 is the most straightforward procedure, for it uses data from the absorption optics system, which has an absolute concentration scale. Start with the scanner coordinate conversion (PI-1026). Then, the main program (PI-1027) is linear least squares fit of ln c vs. r^2, which gives from the slope the weight average molecular weight, M_W, at the midpoint of the r^2 values used. Quadratic least squares (PI-1058) can be used to test for curvature, but a more sensitive check for heterogeneity uses the z-average molecular weight, M_Z, at the same level. To compute this, differentiate the

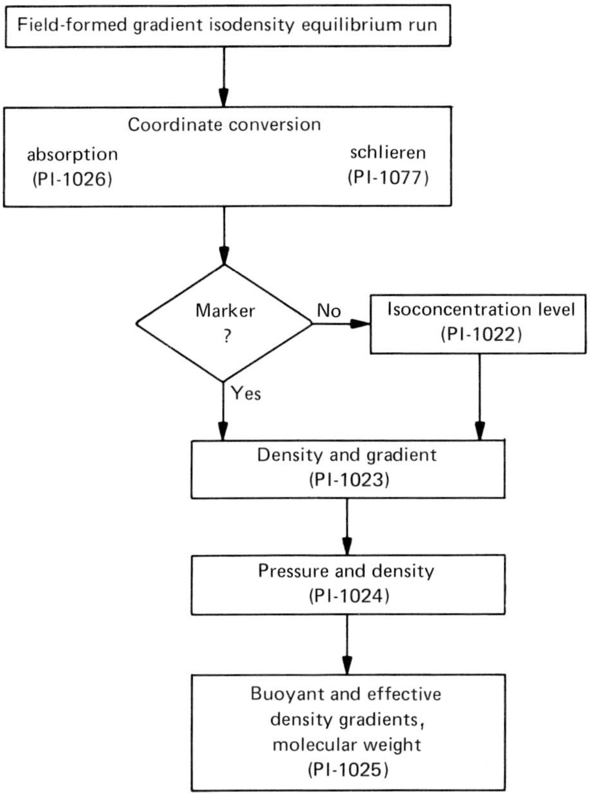

Protocol 3. Isodensity equilibrium method.

concentration profile using either PI-1034 for equally spaced ordinates or PI-1033 for any spacing. Then determine the linear least squares slope with PI-1080 of ln $[(dc/dr)]/r$ vs. r^2, which gives M_Z at the midpoint of the r^2 values used. Inspection of the M_W and M_Z values with their standard errors will reveal if they are significantly different. If so, combine them in PI-1037 to give the heterogeneity parameter with its standard error.

The (high-speed) meniscus depletion method with Rayleigh optics can be processed in the same manner as indicated by the diagonal line from the upper left-hand coordinate conversion box. But, low-speed equilibrium followed by these optics requires additional information. If the initial concentration is known and no solute has precipitated at the bottom meniscus or has been adsorbed to the walls, then PI-1038 yields the coordinate conversion as well as the upper meniscus concentration by integration over the cell. With this information the main route (PI-1027) can be joined as indicated by the diamond at the left. The alternate route down the left-hand side without using c^0 is to differentiate (PI-1034 or PI-1033) and then to compute M_Z (at the mid-value of the concentrations represented) from the slope of the least squares curve fitting of $(dc/dr)/r$ vs. $(c - c_a)$ in PI-1035.

With schlieren optics data and both the knowledge of c^0 and the assumption of no loss of solute, PI-1039 is used to convert the coordinates and to give the meniscus concentration. Then the main route, PI-1027, can be joined as indicated by the diamond at the right. Alternatively, without assumptions, the concentration gradients yield M_Z at the midpoint of the r^2 values from the slope of the least squares fit of $\ln [(dc/dr)/r]$ vs. r^2 indicated in PI-1080, which is redrawn at the lower right to avoid crossing of flow lines.

Additional programs are required for whole cell average molecular weights.

3. *ISODENSITY EQUILIBRIUM METHOD*

Absorption or schlieren optics are used with coordinate conversion programs PI-1026 and PI-1077, respectively, as indicated in Protocol 3. If there is a marker, its density is known and the composition density and density gradient at any other level are determined with PI-1023. This is all that is necessary for buoyant density determinations. If the effect of pressure is to be considered, with or without an overlay of oil, use PI-1024. If additional information about selective binding is known, the effective density gradient can be calculated and then the molecular weight from the bandwidth (PI-1025). If there is no marker, the hinge point or isoconcentration level in the cell that has a density equal to the starting density can be determined from PI-1022. It then constitutes a marker for the main branch of calculations.

PART FOUR

PHOTOBIOLOGY AND MUTATION

19

Effect of Radiation on Viruses

MEYER CHESSIN

Department of Botany
University of Montana
Missoula, Montana

I. INTRODUCTION

In this chapter the term "radiation" is used to include both ionizing and nonionizing wavelengths. The emphasis will be on the direct, mainly in vitro effects, although, because the subject has not been reviewed recently, radiation effects on host plants, both before and after virus infection, will also be considered.

Much of the early work dealing with radiation and viruses employed the more available plant viruses such as the tobacco mosaic virus (Lea, 1947); more recently some animal viruses and bacteriophage have been isolated in highly purified form and have been included in radiation studies. Considerable interest has also developed in radiation effects on virus-infected mammalian cells, as any recent copy of *Virology* bears witness.

Certain aspects of this subject have been reviewed recently (Kleczkowski, 1968*a*; Ginoza, 1968). In his review, Kleczkowski stresses techniques of studying ultraviolet effects and one is referred to some of his earlier reviews mentioned there for further theoretical considerations. I shall therefore summarize the effects of ultraviolet light (UV) on plant viruses. For a more detailed technical analysis of this problem, the reader is referred to the recent review by Kleczkowski (1968*a*). For those wishing an introduction to the field of UV photobiology, a recent publication by Jagger (1967) is recommended.

II. IONIZING RADIATION

A. Physical Considerations

This may be considered to include electromagnetic energy with a wavelength of less than 1000 Å (e.g. X-rays and γ-rays) as well as charged and uncharged particles with energies, generally above 10 ev, sufficient to cause ionization when striking matter (e.g. α-rays and β-rays, protons, and neutrons).

Ginoza (1968) has presented the theory of the physics and biochemistry of ionizing radiation as it pertains to this subject. The following is a briefer summary mostly derived from his material.

Because these radiations act by releasing a large amount of energy in a small volume, biological and chemical effects can be expected to be more profound than is true for radiations such as ultraviolet light, which generally produce excitations only. According to Ginoza the mechanism of action of ionizing radiation involves the ejection of an electron from the target material, with the consequent production of a positively charged site. In biological systems the positive charge tends to migrate to more stable sites such as the ring structures of nucleic acid bases. Unless the charge is neutralized in sufficient time by free or H-bound electrons, profound intramolecular chemical change results. In the case of plant viruses, a single such chemical change anywhere in the nucleic acid chain can be lethal. Since it has been amply demonstrated by now that the nucleic acid is the biologically important component of viruses, it is not surprising that infectivity of viruses can be altered without causing significant changes in other properties such as antigenicity and solubility, which depend upon the protein coat (Bawden, 1964*a*).

In general, the "stopping power" of matter for ionizing radiation is related to the average atomic number. For this reason H_2O, non-bony tissues, and viruses will be traversed to essentially the same extent. Also, charged particles will be less penetrating than uncharged particles or X-rays and γ-rays, but this property can be useful in thickness and sensitive cross-section determinations.

B. Direct vs. Indirect Action

Since water makes up such a large proportion of living cells, and since many chemically reactive products are formed from water upon radiation treatment, it is important to consider whether effects on viruses in aqueous solutions or living cells are due to direct effects of the radiation on the virus particles themselves or whether these effects are mediated, even in part, by reactive chemical species derived from H_2O. Purine and pyrimidine bases of nucleic acids have been shown to be more readily affected by ionizing radiation than the pentose phosphate group in aqueous solution, which may be a reflection of the greater accessibility of the bases to the free radicals produced upon irradiation (Scholes et al., 1960). It is theoretically possible to eliminate indirect effects by irradiating the material while frozen, or by dehydrating the preparation where this is feasible (Ginoza and Norman, 1957).

The presence of oxygen is also important, since oxygen can apparently supply an electron to stabilize a chemical change resulting from electron ejection. The use of nutrient broth and gelatin also reduce indirect effects, presumably by competing with the target for free radicals. In addition, the use of high concentrations of target organisms would be expected to reduce indirect effects (Lea, 1947).

More recently, it has been shown that certain chemicals may actually enhance damage (Ginoza, 1963). Even in frozen preparations, phosphate salts increased damage tenfold. NaCl, Na acetate, and chloroform were also effective.

C. Host-Cell Reactivation

Much literature has accumulated concerning host-cell repair of UV-induced lesions to the cell itself, or to irradiated viruses subsequently inoculated to host cells (Jagger, 1958). Various treatments, especially post-UV exposures to visible light, are effective. Until recently, attempts to demonstrate such repair after exposures to ionizing radiation were unsuccessful (Ginoza, 1968).

Recently, Fluke and Beinecke (1967) reported repair of γ-irradiated bacteriophage. Consequently, it would appear worthwhile to reinvestigate this question using as wide a variety of host cells and viruses as possible to increase the possibility of finding organisms or viruses capable of reversing damage caused by ionizing radiation.

D. Dose-Response Curves (Target Theory)

Generally, where experiments are carefully done (e.g. Ginoza's work with three single-stranded viruses and the free nucleic acids of tobacco mosaic virus and the coliphages R17 and ϕX174) the dose-response curve gives a straight line when the logarithm of survival is plotted directly against dose. That is, first-order kinetics is followed. By most, this result is assumed to be due to inactivation being caused by a single "hit" anywhere within the sensitive volume of the system. In the case of the viruses mentioned, the sensitive volume has been assumed to be identical with the single nucleic acid chain. This "one-hit" hypothesis was first developed as an experimental tool by Lea and colleagues (Lea, 1947).

Target volumes calculated on the basis of "one-hit" inactivations by ionizing radiation agree rather closely with the known proportion of viruses occupied by nucleic acid, supporting the "one-hit" hypothesis. As Kleczkowski has argued, similar assumptions may not be valid for inactivation by ultraviolet light (Kleczkowski, 1960*a*).

III. APPLICATIONS OF IONIZING RADIATION TO PLANT VIRUSES

A. In Vitro Treatment

1. DOSE-RESPONSE CURVES

Apparently without exception, within experimental error, straight-line exponential curves are obtained in survival experiments. Mostly all this work was done with tobacco mosaic virus and its nucleic acid, but Lea (1947) also obtained such curves for intact tobacco necrosis, tomato bushy stunt, and potato X viruses.

2. MODIFICATION OF RADIOSENSITIVITY

Various treatments have been shown to decrease sensitivity of plant viruses to ionizing radiation. Lea (1947) was the first to demonstrate that impurities in preparations of tobacco necrosis and tomato bushy stunt viruses protected them against irradiation damage, so that there was no difference in response whether preparations were wet or dry. Specific compounds known to act as protective agents are gelatin, which increased the resistance of tobacco mosaic virus in aqueous solution essentially to that of a dried preparation of this virus (Lea, 1947), and reduced glutathione, which increased the resistance of tobacco mosaic virus-RNA twofold to X-ray inactivation (Ginoza, 1963).

As mentioned previously, an interesting newer development is the enhancement of sensitivity by the addition of phosphate to the medium, although tobacco mosaic virus-RNA was much less affected than coliphages R17 and ϕ X174 (Ginoza, 1963). Ginoza suggests this difference may be due to the self-protection afforded by the high concentrations (1 mg/ml) of tobacco mosaic virus-RNA required for the assay. This may also explain why tryptone broth failed to protect tobacco mosaic virus in his experiments. However, although concentrations of nucleic acid required for bioassay are well above those which Lea demonstrated resulted in self-protection for tobacco mosaic virus against indirect effects, it is possible, as we have shown, to use sufficiently low concentrations of intact tobacco mosaic virus so that the self-protection effect should be avoided (Chessin et al., 1964). It would appear worthwhile to test further the phosphate effect with concentrations of tobacco mosaic virus such as those used in our experiments.

Ginoza (1963) applies similar reasoning to the absence of a difference between intact tobacco mosaic virus and its infectious RNA when irradiated in tryptone broth. We obtained curves which indicated markedly greater sensitivity of tobacco mosaic virus-RNA as compared to intact virus to high-speed electrons when irradiated in phosphate buffer at room temperature (Chessin et al., 1964). Since this was done with concentrations of RNA which should have afforded considerable self-protection and those of intact virus which should not, even greater differences would presumably have resulted had both been equally self-protected. Our results suggest instead the involvement of indirect effects, in which the protein coat in intact virus acts as an effective protective buffer for the nucleic acid chain contained within. The use of frozen or dried preparations would assist in evaluating this hypothesis.

Temperature might also be expected to influence radiation sensitivity, particularly where indirect action of the medium is important in inactivation. Also, since indirect effects are mediated by diffusible materials, it should be possible to eliminate such effects by using frozen or dried material.

It was shown that inactivation of dry tobacco mosaic virus by deuterons, as might have been expected, was essentially independent of temperature from 20-73° C (Pollard and Dimond, 1956).

It is assumed that oxygen plays an important role in biological radiation damage since it can supply an electron to stabilize the migrated positive charge created by electron ejection (Howard-Flanders and Alper, 1957). I have been unable to find any published reports of the oxygen effect with plant viruses.

3. *CALCULATION OF TARGET VOLUME*

Included among Lea's remarkable early findings was the suggestion that, based upon the application of target theory, the entire virus particle was not the smallest infective unit. Considering when this was done, it is not surprising that Lea found that target diameter was, however, always within a factor of 2 of virus diameter. Apparently the first clear identification of the radiosensitive volume of plant viruses with their nucleic acids was made by Epstein (1953). His results, which are based upon a recalculation of Lea's data, are tabulated in Table 19-1.

TABLE 19-1
A COMPARISON OF NUCLEIC ACID VOLUMES OF SOME PLANT VIRUSES WITH THEIR RADIOSENSITIVE TARGET VOLUMES

Virus	*Target volume* [a] (10^3 nm^3)	*NA volume* [a] (10^3 nm^3)
Tobacco mosaic	2.4	2.5
Tobacco necrosis	1.3	1.2
Tobacco ringspot	1.9	1.2
Tomato bushy stunt	2.0	1.2

[a] Recalculated from Lea (1947).

Possible reasons for the discrepancy in the case of tobacco ringspot and tomato bushy stunt viruses were not given by Epstein (1953).

Ginoza (1968) also includes a similar table. For tobacco mosaic virus and its been demonstrated that acridine orange is intercalated between the base pairs of DNA (Lerman, 1961). A different mechanism is likely for the single-stranded RNA of plant viruses.

In the case of DNA transforming principle, the purine bases appear to be the site of chemical change in photosensitization (Simon and van Venakis, 1962), while the pyrimidine thymine seems to be the site of the biochemical lesion when DNA systems are inactivated by UV light (Setlow and Setlow, 1962).

The photodynamic inactivation of tobacco mosaic virus and its infectious nucleic acid has been studied recently in more detail by several investigators (Tsugita et al., 1965; Sastry and Gordon, 1966; Singer and Fraenkel-Conrat, 1966; Ito, 1967). Also demonstrated was dye-catalyzed photoinactivation of alfalfa and cucumber mosaic viruses and their nucleic acids (Orlob, 1967). Interestingly, both whole cucumber mosaic virus and its RNA were photosensitive in the absence of dye. Since there is no evidence that the virus particle or the nucleic acid absorb in the visible region, Orlob postulated a similar mechanism to the Fe^{3+}-catalyzed photoinactivation demonstrated by Singer and Fraenkel-Conrat (1966). Further studies on this are therefore indicated.

B. In Vivo Studies

1. DIRECT EFFECTS

Light has been shown to affect the establishment and multiplication of plant viruses as well as the expression of symptoms.

a. Viruses favored by post-inoculation illumination. The first work on controlled illumination was reported by Bawden and Roberts (1947). They concluded that most of the profound seasonal effects previously reported were due to light intensity rather than temperature. Similar results were obtained with all the viruses tested, for example, tobacco necrosis, tomato bushy stunt, tobacco mosaic, and the tomato aucuba mosaic. When the fungus *Olpidium brassicae* is used to transmit tobacco necrosis virus to lettuce roots, light also enhances infection after the fungus has penetrated host tissue (Beever and Fry, 1970). However, exposure of the system to light before penetration reduces infection, largely because of zoospore inactivation. Shading leaves before inoculation consistently increased the numbers of local lesions produced; in systemic hosts, the virus content was also greater in shaded leaves than in unshaded controls. Post-inoculation shading in local-lesion hosts gave less consistent results; generally shading decreased lesion numbers.

The mechanism of inhibition by light was unclear; sap extracted from light-grown plants showed no particularly inhibitory effect when added to virus before inoculation.

Wiltshire (1956) studied some biochemical aspects of this problem. CO_2 removal increased susceptibility in light but not in dark, suggesting a photosynthetic involvement. Changes in content of many organic acids were studied as they possibly were related to changes in susceptibility. Only ascorbic acid showed a correlation; however, although it was lost in darkened leaves, ascorbic acid continued to decrease after susceptibility no longer changed. In addition, darkened leaves actually produced more lesions when ascorbic acid was infiltrated. In view of the CO_2 effect, however, the implication of photosynthesis should be tested further.

A promoting effect of post-inoculation illumination on local-lesion production has also been demonstrated by others for potato X and red clover vein mosaic viruses, in addition to tobacco mosaic virus (Goodchild, 1960; Huguelet and Hooker, 1966; Wilcoxson and El-Kandelgy, 1966; Francki, 1960). Whether photosynthesis may be involved has not been demonstrated, although Goodchild showed that for the tobacco mosaic virus-*N. glutinosa* system, the major effect of light was temperature-independent.

b. Light-sensitive viruses. Some viruses are apparently light-sensitive in vivo. The classical example is the tomato spotted wilt virus which produced fewer lesions when leaves were illuminated after inoculation (Best, 1936). On the other hand, lesion production was favored by darkness both before and after inoculation of French bean plants with cucumber mosaic virus (Bhargava, 1951).

Although shading increased the tobacco mosaic virus content of *N. glutinosa* leaves (Bawden and Roberts, 1947) complete darkness markedly reduced virus content (Goodchild, 1960). Under the latter conditions a general breakdown of organic material undoubtedly occurs.

c. Light and "spontaneous" lesions. An interesting sidelight was noted by Francki (1967) in the course of experiments with light and *Gomphrena globosa*. When plants were grown under 2000 ft-c of illumination by fluorescent lights, local lesions were produced even in the absence of virus. Generally, these could be distinguished from those induced by virus. The only other light intensity tested was 180 ft-c which produced no "spontaneous" lesions. Since most experiments with controlled illumination are run at light intensities intermediate between these extremes, it would be of interest to determine the threshold light intensity which produces the "spontaneous" lesions in *Gomphrena globosa*.

d. Photoperiodic effects. In addition to intensity, light duration has been shown to influence susceptibility to infection; for example, a photoperiodic effect on resistance of plants to the potato virus Y (Nienhaus, 1957). Photoperiod has also been shown to affect symptom expression in wheat; when grown under long days only mosaic resulted, while short-day conditions resulted in both mosaic and rosette (McKinney, 1937).

e. Light intensity and symptom expression. Light intensity, as well as duration, has also been shown to influence symptoms caused by certain plant viruses. With some, symptom severity is affected. Perhaps the most interesting effect of reduced light intensity is the decreased severity of mottling caused by the *Abutilon* mosaic, tomato aucuba mosaic, and cucumber 4 viruses. There is no obvious connection between symptoms in these virus diseases and light intensity as there may be for the potato leaf roll or sugar beet yellows diseases, where light enhances symptoms and where infected plants contain more carbohydrate than healthy controls (Bawden, 1952). Since, in diseases such as the *Abutilon* mosaic, increased light is reducing chlorophyll content, it would be of interest to determine whether this is a result of an increased sensitivity of chlorophyll to light. Much evidence has accumulated recently that colored carotenoids, such as B-carotene, protect chlorophyll against photodecomposition (Chessin et al., 1966). Unfortunately, little work has been done on carotenoid content of virus-infected plants. Elmer (1925) found a doubling of the carotene concentration in tobacco mosaic virus infections while the concentration of both chlorophyll and xanthophyll were decreasing. On the other hand, Peterson and McKinney (1938) reported that chlorophyll, carotene, and xanthophyll all decreased by the same percentage. Their results would be in agreement with a carotenoid-protection hypothesis, except that as previously noted (Bawden, 1952), symptoms of mosaic-type diseases in general are enhanced by low light intensity. More recently, Venekamp (1957) has also studied pigment changes in virus-infected plants.

Similar studies with the viruses in which the non-carbohydrate related mottling is greatly enhanced by light should be of considerable interest.

C. Indirect Effects (Photosensitization)

Very little has been reported in this field. Chessin (1960) using a dilute (1.65×10^{-5} M) solution of acridine orange, could demonstrate no photodynamic effect when a dye-tobacco mosaic virus-RNA mixture was rubbed on leaves prior to illumination, in contrast to illumination of the mixture before inoculation. Orlob (1967) showed that free dye not bound to infectious material could reduce infection when applied after inoculation of virus, but this was apparently unrelated to the effect of light.

V. ULTRAVIOLET LIGHT (UV) (2000 Å-4000 Å)

A. Direct Effects

Mulvania was probably the first to deliberately test the sensitivity of tobacco mosaic virus to UV light (Mulvania, 1926). He reported less inactivation relative to bacteria, an observation borne out by later investigators. The first really quantitative work on this problem was done by Hollaender and Duggar (1936). Although initially claiming some correlation between the absorption and inactivation spectra, they ultimately reported considerable discrepancies between the two. This finding has also stood the test of time. A dose-response curve of inactivation of tobacco mosaic virus was published by Price and Gowen (1927) and was shown to be exponential in form.

In most subsequent experiments, tobacco mosaic virus has been the virus used. However, other plant viruses have also been the subject of experimentation, mostly in connection with the phenomenon of photoreactivation. Where dose-response curves have been reported they generally are straight lines, when the logarithm of survival is plotted against dose of UV. Such a response, characteristic of a first-order reaction, is also generally ascribed to a "one-hit" process. However, in view of the large number of quanta (25,000) which a particle of tobacco mosaic virus must absorb before it is inactivated, Kleczkowski has argued against the "one-hit" in favor of a "disequilibrium" hypothesis (Kleczkowski, 1960*a*). Either would be compatible with first-order kinetics. Although I have not made an exhaustive search of the literature I have found the following plant viruses to have been used in UV studies: Tobacco mosaic virus-Rothamsted stock (probably U-1), Datura, U-1, U-2, tomato aucuba, and Lychnis strains; tobacco rattle; barley stripe mosaic; potato X; clover yellow mosaic; cabbage black ringspot; tomato bushy stunt; tobacco necrosis; and the blue-green algal virus LPP-1.

Unfortunately, because of the variety of methods used by the different investigators, including the question of whether lesion numbers are transformed or not for statistical purposes, it is not possible to compare all the viruses used directly for their relative sensitivities to UV. In general, the negative slope of the survival-dose curve (k) could be taken as a measure of sensitivity.

A large body of data have been reported by Bawden and Kleczkowski (1955), so comparisons can be made between the viruses and strains used. However, only four strains of tobacco mosaic virus are strictly comparable in their work, since the other strains of this virus and the other viruses were tested in clarified sap and not in purified form. Relative sensitivities (k value) were 1.6 for the Rothamsted, Datura, and U-1 strains, and 3.3 for the U-2 strain. I have compared the Lychnis strain of tobacco mosaic virus after purification with both the U-1 and U-2 strains, and found its rate of inactivation very similar to that of the U-1 strain (Chessin et al., 1967). Now that many viruses can be obtained in purified form, it would be worth comparing inactivation rates of purified virus. In such work, it would be important to use comparable amounts of virus, since concentration has been shown to influence the rate of inactivation (Bawden and Kleczkowski, 1955; Price, 1965).

B. Photoreactivation

Practically all of the recent work with effects of UV on plant viruses has been done in connection with the phenomenon of photoreactivation (PR), that is, the increased survival of irradiated virus in plants exposed to visible light after inoculation compared to that in plants kept in darkness (usually for 24 hr) after inoculation.

After Kelner's classic rediscovery of this phenomenon in microorganisms (1949), Bawden and Kleczkowski (1953) demonstrated PR in several plant viruses. Bawden (1964*a*) has summarized some comparative studies made by the two investigators. Viruses can be compared directly for extent of PR (which can be taken as the ratio of the negative slopes of survival in darkness to that in light on the semilogarithmic plot), even when they have been tested under different conditions, provided consistent experimental procedures are used for a particular virus. I have made this comparison in Table 19-2 in which I have included all the viruses which have been tested for PR.

Several conclusions can be reached based on the above results:

(1) All plant viruses, with the exception of those with rigid rod-shaped particles, can be at least partly photoreactivated in appropriate hosts.

(2) The infectious nucleic acids of even the rigid rods tested show PR.

(3) DNA virus of photosynthetic plants (blue-green algal virus) exhibits PR, along with the RNA viruses of higher plants.

Unfortunately, the PR indices given in Table 19-2 may not be strictly comparable, mainly because the light intensities to which host plants were exposed after being inoculated with irradiated virus were undoubtedly highly variable, since most received this light under glasshouse conditions.

A likely explanation for the differences shown by viruses of different morphology is the strength of the nucleic acid-protein bond. Although flexuous viruses, in common with those with rigid rods, have a basically helical makeup, the RNA coil is not rigidly held by strong interactions between successive turns of the surface protein because the bonding sites in flexuous viruses are less specific than those of rigid rods; isometric viruses exhibit considerable evidence of loose bonding between the nucleic acid and protein (Caspar and Klug, 1962). Evidence for the importance of nucleic acid-protein bonding in PR is also provided by the photoreactivability of the infectious nucleic acids of the rigid rods of tobacco mosaic virus and tobacco rattle virus when they are irradiated free of their protein, as noted in Table 19-2.

Relatively little has been done on the action spectrum of PR of plant viruses. By using a filter which essentially transmitted no visible light below 5200 Å, no evidence of PR was obtained with potato virus X, while unfiltered white light gave considerable PR (Chessin, 1958). With the DNA blue-green algal virus LPP-1, Wu et al. (1967) demonstrated PR in host cells illuminated with cool white fluorescent or blue light (λ = 360-500 nm), but not with black-light illumination (λ = 300-400 nm). Recently, Hidalgo-Salvatierra and McLaren (1969) have reported a careful study of PR of tobacco mosaic virus-RNA in leaves, and showed an increased PR as wavelength decreased from 550 to 385 nm. The failure of the "black light" to photoreactivate algal virus is probably not due to the fact that it has DNA, since other DNA systems show maximum PR in this region (Jagger, 1967).

TABLE 19-2
PLANT VIRUSES TESTED FOR PHOTOREACTIVATION OF UV DAMAGE

Virus	*Degree of PR*	*Reference*
A. Rigid Rods		
(1) Tobacco mosaic		
(a) Type strain	1.0	Bawden (1964*a*)
(b) Datura strain	1.0	Bawden (1964*a*)
(c) Strain U-2	1.0	Bawden (1964*a*)
(d) Bean form of cowpea strain	1.0	Bawden (1964*a*)
(e) Tomato aucuba strain	1.0	Bawden (1964*a*)
(f) Lychnis strain	1.0	Chessin (unpublished)
(2) Tobacco rattle	1.0	Cadman and Harrison (1959)
(3) Barley stripe mosaic	1.0	Chessin (unpublished)
B. Isometric Particles		
(1) Tomato bushy stunt	1.2	Bawden (1964*a*)
(2) Rothamsted tobacco necrosis	1.2	Bawden (1964*a*)
(3) Cucumber mosaic	1.5	Bawden (1964*a*)
(4) Tobacco ringspot	1.9	Bawden (1964*a*)
(5) Southern bean mosaic	1.5	Price (1965)
C. Bullet-shaped Particles		
(1) Alfalfa mosaic	1.0	Gibbs et al. (1965)
D. Flexuous Threads		
(1) Cabbage black ring	2.0	Bawden (1964*a*)
(2) Potato X	3.0	Bawden (1964*a*)
(3) Clover yellow mosaic	2.0	Chessin (1965)
E. Blue-green Algal Virus	3.0 (7000 lux)	Wu et al. (1967)
LPP-1	2.4 (3000 lux)	Wu et al. (1967)
F. Nucleic Acids		
(1) Tobacco mosaic	2.0	Bawden (1964*a*)
(2) Tobacco rattle	1.9	Harrison and Nixon (1969)

Only one virus has apparently been tested for the temperature coefficient of PR, that of the clover yellow mosaic (Banerjee, 1969). There was no consistent effect on PR in plants held at 15.5-26.6° C for 24 hr after inoculation. Other plant viruses should be tested in this regard.

C. Analytical Photochemistry of Tobacco Mosaic Virus

McLaren (1968) has summarized the voluminous literature on the analytical UV photochemistry of tobacco mosaic virus, and some of his main points follow.

In contrast to experience with bacteria and other cells, UV does not induce mutation in tobacco mosaic virus. Its quantum yield of inactivation, that is, the number of particles inactivated per quantum of UV light, is also quite different from that of cells, a recent calculation being 4.3×10^{-4}. There is an interesting relationship between the quantum yield and particle weight; when enzymes, viruses, and a bacterium were compared, the quantum yield of inactivation was approximately inversely proportional to particle weight to the two-thirds power.

Although most investigators are likely to correct for scattering when studying absorption spectra, McLaren (1968) rightly emphasizes the need to do so for inactivation spectra as well. When such spectra are compared, it is clear that they are different for tobacco mosaic virus, as reported in 1936 by Hollaender and Duggar, and amply confirmed since then but still unexplained. On the other hand, the inactivation spectrum of free infectious tobacco mosaic virus-RNA is essentially identical to its absorption spectrum.

Another important difference between intact virus and infectious RNA is the photoreversibility of the latter when irradiated free of the protein. Apparently, the nature of the protein-nucleic acid bond is critical, since RNA extracted from virus after irradiation is not photoreversible. An interesting variation of this behavior is that when the irradiated free RNA is recombined with protein, it no longer shows PR; however, if reextracted, it recovers photoreversibility. There is considerable evidence for two kinds of UV damage to plant viruses or their infectious nucleic acids, nonphotoreversible and photoreversible. Protein, then, seems to protect the nucleic acid against sustaining photoreversible damage.

There is much interest in the nature of the primary photochemical lesions for both photoreversible and nonreversible damage. Although the U-1 and U-2 strains of tobacco mosaic virus have different spectra for UV inactivation when irradiated wet, when irradiated dry the spectra are identical. This led to the suggestion that the hydration of an amino acid-base linkage is important in inactivation. Streeter and Gordon (1966) have also demonstrated the importance of the protein in determining sensitivity to inactivation by comparing "hybrid" viruses made by combining the nucleic acids of the U-1 and U-2 strains of tobacco mosaic virus to the protein coat of the other. In this case, the sensitivity depended upon the strain which provided the protein coat and not the nucleic acid. That protein is involved in inactivation by UV is suggested by an increased susceptibility of virus to heat denaturation, although solubility and serological properties change little. There is a well-known correlation between the formation of pyrimidine (thymine) dimers and photoreversible damage in several DNA systems. Whether such dimers are involved in the RNA plant viruses is still in dispute, with Carpenter and Kleczkowski (1969) reporting evidence for, and Merriam and Gordon (1967) against, such a correlation.

D. Recent Findings

I conclude this section by referring to some more recent findings.

By irradiating the infectious RNA of tobacco mosaic virus in D_2O Evans and McLaren (1969) found slower inactivation but faster PR in pinto beans as compared to irradiation in H_2O, suggesting a key role for pyrimidine hydrate formation.

Govrier and Kleczkowski (1970) filled an important gap when they reported that the protein of the flexible elongated particles of potato virus X gives some protection to the RNA in the virus against both photoreactivable and nonphotoreactivable UV damage. This behavior is intermediate between that of the rigid rod of tobacco mosaic virus, whose protein protects partly against nonphotoreactivable damage but completely against photoreactivable damage, and the polyhedral particles of tobacco necrosis virus, where both inactivation and photoreactivation are the same whether intact virus or free RNA is irradiated.

Gordon and co-workers have recently demonstrated the importance of the protein in determining UV effects on plant viruses by employing virus "hybrids." Virus reconstituted with the protein of tobacco mosaic virus and the RNA of potato virus X exhibited inactivation and photoreactivation typical of potato virus X (Breck and Gordon, 1970), while combinations of the protein of the polyhedral virus, cowpea chlorotic mottle with RNA of tobacco mosaic virus exhibited photoreactivation and little effect on inactivation (Grouse et al., 1970).

In connection with the evidence for the importance of nucleic acid-protein bond strength in PR, several potential bond "looseners," including low temperature and high pH were tested. These did not render tobacco mosaic virus photoreversible when used during irradiation; however, low temperature consistently reduced UV damage, while high pH (9-9.5) tended to sensitize irradiated virus to inactivation by light in inoculated leaves (Banerjee, 1969). These are new phenomena which should be tested by other investigators.

On the basis of a lesser rate of inactivation by UV of tobacco necrosis virus when assayed in darkened *Chenopodium amaranticolor* as compared to French bean or tobacco, Kleczkowski (1968*b*) concluded that dark reactivation occurs in the *Chenopodium*. Although this is a logical explanation for the results, it would be extremely worthwhile to confirm this by means of cell-free extracts. In general, work on reactivation of UV damage, including PR, in the plant viruses has lagged far behind other systems, and this is nowhere more painfully evident than in the failure to date to demonstrate reactivation of virus in cell-free plant extracts. The photoreactivating enzyme of blue-green algal cells and bean leaves has so far been tested only on transforming DNA of *Hemophilus influenzae* (Werbin and Rupert, 1968; Saito and Werbin, 1969).

20

Tobacco Mosaic Virus: Mutants and Strains

B. HENNIG AND H. G. WITTMANN

Max Planck Institut für Molekulare Genetik
Berlin, Germany

I. INTRODUCTION

Inoculation of a host plant with a virus causes the appearance of more or less striking changes in the infected plant tissue several days later. These changes (symptoms) are the visible expression of the biological activity of the virus.

The symptoms and the chemical and physical properties are characteristic of the virus and remain unchanged in general throughout many virus transmissions under the same conditions (host plant, temperature, light, etc.). Thus, the properties of the virus are hereditary, like those of all cellular organisms.

As with bacteria, plants, and animals, populations of virus with altered hereditary properties arise occasionally from the original virus population. If such variants arise under experimental conditions, so that their descent can be traced back, they are designated as *mutants* of the original virus. On the other hand, virus variants are distinguished as different *strains* if they are isolated from the field, sometimes even on different continents, so that their direct genetic evolution has not been observed. This distinction in terms has mainly practical importance and is without great significance for systematic classification.

All known plant viruses are relatively simple in chemical composition, compared with cellular organisms, in that they are limited to nucleic acid and protein. It has been shown that the nucleic acid carries the genetic information of the virus (Gierer and Schramm, 1956; Fraenkel-Conrat, 1956). All cellular organisms share with viruses the characteristic of hereditary information encoded in the nucleic acid. Thus, much of the knowledge that is acquired relatively easily by studies on viruses helps in understanding the molecular mechanisms taking place in higher organisms.

A. Historical

Early studies of Ivanowsky (1892) and Beijerinck (1898) demonstrated that the mosaic disease of tobacco was caused by a type of infectious particle new at that time. Physical and chemical identification and characterization were not possible, however, until isolation and crystallization of the virus particles by Stanley (1935*b*). In the meantime, genetic determination of the virus became obvious through the isolation of numerous genetically stable variants (McKinney, 1929; Jensen, 1933). These variants attracted attention because of the markedly different symptoms caused in their host plants. Experiments in later years on influencing the spontaneous mutation rate by UV or X-rays, analogous to successful experiments made on cellular organisms, failed to give clear positive results (Gowen, 1941; Mundry, 1957*a*, 1960*b*).

Experiments with tobacco mosaic virus mutants depended, therefore, on spontaneous mutants discovered and isolated by chance—until it was found that virus mutations can be experimentally induced with nitrous acid (Gierer and Mundry, 1958). This finding initiated an intensive production of chemically induced mutants. Chemical analysis of the capsid protein of many such mutants led to confirmation of the genetic code (Wittmann, 1962; Tsugita, 1962*a*; Funatsu and Fraenkel-Conrat, 1964).

II. ISOLATION AND PURIFICATION OF VARIANTS

The purity of a virus preparation is critical, especially when it is used for genetic studies. In studies on cellular organisms, identification and treatment of each individual or of its descendants is, in general, experimentally possible. However, in a single infected host leaf, a virus population can reach a size of about 10^{14} virus particles. Hence, genetic studies on viruses consist of analyses of their population. Reliable interpretation of results requires careful attention to this fact.

A. Methods of Isolation

Evaluation of host symptoms, a property specific for a given virus, is still the simplest and most generally applied method for recognizing variants.

Two principal types of infection occur: (1) Necrotic localized infection: Virus is localized at the primary infected and neighboring cells. These cells die and are visible as a distinct lesion amid the healthy tissue. (2) Systemic infection: The inoculated virus spreads systemically in tissues of the host. Symptoms are visible, and various distinct discolorations and morphological changes of the infected tissue take place.

The necrotic reaction occurs in genetically hypersensitive host plants, for example *Nicotiana glutinosa* L. and *N. tabacum* L. cv. Xanthi-nc. Other hosts become infected systemically almost exclusively, such as *N. tabacum* cv. Samsun, while others are differentiating hosts, for example *N. tabacum* cv. Java, which reacts necrotically to various tobacco mosaic virus variants and systemically to others.

Systemically reacting host plants are especially useful in isolating many of the spontaneous mutants contained in conspicuous yellow spots amid the green mosaic caused by the parent strain vulgare. Because of this symptom, these mutants were named "yellow" strains.

A differential host is preferred for examining and isolating experimentally induced mutants because necrotic mutants derived from a systemic strain are easily visible as lesions on the primary infected leaf. Also, if the initial strain is a necrotic strain, then induced systemic mutants will produce symptoms on the uninoculated leaves.

Special techniques are used for selecting certain mutant types such as conditional lethal mutants. Temperature-sensitive mutants (ts) can be isolated selectively by exposing the infected plants to high temperature for a short time. These are easily recognized, since their lesions enlarge more slowly than the temperature-resistant lesions (Jockusch, 1968*a*).

During experiments in which mutations have been induced chemically, lesions are frequently observed that cannot be transferred by normal extraction and transfer procedures. Such mutants are defective in their protein coat. The RNA is insufficiently coated and is destroyed by leaf nucleases released during the extraction procedures. Transfer of infectivity can be successful when host tissue is extracted with buffer-containing bentonite, which inhibits the ribonuclease activity (Siegel et al., 1962) (see Chapter 1).

B. Methods of Purification

Mutants isolated from a lesion on a differential host plant are likely to be contaminated by other mutants. On transferring the lesion, the contaminating mutants in the population may outgrow the actual virus of interest. It is therefore necessary to purify the mutant by a method which guarantees, at high probability, that the final isolated virus population has arisen from a single virus particle and is thus genetically homogeneous.

Statistical calculations show that infection of a host cell can be initiated by a single infectious virus particle (Lauffer and Price, 1945). To achieve this, the best condition would theoretically be to inoculate the host with a sample of a virus suspension sufficiently diluted so that only one virus particle would come into contact with the host cell. However, there are always uninfectious virus particles in the preparation, and an infectious particle can infect the host tissue only at specific receptor sites. Thus, one might have to inoculate a very great number of plants under this theoretically optimal condition in order to get a single successful infection. Since this is generally impractical, one must compromise, thus running a slight risk of double infection if efficiency of infection is to be acceptable in practice.

If the virus is to be purified with the aid of a systemically reacting host plant, then the optimal dilution of virus suspension to be inoculated is the concentration at which approximately half of all infected plants show the infection. The probability then, according to the Poisson distribution, is sufficiently large that the virus population multiplying in the infected plants originated from a single infection.

More favorable is inoculation into a necrotically reacting host plant. Since each individual lesion develops from a single infection, it is highly probable that the population of each lesion has arisen from a single virus particle, if the virus suspension is sufficiently diluted and aggregation of virus particles with one another is prevented. To increase this probability nearly to certainty, because of the mentioned risk of double infection, the virus extracted from a single lesion or a single systemically infected plant should be purified several times more in the same way. Then the virus preparation resulting from a single lesion or systemically infected leaf of the last purification step can be used as starting material for further experiments.

However, complete purification and constant purity of a virus population to the extent of an absolute genetically homogeneous population is not possible. The spontaneous mutation rate is known to be as high as about 10^{-2} (Sehgal and Krause, 1968). A single lesion is estimated to contain about 10^{7} virus particles, and a single leaf of a systemically reacting host plant contains up to 10^{14} virus particles (Siegel, 1965). Hence, during each multiplication, mutants arise spontaneously. Yet, because of the enormous majority of the initial strain, the chance of outgrowing the initial strain is low unless these mutants have a selective advantage under the conditions of growth. It is necessary to check the fraction of impurities by appropriate tests as a control (see also Section III. A. 1)

III. TOBACCO MOSAIC VIRUS MUTANTS

A. Origin of Mutants

Prerequisite to an intensive analytical study of the genetics of virus is the production of an arbitrarily large number of mutants. The frequency at which mutants arise out of a given population is dependent on environmental conditions.

Mutations occur also under natural conditions without any experimental influence. The frequency of spontaneous mutants is, however, too low. To find conditions under which the mutation rate is increased, numerous physical forces and chemical reagents have been tested for their mutagenic effect. It is important to point out that experimental manipulation can change the balance of multiplication of different preexisting variants in the population through selection or interference. Thus, appropriate measures and controls must be included in the experiment before the appearance of virus mutants can be ascribed to an experimental increase in mutation rate.

1. PROOF AND QUANTITATIVE CONTROL OF EXPERIMENTALLY INDUCED MUTATIONS

a. Mutation Kinetics. If treatment of the genetic material causes phenotypically visible modifications at all, its effect may be either lethal or mutagenic. The number of hits necessary to produce just one definite effect defines a kinetic unit which is treated quantitatively by hit theory. Its simplest model, the one-hit kinetics, is based on most of the actually observed inactivation and mutation processes, as demonstrated in Fig. 20-1 for HNO_2-induced mutants (Mundry and Gierer, 1958; Fraenkel-Conrat, 1961; Brammer, 1963).

If one treats a number of infectious particles (I_0) with an inactivating reagent, then after a certain reaction time (t), only a fraction (I) of the particles are still infectious. This dependence of inactivation on length of treatment can be described by the function

$$I = I_0 e^{-kt}$$

The constant k indicates the average number of hits on the treated virus population per unit time and can be determined experimentally (see below). If not all hits are lethal, but a portion (p) is mutagenic with the frequency pk, then, among the surviving particles (I), the following *absolute number* of mutants (M) is expected

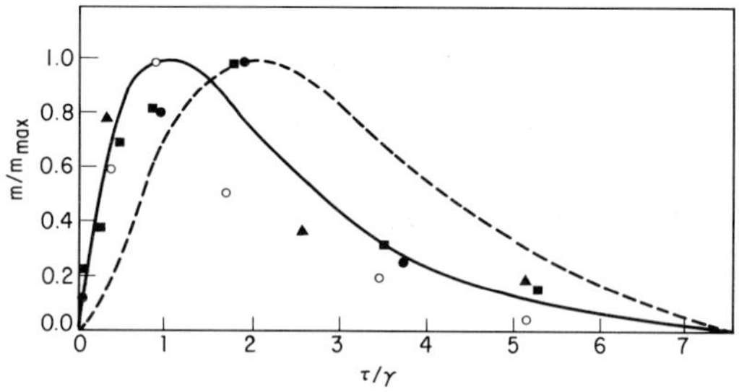

Fig. 20-1. Comparison of mutation kinetics expected from hit theory and actually observed after HNO_2 treatment of intact virus and isolated RNA. (Gierer and Mundry, 1958.)

$$M = pktI = pktI_0e^{-kt}$$

According to this function the absolute number of mutants among the surviving particles increases up to the time $t = I/k$, when it has a maximal value, $M = pI_0/e$, and decreases again with longer treatment.

In contrast, the *relative fraction* of mutants among the surviving particles increases linearly with the time $M/I = pkt$. Of special importance is the time of reaction τ, designated "reduced time," after which each particle on the average has obtained exactly one hit:

(1) After this time, the maximal absolute number of mutants (M) will be observed.

(2) The same value of t/τ for different experiments reflects the same number of hits per particle. The experiments can thus be compared exactly after reduction to the same condition of t/τ. The reduced time can be determined experimentally, and thus also the constant k, as the time of reaction (t) after which infectivity has decreased to 37% (Mundry and Gierer, 1958, see Fig 20-2).

b. Control of effects resulting from contaminants. Naturally the actual appearance of mutants according to theoretical hit kinetics can be expected only if one starts with a genetically homogeneous population. But in practice it is not possible to reach this condition exactly (see also Section II).

Differences between actually observed and theoretically postulated mutation kinetics can be due to preexisting variants: (1) through their greater resistance in relation to inactivation; (2) through their positive selection under the test conditions; or (3) through interference among the different variants during infection.

2. *SPONTANEOUS MUTATIONS*

Since conditions were developed for inducing mutants experimentally, general interest in spontaneous mutations has waned. Further, it is not possible to measure the real spontaneous mutation rate of a virus population, because of the population statistical nature of its analysis. Each mutant is subject to practically uncontrollable selection from the time of its origin until the earliest possible time of measuring the visible symptoms. Measurement of the spontaneous mutation rate of a virus population is, therefore, actually only a measurement of the makeup of the population at a given time and balance of selection. This value must be determined, however, as background for each experimental analysis as a special experimental control. All the experimental conditions should enter into it, except the specific effect of the reagent to be tested. In general, this value lies for different symptoms in the range of about 1 mutated particle to 100-1000 nonmutated particles (Sehgal and Krause, 1968; see also Section IV. B. 4).

This is a relatively high value, but the spontaneous mutation rate of however, is the mutagenic power of HNO_2 on RNA coated by protein subunits in intact virus, on the one hand, and on isolated uncoated RNA, on the other hand (Schuster and Wilhelm, 1963). Uncoated RNA seems to be more sensitive to mutagenic reaction in relation to inactivation; furthermore, the qualitative effect of mutation is different between uncoated RNA and RNA in intact virus.

b. Hydroxylamine. The inactivating and mutagenic effect of hydroxylamine was already well established from studies on bacteriophages when its power to induce tobacco mosaic virus mutants was tested. In contrast to HNO_2, hydroxylamine reacts only on isolated RNA, not on RNA coated with protein (Schuster, 1961; Schuster and Wittmann, 1963). A base analog is produced from cytosine through addition of hydroxylamine to the pyrimidine ring. This is mutagenic if the analog is replaced during RNA replication by uracil as a result of the altered base pairing capacity (Schuster, 1961; Nagata and Martensson, 1968). A lethal pyrimidine ring cleavage is affected in uracil. The reaction speed of hydroxylamine with the pyrimidine bases of RNA is strongly dependent upon pH. Cytosine is attacked more quickly at pH 6 than at pH 9, and uracil is attacked more slowly at pH 6 than pH 9 (Schuster, 1961). Therefore, if one treats tobacco mosaic virus RNA with hydroxylamine at pH 6, the mutagenic reaction with cytosine is about 30 times as fast as the lethal reaction with uracil. At pH 9 the mutagenic reaction is, conversely, 8 times as slow as the lethal reaction.

To perform the reaction, incubate 0.1% solution of isolated RNA in 1 M hydroxylamine at pH 6 and 20° C. To prevent a side reaction of hydroxylamine during inoculation, the RNA must be purified by alcohol precipitation and then inoculated (in phosphate buffer, pH 6) into a host proper for the isolation of mutants.

At the time $t = \tau$, when all RNA molecules have obtained an average of 1 hit each, mutation rate is 5 times the spontaneous rate if necrotic mutants are counted on a differential host (Schuster and Wittmann, 1963).

The measurements satisfy the conditions of 1-hit kinetics, both tor inactivation and mutation, and 1 mutagenic to about 100 lethal hits has been calculated to occur.

c. Alkylation. Alkylation reagents react intensively with virus nucleic acids and inactivate them quickly. The mutagenic effect has been examined after incubation of virus or isolated RNA in vitro with such alkylating substances as mustard gas (pH 7), dimethyl sulfate (DMS, pH 6-9), ethylene oxide (pH 5), and propylene oxide and iodoacetate (pH 4.6) (Fraenkel-Conrat, 1961).

However, mutagenic induction was evident only for dimethyl sulfate. Its mutagenicity is about 1/10-1/5 that of HNO_2. For induction, isolated RNA is incubated for 24 hr at 0° C with 0.5×10^{-3} M DMS at pH 6-9. Following this, the RNA is purified from reagents by alcohol precipitation and inoculated directly, or after reconstitution with native-coat protein, into a proper host plant.

After measuring the number of alkyl groups bound per RNA chain at the time of 37% survival, it was discovered that for the inactivation reaction of all the mentioned alkylating substances there is probably a 1-hit mechanism. It has been established that all reagents mentioned above change guanine, mainly or solely, whereas iodoacetate primarily influences adenine (Fraenkel-Conrat, 1961).

d. Halogenation. The inactivation after reaction of RNA with bromine, N-bromosuccinimide (NBSI), and N-iodosuccinimide is caused by modification of the nucleotide bases (Brammer, 1963). NBSI and bromine alter only guanine and cytosine; N-iodosuccinimide alters only cytosine. A mutagenic effect of these halogenation reactions, therefore, seemed possible.

According to results when isolated RNA (pH 7 in ethylenediaminetetra-acetate or pH 9 in carbonate-bicarbonate buffer) was incubated at 0° C and at a molar ratio of reagent to nucleotides of 1:5-1:10, only NBSI was slightly mutagenic (Brammer, 1963).

e. N-methyl-N'-nitro-N-nitrosoguanidine (MNNG). MNNG proved to be especially good for the induction of mutations in bacteria, so it was interesting to test its effect on tobacco mosaic virus (Singer and Fraenkel-Conrat, 1967). The tests showed that free RNA is quickly inactivated in MNNG-saturated water solution in the pH range of 4.5-9. Moreover, at the time of inactivation of 60-98%, the mutation rate is double the spontaneous rate. The reaction of MNNG in formamide leads to slower inactivation, but also to a 5-fold mutation rate. Mutation rate can be further increased to 10 times that of the control through treatment of intact virus in a pH 5.5 buffer with MNNG.

4. PHYSICAL INDUCTION OF MUTATIONS

In 1927 Muller used X-rays for the first experimental induction of mutations. This technique soon proved equally good on bacteria and multiple-cell organisms. With tobacco mosaic virus also, the treatment of a virus population with radiation was among the first attempts to induce mutations experimentally.

a. Radiation. After treatment of virus preparations with X-rays (149-490 kv of a 1.5 Å radiation), the appearance of 1.2×10^{-8} mutations per Rontgen per particle was measured (Gowen, 1941). Mutants were also isolated after other X-ray experiments (Kausche and Stubbe, 1939). Since those results were quite uncertain and variable and did not take into account the selection of preexisting mutants, an explanation of the mutagenic effect of X-rays on purified tobacco mosaic virus preparations was attempted under particularly careful conditions (Mundry, 1957*a*, 1960*b*).

Some samples of viruses (strain dahlemense in pH 7.5 phosphate buffer) were treated with strong doses of X-rays (10^6 *r* of a 169 kv radiation) and other samples with doses of UV light (253.7 nm), causing equal inactivation. After inactivation to 10^{-6} of the initial infectivity, which corresponds to about 14 hits per virus particle, the appearance of mutants infecting the differential host systemically (instead of necrotically as the original strain) was tested. An increase in mutation rate above the spontaneous rate was apparent. But these "induced mutants" could be fully accounted for as being due to competition phenomena between preexisting mutants and unmutated virus. The same "mutation rate" was reached through appropriate dilution or through properly mixed virus preparations before inoculation (Mundry, 1957*a*). Further investigation has revealed that mutation is very rare in the direction from necrotic to systemic infection, even with the otherwise powerful HNO_2 reagent (Mundry, 1960*a*). Therefore, the experiments were repeated for observation of an increase in mutation rate in the reverse direction of mutation. The results were the same, however, and thus no mutagenic effect of radiation on tobacco mosaic virus could be demonstrated.

Certain dyes and iron salts are supposed to have a catalytic effect on the induction of mutations by exposure to light (Singer and Fraenkel-Conrat, 1965, 1966). Virus or its isolated RNA was incubated in vitro with proflavine (10^{-3}-10^{-5} M) or thiopyronine (10^{-5}-10^{-8} M) in ethylenediaminetetraacetate buffer, pH 7, or borate buffer, pH 9. With proflavine treatment the virus as well as reconstituted RNA showed, after inactivation of more than 99% of the virus, a mutation rate 3.3 times the spontaneous rate for a necrotic infection on a differential host. With thiopyronine the mutation rate was 4 times the background rate under the same conditions.

b. Temperature. In initial studies with tobacco mosaic virus mutants the appearance of yellow spots depended on culture temperature. It was tested whether these mutants can be traced back to mutations induced through high temperatures (Johnson, 1947; Mundry, 1957*b*).

The hereditary change in size of lesions was investigated in infected plants grown at approximately 36° C as compared with a lower temperature. Furthermore the appearance of yellow spot populations arising from the green strain vulgare and green spot populations from the yellow mutant flavum was checked for infected plants grown at temperatures between 17° and 32° C.

Results seemed to indicate that mutation rate increased with temperature, but several isolated spontaneous mutants were shown to reproduce much faster at high temperatures than the initial strain, thereby being positively selected (Kassanis, 1957).

B. Properties of Mutants

Early spontaneous mutants isolated at random (Melchers, 1942; Black and Knight, 1953; Aach, 1957) were characterized intensively in many of their properties (Schramm and Rebensburg, 1942; Aach, 1957; Kramer, 1957; Kramer and Wittmann, 1958; Aach, 1958 *a*, *b*; Wittmann et al., 1965). More general conclusions drawn from these results were quite limited, however, both because of the relatively small number of mutants isolated (the purification procedure was troublesome) and because of their unsystematic production by uncontrolled mutation mechanisms.

An intensive production of chemically induced mutants was first done to gain some insight into the basics of the genetic code. Furthermore, these mutants, with located amino acid replacements in their coat protein, are ideal material for more extensive comparative structural and functional studies. For this purpose the isolated coat protein of these mutants was analyzed by methods of protein chemistry.

1. PROTEIN CHEMICAL STUDIES

The subunits of the protein coat are the only molecules associated with the RNA in the intact tobacco mosaic virus particle. So far there exist difficulties in determining the nucleotide sequence of nucleic acids as large as tobacco mosaic virus RNA. Thus analysis of the amino acid sequence in the coat protein has been the object of chemical studies of mutants.

Evidence of changes in the amino acid sequence, and their localization, can be approached in two somewhat different ways. The technically simplest and quickest procedure for identifying such changes is direct amino acid analysis of the whole polypeptide chains. First, the virus is split into RNA and protein by 67% acetic acid. The protein is isolated and purified and then hydrolyzed by 6 N HCl at 108° C. Quantitative determination of the amino acid composition is done in an automatic amino acid analyzer. However, complementary exchanges in different positions as, for example, (Leu→Phe)+(Phe→Leu), compensate one another in this gross calculation of the amino acid composition of the whole protein. In this case measurements would simulate a protein structure unchanged from that of the initial strain and would reject the mutant as a virus believed to be unmutated in its coat protein gene. Furthermore, in a mutant with two or more exchanged amino acid positions (e.g. with one less serine and threonine and one more leucine and isoleucine) one cannot determine which two types of replacements have actually occurred.

The other way (more troublesome but much more accurate) is the following: The isolated protein is cleaved enzymatically (e.g. by trypsin), or chemically (e.g. with cyanogen bromide, if possible). The resulting peptides are purified by an appropriate chromatographic technique and then hydrolyzed, and the amino acid composition is calculated on the analyzer (Wittmann 1960, 1962; Funatsu, 1964). The risk of rejecting mutants as viruses unmutated in their coat protein by overlooking complementary amino acid replacements is thereby decreased considerably or, in the case of many peptides, completely ruled out. Since then, however, it has been shown that in most cases of chemically induced mutants only one amino acid replacement per mutant occurs, and, even in the case of a double mutant, complementary exchanges are relatively unusual (Wittmann, 1962; Funatsu and Fraenkel-Conrat, 1964; Wittmann and Wittmann-Liebold, 1966).

A change in the amino acid composition must be exactly localized by techniques of sequence analysis (e.g. Edman-degradation and degradation with exopeptidases). Amidation differences are not recognized in normal amino acid analysis, but must be proved electrophoretically if not by sequence analysis.

The coat protein amino acid composition of over 300 spontaneous and chemically induced mutants have been examined chemically (Wittmann 1959*b*, 1960, 1962, 1964; Tsugita and Fraenkel-Conrat, 1960, 1962; Funatsu and Fraenkel-Conrat, 1964; Zaitlin and McCaughey, 1965; Wittmann-Liebold and Wittmann, 1965*a,b;* Rombauts and Fraenkel-Conrat, 1968; Hariharasubramanian and Siegel, 1969). Most of these mutants were isolated after treatment of common tobacco mosaic virus or its spontaneous mutant, A-14, with HNO_2. About one-fourth of these mutants are altered from their original strain in the sequence of coat protein (Table 20-1). More than two-thirds of these have only one amino acid replacement each, and the rest have two or, quite seldomly, three exchanges each (Wittmann, 1962; Tsugita and Fraenkel-Conrat, 1962; Funatsu and Fraenkel-Conrat, 1964; Wittmann-Liebold and Wittmann, 1965*a, b*). As a control, 18 preparations of untreated virus were analyzed (Wittmann, 1962). None of these showed any change in amino acid composition of the isolated tryptic peptides.

A survey of amino acid replacement data reveals the following peculiarities: One particular mutagen induces the substitution of only certain amino acids, and these are exchanged into a limited number of new amino acids each. Conclusions drawn from this are discussed later (see Section V. A). Several mutants are identical with one another in that they have the same type of amino acid replacement in the same sequence position. These mutants, if occurring in the same laboratory, have probably not been induced independently of each other. The first of these mutants isolated probably reappeared several times, since it was contaminating other samples (Funatsu and Fraenkel-Conrat, 1964). One must assume, especially for some "mutants" with an extremely large number of replacements each, that they have derived in a similar way from other strains multiplied in the greenhouse simultaneously with mutants of interest: The coat proteins of "mutants" 328 and 262, claimed to be chemically induced from common tobacco mosaic virus, are each different from that of this common strain in 17 exchanges in the gross amino acid composition (Tsugita, 1962*a*). However, they have, in fact, an amino acid composition identical with that of strain G-TAMV, also handled in that laboratory.

2. *PHYSICO-CHEMICAL STUDIES*

Through comparative studies of physico-chemical properties of the proteins of different coat protein mutants it is possible to draw conclusions concerning

TABLE 20-1

TOBACCO MOSAIC VIRUS MUTANTS WITH LOCATED AMINO ACID EXCHANGES, AND POSSIBLE NUCLEOTIDE ALTERATIONS

Mutant	*Exchange*	*Position*	*References*
Spontaneous			
A 14	Ile → Thr	129	Wittmann-Liebold et al. (1965)
	AUU ACU		
B 13	Asn → Lys	33	Wittmann-Liebold and Wittmann (1965*a*)
	AAU AAA		
CP 415,	Asn → Lys	140	Wittmann et al. (1965)
E 66	AAU AAA		
GK 1	Asn → Ser	73	Wittmann-Liebold and Wittmann (1965*a*)
	AAU AGU		
Flavum	Asp → Ala	19	Wittmann et al. (1965)
	GAU GCU		
Necans	Phe → Leu	10	Wittmann et al. (1965)
	UUU CUU		
	Ala → Val	19	
	GCU GUU		
	Ser → Phe	138	
	UCU UUU		
Reflavescens	Leu → Phe	10	Wittmann et al. (1965)
	CUU UUU		
Revirescens	Leu → Phe	10	Wittmann et al. (1965)
	CUU UUU		
	Val → Asp	19	
	GUU GAU		
430	Pro → Leu	20	Funatsu and Fraenkel-Conrat (1964)
	CCU CUU		
Induced with Nitrous Acid			
Ni 102,	Asp → Gly	66	Wittmann-Liebold and Wittmann (1965*b*)
Ni 116	GAU GGU		
Ni 103	Asp → Gly	19	Wittmann-Liebold and Wittmann (1965*b*)
	GAU GGU		
Ni 109	Glu → Gly	97	Wittmann-Liebold and Wittmann (1965*b*)
	GAA GGA		
Ni 117,	Ser → Phe	138	Wittmann-Liebold and Wittmann (1965*b*)
Ni 252	UCU UUU		Funatsu and Fraenkel-Conrat (1964)
Ni 118,	Pro → Leu	20	Wittmann-Liebold et al. (1965)
Ni 346	CCC CUC		Funatsu and Fraenkel-Conrat (1964)
Ni 445	Ser → Phe	138	Wittmann-Liebold and Wittmann (1965*b*)
	UCU UUU		
Ni 458,	Thr → Ile	59	Wittmann-Liebold et al. (1965)
Ni 470	ACU AUU		

TABLE 20-1 (Continued)

Mutant	*Exchange*	*Position*	*References*
Induced with Fluorouracil			
FU 27	Val → Ala GUU GCU	58	Wittmann-Liebold and Wittmann (1965*a*)
FU 41	Thr → Ala ACU GCU	28	Wittmann-Liebold and Wittmann (1965*a*)
FU 243	Ser → Gly AGU GGU	65	Wittmann-Liebold and Wittmann (1965*a*)
Induced with N-bromosuccinimide			
218, 235	Pro → Leu CCU CUU	20	Funatsu and Fraenkel-Conrat (1964)
342	Ile → Thr AUU ACU	21	Funatsu and Fraenkel-Conrat (1964)
378	Asn → Ser AAU AGU	25	Funatsu and Fraenkel-Conrat (1964)
	Asn → Ser AAU AGU	33	
	Asn → Ser AAU AGU	126	
414	Ser → Gly AGU GGU	65	Funatsu and Fraenkel-Conrat (1964)
Induced with Dimethylsulfate			
215	Ser → Phe UCU UUU	138	Tsugita (1962*a*)
344	Thr → Ala ACU GCU	81	Funatsu and Fraenkel-Conrat (1964)
	Ser → Phe UCU UUU	138	
403	Arg → Gly AGG GGG	46	Funatsu and Fraenkel-Conrat (1964)
Induced by Other Treatment			
483, 487, 498, 539	Gln → His CAPur CAPyr	9	Rombauts and Fraenkel-Conrat (1968)

the relationship between amino acid sequence and the stability of the tertiary and quaternary structure of the coat protein chain.

a. Denaturation. Jockusch (1966*b*) compared the thermal denaturation of several mutants known to differ in temperature-dependent biological behavior. Thus the structures of the coat protein molecules of these mutants must have amino acid replacements crucial in type and location for the biologically functional conformation.

Different thermal stabilities of the tertiary structures of proteins are reflected in their denaturation kinetics. These were measured at 20° and 30° C in the pH range of 5-8. The temperature-dependent mutants are not functionally defective at 20° C but are so at 30° C. In pH 5 buffer, native coat protein molecules aggregate to soluble intermediates of normal virus rods; they are mostly dissociated in pH 8 buffer. In these experiments a protein was designated denatured if, in contrast to its native state, it no longer dissolved in pH 5 buffer and sedimented at low-speed centrifugation. After incubation at 30° C in pH 7 buffer the thermal stability of the temperature-sensitive (ts) mutants was essentially decreased: the half-time of denaturation ranged between 2 and 180 min. In contrast, under these conditions the proteins of the temperature-resistant (tr) strain vulgare and its tr mutants proved relatively stable. The half-time of denaturation was more than 40 hr.

Mutant	*Amino Acid Type*	*Exchange Position*	*Biological Behavior*	*Denaturation* $t_{1/2}(min)$, *pH 7, 30°C*
Vulgare	(initial strain)		tr	> 2500
Ni 109	Glu → Gly	97	tr	> 2500
Ni 103	Asp → Ala	19	ts	20
Ni 1927	Pro → Leu	156	tr	> 2500
Ni 1196	Pro → Ser	63	ts	180
Ni 118	Pro → Leu	20	ts	2
FU 27	Val → Ala	58	tr	> 2500

(Selected from Jockusch, 1966*b*)

In control experiments after lowering the temperature to 20° C, where ts mutants are not thermally unstable, none of the tr and ts proteins showed any clearly measurable denaturation in pH 7 buffer for 40 hr. The result was the same with pH 5 buffer when the temperature then was raised to 30° C. Interpretation of these latter results may be that in pH 5 buffer the native protein subunits aggregate to rods. Completed virus rods are known from biological tests to be no longer temperature-sensitive in the range of physiological temperatures. The same may be true for reaggregated protein rods. Parameters defining the temperature dependency of the thermal stability of tertiary structure can be calculated more exactly. The melting points of denaturation can be measured with the aid of optical rotatory dispersion and were found, for instance, to be 39.5° C for the tr vulgare protein and 27° C for the ts Ni 118 protein (Jockusch et al., 1969). Evaluation of denaturation kinetics enables one to determine the activation energies of denaturation for different proteins, as were derived from the slopes of Arrhenius plots for vulgare protein (170 kcal/mol) and Ni 118 protein (108 kcal/mole) (Jockusch, 1968*b*).

By comparing these results and the known amino acid replacement data, one observes, for instance, the following: The apolar→polar replacement valine→alanine in position 58 (mutant FU 27) has not altered the high thermal stability of the protein compared with that of the initial strain, whereas the polar→apolar replacement threonine→isoleucine in the nearest neighbor position, 59, caused thermal instability. On the other hand, the replacement proline→leucine in position 156 has no effect on temperature dependency, although the same type of replacement in position 20 essentially lowers the denaturation temperature of the protein.

From these examples and further data the following rules are stated (Jockusch, 1968*b*): Both an exchange of polar to apolar amino acids and the substitution of proline by another amino acid can cause a decrease in thermal stability of the tertiary structure of the mutant protein from that of the initial strain. No protein with a replacement apolar to polar amino acid could be detected to be temperature-sensitive. Decisive for a change in thermal stability as a consequence of amino acid replacement of such a crucial type, however, is its location in the sequence.

b. Aggregation. Tobacco mosaic virus consists of about 2130 polypeptide chains aggregated as subunits around the RNA chain and thus forming the protein coat of the virus particle. By incubating the virus rod under proper conditions (alkaline pH or in presence of sodium dodecyl sulfate) the protein subunits can be stripped off from the RNA (Schramm, 1947; May and Knight, 1965). The dissociated native subunits (usually present as trimers, so-called A-protein) reaggregate easily into intermediates of the normal quaternary structure of the virus, stacked discs of 16 subunits each, or even into entire rods at certain pH values (Schramm and Zillig, 1955). Just as one uses denaturation experiments to analyze the importance that certain types of amino acids and their location in the sequence have for the stability of the folding of the polypeptide chain in the tertiary structure, one can investigate with the aid of aggregation experiments their importance for the formation and stability of the quaternary structure. Strain vulgare has in its sequence (position 156), just before the carboxyl terminal end of its polypeptide chain, a proline which is not susceptible to degradation by the usual exopeptidases. In mutant Ni 1927 this proline is replaced by leucine, and thus a protein is available which helps prove that the length of the carboxyl terminal end of the polypeptide chain is not important for normal aggregation. After enzymatic cleavage of the last three amino acids from the native subunit with carboxypeptidase, reaggregation of the treated subunits proceeds normally. This happens also after a previous cycle of denaturation and renaturation of the protein. If one degrades the 15-17 carboxyl terminal amino acids from the denatured polypeptide chain, however, renaturation to the native subunit, and hence aggregation to the normal quaternary structure, is no longer possible (von Sengbusch and Wittmann, 1965).

The importance of an internal position of the coat protein for the aggregation capacity, the proline→leucine replacement in position 20 (mutant Ni 118), of which the thermal stability was known to be decreased, has been analyzed with the aid of sedimentation and optical methods (Jockusch, 1966*b*, Jockusch et al., 1969). When the isolated protein subunits of the tr strain vulgare and of its ts mutant Ni 118 were incubated at 20° C in pH 7 buffer, each aggregated to discs of 2 × 17 subunits each. At 30° C, however, the protein of vulgare still aggregated to form the same discs whereas that of Ni 118 precipitated.

Aggregation of discs to rods could further be induced by dialyzing vulgare protein from pH 7 to pH 5 buffer of low ionic strength both at 20° and 30° C. This was possible with Ni 118 protein only at 20° C. At 30° C the Ni 118 protein precipitated again. Yet, after incubation of Ni 118 protein at 30° C in buffer of high ionic strength, aggregates were obtained which stayed in solution. At first it appeared that the polypeptide chains had been folded into their native tertiary structure, easily soluble and aggregatable. The following experiment, however, shows that this property of Ni 118 protein, demonstrated at 30° C and high ionic strength, is in fact due primarily to a denatured state instead of a distorted aggregation of native subunits to the normal quaternary structure: The dissolved aggregates, if native tertiary structures, should remain in solution during the lowering of temperature, pH, and ionic strength to the values of temperature resistance (below 30° C) and the mentioned aggregation conditions (about pH 5). But the protein of Ni 118 precipitated when incubation at 30° C and high ionic strength was followed by dialysis against pH 5 buffer of low ionic strength at 4° C. Thus, high salt concentration alone must have kept the polypeptide chains in solution in their denatured form.

In vivo, completed virus rods of the mutants tested show the same thermal stability, independent of the temperature-resistant or temperature-sensitive properties of their isolated protein subunits. This was proved by infectivity tests after RNase treatment of virus incubated at 50° C up to 40 hr (Jockusch, 1966*b*). No loss of infectivity occurred with both tr and ts mutants. Whether this high stability of completed virus particles independent of tr or ts properties is due to protein-nucleic acid or to protein-protein interaction in the highly ordered rod should be determinable directly by testing protein rods reaggregated in vitro without RNA, and comparing their stability with that of intact virus and dissociated subunits. This experiment has not yet been performed with clear-cut results.

The most promising results from physico-chemical studies for the goal mentioned are to be expected from experiments with defective mutants. Such mutants can be isolated from host tissue only by special extraction methods (see also Section II and Chapter 11). Even if they are incubated under conditions where protein and RNA isolated from other mutants easily reconstitute to intact RNase-resistant virus particles, an absolute loss of infectivity is obtained after addition of RNase. As shown by electron microscopy with defective mutant PM 2, the protein subunits may actually have a native tertiary structure and can reconstitute with RNA (Zaitlin and Ferris, 1964; Siegel et al., 1966). They have an altered aggregation capacity, however; instead of aggregation to normal compact rods, they form single-stranded or multistranded helices. According to electron micrographs these helices seem to differ from the quaternary structure of the normal tobacco mosaic virus protein in that the individual turns of the helix are not packed densely over one another but are spread apart through a dislocation of the binding sites of the subunits as a result of amino acid exchanges (threonine→isoleucine in position 28, and glutamic acid→aspartic acid in position 95, in mutant PM 2; arginine→cysteine in position 112 in mutant PM 5). The pitch of the helices has been measured to be 23 Å in vulgare and about 280 Å in PM 2 (Siegel et al., 1966).

c. Electrophoresis. Comparing the electrophoretic measurements of different mutants with the known amino acid sequence data of these mutants could provide some information on the three-dimensional structure of the virus rod: (1) about the distribution of charges on the surface of the protein subunit, by comparing the electrophoretic mobilities of denatured and native coat protein; and (2) about the masking of charges in the virus rod, by comparing the electrophoretic mobilities of the native coat protein subunit and reconstituted protein coat.

Before experimental induction of mutations was discovered, a series of spontaneous mutants evolved from the wild strain vulgare seemed to be especially suitable:

vulgare ⟶ flavum ⟶ necans ⟶ virescens

The electrophoretic mobilities of intact virus and dissociated protein subunits were determined for the wild strain and the three mutants in parallel experiments under the same conditions at constant ionic strength (Friedrich-Freksa et al., 1946; Kramer, 1957; Kramer and Wittmann, 1958). These data led to the following conclusions: (1) With the mutation vulgare→flavum, a decrease in acidic groups resulted corresponding to a lower mobility of flavum than of vulgare in the alkaline pH range (pH 7-9); (2) the electrophoretic mobility of the mutants flavum and necans, being identical over the pH range tested, proved that they have an identical number of charged groups; and (3) the mutation necans→revirescens reverted to the charge of the original strain, vulgare.

These interpretations are now supported by analysis of amino acids exchanged (Wittmann et al., 1965):

Position:	10	19	138
Vulgare:	Phe	Asp	Ser
Flavum:	Phe	Ala	Ser
Necans:	Leu	Val	Phe
Revirescens:	Phe	Asp	Phe

A correlation between amino acid exchange and change of electrophoretic mobility was found also by an investigation of chemically induced mutants (von Sengbusch, 1965) (see Table 14-1). The mutant Ni 102 moves more slowly than the wild strain vulgare in the alkaline pH range, because of loss of an acidic group (Asp→Gly in position 102). The mutant CP 415 has an altered mobility in both the acidic and alkaline pH ranges because of an additional basic group (Asn→ Lys in position 140).

In view of these results (i.e. the agreement between mobility and amino acid exchange) it was surprising that the strain vulgare and its mutant Ni 109 showed the same mobility, even though they differed by the amino acid exchange Glu → Gly in position 97, that is, by the loss of an acidic group through mutation. Other experiments indicated this unexpected observation to be due to a steric masking of the charged amino acid in the native vulgare protein. If one destroys the quaternary structure of the virus particle by splitting the protein coat into its individual protein subunits, this previously masked group can then be detected electrophoretically. As expected from the amino acid exchange (Glu→Gly), the protein subunit of Ni 109 moves more slowly than that of vulgare in the alkaline pH range. On the other hand, the protein subunit of A-14 (Ile→Thr in position 129) moves identically with that of vulgare, as is expected from the exchange of uncharged amino acids.

In the native virus rod the polypeptide chain of the coat protein can be shortened at the carboxyl terminal end with carboxypeptidase. In the vulgare rod only the last terminal amino acid, threonine (position 158), is split off, because of the nearby proline (position 156), which blocks the enzymatic function of carboxypeptidase A (Cpase). Further degradation is possible in the rod of the mutant Ni 1927, however, because of an exchange of the proline (position 156) for leucine. This enables one to make electrophoretic comparisons of four slightly different proteins with one another (von Sengbusch and Wittmann, 1965):

vulgare	-Gly-Pro-Ala-Thr
vulgare (Cpase)	-Gly-Pro-Ala
Ni 1927	-Gly-Leu-Ala-Thr
Ni 1927 (Cpase)	-Gly

According to expectations, the untreated native subunits of vulgare and Ni reaction is further intensified after enzymatic cleavage of the last three carboxyl terminal amino acids. Obviously, in native vulgare coat protein an antigenic site is sterically masked by the carboxyl terminal amino acids, which, enzymatically or through a mutated amino acid sequence, can be exposed for binding reaction with antibodies. The different reaction is probably due to free mobility of the polypeptide end in mutant Ni 1927 protein, in contrast to rigid folding of this end in vulgare protein due to the proline in position 156 of its sequence. On the other hand, mutant Ni 1927 protein, and especially the carboxypeptidase-treated Ni 1927 protein, as well as the treated vulgare protein, are not able to precipitate all antibodies out of the vulgare antiserum (von Sengbusch and Wittmann, 1965). This vestigial antibody can no longer be observed in experiments with antiserum against both the treated vulgare and Ni 1927 protein, which is interpreted as reflecting the terminal amino acid threonine taking part in the antigenic site of the coat protein.

Considering the distribution of the amino acid exchanges of the coat protein polypeptide chain according to their serological importance, one gets some information about their position either on the surface or in the interior of the protein molecule, and hence about the three-dimensional folding of the polypeptide chain. These conformational changes, however, could have been induced also by amino acid exchanges in the interior of the molecule, and these, again, could also have altered the serological response.

4. BIOLOGICAL BEHAVIOR

In general, mutants are recognized by an altered symptom which they initiate in the host tissue. Not all mutations, however, alter just that symptom which is under observation. Thus, the possibility must be taken into account of selecting a limited set of mutants out of the whole spectrum of mutants. This especially must be kept in mind for the more general conclusions drawn from these data.

As already discussed (see Section II), there are three ways for host tissues to react toward an infection: obligate systemically, obligate necrotically, or obligate differentially. HNO_2-induced mutants were used in investigating whether these types of reactions are affected either simultaneously or independently of each other in their respective symptoms after inoculation of virus mutants (Sehgal and Krause, 1968). The results indicated that a mutation in the infected virus particle does not lead to a simultaneous change of symptoms in all three host types. The relative frequency of display of symptoms differs greatly and in addition depends on the dose of the mutagen. The data can be interpreted in a way that a single mutagenic hit is already sufficient to cause defective mutants (sterile lesions) with high frequency. Affected very often by a one-step mutation are also changes of symptoms of merely the systemic type of infection and simultaneous alteration of symptoms on the systemic and necrotic hosts. Rarely are symptoms changed only on the necrotic host, and extremely seldom changed is the kind of infection type on the differential host.

By more-step mutations (high number of hits per virus particle after high doses of mutagen) the frequency of defective mutants is increased, as is the frequency of mutants initiating altered symptoms on both systemic and necrotic hosts. On the other hand, the frequency of virus mutants altering symptoms only on the systemic host becomes less than the number altering symptoms solely on the necrotic host.

Especially interesting and important is the rare change in reaction on the differential host as a consequence of virus mutation, since this type of host is used predominantly under test conditions because of its experimental advantages (Mundry and Gierer, 1958; Siegel, 1960; Mundry 1960*a*; Seghal and Krause, 1968). The high percentage of sterile lesions of the mutants had already been observed during the early nitrous acid experiments. Siegel et al. (1962) showed in these lesions an RNase-sensitive infectivity that could be transferred with special conditions of extraction and infection. Further analysis revealed that the function of the coat protein of these mutants was defective, that is, it had lost its capacity to aggregate to compact rods, which protect the RNA.

Several other mutants which are usually transferred without difficulties turned out to be temperature-sensitive. After multiplication in the host tissue at high temperature (30° C instead of 20° C) they also lose the capacity to form intact virus particles. This temperature-sensitive characteristic can be caused by one of at least two different genes (Jockusch, 1966*a*, 1968*a*).

a. ts-I mutants. Most of the ts mutants spread in the host tissue at high temperature as well. Their ts property is therefore concerned neither with a primary event during infection nor with synthesis of RNA. The temperature was changed systematically to determine the time of occurrence of the temperature-sensitive function during the virus multiplication (Jockusch, 1966*a*): The virus was reproduced at 32° C; thereafter the temperature was lowered to 20° C, and infectivity and the formation of virus particles in the tissue were subsequently determined. These experiments, along with reconstitution experiments in vitro, showed that the ts-I property is due to a defective RNA-protein aggregation, the latter being caused by an altered structure of the coat protein.

b. ts-II mutants. Some of the ts mutants do not spread further in the host tissue when the temperature is elevated, and RNA synthesis is not resumed after the temperature is lowered again. Appropriate kinetic experiments showed that the ts function must occur after the primary events at infection and before the formation of complete virus particles; it has been suggested that an enzyme concerned with RNA synthesis is altered (Jockusch, 1966*a*).

Many of the nitrous acid-induced mutants show obvious differences in symptoms from the initial strain but no difference in amino acid sequence of coat protein (Wittmann, 1962, 1964). These mutants support the assumption that the tobacco mosaic virus RNA carries genes in addition to the gene for the coat protein. Localization of such a gene on tobacco mosaic virus RNA has been achieved by inducing mutations with HNO_2 on the virus particles pretreated by a special technique (Kado and Knight, 1966). Protein subunits can be stripped from the virus rod beginning specifically from one end by an appropriate treatment with, for example, sodium dodecyl sulfate. By incubating the virus particles for different times the terminal areas are stripped. Since HNO_2 reacts faster mutagenically on uncoated RNA than on RNA protected by protein (see Section III.A.3), a gene will mutate more frequently when the RNA area of that gene is uncoated. The gene responsible for the necrotic infection on a differential host has been localized by this method. Its distance from the 5′-end is about a quarter of the total RNA length.

IV. TOBACCO MOSAIC VIRUS STRAINS

A. Phylogenetic Relationship of Strains

With many genetic problems, comparison of the properties of different variants is more valuable if the phylogenetic relations of these variants with one another are known. As defined in the introduction to this chapter, the alteration of a mutant from an initial strain has occurred under experimental conditions. Its genetic descent is therefore in general known exactly, including in many cases the number of individual mutation steps through which it has arisen from the original strain, especially with chemically induced mutants. This does not hold true in comparing different strains with one another. Strains also have evolved by mutation, but the number of single mutation steps through which a strain has evolved from another is not known, nor can one trace back experimentally the identity of this original strain. Hence circumstantial evidence must be used in trying to find out the phylogenetic relationship of strains with one another, and hence their evolutionary descent.

Since the hereditary information is coded in the nucleotide base sequence of nucleic acid, the surest evidence on the degree of phylogenetic relationship would be from a comparison of the RNA base sequence of the different strains under investigation. This method has not yet been possible for tobacco mosaic virus RNA. However, the nucleic acid base sequence is directly correlated with the sequence of amino acids in proteins. The method of sequence analysis of proteins has been worked out in detail, and it has already been carried out on the coat protein of some strains of this virus (see Section IV.B.1). Comparison of the amino acid sequence of coat protein therefore seems the best possible criterion for virus classification at this time. This method is still very time-consuming, though, so the criteria generally used are those which are easier to determine, especially for preliminary characterization. According to previous experience obtained with tobacco mosaic virus strains of known amino acid sequences of coat protein, comparison of their amino acid gross compositions, together with a test on serological cross reactions, seems most reliable for preliminary orientation in taxonomy. Moreover, it is necessary to take into account "stable characters," such as exterior morphology, that is, size and form, as well as certain physical properties related to the molecular structure of the virus, whereas host range, symptomatology, type of transmission vector, and other characters of labile nature are not satisfactory (Andrewes, 1968; Gibbs, 1969).

More difficult to determine than the degree of phylogenetic relationship of strains to each other are the limits of taxonomic grouping, that is, when we can say substantially that one virus strain still belongs to the tobacco mosaic virus group whereas another does not. These are problems of general taxonomy, however, and are not discussed here any further (for reference see Knight, 1955; Bellett, 1967; and, more general, Hennig, 1966).

Most of those who work on tobacco mosaic virus consider the following criteria characteristic for strains of the tobacco mosaic virus group:

(1) Morphology: Rod-shaped particles with a length of about 3000 Å and a diameter of about 150-200 Å.

(2) Biochemistry: 5% RNA with the base gross composition A:G:C:U = 1.2:1.0:0.7:1.1, and 95% protein with large amounts of Asp, Thr, Ser, Glu, Ala, Val, and Leu each, and low amounts of Cys, Met, Lys, His, and Trp each.

(3) Physical: Particle weight of the intact virus is about 40×10^6, sedimentation constant is about $s_{20,w} = 200$, molecular weight of the RNA is about 2×10^6, and molecular weight of the coat protein polypeptide chain is about 17,500.

(4) Serology: Cross reaction with antiserum made against the common strain (vulgare) or another strain which is definitely thought to belong to the tobacco mosaic virus group.

(5) Molecular structure: One RNA chain of molecular weight about 2×10^6, consisting of about 6400 nucleotides, is embedded in about 2000 identical and helically arranged coat protein subunits, each of which is folded from one polypeptide chain of molecular weight about 17,500, consisting of about 158 amino acids.

Table 20-2 lists all plant virus strains so far known to belong to the tobacco mosaic group according to the above-stated criteria. Variants isolated in the greenhouse, as, for example, variants M, GA, or J14D1, are considered spontaneous mutants, and thus are not considered here.

Using the extent of differences in values of the coat protein amino acid gross compositions, one can arrange strains so that those with only small differences (2-6 amino acids) from the most similar strain can be placed together in contrast to all others, showing many differences (14 or more) from their most similar strain (see Table 20-3).

Thus, one uses the protein chemical data for a preliminary classification of probable degree of phylogenetic relationship. At this time, however, the degree of phylogenetic relationship is calculated more accurately from the amount of differences among the coat protein amino acid sequences—at least with the

TABLE 20-2
STRAINS OF TOBACCO MOSAIC VIRUS

Abbreviation	*Name*	*Host*	*Geographic Region*	*Reference*
vulgare (U 1)	Common mosaic	tobacco		Iwanowski (1899)
YA	Aucuba mosaic	tomato		Bewley (1924)
CV 3	Cucumber mosaic 3	cucumber		Ainsworth (1935)
CV 4	Cucumber mosaic 4	cucumber		Ainsworth (1935)
Dahlemense		tomato	Germany	Melchers et al. (1940)
HR	Holmes' rib grass	rib grass	U.S.A.	Holmes (1941)
SSM	Southern sunnhemp mosaic	sunn hemp		Capoor (1950)
U 2	Mild mosaic	tobacco	U.S.A.	Singer et al. (1951)
ORSV	Odontoglossum ringspot	orchid	U.S.A.	Jensen and Gold (1951)
U 5		tobacco	U.S.A.	Siegel et al. (1954)
OM		tobacco	Japan	Hiruki et al. (1954)
CPV	Cowpea	cowpea	Nigeria	Lister et al. (1955)
SOV	Sammons' Opuntia	cactus		Sammons et al. (1961)
Y-TAMV	Yellow tomato atypical mosaic	tomato	U.S.A.	Knight et al. (1962)
G-TAMV	Green tomato atypical mosaic	tomato	U.S.A.	Knight et al. (1962)
CV 60		tomato	U.S.A.	Wang and Knight (1967)
CV 61		tomato	U.S.A.	Wang and Knight (1967)
AC-9	American collection No. 9	tomato	U.S.A.	Wang and Knight (1967)
HD	Hall-Davis	tomato	U.S.A.	Wang and Knight (1967)
SJ	San Joaquin	tomato	U.S.A.	Wang and Knight (1967)
VEN	Ventura	tomato	U.S.A.	Wang and Knight (1967)
Aus-II	Australian II	tomato	S. Australia	Wang and Knight (1967)
Dut-I	Dutch I	tomato	Netherlands	Wang and Knight (1967)
YLGP	Yellow leaf G.P.	tomato	Netherlands	Wang and Knight (1967)
K-1		tomato	Japan	Wang and Knight (1967)
PTA		tomato	Argentina	Wang and Knight (1967)
PTV	Philippine	tomato	Philippines	Wang and Knight (1967)
SAF	South African	tomato	S. Africa	Wang and Knight (1967)
CGMMV	Cucumber green mottle mosaic	cucumber	Japan	Inouye et al. (1967)
0 1 through 0 7		orchids	U.S.A.	Kado et al. (1968)
B-TMV 1 and 2		pear tree	Germany	Opel et al. (1969)

TABLE 20-3
AMINO ACID COMPOSITION OF COAT PROTEINS OF TOBACCO MOSAIC VIRUS STRAINS

Strain	*Vulgare*[1]	*OM*[2]	*01/02/03*[3]	*06*[3]	*YA*[4]	*U2*[5]	*G-TAMV*[6]	*Dahlemense*[7]	*Y-TAMV*[8]/*B-TMV1*[9] *AC-9 K-1/PTV*[10] *VC-60/YLPG*[10]	*DUT-1*[10] *AUS-II*[10]	*HD*[10]	*VEN*[10] *VC-61*[10]	*PTA/SJ*[10] *SAF*[10]	*ORSV*[11]	*02/07*[3]	*HR*[12]	*CV4*[13]	*SSM*[14]/*CPV*[15]	*CGMMV*[16]
Asp	18[17]	19	17		19	22		17	18	18	18	19	18	20		17	17	18	20
Thr	16	15	17		17	19		17		16	16	16	16	21		13	12	18/20	10
Ser	16				15	10		16	15			15	15	12		13	24	16/17	24
Glu	16			15		16		19						15		22	10	16	10
Pro	8					10		8						9		8	8	8	6
Gly	6				5	5	4	6						7		4	5	4	9
Ala	14		15	13		17	18	11			12		12	11		18	20	11/12	21
Cys	1					1		1						1		1	—	—	—
Val	14	15	13			12		15			14			9		10	14	11/12	7
Met	—			1		2		1						3		3	—	—	—
Ile	9	8	8	8	8	8		7						8	9	8	6	10	7
Leu	12					11		13						14		11	13	15	18
Tyr	4					6		5						6	7	7	4	8	4
Phe	8					8		8						7		6	11	6	9
Lys	2					1		2						1		2	3	1	4
His	—			1		—		—						—		1	1	1	1
Arg	11			12	12	8		9						10	9	10	10	11	8
Trp	3					2		3						3		2	1	1	2
Total	158	158	158	158	158	158	158	158	158	158	158	158	158	157	158	156	158	155/160	160

[1]Anderer et al. (1965).
[2]Nozu and Okada (1968).
[3]Kado et al. (1968)
[4]Tsugita (1962).
[5]Rentschler (1967*a*).
[6]Knight et al. (1962).
[7]Wittmann-Liebold and Wittmann (1963*b*).
[8]Knight et al. (1962).
[9]Henning (unpublished).
[10]Wang and Knight (1967).
[11]Hennig (unpublished).
[12]Jauregui-Adell et al. (1969).
[13]van Regenmortel (1967).
[14]Rees and Short (1965).
[15]Hennig (unpublished).
[16]Funatsu and Funatsu (1968).
[17]Blank areas under strain indicate that the number of amino acid residue is equal to the given type strain.

strains thus far analyzed. The number of amino acid differences in the individual positions, expressed in percent, is just a measurement of this degree.

Comparison Between		Differences (%) Gross Composition	Sequence
Vulgare	Dahlemense	10	18
Vulgare	U 2	20	26
Dahlemense	U 2	23	30
Dahlemense	HR	22	53
U 2	HR	20	54
Vulgare	HR	22	56

From these data a relationship of these strains can be constructed (Table 20-4). Furthermore, it is demonstrated that mere summation of the amount of differences in the gross compositions is of quite limited application for this purpose. When a certain number of sequence differences is exceeded (about 25%), the amount of differences in the gross compositions have obviously reached a maximum (about 20%), because of the high probability of compensation effects. The mere amount of the differences among the amino acid gross compositions, however, does not take into account qualitative aspects (e.g. presence or absence of cysteine, histidine, or methionine) which may also be important for preliminary classification.

B. Properties of Strains

1. BIOCHEMICAL STUDIES

a. Nucleic acid. Since alterations of the genetic information occur through changes of the nucleic acid base sequence (e.g. through base replacements), it should be theoretically possible to detect genetic differences among different strains even by comparing their relative amounts of the four RNA bases adenine, guanine, cytosine, and uracil. Though numerous comparative measurements have been made of the alkali-hydrolyzed RNA, deviations in measured values of RNA composition between the tobacco mosaic virus strains vulgare, YA, HR, and Y-TAMV remain within the range of error (Knight, 1952; Cooper and Loring, 1954; Knight et al., 1962). Slight differences can be observed by comparing these strains with strains CV 3 and 4. For control, a large difference in nucleotide composition can be shown by comparing these data of tobacco mosaic virus RNA with those on RNA of other plant viruses not belonging to the tobacco mosaic group, for example, the filamentous potato X virus (PVX) and the spherical turnip yellow mosaic virus (TYMV):

TABLE 20-4

TRYPTIC PEPTIDES AND PARTIAL SEQUENCES OF SOME TOBACCO MOSAIC VIRUS STRAINS

I. ORSV (Hennig, unpublished) sequence positions in homology with strain vulgare

T1 (Position 1 - 41):	Acetyl-(Asp$_6$, Thr$_4$, Ser$_6$, Glx$_4$, Pro$_2$, Gly$_1$, Ala$_4$, Cys$_1$, Ile$_2$, Leu$_5$, Tyr$_2$, Phe$_1$, Lys$_1$, Arg$_1$, Trp$_1$)
T2 (Position 42 - 61):	Thr-(Thr, Val, Gln)-(Gln, Gln, Phe, Ser)-Asp-Val-Trp-(Thr, Pro, Ser, Pro, Gln, Leu, Thr, Val, Arg)
T3 (Position 62 - 68):	Phe-Pro-Ala-Gly-Ala-Gly-Tyr-Phe-Arg
T4 (Position 69 - 71):	Val-Tyr-Arg
T5 (Position 72 - 90):	(Asx$_2$, Thr$_1$, Pro$_2$, Ile$_2$, Leu$_3$, Tyr$_1$, Phe$_1$)-Met-Gly-Thr-Phe-Asp-Thr-Arg
T6 (Position 91 - 92):	Asn-Arg
T7 (Position 93 - 112):	(Asx$_3$, Thr$_5$, Glx$_4$, Pro$_2$, Ala$_1$, Val$_1$, Ile$_2$, Leu$_1$, Arg$_1$)
T8 (Position 113 - 122):	Arg-Val-Asp-Asp-Ala-Thr-Val-Ala-Ile-Arg
T9 (Position 123 - 134):	Ser-Ala-Ile-Asn-Asn-Leu-Leu-Asn-Glu-Leu-Ile-Arg
T10 (Position 135 - 158):	(Gly, Thr, Gly)-Met-(Tyr-Asn-Gln-Ser-Thr-Phe-Glu-Val)-Met-Gly-Trp-Thr-Ser-Ser-Leu-Ser-Thr

II. CGMMV (Funatsu and Funatsu, 1968) fragments

T-1:	(Asn, Arg)
T-2:	(Ser$_1$, Pro$_1$, Ala$_2$, Leu$_1$, Phe$_1$, Lys$_1$)
T-3:	(Thr$_1$, Ser$_3$, Gly$_1$, Ile$_1$, Tyr$_1$, Arg$_1$)
T-4:	(Asx$_1$, Thr$_2$, Ser$_4$, Glx$_1$, Ala$_1$, Val$_1$, Leu$_4$, Arg$_1$)
T-5:	(Ser$_1$, Glx$_1$, Arg$_1$)
T-6:	(Asx$_1$, Ile$_1$, Leu$_1$, Arg$_1$)
T-7:	(Asx$_3$, Thr$_2$, Ser$_2$, Glx$_3$, Pro$_2$, Gly$_1$, Ala$_3$, Val$_1$, Ile$_1$, Leu$_1$, Lys$_1$, Arg$_1$)
T-8:	(Asx$_1$, Ser$_1$, Glu$_1$, Pro$_1$, Ala$_1$, Val$_1$, Phe$_2$, Arg$_1$)
T-9:	(Asx$_3$, Ser$_4$, Ala$_4$, Ile$_3$, Leu$_4$, Arg$_1$)

III. B-TMV1 (Hennig, unpublished); Y-TAMV (Funatsu, 1964)

Tryptic peptides Position 42 - 158 identical with dahlemense; 1 amino acid exchange Ser $\longrightarrow$ Asn in peptide 1.

RNA of	Molar Ratio*			
	Adenine	Guanine	Cytosine	Uracil
Vulgare, YA, HR, Y-TAMV	1.20	1.01	0.74	1.06
CV3, CV4	1.03	1.03	0.75	1.21
TYMV	0.91	0.69	1.53	0.89
PVX	1.37	0.87	0.91	0.85

*Data from Knight, 1952; Knight, 1959; Knight et al., 1962.

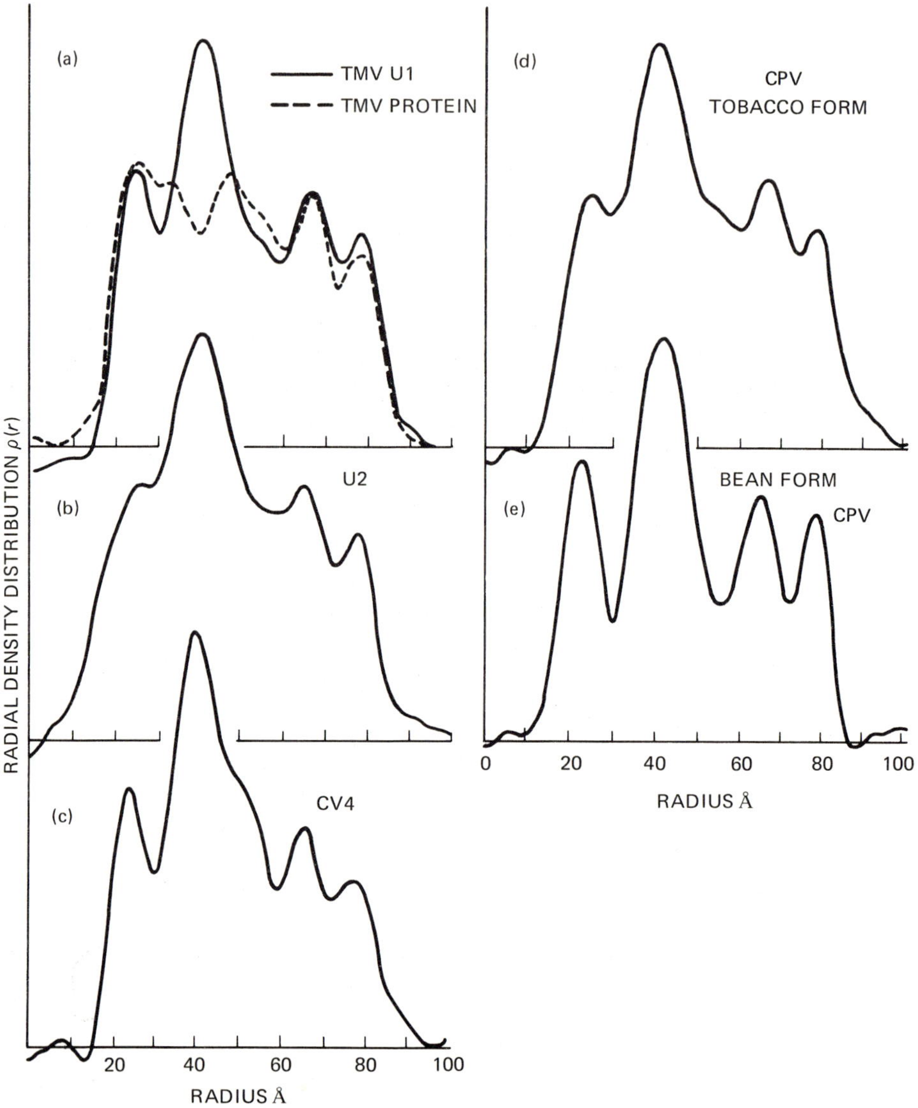

Fig. 20-3. **Comparison of radial electron-density distribution of various tobacco mosaic virus strains. (Holmes and Franklin, 1958.)**

multiplied in tobacco plants, the X-ray pattern of the virus is almost identical with that of strain vulgare, whereas if grown in bean plants it differs from vulgare more than any other strain.

In the protein coat, each of the identical polypeptide chains of a rod is folded together to form a subunit. The number of protein subunits per helical turn of the protein coat varies slightly among the rods of different strains. The value is 16.33 in strain CV4, 16.34 in strain vulgare, and 16.35 in U2 (Franklin et al., 1956*a*). All of the 2130 subunits of a single rod form a uniform surface structure in the vulgare particle. In strain dahlemense, however, four subunits always assemble together more closely, causing deep indentations on the surface of the rod in the interstices between these subunit groups (Caspar, 1963). These differences in assembly strengthen the assumption, already mentioned, that there are small changes in conformation of the individual subunits, which may lead to certain dislocations of the sites responsible for protein-protein interaction between the subunits.

3. *PHYSICO-CHEMICAL STUDIES*

Changes in the structure of the coat protein may lead to changes in radiation absorption, surface charge, or the aggregation process. Conversely, such changes should reveal structural changes.

Ultraviolet-absorption and infrared-absorption spectra have been measured from virus preparations of the strains vulgare, dahlemense, U2, U5, and HR (Dannenberg et al., 1941; Ginoza and Atkinson, 1955), and have been proved to differ only slightly or not at all. This result cannot be surprising, though the content in aromatic amino acids differs somewhat when comparisons are made of the protein chemical data now known for these strains. The nucleic acid content, and thus the absorption due to it, does not vary among the different strains. The experiment should therefore be carried out by using RNA-free protein. Nevertheless, the two strains HR and U2, which have an almost identical content of tyrosine and tryptophan, showed a clearly different spectral curve in the UV region at about 290 nm, known to be due to tyrosine-specific and tryptophan-specific absorption (Ginoza and Atkinson, 1955). This possibly reflects some differences in interaction of these chromophoric residues with their respective environments. Such an interaction must exist since the spectrum of isolated vulgare protein differs from that of an amino acid mixture of composition the same as the vulgare protein (Fraenkel-Conrat and Singer, 1957).

Changes in surface charge will be detected quickly, though roughly, by measuring the position of the isoelectric points. At this pH value the virus suspension becomes intensely turbid. Relative turbidity as an indication of the pH of the suspension has been measured for the strains vulgare, HR, YA, U2, U5, and CV4 (Oster, 1951; Gordon and Price, 1953; Price, 1954; Ginoza and Atkinson, 1955), with maximum turbidity at pH 3.91 for vulgare, pH 4.49 for HR, and as high as pH 4.90 for CV4 (Oster, 1951). More detailed measurements, obtained from electrophoretic studies, were compared for the strains vulgare and dahlemense (Kramer, 1957; Kramer and Wittmann, 1958) and vulgare, U2, and HR (Takahashi, 1955). The native virus rod (with RNA) and the reaggregated coat protein (without RNA) have the same electrophoretic behavior within each of the strains vulgare and dahlemense. Evidently, only the a limited host range, because of resistance genes in plant species not susceptible to this particular virus (Holmes, 1954). In view of this genetically controlled susceptibility, it was surprising when strains known to belong to the tobacco mosaic group were isolated from plants such as orchids (strains 01-07) and cactus (strain SOV). These hosts are phylogenetically quite distantly related to each other and to the plants already known as natural hosts of the earlier isolated strains vulgare, dahlemense, and HR (tobacco, tomato, and other Solanaceae; Plantaginaceae). However, all tobacco mosaic virus strains except CV4, CPV, and SOV can grow in tobacco plants in the greenhouse. Furthermore, the strain vulgare can be multiplied, at least under experimental conditions, in more than 200 different plant species (Holmes, 1938). Susceptibility to tobacco mosaic virus strains thus seems to be scattered throughout the whole plant kingdom.

a. Productivity. The yields of virus of different strains are clearly different among infected plants. This amount can be measured by cutting out separate lesions grown from different strains on an obligate necrotically reacting host, homogenizing them, and inoculating the extracts into a healthy obligate necrotically reacting plant. The number of lesions then formed is counted and the amount of virus in the initial lesion from which the extract has been prepared is calculated. This value is called the *productivity* of the particular strain. It has been measured for some strains under certain identical growing conditions (Veldee and Fraenkel-Conrat, 1962) and found to decrease from about 70×10^3 lesions on the average derived per extracted original lesion from vulgare, down to about 8×10^3 lesions on the average per original lesion for the strain YA.

The productivity of a strain is lowered by inoculating a mixture of different strains. Experiments demonstrated that this interference is due to competition of the different virus particles with one another at each of the limited acceptor sites of the host tissue (Siegel and Wildman, 1956).

In general, only one particle initiates an efficient infection per acceptor site of the host, and this can be a particle from either the one strain or the other. Interference may occur even when one inoculates a host plant with a mixture of two strains only one of which can multiply in this particular host. There is indirect evidence, however, that the insusceptible particles must be able to perform early stages of the infection process and thereby compete with the fully infectious particles of the other strain (Siegel 1959). When the insusceptible particles are inactivated by (for example) UV radiation before being mixed with the susceptible one and the mixture is inoculated, the inactivated insusceptible particles are no longer able to compete with the infectious particles at the acceptor sites, resulting in normal productivity of the susceptible strain.

b. Genes. Nevertheless, uninactivated genes of uninfectious particles can be translated by metabolic mechanisms initiated by infectious particles of another strain inoculated simultaneously. This is known to be the normal way that other plant viruses multiply, as, for example, tobacco necrosis satellite virus, the RNA of which is claimed to have a gene for its coat protein, no genes for its own replication enzymes, but identical 5′-terminal triplets (Reichmann 1964, 1969, 1970; Reichmann et al., 1966). Kado and Knight (1968) used the ability of an infectious tobacco mosaic virus strain to complement defective or missing genes of particles derived from another strain in locating the coat protein gene on tobacco mosaic virus RNA. Particles from the strain vulgare were partially stripped off from their coat protein subunits at the 3′-end of the RNA chain by the method of May and Knight (1965). The uncoated end of the RNA chain was then digested with RNase. The resulting noninfectious particles were mixed with infectious virus from strain HR, and the mixture was inoculated into an obligate systemically reacting host plant. The kinds of coat protein formed during multiplication were identified by immunodiffusion precipitin tests and polyacrylamide gel electrophoresis. Coat protein specific for the strain vulgare could be detected as long as not more than about half of the length of the RNA chain was digested at its 3′-end. This means that the coat protein must be located in the left half of the RNA chain (5′→3′).

V. CONCLUSIONS

A. Genetic code

Evidence that nitrous acid acts as a specific mutagen whose action results in the base change C→U and A→G (Schuster and Schramm, 1958; Mundry and Gierer, 1958) opened the possibility of obtaining information about the genetic code. The rationale of these experiments was the following: Alteration of viral RNA, which is the genetic material and acts as messenger for biosynthesis of the viral coat protein (Gierer and Schramm, 1956; Fraenkel-Conrat, 1956), should give rise to an exchange of amino acids in this protein. Correlation of amino acid exchanges with the type of nucleotide alterations caused by nitrous acid should reveal relationships between amino acid codons involved in the observed exchanges.

There are eight triplets (AAA, AAC, ACA, CAA, ACC, CAC, CCA, and CCC) containing only those bases alterable by nitrous acid. All 64 triplets can be arranged in eight octets, shown in Fig. 20-5. Since UUU codes for phenylalanine (Nirenberg and Matthaei, 1961), the octet containing UUU can be correlated with the amino acids participating in nitrous acid-induced amino acid exchanges (Wittmann, 1962; Wittmann and Wittmann-Liebold, 1966) (Fig. 20-4).

Protein chemical analyses of 117 nitrous acid-induced and 16 spontaneous tobacco mosaic virus mutants led to the following conclusions (Wittmann, 1962):

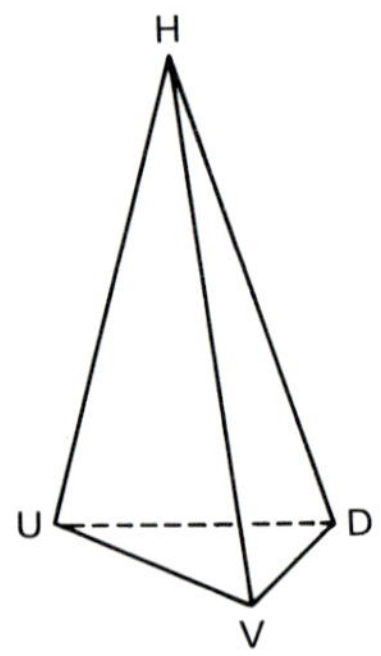

Fig. 20-4. Differences in amino acid sequences between four tobacco mosaic virus strains, shown as a pyramid. (Wittmann et al., 1969.)

(1) The code is nonoverlapping. This conclusion results from the finding that in most cases only one amino acid is exchanged per protein chain. When two exchanges were found, they were never neighboring each other, as would be expected with an overlapping code.

(2) The code is degenerate, because various amino acids (e.g. proline, serine, leucine, and phenylalanine) correspond to more than one triplet. It could further be concluded at that time either that the code is not highly degenerate or that if it is, the triplets coding the same amino acid do not represent a random selection from all 64 triplets but instead are related to one another in nucleotide sequence in a manner that gives rise to the remarkable restrictions found in the pattern of amino acid exchanges. Experiments with the cell-free system of *Escherichia coli* later showed that the latter alternative is true (Nirenberg et al., 1965).

(3) The infecting tobacco mosaic virus RNA strand (or RNA strands with an identical nucleotide sequence), not a complementary strand, serves as template for the synthesis of viral coat protein in vivo. If the complementary RNA strand were the messenger, the direction of amino acid exchanges would be the reverse of that shown in Fig. 20-5. An exchange of phenylalanine to any other amino acid or of any other amino acid to proline has never been found with nitrous acid-induced mutants. It is interesting that such exchanges occurred in spontaneous mutants, for which the restrictions in base alterations do not apply.

These conclusions were confirmed by analyses of many more mutants by Wittmann (1964) and in similar studies carried out independently at the same time (Tsugita, 1962*a*; Funatsu and Fraenkel-Conrat, 1964).

Assuming universality of the code, one should postulate agreement of the data from the mutation experiments with those from the cell-free system. Amino acid exchanges of all mutants should be compatible with an alteration of only one base per triplet. Furthermore, the amino acid exchanges of nitrous

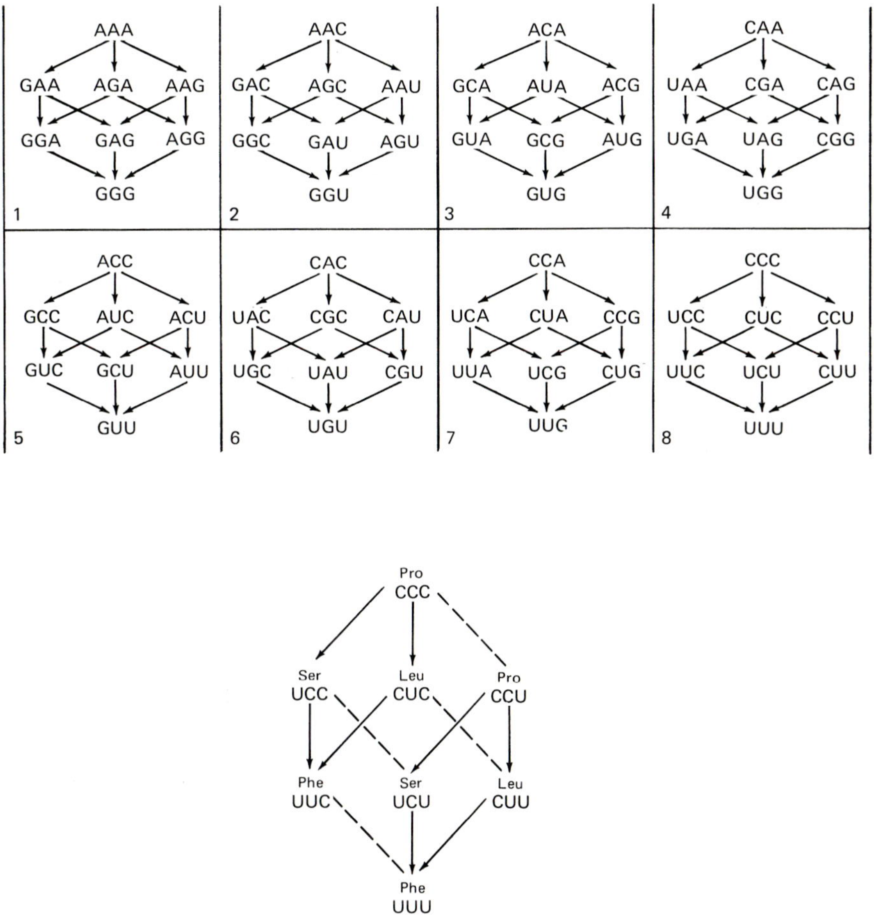

Fig. 20-5. **Vector diagram of HNO_2-induced nucleotide alteration and the correlated amino acid exchanges. (Wittmann, 1962.)**

acid-induced mutants should correspond to an alteration of C→U or A→G in the triplets.

After the codon dictionary was established (through studies on the cell-free system of *E. coli*) it was possible to compare in vivo and in vitro data to see whether the postulations just mentioned were fulfilled (Wittmann and Wittman-Liebold, 1966).

(1) Comparison of the amino acid exchanges with the codons shows that all amino acid exchanges can be correlated with the alteration of only one nucleotide per codon (Table 20-1). If one makes this comparison for amino acids having the same position in the protein sequence of tobacco mosaic virus

strains, only about 50% can be "explained" by alteration of only one nucleotide per codon. As one might expect, more than one nucleotide per codon could have been altered during the evolution of these strains.

(2) A much more critical test is the question whether amino acid exchanges found in nitrous acid-induced mutants correspond only to alterations of C→U or A→G in triplets. As can be seen from Table 20-1, this is the case with all exchanges (except one: glutamic acid→aspartic acid in PM 2, which is probably not a nitrous acid-induced but a spontaneous mutant [A. Siegel, personal communication]). This very good agreement of amino acid exchange data in tobacco mosaic virus with codon assignments from in vitro studies, shows that nitrous acid acts in the manner expected, that the codons based on studies in the "artificial" in vitro system have real biological significance, and that the code is at least very similar in *E. coli* and tobacco.

(3) The probability for each amino acid to be exchanged by a single step to another amino acid has been calculated for spontaneous and nitrous acid-induced mutants of tobacco mosaic virus (Wittmann and Wittmann-Liebold, 1966). Figure 20-6 compares the calculated probability of amino acid exchanges with the frequency of exchanges actually found in nitrous acid-induced mutants. Here again, one sees the very good qualitative agreement,

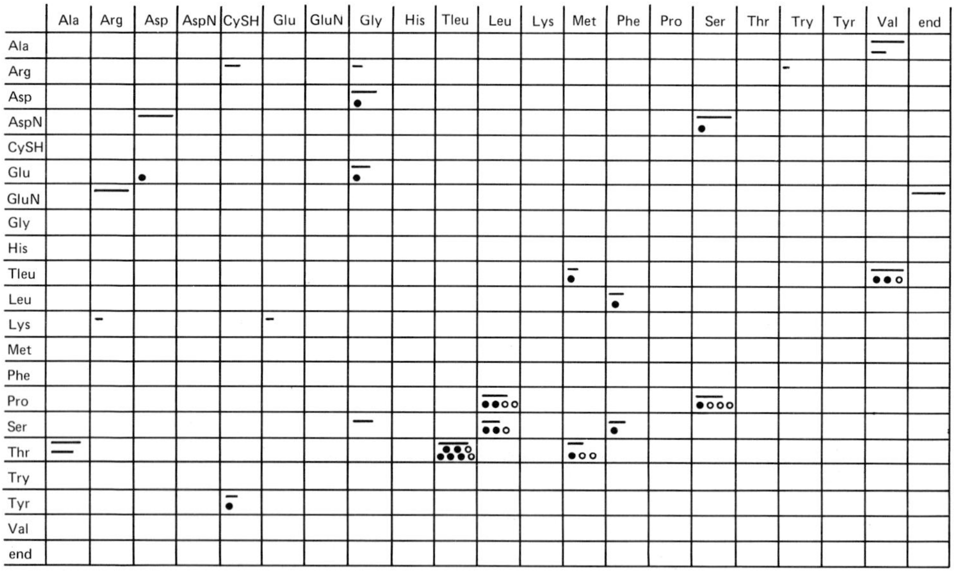

Fig. 20-6. Comparison of amino acid exchanges expected and actually found. The length of each line is proportional to the expected probability of a given amino acid exchange. The number of circles shows how many times a given amino acid exchange has been found. Filled circles indicate exchanges observed in different amino acid positions, and empty circles show the repeats, that is, exchanges at the same position but in different mutants. (Wittmann and Wittmann-Liebold, 1966.)

mentioned above, between "allowed" and observed exchanges. On the other hand, quantitative correlation is not close between the number of amino acid exchanges actually observed and the calculated probabilities for these exchanges, for example, the exchange threonine → alanine is expected with a high probability and has never been observed in nitrous acid-induced mutants, whereas the exchange threonine → isoleucine, which is expected with a lower frequency than threonine → alanine, was found several times and in various parts of the protein chain. This disagreement between calculated and actually observed exchanges is probably a consequence of restrictions caused by requirements of spatial structure of the tobacco mosaic virus rod. This is discussed in more detail in the following section.

(4) It is known that polycations (e.g. streptomycin, spermine, Mg^{2+}), and abnormal temperature induce misreading by bacterial ribosomes, causing incorporation of wrong amino acids. This effect was tested for the tobacco mosaic virus system (Wittmann, 1965*a*). No such exchanges could be detected in the viral coat protein, demonstrating a high fidelity of translation of the genetic message by plant ribosomes. This fidelity has also been shown with other eukaryotic ribosomes (Weinstein et al., 1966).

B. Protein Structure

The nucleotide sequence of tobacco mosaic virus RNA determines the amino acid sequence of the polypeptide chain of the coat protein, as discussed in the previous section. The amino acid sequence, on the other hand, contains all the information necessary for folding the polypeptide chain into its biologically active tertiary structure. After the three-dimensional structure of the coat protein subunit is destroyed by denaturation, the tertiary structure can be renatured merely by incubating the denatured polypeptide chain in a suitable buffer (Anderer, 1959*a*).

The following results lead to the conclusion that different parts of the amino acid sequence of the coat protein differ importantly in formation of the biologically active tertiary structure. Amino acid exchanges in many positions of the coat protein can be present in tobacco mosaic virus strains without changing the tertiary structure or function of the coat protein significantly (Holmes and Franklin, 1958; Wittmann and Wittmann-Liebold, 1966). On the other hand, a single amino acid exchange per subunit may be sufficient to cause extensive alteration of both the tertiary structure and the biological function, as, for example, in the defective mutants (Siegel et al., 1966). Certain regions of the amino acid sequences apparently have to remain unchanged or, at least, are variable in only a certain way, as already known for other proteins (Perutz et al., 1965).

1. *CONSTANT REGIONS OF AMINO ACID SEQUENCE*

Certain amino acids of a protein molecule are absolutely necessary for its biological function, and the three-dimensional structure of the protein requires the side chains of other amino acids. Such amino acid sequences have to be present constantly in homologous sequences of proteins with similar conformation and function; they can be identified by comparing the amino acid sequence of many homologous proteins.

Of the 158 amino acid sequence positions of the coat protein, the amino acids in 47 positions are identical in 4 tobacco mosaic virus strains and 69 coat protein mutants sequenced up to now; that is, 30% of the amino acid sequence is unchanged (see Fig. 20-7). Most of these constant amino acids are distributed

```
              1   2   3   4   5   6   7   8   9   10  11  12  13  14  15  16  17  18
V    Acetyl-Ser-Tyr-Ser-Ile-Thr-Thr-Pro-Ser-Gln-Phe-Val-Phe-Leu-Ser-Ser-Ala-Try-Ala-
D      "    - " - " - " - " - " -Ser- " - " - " - " - " - " - " - " -Val- " - " -
U2    NH2  -Pro- " -Thr- " -Asn- " - " - " - " - " - " -Tyr- " - " - " -Ala-Tyr- " -
H    Acetyl -Ser- " -Asn- " -Thr-Asn-Ser-Asn- " -Tyr-Gln- " -Phe-Ala-Ala-Val-Try- " -

        19  20  21  22  23  24  25  26  27  28  29  30  31  32  33  34  35  36  37
V       Asp-Pro-Ile-Glu-Leu-Ile-Asn-Leu-Cys-Thr-Asn-Ala-Leu-Gly-Asn-Gln-Phe-Gln-Thr-
D        " - " - " - " - " -Leu- " -Val- " - " -Ser-Ser- " - " - " - " - " - " - " -
U2       " - " -Val- " - " -Ile- " -Leu- " - " -Asn-Ala- " - " - " - " - " - " - " -
H       Glu- " -Thr-Pro-Met-Leu- " -Gln- " -Val-Ser- " - " -Ser-Gln-Ser-Tyr- " - " -

        38  39  40  41  42  43  44  45  46  47  48  49  50  51  52  53  54  55  56
V       Gln-Gln-Ala-Arg-Thr-Val-Val-Gln-Arg-Gln-Phe-Ser-Gln-Val-Try-Lys-Pro-Ser-Pro-
D        " - " - " - " - " -Thr- " - " -Gln- " - " - " -Glu- " - " - " - " -Phe- " -
U2       " - " - " - " - " - " - " - " - " - " - " -Ala-Asp-Ala- " - " - " -Ser- " -
H        " -Ala-Gly- " -Asp- " - " -Arg- " - " - " - " -Asn-Leu-Leu-Ser-Thr-Ile-Val-

        57  58  59  60  61  62  63  64  65  66  67  68  69  70  71  72  73  74  75
V       Gln-Val-Thr-Val-Arg-Phe-Pro-Asp-Ser-Asp-Phe-Lys-Val-Tyr-Arg-Tyr-Asn-Ala-Val-
D        " -Ser- " - " - " - " - " -Gly-Asp-Val-Tyr- " - " - " - " - " - " - " - " -
U2      Val-Met- " - " - " - " - " -Ala-Ser-Asp-Phe-Tyr- " - " - " - " - " -Ser-Thr-
H       Ala-Pro-Asn-Gln- " - " - " -Asp-Thr-Gly- " -Arg- " - " -Val-Asn-Ser-Ala-Val-

        76  77  78  79  80  81  82  83  84  85  86  87  88  89  90  91  92  93  94
V       Leu-Asp-Pro-Leu-Val-Thr-Ala-Leu-Leu-Gly-Ala-Phe-Asp-Thr-Arg-Asn-Arg-Ile-Ile-
D        " - " - " - " -Ile- " - " - " - " - " -Thr- " - " - " - " - " - " - " - " -
U2       " - " - " - " - " - " - " - " - " - " -Asn-Ser- " - " - " - " - " - " - " -
H       Ile-Lys- " - " -Tyr-Glu- " - " -Met-Lys- " - " - " - " - " - " - " - " - " -

        95  96  97  98  99  100 101 102 103 104 105 106 107 108 109 110 111 112 113
V       Glu-Val-Glu-Asn-Gln-Ala-Asn-Pro-Thr-Thr-Ala-Glu-Thr-Leu-Asp-Ala-Thr-Arg-Arg-
D        " - " - " - " - " -Gln-Ser- " - " - " - " - " - " - " - " - " - " - " - " -
U2      Glx- " -Asx-Asx-Glx-Ala-Asx- " - " - " - " -(Asx,Thr,Glx,Glx,Pro,Val,Ile) " -
H       Gln-Thr-Glu-Glu-Gln-Ser-Arg- " -Ser-Ala-Ser-Gln-Val-Ala-Asn-Ala-Thr-Gln- " -

        114 115 116 117 118 119 120 121 122 123 124 125 126 127 128 129 130 131 132
V       Val-Asp-Asp-Ala-Thr-Val-Ala-Ile-Arg-Ser-Ala-Ile-Asn-Asn-Leu-Ile-Val-Glu-Leu-
D        " - " - " - " - " - " - " - " - " - " - " - " - " - " - " Val-Asn- " - " -
U2       " - " - " - " - " - " - " - " - " -Ala-Ser- " - " - " - " -Ala- " - " - " -
H        " - " - " - " - " - " - " - " - " -Ser-Gln- " -Gln-Leu- " -Leu- " - " - " -

        133 134 135 136 137 138 139 140 141 142 143 144 145 146 147 148 149 150 151
V       Ile-Arg-Gly-Thr-Gly-Ser-Tyr-Asn-Arg-Ser-Ser-Phe-Glu-Ser-Ser-Ser-Gly-Leu-Val-
D       Val- " - " - " - " -Leu- " - " -Gln-Asn-Thr- " - " - " -Met- " - " - " - " -
U2       " - " - " - " - " -Met-Phe- " - " -Ala-Gly- " - " -Thr-Ala- " - " - " - " -
H       Ser-Asx-His-Gly- " -Tyr-Met- " -Arg- " -Glu- " - " -←—Ala-Ile—→" -Pro-

        152 153 154 155 156 157 158
V       Try-Thr-Ser-Gly-Pro-Ala-Thr
D        " - " - " -Ala- " - " -Ser
U2       " - " -Thr-Thr- " - " -Thr
H        " - " - " -Ala- " - " - "
```

Fig. 20-7. **Comparison of amino acid sequences of the coat protein of tobacco mosaic virus strains. (Wittmann-Liebold and Wittmann, 1967.)**

almost equally along the protein chain. There are, however, differences in the frequency and distribution of apolar and polar amino acids relative to each other: 29 of the constant amino acids are apolar, and most of these are distributed equally along the protein chain. The remaining 18 constant amino acids are polar and are concentrated mainly within three sequence regions. Two of these are discussed again later, since they might be involved in a specific function of the protein. Investigations with other proteins have shown that by appropriate folding of a polypeptide chain, apolar amino acids are placed predominantly in the interior of the folded protein and polar amino acids predominantly at the surface (Perutz et al., 1965). For steric reasons, there is very limited structural variation of the amino acids in the interior of the molecule, that is, predominantly the apolar amino acids. In addition to the result that most of the constant amino acids in the coat protein that are not located in the three clusters of constant polar amino acids are apolar amino acids, those steric reasons might be responsible for the constancy of these amino acids.

There are two longer-sequence regions in the coat protein that are constant in all the tobacco mosaic virus strains and mutants investigated and are composed mainly of polar amino acids: positions 87-94: Phe-Asp-Thr-Arg-Asn-Arg-Ile-Ile, and positions 113-122: Arg-Val-Asp-Asp-Ala-Thr-Val-Ala-Ile-Arg. The primary structure and data gained from X-ray analysis and physicochemical characterization of tobacco mosaic virus indicate that these positions might be responsible for the protein-RNA interaction in the intact virus by ionic bonding between arginine and nucleotide-phosphate, and by the hydrogen bonding of either aspartic acid or threonine to nucleotide base (Hennig, discussed in more detail elsewhere).

2. CONDITIONAL AMINO ACID EXCHANGES

Quantitative analysis of the observed frequencies of the various amino acid exchanges in chemically induced mutants proved that there are significant deviations of the theoretically expected frequencies of amino acid exchanges in the tobacco mosaic virus coat protein (Wittmann and Wittmann-Liebold, 1966; von Sengbusch, 1967). The polar amino acids Asp, Asn, Glu, Gln, Lys, Arg, Ser, and Thr are exchanged less frequently than expected, and some of them not at all. For example, although the exchange Thr→Val should theoretically occur most frequently in the coat protein in HNO_2-induced tobacco mosaic virus mutants, it has never been found. On the other hand, with the apolar amino acids Leu, Ile, Pro, Phe, and Val, all types of exchanges are found in the expected frequencies, even those which are expected in low frequencies, as, for example, Ile→Met and Leu→Phe.

Physicochemical analysis of the coat protein of various mutants showed that, in general, protein molecules with an exchange of a polar amino acid or of proline are thermally less stable than the original molecule and than proteins with an exchange of an apolar amino acid (Jockusch, 1968*b*). The altered stability after substitution of a proline is reasonable, since proline, because of its special structure, forces the polypeptide chain into a rigid bend.

There is no reason to assume that the existence of constant-sequence areas in the polypeptide chain and the difference between theoretically expected and actually observed frequency of amino acid exchanges are caused by a mechanism which acts directly at the RNA. One has rather to assume that those mutations which are expected but not observed do occur, but they are lethal because they lead to a drastic alteration of the protein conformation. This assumption is supported by the occurrence of defective mutants which can be isolated only under certain conditions, and of temperature-sensitive mutants whose proteins are defective after multiplication at high temperature. A model of the three-dimensional molecular structure of tobacco mosaic virus is needed, however, for estimating the value of all these assumptions about the influence of the amino acid sequence on the conformation of its coat protein, but such a model is still to be developed.

ACKNOWLEDGMENT

Thanks are due Drs. L. Bockstahler and G. Nass for help in preparing the English translation of the German manuscript, and for its critical reading.

References*

Aach, H. G. 1957. Serologische untersuchungen an Mutanten des Tabakmosaikvirus. Z. Naturforsch. 12b:614-622.

Aach, H. G. 1958*a*. Spektrophotometrische untersuchungen an Mutanten des Tabakmosaikvirus. Z. Naturforsch. 13b:165-167.

Aach, H. G. 1958*b*. Quantitative aminosäuren-analysen an Mutanten des Tabakmosaikvirus. Z. Naturforsch. 13b:425-433.

Aach, H. G. 1963. Bestimmung der elektrophoretischen Beweglichkeit mittels Zonenelektrophorese im Dichtegradienten. Z. Naturforsch. 18b:337-339.

Abbott, E. V., and J. W. Ingram. 1942. Transmission of chlorotic streak of sugar cane by the leaf hopper *Draeculacephala portola*. Phytopathology 32:99-100.

Abelson, J., and C. A. Thomas. 1966. The anatomy of the T5 bacteriophage DNA molecule. J. Mol. Biol. 18:262-291.

Abu Salih, H. S., A. F. Murant, and M. J. Daft. 1968. Comparison of the passive haemagglutination and bentonite flocculation tests for serological work with plant viruses. J. Gen. Virol. 2:155-166.

Ackers, G. K., and R. L. Steere. 1962. Restricted diffusion of macromolecules through agar-gel membranes. Biochim. Biophys. Acta 59:137-149.

Ackers, G. K., and R. L. Steere. 1967. Molecular sieve methods, pp. 325-365. *In* K. Maramorosch and H. Koprowski [ed.] Methods in virology, vol. 2. Academic Press, New York. 4 vols.

Adams, D. H. 1970. The nature of the scrapie virus: a review of recent progress. Pathol.-Biol. 18:559-577.

Adams, E. T., Jr. 1967. Analysis of self-associating systems by sedimentation equilibrium experiments. Fractions 3:1-18. Spinco Division of Beckman Instruments, Inc., Palo Alto, Calif.

Agrawal, H. O. 1964. Identification of cowpea mosaic virus isolates. Mededel. Landbouwhogeschool. Wageningen 64-5:1-53.

Agrawal, H. O., and G. Bruening. 1966. Isolation of high-molecular weight ^{32}P-labeled influenza virus ribonucleic acid. Proc. Nat. Acad. Sci. U.S. 55:818-825.

Agrawal, H. O., J. W. Kent, and D. M. Mackay. 1965. Rotation technique in electron microscopy of viruses. Science 148:638-640.

Ahmed, M. E., L. M. Black, E. G. Perkins, B. L. Walker, and F. A. Kummerow. 1964. Lipid in potato yellow dwarf virus. Biochem. Biophys. Res. Commun. 17:103-107.

Ainsworth, G. C. 1935. Mosaic diseases of the cucumber. Ann. Appl. Biol. 22:55-67.

Albertsson, P. A. 1960. Partition of cell particles and macromolecules. Almqvist and Wiksells, Uppsala. 231 pp.

Albertsson, P. A. 1967. Two-phase separation of viruses, pp. 303-321. *In* K. Maramorosch and H. Koprowski, [ed.] Methods in virology, vol. 2. Academic Press, New York. 4 vols.

Alberty, R. A. 1948. An introduction to electrophoresis. I. Methods and calculations. II. Analysis and theory. J. Chem. Ed. 25:426-433 and 25:619-625.

---

*The number in parentheses following some of the references is the chapter number in which the reference was cited.

Albrechtová, L., R. Casper and E. Schulze. 1971. Untersuchungen über das Prunus necrotic ring spot virus. Partielle Reinigung, Antiserumherstellung und Gewinnung der -Globulinfraktion. Phytopathol Z. (in press).

Alden, E. 1956. The production of an antiserum against an antigen from an unstable plant virus. Master's Thesis. Univ. Illinois, Urbana.

Alexopoulos, C. J. 1962. Introductory mycology. 2nd ed. Wiley, New York. 613 pp.

Allen, W. R., and J. H. Tremaine. 1965. Characteristics of antigens and antibodies associated with the prunus necrotic ring spot virus immune system. Virology 25:122-132.

Allington, W. B., R. Staples and G. Viehmyer. 1968. Transmission of rose rosette virus by the eriophyid mite *Phyllocoptes fructiphilus*. J. Econ. Entomol. 61:1137-1140.

Almeida, J. D., and A. F. Howatson. 1963. A negative staining method for cell associated virus. J. Cell Biol. 16:616-620.

Amelunxen, F. 1958. Die Virus-Eiweissspindeln der Kakteen. Darstellung elektronenmikroskopische und biochemische Analyse des Virus. Protoplasma 49:140-178.

Amici, A. 1965. Ulteriori richerche sulla diffusione di *Xiphinema index* e sulla presenza di altri nematodi nei vigneti italiani. Riv. Pat. Veg. Ser. IV 1:109-128.

Amos, J., R. G. Hatton, R. C. Knight, and A. M. Massee. 1927. Experiments in the transmission of reversion in black currants. Annu. Rep. East Malling Res. Sta. Kent 13:126-150.

Anderer, F. A. 1959*a*. Reversible Denaturierung des Proteins aus Tabakmosaikvirus. Z. Naturforsch. 14b:642-647.

Anderer, F. A. 1959*b*. Das Molekulargewicht der Peptideinheit im Protein des Tabakmosaikvirus. Z. Naturforsch. 146:24-28.

Anderer, F. A. 1963. Recent studies on the structure of tobacco mosaic virus. Advances Protein Chem. 18:1-35.

Anderer, F. A., and H. D. Schlumberger. 1965*a*. Properties of different artifical antigens immunologically related to tobacco mosaic virus. Biochim. Biophys. Acta 97:503-509. (17)

Anderer, F. A., and H. D. Schlumberger. 1965*b*. Kreuzreaktionen von Antiseren gegen heterologe terminale Aminosäuresequenzen and zwei Stammen des Tabakmosaikvirus. Z. Naturforsch. 20b:564-568.

Anderer, F. A., and H. D. Schlumberger. 1966*a*. Cross-reactions of antisera against the terminal amino acid and dipeptide of tobacco mosaic virus. Biochim. Biophys. Acta 115:222-224. (17)

Anderer, F. A., and H. D. Schlumberger. 1966*b*. Zur immunologischen Spezifitat der Antiseren gegen terminale Aminosäure-Sequenzen von zwei Stammen des Tabakmosaikvirus. Z. Naturforsch. 21b:602-603. (17)

Anderer, F. A., H. D. Schlumberger, and H. Frank. 1967. A serological screening method for the detection of free C-terminal amino acids in virus coat proteins. Biochim. Biophys. Acta 140:80-92.

Anderer, F. A., H. Uhrig, E. Weber and G. Schramm. 1960. Primary structure of the protein of TMV. Nature 186:922.

Anderer, F. A., B. Wittmann-Liebold, and H. G. Wittmann. 1965. Weitere untersuchungen zur amino säuresequenz des proteins in tabakmosaikvirus. Z. Naturforsch. 20b:1203-1213.

Anderson, C. W. 1957. Seed transmission of three viruses in cowpea. Phytopathology 47:515. (Abstr.)

Anderson, N. G. 1956. Techniques for the mass isolation of cellular components, pp. 300-347. *In* G. Oster and A. W. Pollister [ed.] Physical techniques in biological research, vol. 3. Academic Press, New York. 6 vols.

Anderson, N. G. 1966. The development of zonal centrifuges and ancillary systems for tissue fractionation and analysis. National Cancer Institute Monograph 21. U.S. Government Printing Office, Washington, D.C. 526 pp.

Anderson, N. G. 1968. Analytical techniques for cell fractions. VIII. Analytical differential centrifugation in angle head rotors. Anal. Biochem. 23:72-83.

Anderson, N. G., and G. B. Cline. 1967. New centrifugal methods for virus isolation, pp. 137-178. *In* K. Maramorosch and H. Koprowski, [ed.] Methods in virology, vol. 2. Academic Press, New York. 4 vols.

Andrewes, C. H. 1965. Viruses and Noah's ark. Bacteriol. Rev. 29:1-8.

Andrewes, C. H. 1968. Methods of virus classification, pp. 593-613. *In* K. Maramorosch, and H. Koprowski [ed.] Methods in virology, vol. 4. Academic Press, New York. 4 vols.

Andrewes, C. H., and Horstmann. 1949. The susceptibility of viruses to ethyl ether. J. Gen. Microbiol. 3:290-297.

Ansevin, A. T., and M. A. Lauffer. 1959. Native tobacco mosaic virus protein of molecular weight 18,000. Nature 183:1601-1602.

Ansevin, A. T., C. L. Stevens, and M. A. Lauffer. 1964. Polymerization-depolymerization of tobacco mosaic virus protein. III. Changes in ionization and in electrophoretic mobility. Biochemistry 3:1512-1518.

Aoki, S., and I. Takebe. 1969. Infection of tobacco mesophyll protoplasts by tobacco mosaic virus ribonucleic acid. Virology 39:439-448.

Arnott, H. J., S. W. Rosso, and K. M. Smith. 1969. Modifications of plastid ultrastructure in tomato leaf cells infected with tobacco mosaic virus. J. Ultrastructure Res. 27:149-167.

Arnott, H. J., and K. M. Smith. 1967. Electron microscopy of virus-infected sunflower leaves. J. Ultrastructure Res. 19:173-195.

Arnott, H. J., K. M. Smith, and L. L. Fullilove. 1968. Ultrastructure of a cytoplasmic polyhedrosis virus affecting the monarch butterfly *Danaus plexippus*. I. Development of virus and normal polyhedra in the larva. J. Ultrastructure Res. 24:479-507.

Atabekov, J. G., V. K. Novikov, N. A. Kiselev, and A. S. Kaftanova. 1968. Stable intermediate aggregates formed by the polymerization of barley stripe mosaic virus protein. Virology 36:620-638. (15)

Atabekov, J. G., G. A. Popova, N. A. Kiselev, A. S. Kaftanova, and G. V. Petrovsky. 1968. In vitro polymerization of winter wheat mosaic virus antigen. Virology 35:458-472. (15)

Atchison, B. A., R. I. B. Francki, and N. C. Crowley. 1969. Inactivation of lettuce necrotic yellows virus by chelating agents. Virology 37:396-403.

Auclair, J. L. 1963. Aphid feeding and nutrition. Annu. Rev. Entomol. 8:439-490.

Auclair, J. L., and J. J. Cartier. 1963. Pea aphid: rearing on a chemically defined diet. Science 142:1068-1069.

Augier de Montgremier, H. 1963. Libération spontanée d'acide ribonucléique infectieux par le virus de la mosaïque jaune du navet au cours de sa diffusion dans un gel d'agar. Compt. Rend. 257:2201-2203.

Augier de Montgremier, H. 1965. Mise en evidence d'une libération spontanée d'acide ribonucléique infectieux par le virus de la mosaique jaune du navet au cours de sa diffusion dans un gel d'agar. Phytiat. Phytopharm. 13:157-162.

Augier de Montgremier, H., and G. Morel. 1948. Sur la diminution de la teneur en virus (*Marmor tabaci* Holmes) de tissus de tabac cultives in vitro. Compt. Rend. 227:688-689.

Avrameas, S., C. Brailovsky, and T. Ternynck. 1969. Isolation of viruses using specific immunoadsorbents. Virology 37:508-510.

Ayala, A., and M. W. Allen. 1966. Transmission of the California tobacco virus by three species of the nematode genus Trichodorus. Nematologica 12:87. (Abstr.)

Ayala, A., and M. W. Allen. 1968. Transmission of the California tobacco rattle virus (CTRV) by three species of the nematode genus *Trichodorus*. J. Agr. Univ. Puerto Rico 52:101-125.

Babos, P., and B. Kassanis. 1962. Unstable variants of tobacco necrosis virus. Virology 18:206-211.

Bachman, L., and M. M. Salpeter. 1965. Autoradiography with the electron microscope. A quantitative evaluation Lab. Invest. 14:1041-1053, and also *In* G. F. Bahr and E. H. Zeitler [ed.] Symposium on quantitative electron microscopy. Williams and Wilkins, Baltimore. 602 pp.

Backus, R. C., and R. C. Williams. 1950. The use of spraying methods and of volatile suspending media in the preparation of specimens for electron microscopy. J. Appl. Phys. 21:11-15.

Baermann, G. 1917. Eine einfache Methode zur Affindung von Ankylostomum (Nematoden) larven in Erdproben. Meded. Geneesk. Lab. Weltevreden, Feestbundel p. 41.

Bahr, G. F., and E. H. Zeitler. 1965. The determination of magnification in the electron microscope. II. Means for the determination of magnification. Lab. Invest. 14:880-891, and also *In* G. F. Bahr and E. H. Zeitler [ed.] Symposium on quantitative electron microscopy. Williams and Wilkins, Baltimore. 602 pp.

Baker, E. A., and J. T. Martin. 1967. Studies on plant cuticle X. The cuticles of plants of related families. Ann. Appl. Biol. 60:313-319.

Baker, K. F., and S. H. Smith. 1966. Dynamics of seed transmission of plant pathogens. Annu. Rev. Phytopathol. 4:311-334.

Bald, J. G. 1937*a*. The use of numbers of infections for comparing the concentration of plant virus suspensions. I. Dilution experiments with purified suspensions. Ann. Appl. Biol. 24:33-55.

Bald., J. G. 1937*b*. The use of numbers of infections for comparing the concentration of plant virus suspensions. II. Distortion of the dilution series. Ann. Appl. Biol. 24:56-76.

Bald, J. G. 1937*c*. The use of numbers of lesions for comparing the concentrations of plant virus suspensions. IV. Modification of the simple dilution equation. Australian J. Exp. Biol. Med. Sci. 15:211-220.

Bald, J. G. 1948. The development of amoeboid inclusion bodies of tobacco mosaic virus. Australian J. Sci. (Res. 1) 458-463.

Bald, J. G. 1949*a*. A method for the selective staining of viruses in infected plant tissues. Phytopathology 39:395-402.

Bald, J. G. 1949*b*. Additional methods for fixing and staining viruses in infected plant tissues. Amer. J. Bot. 36:335-342.

Bald, J. G. 1964. Cytological evidence for the production of plant virus ribonucleic acid in the nucleus. Virology 22:377-387.

Bald, J. G. 1966. Cytology of plant virus infections. Advances Virus Res. 12:103-125.

Bald, J. G., and G. Samuel. 1934. Some factors affecting the inactivation rate of the virus of tomato spotted wilt. Ann. Appl. Biol. 21:179-190.

Baldwin, R. L. 1959. Equilibrium sedimentation in a density gradient of materials having a continuous distribution of effective densities. Proc. Nat. Acad. Sci. U.S. 45:939-944.

Ball, E. M. 1964. Serology: techniques used in plant virus research, pp. 235-252. *In* M. K. Corbett and H. D. Sisler [ed.] Plant virology. Univ. Florida Press, Gainesville. 527 pp.

Ball, E. M. 1966. A technique for comparing the electrophoretic mobility rates of viruses or virus strains. Arch. Biochem. Biophys. 114:547-556.

Ball, E. M., and M. K. Brakke. 1969. Analysis of antigen-antibody reactions of two plant viruses by density-gradient centrifugation and electron microscopy. Virology 39:746-758.

Bancroft, J. B. 1962. Purification and properties of bean pod mottle virus and associated centrifugal and electrophoretic components. Virology 16:419-427.

Bancroft, J. B. 1968. Plant viruses: defectiveness and dependence. 18th Symp. Soc. Gen. Microbiol. pp. 229-247. Cambridge Univ. Press, London. 372 pp.

Bancroft, J. B. 1970. Brome mosaic. Description of Plant Viruses No. 3. C.M.I./A.A.B.

Bancroft, J. B., and E. Hiebert. 1967. Formation of an infectious nucleoprotein from protein and nucleic acid isolated from a small spherical virus. Virology 32:354-356.

Bancroft, J. B., E. Hiebert, M. W. Rees, and R. Markham. 1968. Properties of cowpea chlorotic mottle virus; its protein and nucleic acid. Virology 34:224-239.

Bancroft, J. B., G. J. Hills, and R. Markham. 1967. A study of the self-assembly process in a small spherical virus; formation of organized structures from protein subunits in vitro. Virology 31:354-379.

Bancroft, J. B., E. L. Moorhead, J. Tuite, and H. P. Liu. 1960. The antigenic characteristics and the relationship among strains of alfalfa mosaic virus. Phytopathology 50:34-39.

Bancroft, J. B., and S. A. Tolin. 1967. Apple latent virus 2 is sowbane mosaic virus. Phytopathology 57:639-640.

Bancroft, J. B., G. W. Wagner, and C. E. Bracker. 1968. The self-assembly of a nucleic acid free pseudo-top component from small spherical virus. Virology 36:146-149.

Banerjee, K. 1969. The mechanism of photoreactivation of some plant viruses. Master's Thesis. Univ. Montana.

Banttari, E. E. 1967. Leafhopper transmission of the oat blue dwarf virus extracted from oats and barley. Phytopathology 57:802.

Banttari, E. E., and M. B. Moore. 1962. Virus cause of blue dwarf of oats and its transmission to barley and flax. Phytopathology 52:897-902.

Banttari, E. E., and R. J. Zeyen. 1969. Chromatographic purification of the oat blue dwarf virus. Phytopathology 59:183-186.

Barber, R. 1966. The chromatographic separation of ribonucleic acids. Biochim. Biophys. Acta 114:422-424.

Barnard, E. A. 1969. Ribonucleases. Annu. Rev. Biochem. 38:677-732.

Barr, D. J. S. 1968. A new species of *Olpidium* parasitic on cucumber roots. Can. J. Bot. 46:1087-1091.

Bartels, W. 1955. Der gegenwartige Stand der Forschung auf dem Gebiet der Inaktivierung pflanzen-pathogener Viren, insbesondere des Tabakmosaikvirus. Phytopathol. Z. 24:117-178.
Baudet, J., O. Croissant, D. G. Dervichian, M. Joly, and J. Mosse. 1951. Variation of the size and the size distribution of tobacco mosaic virus particles depending on the method of preparation. Disc. Faraday Soc. 11:236-246.
Bawden, F. C. 1952. Physiology of virus diseases. Annu. Rev. Plant Physiol. 3:171-185.
Bawden, F. C. 1954. Inhibitors and plant viruses. Advances Virus Res. 2:31-57.
Bawden, F. C. 1959. Effect of nitrous acid on tobacco mosaic virus—mutation or selection. Nature (London) 184:(4688)B.A. 27-29.
Bawden, F C. 1964*a*. Plant viruses and virus diseases. 4th ed. Ronald Press, New York. 361 pp. (1, 2, 11, 19)
Bawden, F. C. 1964*b*. Speculations on the origins and nature of viruses, pp. 365-385. *In* M. K. Corbett and H. D. Sisler [ed.] Plant virology, Univ. Florida Press, Gainesville. 527 pp. (6)
Bawden, F. C., and B. Kassanis. 1945. The suppression of one plant virus by another. Ann. Appl. Biol. 32:52-57.
Bawden, F. C., and B. Kassanis. 1947*a*. *Primula obconica,* a carrier of tobacco necrosis viruses. Ann. Appl. Biol. 34:127-135. (1)
Bawden, F. C., and B. Kassanis. 1947*b*. The behaviour of some naturally occurring strains of potato virus Y. Ann. Appl. Biol. 34:503-516. (5)
Bawden, F. C., and A. Kleczkowski. 1953. The behaviour of some plant viruses after exposure to ultraviolet radiation. J. Gen. Microbiol. 8:145-156.
Bawden, F. C., and A. Kleczkowski. 1955. Studies on the ability of light to counteract the inactivating action of ultraviolet radiation on plant viruses. J. Gen. Microbiol. 13:370-382.
Bawden, F. C., and F. M. Roberts. 1947. The influence of light intensity on the susceptibility of plants to certain viruses. Ann. Appl. Biol. 34:286-296.
Beale, H. P., and C. R. Jones. 1951. Virus diseases of tobacco mosaic and potato yellow dwarf not controlled by certain purified antibiotics. Contrib. Boyce Thompson Inst. 16:395-407.
Beever, J. E., and P. R. Fry. 1970. The effect of light on the transmission of tobacco necrosis virus by *Olpidium brassicae*. Virology 40:357-362.
Behrens, J. 1899. Weiter Beitrage zur Kenntnis der Tabakpflanze. XIV. Die Mauche (Mauke) des Tabaks. Landwirtsch. Vers-Sta. 52:442-447.
Beijerinck, M. W. 1898. Uber ein contagium vivum fluidum als Ursache der Fleckenkrankheit der Tabaksbeatter Verhandee. Koninke. Acad. Witenschap. Amsterdam 6:1-24.
Beirne, B. P. 1956. Leafhoppers (Homoptera: Cicadellidae) of Canada and Alaska. Can. Entomol. 88:Suppl. 2, 180 pp.
Bell, C. D., and G. B. Orlob. 1970. Effects of photosensitization on plant virus protein. Virology 41:12-21.
Bellamy, A. R., and W. K. Joklik. 1967. Studies on reovirus RNA. II. Characterization of reovirus messenger RNA and of the genome RNA segments from which it is transcribed. J. Mol. Biol. 29:19-26.
Bellamy, A. R., L. Shapiro, J. T. August, and W. K. Joklik. 1967. Studies on reovirus RNA. I. Characterization of reovirus genome RNA. J. Mol. Biol. 29:1-17.
Bellett, A. J. D. 1967. Preliminary classification of viruses based on quantitative comparisons of viral nucleic acids. J. Virol. 1:245-259.
Bellin, J., and G. Oster. 1960. Photodynamic inactivation of transforming principle. Biochim. Biophys. Acta 42:533-535.
Benda, G. T. A. 1956. Infection of *Nicotiana glutinosa* L. following injection of two strains of tobacco mosaic virus into a single cell. Virology 2:820-827.
Benda, G. T. A., and C. W. Bennett. 1964. Effect of curly top virus on tobacco seedlings: infection without obvious symptoms. Virology 24:97-101.
Bengtsson, S., and L. Philipson. 1964. Chromatography of animal viruses on pearl-condensed agar. Biochim. Biophys. Acta 79:399-406.
Bengtsson, S., L. Philipson, and P. A. Albertsson. 1962. Counter-current distribution of polio virus. Biochem. Biophys. Res. Commun. 9:318-322.
Benjamini, E., M. Shumuzu, D. J. Young, and C. Y. Leung. 1968. Immunochemical studies on the tobacco mosaic virus protein. VII. The binding of octanoylated peptides of the tobacco mosaic virus protein with antibodies to the whole protein. Biochemistry 7:1261-1264.

Bennett, C. W. 1934. Plant tissue relations of the sugar beet curly-top virus. J. Agr. Res. 48:665-701.
Bennett, C. W. 1935. Studies on properties of the curly-top virus. J. Agr. Res. 50:211-241.
Bennett, C. W. 1936. Further studies on the relation of curly top virus to plant tissue. J. Agr. Res. 53:595-620.
Bennett, C. W. 1940. Acquisition and transmission of viruses by dodder (*Cuscuta subinclusa*). Phytopathology 30:2. (Abstr.)
Bennett, C. W. 1944. Latent virus of dodder and its effect on sugar beet and other plants. Phytopathology 34:77-91.
Bennett, C. W. 1956. Biological relation of plant viruses. Annu. Rev. Plant Physiol. 7:143-170.
Bennett, C. W. 1959. Lychnis ringspot. Phytopathology 49:706-713.
Bennett, C. W. 1962. Curly top virus content of the beet leafhopper influenced by virus concentration in diseased plants. Phytopathology 52:538-541.
Bennett, C. W. 1967*a*. Plant viruses: transmission by dodder, pp. 393-401. *In* K. Maramorosch and H. Koprowski [ed.] Methods in virology, vol. 1. Academic Press, New York. 4 vols. (1)
Bennett, C. W. 1967*b*. Apparent absence of cross protection between strains of the curly top virus in the beet leafhopper *Circulifer tenellus*. Phytopathology 57:207-209.
Bennett, C. W. 1967*c*. Epidemiology of leafhopper-transmitted viruses. Annu. Rev. Phytopathol. 5:87-108.
Bennett, C. W. 1969. Seed transmission of plant viruses. Advances Virus Res. 14:221-261.
Bennett, C. W., E. Carsner, G. H. Coons, and E. W. Brandes. 1946. The Argentine curly-top of sugar beet. J. Agr. Res. 72:19-47.
Bennett, C. W., and A. S. Costa. 1949. The Brazilian curly-top of tomato and tobacco resembling North American and Argentine curly-top of sugar beet. J. Agr. Res. 78:675-693.
Bennett, C. W., and A. S. Costa. 1961. Sowbane mosaic caused by a seed-transmitted virus. Phytopathology 51:546-550.
Bennett, C. W., and K. Esau. 1936. Further studies on the relation of the curly top virus to plant tissues. J. Agr. Res. 53:595-620.
Bennett, C. W., and H. E. Wallace. 1938. Relation of the curly-top virus to the vector, *Eutettix tenellus*. J. Agr. Res. 56:31-51.
Bennett, H. S., and J. H. Luft. 1959. s-Collidine as a basis for buffering fixatives. J. Biophys. Biochem. Cytol. 6:113-114.
Benson, A. P., and R. P. Singh. 1964. Seed transmission of potato spindle tuber in tomato. Amer. Potato J. 41:294.
Beraha, L., M. Varzandeh, and H. H. Thornberry. 1955. Mechanism of the action of abrasives on infection by tobacco mosaic virus. Virology 1:141-151.
Bercks, R. 1960. Serologische Untersuchungen zur Differenzierung von Isolaten des Phaseolus-Virus 2 und ihrer Verwandtschaft mit Phaseolus-Virus 1. Phytopathol. Z. 39:120-128.
Bercks, R. 1963. Untersuchungen über individuelle Unterschiede von Antiseren gegen Kartoffel-X-Virus bei Reaktionen mit verwandten Viren. Phytopathol. Z. 47:301-313.
Bercks, R. 1967*a*. Methodische Untersuchungen über den serologischen Nachweis pflanzenpathogener Viren mit dem Bentonit-Flockungstest, dem Latex-Test und dem Bariumsulfat-Test. Phytopathol. Z. 58:1-17.
Bercks. R. 1967*b*. Vorkommen und Nachweis von Viren in Reben. Weinberg u. Keller 14:151-162.
Bercks, R., and O. Loviosolo. 1965. Serologischer Vergleich von Stammen des Tomatenzwergbusch-Virus (tomato bushy stunt virus). Phytopathol. Z. 52:96-101.
Bercks, R., and G. Querfurth. 1969. Weitere methodische Untersuchungen über den Latextest zum serologischen Nachweis pflanzenpathogener Viren. Phytopathol. Z. 65:243-256.
Bercks, R., and G. Querfurth. 1971. The use of the latex test for the detection of distant serological relationships between plant viruses. J. Gen. Virol. 12:25-32.
Berg, T. M. 1962. A quick and efficient inoculation method for the sap-transmission of viruses from woody to herbaceous hosts. Tijdschr. Plantenziekten 68:231-234.
Bergeson, G. B., and K. L. Athow. 1963. Vector relationship of tobacco ringspot virus (TRSV) and *Xiphinema americanum* and the importance of this vector in TRSV infection of soybean. Phytopathology 53:871. (Abstr.)
Bergoin, M., G. Devauchelle, and C. Vago. 1969. Electron microscopy study of the pox-like virus of *Melolontha melolontha* L. (Coleoptera, Scarabeidae). Arch. Ges. Virusforsch. 28:285-302.

Bergold, G. H., O. M. Suarez, and K. Munz. 1968. Multiplication in and transmission by *Aedes aegypti* of vesicular stomatitis virus. J. Invertebrate Pathol. 11:406-428.

Bernal, J. D., and I. Fankuchen. 1941. X-ray and crystallographic studies of plant virus preparations. I. Introduction and preparation of specimens. II. Modes of aggregation of the virus particles. J. Gen. Physiol. 25:111-165.

Bernardi, G. 1969. Chromatography of nucleic acids on hydroxyapatite III. Chromatography of RNA and polyribonucleotides. Biochim. Biophys. Acta 174:449-457.

Bernhard, S. A. 1968. The structure and function of enzymes, chapters 2 and 3. Benjamin, Inc. New York. 324 pp.

Bernhard, W. 1969. A new staining procedure for electron microscopy. J. Ultrastructure Res. 27:250-265.

Bernhard, W., and E. H. Leduc. 1967*a*. Ultrathin frozen sections. I. Methods and ultrastructural preservation. J. Cell Biol. 34:757-771.

Bernhard, W., and E. H. Leduc. 1967*b*. Ultrathin frozen sections. II. Demonstration of enzyme activity. J. Cell Biol. 34:773-786.

Best, R. J. 1936. The effect of light and temperature on the viruses of tomato spotted wilt and tobacco mosaic. Australian J. Exp. Biol. Med. Sci. 14:223-239.

Best, R. J. 1939. The preservative effect of some reducing systems on tomato spotted wilt virus. Australian J. Exp. Biol. Med. Sci. 17:1-7.

Best, R. J., and G. Samuel. 1936. The effect of various reducing systems on the virus of tomato spotted wilt. Ann. Appl. Biol. 23:759-780.

Betto, E., and D. J. Raski. 1966. Attempts to inoculate *Xiphinema index* with grape fanleaf virus by micro-inoculation. Nematologica 12:453-461.

Bewley, W. F. 1924. Mycological investigations. Chestnut Exp. and Res. Sta., Hertfordshire, Ann. Rep. 9(1923):66-69.

Bhargava, K. S. 1951. Some properties of four strains of cucumber mosaic virus. Ann. Appl. Biol. 38:377-379.

Bils, R. F., and C. E. Hall. 1962. Electron microscopy of wound-tumor virus. Virology 17:123-130.

Bindra, O. S., and E. S. Sylvester. 1961. Effect of insect numbers on aphid transmission of potato leaf-roll virus. Hilgardia 31:279-325.

Birnie, G. D., and D. R. Harvey. 1968. A simple density-gradient engine for loading large capacity zonal ultracentrifuge rotors. Anal. Biochem. 22:171-177.

Black, F. L., and C. A. Knight. 1953. A comparison of some mutants of tobacco mosaic virus. J. Biol. Chem. 202:51-57.

Black, L. M. 1937. A study of potato yellow dwarf in New York. N.Y. (Cornell) Agr. Exp. Sta. Mem. 209. 23 pp.

Black, L. M. 1938. Properties of the potato yellow-dwarf virus. Phytopathology 28:863-874.

Black, L. M. 1941. Further evidence for multiplication of the aster-yellows virus in the aster leafhopper. Phytopathology 31:120-135.

Black, L. M. 1943*a*. Genetic variation in the clover leafhopper's ability to transmit potato yellow-dwarf virus. Genetics 28:200-209.

Black, L. M. 1943*b*. Some properties of aster-yellows virus. Phytopathology 33:2. (Abstr.)

Black, L. M. 1944. Some viruses transmitted by agallian leafhoppers. Proc. Amer. Phil. Soc. 88:132-144.

Black, L. M. 1947. Virus tumors in plants. 6th Growth Symp., Suppl. 79-85.

Black, L. M. 1950. A plant virus that multiplies in its insect vector. Nature 166:852-853.

Black, L. M. 1953*a*. Viruses that reproduce in plants and insects. Ann. N.Y. Acad. Sci. 56:398-413.

Black, L. M. 1953*b*. Loss of vector transmissibility by viruses normally insect transmitted. Phytopathology 43:466. (Abstr.)

Black, L. M. 1953*c*. Occasional transmission of some plant viruses through the eggs of their insect vectors. Phytopathology 43:9-10.

Black, L. M. 1959. Biological cycles of plant viruses in insect vectors, pp. 157-185. *In* F. M. Burnet and W. M. Stanley [ed.] The viruses. Academic Press, New York. 3 vols.

Black, L. M. 1965. Physiology of virus-induced tumors in plants, pp. 236-266. *In* A. Lang [ed.] Handbuck der Pflanzenphysiologie. Springer-Verlag, New York.

Black, L. M. 1969. Insect tissue cultures as tools in plant virus research. Annu. Rev. Phytopathol. 7:73-100.

Black, L. M., and M. K. Brakke. 1952. Multiplication of wound-tumor virus in an insect vector. Phytopathology 42:269-273.
Black, L. M., M. K. Brakke, and A. E. Vatter. 1963. Purification and electron microscopy of tomato spotted-wilt virus. Virology 20:120-130.
Black, L. M., and R. Markham. 1963. Abstracts of communications presented at the joint meeting of the "Nederlandse Planteziekten-kundige Vereniging" and the "Nederlandse Kring Voor Plantevirologie" on 24 May 1962. Wageningen. Neth. J. Plant Pathol. 69:215-217.
Black, L. M., K. M. Smith, G. J. Hills, and R. Markham. 1965. Ultrastructure of potato yellow-dwarf virus. Virology 27:446-449.
Black, L. M., S. Wolcyrz, and R. F. Whitcomb. 1958. A vectorless strain of wound-tumor virus. 7th Int. Congr. Microbiol., Stockholm, Abstr. Commun. p. 255. (Abstr.)
Blanton, M. V., and L. B. Barnett. 1969. Adsorption of ribonucleic acid on bentonite. Anal. Biochem. 32:150-154.
Blum, H. F. 1941. Photodynamic action and diseases caused by light. Reinhold, New York. 309 pp.
Boardman, N. K., R. I. B. Francki, and S. G. Wildman. 1966. Protein synthesis by cell-free extracts of tobacco leaves. III. Comparison of the physical properties and protein synthesizing activity of 70S chloroplast and 80S cytoplasmic ribosomes. J. Mol. Biol. 17:470-489.
Bock, K. R. 1967. Strains of Prunus necrotic ringspot virus in hop (*Humulus lupulus* L.). Ann. Appl. Biol. 59:437-446.
Bock, R. M. 1967. Controlled partial hydrolysis of RNA, pp. 218-221. *In* L. Grossman and K. Moldave [ed.] Methods in enzymology, vol. 12A. Academic Press, New York, 19 vols. (16)
Bockstahler, L. E., and P. Kaesberg. 1962. The molecular weight and other biophysical properties of bromegrass mosaic virus. Biophys. J. 2:1-9.
Boedtker, H. 1968. Molecular weight and conformation of RNA, pp. 429-458. *In* L. Grossman and K. Moldave [ed.] Methods in Enzymology, vol. 12B. Academic Press, New York. 19 vols.
Boman, H. G., and L. E. Westlund. 1956. Protein chromatography on an anion exchange resin. Arch. Biochem. Biophys. 64:217-239.
Bos, L. 1967. Graft transmission of plant viruses, pp. 403-410. *In* K. Maramorosch and H. Koprowski [ed.] Methods in virology, vol. 1. Academic Press, New York. 4 vols.
Bos, L., and P. Granchini. 1968. Peculiar histoid enations in white clover and their relationships to virus diseases and toxic effects of leafhopper feeding. Phytopathol. Z. 61:253-272.
Bos, L., and J. P. H. van der Want. 1962. Early browning of pea, a disease caused by a soil- and seed-borne virus. Tijdschr. Plantenziekten 68:368-390.
Bosch, L., E. M. Bonnet-Smits, and J. van Duin. 1967. *In situ* breakage of turnip yellow mosaic virus RNA and *in situ* aggregation of the fragments. Virology 31:453-460.
Bowen, T. J. 1966. Laboratory centrifuges, pp. 1-23. *In* T. W. Goodwin [ed.] Instrumentation in biochemistry. Academic Press, London. 113 pp.
Boyden, S. V. 1951. The adsorption of proteins on erythrocytes treated with tannic-acid and subsequent haemagglutination by anti-protein sera. J. Exp. Med. 93:107-120.
Bozarth, R. F., and A. F. Ross. 1964. Systemic resistance induced by localized virus infections: extent of changes in uninfected plant parts. Virology 24:446-455.
Bozicevich, J., J. Nason, and D. C. Kayhoe. 1960. Deoxyribonucleic acid (DNA)-bentonite flocculation test for Lupus erythematosus. Proc. Soc. Exp. Biol. Med. 103:636-640.
Bozicevich, J., H. A. Scott, and M. M. Vincent. 1963. The bentonite flocculation test for detection of plant viruses and titration of antibody. Proc. Soc. Exp. Biol. Med. 114:794-798.
Bradbury, S., and G. H. Meek. 1960. A study of potassium permanganate "fixation" for electron microscopy. Quart. J. Microscop. Sci. 101:241-250.
Bradfute, O. E., and L. R. Nault. 1969. Ultrastructure of *Graminae* leaf tissue infected by a mite-borne virus-like agent. Phytopathology 59:1019. (Abstr.)
Bradish, C. J., and L. V. Crawford. 1960. Biophysical studies of the interactions between the viruses of tobacco mosaic and tomato bushy stunt and their rabbit antisera. Virology 11:48-78.
Bradley, D. E. 1962. A study of the negative staining process. J. Gen. Microbiol. 29:503-516.
Bradley, D. E. 1965. The structure of the head, collar and base-plate of "T-even" type bacteriophages. J. Gen. Microbiol. 38:395-408.
Bradley, D. E., and A. Rich. 1956. The fractionation of ribonucleic acids on ecteolacellulose anion exchangers. J. Amer. Chem. Soc. 78:5898-5902.
Bradley, R. H. E. 1962*a*. Different areas of tobacco leaves as sources of potato virus Y for aphids. Virology 16:366-370. (1)

Bradley, R. H. E. 1962*b*. Response of the aphid *Myzus persicae* (Sulz.) to some fluid applied to the mouthparts. Can. Entomol. 94:707-722. (5)
Bradley, R. H. E. 1966. Which of an aphid's stylets carry transmissible virus. Virology 29:396-401.
Bradley, R. H. E., and R. Y. Ganong. 1955. Some effects of formaldehyde on potato virus Y in vitro and on the ability of aphids to transmit virus when their stylets are treated with formaldehyde. Can. J. Microbiol. 1:755-782.
Bradley, R. H. E., C. V. Wade, and F. A. Wood. 1962. Aphid transmission of potato virus Y inhibited by oils. Virology 18:327-328.
Brakke, M. K. 1951. Density gradient centrifugation: a new separation technique. J. Amer. Chem. Soc. 73:1847-1848.
Brakke, M. K. 1953. Zonal separations by density-gradient centrifugation. Arch. Biochem. Biophys. 45:275-290.
Brakke, M. K. 1955. Zone electrophoresis of dyes, proteins and viruses in density-gradient columns of sucrose solutions. Arch. Biochem. Biophys. 55:175-190.
Brakke, M. K. 1956. Stability of potato yellow-dwarf virus. Virology 2:463-476.
Brakke, M. K. 1958. Estimation of sedimentating constants of viruses by density-gradient centrifugation. Virology 6:96-114.
Brakke, M. K. 1960. Density gradient centrifugation and its application to plant viruses. Advances Virus Res. 7:193-224.
Brakke, M. K. 1962. Stability of purified barley stripe mosaic virus. Virology 17:131-142.
Brakke, M. K. 1963. Photometric scanning of centrifuged density gradient columns. Anal. Biochem. 5:271-283.
Brakke, M. K. 1964. Non ideal sedimentation and the capacity of sucrose gradient columns for virus in density-gradient centrifugation. Arch. Biochem. Biophys. 107:388-403.
Brakke, M. K. 1967*a*. Density-gradient centrifugation. *In* K. Maramorosch and H. Koprowski [ed.] Methods in virology, vol. 2, pp. 93-118. Academic Press, New York. 4 vols. (6)
Brakke, M. K. 1967*b*. Miscellaneous problems in virus purification, pp. 119-136. *In* K. Maramorosch and H. Koprowski [ed.] Methods in virology, vol. 2. Academic Press, New York. 4 vols.
Brakke, M. K. 1969*a*. Isolation and purification of vector-borne plant viruses, pp. 527-555. *In* K. Maramorosch [ed.] Viruses, vectors, and vegetation. Interscience, New York. 666 pp.
Brakke, M. K. 1969*b*. Isolation and purification of vector-borne plant viruses, pp. 433-448. *In* K. Maramorosch [ed.] Viruses, vectors, and vegetation. Interscience, New York. 666 pp.
Brakke, M. K. 1970. Systemic infections for the assay of plant viruses. Annu. Rev. Phytopathol. 8:61-84.
Brakke, M. K., and A. P. Estes. 1967. Some factors affecting vector transmission of soil-borne wheat mosaic virus from root washings and soil debris. Phytopathology 57:905-910.
Brakke, M. K., A. P. Estes, and M. L. Schuster. 1965. Transmission of soil-borne wheat mosaic virus. Phytopathology 55:79-86.
Brakke, M. K., A. E. Vatter and L. M. Black. 1954. Size and shape of wound-tumor virus. Abnormal and Pathological Plant Growth Brookhaven Symp. Biol. 6:137-156.
Brammer, K. W. 1963. Chemical modification of viral ribonucleic acid. II. Bromination and iodination. Biochim. Biophys. Acta 72:217-229.
Brandes, J. 1957. Einen elektronenmikroskopische Schnellmethode zum Nachweis faden-und stäbschenförmiger Viren insbesondere in Kartoffeldunkelkeimen. Nachrbe. deut. Pflanzenschutzdienst (Braunschweig) 9:151-152.
Brandes, J. 1964. Identifizierung von getreckten pflanzenpathogenen Viren auf morphologischer Grundlage. Mitt. Biol. Bundesanstalt Land-forstwirtsch. Berlin-Dahlem 110:1-130.
Brandes, J., and R. Bercks. 1965. Gross morphology and serology as a basis for classification of elongated plant viruses. Advances Virus Res. 11:1-24.
Brandes, J., and M. Chessin. 1965. An electron microscope study on the size of Sammons' *Opuntia* virus. Virology 25:673-674.
Brandes, J., and C. Wetter. 1959. Classification of elongated plant viruses on the basis of particle morphology. Virology 8:99-115.
Branton, D., and R. B. Park. 1967. Subunits in chloroplast lamellae. J. Ultrastructure Res. 19:283-303.
Brants, D. H. 1965. Relation between ectodesmata and infection of leaves by C^{14}-labelled tobacco mosaic virus. Virology 26:554-557.

Braun, A. C., and K. Maramorosch. 1951. A method for obtaining saliva from certain leafhoppers. Phytopathology 41:1126-1128.

Breck, L. O. and M. P. Gordon. 1970. Formation, characterization, and photochemical properties of a hybrid plant virus. Virology 40:397-402.

Breece, J. R., and W. H. Hart. 1959. A possible association of nematodes with the spread of peach yellow bud mosaic virus. Plant Dis. Reptr. 43:989-990.

Breed, R. S., E. G. D. Murray, and N. R. Smith. 1957. Bergey's manual of determinative bacteriology. 7th ed. Williams and Wilkins, Baltimore. 333 pp.

Brenner, S., and R. W. Horne. 1959. A negative staining method for high resolution electron microscopy of viruses. Biochim. Biophys. Acta 34:103-110.

Bretz, T. W. 1950. Seed transmission of the elm mosaic virus. Phytopathology 40:3. (Abstr.)

Brinton, C. C., Jr., and M. A. Lauffer. 1959. The electrophoresis of viruses, bacteria and cells, and the microscope method of electrophoresis, pp. 427-492. *In* M. Bier [ed.] Electrophoresis, theory, methods, and applications. Academic Press, New York. 563 pp.

Britten, R. J., and R. B. Roberts. 1960. High resolution density gradient sedimentation analysis. Science 131:32-33.

Broadbent, L. 1965. The epidemiology of tomato mosaic. XI. Seed transmission of TMV. Ann. Appl. Biol. 56:177-205.

Broadbent, L. 1969. Disease control through vector control, pp. 593-630. *In* K. Maramorosch [ed.] Viruses, vectors, and vegetation. Interscience, New York. 666 pp.

Browning, P. M. 1969. Preparative ultracentrifuge applications. Spinco Division of Beckman Instruments, Inc., Palo Alto, Calif.

Browning, P. M. 1970. Ultracentrifuge applications 1969. Spinco Division of Beckman Instruments, Inc., Palo Alto, Calif.

Bruehl, G. W. 1961. Barley yellow dwarf. Amer. Phytopathol. Soc. Monograph No. 1.

Bruehl, G. W., and V. D. Damsteegt. 1961. Soil-borne mosaic of fall-seeded oats in western Washington. Plant Dis. Reptr. 45:884-888.

Brues, C. T., A. L. Melander, and F. M. Carpenter. 1954. Classification of insects. Keys to the living and extinct families of insects, and to the living families of other terrestrial arthropods. Bull. Harvard Mus. Comp. Zool. 108. 917 pp.

Brunt, A. A. 1966. The occurrence of cucumber mosaic virus and four nematode-transmitted viruses in British narcissus crops. Plant Pathol. 15:157-160.

Brunt, A. A., and R. H. Kenten. 1960. Mechanical transmission of cocoa swollen-shoot virus. Virology 12:328-330.

Brunt, A A., and R. H. Kenten. 1963. The use of protein in the extraction of cocoa swollen-shoot virus from cocoa leaves. Virology 19:388-392.

Buchner, P. 1953. Endosymbiose der Tiere mit pflanzlichen Mikroorganismen. Lehrbucher und Monographien aus dem Gebiete der Exakten Wissenschaften. (Birkhauser, Basel) Reich der experimentellen Biologie 12. 771 pp.

Budzynski, A. Z., and H. Fraenkel-Conrat. 1970. Hydrogen exchange as a probe of conformation of tobacco mosaic virus and its coat protein. Biochemistry 9:3301-3309.

Bullivant, S. 1965. Freeze substitution and supporting techniques. Lab. Invest. 14:1178-1195, and also *In* G. F. Bahr and E. H. Zeitler [ed.] 1965. Symposium on quantitative electron Microscopy. Williams and Wilkins, Baltimore. 602 pp.

Burdon, R. H., M. A. Billeter, C. Weissmann, R. C. Warner, S. Ochoa, and C. A. Knight. 1964. Replication of viral RNA. V. Presence of a virus—specific double-stranded RNA in leaves infected with tobacco mosaic virus. Proc. Nat. Acad. Sci. U.S. 52:768-775.

Burness, A. T. H., and F. W. Clothier. 1970. Particle weight and other biophysical properties of encephalomyocarditis virus. J. Gen. Virol. 6:381-393.

Butler, M., and R. H. Leach. 1964. Mycoplasma which induces acidity and cytopathic effect in tissue culture. J. Gen. Microbiol. 34:285-294.

Butler, P. J. G., and A. Klug. 1971. Assembly of the particle of tobacco mosaic virus from RNA disks of protein. Nature New Biology 229:47-50

Buzzell, A., D. Trkula, and M. Lauffer. 1956. X-ray studies on tobacco mosaic virus. Arch. Biochem. Biophys. 63:470-476.

Cadman, C. H. 1959. Some properties of an inhibitor of virus infection from leaves of raspberry. J. Gen. Microbiol. 20:113-128.

Cadman, C. H. 1960. Studies of the relationships between soil-borne viruses of the ringspot type occurring in Britain and continental Europe. Virology 11:653-664.
Cadman, C. H. 1962. Evidence for association of tobacco rattle virus nucleic acid with a cell component. Nature 193:49-52.
Cadman, C. H. 1963. Biology of soil-bone viruses. Annu. Rev. Phytopathol. 1:143-172.
Cadman, C. H. 1965. Filamentous viruses infecting fruit trees and raspberry and their possible mode of spread. Plant Dis. Reptr. 49:230-233.
Cadman, C. H., H. F. Dias, and B. D. Harrison. 1960. Sap-transmissible viruses associated with diseases of grapevines in Europe and North America. Nature 187:577-579.
Cadman, C. H., and B. D. Harrison. 1959. Studies on the properties of soil-borne viruses of the tobacco rattle type occurring in Scotland. Ann. Appl. Biol. 47:542-556.
Cadman, C. H., and B. D. Harrison. 1960. Studies on the behaviour in soils of tomato black ring, raspberry ringspot and arabis mosaic viruses. Virology 10:1-20.
Cadman, C. H., and R. M. Lister. 1961. Relationship between tomato ringspot and peach yellow bud mosaic viruses. Phytopathology 51:29-31.
Caldwell, J. 1934. The physiology of virus diseases in plants. V. The movement of the virus agent in tobacco and tomato. Ann. Appl. Biol. 21:191-205.
Caldwell, J. 1952. Some effects of plant viruses on nuclear division. Ann. Appl. Biol. 39:98-102.
Caldwell, J. 1962. Seed-transmission of viruses. Nature 193:457-459.
Callahan, K. L. 1957. Pollen transmission of elm mosaic virus. Phytopathology 47:5. (Abstr.)
Calvert, E. L., and B. D. Harrison. 1966. Potato mop-top, a soil-borne virus. Plant Pathol. 15:134-139.
Campbell, R. N. 1962. Relationship between the lettuce big-vein virus and its vector, *Olpidium brassicae*. Nature 195:675-677.
Campbell, R. N. 1965. Weeds as reservoir hosts of the lettuce big-vein virus. Can. J. Bot. 43:1141-1149.
Campbell, R. N., and P. R. Fry. 1966. The nature of the associations between *Olpidium brassicae* and lettuce big-vein and tobacco necrosis viruses. Virology 29:222-233.
Campbell, R. N., and R. G. Grogan. 1964. Acquisition and transmission of lettuce big-vein virus by *Olpidium brassicae*. Phytopathology 54:681-690.
Campbell, R. N., R. G. Grogan, and D. E. Purcifull. 1961. Graft transmission of big vein of lettuce. Virology 15:82-85.
Canova, A. 1966. Researches on virus diseases of Gramineae. III. *Polymyxa graminis* vector of wheat mosaic virus. Phytopathol. Mediterranea 5:53-58; Rev. Appl. Mycol. 46:662.
Capoor, S. P. 1950. A mosaic disease of sunn hemp in Bombay. Current Sci. 19:22.
Caro, L. G., and R. P. van Tubergen. 1962. High resolution autoradiography. I. methods. J. Cell Biol. 15:173-188.
Carpenter, J. M., and A. Kleczkowski. 1969. The absence of photoreversible pyrimidine dimers in the ribonucleic acid of ultraviolet-irradiated tobacco mosaic virus. Virology 39:542-547.
Carroll, T. W. 1969. Electron microscopic evidence for the presence of barley stripe mosaic virus in cells of barley embryos. Virology 67:649-657.
Carter, W. 1927. A technique for use with homopterous vectors of plant disease, with special reference to the sugarbeet leafhopper, *Eutettix tenellus* (Baker). J. Agr. Res. 34:449-451.
Carter, W. 1928. An improvement in the technique for feeding homopterous insects. Phytopathology 18:246-247.
Carter, W. 1943. *Peregrinus maidis* (Ashm) and the transmission of corn mosaic. I. Inoculation period and longevity of the virus in the insect. Ann. Entomol. Soc. Amer. 34:551-556.
Carter, W. 1952. Injuries to plants caused by insect toxins. II. Bot. Rev. 18:680-721.
Carter, W. 1962. Insects in relation to plant disease. Interscience, New York. 705 pp.
Carter, W., and J. R. R. Suah. 1964. Studies on the spread of lethal yellowing disease of the coconut palm. FAO Plant Protect. Bull. 12:3-8.
Casals, J. 1968. Filtration of arboviruses through "Millipore" membranes. Nature 217:648-649.
Caspar, D. L. D. 1956. Radial density distribution in the tobacco mosaic virus particle. Nature 177:928-932.
Caspar, D. L. D. 1963. Assembly and stability of the tobacco mosaic virus particle. Advances Protein Chem. 18:37-121.
Caspar, D. L. D. 1966. An analogue for negative staining. J. Mol. Biol. 15:365-371.

Caspar, D. L. D. and A. Klug. 1962. Physical principles in the construction of regular viruses. Cold Spring Harbor Symp. Quant. Biol. 27:1-24.
Cation, D. 1949. Transmission of cherry yellows virus complex through seeds. Phytopathology 39:37-40.
Caudwell, A. 1966. Utilisation de hauts polymeres hydrosoluble pour l'extraction et la purification partielle d'un végétal. Etudes de virologie. Ann. Epiph. 17 (hors-série) :169-178.
Caudwell, A. 1968. The extraction without nicotine of the vine fanleaf virus and its concentration by polyethylene glycol. Ann. Epiph. 19:55-62.
Cech, M. 1962. Some notes on the use of agar gel filtration in the study of plant viruses. Virology 18:487-489.
Ceppellini, R., et al. 1964. Nomenclature for human immunoglobulins. Bull. World Health Organ. 30:447-450.
Chamberlain, E. E., and P. R. Fry. 1950. Influence of method of tomato seed extraction on seed transmission of tobacco-mosaic and tomato-streak. New Zealand J. Sci. Tech. (A) 32:19-23.
Chamberlain, R. W. 1968. Arboviruses, the arthropod-borne animal viruses. Current Topics Microbiol. Immunol. 42:38-58.
Chamberlain, R. W., and W. D. Sudia. 1967. Methods for the study of mosquitoes as virus hosts and vectors, pp. 63-103. *In* K. Maramorosch and H. Koprowski [ed.] Methods in virology, vol. 1. Academic Press, New York. 4 vols.
Chamberlin, M. J. 1965. Comparative properties of DNA, RNA, and hybrid homopolymer pairs. Fed. Proc. 24:1446-1457.
Chambers, T. C., N. C. Crowley, and R. I. B. Francki. 1965. Localization of lettuce necrotic yellows virus in host leaf tissue. Virology 27:320-328.
Chanock, R. M., L. Hayflick, and M. F. Barile. 1962. Growth on artificial medium of an agent associated with atypical pneumonia and its identification as a PPLO. Proc. Nat. Acad. Sci. U.S. 48:41-49.
Charlwood, P. A. 1963. Applications of radioactively labelled marker proteins in density-gradient ultracentrifugation. Anal. Biochem. 5:226-245.
Chen, T. A., and R. R. Granados. 1970. Plant-pathogenic mycoplasma: maintenance in vitro and transmission to *Zea mays* L. Science 167:1633-1636.
Cheo, P. C. 1955. Effect of seed maturation on inhibition of southern bean mosaic virus in bean. Phytopathology 45:17-21.
Chervenka, C. H. 1966. Use of combined schlieren and interference optics for determination of molecular weights from sedimentation equilibrium data. Anal. Chem. 38:356-358.
Chervenka, C. H. 1969. A manual of methods for the analytical ultracentrifuge. Spinco Division of Beckman Instruments, Inc., Palo Alto, Calif.
Chessin, M. 1958. Light quality and photoreactivation of plants and viruses. Ann. Appl. Biol. 46:388-392.
Chessin, M. 1960. Photodynamic inactivation of infectious nucleic acid. Science 132:1840-1841.
Chessin, M. 1965. Photoreactivation of clover yellow mosaic virus. Photochem. Photobiol. 4:709-711.
Chessin, M., R. Livingston, and G. Truscott. 1966. Direct evidence for the sensitized formation of a metastable state of B-Carotene. Trans. Faraday Soc. 62:1519-1524.
Chessin, M., R. A. Solberg, and M. Jakobson. 1964. Differential sensitivity of tobacco mosaic virus and its infectious nucleic acid to high-speed electrons. Nature 202:830-831.
Chessin, M., M. Zaitlin, and R. Solberg. 1967. A new strain of tobacco mosaic virus from *Lychnis alba*. Phytopathology 57:452-453.
Chester, K. S. 1936. Liberation of neutralized virus and antibody from antiserum-virus precipitates. Phytopathology 26:949-964.
Chidlow, J., and J. H. Tremaine. 1971. Limited hydrolysis of cowpea chlorotic mottle virus by trypsin and chymotrypsin. Virology 43:267-278.
Childs, J. F. L. 1956. Transmission experiments and xyloporosis-cachexia relations in Florida. Plant Dis. Reptr. 40:143-145.
Chiu, R. J., and L. M. Black. 1967. Monolayer cultures of insect cell lines and their inoculation with a plant virus. Nature 215:1076-1078.
Chiu, R. J., and L. M. Black. 1969. Assay of wound tumor virus by fluorescent cell counting technique. Virology 37:667-677.

Chiu, R. J., J. H. Jean, M. H. Chen, and T. C. Lo. 1968. Transmission of transitory yellowing virus of rice by two leafhoppers. Phytopathology 58:740-745.

Chiu, R. J., T. C. Lo, C. L. Pi, and M. H. Chen. 1965. Transitory yellowing of rice and its transmission by the leafhopper *Nephotettix apicalis* (Motsch.). Bot. Bull. Acad. Sinica 6:1-18.

Chiu, R. J., D. V. R. Reddy, and L. M. Black. 1966. Inoculation and infection of leafhopper tissue cultures with a plant virus. Virology 30:562-566.

Chiykowski, L. N. 1967. Some host plants of a Canadian isolate of the clover phyllody virus. Can. J. Plant Sci. 47:141-148.

Chiykowski, L. N., and R. K. Chapman. 1965. Migration of the six-spotted leafhopper in central North America. Univ. Wisconsin Res. Bull. 261, 2:21-45.

Christie, J. R. 1952. Some new nematode species of critical importance to Florida growers. Proc. Soil Sci. Soc. Fla. 12:30-39.

Christie, R. G. 1967. Rapid staining procedures for differentiating plant virus inclusions in epidermal strips. Virology 31:268-271.

Clark, J. M., Jr., A. Y. Chang, S. Spiegelman, and M. E. Reichmann. 1965. The in vitro translation of a monocistronic message. Proc. Nat. Acad. Sci. U. S. 54:1193-1197.

Clark, M. F. 1968. Purification and fractionation of alfalfa mosaic virus with polyethylene glycol. J. Gen. Virol. 3:427-432.

Cobb, N. A. 1918. Estimating the nema population of soil. U.S. Dep. Agr. Tech. Circ. No. 1.

Cochran, G. W. 1947. A chromatographic method for the detection of tobacco mosaic virus in juice from diseased Turkish tobacco plants. Phytopathology 37:850-851.

Cochran, G. W., J. L. Chidester, and G. W. Welkie. 1959. Anaerobic column chromatography of nonoxidized components of plant tissue extracts. Biochim. Biophys. Acta 35:190-196.

Cochran, L. C. 1950. Passage of the ringspot virus through peach seed. Phytopathology 40:964. (Abstr.)

Cocking, E. C., and E. Pojnar. 1968. Appearance of tobacco mosaic virus in thin section using different staining procedures. J. Gen. Virol. 2:317-318.

Cohen, R., R. Purcell, and R. L. Steere. 1969. Ultrafiltration of the aster yellows agent. Phytopathology 59:1555. (Abstr.)

Cohen, S. 1942. The isolation and crystallization of plant viruses and other protein macro-molecules by means of hydrophilic colloids. J. Biol. Chem. 144:353-362.

Cohen, S. 1969. In vivo effects in whiteflies of a possible antiviral factor. Virology 37:448-454.

Cohen, S., and F. E. Nitzany, 1966. Transmission and host range of the tomato yellow leaf curl virus. Phytopathology 56:1127-1131.

Cohn, E., and J. Krikun. 1966. A disease of onion associated with ectoparasitic nematode of the genus *Longidorus*. Plant Dis. Reptr. 50:711-712.

Cohn, E., E. Tanne, and F. E. Nitzany. 1970. *Xiphinema italiae*, a new vector of grapevine fanleaf virus. Phytopathology 60:181-182.

Collingwood, C. A., and G. H. L. Decker. 1960. A comparison of various chemicals for the control of the black currant gall mite. Plant Pathol. 9:39-48.

Collingwood, C. A., J. D. R. Vernon, and T. J. Ligowski. 1960. Spraying trials against black currant gall mite. Plant Pathol. 9:135-143.

Commoner, B., J. A. Lippincott, G. B. Shearer, E. E. Richman, and J. H. Wu. 1956. Reconstitution of tobacco mosaic virus components. Nature 178:767-771.

Commoner, B., and F. Mercer. 1951. Inhibition of the biosynthesis of tobacco mosaic virus by thiouracil. Nature 168:113-114.

Commoner, B., and S. D. Rodenberg. 1955. Relationships between tobacco mosaic virus and the non-virus proteins. J. Gen. Physiol. 38:475-492.

Commoner, B., G. B. Shearer, and M. Yamada. 1962. Linear biosynthesis of tobacco mosaic virus. Changes in rod length during the course of infection. Proc. Nat. Acad. Sci. U.S. 48:1788-1795.

Connin, R. V. 1956. The host range of the wheat curl mite, vector of wheat streak mosaic. J. Econ. Entomol. 49:1-4.

Consentino, V., K. Paigen, and R. L. Steere. 1956. Electron microscopy of turnip yellow mosaic virus and the associated abnormal protein. Virology 2:139-148.

Converse, R. H. 1967. Pollen- and seed-borne raspberry viruses. Phytopathology 57:97-98.

Converse, R. H., and R. M. Lister. 1969. The occurrence and some properties of black raspberry latent virus. Phytopathology 59:325-333.

Cook, W. C. 1967. Life history, host plants, and migrations of the beet leafhopper in the western United States. U. S. Dep. Agr. Tech. Bull. 1365. 122 pp.

Cooper, J. I., and P. R. Thomas. 1970. *Trichodorus nanus*, a vector of tobacco rattle virus in Scotland. Plant Pathol. 19:197.

Cooper, W. D., and H. S. Loring. 1954. The purine and pyrimidine composition of the tobacco mosaic virus and the holmes masked strain. J. Biol. Chem. 211:505-515.

Corbett, M. K. 1958. A virus disease of lupines caused by bean yellow mosaic virus. Phytopathology 48:86-91.

Corbett, M. K. 1961. Purification of potato virus X without aggregation. Virology 15:8-15.

Corbett, M. K., and D. A. Roberts. 1962. A rapid method for purifying tobacco ringspot virus and its morphology as determined by electron microscopy and negative staining. Phytopathology 52:902-905.

Costa, A. S. 1944. Quantitative studies with carborundum and its use in local-lesion tests. Phytopathology 34:288-300.

Couch, H. B. 1955. Studies on seed transmission of lettuce mosaic virus. Phytopathology 45:63-70.

Coulter, D. H. 1967. Rapid and improved methods for embedding biological tissues in Epon 812 and Araldite 502. J. Ultrastructure Res. 20:346-355.

Cox, R. A. 1966. Ribonucleic acid from ribonucleoprotein particles. Biochem. Prep. 11:104-109.

Cox, R. A. 1968. The use of guanidinium chloride in the isolation of nucleic acids, pp. 120-129. *In* L. Grossman and K. Moldave [ed.] Methods in enzymology, vol. 12B. Academic Press, New York. 19 vols.

Cox, R. W., and R. W. Horne. 1968. Accurate calibration of the magnification of the A.E.I. EM6B/2 electron microscope using catalase crystals, pp. 579-580. *In* D. S. Bocciarelli [ed.] Electron microscopy 1968, vol. 1. Tipografia Poliglotta Vaticana, Rome.

Cramer, R. 1964. Purification of animal viruses, pp. 145-168. *In* R. J. C. Harris [ed.] Techniques in experimental virology. Academic Press, New York. 450 pp.

Cramer, R., and H. Svensson. 1961. Density-gradient electrophoresis as a new tool in virology. Experientia 17:49-57.

Crawford, L. V. 1963. The physical characteristics of polyoma virus II. The nucleic acid. Virology 19:279-282.

Creeth, J. M., and R. H. Pain. 1967. The determination of molecular weights of biological macromolecules by ultracentrifuge methods. Progr. Biophys. Molec. Biol. 17:217-287.

Cremer, M. G., and G. Kooistra. 1964. Investigations on notched leaf ("kartelblad") of *Gladiolus* and its relation to rattle virus. Nematologica 10:69-70. (Abstr.)

Cremer, M. G., and J. A. van der Veken. 1964. Localization of viral antigen in narcissus leaves infected with yellow stripe virus, determined by means of a fluorescien conjugated antiserum. Neth. J. Plant Pathol. 72:105-113.

Cropley, R. 1961. Cherry leaf-roll virus. Ann. Appl. Biol. 49:524-529.

Cropley, R., A. F. Posnette, and J. M. Thresh. 1964. The effects of black-currant yellows virus and and a strain of reversion virus on yield. Ann. Appl. Biol. 54:177-182.

Crowle, A. J. 1961. Immunodiffusion. Academic Press, New York. 333 pp.

Crowley, N. C. 1955. The effect of seed extracts in the infectivity of plant viruses and its bearing on seed transmission. Australian J. Biol. Sci. 8:56-67.

Crowley, N. C. 1957*a*. The effect of developing embryos on plant viruses. Australian J. Biol. Sci. 10:443-448.

Crowley, N. C. 1957*b*. Studies on the seed transmission of plant virus diseases. Australian J. Biol. Res. 10:449-464.

Crowley, N. C. 1959. Studies on the time of embryo infection by seed-transmitted viruses. Virology 8:116-123.

Crowley, N. C. 1967. Factors affecting the local lesion response of *Nicotiana glutinosa* to lettuce necrotic yellows virus. Virology 31:107-113.

Crowley, N. C., B. D. Harrison, and R. I. B. Francki. 1965. Partial purification of lettuce necrotic yellows virus. Virology 26:290-296.

Cunningham, I. C., T. W. Tinsley, and J. M. Walker. 1966. Hemagglutination with plant and insect viruses. J. Gen. Microbiol. 42:397-401.

Dadd, R. H., and T. E. Mittler. 1965. Studies on the artificial feeding of the aphid *Myzus persicae* (Sulzer). III. Some major nutritional requirements. J. Insect Physiol. 11:717-743.

Daft, M. J. 1965. Some interactions of kinetin and temperature on tobacco leaves infected with tomato aucuba mosaic virus. Ann. Appl. Biol. 55:51-56.

Dahl, D., and C. A. Knight. 1963. Some nitrous acid-induced mutants of tomato atypical mosaic virus. Virology 21:580-586.

Dale, W. T. 1949. Observations on a virus disease of cowpea in Trinidad. Ann. Appl. Biol. 36:327-333.

Dalton, A. J. 1955. A chrome-osmium fixative for electron microscopy. Anat. Rec. 121:281.

Damirdagh, I. S., and A. F. Ross. 1967. A marked synergistic interaction of potato viruses X and Y in inoculated leaves of tobacco. Virology 31:296-307.

Damirdagh, I. S., and R. J. Shepherd. 1970*a*. Purification of the tobacco etch and other viruses of the potato Y group. Phytopathology 60:132-142.

Damirdagh, I. S., and R. J. Shepherd. 1970*b*. Some of the chemical properties of the tobacco etch virus and its protein and nucleic acid components. Virology 40:84-89.

Dannenberg, H., G. Schramm, and H. Flammersfeld. 1948. Die ultraviolettabsorption verschiedener stamme des Tabakmosaik-virus. Z. Naturforsch. 3b:241-245.

Das, C. R., and J. A. Milbrath. 1961. Plant-to-plant transfer of stone fruit ringspot virus in squash by pollination. Phytopathology 51:489-490.

Davis, R. E., and R. F. Whitcomb. 1969. Spectrum of antibiotic sensitivity of aster yellows disease in insects and plants. Phytopathology 59:1556. (Abstr.)

Davis, R. E., and R. F. Whitcomb. 1970. Evidence on possible mycoplasma-like etiology of aster yellows disease: I. Suppression of symptom development in plants by antibiotics. J. Infect. Immunol. 2:201-208.

Davis, R. E., R. F. Whitcomb, and R. L. Steere. 1968. Remission of aster yellows disease by antibiotics. Science 161:793-794.

Davison, E. M. 1969. Cell to cell movement of tobacco ringspot virus. Virology 37:694-696.

Davison, E. M., and R. I. B. Francki. 1969. Observation on negatively stained tobacco ringspot virus and its two RNA-deficient components. Virology 39:235-239.

Day, M. F. 1955*a*. Mechanisms of transmission of viruses by arthropods. Exp. Parasitol. 4:387-418.

Day, M. F. 1955*b*. The mechanism of the transmission of potato leaf roll virus by aphids. Australian J. Biol. Sci. 8:498-513.

Day, M. F., H. Irzykiewiez, and A. McKinnon. 1952. Observations on the feeding of the virus vector *Orosius argentatus* (Evans), and comparisons with certain other jassids. Australian J. Sci. Res., Ser. B, 5:128-142.

Debrot, E. A. 1964. Studies on a strain of raspberry ringspot virus occurring in England. Ann. Appl. Biol. 54:183-191.

Decker, R. V., and J. F. Foster. 1966. The interaction of bovine plasma albumin with detergent anions. Stoichiometry and mechanism of binding of alkylbenzenesulfonates. Biochemistry 5:1242-1254.

deDuve, C., J. Berthet, and H. Beaufay. 1959. Gradient centrifugation of cell particles; theory and application. Progr. Biophys. Chem. 9:326-369.

deFremery, D., and C. A. Knight. 1955. A chemical comparison of three strains of tomato bushy stunt virus. J. Biol. Chem. 214:559-566.

DeLong, D. M. 1948. The leafhoppers, or Cicadellidae, of Illinois (Eurymelinae-Balcluthinae). Bull. Illinois Natur. Hist. Survey 24:376 pp.

DeLong, D. M. 1971. The bionomics of leafhoppers. Annu. Rev. Entomol. 16:179-210.

DeLong, D. M., and H. H. P. Severin. 1950. Spittle-insect vectors of Pierce's disease virus. I. Characters, distribution, and food plants. Hilgardia 19:339-356.

Del Rosario, M. S., and W. H. Sill, Jr. 1958. A method of rearing large colonies of an eriophyid mite, *Aceria tulipae* (Keifer), in pure culture from single eggs or adults. J. Econ. Entomol. 51:303-306.

Del Rosario, M. S., and W. H. Sill, Jr. 1964. Additional biological and ecological characteristics of *Aceria tulipae* (Acarina:Eriophyidae). J. Econ. Entomol. 57:893-896.

Deming, W. E. 1964. Statistical adjustment of data. Dover, New York. 261 pp.

Denby, L. G., and J. M. Wilks. 1963. The effect of tobacco mosaic on the yield of field tomatoes as influenced by sprays of milk and DOSS. Can. J. Plant Sci. 43:457-461

deSequeira, O. A. 1967. Purification and serology of an apple mosaic virus. Virology 31:314-322.

Desjardins, P. R., R. L. Latterell, and J. E. Mitchell. 1954. Seed transmission of tobacco ringspot in Lincoln variety of soybean. Phytopathology 44:86.
deZoeten, G. A. 1966. California tobacco rattle virus, its intracellular appearance and the cytology of the infected cell. Phytopathology 56:744-754.
deZoeten, G. A., and G. Gaard. 1969. Possibilities for inter- and intracellular translocation of some icosahedral plant viruses. J. Cell Biol. 40:814-823.
deZoeten, G. A., and D. E. Schlegel. 1967*a*. Broadbean mottle virus in leaf tissue. Virology 1:173-175. (3)
deZoeten, G. A., and D. E. Schlegel. 1967*b*. Nucleolar and cytoplasmic uridine-^{3}H incorporation in virus-infected plants. Virology 32:416-427. (4)
Dias, H. F. 1963. Host range and properties of grapevine fanleaf and grapevine yellow mosaic viruses. Ann. Appl. Biol. 51:85-95.
Dias, H. F. 1970. Transmission of cucumber necrosis virus by *Olpidium cucurbitacearum* Barr & Dias. Virology 40:828-839.
Dias, H. F., and F. W. Doane. 1968. Evidence for lack of relationship between Canadian cucumber necrosis and tobacco necrosis virus. Can. J. Bot. 46:47-50.
Dias, H. F., and B. D. Harrison. 1963. The relationship between grapevine fanleaf, grapevine yellow mosaic, and arabis mosaic viruses. Ann. Appl. Biol. 51:97-105.
Dias, H. F., and H. E. Waterworth. 1967. The identity of a seed-borne mosaic virus of *Chenopodium amaranticolor* and *C. quinoa*. Can. J. Bot. 45:1285-1295.
DiCamelli, R. F., P. D. Holohan, S. F. Basinger, and J. Lebowitz. 1970. Molecular weight determinations by low-speed sedimentation equilibrium. A combined use of the Nazarian equation and the Griffith rapid equilibrium technique. Anal. Biochem. 36:470-494.
Dieleman-van Zaayen, A., and O. Igesz. 1969. Intracellular appearance of mushroom virus. Virology 39:147-152.
Diener, T. O. 1962. Isolation of an infectious, ribonuclease-sensitive fraction from tobacco leaves recently inoculated with tobacco mosaic virus. Virology 16:140-146.
Diener, T. O. 1971. Potato spindle tuber: a plant virus with properties of a free nucleic acid. III. Subcellular location of PSTV-RNA and the question of whether virions exist in extracts or in situ. Virology 43:75-89.
Diener, T. O., and F. G. Jenifer. 1964. A dependable local lesion assay for turnip yellow mosaic virus. Phytopathology 54:1258-1260.
Diener, T. O., and W. B. Raymer. 1967. Potato spindle tuber virus: a plant virus with properties of a free nucleic acid. Science 158:378-381.
Diener, T. O., and W. B. Raymer. 1969. Potato spindle tuber virus: a plant virus with properties of a free nucleic acid. II. Characterization and partial purification. Virology 37:351-366.
Diener, T. O., and I. R. Schneider. 1968. Virus degradation and nucleic acid release in single-phase phenol systems. Arch. Biochem. Biophys. 124:401-412.
Diener, T. O., H. A. Scott, and J. M. Kaper. 1964. Highly infectious nucleic acid from crude and purified preparations of cucumber mosaic virus (Y strain). Virology 22:131-141.
Diener, T. O., and M. L. Weaver. 1959*a*. A caffeine additive to aid mechanical transmission of necrotic ringspot virus from fruit trees to cucumber. Phytopathology 49:321-322.
Diener, T. O., and M. L. Weaver. 1959*b*. Reversible and irreversible inhibition of necrotic ringspot virus in cucumbers by pancreatic ribonuclease. Virology 7:419-427.
Dixon, W. J., and F. J. Massey. 1957. Introduction to statistical analysis. 2nd ed. McGraw-Hill, New York. 488 pp.
Dobroscky, I. D. 1931*a*. Morphological and cytological studies on the salivary glands and alimentary tract of *Cicadula sexnotata* (Fallen), the carrier of aster yellows virus. Contrib. Boyce Thompson Inst. 3:39-58.
Dobroscky, I. D. 1931*b*. Studies on cranberry false blossom disease and its insect vector. Contrib. Boyce Thompson Inst. 3:59-83.
Doi, Y., M. Terenka, K. Yora, and H. Asuyama. 1967. Mycoplasma- or PLT group-like microorganisms found in the phloem elements of plants infected with mulberry dwarf, potato witches' broom, aster yellows, or paulownia witches' broom. Ann. Phytopathol. Soc. Japan 33:259-266.
Doolittle, S. P., and E. S. Beecher. 1937. Seed transmission of tomato mosaic following the planting of freshly extracted seed. Phytopathology 27:800-801.

Doolittle, S. P., and W. W. Gilbert. 1919. Seed transmission of cucurbit mosaic in the wild cucumber. Phytopathology 9:326-327.

Dorner, R. W., A. Kahn, and S. G. Wildman. 1958. Proteins of green leaves. VIII. The distribution of fraction 1 protein in the plant kingdom as detected by precipitin and ultracentrifugal analysis. Biochim. Biophys. Acta 29:240-245.

Dounin, M. S., and N. N. Popova. 1938. The drop method of virus diagnosis in plant husbandry. Rev. Appl. Mycol. 17:762.

Drake, D. C., and R. K. Chapman. 1965. Evidence for long distance migration of the six-spotted leafhopper into Wisconsin. Univ. Wisconsin Res. Bull. 261, Part 1, 3-20.

Drechsler, C. 1952. Production of zoospores from germinating oospores of *Pythium ultimum* and *Pythium debaryanum*. Bull. Torrey Bot. Club 79:431-450. (Rev. Appl. Mycol. 32:161-162).

Dudman, W. F. 1965. Differentiation of strains of tobacco mosaic virus by immune diffusion in agar plates. Phytopathology 55:635-639.

Dudman, W. F. 1966*a*. Interaction between gelatin solutions and tobacco mosaic virus. Biochim. Biophys. Acta 120:212-221.

Dudman, W. F. 1966*b*. Behaviour of tobacco mosaic virus in agar and gelatin gels. Biochim. Biophys. Acta 126:405-408.

Dudman, W. F. 1966*c*. Precipitation of tobacco mosaic virus by macromolecules; a method for estimating molecular volumes. Nature 211:1049.

Duffus, J. E. 1963. Possible multiplication in the aphid vector of sowthistle yellow vein virus, a virus with an extremely long insect latent period. Virology 21:194-202.

Duffus, J. E., and A. H. Gold. 1965. Transmission of beet western yellows virus by aphids feeding through a membrane. Virology 27:388-390.

Duffus, J. E., and A. H. Gold. 1967. Relationship of tracer-measured aphid feeding to acquisition of beet western yellows virus and to feeding inhibitors in plant extracts. Phytopathology 57:1237-1241.

Duffus, J. E., and A. H. Gold. 1969. Membrane feeding and infectivity neutralization used in a serological comparison of potato leaf roll and beet western yellows viruses. Virology 37: 150-153.

Duggar, B. M. 1930. The problem of seed transmission of the typical mosaic of tobacco. Phytopathology 20:133. (Abstr.)

Dunez, J. 1966. Comparative efficiency of different methods used for purification of fruit tree viruses. Etudes de Virologie. Ann. Epiph. 17 (hors-série):163-168.

Dunn, D. B., and J. H. Hitchborn. 1965. The use of bentonite in the purification of plant viruses. Virology 25:171-192.

Dunnebacke, T. H., and A. K. Kleinschmidt. 1967. Ribonucleic acid from reovirus as seen in protein monolayers by electron microscopy. Z. Naturforsch. 22:159-164.

Durham, A. C. H., J. T. Finch, and A. Klug. 1971*b*. States of aggregation of tobacco mosaic virus protein. Nature New Biology 229:37-42.

Durham, A. C. H., and A. Klug. 1971*a*. Polymerization of tobacco mosaic virus protein and its control. Nature New Biology 229:42-46.

Edelstein, S. J., and H. K. Schachman. 1967. The simultaneous determination of partial specific volumes and molecular weights with microgram quantities. J. Biol. Chem. 242:306-311.

Edwardson, J. R. 1966. Electron microscopy of cytoplasmic inclusions in cell infected with rod-shaped viruses. Amer. J. Bot. 53:359-364.

Eglington, G., and R. J. Hamilton. 1967. Leaf epicuticular waxes. Science 156:1322-1335.

Eigner, J., H. Boedtker, and G. Michaels. 1961. The thermal degradation of nucleic acids. Biochim. Biophys. Acta 51:165-168.

Eisenberg, H. 1969. Sedimentation, partial specific volumes and preferential solvation in deoxyribonucleate solutions. Ann. N.Y. Acad. Sci. 164:25-32.

Elbers, P. F., and J. Pieters. 1964. Accurate determination of magnification in the electron micro-ribonucleate solutions. Ann. N.Y. Acad. Sci. 164:25-32.

Elbers, P. F., and J. Pieters. 1964. Accurate determination of magnification in the electron microscope. J. Ultrastructure Res. 11:25-32.

Elias, H. G. 1964. Ultrazentrifugen. Chemie-Ingenieur-Technik 36:504-522.

Ellen, K. A. O., and S. L. Rhode. 1969. The use of guanidinium thiocyanate to fractionate DNA-like RNA adsorbed on methylated bovine albumin-kieselguhr columns. Biochim. Biophys. Acta 174:117-123.

Elmer, O. H. 1925. Transmissibility and pathological effects of the mosaic disease. Res. Bull. Iowa Agr. Exp. Sta. 82.

Els, H. J., and D. W. Verwoerd. 1969. Morphology of bluetongue virus. Virology 38:213-219.

Englander, S. W., and H. T. Epstein. 1957. Optical methods for measuring nucleoprotein and nucleic acid concentrations. Arch. Biochem. Biophys. 68:144-149.

Epstein, A., and K. R. Barker. 1966. Pathogenicity of *Xiphinema americanum* on *Vinca minor*. Plant Dis. Reptr. 50:420-422.

Epstein, H. T. 1953. Identification of radiosensitive volume with nucleic acid volume. Nature 171:394-395.

Erlandson, R. A. 1964. A new Maraglas, D. E. R. 732, embedment for electron microscopy. J. Cell Biol. 22:704-709.

Esau, K. 1958. Phloem degeneration in celery infected with yellow leafroll virus of peach. Virology 6:348-356.

Esau, K. 1961. Plants, viruses and insects. Harvard Univ. Press, Cambridge. 110 pp.

Esau, K. 1965. Fixation images of sieve element plastids in Beta. Proc. Nat. Acad. Sci. U.S. 54: 429-437.

Esau, K. 1967. Anatomy of plant virus infections. Annu. Rev. Phytopathol. 5:45-76.

Esau, K., and J. Cronshaw. 1967*a*. Tubular components in cells of healthy and tobacco mosaic virus-infected *Nicotiana*. Virology 33:26-35.

Esau, K., and J. Cronshaw. 1967*b*. Relation of tobacco mosaic virus to the host cells. J. Cell Biol. 33:665-678. (10)

Esau, K., J. Cronshaw, and L. L. Hoefert. 1966. Organization of beet yellows-virus inclusions in leaf cells of Beta. Proc. Nat. Acad. Sci. U.S. 55:486-493.

Esau, K., J. Cronshaw, and L. L. Hoeffert. 1967. Relation of beet yellow virus to movement in the sieve tube. J. Cell Biol. 32:71-87.

Eslick, R. F., and M. M. Afanasiev. 1955. Influence of time of infection with barley stripe mosaic on symptoms, plant yield, and seed infection of barley. Plant Dis. Reptr. 39:722-724.

Estes, A. P., and M. K. Brakke. 1966. Correlation of *Polymyxa graminis* with transmission of soil-borne wheat mosaic virus. Virology 28:772-774.

Evans, N. A., and A. D. McLaren. 1969. A kinetic isotope effect in the photoreactivation of U.V.-inactivated tobacco mosaic virus RNA. Photochem. Photobiol. 9:243-246.

Fairbanks, A. 1956. The effects of x-rays on the biosynthesis of tobacco mosaic virus. Ph.D. dissertation. Washington Univ., St. Louis, Mo.

Faivre-Amiot, A., J. P. Moreau, M. T. Cousin and T. Staron. 1970. Essai de mise en culture de l'agent de la phyllodie du trèfle. Ann. Phytopathol. 2:251-258.

Fajardo, T. G. 1930. Studies on the mosaic disease of the bean (*Phaseolus vulgaris* L.). Phytopathology 20:469-493.

Fajardo, T. G., and J. Maranon. 1932. The mosaic disease of sincomas, *Pachyrrhizus erosus* (Linnaeus) Urban. Philippine J. Sci. 48:129-142.

Farkas, G. L., Z. Kiraly, and F. Solymosy. 1960. Role of oxidative metabolism in the localization of plant viruses. Virology 12:408-421.

Fassell, G. N., and S. G. Wildman. 1963. Difference in thermal inactivation kinetics of two strains of tobacco mosaic virus. Biochim. Biophys. Acta 72:230-236.

Fazekas de St. Groth, S. 1962. The neutralization of viruses. Advances Virus Res. 9:1-125.

Feix, G., R. Pollet, and C. Weissmann. 1968. Replication of viral RNA, XVI. Enzymatic synthesis of infectious viral RNA with noninfectious Q_β minus strands as template. Proc. Nat. Acad. Sci. U.S. 59:145-152.

Feldman, J. M., and J. A. Boninsegna. 1968. Antiserum for tomato spotted wilt virus. Nature 219:183-184.

Fellows, H. 1956. Mechanical aids in the study of eriophyid mites in relation to yellow streak mosaic of wheat. Plant Dis. Reptr. 40:601-605.

Fenwick, M. L., R. L. Erikson, and R. M. Franklin. 1964. Replication of the RNA of bacteriophage R17. Science 146:527-530.

Fife, J. M., and V. L. Frampton. 1936. The pH gradient extending from the phloem into the parenchyma of the sugar beet and its relation to the feeding behavior of *Eutettix tenellus*. J. Agr. Res. 53:581-593.

Finch, J. T. 1964. Resolution of the substructure of tobacco mosaic virus in the electron microscope. J. Mol. Biol. 8:872-874.

Finch, J. T., and A. Klug. 1965. The structure of viruses of the papilloma polyoma type. III. Structure of rabbit papilloma virus with an appendix on the topography of contrast in negative-staining for electron microscopy. J. Mol. Biol. 13:1-12.

Finch, J. T., and A. Klug. 1966. Arrangement of protein subunits and the distribution of nucleic acid in turnip yellow mosaic virus. II. Electron microscopic studies. J. Mol. Biol. 15:344-364.

Finch, J. T., and A. Klug. 1967. Structure of broad bean mottle virus. I. Analysis of electron micrographs and comparison with turnip yellow mosaic virus and its top component. J. Mol. Biol. 24:289-302.

Finlay, J. R., and D. S. Teakle. 1969. The effect of pH on the particle stability of a phosphotungstate-stained tobacco necrosis virus. J. Gen. Virol. 5:93-96.

Finter, N. B. 1966. Interferons. North-Holland, Amsterdam. 342 pp.

Fisher, J. M., and D. J. Raski. 1967. Feeding of *Xiphinema index* and *X. diversicaudatum*. Proc. Helminthol. Soc. Wash. D.C. 34:68-72.

Fisher, R. A. 1928. Statistical methods for research workers. 12th ed. Oliver and Boyd, London. 356 pp.

Fisher, R. A., and F. Yates. 1963. Statistical tables. 6th ed. Oliver and Boyd, London. 146 pp.

Flegg, J. J. M. 1967. Extraction of *Xiphinema* and *Longidorus* species from soil by a modification of Cobb's decanting and sieving technique. Ann. Appl. Biol. 60:429-437.

Flemion, F., M. C. Ledbetter, and E. S. Kelley. 1954. Penetration and damage of plant tissues during feeding by the tarnished plant bug (*Lygus lineolaris*). Contrib. Boyce Thompson Inst. 17:347-357.

Flock, R. A., and J. M. Wallace. 1955. Transmission of fig mosaic by the eriophyid mite *Aceria ficus*. Phytopathology 45:52-54.

Fluke, D. J., and B. Beinecke. 1967. Temperature dependence of dry T1 bacteriophage radiation sensitivities on host strains differing in radiation sensitivity. Abstr. Radiat. Res. Soc. May: 7-11.

Forbes, A. R. 1966. Electron microscope evidence for nerves in the mandibular stylets of the green peach aphid. Nature 212:726.

Ford, R. E. 1966. Recovery of pea streak from pea seed parts and its transmission by immature seed. Phytopathology 56:858-859.

Ford, R. E. 1967. Number and size of local lesions of alfalfa mosaic virus reduced by interaction with bean yellow mosaic virus. Phytopathology 57:1274-1275.

Fox, S. W., and J. F. Foster. 1957. Introduction to protein chemistry. Wiley, New York. 459 pp.

Fraenkel-Conrat, H. 1955. The reaction of tobacco mosaic virus with iodine. J. Biol. Chem. 217: 373-381.

Fraenkel-Conrat, H. 1956. The role of the nucleic acid in the reconstitution of active tobacco mosaic virus. J. Amer. Chem. Soc. 78:882-883.

Fraenkel-Conrat, H. 1957. Degradation of tobacco mosaic virus with acetic acid. Virology 4:1-4.

Fraenkel-Conrat, H. 1961. Chemical modification of viral ribonucleic acid. I. Alkylating agents. Biochim. Biophys. Acta 49:169-180.

Fraenkel-Conrat, H. 1965. Structure and function of virus proteins and of viral nucleic acid, pp. 99-151. *In* H. Neurath [ed.] The proteins, vol. 3. Academic Press, New York. 5 vols.

Fraenkel-Conrat, H., 1969. The chemistry and biology of viruses. Academic Press, New York.

Fraenkel-Conrat, H., and M. Colloms. 1967. Reactivity of tobacco mosaic virus and its protein toward acetic anhydride. Biochemistry 6:2740-2745.

Fraenkel-Conrat, H., and K. Narita. 1958. Degradation and structure of tobacco mosaic virus, p. 249. *In* A. Neuberger [ed.] Symposium on protein structure, Paris (1957). Methuen, London. 351 pp.

Fraenkel-Conrat, H., and H. S. Olcott. 1945. Esterification of proteins with alcohols of low molecular weight. J. Biol. Chem. 161:259-268.

Fraenkel-Conrat, H., and M. Sherwood. 1967. The reactivity of the tyrosine residues of tobacco mosaic virus protein with iodine. Arch. Biochem. Biophys. 120:571-577.

Fraenkel-Conrat, H., and B. Singer. 1957. Virus reconstitution. II. Combination of protein and nucleic acid from different strains. Biochim. Biophys. Acta 24:540-548.

Fraenkel-Conrat, H., B. Singer, and A. Tsugita. 1961. Purification of viral RNA by means of bentonite. Virology 14:54-58.

Fraenkel-Conrat, H., S. Veldee, and J. Woo. 1964. The infectivity of tobacco mosaic virus. Virology 22:432-433.

Fraenkel-Conrat, H., and R. C. Williams. 1955. Reconstitution of active tobacco mosaic virus from its inactive protein and nucleic acid components. Proc. Nat. Acad. Sci. U.S. 41:690-698.

Francki, R. I. B. 1962. Infectivity of tobacco mosaic virus from tobacco leaves treated with 2-thiouracil. Virology 17:9-21.

Francki, R. I. B. 1964. Inhibition of cucumber mosaic virus infectivity by leaf extracts. Virology 24:193-199.

Francki, R. I. B. 1966. Some factors affecting particle length distribution in tobacco mosaic virus preparations. Virology 30:388-396.

Francki, R. I. B. 1967. Effect of high light intensities on spontaneous and virus-induced local lesions in *Gomphrena globosa*. Phytopathology 57:329-331.

Francki, R. I. B. 1968. Inactivation of cucumber mosaic virus (Q strain) nucleoprotein by pancreatic ribonuclease. Virology 34:694-700.

Francki, R. I. B., and G. D. McLean. 1968. Purification of potato virus X and preparation of infectious ribonucleic acid by degradation with lithium chloride. Australian J. Biol. Sci. 21:1311-1318.

Francki, R. I. B., and R. E. F. Matthews. 1962. Some effects of 2-thiouracil on the multiplication of turnip yellow mosaic virus. Virology 17:367-380.

Francki, R. I. B., J. W. Randles, T. C. Chambers, and S. B. Wilson. 1966. Some properties of purified cucumber mosaic virus (Q strain). Virology 28:729-741.

Franklin, R. E. 1955. Structure of tobacco mosaic virus. Nature (London) 175:379-381.

Franklin, R. E. 1956*a*. Location of the ribonucleic acid in the tobacco mosaic virus particle. Nature (London) 177:929-930.

Franklin, R. E. 1956*b*. X-ray diffraction studies of cucumber virus 4 and three strains of tobacco mosaic virus. Biochim. Biophys. Acta 19:203-211.

Franklin, R. E., D. L. D. Caspar, and A. Klug. 1959. The structure of viruses as determined by x-ray diffraction, pp. 447-461. *In* C. S. Holton et al. [ed.] Plant pathology, problems and progress. Univ. Wisconsin Press, Madison. 588 pp.

Franklin, R. E., and K. C. Holmes. 1956. The helical arrangement of the protein subunits in tobacco mosaic virus. Biochim. Biophys. Acta 21:405-406.

Franklin, R. E., and K. C. Holmes. 1958. Tobacco mosaic virus: application of the method of isomorphous replacement to the determination of the helical parameters and radical density distribution. Acta Cryst. 11:213-220.

Franklin, R. E., and A. Klug. 1956. The nature of the helical grove on the tobacco mosaic virus particle X-ray diffraction studies. Biochim. Biophys. Acta 19:403-416.

Fraser, K. B. 1967. Defective and delayed myxovirus infections. Brit. Med. Bull. 23:178-184.

Frazier, N. W., and A. F. Posnette. 1957. Transmission and host range studies of strawberry green-petal virus. Ann. Appl. Biol. 45:580-588.

Frazier, N. W., and H. H. P. Severin. 1945. Weed host range of California aster yellows. Hilgardia 16:621-650.

Frazier, N. W., C. E. Yarwood, and A. H. Gold. 1961. Yellow bud virus endemic along California coast. Plant Dis. Reptr. 45:649-651.

Frederiksen, R. A. 1964. Simultaneous infection and transmission of two viruses in flax by *Macrosteles fascifrons*. Phytopathology 54:1028-1030.

Freifelder, D., and A. K. Kleinschmidt. 1965. Single-strand in breaks in duplex DNA of coliphage T7 as demonstrated by electron microscopy. J. Mol. Biol. 14:271-278.

Freitag, J. H. 1936. Negative evidence on multiplication of curly-top virus in the beet leafhopper, *Eutettix tenellus*. Hilgardia 10:305-342.

Freitag, J. H. 1951. Host range of the Pierce's disease virus of grapes as determined by insect transmission. Phytopathology 41:920-934.

Freitag, J. H. 1964. Interaction and mutual suppression among three strains of aster yellows virus. Virology 24:401-413.

Freitag, J. H. 1969. Interactions of plant viruses and virus strains in their insect vectors, pp. 303-325. *In* K. Maramorosch, [ed.] Viruses, vectors, and vegetation. Interscience, New York. 666 pp.

Freitag, J. H., and N. W. Frazier. 1954. Natural infectivity of leafhopper vectors of Pierce's disease virus of grape in California. Phytopathology 44:7-11.

Freitag, J. H., N. W. Frazier, and R. A. Flock. 1952. Six new leafhopper vectors of Pierce's disease virus. Phytopathology 42:533-534.

Freitag, J. H., and H. H. P. Severin. 1936. Ornamental flowering plants experimentally infected with curly top. Hilgardia 10:263-302.

Freitag, J. H., and S. H. Smith. 1969. Effects of tetracyclines on symptom expression and leafhopper transmission of aster yellows. Phytopathology 59:1820-1823.

Frick, G. 1961. Preparation and purification of T2 bacteriophages with an aqueous polymer two-phase system and some properties of the phage suspension obtained. Exp. Cell Res. 23:488-503.

Frick, G., and P. A. Albertsson. 1959. Bacteriophage enrichment in a liquid two-phase system with subsequent treatment with Freon 113. Nature 183:1070-1072.

Fridborg, K., S. Hjertén, S. Höglund, A. Liljas, B. K. S. Lundberg, P. Oxelfelt, L. Philipson, and B. Strandberg. 1965. Purification, electron microscopy and x-ray diffraction studies of the satellite tobacco necrosis virus. Proc. Nat. Acad. Sci. U.S. 54:513-521.

Friedrich-Freksa, H., G. Melcher, and G. Schramm. 1946. Biologischer, chemischer und serologischer Vergleich zweier Parallelmutanten phytopathogener ziren mit ihren Ausgangsformen. Bio. Zentralbl. 65:187-222.

Frigerio, N. A., and M. J. Shaw. 1969. A simple method for determination of glutaraldehyde. J. Histochem. Cytochem. 17:176-181.

Frisch, D. 1969. A photographic reinforcement analysis of neurotubules and cytoplasmic membranes. J. Ultrastructure. Res. 29:357-372.

Fritz, H. G., R. Delhey, and H. Ross. 1964. Zur Gewinning infektioser TMV-RNS aus Pflanzenmaterial mit Hilfe der Gelfiltration. Z. Naturforsch. 12:1165.

Fritzsche, R. 1964. Untersuchungen uber dei Virusubertragung durch Nematoden. Wiss. Z. Univ. Rostock 13:343-347.

Fritzsche, R. 1966. Beitrag zur Okologie von *Xiphinema diversicaudatum* (Micoletzky) Thorne. Nachrichtenbl. Deut. PflSchutzdienst (Berlin) 20:8-11.

Fritzsche, R. 1967. Methoden der Ubertragung pflanzen pathogener Viren durch Nematoden. Biol. Zentralbl. 86:753-759.

Fritzsche, R. 1970. Beitrage zum Ubertragungs mechanismus pflanzenpathogener viren durch Nematoden. Proc. 9th Int. Nematology Symp. Warsaw 1967 (In Press).

Fritzsche, R., and H. Kegler. 1964. Die Ubertragung des Blattrolvirus der Kirsche (Cherry leafroll virus) durch Nematoden. Naturwissenschaften 51:299.

Fritzsche, R., and K. Schmeltzer. 1967. Uber-tragbarkeit des Nelkenringflecken-Virus durch Nematoden. Naturwissenschaften 54:498-499.

Frosheiser, F. I. 1964. Alfalfa mosaic virus transmitted through alfalfa seed. Phytopathology 54:893. (Abstr.)

Frost, R. R., B. D. Harrison, and R. D. Woods. 1967. Apparent symbiotic action between particles of tobacco rattle virus. J. Gen. Virol. 1:57-70.

Fry, P. R. 1958. The relationship of *Olpidium brassicae* (Wor.) Dang. to the big-vein disease of lettuce. New Zealand J. Agr. Res. 1:301-304.

Fry, P. R., and R. N. Campbell. 1966. Transmission of a tobacco necrosis virus by *Olpidium brassicae*. Virology 30:517-527.

Fujisawa, I., T. Hayashi, and C. Matsui. 1967. Electron microscopy of mixed infections with two plant viruses. I. Intracellular interactions between tobacco mosaic virus and tobacco etch virus. Virology 33:70-76. (4)

Fujisawa, I., C. Matsui, and A. Yamaguchi. 1967. Inclusion bodies associated with sugar beet mosaic. Phytopathology 57:210-213.

Fujita, H. 1962. Mathematical theory of sedimentation analysis. Academic Press, New York. 315 pp.

Fukushi, T. 1933. Transmission of virus through the eggs of an insect vector. Proc. Imp. Acad. (Tokyo) 9:457-460.
Fukushi, T. 1934. Plants susceptible to dwarf disease of rice plant. Trans. Sapporo Natur. Hist. Soc. 13:162-166.
Fukushi, T. 1937. An insect vector of the dwarf disease of rice plant. Proc. Imp. Acad. (Tokyo) 13:328-331.
Fukushi, T. 1939. Retention of virus by its insect vectors through several generations. Proc. Imp. Acad. (Tokyo) 15:142-145.
Fukushi, T. 1940. Further studies on the dwarf disease of rice plant. J. Fac. Agr., Hokkaido Imp. Univ. 45:83-154.
Fukushi, T. 1969. Relationships between propagative rice viruses and their vectors, pp. 279-301. *In* K. Maramorosch, [ed.] Viruses, vectors, and vegetation. Interscience, New York. 666 pp.
Fukushi, T., and E. Shikata. 1963. Localization of rice dwarf virus in its insect vector. Virology 21:503-505.
Fukushi, T., E. Shikata, and I. Kimura. 1962. Some morphological characters of rice dwarf virus. Virology 18:192-205.
Fulton, J. R. 1962. Transmission of tobacco ringspot virus by *Xiphinema americanum*. Phytopathology 52:375.
Fulton, R. W. 1952. Mechanical transmission and properties of rose mosaic virus. Phytopathology 42:413-416.
Fulton, R. W. 1955. Curly-top of tobacco in Wisconsin. Plant Dis. Reptr. 39:799-800.
Fulton, R. W. 1957. Properties of certain mechanically transmitted viruses of Prunus. Phytopathology 47:683-687.
Fulton, R. W. 1959. Purification of sour cherry necrotic ringspot and prune dwarf viruses. Virology 9:522-535.
Fulton, R. W. 1964. Transmission of plant viruses by grafting, dodder, seed and mechanical inoculation, pp. 39-67. *In* M. K. Corbett, and H. D. Sisler, [ed.] Plant Virology. Univ. Florida Press, Gainesville. 527 pp.
Fulton, R. W. 1966. Mechanical transmission of viruses of wood plants. Annu. Rev. Phytopathol. 4:79-102.
Fulton, R. W. 1967*a*. Purification and some properties of tobacco streak and tulare apple mosaic viruses. Virology 32:153-162. (1)
Fulton, R. W. 1967*b*. Purification and serology of rose mosaic virus. Phytopathology 57:1197-1201.
Funatsu, G. 1964. Separation of tryptic peptides of tobacco mosaic virus and strain proteins by an improved method of column chromatography. Biochemistry 3:1351-1355.
Funatsu, G., and H. Fraenkel-Conrat. 1964. Location of amino acid exchanges in chemically evoked mutants of tobacco mosaic virus. Biochemistry 3:1356-1362.
Funatsu, G., and M. Funatsu. 1968. Chemical studies on proteins from two tobacco mosaic virus strains; biochemical regulation in diseased plants or injury. Phytopathol. Soc. Japan 1-9.
Funatsu, G., A. Tsugita, and H. Fraenkel-Conrat. 1964. Studies on the amino acid sequence of tobacco mosaic virus protein. V. Amino acid sequences of two peptides from tryptic digests and location of amide group. Arch. Biochem. Biophys. 105:25-41.
Furumoto, W. A., and R. Mickey. 1967*a*. A mathematical model for the infectivity-dilution curve of tobacco mosaic virus: theoretical considerations. Virology 32:216-223.
Furumoto, W. A., and R. Mickey. 1967*b*. A mathematical model for the infectivity-dilution curve of tobacco mosaic virus: experimental tests. Virology 32:224-233.
Furumoto, W. A., and R. Mickey. 1969. Simple graphical methods to substitute for computer assistance in analyzing TMV infectivity data. Virology 39:589-593.
Furumoto, W. A., and S. G. Wildman. 1963*a*. Studies on the mode of attachment of tobacco mosaic virus. Virology 20:45-52.
Furumoto, W. A., and S. G. Wildman. 1963*b*. The specific infectivity of tobacco mosaic virus. Virology 20:53-61. (1)
Gall, J. G. 1967. The light microscope as an optical diffractometer. J. Cell. Sci. 2:163-168.
Galvez, G. E. 1963. Specific virus adsorption by antibodies coupled to a solid matrix. Phytopathology 53:875-876 (Abstr.)
Galvez, G. E. 1964. Loss of virus by filtration through charcoal. Virology 23:307-312.

Galvez, G. E. 1967. The purification of virus-like particles from rice tungro virus-infected plants. Virology 33:357-359.

Galvez, G. E. 1968. Purification and characterization of rice tungro virus by analytical density-gradient centrifugation. Virology 35:418-426.

Galvez, G. E. 1969*a*. Hoja blanca disease of rice, pp. 35-49. *In* The virus diseases of the rice plant. Johns Hopkins Press, Baltimore. 354 pp.

Galvez, G. E. 1969*b*. Transmission of hoja blanca virus of rice, pp. 155-163. *In* The virus diseases of the rice plant. Johns Hopkins Press, Baltimore. 354 pp.

Gamez, R., and L. M. Black. 1968. Particle counts of wound tumor virus during its peak concentration in leafhoppers. Virology 34:444-451.

Gamez, R., L. M. Black, and R. MacLeod. 1967. Reexamination of the serological relationship between wound tumor virus and reovirus. Virology 32:163-165.

Gamez, R., and R. J. Chiu. 1968. The minimum concentration of a plant virus needed for infection of monolayers of vector cells. Virology 34:356-357.

Gamez, R., and M. A. Watson. 1964. Failure of anaesthetized aphids to acquire a transit henbane mosaic virus when their stylets were artificially inserted into leaves of infected or healthy tobacco plants. Virology 22:292-295.

Gardner, W. S. 1967. Electron microscopy of barley stripe mosaic virus: comparative cytology of tissues infected during different stages of maturity. Phytopathology 57:1315-1326.

Garrett, R. G., and J. A. Tomlinson. 1967. Isolate differences in *Olpidium brassicae*. Trans. Brit. Mycol. Soc. 50:429-435.

George, J. A., and T. R. Davidson. 1959. Notes on life-history and rearing of *Colladonus clitellarius* (Say) (Homoptera:Cicadellidae). Can. Entomol. 91:376-379.

George, J. A., and T. R. Davidson. 1963. Pollen transmission of necrotic ringspot and sour cherry yellows from tree-to-tree. Can. J. Plant Sci. 43:276-288.

Gerola, F. M., and M. Bassi. 1966. An electron microscopy study of leaf vein tumors from maize plants experimentally infected with maize rough dwarf virus. Caryologia 19:13-40.

Gerola, F. M., M. Bassi, M. A. Favali, and E. Betto. 1969. An electron microscopy study of the penetration of tobacco mosaic virus into leaves following experimental inoculation. Virology 38:380-386.

Ghabrial, S. A., R. J. Shepherd, and R. G. Grogan. 1967. Chemical properties of three strains of southern bean mosaic virus. Virology 33:17-25.

Gibbs, A. J. 1969. Plant virus classification. Advances Virus Res. 14:263-328.

Gibbs, A. J., and J. C. Gower. 1960. The use of a multiple-transfer method in plant virus transmission studies—some statistical points arising in the analyses of results. Ann. Appl. Biol. 48:75-83.

Gibbs, A. J., and B. D. Harrison. 1964*a*. Nematode transmitted viruses in sugar beet in East Anglia. Plant Pathol. 13:144-150.

Gibbs, A. J., and B. D. Harrison. 1964*b*. A form of pea early browning virus found in Great Britain. Ann. Appl. Biol. 54:1-11.

Gibbs, A. J., E. Hecht-Paivor, and R. D. Woods. 1966. Some properties of three related viruses: Andean potato latent, Dulcamara mottle, and Ononis yellow mosaic. J. Gen. Microbiol. 44:177-193.

Gibbs, A. J., H. L. Nixon, and R. D. Woods. 1965. Properties of purified preparations of lucerne mosaic virus. Virology 19:441-449.

Gicherman, G., and G. Loebenstein. 1968. Competitive inhibition by foreign nucleic acids and induced interference by yeast-RNA with the infection of tobacco mosaic virus. Phytopathology 58:405-409.

Giddings, N. J. 1950. Some interrelationships of virus strains in sugarbeet curly top. Phytopathology 40:377-388.

Gierer, A. 1963. Chemical basis of mutation. Proc. 5th Int. Congr. Biochem., Moscow 1961, vol. 3. pp. 189-197. Macmillan, New York.

Gierer, A., and K. W. Mundry. 1958. Production of mutants of tobacco mosaic virus by chemical alteration of its ribonucleic acid in vitro. Nature 182:1457-1458.

Gierer, A., and G. Schramm. 1956*a*. Die Infektiositat der Nucleinsaure aus Tabakmosaikvirus. Z. Naturforsch. 11b:138-142.

Gierer, A., and G. Schramm. 1956*b*. Infectivity of ribonucleic acid from tobacco mosaic virus. Nature 177:702-703.

Gil-Fernandez, C., and L. M. Black. 1965. Some aspects of the internal anatomy of the leafhopper *Agallia constricta* (Homoptera:Cicadellidae). Ann. Entomol. Soc. Amer. 58:275-284.

Gill, C. C. 1965. Fluorescent metabolites in virus- or rust-infected bean leaves. Can. J. Bot. 43:201-215.

Gilmer, R. M. 1955. Imported Mahaleb seeds as carriers of necrotic ringspot virus. Plant Dis. Reptr. 39:727-728.

Gilmer, R. M., and L. J. Kelts. 1968. Transmission of tobacco mosaic virus in grape seed. Phytopathology 58:277-278.

Gilmer, R. M., D. H. Palmiter, G. A. Schaefers, and F. L. McEwen. 1966. Leafhopper transmission of X-disease virus of stone fruits in New York. N. Y. State Agr. Exp. Sta. Bull. 813:22 pp.

Gilmer, R. M., and R. D. Way. 1960. Pollen transmission of necrotic ringspot, and prune dwarf viruses in sour cherry. Phytopathology 50:624-625.

Gilmer, R. M., and R. D. Way. 1963. Evidence for tree-to-tree transmission of sour cherry yellows virus by pollen. Plant Dis. Reptr. 47:1051-1053.

Gilmer, R. M., and J. M. Wilks. 1967. Seed transmission of tobacco mosaic virus in apple and pear. Phytopathology 57:214-217.

Gilpatrick, J. D., and M. Weintraub. 1952. An unusual type of protection with the carnation mosaic virus. Science 115:701-702.

Ginoza, W. 1963. Radiosensitive molecular weight of single-stranded virus nucleic acids. Nature 199:453-456.

Ginoza, W. 1968. Inactivation of viruses by ionizing radiation and by heat, pp. 139-209. *In* K. Maramorosch, and H. Koprowski [ed.] Methods in virology, vol. 4. Academic Press, New York. 4 vols.

Ginoza, W., and D. E. Atkinson. 1955. Comparison of some physical and chemical properties of eight strains of tobacco mosaic virus. Virology 1:253-260.

Ginoza, W., D. E. Atkinson, and S. G. Wildman. 1954. A differential ability of strains of tobacco mosaic virus to bind host-cell nucleoprotein. Science 119:269-271.

Ginoza, W., C. J. Hoelle, K. B. Vessey, and C. Formack. 1964. Mechanisms of inactivation of single-stranded virus nucleic acids by heat. Nature 203:606-609.

Ginoza, W., and A. Norman. 1957. Radiosensitive molecular weight of TMV nucleic acid. Nature 179:520-521.

Girolami, G. 1955. Comparative anatomical effects of the curly-top and aster-yellows viruses on the flax plant. Bot. Gaz. 116:305-322.

Glauert, A. M., and R. H. Glauert. 1958. Araldite as an embedding medium for electron microscopy. J. Biophys. Biochem. Cytol. 4:191-194.

Glitz, D. G., and C. A. Dekker. 1964. Studies on a ribonuclease from *Ustilago sphaerogena*. I. Purification and properties of the enzyme. Biochemistry 3:1391-1399.

Gold, A. H. 1961. Antihost serum improves plant virus purification. Phytopathology 51:561-565.

Gold, A. H., and J. E. Duffus. 1967. Infectivity neutralization—a serological method as applied to persistent viruses of beets. Virology 31:308-313.

Gold, A. H., C. A. Suneson, B. R. Houston, and J. W. Oswald. 1954. Electron microscopy and seed and pollen transmission of rod-shaped particles associated with the false stripe disease of barley. Phytopathology 44:115-117.

Goldin, M. E. 1963. Virusnye Vklyuchniya v Rastitel noi Kletke i Priroda Virusow. (Viral inclusions in plant cells and the nature of viruses), 2nd ed. Akad. Nauk USSR, Moscow. 204 pp.

Goldstein, B. 1924. Cytological study of living cells of tobacco plants affected with mosaic diseases. Bull. Torrey Bot. Club 51:261-272.

Gomatos, P. J., and I. Tamm. 1963*a*. Animal and plant viruses with double-helical RNA. Proc. Nat. Acad. Sci. U.S. 50:878-885. (6)

Gomatos, P. J., and I. Tamm. 1963*b*. The secondary structure of reovirus RNA. Proc. Nat. Acad. Sci. 49:707-714. (16)

Goodchild, D. 1960. The effect of light on the synthesis of tobacco mosaic virus. Ph.D. Thesis. Univ. Sidney, Australia.

Goodey, J. B. 1963. Laboratory methods for work with plant and soil nematodes. Tech. Bull. 2, Min. Agr., Fish., Fd., 4th ed., London H.M.S.O. 72 pp.

Gooding, G. V., and T. T. Hebert. 1967. A simple technique for purification of tobacco mosaic virus in large quantities. Phytopathology 57:1285.

Gooding, G. V., and W. B. Hewitt. 1962. Grape yellow vein symptomatology, identification, and the association of a mechanically transmissible virus with the disease. Amer. J. Enol. Viticult. 13:196-203.

Gordon, J. A., and W. P. Jencks. 1963. The relationship of structure to the effectiveness of denaturing agents for proteins. Biochemistry 2:47-57.

Gordon, M. P., J. W. Hoff, and J. J. Holland. 1963. Heat inactivation of the infectious ribonucleic acids of polio and tobacco mosaic viruses. Virology 19:416-418.

Gordon, R. B., and W. C. Price. 1953. Differentiation of tobacco mosaic virus strains by differences in pH of maximum optical density. Arch. Biochem. Biophys. 45:117-123.

Govier, D. A. and A. Kleczkowski. 1970. Inactivation by ultraviolet radiation at different wavelengths of the RNA isolated from potato virus X. Photochem. Photobiol. 12:345-354.

Gowen, J. W. 1941. Mutation in *Drosophila*, bacteria and viruses. Cold Spring Harbor Symp. Quant. Biol. 9:187-193.

Grabar, P. 1964. Immunoelectrophoretic analysis, pp. 79-91. *In* I. F. Ackroyd [ed.] Immunological methods. Blackwell, Oxford. 628 pp.

Granados, R. R., and R. K. Chapman. 1968. Identification of some aster-yellows virus strains and their transmission by the aster leafhopper, *Macrosteles fascifrons*. Phytopathology 58:1685-1692.

Granados, R. R., L. S. Ward, and K. Maramorosch. 1968. Insect viremia caused by a plant-pathogenic virus: electron microscopy of vector hemocytes. Virology 34:790-796.

Granboulan, N. 1967. Autoradiographic methods for electron microscopy, pp. 617-637. *In* K. Maramorosch, and H. Koprowski [ed.] Methods in virology, vol. 3. Academic Press, New York. 4 vols.

Granboulan, P. and E. H. Leduc. 1967. Ultrastructural cytochemistry of *Bacillus subtilis*. J. Ultrastructure Res. 20:111-126.

Grant, T. W., and M. K. Corbett. 1960. Mechanical transmission of the infectious variegation virus of citrus. Nature 188:519-520.

Graves, J. H., K. M. Cowan, and R. Trautman. 1968. Immunochemical studies of foot-and-mouth disease. II. Characterization of RNA-free viruslike particles. Virology 34:269-274.

Gray, H. B., Jr., W. A. Bloomfield, and J. E. Hearst. 1967. Sedimentation coefficients of linear and cyclic wormlike coils with excluded-volume effects. J. Chem. Phys. 46:1493-1498.

Gray, R. A. 1955. Activity of an antiviral agent from *Nocardia* on two viruses in intact plants. Phytopathology 45:281-285.

Griffin, G. D., and H. M. Darling. 1964. An ecological study of *Xiphinema americanum* Cobb in an ornamental spruce nursery. Nematologica 10:471-479.

Griffin, G. D., and A. H. Epstein. 1964. Association of a dagger nematode, *Xiphinema americanum* with stunting and winter kill of ornamental spruce. Phytopathology 54:177-180.

Griffin, G. D., J. E. Hoguelet, and J. W. Nelson. 1963. *Xiphinema americanum* as a vector of necrotic ringspot virus of blueberry. Plant Dis. Reptr. 47:703-704.

Grogan, R. G., and R. N. Campbell. 1966. Fungi as vectors and hosts of viruses. Annu. Rev. Phytopathol. 4:29-52.

Grogan, R. G., D. H. Hall, and K. A. Kimble. 1959. Cucurbit mosaic viruses in California. Phytopathology 49:366-376.

Grogan, R. G., and W. C. Schnathorst. 1955. Tobacco ring-spot virus: the cause of lettuce calico. Plant Dis. Reptr. 39:803-805.

Grogan, R. G., R. H. Taylor, and K. A. Kimble. 1964. The effect of placement of reactants on immunodiffusion precipitin patterns. Phytopathology 54:163-166.

Grogan, R. G., J. E. Welch, and R. Bardin. 1952. Common lettuce mosaic and its control by the use of mosaic-free seed. Phytopathology 42:573-578.

Grogan, R. G., F. W. Zink, W. B. Hewitt, and K. A. Kimble. 1958. The association of *Olpidium* with the big-vein disease of lettuce. Phytopathology 48:292-297.

Grossman, L., and K. Moldave [ed.]. 1967, 1968. Methods in enzymology. Vol. 12A and 12B, sections IV and VI. Nucleic acids. Academic Press, New York. 19 vols.

Grouse, L., T. M. Murphy, M. P. Gordon, and E. A. Smuckler. 1970. The photobiology of a spherical phenotypic hybrid of tobacco mosaic virus. Virology 41:385-388.

Gumpf, D. J., and R. I. Hamilton. 1968. Isolation and characterization of barley stripe mosaic virus protein. Virology 35:87-93.

György, B., and M. Nemeth. 1965. The use of sephadex gel-filtration in virus research work on fruit trees. Zastita Bilja, Plant Prot. 85-88; 16:239-242.

Hackett, A. J., E. S. Sylvester, J. Richardson, and P. Wood. 1968. Comparative electron micrographs of sowthistle yellow vein and vesicular stomatitis viruses. Virology 36:693-696.

Hagborg, W. A. F. 1954. Dwarfing of wheat and barley by the barley stripe mosaic (false stripe) virus. Can. J. Bot. 32:24-37.

Hageman, P. C. 1964. Interference of two strains of tobacco mosaic virus. Ph.D. Thesis. Univ. Amsterdam.

Hagley, E. A. C. 1967. Artificial diet for the adult froghopper. Nature 213:414-415.

Haglund, H. 1967. Isoelectric focusing in natural pH gradients—a technique of growing importance for fractionation and characterization of proteins. Science Tools (The LKB Instrument Journal) 14:17-20.

Hake, T. 1965. Studies on the reactions of OsO_4 and $KMnO_4$ with amino acids, peptides and proteins. Lab. Investment. 14:1208-1212, and also *In* G. F. Bahr and E. H. Zeitler, [ed.] Symposium on quantitative electron microscopy. Williams and Wilkins, Baltimore. 602 pp.

Halliwell, R. S., and R. Longston. 1965. Effects of gamma irradiation on symptom expression of barley stripe mosaic virus disease and on two viruses in vivo. Phytopathology 55:1034-1040.

Hamaguchi, K., and E. P. Geiduschek. 1962. The effects of electrolytes on the stability of the deoxyribonucleate helix. J. Amer. Chem. Soc. 84:1329-1338.

Hamilton, R. I. 1965. An embryo test for detecting seed-borne barley stripe mosaic virus in barley. Phytopathology 55:798-799.

Hamilton, R. I., and E. M. Ball. 1966. Antigenic analysis of extracts from barley infected with barley stripe mosaic virus. Virology 30:661-672.

Hampton, R. E., and R. W. Fulton. 1961. The relation of polyphenol oxidase to instability in vitro of prune dwarf and sour cherry necrotic ringspot viruses. Virology 13:44-52.

Hampton, R. O. 1963. Seed transmission of white clover mosaic and clover yellow mosaic viruses in red clover. Phytopathology 53:1139.

Hampton, R. O. 1967. Seed transmission of viruses in red clover. Phytopathology 57:98. (Abstr.)

Hampton, R. O., and E. W. Hanson. 1968. Seed transmission of viruses in red clover; evidence and methodology of detection. Phytopathology 58:914-920.

Hanlon, S. 1966. The importance of London dispersion forces in the maintenance of the deoxyribonucleic acid helix. Biochem. Biophys. Res. Commun. 23:861-867.

Harder, Donald E., and Hugh C. Kirkpatrick. 1966. Cell inclusions in harbaceous hosts of an apple virus. Phytopathology 56:838-841.

Hare, W. W., and G. B. Lucas. 1959. Control of contact transmission of tobacco mosaic virus with milk. Plant Dis. Reptr. 43:152-154.

Hariharasubramanian, U., and A. Siegel. 1969. Characterization of a new defective strain of TMV. Virology 37:203-208.

Harpaz, I. 1959. Needle transmission of a new maize virus. Nature 184:BA 77-78.

Harpaz, I. 1962. *Calligypona marginata*, the vector of maize rough dwarf virus. F.A.O. Plant Prot. Bull. 9:144-147.

Harpaz, I. 1964. On three "leafhopper"—borne plant viruses from England. Proc. Conf. Czechoslovak Plant Virologists, 5th (Prague 1962) 186-194.

Harpaz, I. 1966. Further studies on the vector relations of the maize rough dwarf virus (MRDV). Maydica 11:18-23.

Harrington, W. F., and H. K. Schachman. 1956. Studies on the alkaline degradation of tobacco mosaic virus. I. Ultracentrifugal analysis. Arch. Biochem. Biophys. 65:278-295.

Harris, J. I., and J. Hindley. 1961. The protein subunit of turnip yellow mosaic virus. J. Mol. Biol. 3:117-120.

Harris, J. I., and C. A. Knight. 1955. Studies on the action of carboxypeptidase on tobacco mosaic virus. J. Biol. Chem. 214:215-230.

Harris, W. J. 1962. Holey films for electron microscopy. Nature 196:499-500.

Harrison, B. D. 1956. The infectivity of extracts made from leaves at intervals after inoculation with viruses. J. Gen. Microbiol. 15:210-220.

Harrison, B. D. 1957. Studies on the host range, properties and mode of transmission of beet ringspot virus. Ann. Appl. Biol. 45:462-472.

Harrison, B. D. 1958*a*. Raspberry yellow dwarf, a soil borne virus. Ann. Appl. Biol. 46:221-229. (8)

Harrison, B. D. 1958*b*. Studies on the behavior of potato leaf roll and other viruses in the body of their aphid vector *Myzus persicae* (Sulz). Virology 6:265-277. (6)

Harrison, B. D. 1961. Plant Pathol. Dep., Rothamsted Exp. Sta. Rep. 1960. 118 pp.

Harrison, B. D. 1962. Plant Pathol. Dep., Rothamsted Exp. Sta. Rep. 1961. 105 pp.

Harrison, B. D. 1964*a*. The transmission of plant viruses in soil, pp. 118-147. *In* M. K. Corbett, and H. D. Sisler [ed.] Plant virology. Florida Univ. Press, Gainesville. 527 pp. (9)

Harrison, B. D. 1964*b*. Specific nematode vectors for seriologically distinctive forms of raspberry ringspot and tomato black ring viruses. Virology 22:544-550.

Harrison, B. D. 1967*a*. The transmission of strawberry latent ringspot virus by *Xiphinema diversicaudatum* (Nematoda). Ann. Appl. Biol. 60:405-409.

Harrison, B. D. 1967*b*. Pea early-browning virus (PEBV). Rothamsted Exp. Sta. Rep. 1966, 115 pp.

Harrison, B. D., and C. H. Cadman. 1959. Role of a dagger nematode (*Xiphinema* sp.) in outbreaks of plant disease caused by arabis mosaic virus. Nature (London) 184:1624-1626.

Harrison, B. D., and N. C. Crowley. 1965. Properties and structure of lettuce necrotic yellows virus. Virology 26:297-310.

Harrison, B. D., and R. A. C. Jones. 1970. Host range and some properties of potato mop-top virus. Ann. Appl. Biol. 65:393-402.

Harrison, B. D., W. P. Mowat, and C. E. Taylor. 1961. Transmission of a strain of tomato black ring virus by *Longidorus elongatus* (Nematoda). Virology 14:480-485.

Harrison, B .D., and H. L. Nixon. 1959. Separation and properties of particles of tobacco rattle virus with different lengths. J. Gen. Microbiol. 21:569-581.

Harrison, B. D., and H. L. Nixon. 1969. Some properties of infective preparations made by disrupting tobacco rattle virus with phenol. J. Gen. Microbiol. 21:591-599.

Harrison, B. D., and W. S. Pierpoint. 1963. Relation of polyphenoloxidase in leaf extracts to the instability of cucumber mosaic and other plant viruses. J. Gen. Microbiol. 32:417-427.

Harrison, B. D., and I. M. Roberts. 1968. Association of tobacco rattle virus with mitochondria. J. Gen. Virol. 3:121-124.

Harrison, B. D., and R. D. Winslow. 1961. Laboratory and field studies on the relation of arabis mosaic virus to its nematode vector *Xiphinema diversicaudatum* (Micoletzsky). Ann. Appl. Biol. 49:621-633.

Harrison, B. D., and R. D. Woods. 1966. Serotypes and particle dimensions of tobacco rattle viruses from Europe and America. Virology 28:610-620.

Harshbarger, J. C., and R. L. Taylor. 1968. Neoplasms of insects. Annu. Rev. Entomol. 13:159-190.

Hart, R. G. 1957. On the distribution of purine and pyrimidine bases in the nucleic acid of tobacco mosaic virus. Proc. Nat. Acad. Sci. Wash. 43:457-463.

Hartzell, A. 1937. Movement of intracellular bodies associated with peach yellows. Contrib. Boyce Thompson Inst. 8:375-388.

Haruna, I., and S. Spiegelman. 1966. Selective interference with viral RNA formation in vitro by specific inhibition with synthetic polyribonucleotides. Proc. Nat. Acad. Sci. U.S. 56:1333-1338.

Hatefi, Y., and W. G. Hanstein. 1969. Solubilization of particulate proteins and nonelectrolytes by choatropic agents. Proc. Nat. Acad. Sci. 62:1129-1136.

Havranek, P. 1967. The effect of inoculum concentration on virus multiplication in cucumber cotyledons. I. Lucerne mosaic virus. Acta Virol. 11:444-452.

Hayashi, Y., and S. Kawase. 1964. Base pairing in ribonucleic acid extracted from the cytoplasmic polyhedra of the silkworm. Virology 23:611-614.

Hayashi, T., and C. Matsui. 1967. Electron microscopy of host cells infected with tobacco etch virus. IV. Tritiated uridine and leucine uptake of intracellular virus and intranuclear crystalline inclusions. Virology 33:47-54.

Hayflick, L. [ed.]. 1967. Biology of the mycoplasma. Ann. N. Y. Acad. Sci. 143:1-824.

Hayflick, L. [ed.]. 1969. The mycoplasmatales and the L-phase of bacteria. Appleton-Century-Crofts, New York. 731 pp.

Hearst, J. E., J. B. Ifft, and J. Vinograd. 1961. The effects of pressure on the buoyant behavior of deoxyribonucleic acid and tobacco mosaic virus in a density gradient at equilibrium in the ultracentrifuge. Proc. Nat. Acad. Sci. U.S. 47:1015-1025.

Hearst, J. E., and J. Vinograd. 1961. The net hydration of T-4 bacteriophage deoxyribonucleic acid and the effect of hydration on buoyant behavior in a density gradient at equilibrium in the ultracentrifuge. Proc. Nat. Acad. Sci. U.S. 47:1005-1014.
Hebert, T. T. 1963. Precipitation of plant viruses by polyethylene glycol. Phytopathology 53:362.
Hecht, E. I., and D. F. Bateman. 1964. Nonspecific acquired resistance to pathogens resulting from localized infections by *Thielaviopsis basicola* or viruses in tobacco leaves. Phytopathology 54:523-530.
Hein, A. 1964. Die Wirkung eines Milchfilms auf die Ubertragung eines nichtpersistenten Virus durch Blattlause. Z. PflKrankh. PflSchutz. 71:267-270.
Helms, K. 1965. Interference between two strains of tobacco mosaic virus in leaves of Pinto bean. Virology 27:346-350.
Helms, K., and M. Zaitlin. 1970. Enhancement of infectivity of tobacco mosaic virus particles partially uncoated by alkali. Virology 41:549-557.
Henderson, R. G. 1931. Transmission of tobacco ringspot by seed of petunia. Phytopathology 21:225-229.
Hennig, W. 1966. Phylogenetic systematics. Univ. Illinois Press, Urbana. 263 pp.
Herold, F., G. H. Bergold, and J. Weibel. 1960. Isolation and electron microscopic demonstration of a virus infecting corn (*Zea mays* L.). Virology 12:335-347.
Herold, F., and K. Munz. 1965. Electron microscopic demonstration of virus-like particles in *Peregrinus maidis* following acquisition of maize mosaic virus. Virology 25:412-417.
Herold, F., and K. Munz. 1967. Virus particles in apparently healthy *Peregrinus maidis*. J. Virol. 1:1028-1036.
Herold, F., G. Trujillo, and K. Munz. 1968. Virus-like particles related to hoja blanca disease of rice. Phytopathology 58:546-547.
Herridge, E. A., and D. E. Schlegel. 1962. Autoradiographic studies of tobacco mosaic virus inoculations on host and non-host species. Virology 18:517-523.
Hersh, R. T. 1962. Mutants of TMV and the commaless code. J. Theoret. Biol. 2:326-328.
Herskovits, T. T., and H. Jaillet. 1969. Structural stability and solvent denaturation of myoglobin. Science 163:282-285.
Hewitt, W. B. 1968. Viruses and virus diseases of the grapevine. Rev. Appl. Mycol. 47:433-455.
Hewitt, W. B., and R. G. Grogan. 1967. Unusual vectors of plant viruses. Annu. Rev. Microbiol. 21:205-224.
Hewitt, W. B., B. R. Houston, N. W. Frazier, and J. H. Freitag. 1946. Leafhopper transmission of the virus causing Pierce's disease of grape and dwarf of alfalfa. Phytopathology 36:117-128.
Hewitt, W. B., G. Martelli, H. F. Dias, and R. H. Taylor. 1970. Grapevine fanleaf virus. Description of Plant Viruses No. 28. C.M.I./A.A.B.
Hewitt, W. B., and D. J. Raski. 1962. Nematode vectors of plant viruses, pp. 63-72. *In* K. Maramorosch [ed.] Symposium on biological transmission of disease agents. Academic Press, New York. 192 pp.
Hewitt, W. B., D. J. Raski, and A. C. Goheen. 1958. Nematode vector of soil-borne fanleaf virus of grapevines. Phytopathology 48:586-595.
Hickman, C. J., and H. H. Ho. 1966. Behaviour of zoospores in plant-pathogenic phycomycetes. Annu. Rev. Phytopathol. 4:195-220.
Hidaka, Z. 1960. The behaviour of tobacco stunt virus in soils, particularly supposing *Olpidium brassicae* (Wor.) Dang. as the vector. Proc. Symp. Soil-Borne Viruses, Dundee, Scotland.
Hidaka, Z. 1961. Studies on the tobacco stunt disease. Vector of tobacco stunt disease [in Japanese]. Annu. Rep. Hatano Tobacco Exp. Sta. 1961, p. 110.
Hidaka, Z., K. Nakano, T. Uozumi, T. Shimizu, and C. Hiruki. 1956. Studies on the tobacco stunt disease. Bull. Hatano Tobacco Exp. Sta. 40:1-80.
Hidaka, Z., and A. Tagawa. 1962. The relationship between the occurrence of tobacco stunt disease and *Olpidium brassicae* [in Japanese]. Ann. Phytopathol. Soc. Japan 27:77-78. (Abstr.)
Hidalgo-Salvatierra, O., and A. D. McLaren. 1969. Photoreactivation action spectrum of tobacco mosaic virus-ribonucleic acid inactivated by ultraviolet radiation. Photochem. Photobiol. 9:417-426.
Hiebert, E., J. B. Bancroft, and C. E. Bracker. 1968. The assembly in vitro of some small spherical viruses, hybrid viruses, and other nucleoproteins. Virology 34:492-508.

Higashi, N., A. Matsumoto, K. Tabata, and Y. Nagatomo. 1967. Electron microscope study of development of chickungunya virus in green monkey kidney stable (VERO) cells. Virology 33:55-69.

Hijink, M. J., and K. Kuiper. 1966. Waarnemingen over de Verdeling van Aaltjes in de Grond. Meded. Rijksfaculteit Landbouwwetensschappen Gent. 31:558-571.

Hilborn, M. T., and R. Bonde. 1956. *Datura* spp. as indicator plants for apple and blueberry virus diseases. Phytopathology 46:241. (Abstr.)

Hill, A. V. 1943. Insect transmission and host plants of virescence (big bud of tomato). Australian J. Sci. Ind. Res. 16:85-90.

Hill, J. H., and R. J. Shepherd. 1972. Biochemical properties of turnip mosaic virus. Virology (In Press).

Hills, G. J., and A. Plaskitt. 1968. A protein stain for electron microscopy of small isometric plant virus particles. J. Ultrastructure Res. 25:323-329.

Hino, T. 1962. Studies on the Adzuki-bean mosaic virus. Ann. Phytopathol. Soc. Japan 27:138-142.

Hippel, P. H. von, and T. Schleich. 1969. Ion effects on the solution structure of biological macromolecules. Accounts Chem. Res. 2:257-265.

Hirai, T., and A. Hirai. 1964. Tobacco mosaic virus: cytological evidence of the synthesis in the nucleus. Science 145:589-591.

Hirai, T., A. Hirashima, T. Itah, T. Takahashi, T. Shimomura, and Y. Hayashi. 1966. Inhibitory effect of blasticidin S on tobacco mosaic virus multiplication. Phytopathology 56:1236-1240.

Hirai, T., and T. Shimomura. 1960. The mode of action of some antibiotics in their inhibitory effect on tobacco mosaic virus multiplication. Phytopathol. Z. 40:35-44.

Hirs, C. H. W., S. Moore, and W. H. Stein. 1960. The sequence of the amino acid residues in performic acid-oxidized ribonuclease. J. Biol. Chem. 235:633-647.

Hiruki, C. 1964. Mechanical transmission of tobacco stunt virus. Virology 23:288-290.

Hiruki, C. 1965. Transmission of tobacco stunt virus by *Olpidium brassicae*. Virology 25:541-549.

Hiruki, C. 1967. Host specificity in transmission of tobacco stunt virus by *Olpidium brassicae*. Virology 33:131-136.

Hiruki, C., and Z. Hidaka. (1954) cited after Nozu and Okada, 1968.

Hirumi, H., R. R. Granados, and K. Maramorosch. 1967. Electron microscopy of a plant-pathogenic virus in the nervous system of its insect vector. J. Virol. 1:430-444.

Hirumi, H., and K. Maramorosch. 1964*a*. Insect tissue culture: further studies on the cultivation of embryonic leafhopper tissues in vitro. Contrib. Boyce Thompson Inst. 22:343-352.

Hirumi, H., and K. Maramorosch. 1964*b*. The in vitro cultivation of embryonic leafhopper tissues. Exp. Cell Res. 36:625-631.

Hirumi, H., and K. Maramorosch. 1969. Mycoplasma-like bodies in the salivary glands of insect vectors carrying the aster yellows agent. J. Virol. 3:82-84.

Hitchborn, J. H., and D. B. Dunn. 1965. Differential adsorption of the two components of wild cucumber mosaic virus by bentonite. Virology 26:441-449.

Hitchborn, J. H., and G. J. Hills. 1965. The use of negative staining in the electron microscopic examination of plant viruses in crude extracts. Virology 27:528-540.

Hitchborn, J. H., and G. J. Hills. 1967. Tubular structures associated with turnip yellow mosaic virus in vivo. Science 157:705-706.

Hitchborn, J. H., and G. J. Hills. 1968. A study of tubes produced in plants infected with a strain of turnip yellow mosaic virus. Virology 35:50-70.

Hitchborn, J. H., G. J. Hills, and R. Hull. 1966. Electron microscopy of virus-like particles found in diseased *Plantago lanceolata* in Britian. Virology 28:768-772.

Hjerten, S. 1962. A new method for preparation of agarose for gel electrophoresis. Biochim. Biophys. Acta 62:445-449.

Hodgson, W. A., and J. Munro. 1966. An inhibitor of potato virus X in potato plants infected with *Phytophthora infestans*. Phytopathology 56:560-561.

Holborow, E. J., and G. D. Johnson. 1967. Immunofluorescence, pp. 571-596.*In* D. M. Weir, [ed.] Handbook of experimental immunology, Blackwell, Oxford. 1245 pp.

Hollaender, A., and B. M. Duggar. 1936. Irradiation of plant viruses and of microorganisms with monochromatic light. III. Resistance of the virus of typical tobacco mosaic and *Escherichia coli* to radiation from 3000-2250 Å . Proc. Nat. Acad. Sci. U.S. 22:19-24.

Hollings, M. 1959. Host range studies with fifty two plant viruses. Ann. Appl. Biol. 47:98-108.

Hollings, M. 1962. Studies of pelargonium leaf curl virus. I. Host range, transmission and properties in vitro. Ann. Appl. Biol. 50:189-202.

Hollings, M. 1965*a*. Anemone necrosis, a disease caused by a strain of tobacco ringspot virus. Ann. Appl. Biol. 55:447-457.

Hollings, M. 1965*b*. Disease control through virus-free stock. Annu. Rev. Phytopathol. 3:367-396.

Hollings, M. 1966. Local lesion and other test plants for the identification and culture of viruses, pp. 230-241. *In* A.B.R. Beemster and J. Dijkstra [ed.] Viruses of plants. North-Holland, Amsterdam. 342 pp.

Hollings, M., and O. M. Stone. 1965. Investigation of carnation viruses. II. Carnation ringspot. Ann. Appl. Biol. 56:73-86.

Hollings, M., and O. M. Stone. 1970*a*. Virus purification and production of stock antisera. Rep. Glasshouse Crops Res. Inst. 1969:140.

Hollings, M., and O. M. Stone. 1970*b*. Carnation ringspot virus. Description of Plant Viruses No. 21. C.M.I./A.A.B.

Hollings, M., O. M. Stone, and F. T. Last. 1965. Detection and identification of viruses in mushroom sporophores and mycelium disrupted with ultrasound. Rep. Glasshouse Crops Res. Inst. 1964. 151-154.

Holmes, F. O. 1929*a*. Local lesions in tobacco mosaic. Bot. Gaz. 87:39-55.

Holmes, F. O. 1929*b*. Inoculation methods in tobacco mosaic studies. Bot. Gaz. 87:56-63.

Holmes, F. O. 1931. Local lesions of mosaic in *Nicotiana tabacum* L. Contrib. Boyce Thompson Inst. 3:163-172.

Holmes, F. O. 1938. Taxonomic relationships of plants susceptible to infection by tobacco-mosaic virus. Phytopathology 28:58-66.

Holmes, F. O. 1941. A distinctive strain of tobacco-mosaic virus from plantago. Phytopathology 31:1089-1098.

Holmes, F. O. 1954. Inheritance of resistance to viral diseases in plants. Advances Virus Res. 2:1-30.

Holmes, F. O. 1955. Elimination of spotted wilt from dahlias by propagation of tip cuttings. Phytopathology 45:224-226.

Holmes, F. O. 1956*a*. Elimination of aspermy virus from nightingale chrysanthemum. Phytopathology 56:599-600.

Holmes, F. O. 1956*b*. Elimination of foliage spotting from sweet potatoes. Phytopathology 46:502-504.

Holmes, K. C., and R. E. Franklin. 1958. The radial density distribution in some strains of tobacco mosaic virus. Virology 6:328-336.

Holoubek, V. 1962. Mixed reconstitution between protein from common tobacco mosaic virus and ribonucleic acid from other strains. Virology 18:401-404.

Holoubek, V. 1963. The composition of tobacco mosaic virus protein after the incorporation of 5-fluorouracil into the virus. J. Mol. Biol. 6:164-166.

Honda, Y., and C. Matsui. 1969. Occurrence of tobacco mosaic virus within the nucleus. Virology 39:593-595.

Horne, R. W. 1967. Electron microscopy of isolated virus particles and their components, pp.521-574. *In* K. Maramorosch, and H. Koprowski [ed.] Methods in virology, vol. 3. Academic Press, New York. 4 vols.

Horne, R. W., and P. Wildy. 1961. Symmetry in virus architecture. Virology 15:348-373.

Hosaka, Y., and K. Aizawa. 1964. The fine structure of the cytoplasmic-polyhedrosis virus of the silkworm, *Bombyx mori* (Linnaeus). J. Insect Pathol. 6:53-77.

Hougas, R. W. 1951. Factors affecting sap transmission of the potato yellow-dwarf virus. Phytopathology 41:483-493.

Houston, B. R., K. Esau, and W. B. Hewitt. 1947. The mode of the vector feeding and the tissues involved in the transmission of Pierce's disease virus in grape and alfalfa. Phytopathology 37:247-253.

Howard-Flanders, P., and T. Alper. 1957. The sensitivity of microorganisms to irradiation under controlled gas conditions. Radiation 7:518-540.

Howe, C., C. Morgan, C. de vaux St. Cyr, K. C. Hsu, and H. M. Rose. 1967. Morphogenesis of type 2 parainfluenza virus examined by light and electron microscopy. J. Virol. 1:215-237.

Howell, J. A., H. M. Tsuchiya. and A. G. Fredirckson. 1967. Continuous synchronous culture of photosynthetic micro-organisms. Nature 214:582-584.

Hsiung, G. D. 1965. Use of ultrafiltration for animal virus grouping. Bacteriol. Rev. 29:477-486.

Huguelet, J. E., and W. J. Hooker. 1966. Latent infection of *Gomphrena globosa* by potato virus X. Phytopathology 56:431-437.

Hulett, H. R., and H. S. Loring. 1965. Effect of particle length distribution on infectivity of tobacco mosaic virus. Virology 25:418-430.

Hull, R. 1970. Studies on alfalfa mosaic virus. III. Reversible dissociation and reconstitution studies. Virology 40:34-47.

Hull, R., and A. N. Adams. 1968. Groundnut rosette and its assistor virus. Ann. Appl. Biol. 62:139-145.

Hull, R., M. W. Rees, and M. N. Short. 1969. Studies on alfalfa mosaic virus. I. The protein and nucleic acid. Virology 37:404-415.

Hung, P. P. 1969. Use of DEAE-cellulose for isolation of ribonucleic acids from bacteriophages Q$\beta$and MS2. Biochim. Biophys. Acta 186:220-222.

Hunter, D. E., H. M. Darling, and W. L. Beale. 1969. Seed transmission of potato spindle tuber virus. Amer. Potato J. 46:247-250.

Hurst, R. E., and N. L. Incardona. 1969. Molecular weights of viruses from isopycnic centrifugation with schlieren optics. Virology 37:62-73.

Hurst, R. O., and A. Kuksis. 1958. Degradation of some purine and pyrimidine derivatives by hot alkali. Can. J. Biochem. Physiol. 36:931-936.

Hutchins, L. M. 1933. Identification and control of the phony disease of the peach. Georgia Dep. Entomol. Bull. 78. 55 pp.

Hutchins, L. M. 1939. Apparent localization of phony disease virus in the woody cylinder. Phytopathology 29:12. (Abstr.)

Hutton, E. M., and A. R. Peak. 1951. Inoculation techniques for studying the genetics of virus resistance in plants. J. Australian Inst. Agr. Sci. 17:193-198.

Ifft, J. B., W. R. Martin III, and K. Kinzie. 1970. Density gradient proportionality constants for a number of aqueous binary solutions. Biopolymers 9:597-614.

Ifft, J. B., and J. Vinograd. 1966. The buoyant behavior of bovine serum mercaptalbumin in salt solutions at equilibrium in the ultracentrifuge. II. Net hydration, ion binding, and solvated molecular weight in various salt solutions. J. Phys. Chem. 70:2814-2822.

Ifft, J. B., D. H. Voet, and J. Vinograd. 1961. The determination of density distributions and density gradients in binary solutions at equilibrium in the ultracentrifuge. J. Phys. Chem. 65:1138-1145.

Iglewski, W. J., and R. M. Franklin. 1967. Denaturation and renaturation of viral RNA. III. Purification of the complementary strand of R17 RNA. Proc. Nat. Acad. Sci. U. S. 58:1019-1024.

Ignoffo, C. M. 1968. Viruses—living insects. Current Topics Microbiol. Immunol. 42:129-167.

Iida, T. T., and A. Shinkai. 1969. Transmission of dwarf, yellow dwarf, stripe, and black-streaked dwarf, pp. 125-129. *In* The virus diseases of the rice plant. Johns Hopkins Press, Baltimore. 354 pp.

Inouye, T. 1967. A seed-borne mosaic virus of pea. Ann. Phytopathol. Soc. Japan 33:38-42.

Inouye, T., N. Inouye, M. Asatani, and K. Mitsuhata. 1967. Studies on cucumber green mottle virus I. Jogaku-kenkyu 51:175.

Ishihara, T. 1969. Families and genera of leafhopper vectors, pp. 235-254. *In* K. Maramorosch [ed.] Viruses, vectors, and vegetation. Interscience, New York. 666 pp.

Ishiie, T., Y. Doi, K. Yora, and H. Asuyama. 1967. Suppressive effects of antibiotics of tetracycline group on symptom development of mulberry dwarf disease. Ann. Phytopathol. Soc. Japan 33:267-275.

Israel, H. W., and A. F. Ross. 1967. The fine structure of local lesions induced by tobacco mosaic virus in tobacco. Virology 33:272-286.

Ito, T. 1967. Photoinactivation of acridine-sensitized tobacco mosaic virus ribonucleic acid with monochromatic light. Curr. Mod. Biol. 1:192-195.

Iwanowski, D. 1899. Uber die Mosaikkrankheit der Tabakspflanze. Cbl. Bakteriol. 2F., 5:250-254.

Iwanowski, D. 1903. Uber die Mosaikkrankheit der Tabakspflanze. Z. Pflanzenkrankh. 13:1-41.

Izadpanah, K. and R. J. Shepherd. 1966. Purification and properties of the pea enation mosaic virus. Virology 28:463-476.

Jacobsen, C. F., and K. Linderstrom-Lang. 1940. Method for rapid determination of specific gravity. Acta Physiol. Scand. 1:149-152.

Jacoli, G. G. 1968*a*. Mechanism of inhibition of ribonuclease by bentonite. Can. J. Biochem. 10:1237-1239. (1)

Jacoli, G. G. 1968*b*. Purification of turnip mosaic virus by gel filtration. Biochim. Biophys. Acta 165:299-302.

Jagger, I. C. 1928. The brown blight disease of lettuce. Phytopathology 18:949-950. (Abstr.)

Jagger, I. C. 1940. Brown blight of lettuce. Phytopathology 30:53-64.

Jagger, I. C., and N. Chandler. 1934. Big vein, a disease of lettuce. Phytopathology 24:1253-1256.

Jagger, J. 1958. Photoreactivation. Bacteriol. Rev. 22:99-142.

Jagger, J. 1967. Introduction to research in photobiology. Prentice-Hall, Englewood Cliffs, N. J. 77 pp.

Jaspers, E. M. J., and J. R. Moed. 1966. The complexity of alfalfa mosaic virus, pp. 188-195. *In* A.B.R. Beemster and J. Dijkstra [ed.] Viruses of plants. North-Holland, Amsterdam. 342 pp.

Jauregui-Adell, J., I. Hindennach, and H. G. Wittmann. 1969. Die primare Proteinstruktur von Stammen des Tabakmosaikvirus. Teil V: Aminosaurensequenz (1-61) des Proteins des Tabakmosaikvirus-Stammes Holmes rib grass. Z. Naturforsch. 24b:877-884.

Jeener, R. 1965. Effects on the antigenic determinants of tobacco mosaic virus of a small number of base replacements in the ribonucleic acid. Virology 26:10-15.

Jeener, R., P. Lemoine, and C. Lavand'homme. 1954. Detection et proprietes des formes due virus de la mosaique du tabac depourvues d' acide ribonucleique et non infectieuses. Biochim. Biophys. Acta 14:321-334.

Jeener, R., and J. Rosseels. 1953. Incorporation of 2-thiouracil-^{35}S in the ribose nucleic acid of tobacco mosaic virus. Biochim. Biophys. Acta 11:438.

Jencks, W. P. 1969. Catalysis in chemistry and enzymology. McGraw-Hill, New York. 644 pp.

Jennings, P. R., and A. Pineda. 1971. The effect of the hoja blanca virus on its insect vector. Phytopathology 61:142-143.

Jensen, D. D. 1957. Differential transmission of peach yellow leaf roll virus to peach and celery by the leafhopper, *Colladonus montanus*. Phytopathology 47:575-578.

Jensen, D. D. 1959. A plant virus lethal to its insect vector. Virology 8:164-175.

Jensen, D. D. 1968. Influence of high temperatures on the pathogenicity and survival of Western X-disease virus in leafhoppers. Virology 36:662-667.

Jensen, D. D. 1969. Insect diseases induced by plant-pathogenic viruses, pp. 505-525. *In* K. Maramorosch [ed.] Viruses, vectors, and vegetation. Interscience, New York. 666 pp.

Jensen, D. D., and A. H. Gold. 1951. A virus ring spot of odontoglossum orchid: symptoms, transmission, and electron microscopy. Phytopathology 41:648-653.

Jensen, D. D., W. H. Griggs, C. Q. Gonzales, and H. Schneider. 1964. Pear decline virus transmission by pear psylla. Phytopathology 54:1346-1351.

Jensen, D. D., R. F. Whitcomb, and J. Richardson. 1967. Lethality of injected peach Western X-disease virus to its leafhopper vector. Virology 31:532-538.

Jensen, H. J. 1961. Nematodes affecting Oregon agriculture. Agr. Exp. Sta. Oregon State Univ., Corvallis, Bull. 579:34 pp.

Jensen, H. J., and T. C. Allen. 1964. Transmission of tobacco rattle virus by a stubby-root nematode. *Trichodorus allius*. Plant Dis. Reptr. 48:333-334.

Jensen, J. H. 1933. Isolation of yellow-mosaic viruses from plants infected with tobacco mosaic. Phytopathology 23:964-974.

Jermoljev, E., and L. Albrechtova. 1965. Aplikace suspenze latexu v serologicke diagnoze virovych chorob rostlin. Ochrana Rostlin 2:51-56.

Jha, A., and A. F. Posnette. 1959. Transmission of a virus to strawberry plants by a nematode. Nature (London) 184:962-963.

Jha, A., and A. F. Posnette. 1961. Transmission of arabis mosaic virus by the nematode *Xiphinema diversicaudatum* (Micol.) Virology 13:119-123.

Jockusch, H. 1966*a*. Temperatursensitive Mutanten des Tabakmosaikvirus. I. In vivo-verhalten. Z. Vererbungsl. 98:320-343.

Jockusch, H. 1966*b*. Temperatursensitive Mutanten des Tabakmosaikvirus. II. In vitro-verhalten. Z. Vererbungsl. 98:344-362.

Jockusch, H. 1968*a*. Two mutants of tobacco mosaic virus temperature-sensitive in two different functions. Virology 35:94-101.

Jockusch, H. 1968*b*. Stability and genetic variation of a structural protein. Naturwissenschaften. 55:514-518.

Jockusch, H., R. Koberstein, and R. Jaenicke. 1969. Thermal denaturation equilibria of TMV coat proteins with chemically defined differences in their primary structure. Z. Naturforsch. 24b:613-617.

John, V. T. 1965. A microimmuno-osmophoretic technique for assay of tobacco mosaic virus. Virology 27:121-123.

Johnson, F. 1941. Transmission of plant viruses by dodder. Phytopathology 31:649-656.

Johnson, F. 1942. The complex nature of white clover mosaic. Phytopathology 32:103-116.

Johnson, J. 1947. Virus attenuation and the separation of strains by specific hosts. Phytopathology 37:822-837.

Johnson, J. 1951. Virus particles in various plant species and tissues. Phytopathology 41:78-93.

Johnson, J. H., J. E. Fields, and W. A. Darlington. 1967. Removing viruses from water by poly-electrolytes. Nature 213:665-667.

Joklik, W. K. 1968. The poxviruses. Annu. Rev. Microbiol. 22:359-390.

Joklik, W. K. 1970. The molecular biology of reovirus. J. Cell Physiol. 76:289-301.

Jones, J. C. 1962. Current concepts concerning insect hemocytes. Amer. Zool. 2:209-246.

Jones, R. A. C., and B. D. Harrison. 1969. The behavior of potato mop-top virus in soil, and evidence for its transmission by *Spongospora subterranea* (Wallr.) Lagerh. Ann. Appl. Biol. 63:1-17.

Jones, W. A., M. Jacobson, and R. P. Kahn. 1959. Chemical nature of a plant-virus inhibitor from rice. Nature 184:1146.

Joshi, L., and H. H. Thornberry. 1968. Effect of magnesium trisilicate on the infectivity of tobacco mosaic virus. Virology 35:169-171.

Josse, J., and J. Eigner. 1966. Physical properties of deoxyribonucleic acid. Annu. Rev. Biochem. 35:789-834.

Kabat, E. A., and M. M. Mayer. 1961. Experimental immunochemistry. 2nd ed. Thomas, Springfield, Illinois. 905 pp.

Kado, C. I. 1962. The effect of pH and phosphate on virus infection. Phytopathology 52:362. (Abstr.)

Kado, C. I. 1964. Chenopodium, a sensitive host for TMV-RNA. Virology 24:660-665.

Kado, C. I. 1967. Biological and biochemical characterization of sowbane mosaic virus. Virology 31:217-229.

Kado, C. I. 1968. Purification and properties of a ribonuclease isolated from etiolated cucumber seedlings. Arch. Biochem. Biophys. 125:86-93.

Kado, C. I., and D. R. Black. 1968. The molecular weight of sowbane mosaic virus. Virology 36:137-139.

Kado, C. I., and C. A. Knight. 1966. Location of a local lesion gene in tobacco mosaic virus RNA. Proc. Nat. Acad. Sci. U.S. 55:1276-1283.

Kado, C. I., and C. A. Knight. 1968. The coat protein gene of tobacco mosaic virus. I. Location of the gene by mixed infection. J. Mol. Biol. 36:15-23.

Kado, C. I., and C. A. Knight. 1970. Host specificity of plant viruses. I. Cucumber virus 4. Virology 40:997-1007.

Kado, C. I., M. H. V. van Regenmortel, and C. A. Knight. 1968. Studies on some strains of tobacco mosaic virus in orchids. I. Biological, chemical, and serological studies. Virology 34:17-24.

Kahn, R. P. 1956. Seed transmission of tomato-ringspot virus in the Lincoln variety of soybeans. Phytopathology 46:295.

Kalmakoff, J., and J. H. Tremaine. 1967. Some physical and chemical properties of carnation ringspot virus. Virology 33:10-16.

Kalmus, H., and B. Kassanis. 1945. The use of abrasives in the transmission of plant viruses. Ann. Appl. Biol. 32:230-234.

Kamienska-Zyta, M. 1966. The isolation of virus Y from tobacco leaves by column chromatography in two-phase system of polymers. Acta Biol. Cracoviensia, Ser. Bot. 9:109-113.

Kanervo, V., O. Heinkinheimo, M. Raatikainen, and A. Tinnila. 1957. The leafhopper *Delphacodes pellucida* (F) as the cause and distributor of the damage to oats in Finland. Finnish State Agr. Res. Board 160:1-56.
Kaper, J. M. 1960. Protein components from turnip yellow mosaic virus. Nature 186:219-220.
Kaper, J. M. 1964. Alkaline degradation of turnip yellow mosaic virus. I. The controlled formation of empty protein shells. Biochemistry 3:486-493.
Kaper, J. M. 1968. The small RNA viruses of plants, animals, and bacteria. A. Physical properties, pp. 1-133. *In* H. Fraenkel-Conrat [ed.] Molecular basis of virology. Van Nostrand Reinhold, New York. 656 pp.
Kaper, J. M., T. O. Diener, and H. A. Scott. 1965. Some physical and chemical properties of cucumber mosaic virus (strain Y) and of its isolated ribonucleic acid. Virology 27:54-72.
Karas, J. G., C. L. Hamner, D. J. deZeeuw, and H. M. Sell. 1964. Inhibition of localized virus lesions by N-dimethylamino-succinamic acid. Nature 203:1197.
Karnovsky, M. J. 1965. A formaldehyde-glutaraldehyde fixative of high osmolality for use in electron microscopy. J. Cell Biol. 27:137A-138A.
Kassanis, B. 1939. Intranuclear inclusions in virus infected plants. Ann. Appl. Biol. 26:705-709.
Kassanis, B. 1947. Studies on dandelion yellow mosaic, and other virus diseases of lettuce. Ann. Appl. Biol. 34:412-421.
Kassanis, B. 1949. The transmission of sugar beet yellows virus by mechanical inoculation. Ann. Appl. Biol. 36:270-272.
Kassanis, B. 1952*a*. Some factors affecting the transmission of leaf roll virus by aphids. Ann. Appl. Biol. 39:157-167.
Kassanis, B. 1952*b*. Some effects of high temperature on the susceptibility of plants to infection with viruses. Ann. Appl. Biol. 39:358-369.
Kassanis, B. 1957. Some effects of varying temperature on the quality and quantity of tobacco mosaic virus in infected plants. Virology 4:187-199.
Kassanis, B. 1961. The transmission of potato aucuba mosaic virus by aphids from plants also infected by potato viruses A or Y. Virology 13:93-97.
Kassanis, B. 1962. Properties and behaviour of a virus depending for its multiplication on another. J. Gen. Microbiol. 27:477-488.
Kassanis, B. 1963. Interaction of viruses in plants. Advances Virus Res. 10:219-255.
Kassanis, B. 1964. Properties of tobacco necrosis virus and its association with satellite virus. Ann. Inst. Phytopathol. Benaki NS 6:7-26.
Kassanis, B. 1966. Properties and behaviour of satellite virus, pp. 177-187. *In* A. B. R. Beemster and J. Dijkstra [ed.] Viruses of plants. North-Holland, Amsterdam. 342 pp.
Kassanis, B., and A. Kleczkowski. 1965. Mutual precipitation of two viruses. Nature 205:310.
Kassanis, B., and I. Macfarlane. 1964. Transmission of tobacco necrosis virus by zoospores of *Olpidium brassicae*. J. Gen. Microbiol. 36:79-93.
Kassanis, B., and I. Macfarlane. 1965. Interaction of virus strain, fungus isolate, and host species in the transmission of tobacco necrosis virus. Virology 26:603-612.
Kassanis, B., and I. Macfarlane. 1968. The transmission of satellite viruses of tobacco necrosis virus by *Olpidium brassicae*. J. Gen. Virol. 3:227-232.
Kassanis, B., and T. W. Tinsley. 1958. The freeing of tobacco tissue cultures from potato virus Y by 2-thiouracil. Proc. Third Conf. Potato Virus Dis., Lisse-Wageningen, pp. 153-155.
Kassanis, B., and G. W. Welkie. 1963. The nature and behavior of unstable variants of tobacco necrosis virus. Virology 21:540-550.
Kausche, G. A., and H. Stubbe. 1939. Über die Entstehung einer mit Röntgenstrahlen induzierten 'Mutation' des Tabakmosaikvirus. Naturwissenschaften 27:501-502.
Kausche, G. A., and H. Stubbe. 1940. Zur Frage der Entstehung Rontgenstrahleninduzierten Mutationen beim Tabakmosaikvirus protein. Naturwissenschaften 28:824.
Kay, D. H. [ed.]. 1965. Techniques for electron microscopy. 2nd ed. Blackwell, Oxford. 560 pp.
Kegeles, G., S. Kaplan, and L. Rhodes. 1969. The effects of pressure in high-speed ultracentrifugation of chemically reacting systems. Ann. N. Y. Acad. Sci. 164:183-191.
Keifer, H. H. 1952. The eriophyid mites of California. Bull. Calif. Insect Surv. 2: No. 1.
Keifer, H. H., and N. S. Wilson. 1956. A new species of eriophyid mite responsible for the vection of peach mosaic virus. Calif. Dep. Agr. Bull. 44:145-146.
Keilin, D., and T. Mann. 1938. Polyphenol oxidase: purification, nature and properties. Proc. Roy. Soc. Lond. B 125:187-204.

Kellenberger, E., and W. Arber. 1957. Electron microscopical studies of phage multiplication. I. A method for quantitative analysis of particle suspensions. Virology 3:245-255.

Kellenberger, E., A. Ryter, and J. Sechaud. 1958. Electron microscope study of DNA-containing plasmas. II. Vegetative and mature phage DNA as compared with normal bacterial nucleoids in different physiological states. J. Biophys. Biochem. Cytol. 4:671-678.

Kelly, J. J., and P. Kaesberg. 1962. Preparation of protein subunits from alfalfa mosaic virus under mild conditions. Biochim. Biophys. Acta 55:236-237.

Kelner, A. 1949. Photoreactivation of ultraviolet-irradiated *Escherichia coli* with especial reference to the dose-reduction principle and to ultraviolet-induced mutation. J. Bacteriol. 58:511-522.

Kendrew, J. C. 1959. Structure and function in myoglobin and other proteins. Fed. Proc. 18:740-751.

Kendrick, J. B., and M. W. Gardner. 1924. Soybean mosaic: seed transmission and effect on yield. J. Agr. Res. 27:91-98.

Kennedy, J. S., M. F. Day, and V. F. Eastop. 1962. A conspectus of aphids as vectors of plant viruses. Commonwealth Inst. Entomol. 114 pp.

Kessler, B., E. Swirski, and A. S. Tahori. 1958. Effect of caffeine and other purines upon the ribonucleic acid/deoxyribonucleic acid ratio in leaves, and the suitability of these leaves for aphids. Nature 181:1595-1596.

Keur, J. Y. 1934. Studies of the occurrence and transmission of virus diseases in the genus *Abutilon*. Bull. Torrey Bot. Club 61:53-70.

Khalil, M. T. M., and M. A. Lauffer. 1967. Polymerization depolymerization of tobacco mosaic virus protein. X. Effect of D_2O. Biochemistry 6:2474-2480.

Kickhofen, B., and M. Burger. 1962. Das Verhalten von Proteinen in Phenol I. Ribonuclease. Biochim. Biophys. Acta 65:190-199.

Kieselev, N. A., C. L. Shpitzberg, and B. K. Vainshtein. 1967. Crystallization of catalase in the form of tubes with monomolecular walls. J. Mol. Biol. 25:433-442.

Kimura I., and E. Shikata. 1968. Structural model of rice dwarf virus. Proc. Japan Acad. 44:538-543.

Kiraly, Z., and B. I. Pozsar. 1964. On the inhibition of TMV production by kinetin and adenine in intact tobacco leaves, pp. 61-64. *In* Z. Kiraly and Ubrizsy [ed.] Host-parasite relations in plant pathology. Research Institute for Plant Protection Publication, Budapest, Hungary.

Kirby, K. S. 1968. Isolation of nucleic acids with phenolic solvents, pp. 87-98. *In* L. Grossman and K. Moldave [ed.] Methods in enzymology, vol. 12B. Academic Press, New York. 19 vols.

Kirkaldy, G. W. 1906. Leafhoppers and their natural enemies. Hawaiian Sugar Planters' Assoc. Div. Entomol. Bull. 1, Part IX, 271-479.

Kirkpatrick, H. C., and R. C. Lindner. 1961. Protection of cucumber seedlings from stone fruit ringspot virus by 8-azaguanine. Phytopathology 51:727-730.

Kisimoto, R. 1965. Studies on the polymorphism and its role playing in the population growth of the brown planthopper, *Nilaparvata lugens* Stal. Bull. Shikohu. Agr. Exp. Sta. 13:1-106.

Kisimoto, R. 1967. Genetic variation in the ability of a planthopper vector, *Laodelphax striatellus* (Fallen) to acquire the rice stripe virus. Virology 32:144-152.

Kisimoto, R. 1969. Ecology of insect vectors, forecasting, and chemical control, pp. 243-255. *In* The virus diseases of the rice plant. Johns Hopkins Press, Baltimore. 354 pp.

Kisimoto, R., and M. A. Watson. 1965. Abnormal development of embryos induced by inbreeding in *Delphacodes pellucida* Fabricius and *Delphacodes dubia* Kirschbaum, vectors of European wheat striate mosaic virus. J. Invertebrate Pathol. 7:297-305.

Kitagawa, Y., and E. Shikata. 1969*a*. On some propertiesof rice black-streaked dwarf virus. Mem. Fac. Agr. Hokkaido Univ. 6:439-445.

Kitagawa, Y., and E. Shikata. 1969*b*. Purification of rice black-streaked dwarf virus. Mem. Fac. Agr. Hokkaido Univ. 6:446-451.

Kitajima, E. W. 1965. Electron microscopy of vira cabeca virus (Brazilian tomato spotted wilt virus) within the host cell. Virology 26:89-99.

Kitajima, E. W., and J. A. Lauritis. 1969. Plant virions in plasmodesmata. Virology 37:681-685.

Kitajima, E. W., J. A. Lauritis, and H. Swift. 1969. Morphology and intracellular localisation of a bacilliform latent virus in sweet clover. J. Ultrastructure Res. 29:141-150.

Kleczkowski, A. 1946. Combination between different proteins and between proteins and yeast nucleic acid. Biochem. J. 40:677-687.

Kleczkowski, A. 1949. The transformation of local lesion counts for statistical analysis. Ann. Appl. Biol. 36:139-152.

Kleczkowski, A. 1950. Interpreting relationships between the concentrations of plant viruses and numbers of local lesions. J. Gen. Microbiol. 4:53-69.

Kleczkowski, A. 1959. Aggregation of the protein of tobacco mosaic virus with and without combination with nucleic acid. Virology 7:385-393.

Kleczkowski, A. 1960*a*. Effect of non-ionizing radiation on plant viruses. Ann. N. Y. Acad. Sci. 83:661-669. (19)

Kleczkowski, A. 1960*b*. Serological properties of viruses and their fragments, pp. 196-204. *In* A. B. R. Beemster and J. Dijkstra [ed.] Viruses of plants. North-Holland, Amsterdam. 342 pp. (17)

Kleczkowski, A. 1965. A study of the effect of salt and of pH on precipitation of antigen-antibody compounds. Immunology 8:170-181.

Kleczkowski, A. 1968*a*. Methods of inactivation by ultraviolet radiation, pp. 93-138. *In* K. Maramorosch, and H. Koprowski [ed.] Methods in virology, vol. 4. Academic Press, New York. 4 vols.

Kleczkowski, A. 1968*b*. Dark reactivation of ultraviolet-irradiated tobacco necrosis virus. J. Gen. Virol. 3:19-24. (19)

Kleczkowski, A., and A. van Kammen. 1961. Protection from proteolysis by aggregating the protein isolated from tobacco mosaic virus. Biochim. Biophys. Acta 53:181-185.

Kleinschmidt, A. K., T. H. Dunnebacke, R. S. Spendlove, F. L. Schaffer, and R. F. Whitcomb. 1964. Electron microscopy of RNA from reovirus and wound tumor virus. J. Mol. Biol. 10:282-288.

Klement, Z., Z. Kiraly, and B. I. Pozsar. 1966. Suppression of virus multiplication and local lesion production in tobacco following inoculation with a saprophytic bacterium. Acta Phytopathol. Acad. Sci. Hungary 1:11-18.

Klinkowski, M., and K. Schmelzer. 1960. A necrotic type of potato virus Y. Amer. Potato J. 37:221-228.

Klug, A., and D. L. D. Caspar. 1960. The structure of small viruses. Advances Virus Res. 7:225-325.

Klug, A., and J. T. Finch. 1965. Structure of the viruses of the papilloma-polyoma type. I. Human wart virus. J. Mol. Biol. 11:403-423.

Klug, A., W. Longley, and R. Leberman. 1966. Arrangement of protein subunits and the distribution of nucleic acid in turnip yellow mosaic virus. I. X-ray dlffraction studies. J. Mol. Biol. 15:315-343.

Knight, C. A. 1952. The nucleic acids of some strains of tobacco mosaic virus. J. Biol. Chem. 197:241-249.

Knight, C. A. 1955. Are cucumber viruses 3 and 4 strains of tobacco mosaic virus? A review of the problem. Virology 1:261-267.

Knight, C. A. 1959. Variation and its chemical correlates, pp. 127-156. *In* F. M. Burnet, and W. M. Stanley [ed.] The viruses, vol. 2. Academic Press, New York. 3 vols.

Knight, C. A. 1960. The protein of tobacco mosaic virus. Brookhaven Symp. Biol. 13:232-243.

Knight, C. A. 1963. Chemistry of viruses. Protoplasmatologia 4:1-177.

Knight, C. A. 1964. Preparation and properties of plant virus proteins, pp. 1-48. *In* R. J. C. Harris [ed.] Techniques in experimental virology. Academic Press, New York. 450 pp.

Knight, C. A., D. M. Silva, D. Dahl, and A. Tsugita. 1962. Two distinctive strains of TMV. Virology 16:236-243.

Koch, G., J. M. Biship, and H. Kubinski. 1969. Fractionation of nucleic acids on columns built with methyl esterified albumin kieselguhr, pp. 433-450. *In* K. Habel and N. Salzman [ed.] Fundamental techniques in virology. Academic Press, New York. 527 pp.

Koch, R. 1882. Uber die Milzbrandimpfung. Eine Entgegnung auf den von Pasteur in Genf gehaltenen Vortrag. Reprinted 1912 in Gesammelte Werke von Robert Koch 1:207-231. (Leipzig). Cited in A. J. Riker and R. S. Riker. 1936. Introduction to research on plant diseases. John S. Swift Co., Inc., New York.

Kodama, T., and J. B. Bancroft. 1964. Some properties of infectious ribonucleic acid from broad bean mottle virus. Virology 22:23-32.

Koenig, R. 1969*a*. Analyse serologischer Verwandtschaften innerhalb der potato virus X-Gruppe mit intakten Viren und Virusbruchstücken. Phytopathol. Z. 65:319-386.

Koenig, R. 1969*b*. Behavior of belladonna mottle virus in serological tests in the presence of organic mercury compounds. Virology 38:140-144.

Koenig, R. 1970. Effect of organic mercury compounds in serological tests with some isometric plant viruses. J. Gen. Virol. 7:257-261.

Koenig, R., and R. Bercks. 1968. Anderungen im heterologen Reaktionsvermogen von Antiseren gegen Vertreter der potato virus X-Gruppe im Laufe des Immunisierungsprozesses. Phytopathol. Z. 61:382-398.

Kolehmainen, L., H. Zech, and D. von Wettstein. 1965. The structure of cells during tobacco mosaic virus reproduction. J. Cell Biol. 25:77-97.

Kooistra, G. 1959. Investigations into the chemotherapeutic activity of nitroso hydroxyaryl compounds towards some virus diseases. Acta Bot. Neerl. 8:363-411.

Kosuge, T. 1965. Biochemical aspects of virus inhibitors in vivo and in vitro. Proc. Int. Conf. Virus and Vector on Perennial Hosts, with Special Reference to Vitis. Univ. Calif., Davis. pp. 247-254.

Kotin, L. 1963. On the effect of ionic strength on the melting temperature of DNA. J. Mol. Biol. 7:309-311.

Kramer, E. 1957. Elektrophorestische Untersuchungen an Mutanten des Tabakmosaikvirus. Z. Naturforsch. 12b:609-614.

Kramer, E., and H. G. Wittmann. 1958. Elektrophoretische Untersuchungen der A – Proteine dreier Tabakmosaikvirus – Stämme. Z. Naturforsch. 13b:30-33.

Kramer, G., H. G. Wittmann, and H. Schuster. 1964. Die Erzeugung von Mutanten des tabakmosaikvirus durch den Einbau von Fluoruracil in die Virusnucleinsäure. Z. Naturforsch. 19b:46-51.

Kubo, S., K. Tomaru, T. Nitta, T. Shiroya, and Z. Hidaka. 1965. Chromatography to tobacco mosaic virus, its constituents and nucleic acids extracted from infected tobacco leaf tissues. Virology 26:406-412.

Kuhn, C. W. 1965*a*. Decline of specific infectivity of cowpea chlorotic mottle virus in vivo. Virology 25:9-14. (1)

Kuhn, C. W. 1965*b*. Symptomatology, host range and effect on yield of a seed-transmitted peanut virus. Phytopathology 55:880-884.

Kuhn, C. W., and J. B. Bancroft. 1961. Concentration and specific infectivity changes of alfalfa mosaic virus during systemic infection. Virology 15:281-288.

Kuijt, J. 1968. The biology of parasitic flowering plants. Univ. of California Press, Berkeley. 364 pp.

Kunkel, L. O. 1921. A possible causative agent for the mosaic disease of corn. Bull. Exp. Sta. Hawaiian Sugar Planters Assoc. 3:1-14.

Kunkel, L. O. 1931. Studies on aster yellows in some new host plants. Contrib. Boyce Thompson Inst. 3:85-123.

Kunkel, L. O. 1934. Studies on acquired immunity with tobacco and aucuba mosaics. Phytopathology 24:437-466.

Kunkel, L. O. 1937. Effect of heat on ability of *Cicadula sexnotata* (Fall) to transmit aster yellows. Amer. J. Bot. 24:316-327.

Kunkel, L. O. 1954. Maintenance of yellows-type viruses in plant and insect reservoirs, pp. 150-163. *In* F. W. Hartman, F. L. Horsfall and J. G. Kidd [ed.] The dynamics of virus and rickettsial infections. McGraw-Hill, New York. 461 pp.

Kunkel, L. O. 1955. Cross protection between strains of yellows-type viruses. Advances Virus Res. 3:251-273.

Kurtz, S. M. 1961. A new method of embedding tissues in Vestopal W. J. Ultrastructure Res. 5:468-469.

Kushner, D. J. 1969. Self-assembly of biological structures. Bacteriol. Rev. 33:302-345.

Kutsky, R. J., and T. E. Rawlins. 1950. Inhibition of virus multiplication by naphtalene acetic acid in tobacco tissue cultures as revealed by a spectrophotometric method. J. Bacteriol. 60:763-766.

Labaw, L. W. 1964. Preparation of a 25 Å spacing crystal for magnification calibration above 18,000 × Abstr. 22nd Annu. Meeting Electron Microscope Soc. Amer., Detroit, Michigan.

Lamberti, F., and D. J. Raski. 1964. Transmissione di 'malformazione infectiva' della vite attracerso terreno da *Xiphinema index* Thorne and Allen, in Apulia. Phytopathol. Medit. 3:41-43.

Lamborn, C. R. 1964. An improved method of assaying infectivity of tobacco mosaic virus. Phytopathology 54:898. (Abstr.)

Lamborn, C. R., G. W. Cochran, and J. L. Chidester. 1969. Improved ultrasonic inoculation techniques for tobacco mosaic virus. Phytopathology 59:13 (Abstr.)

Lamborn, C. R., G. W. Cochran, and J. L. Chidester. 1971. An ultrasonic method for tobacco mosaic virus bioassay. Phytopathology 61:1015-1019.

Lang, D., H. Bajard, B. Wolff, and D. Russel. 1967. Electron microscopy of size and shape of viral DNA in solutions of different ionic strengths. J. Mol. Biol. 23:163-181.

Langenberg, W. G., and D. E. Schlegel. 1965. Puromycin inhibition of tobacco mosaic virus multiplication. Phytopathology 55:1065-1066.

Langenberg, W. G., and D. E. Schlegel. 1967. Autoradiography with ^{125}I-labelled antibodies as a means of localizing TMV antigen in plant cells. Virology 32:167-171.

Langenberg, W. G., and D. E. Schlegel. 1969. Localization of tobacco mosaic virus protein in tobacco leaf cells during the early stages of infection. Virology 37:86-93.

Lanni, F. 1964. Viruses and molecular taxonomy, pp. 386-426. *In* M. K. Corbett and H. D. Sisler [ed.] Plant virology. Univ. Florida Press, Gainesville. 527 pp.

Lauffer, M. A. 1948. Interaction between tobacco mosaic virus and bovine serum albumin. J. Biol. Chem. 174:481-488.

Lauffer, M. A., and W. C. Price. 1945. Infection by viruses. Arch. Biochem. 8:449-468.

Lawson, F. R., J. C. Chamberlin, and G. T. York. 1951. Dissemination of the beet leafhopper in California. U. S. Dep. Agr. Tech. Bull. 1030. 59 pp.

Lea, D. E. 1947. Actions of radiation on living cells. MacMillan, New York. 402 pp.

Leberman, R. 1965. Use of uranyl formate as a negative stain. J. Mol. Biol. 13:606.

Leberman, R. 1966. The isolation of plant viruses by means of "simple" coacervates. Virology 30:341-347.

Leduc, E. H., and W. Bernhard. 1967. Recent modifications of the glycol methacrylate embedding procedure. J. Ultrastructure Res. 19:196-199.

Lee, C. L. 1955*a*. Virus-tumor development in relation to lateral-root and bacterial-nodule formation in *Melilotus alba*. Virology 1:152-164.

Lee, C. L. 1955*b*. Anatomical changes in sweet clover shoots infected with wound-tumor virus. Amer. J. Bot. 42:693-698.

Lee, P. E. 1962. Acquisition time and inoculation time in transmission of aster yellows virus by single leafhoppers. Virology 17:394-396.

Lee, P. E. 1963. The extraction of wheat striate mosaic virus from diseased wheat plants. Can. J. Bot. 41:1617-1621.

Lee, P. E. 1965. Viruslike particles in the salivary glands of apparently virus-free leafhoppers. Virology 25:471-472.

Lee, P. E. 1967. Morphology of wheat striate mosaic virus and its localization in infected cells. Virology 33:84-94.

Lee, P. E. 1968. Partial purification of wheat striate mosaic virus and fine structural studies of the virus. Virology 34:583-589.

Lee, P. E., and L. N. Chiykowski. 1963. Infectivity of aster-yellows virus preparations after differential centrifugations of extracts from viruliferous leafhoppers. Virology 21:667-669.

Lee, P. E., and D. D. Jensen. 1963. Crystalline inclusions in *Colladonus montanus* (Van Duzee), a vector of Western X-disease virus. Virology 20:328-332.

Legg, J. T. 1965. Mechanical transmission of hop mosaic virus. Nature 208:1017-1018.

Leiden Symposium on Electron Microscopy and Cytochemistry. 1966. Dutch Philips Organization, Philips Bedrijfsapparatuur, Nederland n.v., Eindhoven.

Lennox, E. S., and M. Cohn. 1967. Immunoglobulins. Annu. Rev. Biochem. Part I. 36:365-402.

Lerman, L. S. 1961. Structural considerations in the interaction of DNA and acridines. J. Mol. Biol. 3:18-30.

Lesnaw, J. A., and M. E. Reichmann. 1969*a*. The structure of tobacco necrosis virus. I. The protein subunit and the nature of the nucleic acid. Virology 39:729-737.

Lesnaw, J. A., and M. E. Reichmann. 1969*b*. The structure of tobacco necrosis virus. II. Terminal amino acid residues of the protein subunit. Virology 39:738-745.

Lesnaw, J. A., and M. E. Reichmann. 1970. Identity of the 5′-terminal RNA nucleotide sequence of the satellite tobacco necrosis virus and its helper virus: possible role of the 5′-terminus in the recognition by virus-specific RNA replicase. Proc. Nat. Acad. Sci. U. S. 66:140-145.

Levin, O. 1958. Chromatography of tobacco mosaic virus and potato virus X. Arch. Biochem. 78:33-45.
Levine, L., J. A. Gordon, and W. P. Jencks. 1963. The relationship of structure to the effectiveness of denaturing agents for deoxyribonucleic acid. Biochemistry 2:168-175.
Levintow, L. 1965. The biochemistry of virus replication. Annu. Rev. Biochem. 34:487-526.
Lima, M. B. 1968. A numerical approach to the *Xiphinema americanum* complex. Compt. Rend. 8th Symp. Int. Nematologie, Antibes, 1965. 30 pp.
Limasset, P., F. Levieil, and M. Sechet. 1948. Influence d'une phytohormone de synthèse sur le develóppement des virus X et Y de la pomme de terre chez le tabac. Compt. Rend. 227:643-645.
Lin, S. C., C. S. Lee, and R. J. Chiu. 1970. Isolation and cultivation of, and inoculation with, a mycoplasma causing white leaf disease of sugarcane. Phytopathology 60:795-797.
Lindberg, G. D., D. H. Hall, and J. C. Walker. 1956. A study of melon and squash mosaic virus. Phytopathology 46:489-495.
Lindner, R. C., P. C. Cheo, H. C. Kirkpatrick and H. C. Govindu. 1960. Some effects of 8-azaguanine on tobacco mosaic virus replication. Phytopathology 50:884-889.
Lindner, R. C., and H. C. Kirkpatrick. 1959. The airbrush as a tool in virus inoculations. Phytopathology 49:507-509.
Linford, M. B., and H. H. McKinney. 1954. Occurrence of *Polymyxa graminis* in roots of small grains in the United States. Plant Dis. Reptr. 38:711-713.
Ling, K. C. 1966. Nonpersistence of the tungro virus of rice in its leafhopper vector, Nephotettix impicticeps. Phytopathology 56:1252-1256.
Ling, K. C. 1969. Nonpropagative leafhopper-borne viruses, pp. 255-277. *In* K. Maramorosch [ed.] Viruses, vectors, and vegetation. Interscience, New York. 666 pp.
Lingens, F., J. Rau, and R. Süssmuth. 1968. II. Produkte der Reaktion von 1-nitroso-3-nitro-1-methyl-guanidin mit Nucleobasin, Nucleosiden, Nucleosidphosphaten und Homopolyribonucleinsäuren. Z. Naturforsch. 23b:1565-1570.
Lister, R. M. 1960. Transmission of soil-borne viruses through seed. Virology 10:547-549.
Lister, R. M. 1964. Strawberry latent ringspot: a new nematode-borne virus. Ann. Appl. Biol. 54:167-176.
Lister, R. M. 1966. Possible relationships of virus-specific products of tobacco rattle virus infection. Virology 28:350-353.
Lister, R. M., and A. F. Murant. 1967. Seed transmission of nematode-borne viruses. Ann. Appl. Biol. 59:49-62.
Lister, R. M., L. C. Raniere, and E. H. Varney. 1963. Relationships of viruses associated with ringspot diseases of blueberry. Phytopathology 53:1031-1035.
Lister, R. M., and J. M. Thresh. 1955. A mosaic disease of leguminous plants caused by a strain of tobacco mosaic virus. Nature (London) 175:1047-1048.
Littau, V. C., and L. M. Black. 1952. Spherical inclusions in plant tumors caused by a virus. Amer. J. Bot. 39:87-95.
Littau, V. C., and K. Maramorosch. 1960. A study of the cytological effects of aster yellows virus on its insect vector. Virology 10:483-500.
Lloyd, D. A., and S. Mandeles. 1970. Partial nucleotide sequences for three unique T1 ribonuclease fragments of tobacco mosaic virus ribonucleic acid. Biochemistry 9:932-938.
Loebenstein, G. 1960. Identifying two sweet potato viruses with paper chromatography. Phytopathology 50:98-99.
Loebenstein, G. 1962. Inducing partial protection in the host plant with native virus protein. Virology 17:574-581.
Loebenstein, G. 1963. Further evidence on systemic resistance induced by localized necrotic virus infections in plants. Phytopathology 53:306-308.
Loebenstein, G., M. Alper, and M. Deutsch. 1964. Preventing aphid-spread cucumber mosaic virus with oils. Phytopathology 54:960-962.
Loebenstein, G., and I. Harpaz. 1960. Virus diseases of sweet potatoes in Israel. Phytopathology 50:100-104.
Loebenstein, G., and N. Linsey. 1963. Effect of virus infection on peroxidase activity and C_6/C_1 ratios. Phytopathology 53:350.
Loebenstein, G., and N. Linsey. 1966. Alterations of peroxidase activity associated with disease symptoms in virus-infected plants. Israel J. Bot. 15:163-167.

Loebenstein, G., and L. Lovrekovich. 1966. Interference with tobacco mosaic virus local lesion formation in tobacco by injecting heat-killed cells of *Pseudomonas syringae*. Virology 30:587-591.

Loebenstein, G., S. Rabina and T. van Praagh. 1966. Induced interference phenomena in virus infections, pp. 150-157. *In* A. B. R. Beemster and J. Dijkstra [ed.] Viruses of plants. North-Holland, Amsterdam. 342 pp.

Loebenstein, G., S. Rabina, and T. van Praagh. 1968. Sensitivity of induced localized acquired resistance to actinomycin D. Virology 34:264-268.

Loebenstein, G., and A. F. Ross. 1963. An extractable agent, induced in uninfected tissues by localized virus infections, that interferes with infection by tobacco mosaic virus. Virology 20:507-517.

Loebenstein, G., B. Sela, and T. van Praagh. 1969. Increase of tobacco mosaic local lesion size and virus multiplication in hypersensitive hosts in the presence of actinomycin D. Virology 37:42-48.

Loebenstein, G., and T. van Praagh. 1964. Extraction of a virus interfering agent induced by localized and systemic infection, pp. 53-58. *In* Z. Kiraly and G. Ubrizsy [ed.] Host-parasite relations in plant pathology. Publ. Research Institute for Plant Protection, Budapest, Hungary.

Loening, U. E. 1967. The fractionation of high-molecular weight ribonucleic acid by polyacrylamide-gel electrophoresis. Biochem. J. 102:251-257.

Lojek, J. S., and G. B. Orlob. 1969. Aphid transmission of tobacco mosaic virus. Science 164:1407-1408.

Lomakina, L. Y., G. M. Razvyaskina, and E. A. Shubnikova. 1963. Cytological and histochemical changes in fat body of Cacada *Psammotettix striatus* Fall infected with winter wheat mosaic virus. Vopr. Virusol. 1963:168-172.

Long, C. [ed.]. 1961. Biochemist's handbook. Van Nostrand Reinhold, New York. 1192 pp.

Longley, W. 1967. The crystal structure of bovine liver catalase. J. Mol. Biol. 30:323-327.

Longley, W., and R. Leberman. 1966. Structural similarity of artificial and natural top components of turnip yellow virus. J. Mol. Biol. 19:223-225.

Longsworth, L. G. 1952. Electrophoresis. Methods Med. Res. 5:63-106.

Longsworth, L. G. 1959. Moving boundary electrophoresis—theory—practice, pp. 91-136, 137-177. *In* M. Bier [ed.] Electrophoresis. Theory, methods and applications. Academic Press, New York. 563 pp.

Longsworth, L. G., and C. F. Jacobsen. 1949. An electrophoretic study of the binding of salt ions by β-lactoglobulin and bovine serum albumin. J. Phys. Colloid Chem. 53:126-135.

Loor, F. 1966. Recherches sur la synthèse d' anticorps anti-TMV non precipitables par l'antigene injecté. Arch. Int. Physiol. Biochim. 74:929-930.

Lopez-Abella, D., F. Jiménez-Millán, and F. Garcia-Hidalgo. 1967. Electron microscope studies of some cephalic structures of *Xiphinema americanum*. Nematologica 13:283-286.

Louie, R., and J. W. Lorbeer. 1966. Mechanical transmission of onion yellow dwarf virus. Phytopathology 56:1020-1023.

Lovisolo, O. 1966. Indagini su virosi di piante ornamentali. Atti del Primo Congresso dell' Unione Fitopatologia mediterranea, Bari-Napoli, 11 pp.

Lovisolo, O., E. Lursoni, M. Conti, and C. Wetter. 1967. Partial purification of maize rough dwarf virus. Naturwissenschaften 54:73-74.

Lovrekovich, L., H. Lovrekovich, and M. A. Stahmann. 1967. The role of peroxidase in resistance of tobacco to *Pseudomonas tabaci*. Phytopathology 57:819-820.

Lownsbery, B. F., and E. F. Serr. 1963. Fruit and nut tree rootstocks as hosts for a root-lesion nematode *Pratylenchus vulnus*. Proc. Amer. Soc. Hort. Sci. 82:250-254.

Lu, Y. T., E. Shikata, and D. Murayama. 1968. Isolation of northern cereal mosaic virus. Mem. Fac. Agr. Hokkaido Univ. 6:335-339.

Ludlum, D. B., and R. C. Warner. 1965. Equilibrium centrifugation in cesium sulfate solutions. J. Biol. Chem. 240:2961-2965.

Luft, J. H. 1956. Permanganate—a new fixative for electron microscopy. J. Biophys. Biochem. Cytol. 2:799-801.

Luft, J. H. 1961. Improvements in epoxy resin embedding methods. J. Biophys. Biochem. Cytol. 9:409-414.

Luftig, R. B. 1967. An accurate measurement of the catalase crystal period and its use as an internal marker for electron microscopy. J. Ultrastructure Res. 20:91-102.

Luftig, R. B. 1968. Further studies on the dimensions of viral and protein structures using the catalase crystal internal marker technique. J. Ultrastructure Res. 23:178-181.

Lund, H. A., A. C. Vatter, and J. B. Hanson. 1958. Biochemical and cytological changes accompanying growth and differentiation in the roots of *Zea mays*. J. Biophys. Biochem. Cytol. 4:87-98.

Lwoff, A. 1965. The specific effectors of viral development. Biochem. J. 96:289-301.

Lyon, H. L. 1921. Three major cane diseases: mosaic, sereh and Fiji disease. Bull. Exp. Sta. Hawaiian Sugar Planters Assoc. 3:17-43.

Lyttleton, J. W., and R. E. F. Matthews. 1958. Release of nucleic acid from turnip yellow mosaic virus under mild conditions. Virology 6:460-471.

McCarthy, B. J., and E. T. Bolton. 1963. An approach to the measurement of genetic relatedness among organisms. Proc. Nat. Acad. Sci. U. S. 50:156-164.

McClean, A. P. D. 1948. Bunchy-top disease of the tomato: additional host plants and the transmission of the virus through the seed of infected plants. Union of South Africa Dep. Agr. Sci. Bull. 256. 28 pp.

McEwen, F. L., and Y. Kawanishi. 1967. Insect transmission of corn mosaic: laboratory studies in Hawaii. J. Econ. Entomol. 60:1413-1417.

McGuire, J. M. 1964. Efficiency of *Xiphinema americanum* as a vector of tobacco ringspot virus. Phytopathology 54:799-801.

Machida, H., T. Hayashi, T. Abe, and Y. Kiho. 1969. In vivo uncoating of tobacco mosaic virus after infection of tobacco leaves. Jap. J. Microbiol. 13:351-357.

McKeen, C. D. 1959. Cucumber necrosis virus. Can. J. Bot. 37:913-925.

McKinney, H. H. 1929. Mosaic diseases in the Canary Islands, West Africa and Gilbraltar. J. Agr. Res. 39:557-578.

McKinney, H. H. 1937. Mosaic diseases of wheat and releated cereals. U. S. Dep. Agr. Circ. 442:1-22.

McKinney, H. H. 1952. Two strains of tobacco mosaic virus, one of which is seed transmitted on etch-immune pungent pepper. Plant Dis. Reptr. 36:184-187.

McKinney, H. H., and L. W. Greeley. 1965. Biological characteristics of barley stripe mosaic virus strains and their evolution. U. S. Dep. Agr. Tech. Bull. 1324. 84 pp.

MacKinnon, J. P. 1965. Aphid-transmitted chlorosis virus of *Physalis floridana* found in a turnip latent virus complex. Can. J. Bot. 43:509-517.

McLaren, A. D. 1968. Photochemistry of a macromolecule: tobacco mosaic virus. Photochem. Photobiol. 8:521-525.

McLean, D. L., and M. S. Kinsey. 1967. Probing behaviour of the pea aphid *Acyrthosphum pisum*. Definitive correlations of electronically recorded wave-forms with aphid-probing activities. Ann. Entomol. Soc. Amer. 60:400-406.

McLean, D. L., and W. A. J. Weight. 1967. An electronic measuring device to record aphid salivation and ingestion. Ann. Entomol. Soc. Amer. 61:180-185.

McLean, G. D., and R. I. B. Francki. 1967. Purification of lettuce necrotic yellows virus by column chromatography on calcium phosphate gel. Virology 31:585-591.

MacLeod, R. 1968. An interpretation of the observed polymorphism of potato yellow dwarf virus. Virology 34:771-777.

MacLeod, R., L. M. Black, and F. H. Moyer. 1966. The fine structure and intracellular localization of potato yellow dwarf virus. Virology 29:540-552.

MacLeod, R., and R. Markham. 1963. Experimental evidence of a relationship between turnip yellow mosaic virus and wild cucumber mosaic virus. Virology 19:190-197.

McWhorter, F. P. 1941*a*. Isometric crystals produced by pisum virus 2 and phaseolus virus 2. Phytopathology 31:760-761.

McWhorter, F. P. 1941*b*. Plant-virus differentiation by trypan blue reactions within infected tissue. Stain Technol. 16:143-149.

McWhorter, F. P. 1957. Chlorazol fast pink B and trypan blue tests for virus and other proteinaceous inclusions. Stain Technol. 32:135-138.

McWhorter, F. P. 1965. Plant virus inclusions. Annu. Rev. Phytopathol. 3:287-312.

Magdoff, B. S. 1960. Electrophoresis of proteins in liquid media, pp. 169-214. *In* P. Alexander and R. J. Block [ed.] A laboratory manual of analytical methods in protein chemistry, including polypeptides, vol. 2. Pergamon Press, London. 3 vols.

Magdoff-Fairchild, B. S. 1967. Electrophoretic and buoyant density variants of southern bean mosaic virus. Virology 31:142-153.

Mahoney, C. H. 1935. Seed transmission of mosaic in inbred lines of muskmelon (*Cucumis melo* L.). Proc. Ann. Soc. Hort. Sci. 32:477-480.

Maia, E. 1966. Induction de synthese d'un facteur antiviral chez le tabac et la Datura par application d'acide phytique. Compt. Rend., Serie D. 262:2099-2101.

Maillet, P., and J. Gouranton. 1971. Etude de cycle biologique du mycoplasme de la phyllodie du trèfle dans l'insecte vecteur, Euscelis lineolatus Brullé (Homoptera, Jassidae). J. Microscopie 11:143-162.

Mallett, M. F., S. Lewis, S. R. Ames, J. M. Nelson, and C. R. Dowson. 1948. Preparation of mushroom tyrosinase. Arch. Biochem. 16:283-289.

Mandeles, S. 1968. Location of unique sequences in tobacco mosaic virus ribonucleic acid. J. Biol. Chem. 243:3671-3674.

Mandeles, S., and G. Bruening. 1968. Tobacco mosaic virus ribonucleic acid. Biochem. Preparations 12:111-117.

Mandell, J. D., and A. D. Hershey. 1960. A fractionating column for analysis of nucleic acids. Anal. Biochem. 1:66-77.

Mandryk, M. 1963. Acquired systemic resistance to tobacco mosaic virus in *Nicotiana tabacum* evoked by stem injection with *Peronospora tabacina* Adam. Australian J. Agr. Res. 14:315-318.

Maramorosch, K. 1950. Influence of temperature on incubation and transmission of the wound tumor virus. Phytopathology 40:1071-1093.

Maramorosch, K. 1952*a*. Direct evidence for the multiplication of aster-yellows virus in its insect vector. Phytopathology 42:59-64.

Maramorosch, K. 1952*b*. Multiplication of aster yellows virus in its vector. Nature 169:194-195.

Maramorosch, K. 1952*c*. Toxicity of cellulose acetate sheets to plants and fish. Science 115:236.

Maramorosch, K. 1953. Incubation period of aster yellows virus. Amer. J. Bot. 40:797-809.

Maramorosch, K. 1955. Mechanical transmission of curly top virus to its insect vector by needle inoculation. Virology 1:286-300.

Maramorosch, K. 1956. Multiplication of aster yellows virus in in vitro preparations of insect tissues. Virology 2:369-376.

Maramorosch, K. 1957. Reversal of virus-caused stunting in plants by gibberellic acid. Science 126:651-652.

Maramorosch, K. 1958*a*. Beneficial effect of virus diseased plants on nonvector insects. Tijdschr. Plantenziekten 64:383-391.

Maramorosch, K. 1958*b*. Viruses that infect and multiply in both plants and insects. Trans. N. Y. Acad. Sci. 20:383-393.

Maramorosch, K. 1963. Arthropod transmission of plant viruses. Annu. Rev. Entomol. 8:369-414.

Maramorosch, K. 1964. Interrelationships between plant pathogenic viruses and insects. Ann. N. Y. Acad. Sci. 118:363-370.

Maramorosch, K. [ed.]. 1969. Viruses, vectors, and vegetation. Interscience, New York. 666 pp.

Maramorosch, K., R. R. Granados, and H. Hirumi. 1970. Mycoplasma diseases of plants and insects. Advances Virus Res. 16:135-193.

Maramorosch, K., and H. Koprowski. [ed.]. 1967. Methods in virology. Academic Press, New York. 4 vols.

Marchis-Mouren, G., and F. Lipman. 1965. On the mechanism of acetylation of fetal and chicken hemoglobins. Proc. Nat. Acad. Sci. U.S. 53:1147-1154.

Marier, J. R., and D. Rose. 1964. Determination of cyanate, and a study of its accumulation in aqueous solutions of urea. Anal. Biochem. 7:304-314.

Markham, R. 1951. Physicochemical studies of the turnip yellow mosaic virus. Disc. Faraday Soc. 11:221-227.

Markham, R. 1959. The biochemistry of plant viruses, pp. 33-125. *In* F. M. Burnet and W. M. Stanley [ed.] The viruses, vol. 2. Academic Press, New York. 3 vols.

Markham, R. 1962. The analytical ultracentrifuge as a tool for the investigation of plant viruses. Advances Virus Res. 9:241-270.

Markham, R. 1966. Separation of viruses. Brit. Med. Bull. 22:153-157.

Markham, R. 1967. The ultracentrifuge, pp. 1-39. *In* K. Maramorosch and H. Koprowski [ed.] Methods in virology, vol. 2. Academic Press, New York. 4 vols.

Markham, R., S. Frey, and G. J. Hills. 1963. Methods for the enhancement of image detail and accentuation of structure in electron microscopy. Virology 22:88-102.

Markham, R., J. H. Hitchborn, G. J. Hills, and S. Frey. 1964. The anatomy of the tobacco mosaic virus. Virology 22:342-359.

Markham, R., and K. M. Smith. 1949. Studies on the virus of turnip yellow mosaic. Parasitology 39:330-342.

Marshall, R. E., C. T. Caskey, and M. Nirenberg. 1967. Fine structure of RNA codewords recognized by bacterial, amphibian, and mammalian transfer RNA. Science 155:820-826.

Martelli, G. P. 1969. Bacilliform particles associated with mottled dwarf of eggplant. (*Solanum melongena* L.). J. Gen. Virol. 5:319-320.

Martelli, G. P., and M. A. Castellano. 1969. The relation of pelargonium leaf curl virus to the nuclei of host cells. Virology 39:610-613.

Martelli, G. P., and D. J. Raski. 1963. Rapporti tra nematodi del Fico e "degenerazione infettiva" della vite. Informatore Fitopatologico 13:416-420.

Martin, R. G., and B. N. Ames. 1961. A method for determining the sedimentation behavior of enzymes: application to protein mixtures. J. Biol. Chem. 236:1372-1379.

Martinez-Palomo, A., and N. Granboulan. 1967. Electron microscopy of adenovirus 12 replication. II. High-resolution autoradiography of infected KB cells labelled with tritiated thymidine. J. Virol. 1:1010-1018.

Massee, A. M. 1928. The life-history of the black currant gall mite, *Eriophyes ribis* (Westw.). Nat. Bull. Entomol. Res. 18:297-309.

Massee, A. M. 1952. Transmission of reversion of black currants. Annu. Rep. East Malling Res. Sta. Kent 1951:162-165.

Massie, H. R., and B. H. Zimm. 1965. The use of hot phenol in preparing DNA. Proc. Nat. Acad. Sci. U.S. 54:1641-1643.

Matsui, C. 1959. Fine structure of X-body. Virology 9:306-313.

Matsui, C., and A. Yamaguchi. 1964. Electron microscopy of host cells infected with tobacco etch virus. I. Fine structures of leaf cells at later stages of infection. Virology 22:40-47.

Matsui, C., and A. Yamaguchi. 1966. Some aspects of plant viruses in situ. Advances Virus Res. 12:127-174.

Matthews, R. E. F. 1957. Plant virus serology. Cambridge Univ. Press, Cambridge. 128 pp.

Matthews, R. E. F. 1960. Properties of nucleoprotein fractions isolated from turnip yellow mosaic virus preparations. Virology 12:521-539.

Matthews, R. E. F. 1966*a*. Reconstitution of RNA from spherical viruses with tobacco mosaic virus protein. Virology 28:165-168.

Matthews, R. E. F. 1966*b*. Reconstitution of turnip yellow mosaic virus RNA with tobacco mosaic virus protein subunits. Virology 30:82-96.

Matthews, R. E. F. 1967. Serological tests for plant viruses, pp. 199-241. *In* K. Maramorosch and H. Koprowski [ed.] Methods in virology, vol. 3. Academic Press, New York. 4 vols.

Matthews, R. E. F. 1970. Serological reactions, pp. 475-506. *In* Plant virology. Academic Press, New York. 778 pp.

Matthews, R. E. F., and J. W. Lyttleton. 1959. Heat inactivation of turnip yellow mosaic virus in vivo. Virology 9:332-342.

Matthews, R. E. F., and J. D. Smith. 1955. The chemotherapy of viruses. Advances Virus Res. 3:49-148.

May, D. S., and C. A. Knight. 1965. Polar stripping of protein subunits from tobacco mosaic virus. Virology 25:502-507.

Mayo, M. A., and E. C. Cocking. 1968. Labelling of tobacco mosaic virus with ^{125}I. J. Gen. Virol. 2:89-97.

Mayor, H. D., and L. E. Jordan. 1968. Nucleic acid molecules: new microdiffusion technique for visualisation. Science 161:1246-1247.

Mazzone, H. M., N. L. Incardona, and P. Kaesberg. 1962. Biochemical and biophysical properties of squash mosaic virus and related macromolecules. Biochim. Biophys. Acta 55:164-175.

Medina, A. C., and R. G. Grogan. 1961. Seed transmission of bean common mosaic viruses. Phytopathology 51:452-456.

Medler, J. T. 1941. The nature of injury to alfalfa caused by *Empoasca fabae* (Harris). Ann. Entomol. Soc. Am. 34:439-450.

Medler, J. T. 1957. Migration of the potato leafhopper—a report on a cooperative study. J. Econ. Entomol. 50:493-497.

Megahed, E. S., and A. Moore. 1969. Inactivation of necrotic ringspot and prune dwarf viruses in seeds of some *Prunus* spp. Phytopathology 59:1758-1760.

Melchers, G., G. Schramm, H. Trusnit, and H. Friedrick-Freska. 1940. Biological, chemical and electron-microscopical investigation of a mosaic virus from tomatoes. Biol. Zentralbl. 60:524-556.

Melchers, G. 1942. Uber einige mutation en des Tabakmosaikvirus und eine Parallelmutation "des tomatenmosaikvirus. Naturwissenschaften 30:48-49.

Melnick, J. L. 1969. Summary of classification of animal viruses. Progr. Med. Virol. 11:451-453.

Melnick, J. L. 1970. Summary of classification of animal viruses. Progr. Med. Virol. 12:337-341.

Melnick, J. L., and R. M. McCombs. 1966. Classification and nomenclature of animal viruses. Progr. Med. Virol. 8:400-409.

Mercer, E. H., and M. S. C. Birbeck. 1966. Electron microscopy, a handbook for biologists. 2nd ed. Blackwell, Oxford. 102 pp.

Merrett, M. J. 1962. The effect of uncoupling agents on the multiplication of tomato-aucuba mosaic virus in tobacco leaves. Physiol. Plant 15:200-205.

Merriam, V., and M. P. Gordon. 1967. Pyrimidine dimer formation in ultraviolet irradiated TMV-RNA. Photochem. Photobiol. 6:309-319.

Meselson, M., and G. M. Nazarian. 1963. The transient state in density-gradient centrifugation, pp. 131-282. *In* J. W. Williams [ed.] Ultracentrifuge analysis in theory and experiment. Academic Press, New York. 280 pp.

Meyer, B. S. 1956. The hydrodynamic system, pp. 596-614. *In* W. Rhuland [ed.] Handbuch der Pflanzenphysiologie, vol. 3. Springer-Verlag, New York.

Miano, M. L., and U. Ferrini. 1969. Acid depurination of deoxyribonucleic acid-polycation complexes. Biochim. Biophys. Acta 174:749-751.

Michaelis, L. 1947. The nature of the interaction of nucleic acids and nuclei with basic dye-stuffs. Cold Spring Harbor Symp. 12:131-142.

Michelson, A. M., 1963. The chemistry of nucleosides and nucleotides. Academic Press, New York. 622 pp.

Miczynski, K. 1968. Detection of free viral nucleic acid in the potato virus X infected tobacco by gel electrophoresis. Phytopathol. Z. 61:309-330.

Middleton, J. T. 1943. The taxonomy, host range and geographic distribution of the genus *Pythium*. Mem. Torrey Bot. Club 20:1-171.

Middleton, J. T., and G. W. Bohn. 1953. Cucumbers, melons, squash. U. S. Rep. Agr. Yearbook of Agriculture, 1953. pp. 483-493.

Miki, T., and C. A. Knight. 1965. Preparation of broad bean mottle virus protein. Virology 25:478-481.

Miki, T., and C. A. Knight. 1967. Some chemical studies on a strain of white clover mosaic virus. Virology 31:55-63.

Miki, T., and C. A. Knight. 1968. The protein subunit of potato virus X. Virology 36:168-173.

Miles, P. W. 1968. Insect secretions in plants. Annu. Rev. Phytopathol. 6:137-164.

Milicić, D. 1963. Viruskoper im Zellsafte. Protoplasma 57:601-610.

Milicić, D., and B. Plavsic. 1956. Eiweisskristalloide in Kakteen-Virustragern. Protoplasma 16:547-555.

Milicić, D., N. Plese, R. Bercks, J. Brandes, R. Casper, and M. Chessin. 1966. Vergleichende serologische und elektronenmikroskopische Untersuchungen an Isolaten des Kakteen-X-Virus. Phytopathol. Z. 55:211-217.

Miller, G. L., and R. H. Golder. 1950. Buffers of pH 2 to 12 for use in electrophoresis. Arch. Biochem. 29:420-423.

Miller, G. L., and W. C. Price. 1946. Physical and chemical studies on southern bean mosaic virus. II. Crystallization by dialysis. Arch. Biochem. 11:329-335.

Millonig, G. 1961. A modified procedure for lead staining of thin sections. J. Biophys. Biochem. Cytol. 11:736-739.

Milne, R. G. 1967. Electron microscopy of leaves infected with sowbane mosaic virus and other small polyhedral viruses. Virology 32:589-600.

Milne, R. G., and G. A. deZoeten. 1967. A comparison of some methods of preparation of thin sections of virus-infected leaves for electron microscopy. J. Ultrastructure Res. 19:398-407.

Milne, R. G., and T. A. Trautner. 1967. Thin sectioning and electron microscopy of SP 50 bacteriophage adsorbed to *Bacillus subtilis*. J. Ultrastructure Res. 20:267-276.
Mims, C. A., M. F. Day, and I. D. Marshall. 1966. Cytopathic effect of Semliki Forest virus in the mosquito *Aedes aegypti*. Amer. J. Trop. Med. Hyg. 15:775-784.
Mitra, S., P. Reichard, R. B. Inman, L. L. Bertsch, and A. Kornberg. 1967. Enzymic synthesis of deoxyribonucleic acid. XXII. Replication of a circular single-stranded DNA template by DNA polymerase of *Escherichia coli*. J. Mol. Biol. 24:429-447.
Mitsuhashi, J. 1967. Infection of leafhopper and its tissues cultivated in vitro with Chilo iridescent virus. J. Invertebrate Pathol. 9:432-434.
Mitsuhashi, J. 1969. Plant-pathogenic viruses in insect vector tissue culture, pp. 475-503. *In* K. Maramorosch [ed.] Viruses, vectors, and vegetation. Interscience, New York. 666 pp.
Mitsuhashi, J., and K. Maramorosch. 1963. Aseptic cultivation of four virus transmitting species of leafhoppers (Cicadellidae). Contrib. Boyce Thompson Inst. 22:165-173.
Mittler, T. E., and R. H. Dadd. 1963. Studies on the artificial feeding of the aphid *Myzus persicae* (Sulzer). I. Relative uptake of water and sucrose solutions. J. Insect Physiol. 9:623-645.
Mittler, T. E., and R. H. Dadd. 1964. Studies on the artificial feeding of the aphid *Myzus persicae*. III. Some major nutritional requirements. J. Insect Physiol. 11:17-43.
Miura, K., I. Fujii, T. Sakkaki, M. Fuke, and S. Kawase. 1968. Double-stranded RNA from cytoplasmic-polyhedrosis virus of the silkworm. J. Virol. 2:1211-1222.
Miura, K., I. Kimura, and N. Suzuki. 1966. Double stranded ribonucleic acid from rice dwarf virus. Virology 28:571-579.
Miyakawa, T., and H. Yoshii. 1951. Transmission of tobacco mosaic virus by dodder, *Cuscuta japonica* and its inhibitory effect on the virus activity. Sci. Bull. Fac. Agr. Kyushu Univ. 12(2):143-154.
Miyamoto, S., and Y. Miyamoto. 1966. Notes on aphid-transmission of potato leafroll virus. Sci. Rep. Hyogo Univ. Agr. Ser. Plant Protect. 7:51-66.
Moed, J. R., and H. Veldstra. 1968. Alfalfa mosaic virus: Comparative investigation of top component A and bottom component by means of fingerprinting and immunological techniques. Virology 36:459-466.
Mohr, W. P., and E. C. Cocking. 1968. A method for preparing highly vacuolated senescent and damaged plant tissue for ultrastructural studies. J. Ultrastructure Res. 21:171-181.
Mollenhauer, H. H. 1964. Plastic embedding mixtures for use in electron microscopy. Stain Technol. 39:111-114.
Monneron, A. 1966*a*. Action of pronase on ultrathin sections of tissue embedded in Epon and glycol methacrylate. Leiden Symposium on Electron Microscopy and Cytochemistry. Dutch Philips Organization, Philips Bedrijfsapparatuur, Nederland n.v., Eindhoven, pp. 57-58.
Monneron, A. 1966*b*. Enzymatic digestion in tissues embedded in Epon. Sixth International Congress of Electron Microscopy, Kyoto, Japan. (Uyeda, S., ed.). p. 27.
Monneron, A., and W. Bernhard. 1966. Action de certaines enzymes sur des tissus inclus en epon. J. Microscopie 5:697-714.
Moor, H., and K. Muhlethaler. 1963. Fine structure in frozen-etched yeast cells. J. Cell Biol. 17:609-628.
Moor, H., K. Muhlethaler, M. Waldner, and A. Frey-Wyssling. 1961. A new freezing ultramicrotome. J. Biophys. Biochem. Cytol. 10:1-13.
Moore, D. H. 1969. Gradient centrifugation, pp. 285-328. *In* Physical techniques and biological research, vol. 2, part B. 2nd ed. Academic Press, New York. 2 vols.
Moorhead, E. L., and W. C. Price. 1953. A new serological test for tobacco mosaic virus. Phytopathology 43:73-77.
Morel, G., and C. Martin. 1952. Guerison de dahlias atteints d'une maladie a virus. Compt. Rend. 235:1324-1325.
Morel, G. W. 1948. Recherches sur la culture associée de parasites obligatoire et de tissus végétaux. Ann. Epiph. (N.S.) 14:123-234.
Morgan, C., and H. M. Rose. 1967. The application of thin sectioning, pp. 576-616. *In* K. Maramorosch and H. Koprowski [ed.] Methods in virology, vol. 3. Academic Press, New York. 4 vols.
Morgan, H. G. 1964. *Xiphinema diversicaudatum* and arabis mosaic virus in south-west England. Nematologica 10:70. (Abstr.)

Morgan, N. G., and B. D. Smith. 1961. The application of lime sulphur for the control of the black currant gall mite. Annu. Rep. Agr. Hort. Res. Sta. Long Ashton Bristol 1960:129-133.

Motoyoshi, F., and N. Oshima. 1968. Multiplication of tobacco mosaic virus in suspension culture of tobacco cells. Jap. J. Microbiol. 12:317-320.

Mowat, W. P. 1968. *Olpidium brassicae:* electrophoretic mobility of zoospores associated with their ability to transmit tobacco necrosis virus. Virology 34:565-568.

Mowat, W. P. 1970. Augusta disease in tulip—a reassessment. Ann. Appl. Biol. 66:17-28.

Mowat, W. P., and C. E. Taylor. 1962. Nematode-borne viruses. Rep. Scot. Hort. Res. Inst. 9:67-68.

Mueller, W. C. 1963. Assay of alfalfa for alfalfa mosaic virus by means of gel diffusion. Plant Dis. Reptr. 47:278-280.

Mueller, W. C., and R. Koenig. 1965. Nuclear inclusions produced by bean yellow mosaic virus as indicators of cross protection. Phytopathology 55:242-243.

Mueller, W. C., and W. F. Rochow. 1961. An aphid injection method for the transmission of barley yellow dwarf virus. Virology 14:253-258.

Muller, H. J. 1927. Artificial transmutation of the gene. Science 66:84-87.

Mulligan, T. 1960. The transmission by mites, host range and properties of ryegrass mosaic virus. Ann. Appl. Biol. 48:575-579.

Mulvania, M. 1926. Studies on the nature of the virus of tobacco mosaic. Phytopathology 16:853-871.

Mundry, K. W. 1957*a*. Zur frage des einflusses von Rontgen-und UV-strahlen auf die Mutationsrate des tabakmosaikvirus nach Bestrahlung reiner praparate. Z. Ind. Abst. Vererbungsl. 88:115-127.

Mundry, K. W. 1957*b*. Die abhängigkeit des auftretens neuer virusstamme von der kulturtemperatur der wirtspflanzen. Z. Ind. Abst. Vererbungsl. 88:407-426.

Mundry, K. W. 1959. The effect of nitrous acid on tobacco mosaic virus: mutation, not selection. Virology 9:722-726.

Mundry, K. W. 1960*a*. Mutationsuntersuchungen am Tabakmosaikvirus in vitro. I. Die abhängigkeit des erfolgs der mutagenese mit HNO_2 von der mutationsrichtung. Z. Vererbungsl. 91:81-86.

Mundry, K. W. 1960*b*. Mutationsuntersuchungen am Tabakmosaikvirus in vitro II. Versuche zum problem der mutagenese durch UV-Bestrahlung reiner viruspropärate. Z. Vererbungsl. 91:87-93.

Mundry, K. W. 1963. Plant virus-host cell relations. Annu. Rev. Phytopathol. 1:173-196.

Mundry, K. W. 1965. A model of the coat protein cistron of tobacco mosaic virus and its biochemical investigation: The model, the experimental approach, and the isolation of a long oligonucleotide from TMV-RNA. Z. Vererbungsl. 97:281-296.

Mundry, K. W. 1967. Strain-specific sequences in TMV-RNA. Mol. Gen. Genet. 100:383-384.

Mundry, K. W., and A. Gierer. 1958. Die erzeugung von mutationen des tabakmosaikvirus durch chemische Behandlung seiner nucleinsäure in vitro. Z. Vererbungsl. 89:614-630.

Munro, J. 1967. Use of molecular gel column to prepare potato virus X inocula for antisera. Amer. Potato J. 44:249-254.

Murakishi, H. H. 1968. Infection of tomato callus cells in suspension with TMV-RNA. Phytopathology 58:993-996.

Murakishi, H. H., J. X. Hartmann, R. N. Beachy and L. E. Pelchers. 1971. Growth curve and yield of tobacco mosaic virus in tobacco callus cells. Virology 43:62-68.

Murakishi, H. H., J. X. Hartmann, L. E. Pelcher and R. N. Beachy. 1970. Improved inoculation of cultured plant cells resulting in high titer and crystal formation. Virology 41:365-367.

Murant, A. F., R. A. Goold, I. M. Roberts, and J. Cathro. 1969. Carrot mottle—a persistent aphid-borne virus with unusual properties and particles. J. Gen. Virol. 4:329-341.

Murant, A. F., and R. M. Lister. 1967. Seed transmission in the ecology of nematode-borne viruses. Ann. Appl. Biol. 59:63-76.

Murant, A. F., and C. E. Taylor. 1965. Treatment of soil with chemicals to prevent transmission of tomato black ring and raspberry ringspot viruses by *Longidorus elongatus* (de Man.) Ann. Appl. Biol. 55:227-237.

Murant, A. F., C. E. Taylor, and J. Chambers. 1968. Properties, relationships and transmission of a strain of raspberry ringspot virus infecting raspberry cultivars immune from the common Scottish strain. Ann. Appl. Biol. 61:175-186.

Murphy, F. A., P. H. Coleman, A. K. Harrison, and G. W. Gary, Jr. 1968. Colorado tick fever virus: an electron microscopic study. Virology 35:28-40.
Mussgay, M., and O. Saurez. 1962. Multiplication of vesicular stomatitis virus in *Aedes aegypti* (L) mosquitoes. Virology 17:202-203.
Myzels, K. J. 1959. Introduction to colloid chemistry. Interscience, New York. 475 pp.
Nagaich, B. B. 1960. Simultaneous transmission of wound-tumor and potato yellow-dwarf viruses of *Agallia constricta* Van Duzee. Science and Culture 25:591-592.
Nagaraj, A. N. 1962. Fluorescent antibody staining of viral antigen in dissociated plant cells. Virology 18:329-332.
Nagaraj, A. N. 1965. Immunofluorescence studies on synthesis and distribution of tobacco mosaic virus antigen in tobacco. Virology 25:133-142.
Nagaraj, A. N., and L. M. Black. 1961. Localization of wound-tumor virus antigen in plant tumors by the use of fluorescent antibodies. Virology 15:289-294.
Nagaraj, A. N., and L. M. Black. 1962. Hereditary variation in the ability of a leafhopper to transmit two unrelated plant viruses. Virology 16:152-162.
Nagaraj, A. N., R. C. Sinha, and L. M. Black. 1961. A smear technique for detecting virus antigen in individual vectors by the use of fluorescent antibodies. Virology 15:205-208.
Nagata, C., and O. Martensson. 1968. On the mechanism of mutagenic action of hydroxylamine. J. Theoret. Biol. 19:133-146.
Nakane, P. K., and G. B. Pierce. 1966. Enzyme labeled antibodies: preparation and application for the localization of antigens. I. Histochem. Cytochem. 14:929-932.
Narita, K. 1958. Isolation of acetylpeptide from enzymic digests of TMV-protein. Biochim. Biophys. Acta 28:184-191.
Narita, K. 1959. Isolation of acetylpeptide from the protein of cucumber virus 4. Biochim. Biophys. Acta 31:372-377.
Narita, K., I. Tsuchida, S. Tsunazawa, and K. Ogata. 1969. Formation of acetylglylpuromycin by the incubation of hen's oviduct minces with puromycin. Biochem. Biophys. Res. Comm. 37:327-332.
Nason, P., V. Schumaker, B. Halsall, and J. Schwedes. 1969. Formation of a streaming convective disturbance which may occur at one gravity during the preparation of samples for zone centrifugation. Biopolymers 7:241-249.
Nasu, S. 1963. Studies on some leafhoppers and planthoppers which transmit virus diseases of rice plant in Japan. Kyushu Agr. Exp. Sta. Bull. 8:153-349.
Nasu, S. 1969*a*. Electron microscopy of transovarial passage of rice dwarf virus, pp. 433-448. *In* K. Maramorosch [ed.] Viruses, vectors, and vegetation. Interscience, New York. 666 pp.
Nasu, S. 1969*b*. Vectors of rice viruses in Asia, pp. 93-109. *In* The virus diseases of the rice plant. Johns Hopkins Press, Baltimore. 354 pp.
Nasu, S., M. Sugiura, T. Wakimoto, and T. Iida. 1967. Rice yellow dwarf virus. Ann. Phytopathol. Soc. Japan 33:343. (Abstr.)
Natti, J. J. 1959. A systemic disease of beans caused by a tobacco necrosis virus. Plant Dis. Reptr. 43:640-644.
Nault, L. R., and W. E. Styer. 1968. Identification of eriophyid mites trapped on silicone grease-coated slides. Phytopathology 58:1061.
Nault, L. R., and W. E. Styer. 1970. Transmission of an eriophyid-borne wheat pathogen by *Aceria tulipae*. Phytopathology 60:1616-1618.
Nelson, D. S., and M. F. Day. 1964. Detection of cauliflower mosaic virus by immune adherence. Phytopathology 54:395-398.
Nelson, M. R., and H. K. Knuhtsen. 1969. Relation of seed transmission to the epidemiology of squash mosaic virus. Phytopathology 54:1042. (Abstr.)
Nelson, R., and E. E. Down. 1933. Influence of pollen and ovule infection in seed transmission of bean mosaic. Phytopathology 23:25. (Abstr.)
Nelson, R. A. 1953. The immune adherence phenomenon. Science 118:733-737.
Nemethy, G., and H. A. Scheraga. 1962. The structure of water and hydrophobic bonding in proteins. III. The thermodynamic properties of hydrophobic bonds in proteins. J. Phys. Chem. 66:1773-1789.
Nene, Y. L. 1960. Studies on the mechanism of inhibition of tobacco-mosaic viral infection by enzymes. Diss. Abstr. 21:1325.

Neurath, A. R., J. T. Stasny, and B. A. Rubin. 1970. Disruption of adenovirus type 7 by lithium iodide resulting in release of viral deoxyribonucleic acid. J. Virol. 5:173-178.

Newhall, A. G. 1923. Seed transmission of lettuce mosaic virus. Phytopathology 13:104-106.

Newmark, P., and R. W. Meyers. 1957. Degradation of tobacco mosaic virus by alkanolamines. Fed. Proc. 16:226.

Newnham, A. G., and H. P. Chu. 1965. An in vitro comparison of the effect of some antibacterial, antifungal and antiprotozoal agents on various strains of mycoplasma (pleuropneumonia-like organisms:PPLO). J. Hyg. 63:1-23.

New York Academy of Sciences. 1969. Advances in ultracentrifugal analysis. Ann. N. Y. Acad. Sci. 164:1-305.

Nichol, L. W., J. L. Bethune, G. Kegeles, and E. L. Hess. 1964. Interacting protein systems, pp. 305-403. *In* H. Neurath [ed.] The proteins, vol. 2. Academic Press, New York. 5 vols.

Nichols, C. W. 1952. The retarding effect of certain plant hormones on tobacco mosaic symptoms. Phytopathology 42:579-580.

Nicolaieff, A., L. Pinck, and L. Hirth. 1969. Examen en microscopie électronique de l'arn replicatif du virus de la mosaique de la luzerne. J. Gen. Virol. 4:283-286.

Nielson, M. W. 1968. The leafhopper vectors of phytopathogenic viruses. Taxonomy, biology, and virus transmission. U.S. Dep. Agr. Tech. Bull. 1382. 386 pp.

Nienhaus, F. 1957. Untersuchungen über den Einflüss von Temperatur und Licht auf die Empfänglichkeit der Pflanzen für das Kartoffel-Y-virus. Phytopathol. Z. 30:189-224.

Nienhaus, F., and B. Stille. 1965. Ubertragung des Kartoffel-X-Virus durch zoosporen von Synchytrium endobioticum. Phytopathol. Z. 54:335-337.

Nilsson-Tillgren, T. 1969. Studies on the biosynthesis of TMV. II. On the RNA synthesis of infected cells. Mol. Gen. Genet. 105:191-202.

Nirenberg, M., P. Leder, M. Bernfield, R. Brimacombe, J. Trupin, F. Rottmann, and C. O'Neal. 1965. RNA codewords and protein synthesis. VII. On the general nature of the RNA code. Proc. Nat. Acad. Sci. U.S. 53:1161-1168.

Nirenberg, M. W., and J. H. Matthaei. 1961. The dependence of cell free protein synthesis in *E. coli* upon naturally occurring or synthetic polyribonucleotides. Proc. Nat. Acad. Sci. U.S. 47:1588-1602.

Nishimura, A., and Y. Hosaka. 1969. Electron microscopic study on RNA of cytoplasmic polyhedrosis virus of the silkworm. Virology 38:550-557.

Nitzany, F. E., and S. Cohen. 1960. A case of interference between alfalfa mosaic virus and cucumber mosaic virus. Virology 11:771-773.

Niu, C. I., V. Shore, and C. A. Knight. 1958. The peptide chains of some plant viruses. Virology 6:226-233.

Nixon, H. L., and H. L. Fisher. 1958. An improved spray droplet technique for quantitative electron microscopy. Brit. J. Appl. Phys. 9:68-70.

Nixon, H. L., and R. D. Woods. 1960. The structure of tobacco mosaic virus protein. Virology 10:157-159.

Noll, H., and E. Stutz. 1968. The use of sodium and lithium dodecyl sulfate in nucleic acid isolation, pp. 129-155. *In* L. Grossman and K. Moldave [ed.] Methods in enzymology, vol. 12B. Academic Press, New York. 19 vols.

Norrby, E. C. J., and P. A. Albertsson. 1960. Concentration of poliovirus by an aqueous polymer two-phase system. Nature 188:1047-1048.

Norton, D. C. 1967. *Xiphinema americanum* as a factor in unthriftiness of red clover. Phytopathology 57:1390-1391.

Novelli, G. D. 1969. The separation of isoaccepting transfer RNA's and the possible role of tRNA in regulation. J. Cell. Physiol. 74, Sup. 1:121-148.

Nozu, Y., and Y. Okada. 1968. Amino acid sequence of a common Japanese strain of tobacco mosaic virus. J. Mol. Biol. 35:643-646.

Nuorteva, P. 1962. Studies on the causes of the phytopathogenicity of *Calligypona pellucida* (F.) (Hom., Araeopidae). Ann. Zool. Soc. 'Vanamo' 23:1-58.

Nyland, G. 1962. Possible virus-induced genetic abnormalities in tree fruits. Science 137:598-599.

Oakley, C. L., and A. J. Fulthorpe. 1953. Antigenic analysis by diffusion. J. Pathol. Bacteriol. 65:49-60.

Oberg, B., P. A. Albertsson, and L. Philipson. 1965. Partition studies on nucleic acids. II. Counter-current distribution of virus RNA. Biochim. Biophys. Acta 108:173-181.

O'Brien, M. J., and W. B. Raymer. 1964. Symptomless hosts of the potato spindle tuber virus. Phytopathology 54:1045-1047.
Oehlen, G. 1967. A comparison of the number of titratable carboxyl groups of TMV wild strains and mutants in the native and denatured states. Mol. Gen. Genetics 99:372-375.
Oertel, C. 1961. Die Verwendung eines Serums gegen normale Pflanzeneiweisse fur die Gweinnung eines hochwertigen Antigens. Phytopathol. Z. 40:272-276.
Oertel, C. 1968. Serologischer Virusnachweis als Routinetest und Bestandteil pflanzen hygienischer Massnahmen im Chrysanthemen jung Pflanzenbetrieb. Z. PflanzenKrankh. 75:605-612.
Ohno, T., Y. Nozu, and Y. Okada. 1971. Polar reconstitution of tobacco mosaic virus (TMV). Virology 44:510-516.
Okuyama, S., K. Yora, and H. Asuyama. 1968. Multiplication of the rice stripe virus in its insect vector. *Laodelphax striatellus* Fallén. Ann. Phytopathol. Soc. Japan 34:255-262.
O'Loughlin, G. T., and T. C. Chambers. 1967. The systemic infection of an aphid by a plant virus. Virology 33:262-271.
Oman, P. W. 1949. The Nearctic leafhoppers (Homoptera:Cicadellidae). A generic classification and check list. Mem. Entomol. Soc. Washington 3:253 pp.
Oman, P. 1969. Criteria of specificity in virus-vector relationships, pp. 1-22. *In* K. Maramorosch [ed.] Viruses, vectors, and vegetation. Interscience, New York. 666 pp.
Oostenbrink, M. 1960. Estimating nematode populations by some selected methods. *In* J. N. Sasser and W. R. Jenkins [ed.] Nematology. Univ. North Carolina Press, Chapel Hill. 480 pp.
Opel, H., H. Kegler, and J. Richter. 1969. Vorkommen und charakterisierung von TMV-stämmen des Kernobstes. Acta Phytopathol. Acad. Sci. Hungaricae 4:1-12.
Orenski, S. W., J. Mitsuhashi, S. M. Ringel, J. F. Martin, and K. Maramorosch. 1965. A presumptive bacterial symbiont from the eggs of the six-spotted leafhopper, *Macrosteles fascifrons* (Stål). Contrib. Boyce Thompson Inst. 23:123-126.
Orlob, G. B. 1966. Feeding and transmission characteristics of *Aceria tulipae* Keifer as vector of wheat streak mosaic virus. Phytopathol. Z. 55:218-238.
Orlob, G. B. 1967. Inactivation of purified plant viruses and their nucleic acids by photosensitizing dyes. Virology 31:402-413.
Oshiro, L. S., H. M. Rose, C. Morgan, and K. C. Hsu. 1967. Electron microscopic study of the development of simian virus 40 by the use of ferritin-labelled antibodies. J. Virol. 1:384-399.
Ossiannilsson, F. 1949. Insect drummers. A study on the morphology and function of the sound producing organ of Swedish Homoptera Auchenorrhyncha, with notes on their sound production. Opuscula Entomol., Suppl. 10. 145 pp.
Ossiannilsson, F. 1966. Insects in the epidemiology of plant viruses. Annu. Rev. Entomol. 11:213-232.
Oster, G. 1951. The isoelectric point of some strains of tobacco mosaic virus. J. Biol. Chem. 190:55-59.
Oster, G., and H. Grimsson. 1949. The interaction of tobacco mosaic virus and of its degradation products with dyes. Arch. Biochem. 24:119-127.
Oster, G., and A. D. MacLaren. 1949. The ultraviolet light and photosensitized inactivation of tobacco mosaic virus. J. Gen. Physiol. 33:215-228.
Oster, G., and N. Yamamoto. 1963. Density-gradient techniques. Chem. Rev. 63:257-268.
Oswald, J. W., and T. Bowman. 1958. Studies on a soil-borne potato virus disease in California. Phytopathology 48:396. (Abstr.)
Oswald, J. W., and B. R. Houston. 1951. A new virus disease of cereals, transmissible by aphids. Plant Dis. Reptr. 35:471-475.
Oswald, J. W., and B. R. Houston. 1953. Host range and epiphytology of the cereal yellow dwarf disease. Phytopathology 43:309-313.
Ouchterlony, O. 1958. Diffusion-in-gel methods for immunological analysis, pp. 1-78. *In* P. Kallós [ed.] Progress in allergy, vol. 5. S. Karger, Basel, Switzerland. 5 vols.
Ouchterlony, O. 1964. Gel-diffusion techniques, pp. 55-78. *In* J. F. Ackroyd [ed.] Immunological methods. Blackwell, Oxford. 628 pp.
Ouchterlony, O. 1967. Immunodiffusion and immunoelectrophoresis, pp. 655-706. *In* D. M. Weir [ed.] Handbook of experimental immunology. 1245 pp. Blackwell, Oxford.
Overbeek, J. Th. G. 1952. Electrokinetic phenomena, pp. 194-244. *In* H. R. Kruyt [ed.] Colloid science, vol. 1. Elsevier, Amsterdam. 2 vols.

Overbeek, J. Th. G., and J. Lÿklema. 1959. Electric potentials in colloidal systems, pp. 1-33. *In* M. Bier [ed.] Electrophoresis. Theory, methods and applications. Academic Press, New York. 563 pp.

Owusu, G. K., N. C. Crowley, and R. I. B. Francki. 1968. Studies of the seed transmission of tobacco ringspot virus. Ann. Appl. Biol. 61:195-202.

Oxelfelt, P. 1966. The use of a sintered-glass filter for concentrating tobacco mosaic virus suspensions. Virology 30:740-741.

Pady, S. M. 1955. The occurrence of the vector of wheat streak mosaic, *Aceria tulipae* on slides exposed in the air. Plant Dis. Reptr. 39:296-297.

Paliwal, Y. C. 1968. Changes in relative virus concentration in *Endria inimica* in relation to its ability to transmit wheat striate mosaic virus. Phytopathology 58:386-87.

Paliwal, Y. C., and R. C. Sinha. 1970. On the mechanism of persistence and distribution of barley yellow dwarf virus in an aphid vector. Virology 42:668-680.

Paliwal, Y. C., and J. T. Slykhuis. 1967. Localization of wheat streak mosaic virus in the alimentary canal of its vector *Aceria tulipae* Keifer. Virology 32:344-353.

Palmiter, D. H., and E. M. Hildebrand. 1943. The yellow-red virosis of peach: its identification and control. N. Y. State Agr. Exp. Sta. Bull. 704. 17 pp.

Parsons, D. F. 1963. Negative staining of thinly spread cells and associated virus. J. Cell Biol. 16:620-626.

Paul, H. L. 1959. Die bestimmung des nucleinsaure-gehaltes pflanzlicher viren mit hilfe einer spektrophotometrischen methode. Z. Naturforsch. 14b:427-432.

Paul, H. L., C. Wetter, H. G. Wittmann, and J. Brandes. 1965. Untersuchungen an odontoglossum ringspot virus, einem verwandten des tabakmosaik-virus I. Physikalische, Chemische, Serologische and Symptomatologischebefunde. Vererbungsl. 97:186-203.

Pavillard, J. 1954. Contribution à l'étude de la croissance des plantes virosées: virus et auxines. Ann. Biol. 30:169-181.

Pease, D. C. 1964. Histological techniques for electron microscopy. 2nd ed. Academic Press, New York. 381 pp.

Perry, V. G. 1958. Parasitism of two species of dagger nematodes (*Xiphinema americanum* and *X. chambersi*) to strawberry. Phytopathology 48:420-423.

Persons, T. D. 1952. Phony peach disease. A review of organized control from 1929 to 1951 and the effect of recent developments on future control programs. Phytopathology 42:286-287.

Perutz, M. F., J. C. Kendrew, and H. C. Watson. 1965. Structure and function of Haemoglobin. II. Some relations between polypeptide chain configuration and amino acid sequence. J. Mol. Biol. 13:669-678.

Peters, D. 1965. The purification of virus-like particles from the aphid *Myzus persicae*. Virology 26:159-161.

Peters, D. 1967. The purification of potato leafroll virus from its vector *Myzus persicae*. Virology 31:46-54.

Peterson, P. D., and H. H. McKinney. 1938. The influence of four mosaic diseases on the plastid pigments and chlorophyllase in tobacco leaves. Phytopathology 28:329-342.

Philipson, L. 1967. Chromatography and membrane separation, pp. 179-233. *In* K. Maramorosch and H. Koprowski [ed.] Methods in virology, vol. 2. Academic Press, New York. 4 vols.

Philipson, L. 1969. Aqueous polymer phase systems in virology, pp. 109-128. *In* K. Habel and N. P. Salzman [ed.] Fundamental techniques in virology. Academic Press, New York. 527 pp.

Philipson, L., P. A. Albertsson, and G. Frick. 1960. The purification and concentration of viruses by aqueous polymer phase systems. Virology 11:553-571.

Philipson, L., and M. Kaufman. 1964. The efficiency of ribonuclease tested with viral ribonucleic acid as substrate. Biochim. Biophys. Acta 80:151-154.

Pickels, E. G. 1943. Sedimentation in the angle centrifuge. J. Gen. Physiol. 26:341-360.

Pierpoint, W. S., and B. D. Harrison. 1963. Copper-dependent and iron-dependent inactivations of cucumber mosaic virus by polyphenols. J. Gen. Microbiol. 32:429-440.

Pinck, L., L. Hirth, and G. Bernardi. 1968. Isolation of replicative RNA from alfalfa mosaic virus-infected plants by chromatography on hydroxyapatite columns. Biochem. Biophys. Res. Commun. 31:481-487.

Pirie, N. W. 1954. The fission of tobacco mosaic virus and some other nucleoproteins by strontium nitrate. Biochem. J. 56:83-86.

Pirone, T. P. 1967. Acquisition and release of infectious tobacco mosaic virus by aphids. Phyto-

pathology 57:463. (Abstr.)
Pirone, T. P. 1969. Mechanism of transmission of stylet-borne viruses, pp. 199-210. *In* K. Maramorosch [ed.] Viruses, vectors, and vegetation. Interscience, New York. 666 pp.
Pirone, T. P., and E. S. Megahed. 1966. Aphid transmissibility of some purified viruses and viral RNA's. Virology 30:631-637.
Pitcher, R. S. 1967. The host-parasite relations and ecology of *Trichodorus viruliferous* on apple roots, as observed from an underground laboratory. Nematologica 13:547-557.
Pitcher, R. S., and J. J. M. Flegg. 1965. Observations of root-feeding by the nematode *Trichodorus viruliferus* (Hooper). Nature (London) 207:317.
Pitcher, R. S., and A. Jha. 1961. On the distribution and infectivity with arabis mosaic virus of a dagger nematode. Plant Pathol. 10:67-71.
Pitcher, R. S., and A. F. Posnette. 1963. Vascular feeding by *Xiphinema diversicaudatum* (Micol.). Nematologica 9:301-304.
Plumb, R. T., and D. A. Vince. 1971. Fixation and electron microscopy of the Rothamsted strain of henbane mosaic virus. J. Gen. Virol. (in press).
Pollard, E., and A. E. Dimond. 1956. The inactivation of tobacco mosaic virus by ionizing radiation. Phytopathology 46:214-218.
Polson, A. 1958*a*. Quantitative gel precipitin tests. Sci. Tools 5:17-20.
Polson, A. 1958*b*. A method for determination of diffusion coefficients by the gel-precipitin technique. Biochim. Biophys. Acta 29:426-428.
Polson, A. 1961. Fractionation of protein mixtures on columns of granulated agar. Biochim. Biophys. Acta 50:565-567.
Polson, A., and R. Cramer. 1958. Zone electrophoresis of type I poliomyelitis virus. Biochim. Biophys. Acta 29:187-192.
Polson, A., and D. Deeks. 1960. The diffusion constants of rabbit, guinea-pig, and rat antibodies to a single common antigen. Biochim. Biophys. Acta 39:208-211.
Polson, A., and B. Russell. 1968. Electrophoresis of viruses, pp. 391-426. *In* K. Maramorosch and H. Koprowski [ed.] Methods in virology, vol. 2. Academic Press, New York. 4 vols.
Poos, F. W., and C. Elliott. 1936. Certain insect vectors of *Aplanobacter stewarti*. J. Agr. Res. 52:585-608.
Porter, C. A., and F. P. McWhorter. 1952. Crystalline inclusions produced by red clover vein mosaic virus. Phytopathology 42:518. (Abstr.)
Posnette, A. F., and C. E. Ellenberger. 1963. Further studies of green petal and other leafhopper-transmitted viruses infecting strawberry and clover. Ann. Appl. Biol. 51:69-83.
Posnette, A. F., and A. H. Strickland. 1948. Virus diseases of Cacao in West Africa. III. Technique of insect transmission. Ann. Appl. Biol. 35:53-63.
Preer, J. R. 1956. A quantitative study of a technique of double-diffusion in agar. J. Immunol. 77:52-60.
Prentice, I. W., and R. V. Harris. 1946. Resolution of strawberry virus complexes by means of the aphid vector *Capitophora fragerial* Theo. Ann. Appl. Biol. 33:50-53.
Prevec, L., Y. Watanabe, C. J. Gauntt, and A. F. Graham. 1968. Transcription of the genomes of Type 1 and Type 3 reoviruses. J. Virol. 2:289-297.
Price, W. C. 1936. Virus concentration in relation to acquired immunity from tobacco ring spot. Phytopathology 26:503-529.
Price, W. C. 1945. Accuracy of the local-lesion method for measuring virus activity. IV. Southern bean mosaic virus. Amer. J. Bot. 32:613-619.
Price, W. C. 1954. Genetic composition in relation to isoelectric point and serological reaction of strains in tobacco mosaic virus. Trans. N. Y. Acad. Sci. 16:196-201.
Price, W. C. 1964. Strains, mutation, acquired immunity and interference, pp. 93-117. *In* M. K. Corbett and H. D. Sisler [ed.] Plant virology. Univ. Florida Press, Gainesville. 527 pp.
Price, W. C. 1965. Inactivation of southern bean mosaic by ultraviolet light. Virology 25:1-8
Price, W. C. 1968. Evidence for restriction of a plant virus to phloem cells. Indian Phytopathol. 21:159-166.
Price, W. C., and W. J. Gowen. 1927. Quantitative studies of tobacco mosaic virus inactivation by ultraviolet light. Phytopathology 27:267-282.
Privalov, P. L., O. B. Ptitsyn, and T. M. Birshtein. 1969. Determination of stability of the DNA double helix in an aqueous medium. Biopolymers 8:559-571.

Privat de Garilhe, M. 1967. Enzymes in nucleic acid research. Holden-Day, San Francisco. 408 pp.

Pryor, D. E. 1944. The big vein disease of lettuce in relation to soil moisture. J. Agr. Res. 68:1-9.

Pryor, D. E. 1946. Exploratory experiments with the big-vein disease of lettuce. Phytopathology 36:264-272.

Purcifull, D. E. 1966. Some properties of tobacco etch virus and its alkaline degradation products. Virology 29:8-14.

Purcifull, D. E., and J. R. Edwardson. 1967. Watermelon mosaic virus: tubular inclusions in pumpkin leaves and aggregates in leaf extracts. Virology 32:393-401.

Putnam, W. L. 1941. The feeding habits of certain leafhoppers. Can. Entomol. 73:39-53.

Quacquarelli, A., G. P. Martelli, and M. Russo. 1967. Richerche sull' agente dell'arricciamento maculato del Carciofo. II. Purificazione, microscopia elettronica e caratteristiche fisicochimiche. Atti 1 Congresso Unione Fitopatologica Mediterranea.

Quantz, L. 1953. Untersuchungen über ein samenübertragbares Mosaikvirus der Ackerbohne (*Vicia faba*). Phytopathol. Z. 20:421-448.

Quantz, L. 1954. Untersuchungen über die Viruskrankheiten der Ackerbohne, Mitt. Biol. Bund. —Anst. Ld.—u. Forstw. 80:171-175.

Quantz, L. 1958. Die Virosen der Leguminosen. *In* M. Klinkowski [ed.] Pflanzliche Virologie 2. Akademie-Verlag, Berlin. 393 pp.

Rader, W. E., H. F. Fitzpatric, and E. M. Hildebrand. 1947. A seed-borne virus of muskmelon. Phytopathology 37:809-816.

Radewald, J. D. 1962. The biology of *Xiphinema index* and the pathological effect of the species on grape. Ph.D. Thesis, Univ. of Calif., Davis. 60 pp.

Radewald, J. D., J. W. Osgood, K. S. Mayberry, A. O. Paulus, and F. Shibuya. 1969. *Longidorus africanus* a pathogen of head lettuce in the Imperial Valley of Southern California. Plant Dis. Reptr. 53:381-384.

Radloff, R., W. Bauer, and J. Vinograd. 1967. A dye-buoyant-density method for the detection and isolation of closed circular duplex DNA: the closed circular DNA in HeLa Cells. Proc. Nat. Acad. Sci. U.S. 57:1514-1521.

Ragetli, H. W. J., Chr., van der Scheer, and J. P. H. van der Want. 1955. Electron-microscopy of fractions obtained by paper chromatography from juices of virus diseased and healthy plants. Tijdschr. Plantenziekten 61:35-46.

Ragetli, H. W. J., and J. P. H. van der Want. 1954. Paper chromatography of plant viruses. Proc. Kon. Ned. Ak. v. Wet. Ser. C 57:621-627.

Ragetli, H. W. J., and M. Weintraub. 1962. Purification and characteristics of a virus inhibitor from *Dianthus caryophyllus* L. I. Purification and activity. Virology 18:232-240.

Ragetli, H. W. J., and M. Weintraub. 1964. Immunoosmophoresis, a rapid and sensitive method for evaluating viruses. Science 144:1023-1024.

Ragetli, H. W. J., and M. Weintraub. 1965. Immunoosmophoresis adapted for weakly charged antigens. Biochim. Biophys. Acta 3:522-528.

Ragetli, H. W. J., and M. Weintraub. 1966. Agar derivatives for gel electrophoresis requiring reduced endosmotic liquid flow. Biochim. Biophys. Acta 112:160-167.

Raine, J. 1967. Leafhopper transmission of witches'-broom and clover phyllody viruses from British Columbia to clover, alfalfa, and potato. Can. J. Bot. 45:441-445.

Ralph, R. K., and P. L. Bergquist. 1967. Separation of viruses into components, pp. 463-545. *In* K. Maramorosch and H. Koprowski [ed.] Methods in virology, vol. 2. Academic Press, New York. 4 vols.

Ralph, R. K., R. E. F. Matthews, and A. I. Matus. 1965. Effects of 2-thiouracil on the formation of double-stranded plant viral ribonucleic acid. Biochim. Biophys. Acta 108:53-66.

Ralph, R. K., R. E. F. Matthews, A. I. Matus, and H. G. Mandel. 1965. Isolation and properties of double-stranded viral RNA from virus-infected plants. J. Mol. Biol. 11:202-212.

Ralph, R. K., and S. J. Wojcik. 1966. Synthesis of double-stranded viral RNA by cell-free extracts from turnip yellow mosaic virus-infected leaves. Biochim. Biophys. Acta 119:347-361.

Rao, A. S. 1968. Biology of *Polymyxa graminis* in relation to soil-borne wheat mosaic virus. Phytopathology 58:1516-1521.

Rappaport, I., and A. Siegel. 1955. Inactivation of tobacco mosaic virus by rabbit antiserum. J. Immunol. 74:106-116.

Rappaport, I., A. Siegel, and R. Haselkorn. 1965. Influence of the state of subunit aggregation on the antigenic specificity of TMV and TYMV. Virology 25:325-328.

Rappaport, I., and J-H. Wu. 1962. Release of inhibited virus infection following irradiation with ultraviolet light. Virology 17:411-419.

Rasa, E. A., and K. Esau. 1961. Anatomic effects of curly top and aster yellows viruses on tomato. Hilgardia 30:469-515.

Raski, D. J., and W. B. Hewitt. 1963. Plant parasitic nematodes as vectors of plant viruses. Phytopathology 53:39-47.

Raski, D. J., W. B. Hewitt, A. C. Goheen, C. E. Taylor, and R. H. Taylor. 1965. Survival of *Xiphinema index* and reservoirs of fanleaf virus in fallowed vineyard soil. Nematologica 11:349-352.

Rawlins, T. E. 1957. A modification of Bald's stain for viruses and for cell inclusions associated with virus infections. Phytopathology 47:307.

Rawlins, T. E. 1962. The inhibitory effect of 2-chloroethyltrimethylammonium chloride treatment on tobacco mosaic virus (TMV) multiplication. Plant Dis. Reptr. 46:170.

Rawlins, T. E., and K. G. Parker. 1934. Influence of rootstocks on the susceptibility of sweet cherry to the buckskin disease. Phytopathology 24:1029-1031.

Rawlins, T. E., and C. M. Tompkins. 1936. Studies on the effect of carborundum as an abrasive in plant virus inoculations. Phytopathology 26:578-587.

Raymer, W. B., and T. O. Diener. 1969. Potato spindle tuber virus: A plant virus with properties of a free nucleic acid. I. Assay, extraction, and concentration. Virology 37:343-350.

Razin, S. 1969. Structure and function in mycoplasma. Annu. Rev. Microbiol. 23:317-356.

Reddi, K. K. 1957. Structural differences in the nucleic acids of some tobacco mosaic virus strains. Biochim. Biophys. Acta 25:528-531.

Reddi, K. K. 1959*a*. Structural differences in the nucleic acids of some tobacco mosaic virus strains. II. Di- and tri-nucleotides in ribonuclease digests. Biochim. Biophys. Acta 32:386-392. (20)

Reddi, K. K. 1959*b*. Studies on tobacco leaf ribonuclease. III. Its role in the synthesis of tobacco mosaic virus nucleic acid. Biochim. Biophys. Acta 33:164-169. (2)

Reddi, K. K., and C. A. Knight. 1956. Structural studies on the nucleic acids of some strains of tobacco mosaic virus. I. The residual material after ribonuclease digestion. J. Biol. Chem. 221:629-633.

Reddick, D., and V. B. Stewart. 1919. Transmission of the virus of bean mosaic in seed and observations on thermal death-point of seed and virus. Phytopathology 9:445-450.

Reddy, D. V. R., and L. M. Black. 1966. Production of wound-tumor virus and wound tumor soluble antigen in the insect vector. Virology 30:551-561.

Reed, L. J., and H. Muench. 1938. A single method of estimating fifty percent endpoints. Amer. J. Hyg. 27:493-497.

Rees, M. W. and M. N. Short. 1965. Variations in the composition of two strains of tobacco mosaic virus in relation to their host. Virology 26:596-602.

Reichmann, M. E. 1959. Potato virus X, part III. Light scattering studies. Can. J. Chem. 37:384-388.

Reichmann, M. E. 1964. The satellite tobacco necrosis virus: a single protein and its genetic code. Proc. Nat. Acad. Sci. U.S. 52:1009-1017.

Reichmann, M. E., A. Y. Chang, L. Faiman, and J. M. Clark, Jr. 1966. The satellite tobacco necrosis virus in studies of genetic coding. Cold Spring Harbor Symp. Quant. Biol. 31:139-144.

Reichmann, M. E., and R. Stace-Smith. 1959. Preparation of infectious ribonucleic acid from potato virus X by means of guanidine denaturation. Virology 9:710-712.

Reid, M. S., and R. E. F. Matthews. 1966. On the origin of the mosaic induced by turnip yellow mosaic viruses. Virology 28:563-570.

Rentschler, L. 1967. Aminosäuresequenzen and physikochemisches Verhalten des Hüllproteins eines Wildstammes des Tabakmosaikvirus. I. Analyse der Primärstruktur (Pos. 62-134) des Hüllproteins vom Wildstamm U2. Mol. Gen. Genet. 100:84-95. (15); II. Aggregationsverhalten und Ladungsverteilung im Vergleich zu den Stämmen vulgare und dahlemense. Mol. Gen. Genet. 100:96-108.

Resconich, E. C. 1963. Movement of tobacco necrosis virus in systemically infected soybeans. Phytopathology 53:913-916.

Reyes, T. T. 1961. Seed transmission of coffee ring spot by Excelsa coffee (*Coffea excelsa*). Plant Dis. Reptr. 45:185.

Reynolds, E. S. 1963. The use of lead citrate at high pH as an electron-opaque stain in electron microscopy. J. Cell Biol. 17:208-212.
Rich, S. 1948. A rapid stain method for detecting certain plant virus inclusions. Stain Technol. 23:19-20.
Richards, E. G., D. C. Teller, and H. K. Schachman. 1968. Ultracentrifuge studies with Rayleigh interference optics. II. Low-speed sedimentation equilibrium of homogeneous systems. Biochemistry 7:1054-1076.
Richardson, J., and E. S. Sylvester. 1965. Aphid honeydew as inoculum for the injection of pea aphids with pea-enation mosaic virus. Virology 25:472-475.
Richardson, J., and E. S. Sylvester. 1968. Further evidence of multiplication of sowthistle yellow-vein virus in its aphid vector, *Hyperomyzus lactucae*. Virology 35:347-355.
Richter, J. 1967. Die Anwendung des Hamagglutinations-Tests nach Boyden zum Nachweis eines pflanzen-pathogenen Virus. Zentralbl. Bakteriol. II. Abt. 121:269-275.
Robb, Sheila M. 1964. Location, structure, and cytochemical staining reactions of the inclusion bodies found in *Dahlia variabilis* infected with dahlia mosaic virus. Virology 23:141-144.
Roberts, J. W., and J. A. Steitz. 1967. The reconstitution of infective bacteriophage R17. Proc. Nat. Acad. Sci. U.S. 58:1416-1421.
Roberts, D. A. 1964. Local-lesion assay of plant viruses, pp. 194-210. *In* M. K. Corbett and H. D. Sisler [ed.] Plant virology. Univ. of Florida Press, Gainesville. 527 pp.
Roberts, D. A., R. G. Christie, and M. C. Archer, Jr. 1970. Infection of apical initials in tobacco shoot meristems by tobacco ringspot virus. Virology 42:217-220.
Roberts, D. A., and W. C. Price. 1967. Infection of apparently uninjured leaves of bean by the viruses of tobacco necrosis and southern bean mosaic. Virology 33:542-545.
Roberts, F. M. 1940. Studies on the feeding methods and penetration rate of *Myzus persicae* Sulz., *Myzus circumflex* Buckt. and *Macrosiphum gei* Koch. Ann. Appl. Biol. 27:348-358.
Roberts, F. M. 1950. The infection of plants by viruses through roots. Ann. Appl. Biol. 37:385-396.
Robinson, D. R., and W. P. Jencks. 1965. The effect of compounds of the urea-guanidinium class on the activity coefficient of acetyltetraglycine ethyl ester and related compounds. J. Amer. Chem. Soc. 87:2462-2470.
Rochow, W. F. 1960. Transmission of barley yellow dwarf virus acquired from liquid extracts by aphids feeding through membranes. Virology 12:223-232.
Rochow, W. F. 1963. Variation within and among aphid-vectors of plant viruses. Ann. New York Acad. Sci. 105:713-729.
Rochow, W. F. 1965. Apparent loss of vector specificity following double infection by two strains of barley yellow dwarf virus. Phytopathology 55:62-68.
Rochow, W. F. 1969. Specificity in aphid transmission of a circulative plant virus, pp. 175-198. *In* K. Maramorosch [ed.] Viruses, vectors, and vegetation. Interscience, New York. 666 pp.
Rochow, W. F., and E. M. Ball. 1967. Serological blocking of aphid transmission of barley yellow dwarf virus. Virology 33:358-362.
Rochow, W. F., and M. W. Brakke. 1964. Purificator of barley yellow dwarf virus. Virology 23:310-332.
Rochow, W. F., and V. F. Eastop. 1966. Variation within *Rhopalosiphum padi* and transmission of barley yellow dwarf virus by clones of some aphid species. Virology 30:286-296.
Roggen, D. R. 1966. On the morphology of *Xiphinema index* reared on grape fanleaf virus infected grapes. Nematologica 12:287-296.
Roggen, D. R., D. J. Raski, and N. Jones. 1967. Further electron microscope observations of *Xiphinema index*. Nematologica 10:489-495.
Rombauts, W. A. 1966. The non-acetylated N-terminal sequence of the green tomato atypical mosaic virus coat protein (G-TAMV). Biochem. Biophys. Res. Commun. 23:549-556.
Rombauts, W., and H. Fraenkel-Conrat. 1968. Artificial histidine-containing mutants of tobacco mosaic virus. Biochemistry 7:3334-3339.
Roodyn, D. B. 1967. Enzyme cytology. Academic Press, New York. 587 pp.
Rosen, L. 1968. Reoviruses. Virol. monogr. 1:73-107.
Rosenbloom, J., and V. Schumaker. 1967. Molecular weight dependence of the rotor speed induced aggregation of deoxyribonucleic acid. Biochemistry 6:276-283.
Rosier, D. J., and A. Klug. 1968. Reconstruction of three-dimensional structures from electron

micrographs. Nature 217:130-134.
Ross, A. F. 1941. The concentration of alfalfa-mosaic virus in tobacco plants at different periods of time after inoculation. Phytopathology 31:410-420.
Ross, A. F. 1959. The interaction of viruses in the host, pp. 511-520. *In* C. S. Holton, G. W. Fisher, R. W. Fulton, H. Hart, and S. E. A. McCallan [ed.] Plant pathology, problems and progress, 1908-1958. Univ. Wisconsin Press, Madison. 588 pp.
Ross, A. F. 1961*a*. Localized acquired resistance to plant virus infection in hypersensitive hosts. Virology 14:329-339.
Ross, A. F. 1961*b*. Systemic acquired resistance induced by localized virus infections in plants. Virology 14:340-358.
Ross, A. F. 1966. Systemic effects of local lesion formation, pp. 127-150. *In* A. B. R. Beemster and J. Dijkstra [ed.] Viruses of plants. North-Holland, Amsterdam. 342 pp.
Rossouw, D. J., and R. W. Fulton. 1963. The effect of certain monovalent and divalent cations on plant virus infection. S. Afr. J. Agr. Sci. 6:193-204.
Rowen, J. W., and W. Ginoza. 1956. Unit particle lengths of strains of tobacco mosaic virus. Biochim. Biophys. Acta 21:416-425.
Roy, D., H. Fraenkel-Conrat, J. Lesnaw, and M. E. Reichmann. 1969. The protein subunit of the satellite tobacco necrosis virus. Virology 38:368-369.
Rubio-Huertos, M. 1950. Estudios sobre inclusiones intracelulares, producidas por virus en las plantas. Microbiol. Espan. 3:207-232.
Rubio-Huertos, M. 1956. Origin and composition of cell inclusions associated with certain tobacco and crucifer viruses. Phytopathology 45:553-556.
Rubio-Huertos, M. 1962. Efectos de la temperatura sobre la sintomatologia del virus "Ringspot" de la petunia. Microbiol. Espan. 15:1-11.
Rubio-Huertos, M., and D. Lopez-Abella. 1966. Ultraestructura de cellulas de pimiento infectadas con un virus y su localizacion en las mismas. Microbiol. Espan. 19:77-86.
Rubio-Huertos, M., C. Matsui, A. Yamaguchi, and T. Kamei. 1968. Electron microscopy of X-body formation in cells of cabbage infected with brassica virus 3. Phytopathology 58:548-549.
Rubio-Huertos, M., and D. H. M. van Slogteren. 1956. Light and electron microscopy of X-bodies associated with broad bean mottle virus and Phaseolus virus 2. Phytopathology 46:401-402.
Rubio-Huertos, M., A. Vela, and D. Lopez-Abella. 1967. Crystalline arrays of spherical particles in turnip yellow mosaic virus infected cells. Virology 32:438-444.
Rueckert, R. R., A. A. Dunker, and C. M. Stoltzfus. 1969. The structure of Mouse-Elberfeld virus. A model. Proc. Nat. Acad. Sci. U.S. 62:912-919.
Rush, M. C. 1970. Serological strains of tobacco ringspot virus transmitted by *Xiphinema americanum*. J. Nematol. 2:265-269.
Rushizky, G. W., A. E. Greco, R. W. Hartley, Jr., and H. A. Sober. 1963. Concentration and desalting of ribonucleoses. Biochem. Biophys. Res. Commun. 10:311-314.
Rushizky, G. W., and C. A. Knight. 1960*a*. An oligonucleotide mapping procedure and its use in the study of tobacco mosaic virus nucleic acid. Virology 11:236-249.
Rushizky, G. W., and C. A. Knight. 1960*b*. Products obtained by digestion of the nucleic acids of some strains of tobacco mosaic virus with pancreatic ribonuclease. Proc. Nat. Acad. Sci. U.S. 46:945-952.
Rushizky, G. W., H. A. Sober, and C. A. Knight. 1962. Products obtained by digestion of the nucleic acids of some strains of tobacco mosaic virus with ribonuclease T1. Biochim. Biophys. Acta 61:56-61.
Russell, G. E. 1962. Sugar beet mild yellowing virus: a persistent aphid transmitted virus. Nature 195:1231.
Ryder, E. J. 1964. Transmission of common lettuce mosaic virus through the gametes of the lettuce plants. Plant Dis. Reptr. 48:522-524.
Ryle, A. P., F. Sanger, L. F. Smith, and R. Kitai. 1955. The disulfide bonds of insulin. Biochem. J. 60:541-556.
Ryter, A., and E. Kellenberger. 1958. L'inclusion au polyester pour l'ultramicrotomie. J. Ultrastructure Res. 2:200-214.

Sabatini, D. D., K. Bensch, and R. J. Barnett. 1963. Cytochemistry and electron microscopy. The preservation of cellular ultrastructure and enzymic activity by aldehyde fixation. J. Cell Biol. 17:19-58.

Sabatini, D. D., F. Miller, and R. J. Barnett. 1964. Aldehyde fixation for morphological and enzyme histochemical studies with the electron microscope. J. Histochem. Cytochem. 12:57-71.

Sagaert, L. 1967. Comparative analysis of three analytical ultracentrifuges. Hung. Sci. Instr. 9:1-5.

Sahtiyanci, S. 1962. Studien über einige wurzelparasitare Olpidiaceen. Arch. Mikrobiol. 41:187-228.

Saito, Y. 1969. Hemagglutination of leafhopper-borne viruses, pp. 463-474. *In* K. Maramorosch [ed.] Viruses, vectors, and vegetation. Interscience, New York. 666 pp.

Saito, Y., and Y. Iwata. 1964. Hemagglutination test for titration of plant virus. Virology 22:426-428.

Saito, Y., K. Takanashi, Y. Iwata, and H. Okamoto. 1964. Studies on the soil-borne virus diseases of wheat and barley. IV. Persistence of virus in the soil and in the roots of diseased plants [in Japanese]. Bull. Nat. Inst. Agr. Sci., Japan, Ser. C. 17:61-74. (Quoted by Estes and Brakke, 1966).

Saito, N., and H. Werbin. 1969. Evidence for a DNA-photoreactivating enzyme in higher plants. Photochem. Photobiol. 9:389-393.

Sakai, F., and I. Takebe. 1970. RNA and protein synthesis in protoplasts isolated from tobacco leaves. Biochim. Biophys. Acta 224:531-540.

Sakurai, Y. 1969. Varietal resistance to stripe, dwarf, yellow dwarf, and black-streaked dwarf, pp. 257-275. *In* The Virus diseases of the rice plant. Johns Hopkins Press, Baltimore. 354 pp.

Salama, F. M., and C. M. Lyman. 1966. Electron microscopy and characterization of internal cork virus of the sweet potato. Phytopathology 56:898. (Abstr.)

Sammons, I. M., and M. Chessin. 1967. Cactus virus in the United States. Nature (London) 191:517-518.

Sanders, F. K. 1964. The infective nucleic acid of animal viruses, pp. 277-304. *In* R. J. C. Harris [ed.] Techniques in experimental virology. Academic Press, New York. 450 pp.

Sänger, H. L. 1961. Untersuchungen über schwer ubertragbare Formen des Rattle-virus. Proc. Conf. Potato Virus Dis. 4th, Braunschweig, Germany 1960. pp. 22-28.

Sänger, H. L. 1968. Characteristics of tobacco rattle virus. I. Evidence that its two particles are functionally defective and mutually complementing. Mol. Gen. Genet. 101:346-367.

Sänger, H. L., M. W. Allen, and A. H. Gold. 1962. Direct recovery of tobacco rattle virus from its nematode vector. Phytopathology 52:750. (Abstr.)

Sänger, H. L., and E. Brandenburg. 1961. Über die Gewinnung von infektiösen Pressaft aus "Wintertype" Pflanzen das Tabak-Rattle-virus durch Phenolextraction: Naturwissenschaften 48:391.

Sänger, H. L., and A. H. Gold. 1962. Transmission of unstable viruses from liquid-nitrogen frozen plant tissue. Phytopathology 52:750. (Abstr.)

Santilli, V., C. M. Nepokroeff, and N. C. Gagliardi. 1962. Susceptibility of pinto bean leaves to tobacco mosaic virus and its relationship to leaf ribonuclease content. Nature 193:656-658.

Santilli, V., J. Piacitelli, and J. H. Wu. 1961. The effect of tobacco mosaic virus protein on virus incubation period and infectivity. Virology 14:109-123.

Sardina, Rodriguez, J., F. Martinez Cordon, and M. Rubio-Huertos. 1948. Tincion de inclusiones en las virosis vegetales. Bot. de Patol Vegetal. y Ent. Agr. 16:311-322.

Sarkar, S. 1960. Interaction and mixed aggregation of proteins from tobacco mosaic virus strains. Z. Naturforsch. 15b:778-786.

Sartory, W. K. 1969. Instability in diffusing fluid layers. Biopolymers 7:251-263.

Sastry, K. S., and M. P. Gordon. 1966. The photodynamic inactivation of tobacco mosaic virus and its ribonucleic acid by acridine orange. Biochim. Biophys. Acta 129:32-41.

Sato, T., Y. Kyogoku, S. Higuchi, Y. Mitsui, Y. Iitaka, M. Tsubori, and K. Miura. 1966. A preliminary investigation on the molecular structure of rice dwarf virus ribonucleic acid. J. Mol. Biol. 16:180-190.

Sauer, N. I. 1966. Simultaneous association of stains of tobacco ringspot virus within *Xiphinema americanum*. Phytopathology 56:862-863.

Schachman, H. K. 1957. Ultracentrifugation, diffusion and viscometry, pp. 32-103. *In* S. P. Colowick and N. O. Kaplan [ed.] Methods in enzymology, vol. 4. Academic Press, New York. 6 vols.

Schachman, H. K. 1959. Ultracentrifugation in biochemistry. Academic Press, New York. 272 pp.

Schade, C. 1967. Untersuchungen zum serologischen Nachweis des Nekrotischen Ringfleckenvirus der Sauerkirsche. I. Der serologische Routinetest bei Steinobst. Phytopathol. Z. 59:352-371.

Scheele, R. B., and M. A. Lauffer. 1967. Acid-base titrations of tobacco mosaic virus and tobacco mosaic virus protein. Biochemistry 6:3076-3081.

Scherrer, K. 1969. Isolation and sucrose gradient analysis of RNA, pp. 413-432. *In* K. Habel and N. P. Salzman [ed.] Fundamental techniques in virology. Academic Press, New York. 527 pp.

Schindler, A. F. 1957. Parasitism and pathogenicity of *Xiphinema diversicaudatum*, anectoparasitic nematode. Nematologica 2:25-31.

Schindler, A. F., and A. J. Braun. 1957. Pathogenicity of an ectoparasitic nematode *Xiphinema diversicaudatum* on strawberries. Nematologica 2:91-93.

Schippers, B. 1963. Transmission of bean common mosaic virus by seed of *Phaseolus vulgaris* L. cultivar Beka. Acta Bot. Neerlandica 12:433-497.

Schito, G. C. 1966. A rapid procedure for the purification of bacterial viruses. Virology 30:157-159.

Schlegel, D. E. 1960*a*. Transmission of several plant viruses by phenol-water extracts of diseased tissues. Phytopathology 50:156-158.

Schlegel, D. E. 1960*b*. Highly infectious phenol extracts from tobacco leaves infected with cucumber mosaic virus. Virology 11:329-338.

Schlegel, D. E., and T. E. Rawlins. 1954. A screening test of the effect of organic compounds on production of tobacco mosaic virus. J. Bacteriol. 67:103-109.

Schmelzer, K. 1956*a*. Beitrage zur kenntnis der Virushemmstaffe in Cuscuta-Arten. Zentralbl. f. Bakteriol., Parasit., Infekt., u. Hygiene 2:482-515.

Schmelzer, K. 1956*b*. Beitrage zur kenntnis der Ubertragbarkeit von viren durch Cuscuta-Arten. Phytopathol. Z. 28:1-56.

Schmelzer, K., R. Bartels, and M. Klinkowski. 1960. Interferenzen zwischen den Viren der Tabakatzmosaik-Gruppe. Phytopathol. Z. 40:52-74.

Schmid, C. W., and J. E.Hearst. 1969. Molecular weights of homogeneous coliphage DNA's from density-gradient sedimentation equilibrium. J. Mol. Biol. 44:143-160.

Schmutterer, H., and P. Ehrahardt. 1964. Untersuchungen uber die Beziehungen zwischen dem Enationen-virus der Erbse und seinen Vektoren II. Versuche zur Ubertragung des Virus durch Injektion von Haemolymphe in verschiedene Aphidenarten. Phytopathol. Z. 50:80-85.

Schneider, F., U. Beiss, and R. Marx. 1964. Experiments on the purification and concentration of the sugar beet. Zucker 17:539-543.

Schneider, I. R. 1965. Introduction, translocation, and distribution of viruses in plants. Advances Virus Res. 11:163-221.

Schneider, I. R., and T. O. Diener. 1968. In vivo and in vitro decline of specific infectivity of tobacco ringspot virus correlated with nucleic acid degradation. Virology 35:150-157.

Scholes, G., J. F. Ward, and J. J. Weiss. 1960. Mechanism of radiation-induced degradation of nucleic acids. J. Mol. Biol. 2:379-391.

Schramm, G. 1947. Uber die spaltung des Tabakmosaikvirus und die Wiedervereinigung der Spaltstücke zu höhermolekularen proteinen. I. Die Spaltungsreaktion. Z. Naturforsch. 2b:112-121.

Schramm, G., and L. Rebensburg. 1942. The comparative characterization of mutants of the tobacco mosaic virus. Naturwissenschaften 30:48-51.

Schramm, G., and B. Röttger. 1959. Untersuchungen über das Tabakmosaikvirus mit fluoreszierenden Antikorpern. Z. Naturforsch. 14b:510-515.

Schramm, G., G. Schumaker, and W. Zillig. 1955. The structure of tobacco mosaic virus: III. Degradation in alkaline solutions. Z. Naturforsch. 10b:481-492.

Schramm, G., and W. Zillig. 1955. Uber die Struktur des Tabakmosaikvirus. IV. Mitt: Die Reaggregation des nucleinsäure freien proteins. Z. Naturforsch. 10b:493-499.

Schumaker, V. N. 1967. Zone centrifugation, pp. 245-339. *In* Advances in biological and medical physics, vol. 11. Academic Press, New York.

Schumaker, V. N., and J. Rosenbloom. 1965. Fundamental mass-transport equations for zone sedimentation velocity. Biochemistry 4:1005-1011.
Schuster, H. 1961. The reaction of tobacco mosaic virus ribonucleic acid with hydroxylamine. J. Mol. Biol. 3:447-457.
Schuster, H. 1964. The infective nucleic acids of plant viruses, pp. 49-78. *In* R. J. C. Harris [ed.] Techniques in experimental virology. Academic Press, New York. 450 pp.
Schuster, H., and G. Schramm. 1958. The effect of chemical changes on the activity of the ribonucleic acid of the tobacco mosaic virus. Z. Naturforsch. 13b:697-704.
Schuster, H., and R. C. Wilhelm. 1963. Reaction differences between tobacco mosaic virus and its free ribonucleic acid with nitrous acid. Biochim. Biophys. Acta 68:554-560.
Schuster, H., and H. G. Wittmann. 1963. The inactivating and mutagenic action of hydroxylamine on tobacco mosaic virus ribonucleic acid. Virology 19:421-430.
Scott, H. A. 1961. Serological detection of barley stripe mosaic virus in single seeds and dehydrated leaf tissues. Phytopathology 51:200-201.
Scott, H. A. 1963. Purification of cucumber mosaic virus. Virology 20:103-106.
Scott, H. A., R. P. Kahn, J. Bozicevich, and M. M. Vincent. 1964. Detection of potato virus X by the bentonite flocculation test. Phytopathology 54:1292-1293.
Scott, H. A., M. Vincent, and W. J. Zaumeyer. 1951. Serological studies of red node and pod mottle viruses. Phytopathology 51:755-758.
Sebanek, J., and J. Hradilik. 1969. Mechanical transmission of beet yellows virus to *Chenopodium quinoa* Willd. and *Chenopodium foliosum* (Moench) Asch. Biol. Plantarum 11:366-369.
Sedat, J., A. Lyon, and R. L. Sinsheimer. 1969. Purification of *Escherichia coli* pulse-labeled RNA by benzoylated DEAE-cellulose chromatography. J. Mol. Biol. 44:415-434.
Seecof, R. L. 1964. Deleterious effects on *Drosophila* development associated with the sigma virus infection. Virology 22:142-148.
Seecof, R. L. 1968. The sigma virus infection of *Drosophila melanogaster*. Current Topics Microbiol. Immunol. 42:59-93.
Sehgal, O. P. 1966. Host range, properties and partial purification of a Missouri isolate of maize dwarf mosaic virus. Plant Dis. Reptr. 50:862-866.
Sehgal, O. P. 1968. Purification, properties and structure of maize dwarf mosaic virus. Phytopathol. Z. 62:232-250.
Sehgal, O. P., and G. F. Krause. 1968. Efficiency of nitrous acid as an inactivating and mutagenic agent of intact tobacco mosaic virus and its isolated nucleic acid. J. Virol. 2:966-971.
Seinhorst, J. W. 1959. A rapid method for the transfer of nematodes from fixative to anhydrous glycerin. Nematologica 4:67-69.
Seinhorst, J. W. 1962. Modifications of the elutriation method for extracting nematodes from soil. Nematologica 8:117-128.
Seinhorst, J. W. 1966. *Longidorus elongatus* on *Fragaria vesca*. Nematologica 12:275-279.
Sela, I., I. Harpaz, and Y. Birk. 1965. Suppression of virus infectivity in diseased plant tissue following treatment with an antiviral factor from virus-infected plants. Virology 25:80-82.
Sela, I., I. Harpaz, and Y. Birk. 1966. Identification of the active component of an antiviral factor isolated from virus-infected plants. Virology 28:71-78.
Selman, I. W. 1964. The effect of kinetin on infection of petunia and tomato leaves with tomato spotted-wilt virus. Ann. Appl. Biol. 53:67-76.
Selman, I. W., and R. G. Milne. 1961. A new local lesion assay method for tomato spotted wilt virus, with a note on cyclical changes in infectivity. Plant Pathol. 10:100-104.
Semal, J. 1966. Study methods for specific virus inhibitors in plant virology. Mededel. Landbouwhogeschool Opzoekingsta Gent. 31:950-955.
Semancik, J. S. 1966*a*. Purification and properties of two isolates of tobacco rattle virus from peppers in California. Phytopathology 56:1190-1193.
Semancik, J. S. 1966*b*. Studies on electrophoretic heterogeneity in isometric plant viruses. Virology 30:698-704.
Semancik, J. S., and J. B. Bancroft. 1965. Stability differences between the nucleoprotein components of bean pod mottle virus. Virology 27:476-483.
Semancik, J. S., and P. A. Reynolds. 1969. Assembly of protein and nucleoprotein particles from extracted tobacco rattle virus protein and RNA. Science 164:559-560.

Sengbusch, P. von. 1965. Aminosaureaustausche und Tertiarstruktur eines Proteins. Verleich von Mutanten des Tabakmosaikvirus mit serologischen und physikochemischen Methoden. Z. Vererbungsl. 96:364-386.

Sengbusch, P. von, and H. G. Wittmann. 1965. Serological and physicochemical properties of the wild strains and two mutants of tobacco mosaic virus with the same amino acid exchange in different positions of the protein chain. Biochem. Biophys. Res. Commun. 18:780-787.

Seth, M. L. 1962. Transmission of pigeon pea sterility by an eriophyid mite. Indian Phytopathol. 15:225-227.

Seth, M. L. 1965. Further observations and studies on pigeon pea sterility. Indian Phytopathol. 18:317-319.

Setlow, R. B., and J. K. Setlow. 1962. Evidence that ultraviolet-induced thymine dimers in DNA cause biological damage. Proc. Nat. Acad. Sci. U.S. 48:1250-1257.

Severin, H. H. P. 1924. Curly leaf transmission experiments. Phytopathology 14:80-93.

Severin, H. H. P. 1945. Evidence of nonspecific transmission of California aster-yellows virus by leafhoppers. Hilgardia 17:21-59.

Severin, H. H. P. 1946. Longevity, or life histories, of leafhopper species on virus-infected and on healthy plants. Hilgardia 17:121-137.

Severin, H. H. P. 1950. Spittle-insect vectors of Pierce's disease virus. II. Life history and virus transmission. Hilgardia 19:357-382.

Severin, H. H. P., and J. H. Freitag. 1933. Some properties of the curly-top virus. Hilgardia 8:1-48.

Severin, H. H. P., F. D. Horn, and N. W. Frazier. 1945. Certain symptoms resembling those of curly top or aster yellows, induced by saliva of *Xerophloea vanduzeei*. Hilgardia 16:337-350.

Severin, H. H. P., and O. Swezy. 1928. Filtration experiments on curly top of sugar beets. Phytopathology 18:681-690.

Shainoff, J. R., and M. A. Lauffer. 1956. Chromatographic purification of southern bean mosaic virus. Arch. Biochem. Biophys. 64:315-318.

Shalla, T. A. 1964. Assembly and aggregation of tobacco mosaic virus in tomato leaflets. J. Cell Biol. 21:253-264.

Shalla, T. A., and A. Amici. 1967. The distribution of viral antigen in cells infected with tobacco mosaic virus as revealed by electron microscopy. Virology 31:78-91.

Shalla, T. A., and J. F. Shepard. 1970. An antigenic analysis of potato virus X and of its degraded protein. II. Evidence for a conformational change associated with the depolymerization of structural protein. Virology 42:835-847.

Sharma, R. D. 1965. Direct damage to strawberry by *Longidorus elongatus* (deMan, 1876). Thorne and Swanger, 1936. Mededel. Landbouwhogeschool Gent. 30:1437-1443.

Sharp, D. G. 1953. Purification and properties of animal viruses. Advances Virus Res. 1:277-313.

Shatkin, A. J. 1968. Viruses containing double-stranded RNA, pp. 351-392. *In* H. Fraenkel-Conrat [ed.] Molecular basis of virology. Van Nostrand Reinhold, New York. 656 pp.

Shaw, J. G. 1969. In vivo removal of protein from tobacco mosaic virus after inoculation of tobacco leaves. II. Some characteristics of the reaction. Virology 37:109-116.

Sheffield, F. M. 1931. The formation of intracellular inclusions in solanaceous hosts infected with aucuba mosaic of tomato. Ann. Appl. Biol. 18:471-493.

Sheffield, F. M. 1934. Experiments bearing on the nature of intracellular inclusions in plant virus diseases. Ann. Appl. Biol. 21:440-453.

Sheffield, F. M. 1939. Micrurgical studies on virus-infected plants. Proc. Roy. Soc. Lond. B. 126:529-538.

Sheffield, F. M. 1941. The cytoplasmic and nuclear inclusions associated with severe etch virus. J. Roy. Microscop. Soc. 61:30-45.

Shepard, J. F. 1968. Electron microscopy of subtilisin-treated tobacco etch virus nuclear and cytoplasmic inclusions. Virology 36:20-29.

Shepard, J. F. 1970. A radial immunodiffusion test for the simultaneous diagnosis of potato viruses S and X. Phytopathology 60:1669-1671.

Shepherd, R. J. 1964. Properties of a mosaic disease of cowpea and its relationship to bean pod mottle virus. Phytopathology 54:466-473.

Shepherd, R. J. 1965. Properties of a mosaic virus of corn and Johnson grass and its relation to the sugarcane mosaic virus. Phytopathology 55:1250-1256.
Shepherd, R. J., G. E. Bruening, and R. J. Wakeman. 1970. Double-stranded DNA from cauliflower mosaic virus. Virology 41:339-347.
Shepherd, R. J., and R. W. Fulton. 1962. Identity of a seed-borne virus of cowpea. Phytopathology 52:489-493.
Shepherd, R. J., and Q. L. Holdeman. 1965. Seed transmission of the Johnson grass strain of the sugarcane mosaic virus in corn. Plant Dis. Reptr. 49:468-469.
Shepherd, R. J., R. J. Wakeman, and S. A. Ghabrial. 1968. Preparation and properties of the protein and nucleic acid components of pea enation mosaic virus. Virology 35:255-267.
Shikata, E. 1962. Observation of rice dwarf virus by means of an electron microscope. Plant Protection 16:313-318.
Shikata, E. 1966. Electron microscopic studies on plant viruses. J. Fac. Agr. Hokkaido Univ. 55 (Part I):1-110.
Shikata, E. 1969. Electron microscopic studies of rice viruses, pp. 223-240. *In* Virus diseases of the rice plant. Johns Hopkins Press, Baltimore. 354 pp.
Shikata, E., and M. Chen. 1969. Electron microscopy of rice transitory yellowing virus. J. Virol. 3:261-264.
Shikata, E., and G. E. Gálvez. 1969. Fine flexuous threadlike particles in cells of plants and insect hosts infected with rice hoja blanca virus. Virology 39:635-641.
Shikata, E., and Y. Lu. 1967. Electron microscopy of northern cereal mosaic virus in Japan. Proc. Japan Acad. 43:918-923.
Shikata, E., and K. Maramorosch. 1965. Electron microscopic evidence for the systemic invasion of an insect host by a plant pathogenic virus. Virology 27:461-475.
Shikata, E., and K. Maramorosch. 1966. Electron microscopy of pea enation mosaic virus in plant cell nuclei. Virology 30:439-454.
Shikata, E., and K. Maramorosch. 1967. Electron microscopy of wound tumor virus assembly sites in insect vectors and plants. Virology 32:363-377.
Shikata, E., and K. Maramorosch. 1969. Electron microscopy of insect-borne viruses in situ, pp. 393-415. *In* K. Maramorosch [ed.] Viruses, vectors, and vegetation. Interscience, New York. 666 pp.
Shikata, E., K. Maramorosch, and R. G. Manadis. 1966. Electron microscopy of pea enation mosaic virus in plants and aphid vectors. Virology 29:426-436.
Shinkai, A. 1956. Differences in rice dwarf virus transmitting ability of leafhoppers from various localities. Ann. Phytopathol. Soc. Japan 21:127. (Abstr.)
Shipp, W., and R. Haselkorn. 1964. Double-stranded RNA from tobacco leaves infected with TMV. Proc. Nat. Acad. Sci. U.S. 52:401-408.
Sibatani, A. 1970. Precipitation and counting of minute quantities of labeled nucleic acids as cetyltrimethylammonium salt. Anal. Biochem. 33:279-285.
Siegel, A. 1959. Mutual exclusion of strains of tobacco mosaic virus. Virology 8:470-477.
Siegel, A. 1960. Studies on the induction of tobacco mosaic virus mutants with nitrous acid. Virology 11:156-167.
Siegel, A. 1965. Artificial production of mutants of tobacco mosaic virus. Advances Virus Res. 11:25-60.
Siegel, A., W. Ginoza, and S. G. Wildman. 1957. The early events of infection with tobacco mosaic virus nucleic acid. Virology 3:554-559.
Siegel, A., G. J. Hills, and R. Markham. 1966. In vitro and in vivo aggregation of the defective PM2 tobacco mosaic virus protein. J. Mol. Biol. 19:140-144.
Siegel, A., and W. Hudson. 1959. Equilibrium centrifugation of two strains of tobacco mosaic virus in density gradients. Biochim. Biophys. Acta 34:254-255.
Siegel, A., and A. Norman. 1958. Action spectra for two strains of tobacco mosaic virus. Virology 6:725-731.
Siegel, A., and S. G. Wildman. 1954. Some natural relationships among strains of tobacco mosaic virus. Phytopathology 44:277-282.
Siegel, A., and S. G. Wildman. 1956. The inactivation of the infectious centers of tobacco mosaic virus by ultraviolet light. Virology 2:69-82.
Siegel, A., and M. Zaitlin. 1964. Infection process in plant virus diseases. Annu. Rev. Phytopathol. 2:179-202.

Siegel, A., and M. Zaitlin. 1965. Defective plant viruses, pp. 113-125. *In* M. Pollard [ed.] Perspectives in virology, vol. 4. Hoeber, New York. 2 vols.

Siegel, A., M. Zaitlin, and O. P. Sehgal. 1962. The isolation of defective tobacco mosaic virus strains. Proc. Nat. Acad. Sci. U.S. 48:1845-1851.

Siegel, B. M. 1964. Modern developments in electron microscopy. Academic Press, New York. 432 pp.

Silberschmidt, K., and J. Caner. 1967. Physiological factors responsible for a specific form of resistance against plant virus infection. Arq. Inst. Biol. S. Paulo 34:127-133.

Sill, W. H., Jr., and J. C. Walker. 1952. A virus inhibitor in cucumber in relation to mosaic resistance. Phytopathology 42:349-352.

Silva, D. M., and C. A. Knight. 1960. Peptic digests of the I-peptide fractions of five strains of tobacco mosaic virus. Virology 12:589-595.

Silverstein, S. C., and S. Dales. 1968. The penetration of reovirus RNA and initiation of its genetic function in L-strain fibroblasts. J. Cell Biol. 36:197-230.

Simon, M. I., and H. van Venakis. 1962. The photodynamic reaction of methylene blue with deoxyribonucleic acid. J. Mol. Biol. 4:488-499.

Simons, J. N. 1954. Vector-virus relationships of pea-enation mosaic and the pea aphid *Macrosiphum pisi* (Kalt.). Phytopathology 44:283-289.

Simons, J. N. 1962. The pseudo-curly top disease in South Florida. J. Econ. Entomol. 55:358-363.

Simons, J. N., and D. M. Coe. 1958. Transmission of pseudo-curly top virus in Florida by a treehopper. Virology 6:43-48.

Simpson, R. W., and R. E. Hauser. 1966. Structural components of vesicular stomatitis virus. Virology 29:654-667.

Singer, B., and H. Fraenkel-Conrat. 1962. Enzyme resistance of complexes of ribonucleic acid with metals. Biochemistry 1:852-858.

Singer, B., and H. Fraenkel-Conrat. 1965. Effects of light in the presence of iron salts on ribonucleic acid and model compounds. Biochemistry 4:226-233.

Singer, B., and H. Fraenkel-Conrat. 1966. Dye catalyzed photoinactivation of tobacco mosaic virus ribonucleic acid. Biochemistry 5:2446-2450.

Singer, B., and H. Fraenkel-Conrat. 1967. Chemical modification of viral RNA, VI. The action of N-methyl-N′ nitro-N-nitrosoguanidine. Proc. Nat. Acad. Sci. U.S. 58:234-239.

Singer, B., and H. Fraenkel-Conrat. 1969*a*. Chemical modification of viral ribonucleic acid. VII. The action of methylating agents and nitrosoguanidine on polynucleotides including tobacco mosaic virus ribonucleic acid. Biochemistry 8:3260-3266.

Singer, B., and H. Fraenkel-Conrat. 1969*b*. Chemical modification of viral ribonucleic acid. VIII. The chemical and biological effects of methylating agents and nitrosoguanidine on tobacco mosaic virus. Biochemistry 8:3266-3269.

Singer, J. M., and C. M. Plotz. 1956. The latex fixation test. I. Application to the serologic diagnosis of rheumatoid arthritis. Amer. J. Med. 21:888-982.

Singer, S. J., J. G. Bald, S. G. Wildman, and R. D. Owen. 1951. The detection and isolation of naturally occurring strains of tobacco mosaic virus by electrophoresis. Science 114:463-465.

Singer, S. J., and A. F. Schick. 1961. The properties of specific stains for electron microscopy prepared by the conjugation of antibody molecules with ferritin. J. Biophys. Biochem. Cytol. 9:519-537.

Singh, G. P., D. C. Arny, and G. S. Pound. 1960. Studies on the stripe mosaic of barley, including effects of temperature and age of host on disease development and seed infection. Phytopathology 50:290-296.

Singh, R. P., A. P. Benson, and F. M. Salama. 1966. Purification and electron microscopy of potato spindle tuber virus. Phytopathology 56:901-902. (Abstr.)

Sinha, R. C. 1960. Comparison of the ability of nymph and adult *Delphacodes pellucida* Fabricius to transmit European wheat striate mosaic virus. Virology 10:344-352.

Sinha, R. C. 1965*a*. Recovery of potato yellow dwarf virus from hemolymph and internal organs of an insect vector. Virology 27:118-119.

Sinha, R. C. 1965*b*. Sequential infection and distribution of wound-tumor virus in the internal organs of a vector after ingestion of virus. Virology 26:673-686.

Sinha, R. C. 1967. Response of wound-tumor virus infection in insects to vector age and temperature. Virology 31:746-748.

Staples, R., and W. B. Allington. 1956. Streak mosaic of wheat in Nebraska and its control. Nebraska Univ. Agr. Exp. Sta. Res. Bull. 178. 41 pp.

Steere, R. L. 1956. Purification and properties of tobacco ringspot virus. Phytopathology 46:60-69.

Steere, R. L. 1957. Electron microscopy of structural detail in frozen biological specimens. J. Biophys. Biochem. Cytol. 3:45-60.

Steere, R. L. 1959. The purification of plant viruses. Advances Virus Res. 6:1-73.

Steere, R. L. 1963. Tobacco mosaic virus: purifying and sorting associated particles according to length. Science 140:1089-1090.

Steere, R. L. 1964. Purification, pp. 211-234. *In* M. K. Corbett and H. D. Sisler [ed.] Plant virology. Univ. Florida Press, Gainesville. 527 pp.

Steere, R. L. 1967. Gel filtration of aster yellows virus. Phytopathology 57:832-833. (Abstr.)

Steere, R. L., and G. K. Ackers. 1962*a*. Purification and separation of tobacco mosaic virus and southern bean mosaic virus by agar-gel filtration. Nature 194:114-116.

Steere, R. L., and G. K. Ackers. 1962*b*. Restricted-diffusion chromatography through calibrated columns of granulated agar gel: a simple method for particle-size determination. Nature 196:475-476.

Steere, R. L., and R. C. Williams. 1953. Identification of crystalline inclusion bodies extracted intact from plant cells infected with tobacco mosaic virus. Amer. J. Bot. 40:81-84.

Stegwee, D., and M. B. Ponsen. 1958. Multiplication of potato leaf roll virus in the aphid *Myzus persicae* (Sulz.). Entomol. Exp. Appl. 1:291-300.

Steinberg, I. Z., and H. K. Schachman. 1966. Ultracentrifugation studies with adsorption optics. V. Analysis of interacting systems involving macromolecules and small molecules. Biochemistry 5:3728-3747.

Stevens, C. L., and M. A. Lauffer. 1965. Polymerization depolymerization of tobacco mosaic virus protein. IV. The role of water. Biochemistry 4:31-37.

Storey, H. H. 1932. The inheritance by an insect vector of the ability to transmit a plant virus. Proc. Roy. Soc. (London), Ser. B. 112:46-60.

Storey, H. H. 1933. Investigations of the mechanism of the transmission of plant viruses by insect vectors. I. Proc. Roy. Soc. (London), Ser. B. 113:463-485.

Storey, H. H. 1938. Investigations of the mechanism of the transmission of plant viruses by insect vectors. II. The part played by puncture in transmissions. Proc. Roy. Soc. (London), Ser. B. 125:455-477.

Storey, H. H. 1939. Investigations of the mechanism of the transmission of plant viruses by insect vectors. III. The insect's saliva. Proc. Roy. Soc. (London), Ser. B. 127:526-643.

Storey, H. H., and A. K. Ryland. 1955. Transmission of groundnut rosette virus. Ann. Appl. Biol. 43:423-432.

Strauss, J. H., Jr. 1964. A modified method of MS2-RNA Assay. J. Mol. Biol. 10:422.

Strauss, J. H., Jr., and R. L. Sinsheimer. 1967. Characterization of an infectivity assay for the ribonucleic acid of bacteriophage MS2. J. Virol. 1:711-716.

Strauss, J. H., Jr., and R. L. Sinsheimer. 1968. Initial kinetics of degradation of MS2 ribonucleic acid by ribonuclease, heat and alkali and the presence of configurational restraints in this ribonucleic acid. J. Mol. Biol. 34:453-465.

Streeter, D. G., and M. P. Gordon. 1966. Separation of U1 and U2 strains of tobacco mosaic virus by continuous freeflow electrophoresis. Phytopathology 56:419-421.

Streissle, G., and R. R. Granados. 1968. The fine structure of wound tumor and reovirus. Arch. Ges. Virusforsch. 25:369-372.

Streissle, G., and K. Maramorosch. 1963. Reovirus and wound-tumor virus: serological cross reactivity. Science 140:996-997.

Streissle, G., L. Rosen, and T. Tokumitsu. 1968. Host specificity of wound tumor virus and reovirus and their serologic relationships. Arch. Ges. Virusforsch. 22:409-416.

Strohmaier, K. 1967. Virus concentration by ultrafiltration, pp. 245-274. *In* K. Maramorosch and H. Koprowski [ed.] Methods in virology, vol. 2. Academic Press, New York. 4 vols.

Stubbs, J. D., and P. Kaesberg. 1964. A protein subunit of bromegrass mosaic virus. J. Mol. Biol. 8:314-323.

Studier, F. W. 1965. Sedimentation studies of the size and shape of DNA. J. Mol. Biol. 11:373-390.

Studier, F. W. 1969. Conformational changes of single-stranded DNA. J. Mol. Biol. 41:189-197.

Sueoka, N. 1959. A statistical analysis of deoxyribonucleic acid distribution in density-gradient centrifugation. Proc. Nat. Acad. Sci. U.S. 45:1480-1490.

Sugiyama, T. 1966. Tobacco mosaic virus like rods formed by "mixed reconstitution" between MS2 ribonucleic acid and tobacco mosaic virus protein. Virology 28:488-492.

Sugiyama, T., and H. Fraenkel-Conrat. 1963. The end-groups of tobacco mosaic virus RNA. II. Nature of the 3′-linked chain end in TMV and of both ends in four strains. Biochemistry 2:332-334.

Sugiyama, T., R. R. Herbert, and R. A. Hartman. 1967. Ribonucleoprotein complexes formed between bacteriophage MS2 RNA and MS2 protein in vitro. J. Mol. Biol. 25:455-463.

Sukhov, K. S. 1943*a*. A purified protein preparation of winter wheat mosaic virus. Compt. Rend. (Dokl.) Acad. Sci. USSR 27:377-379.

Sukhov, K. S. 1943*b*. Proof of infectiousness of a purified protein preparation of oat mosaic virus (Zakuklivanie in oats). Compt. Rend. (Dokl.) Acad. Sci. USSR 40:167-169.

Sukhov, K. S., and M. N. Sukhova. 1940. Interrelations between the virus of a new grain mosaic disease (Zakuklivanie) and its carrier *Delphax striatella* Fallen. Compt. Rend. (Dokl.) Acad. Sci. USSR 26:479-482.

Sukhov, K. S., A. M. Vovk, and T. S. Alexeeva. 1943. Purified protein preparation from the virus of oat mosaic (Zakuklivanie). Compt. Rend. (Dokl.) Acad. Sci. USSR 41:344-346.

Süss, R., E. Sander, B. Röttger, and H. Senger. 1965. Versuche zur vermehrung infektioser ribonukleinsäure aus Tabakmosaikvirus in *Chlorella*. Biochim. Biophys. Acta 95:388-397.

Sutic, D., and B. Djordjevic. 1964. Effect of 5-fluorouracil on antigenic properties of tobacco mosaic virus. Nature 203:434-435.

Suzuki, N. 1969. Purification of single and double-stranded vector-borne RNA viruses, pp. 557-578. *In* K. Maramorosch [ed.] viruses, vectors, and vegetation. Interscience, New York. 666 pp.

Suzuki, N., and I. Kimura. 1969. Purification, bioassay, properties, and serology of rice viruses, pp. 207-221. *In* The virus diseases of the rice plant. Johns Hopkins Press, Baltimore. 354 pp.

Svedberg, T., and K. O. Pedersen. 1940. The ultracentrifuge. Clarendon Press, Oxford. (Also reprinted by the Johnson Reprint Corp., N.Y.) 478 pp.

Svensson, H. 1960. Zonal density-gradient electrophoresis, pp. 195-244. *In* P. Alexander and R. J. Block [ed.] A laboratory manual of analytical methods in protein chemistry including polypeptides, vol. 1. Pergamon Press, London. 3 vols.

Svensson, H., and T. E. Thompson. 1961. Translational diffusion methods in protein chemistry, pp. 58-118. *In* P. Alexander and R. J. Block [ed.] A laboratory manual of analytical methods in protein chemistry, including polypeptides, vol. 3. Pergamon Press, London. 3 vols.

Svensson, H., and E. Valmet. 1955. Density-gradient electrophoresis — a new method of separating electrically charged compounds. Science Tools 2:11-12.

Swenson, K. G. 1967. Plant virus transmission by insects, pp. 267-307. *In* K. Maramorosch and H. Koprowski [ed.] Methods in virology, vol. 1. Academic Press, New York. 4 vols.

Swenson, K. G. 1969. Plant susceptibility to virus infection by insect transmission, pp. 143-157. *In* K. Maramorosch [ed.] Viruses, vectors, and vegetation. Interscience, New York. 666 pp.

Swenson, K. G., and J. A. Milbrath. 1964. Insect and mite transmission tests with the *Prunus* ringspot virus. Phytopathology 54:399-404.

Swezy, O., and H. H. P. Severin. 1930. A rickettsialike microorganism in *Eutettix tenellus* (Baker), the carrier of curly top of sugar beets. Phytopathology 20:169-178.

Swingle, R. U. 1942. Phloem necrosis — a virus disease of the American elm. U.S. Dep. Agr. Circ. 640, 8 pp.

Sylvester, E. S. 1958. Latent period phenomena in transmission of sugar beet yellow-net virus by green peach aphids. J. Econ. Entomol. 51:812-818.

Sylvester, E. S. 1965. The latent period of pea-enation mosaic virus in the pea aphid, *Acyrtosiphon pisum* (Harris)—an approach to its estimation. Virology 25:62-67.

Sylvester, E. S. 1967. Retention of inoculativity in the transmission of pea enation mosaic virus by pea aphids as associated with virus isolates, aphid reproduction and excretion. Virology 32:524-531.

Sylvester, E. S. 1969. Virus transmission by aphids — a viewpoint, pp. 159-173. *In* K. Maramorosch [ed.] Viruses, vectors, and vegetation. Interscience, New York. 666 pp.

Sylvester, E. S., and R. H. Bradley. 1962. Effect of treating stylets of aphids with formalin on the transmission of sugarbeet yellow virus. Virology 17:381-386.

Sylvester, E. S., and J. Richardson. 1964. Transmission of cabbage mosaic virus by green peach aphids. Stylet transmission studies. Virology 22:520-538.

Sylvester, E. S., and J. Richardson. 1966. Recharging pea aphids with pea enation mosaic virus. Virology 30:592-597.

Symington, J., B. Commoner, and M. Yamada. 1962. Linear biosynthesis of tobacco mosaic virus: evidence that short virus rods are natural products of TMV biosynthesis. Proc. Nat. Acad. Sci. U.S. 48:1675-1683.

Symons, R. H., M. W. Rees, M. N. Short, and R. Markham. 1963. Relationship between the ribonucleic acid and protein of some plant viruses. J. Mol. Biol. 6:1-15.

Szybalski, W. 1967. Effects of elevated temperatures on DNA and on some polynucleotides: denaturation, renaturation and cleavage of glycosidic and phosphate ester bonds, pp. 73-122. *In* A. H. Rose [ed.] Thermobiology. Academic Press, New York. 653 pp.

Szybalski, W. 1968. Use of cesium sulfate for equilibrium density-gradient centrifugation, pp. 330-360. *In* L. Grossman and K. Moldave [ed.] Methods in enzymology, part II, vol. 12. Academic Press, New York. 19 vols.

Tabor, D., and R. H. S. Winterton. 1968. Surface forces: direct measurement of normal and retarded van der Waals forces. Nature 219:1120-1121.

Taconis, P. J., and K. Kuiper. 1963. Overdracht van het *Nicotiana* virus 5 door aaltjes van het geslacht *Trichodorus* in zaailingen van 5 ges wassen. Versl. Meded. Plziektenk. Dienst. Wageningen. 141:177-178.

Takahashi, W. N. 1955. Anomalous proteins associated with three strains of tobacco mosaic virus. Virology 1:393-396.

Takahashi, W. N. 1956. Increasing the sensitivity of the local-lesion method of virus assay. Phytopathology 46:654-656.

Takahashi, W. M. 1959. The role of an anomalous noninfectious protein in virus synthesis. Virology 9:437-445.

Takahashi, W. N. 1964. The biochemistry of virus infection, pp. 315-329. *In* M. K. Corbett and H. D. Sisler [ed.] Plant virology. Univ. Florida Press, Gainesville. 527 pp.

Takahashi, W. N., and M. Ishii. 1952. An abnormal protein associated with tobacco mosaic virus infection. Nature 169:419-420.

Takahashi, W. N., and M. Ishii. 1953. A macromolecular protein associated with tobacco mosaic virus infection: its isolation and proteins. Amer. J. Bot. 40:85-90.

Takahashi, Y., and I. Sekiya. 1962. Adipose tissue of the green rice leafhopper, *Nephotettix cincticeps* Uhler, infected with the virus of the yellow dwarf disease of the rice plant. Jap. J. Appl. Entomol. Zool. 6:90-94.

Takebe, I., and Y. Otsuki. 1969. Infection of tobacco mesophyll protoplasts by tobacco mosaic virus. Proc. Nat. Acad. Sci. U.S. 64:843-848.

Takebe, I., Y. Otsuki, and S. Aoki. 1968. Isolation of tobacco mesophyll cells in intact and active state. Plant and Cell Physiol. (Tokyo) 9:115-124.

Tanford, C. 1961. Physical chemistry of macromolecules. Wiley, New York. 710 pp.

Tanford, C. 1968. Protein denaturation. Advances Protein Chem. 23:121-282.

Tanford, C. 1970. Protein denaturation. Part C. Theoretical models for the mechanism of denaturation. Advances Protein Chem. 24:2-95.

Taniguchi, T. 1962*a*. A rapid method for microanalytical determination of the amount of tobacco mosaic virus in plant tissues. Nature 194:708. (2)

Taniguchi, T. 1962*b*. Determination of the limiting amount of tobacco mosaic virus attaching to ion exchange resin and the dissociation constant of their complex. Virology 18:646-647. (13)

Taniguchi, T. 1964. The accumulation in plants of chromatographically separated components of tobacco mosaic virus. Virology 22:245-252.

Taylor, C. E. 1962. Transmission of raspberry ringspot virus in *Longidorus elongatus* (deMan) (Nematoda: Dorylaimoidea). Virology 17:493-494.

Taylor, C. E. 1964. Nematode-borne viruses. Rep. Scot. Hort. Res. Inst. 11:64-66.

Taylor, C. E. 1967. The multiplication of *Longidorus elongatus* (deMan) on different host plants with reference to virus transmission. Ann. Appl. Biol. 59:275-281.

Taylor, C. E. 1968*a*. The association of ringspot viruses with their nematode vectors. Compt. Rend. 8th Symp. Int. Nematologie, Antibes 1965. pp. 109-110.

Taylor, C. E. 1968*b*. Association of virus and vectors. Rep. Scot. Hort. Res. Inst. 14:66.

Taylor, C. E. 1970. The association of *Longidorus elongatus* with raspberry ringspot and tomato black ring viruses. Proc. 9th Int. Nematol. Symp., Warsaw, 1967. pp. 283-289.

Taylor, C. E., and A. F. Murant. 1966. The use of quintozene (PCNB) as a nematocide. Proc. 3rd Brit. Insect. and Fungicide Conf. 1965. pp. 514-520.

Taylor, C. E., and A. F. Murant. 1968. Chemical control of raspberry ringspot and tomato black ring viruses in strawberry. Plant Pathol. 17:171-178.

Taylor, C. E., and A. F. Murant. 1969. Transmission of strains of raspberry ringspot and tomato black ring viruses by *Longidorus elongatus* (deMan). Ann. Appl. Biol. 64:43-48.

Taylor, C. E., and D. J. Raski. 1964. On the transmission of grape fanleaf by *Xiphinema index*. Nematologica 10:489-495.

Taylor, C. E., and W. M. Robertson. 1969. The location of raspberry ringspot and tomato black ring viruses in the nematode vector, *Longidorus elongatus* (deMan). Ann. Appl. Biol. 64:233-237.

Taylor, C. E., and W. M. Robertson. 1970*a*. Location of tobacco rattle virus in the nematode vector, *Trichodorus pachydermus* Seinhorst. J. Gen. Virol. 6:179-182.

Taylor, C. E., and W. M. Robertson. 1970*b*. Electron microscopy. Rep. Scot. Hort. Inst. 16 (in press).

Taylor, C. E., and P. R. Thomas. 1968. The association of *Xiphinema diversicaudatum* (Micoletzky) with strawberry latent ringspot and arabis mosaic viruses in a raspberry plantation. Ann. Appl. Biol. 62:147-157.

Taylor, C. E., P. R. Thomas, W. M. Robertson, and I. M. Roberts. 1970. An electron microscope study of the oesophageal region of *Longidorus elongatus* (deMan). Nematologica 16:6-12.

Taylor, R. H., R. G. Grogan, and K. A. Kimble. 1961. Transmission of tobacco mosaic virus in tomato seed. Phytopathology 51:837-842.

Teakle, D. S. 1960. Association of *Olpidium brassicae* and tobacco necrosis virus. Nature 188:431-432.

Teakle, D. S. 1962*a*. Necrotic symptoms of tobacco necrosis virus in roots. Phytopathology 52:1037-1040.

Teakle, D. S. 1962*b*. Transmission of tobacco necrosis virus by a fungus, *Olpidium brassicae*. Virology 18:224-231.

Teakle, D. S. 1964. Transmission of tobacco necrosis virus by *Olpidium brassicae*. Diss. Abstr. 24:4904.

Teakle, D. S. 1967. Fungus transmission of plant viruses, pp. 369-391. *In* K. Maramorosch and H. Koprowski [ed.] Methods in virology, vol. 1. Academic Press, New York. 4 vols.

Teakle, D. S. 1969. Fungi as vectors and hosts of viruses, pp. 23-54. *In* K. Maramorosch [ed.] Viruses, vectors, and vegetation. Interscience, New York. 666 pp.

Teakle, D. S., and A. H. Gold. 1963. Further studies of *Olpidium* as a vector of tobacco necrosis virus. Virology 19:310-315.

Teakle, D. S., and A. H. Gold. 1964. Prolonging the motility and virus-transmitting ability of *Olpidium* zoospores with chemicals. Phytopathology 54:29-32.

Teakle, D. S., and C. Hiruki. 1964. Vector specificity in *Olpidium*. Virology 24:539-544.

Teakle, D. S., and D. R. L. Steindl. 1969. Virus-like particles in galls on sugarcane plants affected by Fiji disease. Virology 37:139-145.

Teakle, D. S., and C. E. Yarwood. 1962. Improved recovery to tobacco necrosis virus from roots by means of *Olpidium brassicae*. Phytopathology 52:366. (Abstr.)

Teliz, D. 1967. Effects of nematode extraction method, soil mixture and nematode numbers on the transmission of tobacco ringspot virus by *Xiphinema americanum*. Nematologica 13:177-185.

Teliz, D., R. G. Grogan, and B. F. Lownsbery. 1966. Transmission of tomato ringspot, peach yellow bud mosaic and grape yellow vein viruses by *Xiphinema americanum*. Phytopathology 56:658-663.

Teller, D., T. A. Horbett, E. G. Richards, and H. K. Schachman. 1969. Ultracentrifuge studies with Rayleigh interference optics. III. Computational methods applied to high-speed sedimentation equilibrium experiments. Ann. N. Y. Acad. Sci. 164:66-100.

Temmink, J. H. M., and R. N. Campbell. 1969. Specific adsorption of tobacco necrosis virus particles onto zoospores of *Olpidium brassicae*. Phytopathology 59:1053 (Abstr.)

Temmink, J. H. M., R. N. Campbell, and P. R. Smith. 1970. Specificity and site of in vitro acquisition of tobacco necrosis virus by zoospores of *Olpidium brassicae*. J. Gen. Virol. 9:201-213.

Thomas, P. E. and R. W. Fulton. 1968. Correlation of ectodesmata number with nonspecific resistance to initial virus infection. Virology 34:459-469.

Thomas, P. R. 1960. Crop and weed plants compared as hosts of viruliferous *Longidorus elongatus* (deMan). Plant Pathol. 18:23-28.

Thomas, P. R. 1970. Host status of some plants for *Xiphinema diversicaudatum* (Micol.) and their susceptibility to viruses transmitted by this species. Ann. Appl. Biol. 65:169-178.

Thomas, W. D., Jr., and R. W. Graham. 1951. Seed transmission of red node virus in Pinto bean. Phytopathology 41:959-962.

Thompson, R. C., and S. P. Doolittle. 1942. Influence of temperature on the expression of big-vein symptoms in lettuce. Phytopathology 32:542-544.

Thornberry, H. H. 1935*a*. Effect of phosphate buffers on infectivity of tobacco mosaic virus. Phytopathology 25:618-627.

Thornberry, H. H. 1935*b*. Effect of tannic acid on the infectivity of tobacco-mosaic virus. Phytopathology 25:931-937.

Thottappilly, G. 1968. Untersuchungen uber die Beziehungen zwischen dem Erbsenblattrollvirus und seinen Vektoren sowie uber ein neuss Pilzund Blattlausubertragbares Virus der Erbse. Doctoral Thesis, Justus Liebig University, Giessen.

Thottappilly, G., and H. Schmutterer. 1968. Zur Kenntnis eines mechanisch, samen-, pilz- und insektenubertragbaren neuen Virus der Erbse. Z. Pflanzenkr. Pflanzenschutz 75:1-8.

Thresh, J. M. 1964*a*. Increased susceptibility to the mite vector (*Phytoptus ribis* Nal.) caused by infection with black currant reversion virus. Nature 202:1028.

Thresh, J. M. 1964*b*. Association between black currant reversion virus and its gall mite vector. Nature 202:1085-1087.

Thresh, J. M. 1964*c*. Warm water treatments to eliminate the gall mite *Phytoptus ribis* Nal. from black currant cuttings. Jubilee Ann. Rep. East Malling Res. Sta. Kent 1963:131-132.

Thung, T. H., and J. P. H. van der Want. 1951. Viren en looistoffen. Tijdschr. Plantenziekten 57:173-174.

Timian, R. G. 1967. Barley stripe mosaic seed transmission and barley yield as influenced by time of infection. Phytopathology 57:1375-1377.

Timian, R. G., W. B. Lunsetter, and H. J. Klosterman. 1962. Liquid-liquid phase separation in concentrating barley stripe mosaic virus. Phytopathology 52:755. (Abstr.)

Timian, R. G., and S. M. Savage. 1966. Purification of barley stripe mosaic virus with chloroform and charcoal. Phytopathology 56:1233-1235.

Tiselius, A. 1954. Chromatography of proteins on calcium phosphate columns. Arkiv. for Kemi 7:433.

Tiselius, A., S. Hjertén, and S. Jerstedt. 1965. Particle-sieve electrophoresis of viruses in polyacrylamide gels, exemplified by purification of turnip yellow mosaic virus. Archiv ges. Virusforschung 17:512-521.

Tiselius, A., S. Hjertén, and O. Levin. 1956. Protein chromatography on calcium phosphate columns. Arch. Biochem. Biophys. 65:132-155.

Todd, J. M. 1965. Soil-borne virus diseases of potato. Scottish Plant Breed. Sta. Rec., 1965:209-235.

Toler, R. W. and T. T. Hebert. 1964. Properties and transmission of soil-borne oat mosaic virus. Phytopathology 54:428-433.

Tomita, K., and A. Rich. 1964. X-ray diffraction investigations of complementary RNA. Nature 201:1160-1163.

Tomlinson, J. A. 1958. Crook root of watercress. III. The causal organism *Spongospora subterranea* (Wallr.) Lagerh. f. sp. nasturtii f. sp. nov. Trans. Brit. Mycol. Soc. 41:491-498.

Tomlinson, J. A. 1963. Effect of phosphate and borate on the infectivity of some viruses during purification. Nature 200:93-94.

Tomlinson, J. A. 1964. Purification and properties of lettuce mosaic virus. Ann. Appl. Biol. 53:95-102.

Tomlinson, J. A., and R. G. Garrett. 1962. Role of *Olpidium* in the transmission of big-vein disease of lettuce. Nature 194:249-250.

Tomlinson, J. A., and R. G. Garrett. 1964. Studies on the lettuce big-vein virus and its vector *Olpidium brassicae* (Wor.) Dang. Ann. Appl. Biol. 54:45-61.

Tomlinson, J. A., R. J. Shepherd, and J. C. Walker. 1959. Purification, properties, and serology of cucumber mosaic virus. Phytopathology 49:293-299.

Tomlinson, J. A., B. R. Smith, and R. G. Garrett. 1962. Graft transmission of lettuce big vein. Nature 193:599-600.
Tomlinson, J. A., S. H. Smith, and H. E. Williams. 1968. Purification and properties of a mechanically transmitted virus from pear pollen and corolla. Phytopathology 58:1026-1031.
Tomlinson, J. A., and D. G. A. Walkey. 1967*a*. The isolation and identification of rhubarb viruses occurring in Britain. Ann. Appl. Biol. 59:415-427. (10)
Tomlinson, J. A., and D. G. A. Walkey. 1967*b*. Effects of ultrasonic treatment on turnip mosaic virus and potato virus X. Virology 32:267-278. (11, 17)
Townshend, J. L. 1967. Plant parasitic nematodes in grape and raspberry soils of Ontario and a comparison of extraction techniques. Can. Plant Dis. Surv. 47:83-86.
Townsley, P. M. 1959*a*. The electrophoretic analysis of plant virus preparations. Can. J. Biochem. Physiol. 37:119-126.
Townsley, P. M. 1959*b*. Zone electrophoresis of tobacco mosaic virus. Can. J. Biochem. Physiol. 37:1025-1031.
Townsley, P. M. 1961. Chromatography of tobacco mosaic virus (TMV) on chitin columns. Nature 191:626.
Toyoda, L. S., I. Kimura, and N. Suzuki. 1965. Purification of rice dwarf virus. Ann. Phytopathol. Soc. Japan 30:225-230.
Trautman, J. L., W. K. Bailey, and C. A. Thomas. 1967. Seed transmission of peanut stunt virus. Phytopathology 57:1280-1281.
Trautman, R. 1956. Operating and comparing procedures facilitating schlieren pattern analysis in analytical ultracentrifugation. J. Phys. Chem. 60:1211-1217.
Trautman, R. 1960. Determination of density gradients in isodensity equilibrium ultracentrifugation. Arch. Biochem. Biophys. 87:289-292.
Trautman, R. 1964. Ultracentrifugation, pp. 211-297. *In* D. W. Newman [ed.] Instrumental methods of experimental biology. Macmillan, New York. 560 pp.
Trautman, R. 1966. The impact of computers on ultracentrifugation. Fractions. Spinco Division of Beckman Instruments, Inc., Palo Alto, California. 2:2-14
Trautman, R. 1969. The impact of the desk-top computer on ultracentrifugation. Ann. N.Y. Acad. Sci. 164:52-65.
Trautman, R., S. P. Spragg, and H. B. Halsall. 1969. Absorption optics data processing for sedimentation and diffusion coefficients from moving boundary ultracentrifugation. Anal. Biochem. 28:396-415.
Tremaine, J. H. 1961. Removal of host antigens from plant virus preparations by ion exchange chromatography. Can. J. Bot. 39:1705-1709.
Tremaine, J. H. 1966. The amino acid and nucleotide composition of the bean and cowpea strains of southern bean mosaic virus. Virology 30:348-354.
Tremaine, J. H., and D. E. Goldsack. 1968. The structure of regular viruses in relation to their subunit amino acid composition. Virology 35:227-237.
Tremaine, J. H., and R. Stace-Smith. 1968. Chemical composition and biophysical properties of tomato ringspot virus. Virology 35:102-107.
Tremaine, J. H., and R. S. Willison. 1962. A merthiolate-induced precipitation in gel-diffusion tests with plant sap. Phytopathology 52:179-180.
Tremaine, J. H., and N. S. Wright. 1967. Cross reactive antibodies in antisera to two strains of southern bean mosaic virus. Virology 31:481-488.
Tsao, P. W. 1963. Intranuclear inclusion bodies in the leaves of cotton plants infected with leaf crumple virus. Phytopathology 53:243-244.
Tsugita, A. 1962*a*. The proteins of mutants of TMV: composition and structure of chemically evoked mutants of TMV-RNA. J. Mol. Biol. 5:284-292. (20, 15)
Tsugita, A. 1962*b*. The proteins of mutants of TMV: classification of spontaneous and chemically evoked strains. J. Mol. Biol. 5:293-300. (20, 15)
Tsugita, A., and H. Fraenkel-Conrat. 1960. The amino acid composition and C-terminal sequence of chemically evoked mutant of tobacco mosaic virus. Proc. Nat. Acad. Sci. U.S. 46:636-642.
Tsugita, A., and H. Fraenkel-Conrat. 1962. The composition of proteins of chemically evoked mutants of TMV-RNA. J. Mol. Biol. 4:73-82.
Tsugita, A., and H. Fraenkel-Conrat. 1963. Contributions from TMV studies to the problem of genetic information transfer and coding, pp. 477-520. *In* J. H. Taylor [ed.] Molecular genetics, Part I. Academic Press, New York. 544 pp.

Tsugita, A., D. T. Gish, J. Young, H. Fraenkel-Conrat, C. A. Knight, and W. M. Stanley. 1960. The complete amino acid sequence of the protein of tobacco mosaic virus. Proc. Nat. Acad. Sci. U.S. 46:1463-1469.

Tsugita, A., Y. Okada, and K. Uehara. 1965. Photosensitized inactivation of ribonucleic acids in the presence of riboflavin. Biochim. Biophys. Acta 103:360-363.

Tsung, C. M., G. Funatsu, and J. D. Young. 1964. Studies on the amino acid sequence of tobacco mosaic virus protein. Arch. Biochem. Biophys. 105:42-50.

Tswett, M. 1906*a*. Physikalisch-chemische Studien uber das Chlorophyll. Die Adsorptionen. Ber. dtsch. bot. Ges. bis. 24:316-323.

Tswett, M. 1906*b*. Adsorptionsanalyse und chromatographische Methode. Anwendung aut die Chemie des Chlorophylls. Ber. dtsch. bot. Ges. bis. 24:384-393.

Tuite, J. 1960. The natural occurence of tobacco ringspot virus. Phytopathology 50:296-298.

Turner, W. F. 1949. Insect vectors of phony peach disease. Science 109:87-88.

Upreti, G. C., B. B. Nagaich, and S. B. Lal. 1964. Inhibition of potato viruses X and Y by some chemicals. Indian Potato J. 6:19-23.

Uyemoto, J. K., and R. G. Grogan. 1969. Chemical characterization of tobacco necrosis and satellite viruses. Virology 39:79-89.

Vago, C., and O. Flandre. 1963. Culture prolongée de tissus d'insectes et de vecteurs de maladies en coagulum plasmatique. Ann. Epiphyties 14(III):127-139.

Valenta, V. 1959*a*. Interference studies with yellows-type plant viruses. I. Cross protection tests with European viruses. Acta Virol. 3:65-72.

Valenta, V. 1959*b*. Interference studies with yellows-type plant viruses. II. Cross protection tests with European and American viruses. Acta Virol. 3:145-152.

Valentine, R. C. 1961. Contrast enhancement in the electron microscopy of viruses. Advances Virus Res. 8:287-318.

Valleau, W. D. 1932. Seed transmission and sterility studies of two strains of tobacco ringspot. Ky. Agr. Exp. Sta. Res. Bull. 327:43-80.

Valleau, W. D. 1935. Localization and resistance of tobacco mosaic in *Nicotiana*. Ky. Agr. Exp. Sta. Res. Bull. 360:202-230.

Valleau, W. D. 1939. Symptoms of yellow ringspot and longevity of the virus in tobacco seed. Phytopathology 29:549-551.

van Boven, C. P. A., H. L. Ensering, and W. Hijmans. 1968. Size determination by the filtration method of the reproductive elements of group A streptococcal L-forms. J. Gen. Microbiol. 52:403-412.

van Bruggen, E. F. J., H. Wiebenga, and M. Gruber. 1962. Structure and properties of hemocyanins. I. Electron micrographs of hemocyanin and apohemocyanin from *Helix pomatia* at different pH values. J. Mol. Biol. 4:1-7.

van der Meer, F. A. 1965. Investigations of currant viruses in the Netherlands. II. Further observations on spoon leaf virus, a soil-borne virus transmitted by the nematode *Longidorus elongatus*. Neth. J. Plant Pathol. 71:33-46.

van der Veken, J. A., and J. Semal. 1966. Aphid transmission of beet yellows virus inhibited by mineral oil. Phytopathology 56:1210-1211.

van der Veken, J. A., D. H. M. van Slogteren, and J. P. H. van der Want. 1962. Immunological methods, pp. 422-463. *In* H. F. Linskens and M. V. Tracey [ed.] Moderne Methoden der Pflanzenanalyse, vol. 5.

van Griensven, L. J. L. D., and A. van Kammen. 1969. The isolation of cowpea mosaic virus-induced ribonuclease-resistant RNA. Evidence for two double-stranded RNA components. J. Gen. Virol. 4:423-428.

van Holde, K. E. 1967. Sedimentation equilibrium. Fractions 1:1-10. Spinco Division of Beckman Instruments, Inc., Palo Alto, California.

van Holde, K. E., and R. L. Baldwin. 1958. Rapid attainment of sedimentation equilibrium. J. Phys. Chem. 62:734-743.

van Hoof, H. A. 1962. *Trichodorus pachydermus* and *T. teres*, vectors of the early browning virus of peas. Tijdschr. Plziekt. 68:391-396.

van Hoof, H. A. 1964*a*. *Trichodorus teres* a vector of rattle virus. Neth. J. Plant Pathol. 70:187.

van Hoof, H. A. 1964*b*. Serial transmission of rattle virus by a single male of *Trichodorus pachydermus* Seinhorst. Nematologica 10:141-144.

van Hoof, H. A. 1966. Nematode populations active and inactive with regard to transmission of NEPO viruses. Nematologica 12:615-618.

van Hoof, H. A. 1968. Transmission of tobacco rattle by *Trichodorus* species. Nematologica 14:20-24.

van Hoof, H. A., D. Z. Maat, and J. W. Seinhorst. 1966. Viruses of the tobacco rattle virus group in northern Italy: their vectors and serological relationships. Neth. J. Plant Pathol. 72:253-258.

van Kammen, A. 1967. Purification and properties of the components of cowpea mosaic virus. Virology 31:633-642.

van Kammen, A. 1968. The relationship between the components of cowpea mosaic virus. I. Two ribonucleoprotein particles necessary for the infectivity of CPMV. Virology 34:312-318.

van Kammen, A., D. Noordam, and T. H. Thung. 1961. The mechanism of inhibition of infection with tobacco mosaic virus by an inhibitor from carnation sap. Virology 14:100-108.

van Ravenswaay-Claasen, J. C., A. B. J. van Leeuwen, G. A. H. Duijts, and L. Bosch. 1967. In vitro translation of alfalfa mosaic virus RNA. J. Mol. Biol. 23:535-544.

van Regenmortel, M. H. V. 1962*a*. Fraction-1 protein: a common contaminant in plant virus preparations. S. Afr. J. Lab. Clin. Med. 8:164. (17)

van Regenmortel, M. H. V. 1962*b*. Purification of a plant virus by filtration through granulated agar (preliminary report). Virology 17:601-602. (13)

van Regenmortel, M. H. V. 1963. Serologically related plant contaminants in preparations of partially purified plant viruses. Virology 21:657-658.

van Regenmortel, M. H. V. 1964*a*. Purification of plant viruses by zone electrophoresis. Virology 23:495-502. (11)

van Regenmortel, M. H. V. 1964*b*. Separation of an antigenic plant protein from preparations of plant viruses. Phytopathology 54:282-289. (13)

van Regenmortel, M. H. V. 1966. Plant virus serology. Advances Virus Res. 12:207-271.

van Regenmortel, M. H. V. 1967*a*. Biochemical and biophysical properties of cucumber mosaic virus. Virology 31:391-396. (15)

van Regenmortel, M. H. V. 1967*b*. Serological studies on naturally occurring strains and chemically induced mutants of tobacco mosaic virus. Virology 31:467-480. (17)

van Regenmortel, M. H. V., J. Brandes, and R. Bercks. 1962. Investigations on the properties of watermelon mosaic virus. Phytopathol. Z. 45:205-216.

van Regenmortel, M. H. V., A. C. Nel, and J. S. Hahn. 1964. Separation of three plant viruses by zone electrophoresis. S. Afr. J. Agr. Sci. 7:817-822.

van Slogteren, D. H. M. 1955. Serological microreactions with plant viruses under paraffin oil. Proc. 2nd Conf. Potato Virus Dis. Lisse-Wageningen 1954, pp. 51-54.

van Slogteren, D. H. M., and N. P. deVos. 1966. Tulip breaking virus, its serological behaviour and serological relationship to a virus isolated from lily, pp. 320-323. *In* A. B. R. Beemster and J. Dijkstra [ed.] Viruses of plants. North-Holland, Amsterdam. 342 pp.

van Slogteren, D. H. M., and H. R. Visscher. 1967. Transmission of a tobacco necrosis virus, causing "Augusta disease" to the roots of tulip by zoospores of the fungus *Olpidium brassicae* (Wor.) Dang. Med. Rijksfaculteit Landbouw-Wet. (Gent) 32:927-938.

van Slogteren, E. 1952. The serological diagnosis of plant diseases caused by virus. Rep. 13th Int. Hort. Congr. 1952, 1:564-569.

van Slogteren, E., and D. H. M. van Slogteren. 1957. Serological identification of plant viruses and serological diagnosis of virus diseases of plants. Annu. Rev. Microbiol. 11:149-164.

Varma, P., and A. J. Gibbs. 1967. Preliminary studies on sap-transmissible viruses of red clover (*Trifolium pratense* L.) in England and Wales. Ann. Appl. Biol. 59:23-30.

Varma, A., A. J. Gibbs, R. D. Woods, and J. T. Finch. 1968. Some observations on he structure of the filamentous particles of several plant viruses. J. Gen. Virol. 2:107-114.

Vasquez, C., and P. Tournier. 1964. New interpretation of the reovirus structure. Virology 24:128-130.

Veldee, S., and H. Fraenkel-Conrat. 1962. The characterization of tobacco mosaic virus strains by their productivity. Virology 18:56-63.

Venekamp, J. H. 1957. Onderzoek naar Biochemische Verschillen Tussen Viruszieke en Gezonde Planten. Jaarverslag Inst. Plantenziektenk. Onderz., Wageningen 87-90.

Venekamp, J. H., and W. H. M. Mosch. 1963. Chromatographic studies on plant viruses. I. The isolation of potato virus X by means of various systems of adsorption chromatrography. Virology 19:316-321.

Venekamp, J. H., and W. H. M. Mosch. 1964*a*. Chromatographic studies on plant viruses. II. The use of polyethylene glycol in clarification and purification of tobacco mosaic virus. Virology 22:503-507.

Venekamp, J. H., and W. H. M. Mosch. 1964*b*. Chromatographic studies on plant viruses. III. The purification of potato virus X, potato virus Y, tobacco mosaic virus and potato stem mottle virus by chromatography on cellulose columns with polyethylene glycol-containing solutions as solvents. Virology 23:394-402.

Venekamp, J. H., and W. H. M. Mosch. 1964*c*. Chromatographic purification of plant viruses on cellulose columns with polyethylene glycol-containing solutions as solvents. Neth. J. Plant Pathol. 70:85-89.

Venekamp, J. H., and W. H. M. Mosch. 1968. The chromatographic purification of plant viruses dependent on the properties of the virus particles. Proc. Conf. Polish Virol. Krakow, Poland.

Venekamp, J. H., W. H. M. Mosch, and K. E. Erkelens-Nanninga. 1964. Chromatographic purification of the carnation ringspot, carnation mottle, and tobacco necrosis viruses. Phytopathology 54:608-609.

Venekamp, J. H., W. H. M. Mosch, and J. P. W. Noordink. 1965. Chromatographic purification of plant viruses. Proc. Int. Conf. Plant Viruses, Wageningen, July 1965. pp. 108-124.

Venner, H. 1966. Stabilitat der N-Glykosidbindung von Nuckleotiden. Z. Physiol. Chem. 344:189-196.

Verduin, B. J. M., and J. B. Bancroft. 1969. The infectivity of tobacco mosaic virus RNA in coat proteins from spherical viruses. Virology 37:501-507.

Verwoerd, D. W. 1969. Purification and characterization of blue tongue virus. Virology 38:203-212.

Vidano, C. 1970. Phases of maize rough dwarf virus multiplication in the vector *Laodelphax striatellus* Fallen. Virology 41:218-232.

Vinograd, J. 1963. Sedimentation equilibrium in a density gradient, pp. 854-870. *In* S. P. Colowick and N. O. Kaplan [ed.] Methods in Enzymology, vol. 6. Academic Press, New York. 7 vols.

Vinograd, J., and R. Bruner. 1966. Band centrifugation in self-generating density gradients. Fractions 1:2-9. Spinco Division of Beckman Instruments, Inc. Palo Alto, Calif.

Vinograd, J., and J. E. Hearst. 1962. Equilibrium sedimentation of macromolecules and viruses in a density gradient. Progr. Chem. Org. Nat. Products 20:372-422.

Vinson, C. G. and A. W. Petri. 1941. Mosaic disease of tobacco. II. Activity of the virus precipitated by lead acetate. Contrib. Boyce Thompson Inst. 3:131-145.

Vuittenez, A. 1961. Les Nématodes vecteurs de virus et le probleme de la dégénérescence infectieuse de la vigne. Les Nématodes. C.N.R.A. Versailles, 1961. 55:78.

Vuittenez, A., and R. Legin. 1964. Confirmation de la transmission des virus de la dégénérescence infectieuse par *Xiphinema index* (Thorne et Allen) et de l'activité parasitaire de ce nématode sur la vigne. Etudes de Virol. Appl. 5:59-62.

Vulic, M., and B. Arenz. 1962. Uber eine verbesserte Arbeitsweise zur serologischen Bestimmung des Y-virus in der Kartoffelphlanze. Nachr. Deutsch. Pflanzenschutzdienst (Braunschweig) 14:65-69.

Wagner, R. R., and A. H. Levy. 1960. Interferon as a chemical intermediary in viral interference. Ann. N. Y. Acad. Sci. 88:1308-1318.

Wagnon, H. K., and J. R. Breece. 1955. Evidence of retention of peach yellow bud mosaic virus in soil. Phytopathology 45:696. (Abstr.)

Wakimoto, S. 1960. An anti-bacterial serum method for the purification of phages which attack phytopathogenic bacteria. Ann. Phytopathol. Soc. Japan 25:205-208.

Walkinshaw, C. H., G. D. Griffin, and R. H. Larson. 1961. *Trichodorus christiei* as a vector of potato corky ringspot (tobacco rattle) virus. Phytopathology 51:806-808.

Wallace, J. M. 1944. Acquired immunity from curly top in tobacco and tomato. J. Agr. Res. 69:187-214.

Wallace, J. M. 1965. Protection in citrus after recovery from seedling yellows and loss of the yellows-inducing virus. Phytopathology 55:1081. (Abstr.)

Wallace, J. M., and R. J. Drake. 1962. A high rate of seed transmission of avocado sun-blotch virus from symptomless trees and the origin of such trees. Phytopathology 52:237-241.

Walters, H. J., and M. A. Bryan. 1965. Purification of bean pod mottle virus using the agar column technique. Phytopathology 55:502. (Abstr.)

Wang, A. L., and C. A. Knight. 1967. Analysis of protein components of tomato strains of tobacco mosaic virus. Virology 31:101-106.

Warmke, H. E., and R. G. Christie. 1967. Use of dilute osmium tetroxide for preservation of three-dimensional crystals of tobacco mosaic virus. Virology 32:534-537.

Warmke, H. E., and J. R. Edwardson. 1966*a*. Electron microscopy of crystalline inclusions of tobacco mosaic virus in leaf tissue. Virology 30:45-57.

Warmke, H. E., and J. R. Edwardson. 1966*b*. Use of potassium permanganate as a fixative for virus particles in plant tissues. Virology 28:693-700.

Wasserman, E., and L. Levine. 1961. Quantitative microcomplement fixation and its use in the study of antigenic structures by specific antigen-antibody inhibition. J. Immunol. 87:290-295.

Watanabe, Y., S. Millward, and A. F. Graham. 1968. Regulation of transcription of the reovirus genome. J. Mol. Biol. 36:107-123.

Watanabe, Y., L. Prevec, and A. F. Graham. 1967. Specificity in transcription of the reovirus genome. Proc. Nat. Acad. Sci. U.S. 58:1040-1046.

Watson, M. A. 1936. Factors affecting the amount of infection obtained by aphid transmission of the Hyoscymus virus 3. Phil. Trans. Roy. Soc. London, B. 226:457-489.

Watson, M. A. 1940. Studies in the transmission of sugar beet yellows virus by the aphids *Myzus persicae* (Sulz). Proc. Roy. Soc. London, B. 125:144-170.

Watson, M. A. 1960. Evidence for interaction or genetic recombination between potato viruses Y and C in infected plants. Virology 10:211-232.

Watson, M. A. 1967. Epidemiology of aphid transmitted viruses. Outl. Agr. 5:155-166.

Watson, M. A., and T. Mulligan. 1960. The manner of transmission of some barley yellow dwarf viruses by different aphid species. Ann. Appl. Biol. 48:711-720.

Watson, M. A., and B. A. M. Okusanya. 1967. Studies on the transmission of groundnut rosette virus by *Aphis craccivora* (Koch). Ann. Appl. Biol. 66:199-208.

Watson, M. A., and F. M. Roberts. 1939. A comparative study of the transmission of Hyoscymus virus 3, potato virus Y and cucumber virus by the vector *Myzus persicae* (Sulz). Proc. Roy. Soc. London, B. 127:543-576.

Watson, M. A., E. P. Sergeant, and E. A. Lennon. 1964. Carrot motley dwarf and parsnip mottle viruses. Ann. Appl. Biol. 54:153-166.

Weathers, L. G., and E. C. Calavan. 1959. Nucellar embryony—a means of freeing citrus clones of viruses, pp. 197-201. *In* J. M. Wallace [ed.] Citrus virus diseases. Univ. Calif., Div. Agr. Sci. 243 pp.

Wechmar, M. B. von. 1967. Ph.D. dissertation. University of Stellenbosch.

Wechmar, M. B. von, and M. H. V. van Regenmortel. 1966. Virus diseases of cereals in South Africa. I. Bromegrass mosaic virus. S. Afr. J. Agr. Sci. 9:443-452.

Wechmar, M. B. von, and M. H. V. van Regenmortel. 1968. Serological studies on bromegrass mosaic virus and its protein fragments. Virology 34:36-45.

Weinstein, J. B., S. M. Friedman, and M. Ochoa. 1966. Fidelity during translation of the genetic code. Cold Spring Harbor Symp. Quant. Biol. 37:671-681.

Weintraub, M., and H. W. J. Ragetli. 1966. Fine structure of inclusions and organelles in *Vicia faba* infected with bean yellow mosaic virus. Virology 28:290-302.

Weintraub, M., and H. W. J. Ragetli. 1968. Intracellular characterization of bean yellow mosaic virus-induced inclusions by differential enzyme digestion. J. Cell Biol. 38:316-328.

Weintraub, M., H. W. J. Ragetli, and P. M. Townsley. 1962. The use of agar gel electrophoresis for the electron microscopic analysis of virus particles. Virology 17:196-198.

Weischer, B. 1966. Ein Beitrag zur geographischen Verbreitung und Okologie von Arten der Gattungen *Xiphinema* und *Longidorus*. Mitt. biol. Bund Anst Landwirt. u. Forstwirt. 118:100-106.

Weiser, J. 1965. Vagoiavirus, gen. n., a virus causing disease in insects. J. Invertebrate Pathol. 7:82-85.

Weissmann, C., M. A. Billeter, E. Vinuela, and M. Libonati. 1966. Double-stranded RNA and the replication of viral RNA, pp. 249-274. *In* A. B. R. Beemster and J. Dijkstra [ed.] Viruses of plants. North-Holland, Amsterdam. 342 pp.

Weller, D. L., and R. S. Morgan. 1967. A new method for preparing ribosomal ribonucleic acid. Biochemistry 6:983-988.

Wells, J. M., and H. D. Sisler. 1968. Preparation and properties of infectious ribonucleic acid from southern bean mosaic virus. Phytopathoggy 58:1588-1592.

Werbin, H., and C. S. Rupert. 1968. Presence of photoreactivating enzyme in blue-green algal cells. Photochem. Photobiol. 7:225-230.

Wessel-Riemens, P. C. 1965. The use of proteins and alkaloids in mechanical transmission of Nigerian isolates of cocoa swollen-shoot virus. Virology 27:566-570.

Westley, F., and I. Cohen. 1966. Tables of values relating the axial ratio to the frictional ratio of an ellipsoid of revolution. Biopolymers 4:201-204.

Wetter, C. 1965. Serology in virus-disease diagnosis. Annu. Rev. Phytopathol. 3:19-42.

Wetter, C. 1967*a*. Immunodiffusion of tobacco mosaic virus and its interaction with agar. Virology 31:498-507.

Wetter, C. 1967*b*. Der Einfluss von Agar- und Elektrolytkonzentration auf die Immunodiffusion von Tabakmosaic- und carnation latent virus. Z. Naturforsch. 22b:1008-1013.

Wetter, C., and H. L. Paul. 1967. Serologische und physikalische Eigenschaften von Sammons Opuntia Virus. Mit 2 Abb. Phytopathol. Z. 60:92-95.

Whitcomb, R. F. 1964. A comparison of serological scoring with test plant scoring of leafhoppers infected by wound tumor virus. Virology 24:488-492.

Whitcomb, R. F. 1965. Clarification of extracts containing potato yellow dwarf virus. Phytopathology 55:746-748.

Whitcomb, R. F. 1967. Infectivity dilution curve of a leafhopper-borne virus on plant leaves. Phytopathology 57:836. (Abstr.)

Whitcomb, R. F. 1969. Bioassay of plant viruses transmitted persistently by their vectors, pp. 449-462. *In* K. Maramorosch [ed.] Viruses, vectors, and vegetation. Interscience, New York. 666 pp.

Whitcomb, R. F. 1971. Concepts and problems in bioassay of the agents of clover wound tumor, peach Western X, and aster yellows diseases. U.S. Dep. Agr. Tech. Bull. (in press).

Whitcomb, R. F., and L. M. Black. 1961. Synthesis and assay of wound-tumor soluble antigen in an insect vector. Virology 15:136-145.

Whitcomb, R. F., and R. E. Davis. 1969. Properties of the aster yellows agent. Phytopathology 59:1561. (Abstr.)

Whitcomb, R. F., and R. E. Davis. 1970*a*. Evidence on possible mycoplasma-like etiology of aster yellows disease: II. Suppression of aster yellows in insect vectors. J. Infect. Immunol. 2:209-215.

Whitcomb, R. F., and R. E. Davis. 1970*b*. Mycoplasma and phytarboviruses as plant pathogens persistently transmitted by insects. Annu. Rev. Entomol. 15:405-464.

Whitcomb, R. F., R. E. Davis, R. Purcell, R. Cohen, and R. L. Steere. 1970. Aster yellows disease: mycoplasma-like properties of the infectious agent. Science (in press).

Whitcomb, R. F., and D. D. Jensen. 1968. Proliferative symptoms in leafhoppers infected with Western X-disease virus. Virology 35:174-177.

Whitcomb, R. F., D. D. Jensen, and J. Richardson. 1966*a*. The infection of leafhoppers by Western X-disease virus. I. Frequency of transmission after injection or acquisition feeding. Virology 28:448-453.

Whitcomb, R. F.. D. D. Jensen, and J. Richardson. 1966*b*. The infection of leafhoppers by Western X-disease virus. II. Fluctuation of virus concentration in the hemolymph after injection. Virology 28:454-458.

Whitcomb, R. F., D. D. Jensen, and J. Richardson. 1967. The infection of leafhoppers by Western X-disease virus. III. Salivary, neural, and adipose histopathology. Virology 31:539-549.

Whitcomb, R. F.. D. D. Jensen, and J. Richardson. 1968*a*. The infection of leafhoppers by Western X-disease virus. IV. Pathology in the alimentary tract. Virology 34:69-78.

Whitcomb, R. F., D. D. Jensen, and J. Richardson. 1968*b*. The infection of leafhoppers by Western X-disease virus. V. Properties of the infectious agent. J. Invertebrate Pathol. 12:192-201.

Whitcomb, R. F., D. D. Jensen, and J. Richardson. 1968*c*. The infection of leafhoppers by Western X-disease virus. VI. Cytopathological interrelationships. J. Invertebrate Pathol. 12:202-221.

Whitcomb, R. F., J. Richardson, and D. D. Jensen. 1971. Anatomical and cytological notes on the leafhopper *Colladonus montanus* Van Duzee. (In preparation).

Whitcomb, R. F., M. Shapiro, and J. Richardson. 1966. An Erwinia-like bacterium pathogenic to leafhoppers. J. Invertebrate Pathol. 8:299-307.

Whitcomb, R. F., and R. C. Sinha. 1964. Effect of host components and sucrose on infection by potato yellow dwarf virus. Phytopathology 54:142-146.

Whitcomb, R. F., and R. S. Spendlove. 1966. Density gradient centrifugation of virus-antibody complexes: A sensitive serological method. Virology 30:752-754.

Whitehead, A. G. 1965. Nematodes associated with "Docking Disorders" of sugar beet. Brit. Sugar Beet Rev. 34:77-78, 83-84, 92.

Whitehead, A. G., and J. R. Hemming. 1965. A comparison of some quantitative methods of extracting small vermiform nematodes from soil. Ann. Appl. Biol. 55:25-38.

Whitfeld, P. R., and S. Williams. 1963. On the ribonuclease activity associated with tobacco mosaic virus preparations. Virology 21:156-161.

Whitney, D. L., and C. Tanford. 1962. Solubility of amino acids in aqueous urea solution and its implications for the denaturation of proteins by urea. J. Biol. Chem. 237:1735-1737.

Wicker, R., and S. Avrameas. 1969. Localization of virus antigens by enzyme labelled antibodies. J. Gen. Virol. 4:465-471.

Wilcoxson, R. D., and S. M. El-Kandelgy. 1966. Effect of light on formation of lesions in *G. globosa* by red clover vein mosaic virus. Phytopathology 56:364.

Wildman, S. G. 1959. The process of infection and virus synthesis with tobacco mosaic virus and other plant viruses, pp. 1-31. *In* F. M. Burnet and W. M. Stanley [ed.] The viruses, vol. 2. Academic Press, New York, 3 vols.

Williams, J. W. [ed.] 1963. Ultracentrifugal analysis in theory and experiments. Academic Press, New York. 282 pp.

Williams, J. W., K. E. Van Holde, R. L. Baldwin, and H. Fujita. 1958. The theory of sedimentation analysis. Chem. Rev. 58:715-806.

Williams, P. R., and B. E. Juniper. 1968. An improved replica technique for examining plant surfaces in the electron microscope. Proc. Roy. Microscop. Soc. 3:129-131.

Williams, R. C. 1953. A method of freeze-drying for electron microscopy. Exp. Cell Res. 4:188-201.

Williams, R. C., and R. C. Backus. 1949. Macromolecular weights determined by direct particle counting. I. The weight of the bushy stunt virus particle. J. Amer. Chem. Soc. 71:4052-4057.

Williams, S., and P. R. Whitfeld. 1966. Fractionation of partially degraded tobacco mosaic virus ribonucleic acid. Nature 211:102-104.

Willison, J. H. M., and E. C. Cocking. 1969. Freeze-etching observation on tobacco leaves infected with tobacco mosaic virus. J. Gen. Virol. 4:229-233.

Willison, R. S., M. Weintraub, and J. D. Ferguson. 1956. Purification and electron microscopy of viruses causing cherry yellows and related diseases. Can J. Bot. 34:86-103.

Wilson, M. W., and B. H. Pringle. 1955. Interpretation of the Ouchterlony precipitin test. J. Immunol. 75:460-469.

Wilson, M. W., and B. H. Pringle. 1956. Cross-reactions in the Ouchterlony plate analysis of native and halogenated bovine serum albumins. J. Immunol. 77:325-331.

Wilson, N. S., and L. S. Jones. 1957. Transmission studies with *Eriophyes insidiosus*. Unpublished Rep., Fruit Insects Sect., Entomol. Res. Div., U.S. Dep. Agr.

Wilson, N. S., L. S. Jones, and L. C. Cochran. 1955. An eriophyid mite vector of the peach mosaic virus. Plant Dis. Reptr. 39:889-892.

Wiltshire, G. H. 1956. The effect of darkening on the susceptibility of plants to infection with viruses. Ann. Appl. Biol. 44:233-255.

Winkler, A. J. 1949. Pierce's disease investigations. Hilgardia 19:207-264.

Wischnitzer, S. 1967. Current techniques in biomedical electron microscopy. Int. Rev. Cytol. 22:2-61.

Wittmann, H. G. 1959*a*. Darstellung und physikalische charakterisierung des peptideketten des tabakmosaikvirus. Experimentia 15:174-175. (15)

Wittmann, H. G. 1959*b*. Vergleich der proteine des normalstammes und einer nitritmutante des tabakmosaikvirus. Z. Vererbungsl. 90:463-475. (20)

Wittmann, H. G. 1960. Comparison of the tryptic peptides of chemically induced and spontaneous mutants of tobacco mosaic virus. Virology 12:609-612.

Wittmann, H. G. 1962. Proteinuntersuchungen an mutanten des tabakmosaikvirus als beitrag zum problem des genetischen codes. Z. Vererbungsl. 93:491-530.

Wittmann, H. G. 1964. Proteinanalysen von chemisch induzierten mutanten des tabakmosaikvirus. Z. Vererbungsl. 95:333-344.

Wittmann, H. G. 1965*a*. Uber die frage der mehrdeutigkeit des genetischen codes, untersucht am Hüllenprotein des Tabakmosaikvirus. Z. Vererbungsl. 97:138-149.

Wittmann, H. G. 1965*b*. Die proteinstruktur der defektmutante PM2 des Tabakmosaikvirus. Z. Vererbungsl. 97:297-304.

Wittmann, H. G. 1965*c*. Einstamm des Tabakmosaikvirus mit freier N-Terminaler aminosäure in der Hüllproteinkette. Z. Vererbungsl. 97:204-208.

Wittmann, H. G., I. Hindennach, and B. Wittmann-Liebold. 1969. Die primäre Proteinstruktur von Stämmen des Tabakmosaikvirus. Teil V: Aminosäurensequenz (Positionen 62-156) des Proteins des Tabakmosaikvirusstammes Holmes rib grass. Z. Naturforsch. 24b:870-876.

Wittman, H. G., and H. L. Sanger. 1965. I. Phytopathogene Viren. Fortschritte Bot. 27:250-290.

Wittman, H. G., and B. Wittmann-Liebold. 1966. Protein chemical studies of the RNA viruses and their mutants. Cold Spring Harbor Symp. Quant. Biol. 31:163-172.

Wittmann, H. G., B. Wittmann-Liebold, and Jauregui-adell. 1965. Die primäre Proteinstruktur Temperatur-sensitiver Mutanten des Tabakmosaikvirus. Z. Naturforsch. 20b:1224-1234.

Wittmann-Liebold, B., and H. G. Wittmann. 1964. Die primäre Proteinstruktur von Stämmen des Tabakmosaikvirus. Aminosäuresequenzen des Proteins des Tabakmosaikvirusstammes dahlemense. Teil II (Position 1-41 and 135-158). Z. Naturforsch. 18b:1032-1049.

Wittmann-Liebold, B., and H. G. Wittmann. 1964. Die primaře Proteinstruktur von Stämmen des Tabakmosaikvirus Aminosäuresequenzen des proteins des Tabakmosaikvirusstammes dahlemense Teil I (Position 42-134). Hoppe-Seyler Ztschr. Physiol. Chem. 335:69-116.

Wittmann-Liebold, B., and H. G. Wittmann. 1965*a*. Lokalisierung von aminosäureaustauschen bei spontanmutanten und nach fluoruracil-einbau isolierten Mutanten des Tabakmosaikvirus. Z. Vererbungsl. 97:218-225. (20)

Wittmann-Liebold, B., and H. G. Wittmann. 1965*b*. Lokalisierung von aminosäureaustauschen bei Nitritmutanten des Tabakmosaikvirus. Z. Vererbungsl. 97:305-326. (20)

Wittmann-Liebold, B., and H. G. Wittmann. 1967. Coat proteins of strains of two RNA viruses: comparison of their amino acid sequences. Mol. Gen. Genet. 100:358-363.

Wolanski, B. S., and R. I. B. Francki. 1969. Structure of lettuce necrotic yellows virus. II. Electron microscopic studies on the effect of pH of phosphotungstic acid stain on the morphology of the virus. Virology 37:437-447.

Wolanski, B. S., R. I. B. Francki, and T. C. Chambers. 1967. Structure of lettuce necrotic yellows virus. I. Electron microscopy of negatively stained preparations. Virology 33:287-296.

Wolf, G., and R. Casper. 1971. Disc electrophoretic separation of elongated plant viruses in polyacrylamide-agarose gels. J. Gen. Virol. (in press).

Wolstenholme, D. R., and L. E. Bockstahler. 1967. Electron microscopy of double-stranded RNA induced by turnip yellow mosaic virus and tobacco mosaic virus. Mol. Gen. Genet. 100:349-357.

Wood, R. L., and J. H. Luft. 1965. The influence of buffer systems on fixation with osmium tetroxide. J. Ultrastructure Res. 12:22-45.

Woods, M. W., and H. G. duBuy. 1951. The action of mutant chondriogenes and viruses on plant cells with special reference to the plastids. Am. J. Bot. 38:419-434.

Woods, M. W., and R. V. Eck. 1948. Nuclear inclusions produced by a strain of tobacco mosaic virus. Phytopathology 38:852-856.

Woody, B. R., and C. A. Knight. 1959. Peptide maps obtained with tryptic digests of the proteins of some strains of tobacco mosaic virus. Virology 9:359-374.

Worley, J. F. 1969. Chains, filamentous forms and budding of mycoplasma-like organisms in aster yellows-infected plants. Phytopathology 59:1561-1562. (Abstr.)

Worley, J. F. 1970. Possible replicative forms of a mycoplasma like organism and their location in aster yellows diseased *Nicotiana* and aster. Phytopathology 60:284-292.

Worley, J. F., and I. R. Schneider. 1963. Progressive distribution of southern bean mosaic virus antigen in bean leaves determined with a fluorescent antibody stain. Phytopathology 53:1255-1257.

Wright, K. A. 1965. The histology of the oesophageal region of *Xiphinema index* Thorne and Allen, 1950, as seen with the electron microscope. Can. J. Zool. 43:319-328.

Wrigley, N. G. 1968. The lattice spacing of crystalline catalase as an internal standard of length in electron microscopy. J. Ultrastructure Res. 24:454-464.

Wu, J.-H., and J. E. Dimitman. 1970. Leaf structure and callus formation as determinants of TMV movement in bean leaves as revealed by UV irradiation studies. Virology 40:820-827.

Wu, J.-H., W. Hudson, and S. G. Wildman. 1962. Quantitative analysis of interference produced by related strains of TMV compared with noninfectious TMV components. Phytopathology 52:1264-1266.

Wu, J.-H., R. A. Lewin, and H. Werbin. 1967. Photoreactivation of ultraviolet-irradiated blue-green algal virus LPP-1. Virology 31:657-664.

Wu, J.-H. and I. Rappaport. 1962. Competition between infectious tobacco mosaic virus nucleic acid and intact virus on *Nicotiana glutinosa*. Nature 193:908.

Yamazaki, H., J. B. Bancroft, and P. Kaesberg. 1961. Biophysical studies of broad bean mottle virus. Proc. Nat. Acad. Sci. U.S. 47:979-983.

Yamazaki, H., and P. Kaesberg. 1961. The preparation and some properties of protein subunits obtained from wild cucumber mosaic virus. Biochim. Biophys. Acta 53:173-180.

Yamazaki, H., and P. Kaesberg. 1963*a*. Degradation of bromegrass mosaic virus with calcium chloride and the isolation of its protein and nucleic acid. J. Mol. Biol. 7:760-762. (15)

Yamazaki, H., and P. Kaesberg. 1963*b*. Isolation and characterization of a protein subunit of broad bean mottle virus. J. Mol. Biol. 6:465-473. (15)

Yarwood, C. E. 1952. The phosphate effect in plant virus inoculations. Phytopathology 42:137-143.

Yarwood, C. E. 1953. Quick virus inoculation by rubbing with fresh leaf discs. Plant Dis. Reptr. 37:501-502.

Yarwood, C. E. 1955. Mechanical transmission of an apple mosaic virus. Hilgardia 23:613-628.

Yarwood, C. E. 1957*a*. Mechanical transmission of fruit tree viruses. Proc. Can. Phytopathol. Soc. 25:19-20. (Abstr.)

Yarwood, C. E. 1957*b*. Mechanical transmission of plant viruses. Advances Virus Res. 4:243-278.

Yarwood, C. E. 1959. Tobacco necrosis virus in *Cleome*. Plant Dis. Reptr. 43:390-393.

Yarwood, C. E. 1960*a*. Localized acquired resistance to tobacco mosaic virus. Phytopathology 50:741-744.

Yarwood, C. E. 1960*b*. Topical susceptibility of plants to viruses. Virology 12:245-257. (1)

Yarwood, C. E. 1962. Further phosphate effects in virus inoculations. Phytopathology 52:206-209.

Yarwood, C. E. 1964. Assay of infectivity, pp. 79-110. *In* R. J. C. Harris [ed.] Techniques in experimental virology. Academic Press, New York. 450 pp.

Yarwood, C. E. 1966. Bentonite aids virus transmission. Virology 28:459-462.

Yarwood, C. E. 1968. Sequence of supplements in virus inoculations. Phytopathology 58:132-136.

Yarwood, C. E. 1969*a*. Charcoal in virus inoculations. Phytopathology 59:71-75. (1)

Yarwood, C. E. 1969*b*. Sulfite in plant virus inoculations. Virology 39:74-78. (1)

Yarwood, C. E. and R. W. Fulton. 1967. Mechanical transmission of plant viruses, pp. 237-266. *In* K. Maramorosch and H. Koprowski [ed.] Methods in virology, vol. 1. Academic Press, New York. 4 vols.

Yarwood, C. E., and E. Hecht-Poinar. 1970. Magnesium silicate in virus transmission. Virology 41:436-443.

Yates, F. 1937. The design and analysis of factorial experiments. Imp. Bur. Soil Tech. Commun. No. 35.

Youden, W. J., and H. P. Beale. 1934. A statistical study of the local lesion method for estimating tobacco mosaic virus. Contrib. Boyce Thompson Inst. 6:437-454.

Youden, W. J., H. P. Beale, and J. D. Guthrie. 1935. Relation of virus concentration to the number of lesions produced. Contrib. Boyce Thompson Inst. 7:37-53.

Young, D. A. 1968. Taxonomic study of the *Cicadellinae* (Homoptera:Cicadellidae). Part 1. Proconiini. U.S. Nat. Mus. Bull. 261. 287 pp.

Yphantis, D. A. 1960. Rapid determination of molecular weights of peptides and proteins. Ann. N.Y. Acad. Sci. 88:586-601.

Yphantis, D. A. 1964. Equilibrium ultracentrifugation of dilute solutions. Biochemistry 3:297-317.

Zaitlin, M. 1958. Continuous filter-paper electrophoresis of tobacco mosaic virus. J. Chromatogr. 1:186-199.

Zaitlin, M., and W. R. Ferris. 1964. Unusual aggregation of a non-functional tobacco mosaic virus protein. Science 143:1451-1452.

Zaitlin, M. and C. L. Keswani. 1964. Relative infectivities of tissue containing a defective tobacco mosaic virus strain. Virology 24:495-498.

Zaitlin, M., and W. F. McCaughey. 1965. Amino acid composition of a nonfunctional tobacco mosaic virus protein. Virology 26:500-503.

Zaitlin, M., and A. Siegel. 1963. A virus inhibitor from tobacco. Phytopathology 53:224-227.

Zalokar, M. 1965. A simple freeze-substitution method for electron microscopy. J. Ultrastructure Res. 15:469-479.

Zazhurilo, V. K., and G. M. Sitnikova. 1941. Interrelations between mosaic disease virus of winter wheat and its vector, *Deltocephalus striatus* L. Dokl. Vses. Akad. Sel'Skokhoz. Mauk. 11:27-29. (Proc. Lenin Acad. Agr. Sci., USSR).

Zech, H. 1952. Infektionsvorggang und Wanderung des Tabakmosaik-virus. Planta 40:461-514.

Zech, H., and L. Vogt-Kohne. 1956. Untersuchungen zur reproduktion des Tabakmosaikvirus I. Elektronenmikroskopische Beobachtungen. Exp. Cell Res. 10:458-475.

Zimmer, C., and H. Triebel. 1969. Studies on conformational changes in the DNA structure induced by protonation: reversible and irreversible acid titrations and sedimentation measurements. Biopolymers 8:573-593.

Zink, F. W., R. G. Grogan, and J. E. Welch. 1956. The effect of the percentage of seed transmission upon subsequent spread of lettuce mosaic virus. Phytopathology 46:662-664.

Zobel, C. R., and M. Beer. 1965. The use of heavy metal salts as electron strains. Rev. Cytol. 18:363-400.

Index